Basic College Mathematics

Basic College Mathematics

Fifth Edition

Elayn Martin-Gay

University of New Orleans

PEARSON

Boston   Columbus   Hoboken   Indianapolis   New York   San Francisco
Amsterdam   Cape Town   Dubai   London   Madrid   Milan   Munich   Paris   Montréal   Toronto
Delhi   Mexico City   São Paulo   Sydney   Hong Kong   Seoul   Singapore   Taipei   Tokyo

Editorial Director, Mathematics: *Christine Hoag*
Editor-in-Chief: *Michael Hirsch*
Acquisitions Editor: *Mary Beckwith*
Project Manager: *Lauren Morse*
Project Team Lead: *Christina Lepre*
Assistant Editor: *Matthew Summers*
Editorial Assistant: *Megan Tripp*
Executive Development Editor: *Dawn Nuttall*
Program Team Lead: *Karen Wernholm*
Program Manager: *Patty Bergin*
Cover and Illustration Design: *Tamara Newnam*
Program Design Lead: *Heather Scott*
Director, Course Production: *Ruth Berry*
Executive Content Manager, MathXL: *Rebecca Williams*
Senior Content Developer, TestGen: *John Flanagan*
Media Producer: *Audra Walsh*
Director of Marketing, Mathematics: *Michelle Renda*
Senior Marketing Manager: *Rachel Ross*
Marketing Assistant: *Kelly Cross*
Senior Author Support/Technology Specialist: *Joe Vetere*
Senior Procurement Specialist: *Carol Melville*
Interior Design, Production Management, Answer Art, and Composition:
  Integra Software Services Pvt. Ltd.
Text Art: *Scientific Illustrators*

Acknowledgments of third party content appear on page xvi, which constitutes an extension
of this copyright page.

PEARSON, ALWAYS LEARNING, and MYMATHLAB are exclusive trademarks in the
U.S. and/or other countries owned by Pearson Education, Inc. or its affiliates.

Unless otherwise indicated herein, any third-party trademarks that may appear in this work
are the property of their respective owners and any references to third-party trademarks,
logos or other trade dress are for demonstrative or descriptive purposes only. Such references are not intended to imply any sponsorship, endorsement, authorization, or promotion
of Pearson's products by the owners of such marks, or any relationship between the owner
and Pearson Education, Inc. or its affiliates, authors, licensees or distributors.

Library of Congress Cataloging-in-Publication Data
Martin-Gay, K. Elayn
  Basic college mathematics / Elayn Martin-Gay, University of New Orleans. – 5th edition.
    pages cm
  ISBN-13: 978-0-321-95097-0 (alk. paper)
  ISBN-10: 0-321-95097-6 (alk. paper)
1. Mathematics–Textbooks.   I. Title.
  QA39.3.M37 2015
  510–dc23   2013043430

2 3 4 5 6 7 8 9 10—CRK—18 17 16 15

www.pearsonhighered.com

ISBN-10: 0-321-95097-6 (Student Edition)
ISBN-13: 978-0-321-95097-0

In Memory of
Alvin Numa Cahall, Jr.

A wonderful friend who used mathematics in his own way—through
a thorough knowledge of electronics and sound technology.
He is sorely missed by family and friends.

Contents

Preface

Basic College Mathematics, **Fifth Edition** was written to provide a solid foundation in the basics of college mathematics, including the topics of whole numbers, fractions, decimals, ratio and proportion, percent, and measurement as well as introductions to geometry, statistics and probability, and algebra topics. Specific care was taken to make sure students have the most up-to-date relevant text preparation for their next mathematics course or for nonmathematical courses that require an understanding of basic mathematical concepts. I have tried to achieve this by writing a user-friendly text that is keyed to objectives and contains many worked-out examples. As suggested by AMATYC and the NCTM Standards (plus Addenda), real-life and real-data applications, data interpretation, conceptual understanding, problem solving, writing, cooperative learning, appropriate use of technology, mental mathematics, number sense, estimation, critical thinking, and geometric concepts are emphasized and integrated throughout the book.

The many factors that contributed to the success of the previous editions have been retained. In preparing the Fifth Edition, I considered comments and suggestions of colleagues, students, and many users of the prior edition throughout the country.

What's New in the Fifth Edition?

- **The Martin-Gay Program** has been revised and enhanced with a new design in the text and MyMathLab® to actively encourage students to use the text, video program, Video Organizer, and Student Organizer as an integrated learning system.

- **The New Video Organizer** is designed to help students take notes and work practice exercises while watching the Interactive Lecture Series videos (available in MyMathLab and on DVD). All content in the Video Organizer is presented in the same order as it is presented in the videos, making it easy for students to create a course notebook and build good study habits.

  — Covers all of the video examples in order.

  — Provides ample space for students to write down key definitions and properties.

  — Includes "Play" and "Pause" button icons to prompt students to follow along with the author for some exercises while they try others on their own.

  The Video Organizer is available in a loose-leaf, notebook-ready format. It is also available for download in MyMathLab.

- **Vocabulary, Readiness & Video Check** questions have been added prior to every section exercise set. These exercises quickly check a student's understanding of new vocabulary words. The **readiness** exercises center on a student's understanding of a concept that is necessary in order to continue to the exercise set. **New Video check questions for the Martin-Gay Interactive Lecture videos** are now included in every section for each learning objective. **These exercises are all available for assignment in MyMathLab** and are a great way to assess whether students have viewed and understood the key concepts presented in the videos.

- **New Student Success Tips Videos** are 3- to 5-minute video segments designed to be daily reminders to students to continue practicing and maintaining good organizational and study habits. They are organized in three categories and are available in MyMathLab and the Interactive Lecture Series. The categories are:

1. Success Tips that apply to any course in college in general, such as Time Management.

2. Success Tips that apply to any mathematics course. One example is based on understanding that mathematics is a course that requires homework to be completed in a timely fashion.

3. Section- or Content-specific Success Tips to help students avoid common mistakes or to better understand concepts that often prove challenging. One example of this type of tip is how to apply the order of operations to simplify an expression.

- **Interactive DVD Lecture Series**, featuring your text author (Elayn Martin-Gay), provides students with active learning at their own pace. The videos offer the following resources and more:

  A complete lecture for each section of the text highlights key examples and exercises from the text. "Pop-ups" reinforce key terms, definitions, and concepts.

  An interface with menu navigation features allows students to quickly find and focus on the examples and exercises they need to review.

  Interactive Concept Check exercises measure students' understanding of key concepts and common trouble spots.

  New Student Success Tips Videos.

- **The Interactive DVD Lecture Series** also includes the following resources for test prep:

  The Chapter Test Prep Videos help students during their most teachable moment—when they are preparing for a test. This innovation provides step-by-step solutions for the exercises found in each Chapter Test. For the Fifth Edition, the chapter test prep videos are also available on YouTube™. The videos are captioned in English and Spanish.

  The Practice Final Exam Videos help students prepare for an end-of-course final. Students can watch full video solutions to each exercise in the Practice Final Exam at the end of this text.

- **The Martin-Gay MyMathLab** course has been updated and revised to provide more exercise coverage, including assignable video check questions and an expanded video program. There are section lecture videos for every section, which students can also access at the specific objective level; Student Success Tips videos; and an increased number of watch clips at the exercise level to help students while doing homework in MathXL. Suggested homework assignments have been premade for assignment at the instructor's discretion.

- **New MyMathLab Ready to Go Courses** (access code required) provide students with all the same great MyMathLab features that you're used to, but make it easier for instructors to get started. Each course includes preassigned homework and quizzes to make creating your course even simpler. Ask your Pearson representative about the details for this particular course or to see a copy of this course.

Key Pedagogical Features

The following key features have been retained and/or updated for the Fifth Edition of the text:

Problem-Solving Process   This is formally introduced in Chapter 1 with a four-step process that is integrated throughout the text. The four steps are **Understand, Translate, Solve,** and **Interpret.** The repeated use of these steps in a variety of examples shows their wide applicability. Reinforcing the steps can increase students' comfort level and confidence in tackling problems.

Exercise Sets Revised and Updated   The exercise sets have been carefully examined and extensively revised. Special focus was placed on making sure that even- and odd-numbered exercises are paired and that real-life applications were updated.

Examples   Detailed, step-by-step examples were added, deleted, replaced, or updated as needed. Many examples reflect real life. Additional instructional support is provided in the annotated examples.

Practice Exercises   Throughout the text, each worked-out example has a parallel Practice exercise. These invite students to be actively involved in the learning process. Students should try each Practice exercise after finishing the corresponding example. Learning by doing will help students grasp ideas before moving on to other concepts. Answers to the Practice exercises are provided at the bottom of each page.

Helpful Hints   Helpful Hints contain practical advice on applying mathematical concepts. Strategically placed where students are most likely to need immediate reinforcement, Helpful Hints help students avoid common trouble areas and mistakes.

Concept Checks   This feature allows students to gauge their grasp of an idea as it is being presented in the text. Concept Checks stress conceptual understanding at the point-of-use and help suppress misconceived notions before they start. Answers appear at the bottom of the page. Exercises related to Concept Checks are included in the exercise sets.

Mixed Practice Exercises   In the section exercise sets, these exercises require students to determine the problem type and strategy needed to solve it just as they would need to do on a test.

Integrated Reviews   This unique, mid-chapter exercise set (and notes where appropriate) helps students assimilate new skills and concepts that they have learned separately over several sections. These reviews provide yet another opportunity for students to work with "mixed" exercises as they master the topics.

Vocabulary Check   This feature provides an opportunity for students to become more familiar with the use of mathematical terms as they strengthen their verbal skills. These appear at the end of each chapter before the Chapter Highlights. Vocabulary, Readiness & Video exercises provide practice at the section level.

Chapter Highlights   Found at the end of every chapter, these contain key definitions and concepts with examples to help students understand and retain what they have learned and help them organize their notes and study for tests.

Chapter Review   The end of every chapter contains a comprehensive review of topics introduced in the chapter. The Chapter Review offers exercises keyed to every section in the chapter, as well as Mixed Review exercises that are not keyed to sections.

Chapter Test and Chapter Test Prep Videos   The Chapter Test is structured to include those exercises that involve common student errors. The **Chapter Test Prep Videos** gives students instant access to a step-by-step video solution of each exercise in the Chapter Test.

Cumulative Review   This review follows every chapter in the text (except Chapter 1). Each odd-numbered exercise contained in the Cumulative Review is an earlier worked example in the text that is referenced in the back of the book along with the answer.

Writing Exercises ✎ These exercises occur in almost every exercise set and require students to provide a written response to explain concepts or justify their thinking.

Applications Real-world and real-data applications have been thoroughly updated, and many new applications are included. These exercises occur in almost every exercise set and show the relevance of mathematics and help students gradually and continuously develop their problem-solving skills.

Review Exercises These exercises occur in each exercise set (except in Chapter 1) and are keyed to earlier sections. They review concepts learned earlier in the text that will be needed in the next section or chapter.

Exercise Set Resource Icons Located at the opening of each exercise set, these icons remind students of the resources available for extra practice and support:

See Student Resources descriptions on page xv for details on the individual resources available.

Exercise Icons These icons facilitate the assignment of specialized exercises and let students know what resources can support them.

- ▶ DVD Video icon: exercise worked on the Interactive DVD Lecture Series.
- △ Triangle icon: identifies exercises involving geometric concepts.
- ✎ Pencil icon: indicates a written response is needed.
- ▦ Calculator icon: optional exercises intended to be solved using a scientific or graphing calculator.

Group Activities Found at the end of each chapter, these activities are for individual or group completion, and are usually hands-on or data-based activities that extend the concepts found in the chapter, allowing students to make decisions and interpretations and to think and write about algebra.

Optional: Calculator Exploration Boxes and Calculator Exercises The optional Calculator Explorations provide keystrokes and exercises at appropriate points to give students an opportunity to become familiar with these tools. Section exercises that are best completed by using a calculator are identified by ▦ for ease of assignment.

Tara LaFrance, *Delgado Community College*
John LaMaster, *Indiana Purdue University Fort Wayne*
Nancy Lange, *Inver Hills Community College*
Judy Langer, *Westchester Community College*
Lisa Lindloff, *McLinnan Community College*
Sandy Lofstock, *St. Petersburg College*
Kathy Lovelle, *Westchester Community College*
Nicole Mabine, *North Lake College*
Jean McArthur, *Joliet Junior College*
Kevin McCandless, *Evergreen Valley College*
Ena Michael, *State College of Florida*
Daniel Miller, *Niagara County Community College*
Marica Molle, *Metropolitan Community College*
Carol Murphy, *San Diego Miramar College*
Charlotte Newsom, *Tidewater Community College*
Cao Nguyen, *Central Piedmont Community College*
Greg Nguyen, *Fullerton College*
Eric Oilila, *Jackson Community College*

Linda Padilla, *Joliet Junior College*
Armando Perez, *Laredo Community College*
Davidson Pierre, *State College of Florida*
Marilyn Platt, *Gaston College*
Chris Riola, *Moraine Valley College*
Carole Shapero, *Oakton Community College*
Janet Sibol, *Hillsborough Community College*
Anne Smallen, *Mohawk Valley Community College*
Barbara Stoner, *Reading Area Community College*
Jennifer Strehler, *Oakton Community College*
Ellen Stutes, *Louisiana State University Eunice*
Tanomo Taguchi, *Fullerton College*
Robyn Toman, *Anne Arundel Community College*
MaryAnn Tuerk, *Elsin Community College*
Walter Wang, *Baruch College*
Leigh Ann Wheeler, *Greenville Technical Community College*
Darlene Williams, *Delgado Community College*
Valerie Wright, *Central Piedmont Community College*

A special thank you to those students who participated in our design review: Katherine Browne, Mike Bulfin, Nancy Canipe, Ashley Carpenter, Jeff Chojnachi, Roxanne Davis, Mike Dieter, Amy Dombrowski, Kay Herring, Todd Jaycox, Kaleena Levan, Matt Montgomery, Tony Plese, Abigail Polkinghorn, Harley Price, Eli Robinson, Avery Rosen, Robyn Schott, Cynthia Thomas, and Sherry Ward.

Elayn Martin-Gay

About the Author

Elayn Martin-Gay has taught mathematics at the University of New Orleans for more than 25 years. Her numerous teaching awards include the local University Alumni Association's Award for Excellence in Teaching, and Outstanding Developmental Educator at University of New Orleans, presented by the Louisiana Association of Developmental Educators.

Prior to writing textbooks, Elayn Martin-Gay developed an acclaimed series of lecture videos to support developmental mathematics students in their quest for success. These highly successful videos originally served as the foundation material for her texts. Today, the videos are specific to each book in the Martin-Gay series. The author has also created Chapter Test Prep Videos to help students during their most "teachable moment"—as they prepare for a test—along with Instructor-to-Instructor videos that provide teaching tips, hints, and suggestions for each developmental mathematics course, including basic mathematics, prealgebra, beginning algebra, and intermediate algebra.

Elayn is the author of 12 published textbooks as well as multimedia, interactive mathematics, all specializing in developmental mathematics courses. She has also published series in Algebra 1, Algebra 2, and Geometry. She has participated as an author across the broadest range of educational materials: textbooks, videos, tutorial software, and courseware. This provides an opportunity of various combinations for an integrated teaching and learning package offering great consistency for the student.

Applications Index

The Whole Numbers

A Selection of Resources for Success in This Mathematics Course

Textbook

Instructor

MyMathLab and MathXL

Video Organizer

Student Organizer

Interactive Lecture Series

For more information about the resources illustrated above, read Section 1.1.

Whole numbers are the basic building blocks of mathematics. The whole numbers answer the question "How many?"

This chapter covers basic operations on whole numbers. Knowledge of these operations provides a good foundation on which to build further mathematical skills.

1.1 Study Skill Tips for Success in Mathematics

Objectives

A Get Ready for This Course.

B Understand Some General Tips for Success.

C Know How to Use This Text.

D Know How to Use Video and Notebook Organizer Resources.

E Get Help as Soon as You Need It.

F Learn How to Prepare for and Take an Exam.

G Develop Good Time Management.

Before reading Section 1.1, you might want to ask yourself a few questions.

1. When you took your last math course, were you organized? Were your notes and materials from that course easy to find, or were they disorganized and hard to find—if you saved them at all?

2. Were you satisfied—really satisfied—with your performance in that course? In other words, do you feel that your outcome represented your best effort?

If the answer is "no" to these questions, then it is time to make a change. Changing to or resuming good study skill habits is not a process you can start and stop as you please. It is something that you must remember and practice each and every day. To begin, continue reading this section.

Objective A Getting Ready for This Course

Now that you have decided to take this course, remember that a *positive attitude* will make all the difference in the world. Your belief that you can succeed is just as important as your commitment to this course. Make sure you are ready for this course by having the time and positive attitude that it takes to succeed.

Make sure that you are familiar with the way that this course is being taught. Is it a traditional course, in which you have a printed textbook and meet with an instructor? Is it taught totally online, and your textbook is electronic and you e-mail your instructor? Or is your course structured somewhere in between these two methods? (Not all of the tips that follow will apply to all forms of instruction.)

Also make sure that you have scheduled your math course for a time that will give you the best chance for success. For example, if you are also working, you may want to check with your employer to make sure that your work hours will not conflict with your course schedule.

On the day of your first class period, double-check your schedule and allow yourself extra time to arrive on time in case of traffic problems or difficulty locating your classroom. Make sure that you are aware of and bring all necessary class materials.

Objective B General Tips for Success

Below are some general tips that will increase your chance for success in a mathematics class. Many of these tips will also help you in other courses you may be taking.

Most important! Organize your class materials. In the next couple pages, many ideas will be presented to help you organize your class materials—notes, any handouts, completed homework, previous tests, etc. In general, you MUST have these materials organized. All of them will be valuable references throughout your course and when studying for upcoming tests and the final exam. One way to make sure you can locate these materials when you need them is to use a three-ring binder. This binder should be used solely for your mathematics class and should be brought to each and every class and/or lab. This way, any material can be immediately inserted in a section of this binder and will be there when you need it.

Form study groups and/or exchange names and e-mail addresses. Depending on how your course is taught, you may want to keep in contact with your fellow students. Some ways of doing this are to form a study group—whether in person or through the Internet. Also, you may want to ask if anyone is interested in exchanging e-mail addresses or any other form of contact.

> **Helpful Hint**
>
> **MyMathLab® and MathXL®**
> When assignments are turned in online, keep a hard copy of your complete written work. You will need to refer to your written work to be able to ask questions and to study for tests later.

Choose to attend all class periods. If possible, sit near the front of the classroom. This way, you will see and hear the presentation better. It may also be easier for you to participate in classroom activities.

Do your homework. You've probably heard the phrase "practice makes perfect" in relation to music and sports. It also applies to mathematics. You will find that the more time you spend solving mathematics exercises, the easier the process becomes. Be sure to schedule enough time to complete your assignments before the due date assigned by your instructor.

Check your work. Review the steps you took while working a problem. Learn to check your answers in the original exercises. You may also compare your answers with the "Answers to Selected Exercises" section in the back of the book. If you have made a mistake, try to figure out what went wrong. Then correct your mistake. If you can't find what went wrong, **don't** erase your work or throw it away. Show your work to your instructor, a tutor in a math lab, or a classmate. It is easier for someone to find where you had trouble if he or she looks at your original work.

Learn from your mistakes and be patient with yourself. Everyone, even your instructor, makes mistakes. (That definitely includes me—Elayn Martin-Gay.) Use your errors to learn and to become a better math student. The key is finding and understanding your errors.

Was your mistake a careless one, or did you make it because you can't read your own math writing? If so, try to work more slowly or write more neatly and make a conscious effort to carefully check your work.

Did you make a mistake because you don't understand a concept? Take the time to review the concept or ask questions to better understand it.

Did you skip too many steps? Skipping steps or trying to do too many steps mentally may lead to preventable mistakes.

Know how to get help if you need it. It's all right to ask for help. In fact, it's a good idea to ask for help whenever there is something that you don't understand. Make sure you know when your instructor has office hours and how to find his or her office. Find out whether math tutoring services are available on your campus. Check on the hours, location, and requirements of the tutoring service.

Don't be afraid to ask questions. You are not the only person in class with questions. Other students are normally grateful that someone has spoken up.

Turn in assignments on time. This way, you can be sure that you will not lose points for being late. Show every step of a problem and be neat and organized. Also be sure that you understand which problems are assigned for homework. If allowed, you can always double-check the assignment with another student in your class.

Objective **C** Knowing and Using Your Text ▶

Flip through the pages of this text or view the e-text pages on a computer screen. Start noticing examples, exercise sets, end-of-chapter material, and so on. Every text is organized in some manner. Learn the way this text is organized by reading about and then finding an example in your text of each type of resource listed below. Finding and using these resources throughout your course will increase your chance of success.

- *Practice Exercises.* Each example in every section has a parallel Practice exercise. As you read a section, try each Practice exercise after you've finished the corresponding example. Answers are at the bottom of the page. This "learn-by-doing" approach will help you grasp ideas before you move on to other concepts.

- *Symbols at the Beginning of an Exercise Set.* If you need help with a particular section, the symbols listed at the beginning of each exercise set will remind you of the resources available.

Helpful Hint

MyMathLab® and MathXL®
If you are doing your homework online, you can work and re-work those exercises that you struggle with until you master them. Try working through all the assigned exercises twice before the due date.

Helpful Hint

MyMathLab® and MathXL®
If you are completing your homework online, it's important to work each exercise on paper before submitting the answer. That way, you can check your work and follow your steps to find and correct any mistakes.

Helpful Hint

MyMathLab® and MathXL®
Be aware of assignments and due dates set by your instructor. Don't wait until the last minute to submit work online.

- *Objectives.* The main section of exercises in each exercise set is referenced by an objective, such as **A** or **B**, and also an example(s). There is also often a section of exercises entitled "Mixed Practice," which is referenced by two or more objectives or sections. These are mixed exercises written to prepare you for your next exam. Use all of this referencing if you have trouble completing an assignment from the exercise set.

- *Icons (Symbols).* Make sure that you understand the meaning of the icons that are beside many exercises. ▶ tells you that the corresponding exercise may be viewed on the video Lecture Series that corresponds to that section. ✎ tells you that this exercise is a writing exercise in which you should answer in complete sentences. △ tells you that the exercise involves geometry.

- *Integrated Reviews.* Found in the middle of each chapter, these reviews offer you a chance to practice—in one place—the many concepts that you have learned separately over several sections.

- *End-of-Chapter Opportunities.* There are many opportunities at the end of each chapter to help you understand the concepts of the chapter.

  Vocabulary Checks contain key vocabulary terms introduced in the chapter.

  Chapter Highlights contain chapter summaries and examples.

  Chapter Reviews contain review problems. The first part is organized section by section and the second part contains a set of mixed exercises.

  Chapter Tests are sample tests to help you prepare for an exam. The Chapter Test Prep Videos found in the Interactive Lecture Series, MyMathLab, and YouTube provide the video solution to each question on each Chapter Test.

  Cumulative Reviews start at Chapter 2 and are reviews consisting of material from the beginning of the book to the end of that particular chapter.

- *Student Resources in Your Textbook.* You will find a **Student Resources** section at the back of this textbook. It contains the following to help you study and prepare for tests:

  Study Skill Builders contain study skills advice. To increase your chance for success in the course, read these study tips and answer the questions.

  Bigger Picture—Study Guide Outline provides you with a study guide outline of the course, with examples.

  Practice Final provides you with a Practice Final Exam to help you prepare for a final.

- *Resources to Check Your Work.* The **Answers to Selected Exercises** section provides answers to all odd-numbered section exercises and to all integrated review, chapter review, chapter test, and cumulative review exercises. Use the **Solutions to Selected Exercises** to see the worked-out solution to every other odd-numbered exercise.

Objective **D** Knowing and Using Video and Notebook Organizer Resources ▶

Video Resources

Below is a list of video resources that are all made by me—the author of your text, Elayn Martin-Gay. By making these videos, I can be sure that the methods presented are consistent with those in the text.

- *Interactive DVD Lecture Series.* Exercises marked with a ▶ are fully worked out by the author on the DVDs and within MyMathLab. The lecture series provides approximately 20 minutes of instruction per section and is organized by Objective.

> **Helpful Hint**
>
> **MyMathLab®**
>
> In MyMathLab, you have access to the following video resources:
> - Lecture Videos for each section
> - Chapter Test Prep Videos
>
> Use these videos provided by the author to prepare for class, review, and study for tests.

- *Chapter Test Prep Videos.* These videos provide solutions to all of the Chapter Test exercises worked out by the author. They can be found in MyMathLab, the Interactive Lecture series, and You Tube. This supplement is very helpful before a test or exam.
- *Student Success Tips.* These video segments are about 3 minutes long and are daily reminders to help you continue practicing and maintaining good organizational and study habits.
- *Final Exam Videos.* These video segments provide solutions to each question. These videos can be found within MyMathLab and the Interactive Lecture Series.

Notebook Organizer Resources

The resources below are in three-ring notebook ready form. They are to be inserted in a three-ring binder and completed. Both resources are numbered according to the sections in your text to which they refer.

- *Video Organizer.* This organizer is closely tied to the Interactive Lecture (Video) Series. Each section should be completed while watching the lecture video on the same section. Once completed, you will have a set of notes to accompany the Lecture (Video) Series section by section.
- *Student Organizer.* This organizer helps you study effectively through note-taking hints, practice, and homework while referencing examples in the text and examples in the Lecture Series.

Objective E Getting Help

If you have trouble completing assignments or understanding the mathematics, get help as soon as you need it! This tip is presented as an objective on its own because it is so important. In mathematics, usually the material presented in one section builds on your understanding of the previous section. This means that if you don't understand the concepts covered during a class period, there is a good chance that you will not understand the concepts covered during the next class period. If this happens to you, get help as soon as you can.

Where can you get help? Many suggestions have been made in this section on where to get help, and now it is up to you to get it. Try your instructor, a tutoring center, or a math lab, or you may want to form a study group with fellow classmates. If you do decide to see your instructor or go to a tutoring center, make sure that you have a neat notebook and are ready with your questions.

Helpful Hint

MyMathLab® and MathXL®

- Use the **Help Me Solve This** button to get step-by-step help for the exercise you are working. You will need to work an additional exercise of the same type before you can get credit for having worked it correctly.
- Use the **Video** button to view a video clip of the author working a similar exercise.

Objective F Preparing for and Taking an Exam

Make sure that you allow yourself plenty of time to prepare for a test. If you think that you are a little "math anxious," it may be that you are not preparing for a test in a way that will ensure success. The way that you prepare for a test in mathematics is important. To prepare for a test:

1. Review your previous homework assignments.
2. Review any notes from class and section-level quizzes you have taken. (If this is a final exam, also review chapter tests you have taken.)
3. Review concepts and definitions by reading the Chapter Highlights at the end of each chapter.
4. Practice working out exercises by completing the Chapter Review found at the end of each chapter. (If this is a final exam, go through a Cumulative Review. There is one found at the end of each chapter except Chapter 1. Choose the review found at the end of the latest chapter that you have covered in your course.) *Don't stop here!*

Helpful Hint

MyMathLab® and MathXL®

Review your written work for previous assignments. Then, go back and re-work previous assignments. Open a previous assignment, and click **Similar Exercise** to generate new exercises. Re-work the exercises until you fully understand them and can work them without help features.

5. It is important that you place yourself in conditions similar to test conditions to find out how you will perform. In other words, as soon as you feel that you know the material, get a few blank sheets of paper and take a sample test. There is a Chapter Test available at the end of each chapter, or you can work selected problems from the Chapter Review. Your instructor may also provide you with a review sheet. During this sample test, do not use your notes or your textbook. Then check your sample test. If your sample test is the Chapter Test in the text, don't forget that the video solutions are in MyMathLab, the Interactive Lecture Series, and YouTube. If you are not satisfied with the results, study the areas that you are weak in and try again.

6. On the day of the test, allow yourself plenty of time to arrive at where you will be taking your exam.

When taking your test:

1. Read the directions on the test carefully.

2. Read each problem carefully as you take the test. Make sure that you answer the question asked.

3. Watch your time and pace yourself so that you can attempt each problem on your test.

4. If you have time, check your work and answers.

5. Do not turn your test in early. If you have extra time, spend it double-checking your work.

Objective G Managing Your Time

As a college student, you know the demands that classes, homework, work, and family place on your time. Some days you probably wonder how you'll ever get everything done. One key to managing your time is developing a schedule. Here are some hints for making a schedule:

1. Make a list of all of your weekly commitments for the term. Include classes, work, regular meetings, extracurricular activities, etc. You may also find it helpful to list such things as laundry, regular workouts, grocery shopping, etc.

2. Next, estimate the time needed for each item on the list. Also make a note of how often you will need to do each item. Don't forget to include time estimates for the reading, studying, and homework you do outside of your classes. You may want to ask your instructor for help estimating the time needed.

3. In the exercise set that follows, you are asked to block out a typical week on the schedule grid given. Start with items with fixed time slots like classes and work.

4. Next, include the items on your list with flexible time slots. Think carefully about how best to schedule items such as study time.

5. Don't fill up every time slot on the schedule. Remember that you need to allow time for eating, sleeping, and relaxing! You should also allow a little extra time in case some items take longer than planned.

6. If you find that your weekly schedule is too full for you to handle, you may need to make some changes in your workload, classload, or other areas of your life. You may want to talk to your advisor, manager or supervisor at work, or someone in your college's academic counseling center for help with such decisions.

1.1 Exercise Set MyMathLab® ▶

1. What is your instructor's name?

2. What are your instructor's office location and office hours?

3. What is the best way to contact your instructor?

4. Do you have the name and contact information of at least one other student in class?

5. Will your instructor allow you to use a calculator in this class?

6. Why is it important that you write step-by-step solutions to homework exercises and keep a hard copy of all work submitted online?

7. Is there a tutoring service available on campus? If so, what are its hours? What services are available?

8. Have you attempted this course before? If so, write down ways that you might improve your chances of success during this attempt.

9. List some steps that you can take if you begin having trouble understanding the material or completing an assignment. If you are completing your homework in MyMathLab® and MathXL®, list the resources you can use for help.

10. How many hours of studying does your instructor advise for each hour of instruction?

11. What does the ╲ icon in this text mean?

12. What does the △ icon in this text mean?

13. What does the ▶ icon in this text mean?

14. Search the minor columns in your text. What are Practice exercises?

15. When might be the best time to work a Practice exercise?

16. Where are the answers to Practice exercises?

17. What answers are contained in this text and where are they?

18. What are Study Skill Tips of the Day and where are they?

19. What and where are Integrated Reviews?

20. How many times is it suggested that you work through the homework exercises in MathXL® before the submission deadline?

21. How far in advance of the assigned due date is it suggested that homework be submitted online? Why?

22. Chapter Highlights are found at the end of each chapter. Find the Chapter 1 Highlights and explain how you might use it and how it might be helpful.

23. Chapter Reviews are found at the end of each chapter. Find the Chapter 1 Review and explain how you might use it and how it might be helpful.

24. Chapter Tests are found at the end of each chapter. Find the Chapter 1 Test and explain how you might use it and how it might be helpful when preparing for an exam on Chapter 1. Include how the Chapter Test Prep Videos may help. If you are working in MyMathLab® and MathXL®, how can you use previous homework assignments to study?

25. What is the Video Organizer? Explain the contents and how it might be used.

26. What is the Student Organizer? Explain the contents and how it might be used.

27. Read or reread objective **G** and fill out the schedule grid on the next page.

	Monday	Tuesday	Wednesday	Thursday	Friday	Saturday	Sunday
4:00 a.m.							
5:00 a.m.							
6:00 a.m.							
7:00 a.m.							
8:00 a.m.							
9:00 a.m.							
10:00 a.m.							
11:00 a.m.							
12:00 p.m.							
1:00 p.m.							
2:00 p.m.							
3:00 p.m.							
4:00 p.m.							
5:00 p.m.							
6:00 p.m.							
7:00 p.m.							
8:00 p.m.							
9:00 p.m.							
10:00 p.m.							
11:00 p.m.							
Midnight							
1:00 a.m.							
2:00 a.m.							
3:00 a.m.							

1.2 Place Value, Names for Numbers, and Reading Tables

Objectives

A Find the Place Value of a Digit in a Whole Number.

B Write a Whole Number in Words and in Standard Form.

C Write a Whole Number in Expanded Form.

D Read Tables.

The **digits** 0, 1, 2, 3, 4, 5, 6, 7, 8, and 9 can be used to write numbers. For example, the **whole numbers** are

0, 1, 2, 3, 4, 5, 6, 7, 8, 9, 10, 11, . . .

and the **natural numbers** are 1, 2, 3, 4, 5, 6, 7, 8, 9, 10, 11, . . .

The three dots (. . .) after each 11 means that these lists continue indefinitely. That is, there is no largest whole number. The smallest whole number is 0. Also, there is no largest natural number. The smallest natural number is 1.

Objective A Finding the Place Value of a Digit in a Whole Number

The position of each digit in a number determines its **place value.** For example, the distance (in miles) between the planet Mercury and the planet Earth can be represented by the whole number 48,337,000. Next is a place-value chart for this whole number.

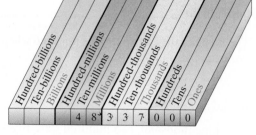

The two 3s in 48,337,000 represent different amounts because of their different placements. The place value of the 3 on the left is hundred-thousands. The place value of the 3 on the right is ten-thousands.

Examples   Find the place value of the digit 3 in each whole number.

1. 396,418
↑
hundred-thousands

2. 93,192
↑
thousands

3. 534,275,866
↑
ten-millions

🔲 Work Practice 1–3

Practice 1–3

Find the place value of the digit 8 in each whole number.
1. 38,760,005
2. 67,890
3. 481,922

Objective B Writing a Whole Number in Words and in Standard Form ▶

A whole number such as 1,083,664,500 is written in **standard form.** Notice that commas separate the digits into groups of three, starting from the right. Each group of three digits is called a **period.** The names of the first four periods are shown in red.

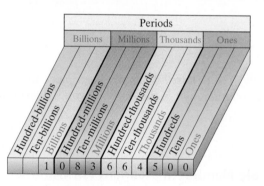

Writing a Whole Number in Words

To write a whole number in words, write the number in each period followed by the name of the period. (The ones period is usually not written.) This same procedure can be used to read a whole number.

For example, we write 1,083,664,500 as

one **billion,**
eighty-three **million,**
six hundred sixty-four **thousand,**
five hundred

> **Helpful Hint**   Notice the commas after the name of each period.

Answers
1. millions   **2.** hundreds
3. ten-thousands

Helpful Hint

The name of the ones period is not used when reading and writing whole numbers. For example,

9,265

is read as

"nine **thousand,** two hundred sixty-five."

Practice 4–6

Write each number in words.

4. 67

5. 395

6. 12,804

Examples Write each number in words.

4. 85   eighty-five

5. 126   one hundred twenty-six

6. 27,034   twenty-seven thousand, thirty-four

■ Work Practice 4–6

Helpful Hint   The word "and" is *not* used when reading and writing whole numbers. It is used when reading and writing mixed numbers and some decimal values, as shown later in this text.

Practice 7

Write 321,670,200 in words.

Example 7   Write 106,052,447 in words.

Solution: 106,052,447 is written as

one hundred six **million**, fifty-two **thousand**, four hundred forty-seven

■ Work Practice 7

✓**Concept Check** True or false? When writing a check for $2600, the word name we write for the dollar amount of the check is "two thousand sixty." Explain your answer.

Practice 8–11

Write each number in standard form.

8. twenty-nine

9. seven hundred ten

10. twenty-six thousand, seventy-one

11. six million, five hundred seven

Answers

4. sixty-seven **5.** three hundred ninety-five **6.** twelve thousand, eight hundred four **7.** three hundred twenty-one million, six hundred seventy thousand, two hundred **8.** 29 **9.** 710 **10.** 26,071 **11.** 6,000,507

✓**Concept Check Answer**

false

Writing a Whole Number in Standard Form

To write a whole number in standard form, write the number in each period, followed by a comma.

Examples   Write each number in standard form.

8. sixty-one   61   **9.** eight hundred five   805

10. nine thousand, three hundred eighty-six

9,386   or   9386

11. two million, five hundred sixty-four thousand, three hundred fifty

2,564,350

■ Work Practice 8–11

Helpful Hint

A comma may or may not be inserted in a four-digit number. For example, both

9,386   and   9386

are acceptable ways of writing nine thousand, three hundred eighty-six.

Objective C Writing a Whole Number in Expanded Form

The place value of a digit can be used to write a number in expanded form. The **expanded form** of a number shows each digit of the number with its place value. For example, 5672 is written in expanded form as

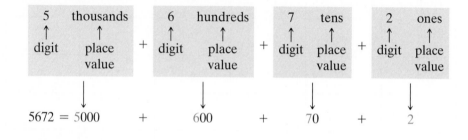

Example 12   Write 2,706,449 in expanded form.

Solution: 2,000,000 + 700,000 + 6000 + 400 + 40 + 9

■ Work Practice 12

Practice 12

Write 1,047,608 in expanded form.

We can visualize whole numbers by points on a line. The line below is called a **number line.** This number line has equally spaced marks for each whole number. The arrow to the right simply means that the whole numbers continue indefinitely. In other words, there is no largest whole number.

Number Line

We will study number lines further in Section 1.5.

Objective D Reading Tables

Now that we know about place value and names for whole numbers, we introduce one way that whole numbers may be presented. **Tables** are often used to organize and display facts that involve numbers. The table on the next page shows the ten countries with the most Nobel Prize winners since the inception of the Nobel Prize in 1901, and the categories of the prizes. The numbers for the Economics prize reflect the winners since 1969, when this category was established. (The numbers may seem large for two reasons: first, the annual Nobel Prize is often awarded to more than one individual, and second, several award winners hold dual citizenship, so they are counted in two countries.)

Answer
12. 1,000,000 + 40,000 + 7000 + 600 + 8

Countries with Most Nobel Prize Winners, 1901–2013

Country	Chemistry	Economics	Literature	Peace	Physics	Physiology & Medicine	Total
United States	68	54	11	22	90	97	342
United Kingdom	27	8	8	9	23	31	106
Germany	27	1	8	4	22	17	79
France	7	1	15	8	13	11	55
Sweden	4	2	9	5	4	8	32
Switzerland	6	0	2	3	4	6	21
Russia (USSR)	1	1	4	2	10	2	20
Japan	6	0	2	1	6	2	17
Netherlands	3	1	0	1	9	2	16
Italy	1	0	6	1	4	3	15

Source: Based on data from Encyclopaedia Britannica, Inc.

For example, by reading from left to right along the row marked "United States," we find that the United States has 68 Chemistry, 54 Economics, 11 Literature, 22 Peace, 90 Physics, and 97 Physiology and Medicine Nobel Prize winners.

Practice 13

Use the Nobel Prize Winner table to answer the following questions:

a. How many Nobel Prize winners in Literature come from France?

b. Which countries shown have more than 60 Nobel Prize winners?

Answers

13. a. 15   **b.** United States, United Kingdom, and Germany

Example 13   Use the Nobel Prize Winner table to answer each question.

a. How many total Nobel Prize winners are from Sweden?

b. Which countries shown have fewer Nobel Prize winners than Russia?

Solution:

a. Find "Sweden" in the left column. Then read from left to right until the "Total" column is reached. We find that Sweden has 32 Nobel Prize winners.

b. Russia has 20 Nobel Prize winners. Japan has 17, Netherlands has 16, and Italy has 15, so they have fewer Nobel Prize winners than Russia.

 Work Practice 13

Vocabulary, Readiness & Video Check

Use the choices below to fill in each blank.

standard form	period	whole
expanded form	place value	words

1. The numbers 0, 1, 2, 3, 4, 5, 6, 7, 8, 9, 10, 11, 12, … are called _____ numbers.

2. The number 1,286 is written in _____.

3. The number "twenty-one" is written in _____.

4. The number 900 + 60 + 5 is written in _____.

5. In a whole number, each group of three digits is called a(n) _____.

6. The _____ of the digit 4 in the whole number 264 is ones.

Martin-Gay Interactive Videos   *Watch the section lecture video and answer the following questions.*

Objective A 7. In ▣ Example 1, what is the place value of the digit 6? ▶

Objective B 8. Complete this statement based on ▣ Example 3: To read (or write) a number, read from _____ to _____. ▶

Objective C 9. In ▣ Example 5, what is the expanded-form value of the digit 8? ▶

Objective D 10. Use the table given in ▣ Example 6 to determine which mountain in the table is the shortest. ▶

See Video 1.2

1.2 **Exercise Set** MyMathLab® ▶

Objective **A** *Determine the place value of the digit 5 in each whole number. See Examples 1 through 3.*

▶ **1.** 657

2. 905

▶ **3.** 5423

4. 6527

5. 43,526,000

6. 79,050,000

7. 5,408,092

8. 51,682,700

Objective **B** *Write each whole number in words. See Examples 4 through 7.*

9. 354

10. 316

11. 8279

12. 5445

▶ **13.** 26,990

14. 42,009

15. 2,388,000

16. 3,204,000

17. 24,350,185

18. 47,033,107

Write each number in the sentence in words. See Examples 4 through 7.

19. As of July 2013, the population of Iceland was 315,281. (*Source:* CIA World Factbook)

20. The land area of Belize is 22,806 square kilometers. (*Source:* CIA World Factbook)

21. The Burj Khalifa, in Dubai, United Arab Emirates, a hotel and office building, is the world's tallest building at a height of 2717 feet. (*Source:* Council on Tall Buildings and Urban Habitat)

22. As of October 2013, there were 119,948 patients in the United States waiting for an organ transplant. (*Source:* Organ Procurement and Transplantation Network)

23. In 2012, UPS received an average of 32,100,000 online tracking requests each day. (*Source:* UPS)

24. Each year, an estimated 500,000,000 Americans visit carnivals, fairs, and festivals. (*Source:* Outdoor Amusement Business Association)

25. The highest point in Colorado is Mount Elbert, at an elevation of 14,433 feet. (*Source:* U.S. Geological Survey)

26. The highest point in Oregon is Mount Hood, at an elevation of 11,239 feet. (*Source:* U.S. Geological Survey)

27. In 2013, the Great Internet Mersenne Prime Search, a cooperative computing project, helped find a prime number that has over 17,000,000 digits. (*Source:* Mersenne Research, Inc.)

28. The Goodyear blimp *Eagle* holds 202,700 cubic feet of helium. (*Source:* The Goodyear Tire & Rubber Company)

Write each whole number in standard form. See Examples 8 through 11.

29. Six thousand, five hundred eighty-seven

30. Four thousand, four hundred sixty-eight

31. Fifty-nine thousand, eight hundred

32. Seventy-three thousand, two

33. Thirteen million, six hundred one thousand, eleven

34. Sixteen million, four hundred five thousand, sixteen

35. Seven million, seventeen

36. Two million, twelve

37. Two hundred sixty thousand, nine hundred ninety-seven

38. Six hundred forty thousand, eight hundred eighty-one

Write the whole number in each sentence in standard form. See Examples 8 through 11.

39. After an orbit correction in October 2013, the International Space Station orbited Earth at an average altitude of about four hundred eighteen kilometers. (*Source:* Heavens Above)

40. The average distance between the surfaces of Earth and the Moon is about two hundred thirty-four thousand miles.

41. La Rinconada, Peru, is the highest town in the world. It is located sixteen thousand, seven hundred thirty-two feet above sea level. (*Source:* Russell Ash: *Top 10 of Everything*)

42. The world's tallest freestanding tower is the Tokyo Sky Tree in Japan. Its height is two thousand eighty feet tall. (*Source:* Council on Tall Buildings and Urban Habitat)

43. The Warner Bros. film *Harry Potter and the Deathly Hallows Part 2* holds the record for U.S./ Canada opening day box office gross when it took in approximately ninety-one million, seventy-one thousand dollars on its opening day in 2011. (*Source:* Box Office Mojo)

44. The Buena Vista film *Marvel's The Avengers* set the U.S./Canada record for second-highest opening day box office gross when it took in approximately eighty million, eight hundred fourteen thousand dollars on its opening day in 2012. (*Source:* Box Office Mojo)

45. In 2012, the UPS delivery fleet consisted of one hundred one thousand vehicles. (*Source:* UPS)

46. Morten Andersen, who played football for New Orleans, Atlanta, N.Y. Giants, Kansas City, and Minnesota between 1982 and 2007, holds the record for the most points scored in a career. Over his 25-year career he scored two thousand, five hundred forty-four points. (*Source:* NFL.com)

Objective C *Write each whole number in expanded form. See Example 12.*

47. 406   **48.** 789   **49.** 3470   **50.** 6040

51. 80,774   **52.** 20,215   **53.** 66,049   **54.** 99,032

55. 39,680,000   **56.** 47,703,029

Objectives B C D Mixed Practice *The table shows the six tallest mountains in New England and their elevations. Use this table to answer Exercises 57 through 62. See Example 13.*

Mountain (State)	Elevation (in feet)
Boott Spur (NH)	5492
Mt. Adams (NH)	5774
Mt. Clay (NH)	5532
Mt. Jefferson (NH)	5712
Mt. Sam Adams (NH)	5584
Mt. Washington (NH)	6288
Source: U.S. Geological Survey	

Elevation in feet

57. Write the elevation of Mt. Clay in standard form and then in words.

58. Write the elevation of Mt. Washington in standard form and then in words.

59. Write the height of Boott Spur in expanded form.

60. Write the height of Mt. Jefferson in expanded form.

61. Which mountain is the tallest in New England?

62. Which mountain is the second tallest in New England?

The table shows the top ten museums in the world in 2012. Use this table to answer Exercises 63 through 68. See Example 13.

Top 10 Museums Worldwide in 2012		
Museum	**Location**	**Visitors**
Louvre	Paris, France	9,720,000
National Museum of Natural History	Washington, DC, United States	7,600,000
National Air and Space Museum	Washington, DC, United States	6,800,000
The Metropolitan Museum of Art	New York, NY, United States	6,116,000
British Museum	London, United Kingdom	5,576,000
Tate Modern	London, United Kingdom	5,319,000
National Gallery	London, United Kingdom	5,164,000
Vatican Museums	Vatican City	5,065,000
American Museum of Natural History	New York, NY, United States	5,000,000
Natural History Museum	London, United Kingdom	4,936,000

(*Source:* Themed Entertainment Association)

63. Which museum had fewer visitors, the National Gallery in London or the National Air and Space Museum in Washington, DC?

64. Which museum had more visitors, the British Museum in London or the Louvre in Paris?

65. How many people visited the Vatican Museums? Write the number of visitors in words.

66. How many people visited The Metropolitan Museum of Art? Write the number of visitors in words.

67. How many of 2012's top ten museums in the world were located in the United States?

68. How many of 2012's top ten museums in the world were visited by fewer than 6,000,000 people?

Concept Extensions

69. Write the largest four-digit number that can be made from the digits 1, 9, 8, and 6 if each digit must be used once. ____ ____ ____ ____

70. Write the largest five-digit number that can be made using the digits 5, 3, and 7 if each digit must be used at least once. ____ ____ ____ ____ ____

Check to see whether each number written in standard form matches the number written in words. If not, correct the number in words. See the Concept Check in this section.

71.

60–8124/7233 1000613331	1401
DATE _____	
PAY TO THE ORDER OF	$ 105.00
One Hundred Fifty and 00/100 ——— DOLLARS	
FIRST STATE BANK OF FARTHINGTON FARTHINGTON, IL 64422	
MEMO _____	_____
⑆621497260⑆ 1000613331⑆ 1401	

72.

60–8124/7233 1000613331	1402
DATE _____	
PAY TO THE ORDER OF	$ 7030.00
Seven Thousand Thirty and 00/100 ——— DOLLARS	
FIRST STATE BANK OF FARTHINGTON FARTHINGTON, IL 64422	
MEMO _____	_____
⑆621497260⑆ 1000613331⑆ 1402	

73. If a number is given in words, describe the process used to write this number in standard form.

74. If a number is written in standard form, describe the process used to write this number in expanded form.

75. In June 2013, the MilkyWay-2, a high-speed computer built by China's National University of Defense Technology, was ranked as the world's fastest computer. Its speed was clocked at nearly 34 petaflops, or more than 34 quadrillion arithmetic operations per second. Look up "quadrillion" (in the American system) and use the definition to write this number in standard form. (*Source:* top500.org)

76. As of December 2012, the national debt of France was approximately $5 trillion. Look up "trillion" (in the American system) and use the definition to write 5 trillion in standard form. (*Source:* CIA World Factbook)

77. The Pro Football Hall of Fame was established on September 7, 1963, in this town. Use the information and the diagram to the right to find the name of the town.

- Alliance is east of Massillon.
- Dover is between Canton and New Philadelphia.
- Massillon is not next to Alliance.
- Canton is north of Dover.

N W E S

Pro Football Hall of Fame

OHIO

Adding Whole Numbers and Perimeter

Objective A Adding Whole Numbers

According to Gizmodo, the iPod nano (currently in its seventh generation) is still the best overall MP3 player.

Suppose that an electronics store received a shipment of two boxes of iPod nanos one day and an additional four boxes of iPod nanos the next day. The **total** shipment in the two days can be found by adding 2 and 4.

2 boxes of iPod nanos + 4 boxes of iPod nanos = 6 boxes of iPod nanos

The **sum** (or total) is 6 boxes of iPod nanos. Each of the numbers 2 and 4 is called an **addend,** and the process of finding the sum is called **addition.**

$$\underset{\text{addend}}{2} \; + \; \underset{\text{addend}}{4} \; + \; \underset{\text{sum}}{6}$$

To add whole numbers, we add the digits in the ones place, then the tens place, then the hundreds place, and so on. For example, let's add 2236 + 160.

$$\begin{array}{r} 2236 \\ + \; 160 \\ \hline 2396 \end{array}$$

Line up numbers vertically so that the place values correspond. Then add digits in corresponding place values, starting with the ones place.

— sum of ones
— sum of tens
— sum of hundreds
— sum of thousands

| **Example 1** | Add: 23 + 136 |

Solution:
$$\begin{array}{r} 23 \\ + \; 136 \\ \hline 159 \end{array}$$

■ Work Practice 1

When the sum of digits in corresponding place values is more than 9, **carrying** is necessary. For example, to add 365 + 89, add the ones-place digits first.

Carrying
$$\begin{array}{r} \overset{1}{3}65 \\ + \quad 89 \\ \hline 4 \end{array}$$
5 ones + 9 ones = **14 ones** or **1 ten** + **4 ones**
Write the 4 ones in the ones place and carry the 1 ten to the tens place.

Next, add the tens-place digits.
$$\begin{array}{r} \overset{1}{\overset{1}{3}}65 \\ + \quad 89 \\ \hline 54 \end{array}$$
1 ten + 6 tens + 8 tens = **15 tens** or **1 hundred** + **5 tens**
Write the 5 tens in the tens place and carry the 1 hundred to the hundreds place.

Next, add the hundreds-place digits.
$$\begin{array}{r} \overset{1\,1}{3}65 \\ + \quad 89 \\ \hline 454 \end{array}$$
1 hundred + 3 hundreds = **4 hundreds**
Write the 4 hundreds in the hundreds place.

Objectives

A Add Whole Numbers.

B Find the Perimeter of a Polygon.

C Solve Problems by Adding Whole Numbers.

Practice 1

Add: 7235 + 542

Answer
1. 7777

17

Practice 2

Add: 27,364 + 92,977

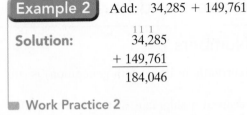

Example 2 Add: 34,285 + 149,761

Solution:

$$
\begin{array}{r}
\overset{1\ 1\ \ 1}{34{,}285} \\
+\ 149{,}761 \\
\hline
184{,}046
\end{array}
$$

■ Work Practice 2

✓**Concept Check** What is wrong with the following computation?

$$
\begin{array}{r}
394 \\
+\ 283 \\
\hline
577
\end{array}
$$

Before we continue adding whole numbers, let's review some properties of addition that you may have already discovered. The first property that we will review is the **addition property of 0.** This property reminds us that the sum of 0 and any number is that same number.

Addition Property of 0

The sum of 0 and any number is that number. For example,

$$7 + 0 = 7$$
$$0 + 7 = 7$$

Next, notice that we can add any two whole numbers in any order and the sum is the same. For example,

$$4 + 5 = 9 \quad \text{and} \quad 5 + 4 = 9$$

We call this special property of addition the **commutative property of addition.**

Commutative Property of Addition

Changing the **order** of two addends does not change their sum. For example,

$$2 + 3 = 5 \quad \text{and} \quad 3 + 2 = 5$$

Another property that can help us when adding numbers is the **associative property of addition.** This property states that when adding numbers, the grouping of the numbers can be changed without changing the sum. We use parentheses to group numbers. They indicate which numbers to add first. For example, let's use two different groupings to find the sum of 2 + 1 + 5.

$$(2 + 1) + 5 = 3 + 5 = 8$$

Also,

$$2 + (1 + 5) = 2 + 6 = 8$$

Both groupings give a sum of 8.

Answer

2. 120,341

✓**Concept Check Answer**

forgot to carry 1 hundred to the hundreds place

Associative Property of Addition

Changing the **grouping** of addends does not change their sum. For example,

$$3 + (5 + 7) = 3 + 12 = 15 \quad \text{and} \quad (3 + 5) + 7 = 8 + 7 = 15$$

The commutative and associative properties tell us that we can add whole numbers using any order and grouping that we want.

When adding several numbers, it is often helpful to look for two or three numbers whose sum is 10, 20, and so on. Why? Adding multiples of 10 such as 10 and 20 is easier.

Example 3   Add: $13 + 2 + 7 + 8 + 9$

Solution:

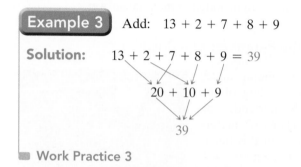

$13 + 2 + 7 + 8 + 9 = 39$

$20 + 10 + 9$

39

Practice 3

Add: $11 + 7 + 8 + 9 + 13$

■ Work Practice 3

Feel free to use the process of Example 3 anytime when adding.

Example 4   Add: $1647 + 246 + 32 + 85$

Solution:

$$
\begin{array}{r}
\overset{1\,2\,2}{1647} \\
246 \\
32 \\
+\ \ 85 \\
\hline
2010
\end{array}
$$

Practice 4

Add: $19 + 5042 + 638 + 526$

■ Work Practice 4

Objective B Finding the Perimeter of a Polygon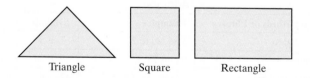

In geometry addition is used to find the perimeter of a polygon. A **polygon** can be described as a flat figure formed by line segments connected at their ends. (For more review, see Appendix A.3.) Geometric figures such as triangles, squares, and rectangles are called polygons.

Triangle   Square   Rectangle

The **perimeter** of a polygon is the *distance around* the polygon. This means that the perimeter of a polygon is the sum of the lengths of its sides.

Answers

3. 48   **4.** 6225

Practice 5

Find the perimeter of the polygon shown. (A centimeter is a unit of length in the metric system.)

2 centimeters
5 centimeters   8 centimeters
15 centimeters

Example 5   Find the perimeter of the polygon shown.

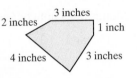

3 inches
2 inches
1 inch
4 inches   3 inches

Solution: To find the perimeter (distance around), we add the lengths of the sides.

2 in. + 3 in. + 1 in. + 3 in. + 4 in. = 13 in.

The perimeter is 13 inches.

▪ Work Practice 5

To make the addition appear simpler, we will often not include units with the addends. If you do this, make sure units are included in the final answer.

Practice 6

A park is in the shape of a triangle. Each of the park's three sides is 647 feet. Find the perimeter of the park.

Example 6   Calculating the Perimeter of a Building

The world's largest commercial building under one roof is the flower auction building of the cooperative VBA in Aalsmeer, Netherlands. The floor plan is a rectangle that measures 776 meters by 639 meters. Find the perimeter of this building. (A meter is a unit of length in the metric system.) (*Source: The Handy Science Answer Book,* Visible Ink Press)

Solution: Recall that opposite sides of a rectangle have the same length. To find the perimeter of this building, we add the lengths of the sides. The sum of the lengths of its sides is

776 meters

639 meters   639 meters

776 meters

```
   639
   639
   776
 + 776
  2830
```

The perimeter of the building is 2830 meters.

▪ Work Practice 6

Objective C Solving Problems by Adding ▶

Often, real-life problems occur that can be solved by adding. The first step in solving any word problem is to *understand* the problem by reading it carefully.

Descriptions of problems solved through addition *may* include any of these key words or phrases:

Addition		
Key Words or Phrases	**Examples**	**Symbols**
added to	5 added to 7	7 + 5
plus	0 plus 78	0 + 78
increased by	12 increased by 6	12 + 6
more than	11 more than 25	25 + 11
total	the total of 8 and 1	8 + 1
sum	the sum of 4 and 133	4 + 133

Answers

5. 30 cm   **6.** 1941 ft

To solve a word problem that involves addition, we first use the facts given to write an addition statement. Then we write the corresponding solution of the real-life problem. It is sometimes helpful to write the statement in words (brief phrases) and then translate to numbers.

Example 7 Finding the Number of Vehicles Sold in the United States

In 2011, a total of 12,734,424 passenger vehicles were sold in the United States. In 2012, total passenger vehicle sales in the United States had increased by 1,705,636 vehicles. Find the total number of passenger vehicles sold in the United States in 2012. (*Source:* Alliance of Automobile Manufacturers)

Solution: The key phrase here is "had increased by," which suggests that we add. To find the number of vehicles sold in 2012, we add the increase, 1,705,636, to the number of vehicles sold in 2011.

In Words		Translate to Numbers
Number sold in 2011	⟶	12,734,424
+ increase	⟶	+ 1,705,636
Number sold in 2012	⟶	14,440,060

The number of passenger vehicles sold in the United States in 2012 was 14,440,060.

▪ Work Practice 7

Graphs can be used to visualize data. The graph shown next is called a **bar graph.** For this bar graph, the height of each bar is labeled above the bar. To check this height, follow the top of each bar to the vertical line to the left. For example, the first bar is labeled 185. Follow the top of that bar to the left until the vertical line is reached, between 180 and 200, but closer to 180, or 185.

Example 8 Reading a Bar Graph

In the following graph, each bar represents a country and the height of each bar represents the number of threatened mammal species identified in that country.

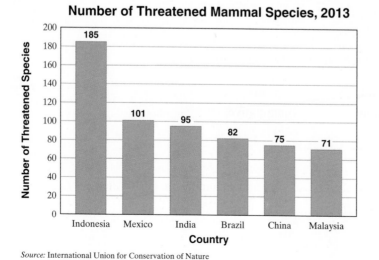

Number of Threatened Mammal Species, 2013

Source: International Union for Conservation of Nature

(Continued on next page)

Practice 7

Georgia produces 70 million pounds of freestone peaches per year. The second largest U.S. producer of peaches, South Carolina, produces 50 million more freestone peaches than Georgia. How much does South Carolina produce? (*Source:* farms.com)

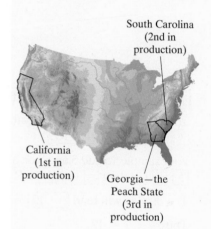

Practice 8

Use the graph in Example 8 to answer the following:

a. Which country shown has the fewest threatened mammal species?

b. Find the total number of threatened mammal species for Brazil, India, and Mexico.

Answers

7. 120 million lb

8. a. Malaysia   **b.** 278

a. Which country shown has the greatest number of threatened mammal species?

b. Find the total number of threatened mammal species for Malaysia, China, and Indonesia.

Solution:

a. The country with the greatest number of threatened mammal species corresponds to the tallest bar, which is Indonesia.

b. The key word here is "total." To find the total number of threatened mammal species for Malaysia, China, and Indonesia, we add.

In Words		Translate to Numbers
Malaysia	$\longrightarrow$	71
China	$\longrightarrow$	75
Indonesia	$\longrightarrow$	+ 185
	Total	331

The total number of threatened mammal species for Malaysia, China, and Indonesia is 331.

▶ Work Practice 8

🖩 Calculator Explorations Adding Numbers

To add numbers on a calculator, find the keys marked
$+$ and $=$ or ENTER .

For example, to add 5 and 7 on a calculator, press the keys
5 $+$ 7 then $=$ or ENTER.

The display will read ☐ 12 .

Thus, 5 + 7 = 12.

To add 687, 981, and 49 on a calculator, press the keys
687 $+$ 981 $+$ 49 then $=$ or ENTER .

The display will read ☐ 1717 .

Thus, 687 + 981 + 49 = 1717. (Although entering 687, for example, requires pressing more than one key, here numbers are grouped together for easier reading.)

Use a calculator to add.

1. 89 + 45

2. 76 + 97

3. 285 + 55

4. 8773 + 652

5.
```
    985
   1210
    562
 +   77
```

6.
```
    465
   9888
    620
 + 1550
```

Vocabulary, Readiness & Video Check

Use the choices below to fill in each blank. Some choices may be used more than once.

sum	order	addend	associative
perimeter	number	grouping	commutative

1. The sum of 0 and any number is the same _____.

2. The sum of any number and 0 is the same _____.

3. In 35 + 20 = 55, the number 55 is called the _____ and 35 and 20 are each called a(n) _____.

4. The distance around a polygon is called its _____.

5. Since $(3 + 1) + 20 = 3 + (1 + 20)$, we say that changing the _____ in addition does not change the sum. This property is called the _____ property of addition.

6. Since $7 + 10 = 10 + 7$, we say that changing the _____ in addition does not change the sum. This property is called the _____ property of addition.

Martin-Gay Interactive Videos   *Watch the section lecture video and answer the following questions.*

Objective A 7. Complete this statement based on the lecture before ⊞Example 1: To add whole numbers, we line up _____ values and add from _____ to _____. ▷

Objective B 8. In ⊞Example 4, the perimeter of what type of polygon is found? How many addends are in the resulting addition problem? ▷

Objective C 9. In ⊞Example 6, what key word or phrase indicates addition? ▷

See Video 1.3

1.3 **Exercise Set** MyMathLab® ▷

Objective A *Add. See Examples 1 through 4.*

1.
```
  14
+ 22
```

2.
```
  27
+ 31
```

3.
```
   62
+ 230
```

4.
```
   37
+ 542
```

5.
```
  12
  13
+ 24
```

6.
```
  23
  45
+ 30
```

▷ 7.
```
  5267
+  132
```

8.
```
   236
+ 6243
```

9. $53 + 64$

10. $41 + 74$

11. $22 + 490$

12. $35 + 470$

13. $22{,}781 + 186{,}297$

14. $17{,}427 + 821{,}059$

▷ 15.
```
  8
  9
  2
  5
+ 1
```

16.
```
  3
  5
  8
  5
+ 7
```

17.
```
   6
  21
  14
   9
+ 12
```

18.
```
  12
   4
   8
  26
+ 10
```

19.
```
  81
  17
  23
  79
+ 12
```

20.
```
  64
  28
  56
  25
+ 32
```

21. $62 + 18 + 14$

22. $23 + 49 + 18$

23. $40 + 800 + 70$

24. $30 + 900 + 20$

25. $7542 + 49 + 682$

26. $1624 + 32 + 976$

▶ **27.** 24 + 9006 + 489 + 2407   **28.** 16 + 1056 + 748 + 7770

29.
```
  627
  628
+ 629
```

30.
```
  427
  383
+ 229
```

31.
```
 6820
 4271
+5626
```

32.
```
 6789
 4321
+5555
```

33.
```
 507
 593
+ 10
```

34.
```
 864
  33
+356
```

35.
```
 4200
 2107
+2692
```

36.
```
 5000
 1400
+3021
```

37.
```
    49
   628
  5 762
+29,462
```

38.
```
    26
   582
  4 763
+62,511
```

39.
```
 121,742
  57,279
  26,586
+426,782
```

40.
```
 504,218
 321,920
  38,507
+594,687
```

Objective B *Find the perimeter of each figure. See Examples 5 and 6.*

△ **41.** 9 inches, 12 inches, 12 inches, 9 inches

△ **42.** 3 kilometers, 3 kilometers, 5 kilometers, 5 kilometers

▶ △ **43.** 7 feet, 8 feet, 10 feet

△ **44.** 3 centimeters, 5 centimeters, 4 centimeters

△ **45.** 4 inches, Rectangle, 8 inches

△ **46.** 8 miles, Rectangle, 4 miles

△ **47.** 2 yards, 2 yards, Square

△ **48.** 23 centimeters, 23 centimeters, Square

⊛ **49.** 8 inches, 1 inch, 3 inches, 5 inches, 5 inches, 7 inches

△ **50.** 6 inches, 5 inches, 5 inches, 7 inches, 7 inches, 3 inches, 4 inches

△ **51.**

△ **52.**

Objectives A B C Mixed Practice–Translating *Solve. See Examples 1 through 8.*

▶ **53.** Find the sum of 297 and 1796.

54. Find the sum of 802 and 6487.

55. Find the total of 76, 39, 8, 17, and 126.

56. Find the total of 89, 45, 2, 19, and 341.

57. What is 452 increased by 92?

58. What is 712 increased by 38?

59. What is 2686 plus 686 plus 80?

60. What is 3565 plus 565 plus 70?

61. The estimated population of Florida was 19,318 thousand in 2012. If is projected to increase by 1823 thousand by 2020. What is Florida's projected population in 2020? (*Source:* U.S. Census Bureau, Florida Office of Economic & Demographic Research)

62. The estimated population of California was 38,041 thousand in 2012. It is projected to increase by 2603 thousand by 2020. What is California's projected population in 2020? (*Source:* U.S. Census Bureau, California Department of Finance)

▶ **63.** The highest point in South Carolina is Sassafras Mountain at 3560 feet above sea level. The highest point in North Carolina is Mt. Mitchell, whose peak is 3124 feet increased by the height of Sassafras Mountain. Find the height of Mt. Mitchell. (*Source:* U.S. Geological Survey)

64. The distance from Kansas City, Kansas, to Hays, Kansas, is 285 miles. Colby, Kansas, is 98 miles farther from Kansas City than Hays. Find the total distance from Kansas City to Colby.

△ **65.** Leo Callier is installing an invisible fence in his backyard. How many feet of wiring are needed to enclose the yard below?

△ **66.** A homeowner is considering installing gutters around her home. Find the perimeter of her rectangular home.

67. The tallest waterfall in the United States is Yosemite Falls in Yosemite National Park in California. Yosemite Falls is made up of three sections, as shown in the graph. What is the total height of Yosemite Falls? (*Source:* U.S. Department of the Interior)

Tallest U.S. Waterfall
(by Sections)

68. Jordan White, a nurse at Mercy Hospital, is recording fluid intake on a patient's medical chart. During his shift, the patient had the following types and amounts of intake measured in cubic centimeters (cc). What amount should Jordan record as the total fluid intake for this patient?

Oral	Intravenous	Blood
240	500	500
100	200	
355		

69. In 2012, Harley-Davidson sold 172,251 of its motorcycles domestically. In addition, 77,598 Harley-Davidson motorcycles were sold internationally. What was the total number of Harley-Davidson motorcycles sold in 2012? (*Source:* Harley-Davidson, Inc.)

70. Hank Aaron holds Major League Baseball's record for the most runs batted in over his career. He batted in 1305 runs from 1954 to 1965. He batted in another 992 runs from 1966 until he retired in 1976. How many total runs did Hank Aaron bat in during his career in professional baseball?

71. During August 2013, a total of 999,040 vehicles were produced in the United States. During the same period, a total of 475,671 vehicles were produced in Canada and Mexico. What was the total number of vehicles produced in North America in August 2013? (*Source:* WardsAuto InfoBank)

72. In 2012, the country of New Zealand had 26,767,000 more sheep than people. If the human population of New Zealand in 2012 was 4,433,000, what was the sheep population? (*Source:* Statistics New Zealand)

73. The largest permanent Monopoly board is made of granite and located in San Jose, California. Find the perimeter of the square playing board.

31 ft

31 ft

74. The smallest commercially available jigsaw puzzle (with a minimum of 1000 pieces) is manufactured in Hong Kong, China. (*Source: Guinness World Records*) Find the exact perimeter of this rectangular-shaped puzzle in millimeters.

182 millimeters
(about 7 in.)

257 millimeters
(about 10 in.)

75. In 2013, there were 2657 Gap Inc. (Gap, Banana Republic, Old Navy) stores located in the United States and 438 located outside the United States. How many Gap Inc. stores were located worldwide? (*Source:* Gap Inc.)

76. Wilma Rudolph, who won three gold medals in track and field events in the 1960 Summer Olympics, was born in 1940. Allyson Felix, who also won three gold medals in track and field events but in the 2012 Summer Olympics, was born 45 years later. In what year was Allyson Felix born?

The table shows the number of Target stores in ten states. Use this table to answer Exercises 77 through 82.

The Top States for Target Stores in 2012	
State	**Number of Stores**
California	257
Florida	123
Illinois	89
Michigan	59
Minnesota	75
New York	67
Ohio	64
Pennsylvania	63
Texas	149
Virginia	57
(*Source:* Target Corporation)	

77. Which state has the most Target stores?

78. Which of the states listed in the table has the fewest Target stores?

79. What is the total number of Target stores located in the three states with the most Target stores?

80. How many Target stores are located in the ten states listed in the table?

81. Which pair of neighboring states has more Target stores combined, New York and Pennsylvania or Michigan and Ohio?

82. Target operates stores in 49 states. There are 775 Target stores located in the states not listed in the table. How many Target stores are in the United States?

83. The state of Delaware has 2997 miles of urban high-ways and 3361 miles of rural highways. Find the total highway mileage in Delaware. (*Source:* U.S. Federal Highway Administration)

84. The state of Rhode Island has 5260 miles of urban highways and 1225 miles of rural highways. Find the total highway mileage in Rhode Island. (*Source:* U.S. Federal Highway Administration)

Concept Extensions

85. In your own words, explain the commutative property of addition.

86. In your own words, explain the associative property of addition.

87. Give any three whole numbers whose sum is 100.

88. Give any four whole numbers whose sum is 25.

89. Add: 56,468,980 + 1,236,785 + 986,768,000

90. Add: 78,962 + 129,968,350 + 36,462,880

Check each addition below. If it is incorrect, find the correct answer. See the Concept Check in this section.

91.	**92.**	**93.**	**94.**
566	773	14	19
932	659	173	214
+871	+481	86	49
2369	1913	+257	+651
		520	923

1.4 Subtracting Whole Numbers

Objectives

A Subtract Whole Numbers.

B Solve Problems by Subtracting Whole Numbers.

Objective A Subtracting Whole Numbers

If you have $5 and someone gives you $3, you have a total of $8, since 5 + 3 = 8. Similarly, if you have $8 and then someone borrows $3, you have $5 left. **Subtraction** is finding the **difference** of two numbers.

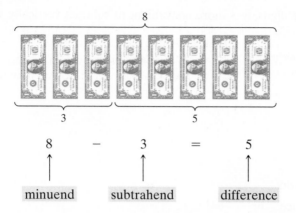

$$8 \quad - \quad 3 \quad = \quad 5$$

minuend   subtrahend   difference

In this example, 8 is the **minuend,** and 3 is the **subtrahend.** The **difference** between these two numbers, 8 and 3, is 5.

Notice that addition and subtraction are very closely related. In fact, subtraction is defined in terms of addition.

$8 - 3 = 5$ because $5 + 3 = 8$

This means that subtraction can be *checked* by addition, and we say that addition and subtraction are reverse operations.

Example 1   Subtract. Check each answer by adding.

a. $12 - 9$ **b.** $22 - 7$ **c.** $35 - 35$ **d.** $70 - 0$

Solution:

a. $12 - 9 = 3$ because $3 + 9 = 12$
b. $22 - 7 = 15$ because $15 + 7 = 22$
c. $35 - 35 = 0$ because $0 + 35 = 35$
d. $70 - 0 = 70$ because $70 + 0 = 70$

■ **Work Practice 1**

Look again at Examples 1(c) and 1(d).

1(c) $35 - 35 = 0$ 1(d) $70 - 0 = 70$

same   difference   a number   difference is the
number   is 0   minus 0   same number

These two examples illustrate the subtraction properties of 0.

Subtraction Properties of 0

The difference of any number and that same number is 0. For example,

$11 - 11 = 0$

The difference of any number and 0 is that same number. For example,

$45 - 0 = 45$

To subtract whole numbers we subtract the digits in the ones place, then the tens place, then the hundreds place, and so on. When subtraction involves numbers

Practice 1

Subtract. Check each answer by adding.

a. $14 - 6$
b. $20 - 8$
c. $93 - 93$
d. $42 - 0$

Answers
1. a. 8 **b.** 12 **c.** 0 **d.** 42

of two or more digits, it is more convenient to subtract vertically. For example, to subtract 893 − 52,

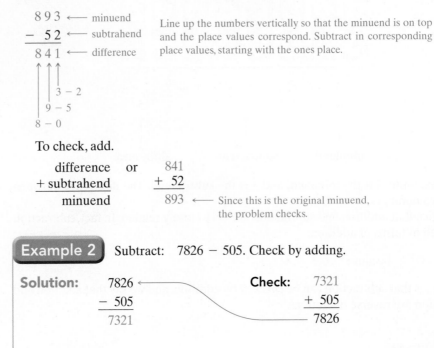

893 ←— minuend

− 52 ←— subtrahend

841 ←— difference

Line up the numbers vertically so that the minuend is on top and the place values correspond. Subtract in corresponding place values, starting with the ones place.

3 − 2
9 − 5
8 − 0

To check, add.

difference   or   841
+ subtrahend   + 52
   minuend   893 ←— Since this is the original minuend, the problem checks.

> **Example 2** Subtract: 7826 − 505. Check by adding.
>
> **Solution:**   7826 ←   **Check:**   7321
>   − 505   + 505
>   7321   7826

■ Work Practice 2

Subtracting by Borrowing

When subtracting vertically, if a digit in the second number (subtrahend) is larger than the corresponding digit in the first number (minuend), **borrowing** is necessary. For example, consider

8 1
− 6 3

Since the 3 in the ones place of 63 is larger than the 1 in the ones place of 81, borrowing is necessary. We borrow 1 ten from the tens place and add it to the ones place.

Borrowing

           7 11
8 − 1 = 7 → 8 1̶ ←— 1 ten + 1 one = 11 ones
tens ten tens
                − 6 3

Now we subtract the ones-place digits and then the tens-place digits.

7 11
8 1̶ ←   **Check:**   18
−6 3   +63
 1 8 ←— 11 − 3 = 8   81   The original minuend.
           7 − 6 = 1

Practice 3

Subtract. Check by adding.

a.   697
        − 49

b.   326
        −245

c.   1234
        − 822

> **Example 3** Subtract: 543 − 29. Check by adding.
>
>   3 13
> **Solution:**   5 4̶ 3̶ ←   **Check:**   514
>   − 29   + 29
>   5 1 4   543

■ Work Practice 3

Sometimes we may have to borrow from more than one place. For example, to subtract 7631 − 152, we first borrow from the tens place.

$$
\begin{array}{r}
76\overset{2\ 11}{\cancel{3}\cancel{1}} \\
-\ 15\ 2 \\
\hline
9 \leftarrow 11 - 2 = 9
\end{array}
$$

In the tens place, 5 is greater than 2, so we borrow again. This time we borrow from the hundreds place.

6 hundreds − **1 hundred** = 5 hundreds

1 hundred + 2 tens
or
10 tens + 2 tens = 12 tens

$$
\begin{array}{r}
7\overset{5\ \overset{12}{2}\ 11}{\cancel{6}\cancel{3}\cancel{1}} \\
-\ 1\ 5\ 2 \\
\hline
7\ 4\ 7\ 9
\end{array}
$$

Check:
$$
\begin{array}{r}
7479 \\
+\ 152 \\
\hline
7631
\end{array}
$$
 The original minuend.

Example 4 Subtract: 900 − 174. Check by adding.

Solution: In the ones place, 4 is larger than 0, so we borrow from the tens place. But the tens place of 900 is 0, so to borrow from the tens place we must first borrow from the hundreds place.

$$
\begin{array}{r}
\overset{8\ 10}{\cancel{9}\cancel{0}}\ 0 \\
-\ 1\ 7\ 4
\end{array}
$$

Now borrow from the tens place.

$$
\begin{array}{r}
\overset{8\ \overset{9}{10}\ 10}{\cancel{9}\cancel{0}\cancel{0}} \\
-\ 1\ 7\ 4 \\
\hline
7\ 2\ 6
\end{array}
$$

Check:
$$
\begin{array}{r}
726 \\
+174 \\
\hline
900
\end{array}
$$

■ Work Practice 4

Objective B Solving Problems by Subtracting ▶

Often, real-life problems occur that can be solved by subtracting. The first step in solving any word problem is to *understand* the problem by reading it carefully.

Descriptions of problems solved through subtraction *may* include any of these key words or phrases:

Subtraction		
Key Words or Phrases	**Examples**	**Symbols**
subtract	subtract 5 from 8	8 − 5
difference	the difference of 10 and 2	10 − 2
less	17 less 3	17 − 3
less than	2 less than 20	20 − 2
take away	14 take away 9	14 − 9
decreased by	7 decreased by 5	7 − 5
subtracted from	9 subtracted from 12	12 − 9

Practice 4

Subtract. Check by adding.
a.
$$
\begin{array}{r}
400 \\
-164
\end{array}
$$
b.
$$
\begin{array}{r}
1000 \\
-\ 762
\end{array}
$$

Helpful Hint Be careful when solving applications that suggest subtraction. Although order *does not* matter when adding, order *does* matter when subtracting. For example, 20 − 15 and 15 − 20 do not simplify to the same number.

✓**Concept Check** In each of the following problems, identify which number is the minuend and which number is the subtrahend.

a. What is the result when 6 is subtracted from 40?

b. What is the difference of 15 and 8?

c. Find a number that is 15 fewer than 23.

To solve a word problem that involves subtraction, we first use the facts given to write a subtraction statement. Then we write the corresponding solution of the real-life problem. It is sometimes helpful to write the statement in words (brief phrases) and then translate to numbers.

Practice 5

The radius of Uranus is 15,759 miles. The radius of Neptune is 458 miles less than the radius of Uranus. What is the radius of Neptune? (*Source:* National Space Science Data Center)

Example 5   Finding the Radius of a Planet

The radius of Jupiter is 43,441 miles. The radius of Saturn is 7257 miles less than the radius of Jupiter. Find the radius of Saturn. (*Source:* National Space Science Data Center)

43,441 miles   ?

Jupiter   Saturn

Solution:

In Words		Translate to Numbers
radius of Jupiter	⟶	$\overset{\overset{13}{3\,13\;3\;\cancel{3}11}}{4\cancel{3},4\cancel{4}\cancel{1}}$
− 7257	⟶	− 7 257
radius of Saturn	⟶	36,184

The radius of Saturn is 36,184 miles.

> **Helpful Hint**   Since subtraction and addition are reverse operations, don't forget that a subtraction problem can be checked by adding.

■ Work Practice 5

Practice 6

During a sale, the price of a new suit is decreased by $47. If the original price was $92, find the sale price of the suit.

Example 6   Calculating Miles per Gallon

A subcompact car gets 42 miles per gallon of gas. A full-size car gets 17 miles per gallon of gas. Find the difference between the subcompact car miles per gallon and the full-size car miles per gallon.

Solution:

In Words		Translate to Numbers
subcompact miles per gallon	⟶	$\overset{3\;12}{\cancel{4}\cancel{2}}$
− full-size miles per gallon	⟶	−1 7
difference in miles per gallon		2 5

The difference in the subcompact car miles per gallon and the full-size car miles per gallon is 25 miles per gallon.

■ Work Practice 6

Answers

5. 15,301 miles   **6.** $45

✓**Concept Check Answers**

a. minuend: 40; subtrahend: 6

b. minuend: 15; subtrahend: 8

c. minuend: 23; subtrahend: 15

Helpful Hint

Once again, because subtraction and addition are reverse operations, don't forget that a subtraction problem can be checked by adding.

Calculator Explorations Subtracting Numbers

To subtract numbers on a calculator, find the keys marked $-$ and $=$ or ENTER.

For example, to find $83 - 49$ on a calculator, press the keys 83 $-$ 49 then $=$ or ENTER .

The display will read ⎡___34___⎤.

Thus, $83 - 49 = 34$.

Use a calculator to subtract.

1. $865 - 95$ **2.** $76 - 27$

3. $147 - 38$ **4.** $366 - 87$

5. $9625 - 647$ **6.** $10,711 - 8925$

Vocabulary, Readiness & Video Check

Use the choices below to fill in each blank.

   0   minuend   difference

   number   subtrahend

1. The difference of any number and that same number is _____.

2. The difference of any number and 0 is the same _____.

3. In $37 - 19 = 18$, the number 37 is the _____, and the number 19 is the _____.

4. In $37 - 19 = 18$, the number 18 is called the _____.

Find each difference.

5. $6 - 6$ **6.** $93 - 93$ **7.** $600 - 0$ **8.** $5 - 0$

Martin-Gay Interactive Videos

Watch the section lecture video and answer the following questions.

Objective A   **9.** In ▣ Example 2, explain how we end up subtracting 7 from 12 in the ones place. ▶

Objective B   **10.** Complete this statement based on ▣ Example 4: Order does not matter when _____, but order does matter when _____. ▶

See Video 1.4 🍎

1.4 Exercise Set MyMathLab ▶

Objective A *Subtract. Check by adding. See Examples 1 and 2.*

1. 67
− 23

2. 72
− 41

3. 389
− 124

4. 572
− 321

5. 167
− 32

6. 286
− 45

7. 2677 − 423

8. 5766 − 324

9. 6998 − 1453

10. 4912 − 2610

▶ **11.** 749
− 149

12. 257
− 257

Subtract. Check by adding. See Examples 1 through 4.

▶ **13.** 62
− 37

14. 55
− 29

15. 70
− 25

16. 80
− 37

17. 938
− 792

18. 436
− 275

19. 922
− 634

20. 674
− 299

21. 600
− 432

22. 300
− 149

23. 142
− 36

24. 773
− 29

25. 923
− 476

26. 813
− 227

27. 6283
− 560

28. 5349
− 720

29. 533
− 29

30. 724
− 16

31. 200
− 111

32. 300
− 211

33. 1983
− 1904

34. 1983
− 1914

35. 56,422
− 16,508

36. 76,652
− 29,498

37. 50,000 − 17,289

38. 40,000 − 23,582

39. 7020 − 1979

40. 6050 − 1878

▶ **41.** 51,111 − 19,898

42. 62,222 − 39,898

Objective B *Solve. See Examples 5 and 6.*

▶ **43.** Subtract 5 from 9.

44. Subtract 9 from 21.

▶ **45.** Find the difference of 41 and 21.

46. Find the difference of 16 and 5.

47. Subtract 56 from 63.

48. Subtract 41 from 59.

49. Find 108 less 36.

50. Find 25 less 12.

51. Find 12 subtracted from 100.

52. Find 86 subtracted from 90.

53. Professor Graham is reading a 503-page book. If she has just finished reading page 239, how many more pages must she read to finish the book?

54. When a couple began a trip, the odometer read 55,492. When the trip was over, the odometer read 59,320. How many miles did they drive on their trip?

55. In 2008, the hole in the Earth's ozone layer over Antarctica was about 25 million square kilometers in size. By 2012, the hole had shrunk to about 18 million square kilometers. By how much did the hole shrink from 2008 to 2012? (*Source:* NASA Ozone Watch)

56. Bamboo can grow to 98 feet while Pacific giant kelp (a type of seaweed) can grow to 197 feet. How much taller is the kelp than the bamboo?

Bamboo   Kelp

A river basin is the geographic area drained by a river and its tributaries. The Mississippi River Basin is the third largest in the world and is divided into six sub-basins, whose areas are shown in the following bar graph. Use this graph for Exercises 57 through 60.

57. Find the total U.S. land area drained by the Upper Mississippi and Lower Mississippi sub-basins.

58. Find the total U.S. land area drained by the Ohio and Tennessee sub-basins.

59. How much more land is drained by the Missouri sub-basin than the Arkansas Red-White sub-basin?

60. How much more land is drained by the Upper Mississippi sub-basin than the Lower Mississippi sub-basin?

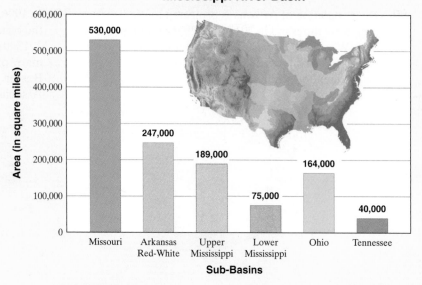

Mississippi River Basin

Area (in square miles) vs Sub-Basins

- Missouri: 530,000
- Arkansas Red-White: 247,000
- Upper Mississippi: 189,000
- Lower Mississippi: 75,000
- Ohio: 164,000
- Tennessee: 40,000

61. The peak of Mt. McKinley in Alaska is 20,320 feet above sea level. The peak of Long's Peak in Colorado is 14,255 feet above sea level. How much higher is the peak of Mt. McKinley than Long's Peak? (*Source:* U.S. Geological Survey)

Mt. McKinley, Alaska   Long's Peak, Colorado

62. On January 12, 1916, the city of Indianapolis, Indiana, had the greatest temperature change in a day. It dropped 58 degrees. If the high temperature was 68° Fahrenheit, what was the low temperature?

63. The Oroville Dam, on the Feather River, is the tallest dam in the United States at 754 feet. The Hoover Dam, on the Colorado River, is 726 feet high. How much taller is the Oroville Dam than the Hoover Dam? (*Source:* U.S. Bureau of Reclamation)

64. A new iPhone with 32 GB costs $299. Jocelyn Robinson has $713 in her savings account. How much will she have left in her savings account after she buys the iPhone? (*Source:* Apple, Inc.)

65. The distance from Kansas City to Denver is 645 miles. Hays, Kansas, lies on the road between the two and is 287 miles from Kansas City. What is the distance between Hays and Denver?

66. Pat Salanki's blood cholesterol level is 243. The doctor tells him it should be decreased to 185. How much of a decrease is this?

67. A new DVD player with remote control costs $295. A college student has $914 in her savings account. How much will she have left in her savings account after she buys the DVD player?

68. A stereo that regularly sells for $547 is discounted by $99 in a sale. What is the sale price?

69. The population of Arizona is projected to grow from 6499 thousand in 2012 to 9129 thousand in 2032. What is Arizona's projected population increase over this time? (*Source:* Arizona Department of Administration)

70. In 1996, the centennial of the Boston Marathon, the official number of participants was 38,708. In 2013, there were 11,869 fewer participants. How many official participants were there for the 2013 Boston Marathon? (*Source:* Boston Athletic Association)

The decibel (dB) is a unit of measurement for sound. Every increase of 10 dB is a tenfold increase in sound intensity. The bar graph below shows the decibel levels for some common sounds. Use this graph for Exercises 71 through 74.

71. What is the dB rating for live rock music?

72. Which is the quietest of all the sounds shown in the graph?

73. How much louder is the sound of snoring than normal conversation?

74. What is the difference in sound intensity between live rock music and loud television?

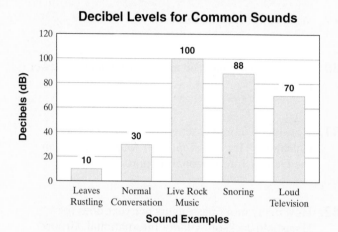

Decibel Levels for Common Sounds

75. The 113th Congress has 535 senators and representatives. Of these, 189 were registered Boy Scouts at some time in their lives. How many members of the 113th Congress were never Boy Scouts? (*Source:* Boy Scouts of America)

76. In the United States, there were 28,799 tornadoes from 1990 through 2012. In all, 13,205 of these tornadoes occurred from 1990 through 2000. How many tornadoes occurred during the period after 2000? (*Source:* Storm Prediction Center, National Weather Service)

77. Until recently, the world's largest permanent maze was located in Ruurlo, Netherlands. This maze of beech hedges covers 94,080 square feet. A new hedge maze using hibiscus bushes at the Dole Plantation in Wahiawa, Hawaii, covers 100,000 square feet. How much larger is the Dole Plantation maze than the Ruurlo maze? (*Source: The Guinness Book of Records*)

78. There were only 27 California condors in the entire world in 1987. To date, the number has increased to an estimated 223 living in the wild. How much of an increase is this? (*Source:* California Department of Fish and Wildlife)

The bar graph shows the top five U.S. airports according to number of passengers arriving and departing in 2013. Use this graph to answer Exercises 79 through 82.

79. Which airport is the busiest?

80. Which airports have 60 million passengers or fewer per year?

81. How many more passengers per year does the Chicago O'Hare International Airport have than the Denver International Airport?

82. How many more passengers per year does the Hartsfield-Jackson Atlanta International Airport have than the Dallas/Ft. Worth International Airport?

Top Five Airports in U.S.

Source: Airports Council International

Solve.

83. Two seniors, Jo Keen and Trudy Waterbury, were candidates for student government president. Who won the election if the votes were cast as follows? By how many votes did the winner win?

Class	Candidate	
	Jo	**Trudy**
Freshman	276	295
Sophomore	362	122
Junior	201	312
Senior	179	18

84. Two students submitted advertising budgets for a student government fund-raiser.

	Student A	**Student B**
Radio ads	$600	$300
Newspaper ads	$200	$400
Posters	$150	$240
Handbills	$120	$170

If $1200 is available for advertising, how much excess would each budget have?

Mixed Practice (Sections 1.3 and 1.4) *Add or subtract as indicated.*

85.
$$986 + 48$$

86.
$$986 - 48$$

87. $76 - 67$

88. $80 + 93 + 17 + 9 + 2$

89.
$$9000 - 482$$

90.
$$10{,}000 - 1786$$

91.
$$10{,}962 \\ 4851 \\ + 7063$$

92.
$$12{,}468 \\ 3211 \\ + 1988$$

Concept Extensions

For each exercise, identify which number is the minuend and which number is the subtrahend. See the Concept Check in this section.

93.
$$48 - 1$$

94.
$$2863 - 1904$$

95. Subtract 7 from 70.

96. Find 86 decreased by 25.

Identify each answer as correct or incorrect. Use addition to check. If the answer is incorrect, then write the correct answer.

97.
$$\begin{array}{r} 741 \\ -\ 56 \\ \hline 675 \end{array}$$

98.
$$\begin{array}{r} 478 \\ -\ 89 \\ \hline 389 \end{array}$$

99.
$$\begin{array}{r} 1029 \\ -\ 888 \\ \hline 141 \end{array}$$

100.
$$\begin{array}{r} 7615 \\ -\ 547 \\ \hline 7168 \end{array}$$

Fill in the missing digits in each problem.

101.
$$\begin{array}{r} 526\_ \\ -2\_85 \\ \hline 28\_4 \end{array}$$

102.
$$\begin{array}{r} 10,\_4\_ \\ -\ 85\_4 \\ \hline \_710 \end{array}$$

103. Is there a commutative property of subtraction? In other words, does order matter when subtracting? Why or why not?

104. Explain why the phrase "Subtract 7 from 10" translates to "$10 - 7$."

105. The local college library is having a Million Pages of Reading promotion. The freshmen have read a total of 289,462 pages; the sophomores have read a total of 369,477 pages; the juniors have read a total of 218,287 pages; and the seniors have read a total of 121,685 pages. Have they reached a goal of one million pages? If not, how many more pages need to be read?

1.5   Rounding and Estimating

Objective A Rounding Whole Numbers

Objectives

A Round Whole Numbers.

B Use Rounding to Estimate Sums and Differences.

C Solve Problems by Estimating.

Rounding a whole number means approximating it. A rounded whole number is often easier to use, understand, and remember than the precise whole number. For example, instead of trying to remember the Minnesota state population as 5,197,621, it is much easier to remember it rounded to the nearest million: 5,000,000, or 5 million people.(*Source: World Almanac*)

Recall from Section 1.2 that the line below is called a number line. To **graph** a whole number on this number line, we darken the point representing the location of the whole number. For example, the number 4 is graphed below.

On a number line, the whole number 36 is closer to 40 than 30, so 36 rounded to the nearest ten is 40.

The whole number 52 is closer to 50 than 60, so 52 rounded to the nearest ten is 50.

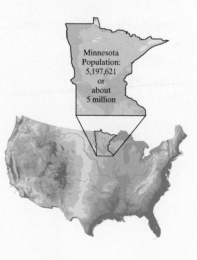

Minnesota Population: 5,197,621 or about 5 million

In trying to round 25 to the nearest ten, we see that 25 is halfway between 20 and 30. It is not closer to either number. In such a case, we round to the larger ten, that is, to 30.

Here, we round "up."

To round a whole number without using a number line, follow these steps:

Rounding Whole Numbers to a Given Place Value

Step 1: Locate the digit to the right of the given place value.

Step 2: If this digit is 5 or greater, add 1 to the digit in the given place value and replace each digit to its right by 0.

Step 3: If this digit is less than 5, replace it and each digit to its right by 0.

Practice 1

Round to the nearest ten.

a. 57

b. 641

c. 325

Example 1 Round 568 to the nearest ten.

Solution: 5 6⑧   The digit to the right of the tens place is the ones place, which is circled.
            ↑
        tens place

            5 6⑧   Since the circled digit is 5 or greater, add 1 to the 6 in the tens place and replace the digit to the right by 0.
          ↑   ↖
      Add 1.   Replace
              with 0.

We find that 568 rounded to the nearest ten is 570.

■ Work Practice 1

Practice 2

Round to the nearest thousand.

a. 72,304

b. 9222

c. 671,800

Example 2 Round 278,362 to the nearest thousand.

Solution: Thousands place
                ↓ ┌─ 3 is less than 5.
            278,③62
              ↑   ↑
      Do not add 1. Replace with zeros.

The number 278,362 rounded to the nearest thousand is 278,000.

■ Work Practice 2

Answers

1. a. 60 **b.** 640 **c.** 330

2. a. 72,000 **b.** 9000 **c.** 672,000

Example 3 Round 248,982 to the nearest hundred.

Solution:

Hundreds place

8 is greater than or equal to 5.

248,9⑧2

Add 1.   9 + 1 = 10, so replace the digit 9 by 0 and carry 1 to the place value to the left.

2 4 8, 9 8 2

Add 1.   Replace with zeros.

The number 248,982 rounded to the nearest hundred is 249,000.

▪ **Work Practice 3**

✓**Concept Check** Round each of the following numbers to the nearest *hundred*. Explain your reasoning.

a. 59   **b.** 29

Objective B Estimating Sums and Differences

By rounding addends, minuends, and subtrahends, we can estimate sums and differences. An estimated sum or difference is appropriate when the exact number is not necessary. Also, an estimated sum or difference can help us determine if we made a mistake in calculating an exact amount. To estimate the sum below, round each number to the nearest hundred and then add.

768	rounds to	800
1952	rounds to	2000
225	rounds to	200
+ 149	rounds to	+ 100
		3100

The estimated sum is 3100, which is close to the **exact** sum of 3094.

Example 4 Round each number to the nearest hundred to find an estimated sum.

```
  294
  625
 1071
+ 349
```

Solution:

Exact:		**Estimate:**
294	rounds to	300
625	rounds to	600
1071	rounds to	1100
+ 349	rounds to	+ 300
		2300

The estimated sum is 2300. (The exact sum is 2339.)

▪ **Work Practice 4**

Practice 3

Round to the nearest hundred.
a. 3474
b. 76,243
c. 978,965

Practice 4

Round each number to the nearest ten to find an estimated sum.

```
  49
  25
  32
  51
+ 98
```

Answers
3. a. 3500   **b.** 76,200   **c.** 979,000
4. 260

✓**Concept Check Answers**
a. 100   **b.** 0

Practice 5

Round each number to the nearest thousand to find an estimated difference.

$$3785$$
$$-\,2479$$

Example 5 Round each number to the nearest hundred to find an estimated difference.

$$4725$$
$$-\,2879$$

Solution:

Exact:		**Estimate:**
4725	rounds to	4700
− 2879	rounds to	− 2900
		1800

The estimated difference is 1800. (The exact difference is 1846.)

■ Work Practice 5

Objective C Solving Problems by Estimating ▶

Making estimates is often the quickest way to solve real-life problems when solutions do not need to be exact.

Practice 6

Tasha Kilbey is trying to estimate how far it is from Gove, Kansas, to Hays, Kansas. Round each given distance on the map to the nearest ten to estimate the total distance.

Example 6 Estimating Distances

A driver is trying to quickly estimate the distance from Temple, Texas, to Brenham, Texas. Round each distance given on the map to the nearest ten to estimate the total distance.

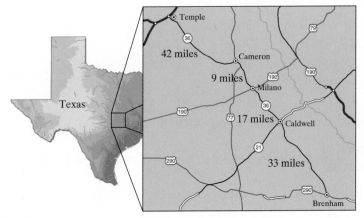

Solution:

Exact Distance:		**Estimate:**
42	rounds to	40
9	rounds to	10
17	rounds to	20
+ 33	rounds to	+ 30
		100

It is approximately 100 miles from Temple to Brenham. (The exact distance is 101 miles.)

■ Work Practice 6

Answers

5. 2000   **6.** 80 mi

Example 7 Estimating Data

In three recent months, the numbers of tons of mail that went through Hartsfield-Jackson Atlanta International Airport were 635, 687, and 567. Round each number to the nearest hundred to estimate the tons of mail that passed through this airport.

Solution:

Exact Tons of Mail:		**Estimate:**
635	rounds to	600
687	rounds to	700
+567	rounds to	+600
		1900

The approximate tonnage of mail that moved through Atlanta's airport over this period was 1900 tons. (The exact tonnage was 1889 tons.)

▧ Work Practice 7

Practice 7

In 2012, Ecuador topped the International Union for Conservation of Nature's Red List of Threatened Species. At that time, Ecuador was home to 139 threatened bird and mammal species, 316 threatened other animal species, and 1842 threatened plant species. Round each number to the nearest hundred to estimate the total number of threatened species in Ecuador. (*Source:* International Union for Conservation of Nature)

Answer

7. 2200 total threatened species

Vocabulary, Readiness & Video Check

Use the choices below to fill in each blank.

60	rounding	exact
70	estimate	graph

1. To _____ a number on a number line, darken the point representing the location of the number.

2. Another word for approximating a whole number is _____.

3. The number 65 rounded to the nearest ten is _____, but the number 61 rounded to the nearest ten is _____.

4. A(n) _____ number of products is 1265, but a(n) _____ is 1000.

Martin-Gay Interactive Videos

See Video 1.5 🍎

Watch the section lecture video and answer the following questions.

Objective A **5.** In ▤ Example 1, when rounding the number to the nearest ten, why do we replace the digit 3 with a 4? ▶

Objective B **6.** As discussed in ▤ Example 3, explain how a number line can help us understand how to round 22 to the nearest ten. ▶

Objective C **7.** What is the significance of the circled digit in each height value in ▤ Example 5? ▶

1.5 Exercise Set MyMathLab ▶

Objective A *Round each whole number to the given place. See Examples 1 through 3.*

1. 423 to the nearest ten

2. 273 to the nearest ten

▶ **3.** 635 to the nearest ten

4. 846 to the nearest ten

5. 2791 to the nearest hundred

6. 8494 to the nearest hundred

7. 495 to the nearest ten

8. 898 to the nearest ten

9. 21,094 to the nearest thousand

10. 82,198 to the nearest thousand

11. 33,762 to the nearest thousand

12. 42,682 to the nearest ten-thousand

13. 328,495 to the nearest hundred

14. 179,406 to the nearest hundred

▶ **15.** 36,499 to the nearest thousand

16. 96,501 to the nearest thousand

17. 39,994 to the nearest ten

18. 99,995 to the nearest ten

19. 29,834,235 to the nearest ten-million

20. 39,523,698 to the nearest million

Complete the table by estimating the given number to the given place value.

		Tens	Hundreds	Thousands
21.	5281			
22.	7619			
23.	9444			
24.	7777			
25.	14,876			
26.	85,049			

Round each number to the indicated place.

27. The University of California, Los Angeles, had a total undergraduate enrollment of 27,941 students in fall 2012. Round this number to the nearest thousand. (*Source:* UCLA)

28. In 2012, there were 12,997 Burger King restaurants worldwide. Round this number to the nearest thousand. (*Source:* Burger King Worldwide, Inc.)

29. Kareem Abdul-Jabbar holds the NBA record for points scored, a total of 38,387 over his NBA career. Round this number to the nearest thousand. (*Source:* National Basketball Association)

30. It takes 60,149 days for Neptune to make a complete orbit around the Sun. Round this number to the nearest hundred. (*Source:* National Space Science Data Center)

31. In 2013, the most valuable brand in the world was Apple, having just overtaken the longtime leader, Coca-Cola. The estimated brand value of Apple was $98,316,000,000. Round this to the nearest billion. (*Source:* Interbrand)

32. According to the 2013 Population Clock, the population of the United States was 316,797,189 in October 2013. Round this population figure to the nearest million. (*Source:* U.S. Census population clock)

33. The average salary for a Boston Red Sox baseball player during the 2013 season was $5,021,850. Round this average salary to the nearest thousand. (*Source:* CBSSports.com)

34. In FY 2013, the Procter & Gamble Company had $84,167,000,000 in sales. Round this sales figure to the nearest billion. (*Source:* Procter & Gamble)

35. In the United States in 2012, the travel industry generated $128,800,000,000 in tax revenue for local, state, and federal governments. Round this travel-related tax revenue figure to the nearest billion. (*Source:* U.S. Travel Association)

36. U.S. farms produced 3,149,166,000 bushels of soybeans in 2013. Round the soybean production figure to the nearest ten-million. (*Source:* U.S. Department of Agriculture)

Objective B *Estimate the sum or difference by rounding each number to the nearest ten. See Examples 4 and 5.*

37.
$$\begin{array}{r} 39 \\ 45 \\ 22 \\ +\,17 \\ \hline \end{array}$$

38.
$$\begin{array}{r} 52 \\ 33 \\ 15 \\ +\,29 \\ \hline \end{array}$$

39.
$$\begin{array}{r} 449 \\ -\,373 \\ \hline \end{array}$$

40.
$$\begin{array}{r} 555 \\ -\,235 \\ \hline \end{array}$$

Estimate the sum or difference by rounding each number to the nearest hundred. See Examples 4 and 5.

41.
$$\begin{array}{r} 1913 \\ 1886 \\ +\,1925 \\ \hline \end{array}$$

42.
$$\begin{array}{r} 4050 \\ 3133 \\ +\,1220 \\ \hline \end{array}$$

43.
$$\begin{array}{r} 1774 \\ -\,1492 \\ \hline \end{array}$$

44.
$$\begin{array}{r} 1989 \\ -\,1870 \\ \hline \end{array}$$

45.
$$\begin{array}{r} 3995 \\ 2549 \\ +\,4944 \\ \hline \end{array}$$

46.
$$\begin{array}{r} 799 \\ 1655 \\ +\,271 \\ \hline \end{array}$$

Three of the given calculator answers below are incorrect. Find them by estimating each sum.

47. 463 + 219 602

48. 522 + 785 1307

49. 229 + 443 + 606 1278

50. 542 + 789 + 198 2139

51. 7806 + 5150 12,956

52. 5233 + 4988 9011

Helpful Hint
Estimation is useful to check for incorrect answers when using a calculator. For example, pressing a key too hard may result in a double digit, while pressing a key too softly may result in the digit not appearing in the display.

Objective C *Solve each problem by estimating. See Examples 6 and 7.*

53. An appliance store advertises three refrigerators on sale at $899, $1499, and $999. Round each cost to the nearest hundred to estimate the total cost.

54. Suppose you scored 89, 97, 100, 79, 75, and 82 on your biology tests. Round each score to the nearest ten to estimate your total score.

55. The distance from Kansas City to Boston is 1429 miles and from Kansas City to Chicago is 530 miles. Round each distance to the nearest hundred to estimate how much farther Boston is from Kansas City than Chicago is.

56. The Gonzales family took a trip and traveled 588, 689, 277, 143, 59, and 802 miles on six consecutive days. Round each distance to the nearest hundred to estimate the distance they traveled.

▷ 57. The peak of Mt. McKinley, in Alaska, is 20,320 feet above sea level. The top of Mt. Rainier, in Washington, is 14,410 feet above sea level. Round each height to the nearest thousand to estimate the difference in elevation of these two peaks. (*Source:* U.S. Geological Survey)

58. A student is pricing new car stereo systems. One system sells for $1895 and another system sells for $1524. Round each price to the nearest hundred dollars to estimate the difference in price of these systems.

59. In 2012, the United States Postal Service delivered 159,859,000,000 pieces of mail. In 2011, it delivered 168,297,000,000 pieces of mail. Round each number to the nearest billion to estimate how much the mail volume decreased from 2011 to 2012. (*Source:* United States Postal Service)

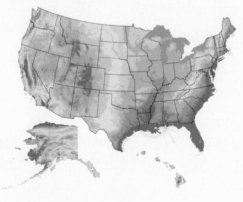

60. Round each distance given on the map to the nearest ten to estimate the total distance from North Platte, Nebraska, to Lincoln, Nebraska.

61. Head Start is a national program that provides developmental and social services for America's low-income preschool children ages three to five. Enrollment figures in Head Start programs showed an increase from 1,125,209 in 2011 to 1,128,030 in 2012. Round each number of children to the nearest thousand to estimate this increase. (*Source:* U.S. Department of Health and Human Services)

62. Enrollment figures at a local community college showed an increase from 49,713 credit hours in 2012 to 51,746 credit hours in 2013. Round each number to the nearest thousand to estimate the increase.

Mixed Practice (Sections 1.2 and 1.5) *The following table shows the top five leading U.S. advertisers in 2012 and the amount of money spent that year on advertising. Complete this table. The first line is completed for you. (Source: Ad Age DataCenter)*

	Advertiser	Amount Spent on Advertising in 2012 (in millions of dollars)	Amount Written in Standard Form	Standard Form Rounded to Nearest Ten-Million	Standard Form Rounded to Nearest Hundred-Million
	Procter & Gamble Co.	$4829	$4,829,000,000	$4,830,000,000	$4,800,000,000
63.	General Motors Co.	$3067			
64.	Comcast Corp.	$2989			
65.	AT&T	$2910			
66.	Verizon Communications	$2381			

Concept Extensions

67. Find one number that when rounded to the nearest hundred is 5700.

68. Find one number that when rounded to the nearest ten is 5700.

69. A number rounded to the nearest hundred is 8600.
   a. Determine the smallest possible number.
   b. Determine the largest possible number.

70. On August 23, 1989, it was estimated that 1,500,000 people joined hands in a human chain stretching 370 miles to protest the fiftieth anniversary of the pact that allowed what was then the Soviet Union to annex the Baltic nations in 1939. If the estimate of the number of people is to the nearest hundred-thousand, determine the largest possible number of people in the chain.

71. In your own words, explain how to round a number to the nearest thousand.

72. In your own words, explain how to round 9660 to the nearest thousand.

△**73.** Estimate the perimeter of the rectangle by first rounding the length of each side to the nearest ten.

54 meters

Rectangle   17 meters

△**74.** Estimate the perimeter of the triangle by first rounding the length of each side to the nearest hundred.

5950 miles   7693 miles

8203 miles

Objectives

A Use the Properties of Multiplication.

B Multiply Whole Numbers.

C Multiply by Whole Numbers Ending in Zero(s).

D Find the Area of a Rectangle.

E Solve Problems by Multiplying Whole Numbers.

Multiplication Shown as Repeated Addition Suppose that we wish to count the number of laptops provided in a computer class. The laptops are arranged in 5 rows, and each row has 6 laptops.

6 laptops in each row

Adding 5 sixes gives the total number of laptops. We can write this as $6 + 6 + 6 + 6 + 6 = 30$ laptops. When each addend is the same, we refer to this as **repeated addition.**

Multiplication is repeated addition but with different notation.

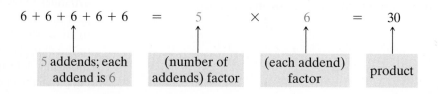

$6 + 6 + 6 + 6 + 6$	=	5	×	6	=	30
5 addends; each addend is 6		(number of addends) factor		(each addend) factor		product

The × is called a **multiplication sign.** The numbers 5 and 6 are called **factors.** The number 30 is called the **product.** The notation 5×6 is read as "five times six." The symbols · and ( ) can also be used to indicate multiplication.

$$5 \times 6 = 30, \quad 5 \cdot 6 = 30, \quad (5)(6) = 30, \quad \text{and} \quad 5(6) = 30$$

✓**Concept Check**

a. Rewrite $5 + 5 + 5 + 5 + 5 + 5 + 5$ using multiplication.

b. Rewrite 3×16 as repeated addition. Is there more than one way to do this? If so, show all ways.

Objective A Using the Properties of Multiplication

As with addition, we memorize products of one-digit whole numbers and then use certain properties of multiplication to multiply larger numbers. (If necessary, review the multiplication of one-digit numbers in Appendix A.2)

Notice that when any number is multiplied by 0, the result is always 0. This is called the **multiplication property of 0.**

✓ Concept Check Answers
a. $7 \times 5 = 35$
b. $16 + 16 + 16 = 48$; yes,
$3 + 3 + 3 + 3 + 3 + 3 + 3 + 3 + 3 + 3 + 3 + 3 + 3 + 3 + 3 + 3 = 48$

Multiplication Property of 0

The product of 0 and any number is 0. For example,

$$5 \cdot 0 = 0 \quad \text{and} \quad 0 \cdot 8 = 0$$

Also notice in Appendix A.2 that when any number is multiplied by 1, the result is always the original number. We call this result the **multiplication property of 1.**

Multiplication Property of 1

The product of 1 and any number is that same number. For example,

$$1 \cdot 9 = 9 \quad \text{and} \quad 6 \cdot 1 = 6$$

Example 1 Multiply.

a. 6×1 **b.** $0(18)$ **c.** $1 \cdot 45$ **d.** $(75)(0)$

Solution:

a. $6 \times 1 = 6$ **b.** $0(18) = 0$

c. $1 \cdot 45 = 45$ **d.** $(75)(0) = 0$

■ Work Practice 1

Practice 1

Multiply.

a. 3×0

b. $4(1)$

c. $(0)(34)$

d. $1 \cdot 76$

Like addition, multiplication is commutative and associative. Notice that when multiplying two numbers, the order of these numbers can be changed without changing the product. For example,

$$3 \cdot 5 = 15 \quad \text{and} \quad 5 \cdot 3 = 15$$

This property is the **commutative property of multiplication.**

Commutative Property of Multiplication

Changing the **order** of two factors does not change their product. For example,

$$9 \cdot 2 = 18 \quad \text{and} \quad 2 \cdot 9 = 18$$

Another property that can help us when multiplying is the **associative property of multiplication.** This property states that when multiplying numbers, the grouping of the numbers can be changed without changing the product. For example,

$$(2 \cdot 3) \cdot 4 = 6 \cdot 4 = 24$$

Also,

$$2 \cdot (3 \cdot 4) = 2 \cdot 12 = 24$$

Both groupings give a product of 24.

Associative Property of Multiplication

Changing the **grouping** of factors does not change their product. From the previous page, we know that for example,

$$(2 \cdot 3) \cdot 4 = 2 \cdot (3 \cdot 4)$$

With these properties, along with the **distributive property,** we can find the product of any whole numbers. The distributive property says that multiplication **distributes** over addition. For example, notice that $3(2 + 5)$ simplifies to the same number as $3 \cdot 2 + 3 \cdot 5$.

$$3(2 + 5) = 3(7) = 21$$

$$3 \cdot 2 + 3 \cdot 5 = 6 + 15 = 21$$

Since $3(2 + 5)$ and $3 \cdot 2 + 3 \cdot 5$ both simplify to 21, then

$$3(2 + 5) = 3 \cdot 2 + 3 \cdot 5$$

Notice in $3(2 + 5) = 3 \cdot 2 + 3 \cdot 5$ that each number inside the parentheses is multiplied by 3.

Distributive Property

Multiplication distributes over addition. For example,

$$2(3 + 4) = 2 \cdot 3 + 2 \cdot 4$$

Practice 2

Rewrite each using the distributive property.

a. $5(2 + 3)$

b. $9(8 + 7)$

c. $3(6 + 1)$

Example 2 Rewrite each using the distributive property.

a. $3(4 + 5)$ **b.** $10(6 + 8)$ **c.** $2(7 + 3)$

Solution: Using the distributive property, we have

a. $3(4 + 5) = 3 \cdot 4 + 3 \cdot 5$

b. $10(6 + 8) = 10 \cdot 6 + 10 \cdot 8$

c. $2(7 + 3) = 2 \cdot 7 + 2 \cdot 3$

▬ Work Practice 2

Objective B Multiplying Whole Numbers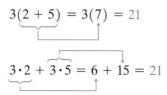

Let's use the distributive property to multiply $7(48)$. To do so, we begin by writing the expanded form of 48 (see Section 1.2) and then applying the distributive property.

$$7(48) = 7(40 + 8) \qquad \text{Write 48 in expanded form.}$$
$$= 7 \cdot 40 + 7 \cdot 8 \qquad \text{Apply the distributive property.}$$
$$= 280 + 56 \qquad \text{Multiply.}$$
$$= 336 \qquad \text{Add.}$$

Answers

2. a. $5(2 + 3) = 5 \cdot 2 + 5 \cdot 3$
 b. $9(8 + 7) = 9 \cdot 8 + 9 \cdot 7$
 c. $3(6 + 1) = 3 \cdot 6 + 3 \cdot 1$

This is how we multiply whole numbers. When multiplying whole numbers, we will use the following notation.

First:

$$\begin{array}{r} \overset{5}{4}8 \\ \times\ 7 \\ \hline 336 \end{array}$$

$7 \cdot 8 = 56$ Write 6 in the ones place and carry 5 to the tens place.

Next:

$$\begin{array}{r} \overset{5}{4}8 \\ \times\ 7 \\ \hline 336 \end{array}$$

$7 \cdot 4 + 5 = 28 + 5 = 33$

The product of 48 and 7 is 336.

Example 3 Multiply:

a. $\begin{array}{r} 25 \\ \times\ 8 \end{array}$ **b.** $\begin{array}{r} 246 \\ \times\ 5 \end{array}$

Solution:

a. $\begin{array}{r} \overset{4}{2}5 \\ \times\ 8 \\ \hline 200 \end{array}$ **b.** $\begin{array}{r} \overset{2\,3}{2}46 \\ \times\ 5 \\ \hline 1230 \end{array}$

▨ Work Practice 3

To multiply larger whole numbers, use the following similar notation. Multiply 89×52.

Step 1

$$\begin{array}{r} \overset{1}{8}9 \\ \times\ 52 \\ \hline 178 \end{array}$$
← Multiply 89×2.

Step 2

$$\begin{array}{r} \overset{4}{8}9 \\ \times\ 52 \\ \hline 178 \\ 4450 \end{array}$$
← Multiply 89×50.

Step 3

$$\begin{array}{r} 89 \\ \times\ 52 \\ \hline 178 \\ 4450 \\ \hline 4628 \end{array}$$
← Add.

The numbers 178 and 4450 are called **partial products.** The sum of the partial products, 4628, is the product of 89 and 52.

Example 4 Multiply: 236×86

Solution:

$$\begin{array}{r} 236 \\ \times\ 86 \\ \hline 1416 \\ 18880 \\ \hline 20{,}296 \end{array}$$

← 6(236)
← 80(236)
Add.

▨ Work Practice 4

Example 5 Multiply: 631×125

Solution:

$$\begin{array}{r} 631 \\ \times\ 125 \\ \hline 3155 \\ 12620 \\ 63100 \\ \hline 78{,}875 \end{array}$$

← 5(631)
← 20(631)
← 100(631)
Add.

▨ Work Practice 5

Practice 3

Multiply.

a. $\begin{array}{r} 36 \\ \times\ 4 \end{array}$ **b.** $\begin{array}{r} 132 \\ \times\ 9 \end{array}$

Practice 4

Multiply.

a. $\begin{array}{r} 594 \\ \times\ 72 \end{array}$ **b.** $\begin{array}{r} 306 \\ \times\ 81 \end{array}$

Practice 5

Multiply.

a. $\begin{array}{r} 726 \\ \times\ 142 \end{array}$ **b.** $\begin{array}{r} 288 \\ \times\ 4 \end{array}$

Answers

3. a. 144   **b.** 1188
4. a. 42,768   **b.** 24,786
5. a. 103,092   **b.** 1152

✓**Concept Check** Find and explain the error in the following multiplication problem.

$$
\begin{array}{r}
102 \\
\times\ 33 \\
\hline
306 \\
306 \\
\hline
612
\end{array}
$$

Objective C Multiplying by Whole Numbers Ending in Zero(s)

Interesting patterns occur when we multiply by a number that ends in zeros. To see these patterns, let's multiply a number, say 34, by 10, then 100, then 1000.

$$34 \cdot 10 = 340 \qquad \text{1 zero attached to 34.}$$

$$34 \cdot 100 = 3400 \qquad \text{2 zeros attached to 34.}$$

$$34 \cdot 1000 = 34{,}000 \qquad \text{3 zeros attached to 34.}$$

These patterns help us develop a shortcut for multiplying by whole numbers ending in zeros.

> To multiply by 10, 100, 1000, and so on,
>   Form the product by attaching the number of zeros in that number to the other factor.
>   For example, $41 \cdot 100 = 4100$.
>   2 zeros

Examples Multiply.

6. $176 \cdot 1000 = 176{,}000$ Attach 3 zeros.

7. $2041 \cdot 100 = 204{,}100$ Attach 2 zeros.

■ Work Practice 6–7

We can use a similar format to multiply by any whole number ending in zeros. For example, since

$$15 \cdot 500 = 15 \cdot 5 \cdot 100,$$

we find the product by multiplying 15 and 5, then attaching two zeros to the product.

$$
\begin{array}{r}
\overset{2}{15} \\
\times\ 5 \\
\hline
75
\end{array}
\qquad 15 \cdot 500 = 7500
$$

Practice 6–7

Multiply.
6. $75 \cdot 100$
7. $808 \cdot 1000$

Answers
6. 7500   **7.** 808,000

✓**Concept Check Answer**

$$
\begin{array}{r}
102 \\
\times\ 33 \\
\hline
306 \\
3060 \\
\hline
3366
\end{array}
$$

Multiply.

8. $25 \cdot 9000 = 225{,}000$ $\begin{array}{r} 4 \\ 25 \\ \times\, 9 \\ \hline 225 \end{array}$ Attach 3 zeros.

9. $20 \cdot 7000 = 140{,}000$ Attach 4 zeros.
$\underline{2 \cdot 7}$

■ Work Practice 8–9

Practice 8–9

Multiply.
8. $35 \cdot 3000$
9. $600 \cdot 600$

Objective D Finding the Area of a Rectangle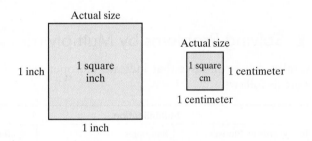

A special application of multiplication is finding the **area** of a region. Area measures the amount of surface of a region. For example, we measure a plot of land or the living space of a home by its area. The figures below show two examples of units of area measure. (A centimeter is a unit of length in the metric system.)

For example, to measure the area of a geometric figure such as the rectangle below, count the number of square units that cover the region.

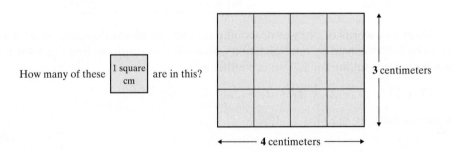

This rectangular region contains 12 square units, each 1 square centimeter. Thus, the area is 12 square centimeters. This total number of squares can be found by counting or by multiplying $4 \cdot 3$ (length · width).

Area of a rectangle = length · width
= (4 centimeters)(3 centimeters)
= 12 square centimeters

In this section, we find the areas of rectangles only. In later sections, we will find the areas of other geometric regions.

Helpful Hint

Notice that area is measured in **square** units while perimeter is measured in units.

Answers
8. 105,000 **9.** 360,000

Practice 10

The state of Wyoming is in the shape of a rectangle whose length is 360 miles and whose width is 280 miles. Find its area.

Example 10 Finding the Area of a State

The state of Colorado is in the shape of a rectangle whose length is 380 miles and whose width is 280 miles. Find its area.

Solution: The area of a rectangle is the product of its length and its width.

Area = length · width
   = (380 miles)(280 miles)
   = 106,400 square miles

The area of Colorado is 106,400 square miles.

🔳 Work Practice 10

Objective E Solving Problems by Multiplying 🔘

There are several words or phrases that indicate the operation of multiplication. Some of these are as follows:

Multiplication		
Key Words or Phrases	**Examples**	**Symbols**
multiply	multiply 5 by 7	5 · 7
product	the product of 3 and 2	3 · 2
times	10 times 13	10 · 13

Many key words or phrases describing real-life problems that suggest addition might be better solved by multiplication instead. For example, to find the **total** cost of 8 shirts, each selling for $27, we can either add

27 + 27 + 27 + 27 + 27 + 27 + 27 + 27

or we can multiply 8(27).

Practice 11

A particular computer printer can print 16 pages per minute in color. How many pages can it print in 45 minutes?

Example 11 Finding DVD Space

A digital video disc (DVD) can hold about 4800 megabytes (MB) of information. How many megabytes can 12 DVDs hold?

Solution: Twelve DVDs will hold 12 × 4800 megabytes.

In Words	**Translate to Numbers**
megabytes per disc →	4800
× DVDs →	× 12
	9600
	48000
total megabytes	57,600

Twelve DVDs will hold 57,600 megabytes.

🔳 Work Practice 11

Example 12 Budgeting Money

Suzanne Scarpulla and a friend plan to take their children to the Georgia Aquarium in Atlanta, the world's largest aquarium. The peak hour ticket price for each child is $29 and for each adult, $35. If five children and two adults plan to go, how much money is needed for admission? (*Source:* GeorgiaAquarium.org)

Solution: If the price of one child's ticket is $29, the cost for 5 children is $5 \times 29 = \$145$. The price of one adult ticket is $35, so the cost for two adults is $2 \times 35 = \$70$. The total cost is:

In Words		**Translate to Numbers**
cost for 5 children	$\rightarrow$	145
+ cost for 2 adults	$\rightarrow$	+ 70
total cost		215

The total cost is $215.

▪ Work Practice 12

Practice 12

Ken Shimura purchased DVDs and CDs through a club. Each DVD was priced at $11, and each CD cost $9. Ken bought eight DVDs and five CDs. Find the total cost of the order.

Example 13 Estimating Word Count

The average page of a book contains 259 words. Estimate, rounding each number to the nearest hundred, the total number of words contained on 212 pages.

Solution: The exact number of words is 259×212. Estimate this product by rounding each factor to the nearest hundred.

$$259 \quad \text{rounds to} \quad 300$$
$$\times 212 \quad \text{rounds to} \quad \times 200,$$

$$300 \times 200 = 60,000$$
$$3 \cdot 2 = 6$$

There are approximately 60,000 words contained on 212 pages.

▪ Work Practice 13

Practice 13

If an average page in a book contains 163 words, estimate, rounding each number to the nearest hundred, the total number of words contained on 391 pages.

🖩 **Calculator Explorations** **Multiplying Numbers**

To multiply numbers on a calculator, find the keys marked $\boxed{\times}$ and $\boxed{=}$ or $\boxed{\text{ENTER}}$. For example, to find $31 \cdot 66$ on a calculator, press the keys $\boxed{31}$ $\boxed{\times}$ $\boxed{66}$ then $\boxed{=}$ or $\boxed{\text{ENTER}}$. The display will read $\boxed{\quad\quad 2046}$. Thus, $31 \cdot 66 = 2046$.

Use a calculator to multiply.

1. 72×48 **2.** 81×92

3. $163 \cdot 94$ **4.** $285 \cdot 144$

5. $983(277)$ **6.** $1562(843)$

Answers
12. $133   **13.** 80,000 words

Vocabulary, Readiness & Video Check

Use the choices below to fill in each blank.

area	grouping	commutative	1	product	length
factor	order	associative	0	distributive	number

1. The product of 0 and any number is _____.

2. The product of 1 and any number is the _____.

3. In $8 \cdot 12 = 96$, the 96 is called the _____ and 8 and 12 are each called a(n) _____.

4. Since $9 \cdot 10 = 10 \cdot 9$, we say that changing the _____ in multiplication does not change the product. This property is called the _____ property of multiplication.

5. Since $(3 \cdot 4) \cdot 6 = 3 \cdot (4 \cdot 6)$, we say that changing the _____ in multiplication does not change the product. This property is called the _____ property of multiplication.

6. _____ measures the amount of surface of a region.

7. Area of a rectangle = _____ $\cdot$ width.

8. We know $9(10 + 8) = 9 \cdot 10 + 9 \cdot 8$ by the _____ property.

Martin-Gay Interactive Videos

Watch the section lecture video and answer the following questions.

See Video 1.6

Objective A   9. The expression in ▤ Example 3 is rewritten using what property? ▶

Objective B   10. During the multiplication process for ▤ Example 5, why is a single zero placed at the end of the second partial product? ▶

Objective C   11. Explain two different approaches to solving the multiplication problem $50 \cdot 900$ in ▤ Example 7. ▶

Objective D   12. Why are the units to the answer to ▤ Example 8 not just meters? What are the correct units? ▶

Objective E   13. In ▤ Example 9, why can "total" imply multiplication as well as addition? ▶

1.6   Exercise Set MyMathLab® ▶

Objective A *Multiply. See Example 1.*

▶ **1.** $1 \cdot 24$ **2.** $55 \cdot 1$ ▶ **3.** $0 \cdot 19$ **4.** $27 \cdot 0$

5. $8 \cdot 0 \cdot 9$ **6.** $7 \cdot 6 \cdot 0$ **7.** $87 \cdot 1$ **8.** $1 \cdot 41$

Use the distributive property to rewrite each expression. See Example 2.

9. $6(3 + 8)$ **10.** $5(8 + 2)$ **11.** $4(3 + 9)$

12. $6(1 + 4)$ ▶ **13.** $20(14 + 6)$ **14.** $12(12 + 3)$

Objective B *Multiply. See Example 3.*

15. 64
　　　× 8

16. 79
　　　× 3

17. 613
　　　× 6

18. 638
　　　× 5

▶ **19.** 277 × 6

20. 882 × 2

21. 1074 × 6

22. 9021 × 3

Objectives A B Mixed Practice *Multiply. See Examples 1 through 5.*

23. 89
　　　× 13

24. 91
　　　× 72

25. 421
　　　× 58

26. 526
　　　× 23

27. 306
　　　× 81

28. 708
　　　× 21

29. (780)(20)

30. (720)(80)

31. (495)(13)(0)

32. (593)(47)(0)

33. (640)(1)(10)

34. (240)(1)(20)

35. 1234 × 39

36. 1357 × 79

37. 609 × 234

38. 807 × 127

▶ **39.** 8649
　　　× 274

40. 1234
　　　× 567

41. 589
　　　× 110

42. 426
　　　× 110

43. 1941
　　　× 2035

44. 1876
　　　× 1407

Objective C *Multiply. See Examples 6 through 9.*

▶ **45.** 8 × 100

46. 6 × 100

47. 11 × 1000

48. 26 × 1000

49. 7406 · 10

50. 9054 · 10

51. 6 · 4000

52. 3 · 9000

▶ **53.** 50 · 900

54. 70 · 300

55. 41 · 80,000

56. 27 · 50,000

Objective D Mixed Practice (*Section 1.3*) *Find the area and the perimeter of each rectangle. See Example 10.*

▶ △ **57.**　9 meters
　　　　　7 meters

△ **58.** 3 inches
　　　　13 inches

△ **59.** 17 feet
　　　　40 feet

△ **60.** 25 centimeters
　　　　20 centimeters

Objective E Mixed Practice (Section 1.5) *Estimate the products by rounding each factor to the nearest hundred. See Example 13.*

61. 576×354　　**62.** 982×650　　**63.** 604×451　　**64.** 111×999

Without actually calculating, mentally round, multiply, and choose the best estimate.

65. $38 \times 42 =$
 a. 16
 b. 160
 c. 1600
 d. 16,000

66. $2872 \times 12 =$
 a. 2872
 b. 28,720
 c. 287,200
 d. 2,872,000

67. $612 \times 29 =$
 a. 180
 b. 1800
 c. 18,000
 d. 180,000

68. $706 \times 409 =$
 a. 280
 b. 2800
 c. 28,000
 d. 280,000

Objectives D E Mixed Practice–Translating *Solve. See Examples 10 through 13.*

69. Multiply 80 by 11.

70. Multiply 70 by 12.

71. Find the product of 6 and 700.

72. Find the product of 9 and 900.

73. Find 2 times 2240.

74. Find 3 times 3310.

75. One tablespoon of olive oil contains 125 calories. How many calories are in 3 tablespoons of olive oil? (*Source: Home and Garden Bulletin No. 72*, U.S. Department of Agriculture)

76. One ounce of hulled sunflower seeds contains 14 grams of fat. How many grams of fat are in 8 ounces of hulled sunflower seeds? (*Source: Home and Garden Bulletin No. 72*, U.S. Department of Agriculture)

77. The textbook for a course in biology costs $94. There are 35 students in the class. Find the total cost of the biology books for the class.

78. The seats in a lecture hall are arranged in 14 rows with 34 seats in each row. Find how many seats are in this room.

79. Cabot Creamery is packing a pallet of 20-lb boxes of cheddar cheese to send to a local restaurant. There are five layers of boxes on the pallet, and each layer is four boxes wide by five boxes deep.
 a. How many boxes are in one layer?
 b. How many boxes are on the pallet?
 c. What is the weight of the cheese on the pallet?

80. An apartment building has *three floors*. Each floor has five rows of apartments with four apartments in each row.
 a. How many apartments are on 1 floor?
 b. How many apartments are in the building?

81. A plot of land measures 80 feet by 110 feet. Find its area.

 82. A house measures 45 feet by 60 feet. Find the floor area of the house.

△ **83.** The largest hotel lobby can be found at the Hyatt Regency in San Francisco, CA. It is in the shape of a rectangle that measures 350 feet by 160 feet. Find its area.

△ **84.** Recall from an earlier section that the world's largest commercial building under one roof is the flower auction building of the cooperative VBA in Aalsmeer, Netherlands. The floor plan is a rectangle that measures 776 meters by 639 meters. Find the area of this building. (*Source: The Handy Science Answer Book,* Visible Ink Press)

776 meters

639 meters

85. A pixel is a rectangular dot on a graphing calculator screen. If a graphing calculator screen contains 62 pixels in a row and 94 pixels in a column, find the total number of pixels on a screen.

86. A certain compact disc (CD) can hold 700 megabytes (MB) of information. How many MB can 17 discs hold?

87. A line of print on a computer contains 60 characters (letters, spaces, punctuation marks). Find how many characters there are in 35 lines.

88. An average cow eats 3 pounds of grain per day. Find how much grain a cow eats in a year. (Assume 365 days in 1 year.)

89. One ounce of Planters® Dry Roasted Peanuts has 170 calories. How many calories are in 8 ounces? (*Source:* Kraft Foods)

90. One ounce of Planters® Dry Roasted Peanuts has 14 grams of fat. How many grams of fat are in 16 ounces? (*Source:* Kraft Foods)

91. The Thespian club at a local community college is ordering T-shirts. T-shirts size S, M, or L cost $10 each and T-shirts size XL or XXL cost $12 each. Use the table below to find the total cost. (The first row is filled in for you.)

T-Shirt Size	Number of Shirts Ordered	Cost per Shirt	Cost per Size Ordered
S	4	$10	$40
M	6		
L	20		
XL	3		
XXL	3		

92. The student activities group at North Shore Community College is planning a trip to see the local minor league baseball team. Tickets cost $5 for students, $7 for nonstudents, and $2 for children under 12. Use the following table to find the total cost.

Person	Number of Persons	Cost per Person	Cost per Category
Student	24	$5	$120
Nonstudent	4		
Children under 12	5		

93. Celestial Seasonings of Boulder, Colorado, is a tea company that specializes in herbal teas, accounting for over $100,000,000 in herbal tea blend sales in the United States annually. Their plant in Boulder has bagging machines capable of bagging over 1000 bags of tea per minute. If the plant runs 24 hours a day, how many tea bags are produced in one day? (*Source:* Celestial Seasonings)

94. The number of "older" Americans (ages 65 and older) has increased fourteenfold since 1900. If there were 3 million "older" Americans in 1900, how many were there in 2012? (*Source:* U.S. Census Bureau)

Mixed Practice (Sections 1.3, 1.4, 1.6) *Perform each indicated operation.*

95. $\begin{array}{r} 128 \\ +\ 7 \\ \hline \end{array}$

96. $\begin{array}{r} 126 \\ -\ 8 \\ \hline \end{array}$

97. $\begin{array}{r} 134 \\ \times\ 16 \\ \hline \end{array}$

98. $47 + 26 + 10 + 231 + 50$

99. Find the sum of 19 and 4.

100. Find the product of 19 and 4.

101. Find the difference of 19 and 4.

102. Find the total of 19 and 4.

Concept Extensions

Solve. See the first Concept Check in this section.

103. Rewrite $7 + 7 + 7 + 7$ using multiplication.

104. Rewrite $11 + 11 + 11 + 11 + 11 + 11$ using multiplication.

105. a. Rewrite $3 \cdot 5$ as repeated addition.
   b. Explain why there is more than one way to do this.

106. a. Rewrite $4 \cdot 5$ as repeated addition.
   b. Explain why there is more than one way to do this.

Find and explain the error in each multiplication problem. See the second Concept Check in this section.

107. $\begin{array}{r} 203 \\ \times\ 14 \\ \hline 812 \\ 203 \\ \hline 1015 \\ \hline \end{array}$

108. $\begin{array}{r} 31 \\ \times\ 50 \\ \hline 155 \\ \hline \end{array}$

Fill in the missing digits in each problem.

109. $\begin{array}{r} 4\_ \\ \times\ \_3 \\ \hline 126 \\ 3780 \\ \hline 3906 \\ \hline \end{array}$

110. $\begin{array}{r} \_7 \\ \times\ 6\_ \\ \hline 171 \\ 3420 \\ \hline 3591 \\ \hline \end{array}$

111. Explain how to multiply two 2-digit numbers using partial products.

112. In your own words, explain the meaning of the area of a rectangle and how this area is measured.

113. A window washer in New York City is bidding for a contract to wash the windows of a 23-story building. To write a bid, the number of windows in the building is needed. If there are 7 windows in each row of windows on 2 sides of the building and 4 windows per row on the other 2 sides of the building, find the total number of windows.

114. During the 2012–2013 regular season, Kevin Durant of the Oklahoma City Thunder led the NBA in total points scored. He scored 139 three-point field goals, 592 two-point field goals, and 679 free throws (worth one point each). How many points did Kevin Durant score during the 2012–2013 regular season? (*Source:* NBA)

1.7 Dividing Whole Numbers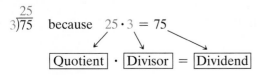

Suppose three people pooled their money and bought a raffle ticket at a local fundraiser. Their ticket was the winner, and they won a $75 cash prize. They then divided the prize into three equal parts so that each person received $25.

Objectives

A Divide Whole Numbers.

B Perform Long Division.

C Solve Problems That Require Dividing by Whole Numbers.

D Find the Average of a List of Numbers.

Objective A Dividing Whole Numbers

The process of separating a quantity into equal parts is called **division.** The division above can be symbolized by several notations.

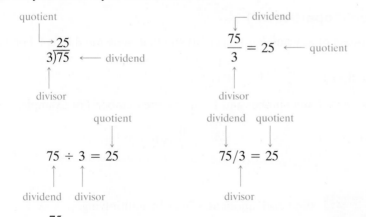

(In the notation $\dfrac{75}{3}$, the bar separating 75 and 3 is called a **fraction bar.**) Just as subtraction is the reverse of addition, division is the reverse of multiplication. This means that division can be checked by multiplication.

$3\overline{)75}^{25}$ because $25 \cdot 3 = 75$

$\boxed{\text{Quotient}} \cdot \boxed{\text{Divisor}} = \boxed{\text{Dividend}}$

Since multiplication and division are related in this way, you can use your knowledge of multiplication facts (or study Appendix A.2) to review quotients of one-digit divisors if necessary.

Practice 1

Find each quotient. Check by multiplying.

a. $9\overline{)72}$

b. $40 \div 5$

c. $\dfrac{24}{6}$

Practice 2

Find each quotient. Check by multiplying.

a. $\dfrac{7}{7}$ **b.** $5 \div 1$

c. $1\overline{)11}$ **d.** $4 \div 1$

e. $\dfrac{10}{1}$ **f.** $21 \div 21$

Practice 3

Find each quotient. Check by multiplying.

a. $\dfrac{0}{7}$ **b.** $8\overline{)0}$

c. $7 \div 0$ **d.** $0 \div 14$

Answers

1. a. 8 **b.** 8 **c.** 4 **2. a.** 1 **b.** 5
c. 11 **d.** 4 **e.** 10 **f.** 1 **3. a.** 0
b. 0 **c.** undefined **d.** 0

Example 1 Find each quotient. Check by multiplying.

a. $42 \div 7$ **b.** $\dfrac{64}{8}$ **c.** $3\overline{)21}$

Solution:

a. $42 \div 7 = 6$ because $6 \cdot 7 = 42$

b. $\dfrac{64}{8} = 8$ because $8 \cdot 8 = 64$

c. $3\overline{)21}^{\,7}$ because $7 \cdot 3 = 21$

■ Work Practice 1

Example 2 Find each quotient. Check by multiplying.

a. $1\overline{)7}$ **b.** $12 \div 1$ **c.** $\dfrac{6}{6}$ **d.** $9 \div 9$ **e.** $\dfrac{20}{1}$ **f.** $18\overline{)18}$

Solution:

a. $1\overline{)7}^{\,7}$ because $7 \cdot 1 = 7$ **b.** $12 \div 1 = 12$ because $12 \cdot 1 = 12$

c. $\dfrac{6}{6} = 1$ because $1 \cdot 6 = 6$ **d.** $9 \div 9 = 1$ because $1 \cdot 9 = 9$

e. $\dfrac{20}{1} = 20$ because $20 \cdot 1 = 20$ **f.** $18\overline{)18}^{\,1}$ because $1 \cdot 18 = 18$

■ Work Practice 2

Example 2 illustrates the important properties of division described next:

Division Properties of 1

The quotient of any number (except 0) and that same number is 1. For example,

$$8 \div 8 = 1 \qquad \dfrac{5}{5} = 1 \qquad 4\overline{)4}^{\,1}$$

The quotient of any number and 1 is that same number. For example,

$$9 \div 1 = 9 \qquad \dfrac{6}{1} = 6 \qquad 1\overline{)3}^{\,3} \qquad \dfrac{0}{1} = 0$$

Example 3 Find each quotient. Check by multiplying.

a. $9\overline{)0}$ **b.** $0 \div 12$ **c.** $\dfrac{0}{5}$ **d.** $\dfrac{3}{0}$

Solution:

a. $9\overline{)0}^{\,0}$ because $0 \cdot 9 = 0$ **b.** $0 \div 12 = 0$ because $0 \cdot 12 = 0$

c. $\dfrac{0}{5} = 0$ because $0 \cdot 5 = 0$

d. If $\dfrac{3}{0}$ = a *number,* then the *number* times $0 = 3$. Recall from Section 1.6 that any number multiplied by 0 is 0 and not 3. We say, then, that $\dfrac{3}{0}$ is **undefined.**

◼ Work Practice 3

Example 3 illustrates important division properties of 0.

Division Properties of 0

The quotient of 0 and any number (except 0) is 0. For example,

$$0 \div 9 = 0 \qquad \frac{0}{5} = 0 \qquad 14\overline{)0}^{\,0}$$

The quotient of any number and 0 is not a number. We say that

$$\frac{3}{0}, \quad 0\overline{)3}, \quad \text{and} \quad 3 \div 0$$

are **undefined.**

Objective B Performing Long Division ▶

When dividends are larger, the quotient can be found by a process called **long division.** For example, let's divide 2541 by 3.

$$\text{divisor} \rightarrow 3\overline{)2541}$$
$$\uparrow$$
$$\text{dividend}$$

We can't divide 3 into 2, so we try dividing 3 into the first two digits.

$$\begin{array}{r} 8 \\ 3\overline{)2541} \end{array} \qquad 25 \div 3 = 8 \text{ with 1 left, so our best estimate is 8. We place 8 over the 5 in 25.}$$

Next, multiply 8 and 3 and subtract this product from 25. Make sure that this difference is less than the divisor.

$$\begin{array}{r} 8 \\ 3\overline{)2541} \\ -24 \\ \hline 1 \end{array} \qquad \begin{array}{l} 8(3) = 24 \\ 25 - 24 = 1, \text{ and 1 is less than the divisor 3.} \end{array}$$

Bring down the next digit and go through the process again.

$$\begin{array}{r} 84 \\ 3\overline{)2541} \\ -24\downarrow \\ \hline 14 \\ -12 \\ \hline 2 \end{array} \qquad \begin{array}{l} 14 \div 3 = 4 \text{ with 2 left} \\ \\ 4(3) = 12 \\ 14 - 12 = 2 \end{array}$$

Once more, bring down the next digit and go through the process.

$$\begin{array}{r} 847 \\ 3\overline{)2541} \\ -24 \\ \hline 14 \\ -12 \\ \hline 21 \\ -21 \\ \hline 0 \end{array} \qquad \begin{array}{l} 21 \div 3 = 7 \\ \\ \\ \\ 7(3) = 21 \\ 21 - 21 = 0 \end{array}$$

The quotient is 847. To check, see that $847 \times 3 = 2541$.

Practice 4

Divide. Check by multiplying.

a. $4908 \div 6$

b. $2212 \div 4$

c. $753 \div 3$

Helpful Hint   Since division and multiplication are reverse operations, don't forget that a division problem can be checked by multiplying.

Example 4   Divide: $3705 \div 5$. Check by multiplying.

Solution:

$$
\begin{array}{r}
7 \\
5\overline{)3705} \\
-35\downarrow \\
\hline
20
\end{array}
$$

$37 \div 5 = 7$ with 2 left. Place this estimate, 7, over the 7 in 37.

$7(5) = 35$

$37 - 35 = 2$, and 2 is less than the divisor 5.

⌐ Bring down the 0.

$$
\begin{array}{r}
74 \\
5\overline{)3705} \\
-35 \\
\hline
20 \\
-20\downarrow \\
\hline
05
\end{array}
$$

$20 \div 5 = 4$

$4(5) = 20$

$20 - 20 = 0$. and 0 is less than the divisor 5.

⌐ Bring down the 5.

$$
\begin{array}{r}
741 \\
5\overline{)3705} \\
-35 \\
\hline
20 \\
-20\downarrow \\
\hline
5 \\
-5 \\
\hline
0
\end{array}
$$

$5 \div 5 = 1$

$1(5) = 5$

$5 - 5 = 0$

Check:

$$
\begin{array}{r}
741 \\
\times \quad 5 \\
\hline
3705
\end{array}
$$

■ Work Practice 4

Practice 5

Divide and check by multiplying.

a. $7\overline{)2128}$

b. $9\overline{)45{,}900}$

Example 5   Divide and check: $1872 \div 9$

Solution:

$$
\begin{array}{r}
208 \\
9\overline{)1872} \\
-18\downarrow\downarrow \\
\hline
07 \\
-0\downarrow \\
\hline
72 \\
-72 \\
\hline
0
\end{array}
$$

$2(9) = 18$

$18 - 18 = 0$; bring down the 7.

$0(9) = 0$

$7 - 0 = 7$; bring down the 2.

$8(9) = 72$

$72 - 72 = 0$

Check: $208 \cdot 9 = 1872$

■ Work Practice 5

Answers

4. a. 818   **b.** 553   **c.** 251

5. a. 304   **b.** 5100

Naturally, quotients don't always "come out even." Making 4 rows out of 26 chairs, for example, isn't possible if each row is supposed to have exactly the same number of chairs. Each of 4 rows can have 6 chairs, but 2 chairs are still left over.

4 rows — 6 chairs in each row

2 chairs left over

We signify "leftovers" or **remainders** in this way:

$$\begin{array}{r} 6 \ \ \text{R } 2 \\ 4)\overline{26} \end{array}$$

The **whole number part of the quotient** is 6; the **remainder part of the quotient** is 2. Checking by multiplying,

whole number part	·	divisor	+	remainder part	=	dividend
↓		↓		↓		↓
6	·	4	+	2		
		24	+	2	=	26

Example 6   Divide and check: $2557 \div 7$

Solution:

$$\begin{array}{r} 365 \ \ \text{R } 2 \\ 7)\overline{2557} \\ -21 \\ \hline 45 \\ -42 \\ \hline 37 \\ -35 \\ \hline 2 \end{array}$$

$3(7) = 21$

$25 - 21 = 4$; bring down the 5.

$6(7) = 42$

$45 - 42 = 3$; bring down the 7.

$5(7) = 35$

$37 - 35 = 2$; the remainder is 2.

Check: 365 · 7 + 2 = 2557

whole number part · divisor + remainder part = dividend

 Work Practice 6

Practice 6
Divide and check.
a. $4)\overline{939}$
b. $5)\overline{3287}$

Answers
6. a. 234 R 3   **b.** 657 R 2

Practice 7

Divide and check.

a. $9\overline{)81,605}$

b. $4\overline{)23,310}$

Example 7 Divide and check: $56,717 \div 8$

Solution:

$$
\begin{array}{r}
7089 \;\; R\,5 \\
8\overline{)56717} \\
\end{array}
$$

$$
\begin{array}{rl}
-56\!\downarrow\!\downarrow\!\downarrow & \quad 7(8) = 56 \\
\hline
07 & \quad \text{Subtract and bring down the 7.} \\
-0\!\downarrow & \quad 0(8) = 0 \\
\hline
71 & \quad \text{Subtract and bring down the 1.} \\
-64\!\downarrow & \quad 8(8) = 64 \\
\hline
77 & \quad \text{Subtract and bring down the 7.} \\
-72 & \quad 9(8) = 72 \\
\hline
5 & \quad \text{Subtract. The remainder is 5.}
\end{array}
$$

Check: $7089 \quad \cdot \quad 8 \quad + \quad 5 \quad = \quad 56{,}717$

whole number part	·	divisor	+	remainder part	=	dividend

Work Practice 7

When the divisor has more than one digit, the same pattern applies. For example, let's find $1358 \div 23$.

$$
\begin{array}{r}
5 \\
23\overline{)1358} \\
-115\!\downarrow \\
\hline
208
\end{array}
$$

$135 \div 23 = 5$ with 20 left over. Our estimate is 5.

$5(23) = 115$

$135 - 115 = 20.$ Bring down the 8.

Now we continue estimating.

$$
\begin{array}{r}
59 \;\; R\,1 \\
23\overline{)1358} \\
-115 \\
\hline
208 \\
-207 \\
\hline
1
\end{array}
$$

$208 \div 23 = 9$ with 1 left over.

$9(23) = 207$

$208 - 207 = 1.$ The remainder is 1.

To check, see that $59 \cdot 23 + 1 = 1358$.

Practice 8

Divide: $8920 \div 17$

Example 8 Divide: $6819 \div 17$

Solution:

$$
\begin{array}{r}
401 \;\; R\,2 \\
17\overline{)6819} \\
\end{array}
$$

$$
\begin{array}{rl}
-68\!\downarrow\!\downarrow & \quad 4(17) = 68 \\
\hline
01 & \quad \text{Subtract and bring down the 1.} \\
-0\!\downarrow & \quad 0(17) = 0 \\
\hline
19 & \quad \text{Subtract and bring down the 9.} \\
-17 & \quad 1(17) = 17 \\
\hline
2 & \quad \text{Subtract. The remainder is 2.}
\end{array}
$$

To check, see that $401 \cdot 17 + 2 = 6819$.

Work Practice 8

Answers

7. a. 9067 R 2   **b.** 5827 R 2

8. 524 R 12

Example 9 Divide: 51,600 ÷ 403

Solution:

```
          128  R 16
   403)51600
       −403↓
        1130
        −806↓
        3240
       −3224
          16
```

1(403) = 403
Subtract and bring down the 0.
2(403) = 806
Subtract and bring down the 0.
8(403) = 3224
Subtract. The remainder is 16.

To check, see that $128 \cdot 403 + 16 = 51{,}600$.

🔲 **Work Practice 9**

Practice 9

Divide: 33,282 ÷ 678

Division Shown as Repeated Subtraction To further understand division, recall from Section 1.6 that addition and multiplication are related in the following manner:

$$\underbrace{3 + 3 + 3 + 3}_{\text{4 addends; each addend is 3}} = 4 \times 3 = 12$$

In other words, multiplication is repeated addition. Likewise, division is repeated subtraction.

For example, let's find

$$35 \div 8$$

by repeated subtraction. Keep track of the number of times 8 is subtracted from 35. We are through when we can subtract no more because the difference is less than 8.

35 ÷ 8:   Repeated Subtraction

```
  35 ⎫
  −8 ⎬ 1 time
  ───
  27 ⎫
  −8 ⎬ 2 times
  ───
  19 ⎫
  −8 ⎬ 3 times
  ───
  11 ⎫
  −8 ⎬ 4 times
  ───
   3  ⟵ Remainder
```
(We cannot subtract 8 again.)

Thus, $35 \div 8 = 4 \text{ R } 3$.

To check, perform the same multiplication as usual and finish by adding in the remainder.

whole number part of quotient	·	divisor	+	remainder	=	dividend
↓		↓		↓		↓
4	·	8	+	3	=	35

Answer

9. 49 R 60

Objective C Solving Problems by Dividing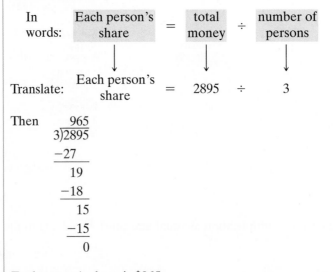

Below are some key words and phrases that may indicate the operation of division:

Division		
Key Words or Phrases	**Examples**	**Symbols**
divide	divide 10 by 5	$10 \div 5$ or $\dfrac{10}{5}$
quotient	the quotient of 64 and 4	$64 \div 4$ or $\dfrac{64}{4}$
divided by	9 divided by 3	$9 \div 3$ or $\dfrac{9}{3}$
divided or shared equally among	$100 divided equally among five people	$100 \div 5$ or $\dfrac{100}{5}$
per	100 miles per 2 hours	$\dfrac{100 \text{ miles}}{2 \text{ hours}}$

✔**Concept Check** Determine whether each of the following is the correct way to represent "the quotient of 60 and 12." Explain your answer.

a. $12 \div 60$

b. $60 \div 12$

Practice 10

Three students bought 171 blank CDs to share equally. How many CDs did each person get?

Example 10   Finding Shared Earnings

Three college students share a paper route to earn money for expenses. The total in their fund after expenses was $2895. How much is each person's equal share?

Solution:

In words: Each person's share   =   total money   ÷   number of persons

Translate: Each person's share   =   2895   ÷   3

Then

$$\begin{array}{r} 965 \\ 3\overline{)2895} \\ \underline{-27} \\ 19 \\ \underline{-18} \\ 15 \\ \underline{-15} \\ 0 \end{array}$$

Each person's share is $965.

■ Work Practice 10

Answer

10. 57 CDs

✔**Concept Check Answers**

a. incorrect   **b.** correct

Example 11 Dividing Number of Downloads

As part of a promotion, an executive receives 238 cards, each good for one free song download. If she wants to share them evenly with 19 friends, how many download cards will each friend receive? How many will be left over?

Solution:

$$\begin{array}{r} 12 \text{ R } 10 \\ 19\overline{)238} \\ -19 \\ \hline 48 \\ -38 \\ \hline 10 \end{array}$$

Each friend will receive 12 download cards. The cards cannot be divided equally among her friends since there is a nonzero remainder. There will be 10 download cards left over.

■ Work Practice 11

Objective D Finding Averages ▶

A special application of division (and addition) is finding the average of a list of numbers. The **average** of a list of numbers is the sum of the numbers divided by the *number* of numbers.

$$\text{average} = \frac{\text{sum of numbers}}{\textit{number of numbers}}$$

Example 12 Averaging Scores

A mathematics instructor is checking a simple program she wrote for averaging the scores of her students. To do so, she averages a student's scores of 75, 96, 81, and 88 by hand. Find this average score.

Solution: To find the average score, we find the sum of the student's scores and divide by 4, the number of scores.

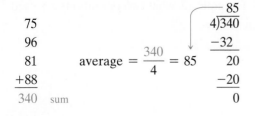

$$\begin{array}{r} 75 \\ 96 \\ 81 \\ +88 \\ \hline 340 \quad \text{sum} \end{array}$$

$$\text{average} = \frac{340}{4} = 85$$

$$\begin{array}{r} 85 \\ 4\overline{)340} \\ -32 \\ \hline 20 \\ -20 \\ \hline 0 \end{array}$$

The average score is 85.

■ Work Practice 12

Practice 11

Printers can be packed 12 to a box. If 532 printers are to be packed but only full boxes are shipped, how many full boxes will be shipped? How many printers are left over and not shipped?

Practice 12

To compute a safe time to wait for reactions to occur after allergy shots are administered, a lab technician is given a list of elapsed times between administered shots and reactions. Find the average of the times 4 minutes, 7 minutes, 35 minutes, 16 minutes, 9 minutes, 3 minutes, and 52 minutes.

Answers

11. 44 full boxes; 4 printers left over
12. 18 minutes

Calculator Explorations   Dividing Numbers

To divide numbers on a calculator, find the keys marked ÷ and = or ENTER . For example, to find 435 ÷ 5 on a calculator, press the keys 435 ÷ 5 then = or ENTER . The display will read 87 . Thus, 435 ÷ 5 = 87.

Use a calculator to divide.

1. 848 ÷ 16

2. 564 ÷ 12

3. 95)‾5890

4. 27)‾1053

5. $\dfrac{32,886}{126}$

6. $\dfrac{143,088}{264}$

7. 0 ÷ 315

8. 315 ÷ 0

Vocabulary, Readiness & Video Check

Use the choices below to fill in each blank. Some choices may be used more than once.

1	number	divisor	dividend
0	undefined	average	quotient

1. In 90 ÷ 2 = 45, the answer 45 is called the _____ , 90 is called the _____ , and 2 is called the

_____ .

2. The quotient of any number and 1 is the same _____ .

3. The quotient of any number (except 0) and the same number is _____ .

4. The quotient of 0 and any number (except 0) is _____ .

5. The quotient of any number and 0 is _____ .

6. The _____ of a list of numbers is the sum of the numbers divided by the _____ of numbers.

Martin-Gay Interactive Videos

See Video 1.7 🍎

Watch the section lecture video and answer the following questions.

Objective A   **7.** Look at ⊞ Examples 6–8. What number can never be the divisor in division? ▶

Objective B   **8.** In ⊞ Example 10, how many 102s are in 21? How does this result affect the quotient? ▶

          9. What calculation would you use to check the answer in ⊞ Example 10? ▶

Objective C   **10.** In ⊞ Example 11, what is the importance of knowing that the distance to each hole is the same? ▶

Objective D   **11.** As shown in ⊞ Example 12, what two operations are used when finding an average? ▶

1.7 Exercise Set MyMathLab® ▶

Objective A *Find each quotient. See Examples 1 through 3.*

1. $54 \div 9$ **2.** $72 \div 9$ **▶ 3.** $36 \div 3$ **4.** $24 \div 3$ **5.** $0 \div 8$

6. $0 \div 4$ **▶ 7.** $31 \div 1$ **8.** $38 \div 1$ **▶ 9.** $\dfrac{18}{18}$ **10.** $\dfrac{49}{49}$

11. $\dfrac{24}{3}$ **12.** $\dfrac{45}{9}$ **▶ 13.** $26 \div 0$ **14.** $\dfrac{12}{0}$ **15.** $26 \div 26$

16. $6 \div 6$ **▶ 17.** $0 \div 14$ **18.** $7 \div 0$ **19.** $18 \div 2$ **20.** $18 \div 3$

Objectives A B Mixed Practice *Divide and then check by multiplying. See Examples 1 through 5.*

21. $3\overline{)87}$ **22.** $5\overline{)85}$ **23.** $3\overline{)222}$ **24.** $8\overline{)640}$ **25.** $3\overline{)1014}$ **26.** $4\overline{)2104}$

27. $\dfrac{30}{0}$ **28.** $\dfrac{0}{30}$ **29.** $63 \div 7$ **30.** $56 \div 8$ **31.** $150 \div 6$ **32.** $121 \div 11$

Divide and then check by multiplying. See Examples 6 and 7.

33. $7\overline{)479}$ **34.** $7\overline{)426}$ **35.** $6\overline{)1421}$ **36.** $3\overline{)1240}$

37. $305 \div 8$ **38.** $167 \div 3$ **39.** $2286 \div 7$ **40.** $3333 \div 4$

Divide and then check by multiplying. See Examples 8 and 9.

▶ 41. $55\overline{)715}$ **42.** $23\overline{)736}$ **43.** $23\overline{)1127}$ **44.** $42\overline{)2016}$ **45.** $97\overline{)9417}$

46. $44\overline{)1938}$ **47.** $3146 \div 15$ **48.** $7354 \div 12$ **49.** $6578 \div 13$ **50.** $5670 \div 14$

51. $9299 \div 46$ **52.** $2505 \div 64$ **53.** $\dfrac{12{,}744}{236}$ **54.** $\dfrac{5781}{123}$ **55.** $\dfrac{10{,}297}{103}$

56. $\dfrac{23{,}092}{240}$ **▶ 57.** $20{,}619 \div 102$ **58.** $40{,}853 \div 203$ **59.** $244{,}989 \div 423$ **60.** $164{,}592 \div 543$

Divide. See Examples 1 through 9.

61. $7\overline{)119}$

62. $8\overline{)104}$

63. $7\overline{)3580}$

64. $5\overline{)3017}$

65. $40\overline{)85,312}$

66. $50\overline{)85,747}$

67. $142\overline{)863,360}$

68. $214\overline{)650,560}$

Objective **C** Translating *Solve. See Examples 10 and 11.*

69. Find the quotient of 117 and 5.

70. Find the quotient of 94 and 7.

71. Find 200 divided by 35.

72. Find 116 divided by 32.

73. Find the quotient of 62 and 3.

74. Find the quotient of 78 and 5.

75. Martin Thieme teaches American Sign Language classes for $65 per student for a 7-week session. He collects $2145 from the group of students. Find how many students are in the group.

76. Kathy Gomez teaches Spanish lessons for $85 per student for a 5-week session. From one group of students, she collects $4930. Find how many students are in the group.

77. The gravity of Jupiter is 318 times as strong as the gravity of Earth, so objects on Jupiter weigh 318 times as much as they weigh on Earth. If a person would weigh 52,470 pounds on Jupiter, find how much the person weighs on Earth.

78. Twenty-one people pooled their money and bought lottery tickets. One ticket won a prize of $5,292,000. Find how many dollars each person received.

79. An 18-hole golf course is 5580 yards long. If the distance to each hole is the same, find the distance between holes.

80. A truck hauls wheat to a storage granary. It carries a total of 5768 bushels of wheat in 14 trips. How much does the truck haul each trip if each trip it hauls the same amount?

81. There is a bridge over highway I-35 every three miles. The first bridge is at the beginning of a 265-mile stretch of highway. Find how many bridges there are over 265 miles of I-35.

82. The white stripes dividing the lanes on a highway are 25 feet long, and the spaces between them are 25 feet long. Let's call a "lane divider" a stripe followed by a space. Find how many whole "lane dividers" there are in 1 mile of highway. (A mile is 5280 feet.)

83. Ari Trainor is in the requisitions department of Central Electric Lighting Company. Light poles along a highway are placed 492 feet apart. The first light pole is at the beginning of a 1-mile strip. Find how many poles he should order for the 1-mile strip of highway. (A mile is 5280 feet.)

84. Professor Lopez has a piece of rope 185 feet long that she wants to cut into pieces for an experiment in her physics class. Each piece of rope is to be 8 feet long. Determine whether she has enough rope for her 22-student class. Determine the amount extra or the amount short.

85. Broad Peak in Pakistan is the twelfth-tallest mountain in the world. Its elevation is 26,400 feet. A mile is 5280 feet. How many miles tall is Broad Peak? (*Source:* National Geographic Society)

86. Arian Foster of the Houston Texans led the NFL in touchdowns during the 2012 regular football season, scoring a total of 102 points from touchdowns. If a touchdown is worth 6 points, how many touchdowns did Foster make during 2012? (*Source:* National Football League)

87. Find how many yards are in 1 mile. (A mile is 5280 feet; a yard is 3 feet.)

88. Find how many whole feet are in 1 rod. (A mile is 5280 feet; 1 mile is 320 rods.)

Objective **D** *Find the average of each list of numbers. See Example 12.*

89. 10, 24, 35, 22, 17, 12

90. 37, 26, 15, 29, 51, 22

91. 205, 972, 210, 161

92. 121, 200, 185, 176, 163

93. 86, 79, 81, 69, 80

94. 92, 96, 90, 85, 92, 79

The normal monthly temperatures in degrees Fahrenheit for Salt Lake City, Utah, is given in the graph. Use this graph to answer Exercises 95 and 96. (Source: National Climatic Data Center)

Normal Monthly Temperatures
(in Fahrenheit) **for**
Salt Lake City, Utah

95. Find the average temperature for June, July, and August.

96. Find the average temperature for October, November, and December.

Mixed Practice (*Sections 1.3, 1.4, 1.6, 1.7*) *Perform each indicated operation. Watch the operation symbol.*

97. 82 + 463 + 29 + 8704

98. 23 + 407 + 92 + 7011

99. 546
    × 28

100. 712
    × 54

101. 722
    − 43

102. 712
    − 54

103. $\dfrac{45}{0}$

104. $\dfrac{0}{23}$

105. 228 ÷ 24

106. 304 ÷ 31

Concept Extensions

Match each word phrase to the correct translation. (Not all letter choices will be used.) See the Concept Check in this section.

107. The quotient of 40 and 8

108. The quotient of 200 and 20

109. 200 divided by 20

110. 40 divided by 8

    a. $20 \div 200$ **b.** $200 \div 20$
    c. $40 \div 8$ **d.** $8 \div 40$

The following table shows the top eight countries with the most Nobel Prize winners through 2013. Use this table to answer Exercises 111 and 112. (Source: Based on data from Encyclopaedia Britannica, Inc.)

111. Find the average number of Nobel Prize winners for the countries shown.

112. Find the average number of Nobel Prize winners per category for the United States.

Countries with Most Nobel Prize Winners, 1901–2013							
Country	Chemistry	Economics	Literature	Peace	Physics	Physiology & Medicine	Total
United States	68	54	11	22	90	97	342
United Kingdom	27	8	8	9	23	31	106
Germany	27	1	8	4	22	17	79
France	7	1	15	8	13	11	55
Sweden	4	2	9	5	4	8	32
Switzerland	6	0	2	3	4	6	21
Russia (USSR)	1	1	4	2	10	2	20
Japan	6	0	2	1	6	2	17

In Example 12 in this section, we found that the average of 75, 96, 81, and 88 is 85. Use this information to answer Exercises 113 and 114.

113. If the number 75 is removed from the list of numbers, does the average increase or decrease? Explain why.

114. If the number 96 is removed from the list of numbers, does the average increase or decrease? Explain why.

115. Without computing it, tell whether the average of 126, 135, 198, 113 is 86. Explain why it is possible or why it is not.

116. Without computing it, tell whether the average of 38, 27, 58, and 43 is 17. Explain why it is possible or why it is not.

117. If the area of a rectangle is 60 square feet and its width is 5 feet, what is its length?

118. If the area of a rectangle is 84 square inches and its length is 21 inches, what is its width?

119. Write down any two numbers whose quotient is 25.

120. Write down any two numbers whose quotient is 1.

121. Find $26 \div 5$ using the process of repeated subtraction.

122. Find $86 \div 10$ using the process of repeated subtraction.

Operations on Whole Numbers

1. 23
46
+ 79

2. 7006
− 451

3. 36
× 45

4. $8\overline{)4496}$

5. $1 \cdot 79$

6. $\dfrac{36}{0}$

7. $9 \div 1$

8. $9 \div 9$

9. $0 \cdot 13$

10. $7 \cdot 0 \cdot 8$

11. $0 \div 2$

12. $12 \div 4$

13. $4219 - 1786$

14. $1861 + 7965$

15. $5\overline{)1068}$

16. 1259
× 63

17. $3 \cdot 9$

18. $45 \div 5$

19. 207
− 69

20. 207
+ 69

21. $7\overline{)7695}$

22. $9\overline{)1000}$

23. $32\overline{)21,222}$

24. $65\overline{)70,000}$

25. $4000 - 2976$

26. $10,000 - 101$

27. 303
× 101

28. $(475)(100)$

29. Find the total of 57 and 8.

30. Find the product of 57 and 8.

Answers

1. _____
2. _____
3. _____
4. _____
5. _____
6. _____
7. _____
8. _____
9. _____
10. _____
11. _____
12. _____
13. _____
14. _____
15. _____
16. _____
17. _____
18. _____
19. _____
20. _____
21. _____
22. _____
23. _____
24. _____
25. _____
26. _____
27. _____
28. _____
29. _____
30. _____

31. _____

32. _____

33. _____

34. _____

35. _____

36. _____

37. _____

38. _____

39. _____

40. _____

41. _____

42. _____

43. _____

44. _____

45. _____

46. _____

31. Find the quotient of 62 and 9.

32. Find the difference of 62 and 9.

33. Subtract 17 from 200.

34. Find the difference of 432 and 201.

Complete the table by rounding the given number to the given place value.

		Tens	Hundreds	Thousands
35.	9735			
36.	1429			
37.	20,801			
38.	432,198			

Find the perimeter and area of each figure.

△ **39.**

Square   6 feet

△ **40.**

14 inches

Rectangle   7 inches

Find the perimeter of each figure.

△ **41.**

13 miles   9 miles

6 miles

△ **42.**

3 meters

4 meters

3 meters

3 meters

Find the average of each list of numbers.

43. 19, 15, 25, 37, 24

44. 108, 131, 98, 159

45. The Mackinac Bridge is a suspension bridge that connects the lower and upper peninsulas of Michigan across the Straits of Mackinac. Its total length is 26,372 feet. The Lake Pontchartrain Bridge is a twin concrete trestle bridge in Slidell, Louisiana. Its total length is 28,547 feet. Which bridge is longer and by how much? (*Sources:* Mackinac Bridge Authority and Federal Highway Administration, Bridge Division)

Straits of Mackinac, MI

Slidell, LA

46. The average teenage male American consumes 2 quarts of carbonated soft drinks per day. On average, how many quarts of carbonated soft drinks would be consumed in a year? (Use 365 for the number of days.) (*Source:* American Beverage Association)

1.8 An Introduction to Problem Solving

Objective A Solving Problems Involving Addition, Subtraction, Multiplication, or Division

In this section, we decide which operation to perform in order to solve a problem. Don't forget the key words and phrases that help indicate which operation to use. Some of these are listed below and were introduced earlier in the chapter. Also included are several words and phrases that translate to the symbol " = ":

Addition (+)	Subtraction (−)	Multiplication (·)	Division (÷)	Equality (=)
sum	difference	product	quotient	equals
plus	minus	times	divide	is equal to
added to	subtract	multiply	shared equally	is/was
more than	less than	multiply by	among	yields
increased by	decreased by	of	divided by	
total	less	double/triple	divided into	

The following problem-solving steps may be helpful to you:

Problem-Solving Steps

1. **UNDERSTAND** the problem. Some ways of doing this are to read and reread the problem, construct a drawing, and look for key words to identify an operation.
2. **TRANSLATE** the problem. That is, write the problem in short form using words, and then translate to numbers and symbols.
3. **SOLVE** the problem. It is helpful to estimate the solution by rounding. Then carry out the indicated operation from step 2.
4. **INTERPRET** the results. *Check* the proposed solution in the stated problem and *state* your conclusions. Write your results with the correct units attached.

Example 1 Calculating the Length of a River

The Hudson River in New York State is 306 miles long. The Snake River in the northwestern United States is 732 miles longer than the Hudson River. How long is the Snake River? (*Source:* U.S. Department of the Interior)

Solution:

1. **UNDERSTAND.** Read and reread the problem, and then draw a picture. Notice that we are told that Snake River is 732 miles longer than the Hudson River. The phrase "longer than" means that we add.

```
|←— 306 miles —→|
     Hudson River        |←———— 732 miles ————→|

                    Snake River
```

(Continued on next page)

Objectives

A Solve Problems by Adding, Subtracting, Multiplying, or Dividing Whole Numbers.

B Solve Problems That Require More Than One Operation.

Practice 1

The building called 555 California Street is the second-tallest building in San Francisco, California, at 779 feet. The tallest building in San Francisco is the Transamerica Pyramid, which is 74 feet taller than 555 California Street. How tall is the Transamerica Pyramid? (*Source: The World Almanac*)

555 California Street   Transamerica Pyramid

Answer
1. 853 ft

2. TRANSLATE.

In words:	Snake River	is	732 miles	longer than	the Hudson River
	↓	↓	↓	↓	↓
Translate:	Snake River	=	732	+	306

3. SOLVE: Let's see if our answer is reasonable by also estimating. We will estimate each addend to the nearest hundred.

$$\begin{array}{rl} 732 & \text{rounds to} \\ +306 & \text{rounds to} \\ \hline 1038 & \text{exact} \end{array} \quad \begin{array}{rl} 700 \\ +300 \\ \hline 1000 & \text{estimate} \end{array}$$

4. INTERPRET. *Check* your work. The answer is reasonable since 1038 is close to our estimated answer of 1000. *State* your conclusion: The Snake River is 1038 miles long.

⬛ Work Practice 1

Practice 2

Four friends bought a lottery ticket and won $65,000. If each person is to receive the same amount of money, how much does each person receive?

Example 2 Filling a Shipping Order

How many cases can be filled with 9900 cans of jalapeños if each case holds 48 cans? How many cans will be left over? Will there be enough cases to fill an order for 200 cases?

Solution:

1. UNDERSTAND. Read and reread the problem. Draw a picture to help visualize the situation.

48 cans   48 cans   48 cans

9900 cans

JALAPEÑOS   JALAPEÑOS   JALAPEÑOS   JALAPEÑOS

Since each case holds 48 cans, we want to know how many 48s there are in 9900. We find this by dividing.

2. TRANSLATE.

In words:	Number of cases	is	9900	divided by	48
	↓	↓	↓	↓	↓
Translate:	Number of cases	=	9900	÷	48

3. SOLVE: Let's estimate a reasonable solution before we actually divide. Since 9900 rounded to the nearest thousand is 10,000 and 48 rounded to the nearest ten is 50, $10{,}000 \div 50 = 200$. Now find the exact quotient.

$$\begin{array}{r} 206 \ \text{R}12 \\ 48\overline{)9900} \\ -96 \\ \hline 300 \\ -288 \\ \hline 12 \end{array}$$

Answer

2. $16,250

4. INTERPRET. *Check* your work. The answer is reasonable since 206 R 12 is close to our estimate of 200. *State* your conclusion: 206 cases will be filled, with 12 cans left over. There will be enough cases to fill an order for 200 cases.

■ Work Practice 2

Example 3   Calculating Budget Costs

The director of a learning lab at a local community college is working on next year's budget. Thirty-three new DVD players are needed at a cost of $187 each. What is the total cost of these DVD players?

Solution:

1. UNDERSTAND. Read and reread the problem, and then draw a diagram.

33 DVD Players

$187 $187   ... $187

From the phrase "total cost," we might decide to solve this problem by adding. This would work, but repeated addition, or multiplication, would save time.

2. TRANSLATE.

In words:	Total cost	is	number of DVD players	times	cost of a DVD player
	↓	↓	↓	↓	↓
Translate:	Total cost	=	33	×	$187

3. SOLVE: Once again, let's estimate a reasonable solution.

$$
\begin{array}{ll}
187 & \text{rounds to} \quad 200 \\
\underline{\times\ 33} & \text{rounds to} \quad \underline{\times\ 30} \\
561 & 6000 \quad \text{estimate} \\
\underline{5610} & \\
6171 & \text{exact}
\end{array}
$$

4. INTERPRET. *Check* your work. *State* your conclusion: The total cost of the DVD players is $6171.

■ Work Practice 3

Example 4   Calculating a Public School Teacher's Salary

In 2012, the average salary for a public school teacher in California was $69,324. For the same year, the average salary for a public school teacher in Iowa was $17,796 less than this. What was the average public school teacher's salary in Iowa? (*Source:* National Education Association)

Solution:

1. UNDERSTAND. Read and reread the problem. Notice that we are told that the Iowa salary is $17,796 less than the California salary. The phrase "less than" indicates subtraction.

(Continued on next page)

Practice 3

The director of the learning lab also needs to include in the budget a line for 425 blank CDs at a cost of $4 each. What is this total cost for the blank CDs?

Practice 4

In 2012, the average salary for a public school teacher in Alaska was $65,468. For the same year, the average salary for a public school teacher in Hawaii was $11,168 less than this. What was the average public school teacher's salary in Hawaii? (*Source:* National Education Association)

Answers
3. $1700 **4.** $54,300

2. TRANSLATE. Remember that order matters when subtracting, so be careful when translating.

In words: | Iowa salary | is | California salary | minus | $17,796 |

Translate: Iowa salary   =   69,324   −   17,796

3. SOLVE. This time, instead of estimating, let's check by adding.

$$\begin{array}{r} 69{,}324 \\ -17{,}796 \\ \hline 51{,}528 \end{array} \qquad \textbf{Check:} \quad \begin{array}{r} 51{,}528 \\ +17{,}796 \\ \hline 69{,}324 \end{array}$$

4. INTERPRET. *Check* your work. The check is above. *State* your conclusion: The average Iowa teacher's salary in 2012 was $51,528.

■ Work Practice 4

Objective B Solving Problems That Require More Than One Operation ▶

We must sometimes use more than one operation to solve a problem.

Practice 5

A gardener is trying to decide how much fertilizer to buy for his yard. He knows that his lot is in the shape of a rectangle that measures 90 feet by 120 feet. He also knows that the floor of his house is in the shape of a rectangle that measures 45 feet by 65 feet. How much area of the lot is not covered by the house?

65 feet
45 feet
120 feet
90 feet

Answer

5. 7875 sq ft

Example 5   Planting a New Garden

A gardener bought enough plants to fill a rectangular garden with length 30 feet and width 20 feet. Because of shading problems from a nearby tree, the gardener changed the width of the garden to 15 feet. If the area is to remain the same, what is the new length of the garden?

Solution:

1. UNDERSTAND. Read and reread the problem. Then draw a picture to help visualize the problem.

Old garden
New garden
20 feet
30 feet
15 feet

2. TRANSLATE. Since the area of the new garden is to be the same as the area of the old garden, let's find the area of the old garden. Recall that

Area = length × width = 30 feet × 20 feet = 600 square feet

Since the area of the new garden is to be 600 square feet also, we need to see how many 15s there are in 600. This means division. In other words,

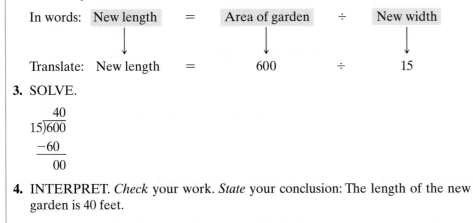

In words: New length　=　Area of garden　÷　New width

Translate: New length　=　　　600　　　÷　　　15

3. SOLVE.

$$\begin{array}{r} 40 \\ 15{\overline{\smash{\big)}\,600}} \\ \underline{-60} \\ 00 \end{array}$$

4. INTERPRET. *Check* your work. *State* your conclusion: The length of the new garden is 40 feet.

■ Work Practice 5

Vocabulary, Readiness & Video Check

Martin-Gay Interactive Videos　　*Watch the section lecture video and answer the following questions.*

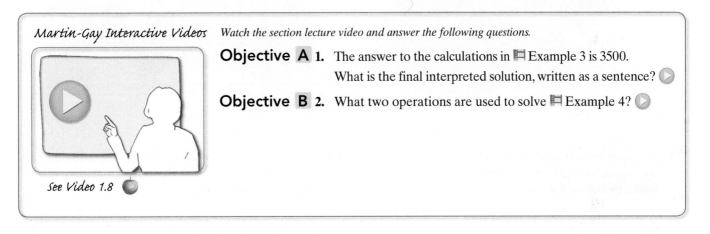

Objective A 1. The answer to the calculations in ▤ Example 3 is 3500. What is the final interpreted solution, written as a sentence? ▶

Objective B 2. What two operations are used to solve ▤ Example 4? ▶

See Video 1.8 🍎

1.8 Exercise Set MyMathLab® ▶

Objective A *Solve. Exercises 1, 2, 11, and 12 have been started for you. See Examples 1 through 4.*

1. 41 increased by 8 is what number?

Start the solution:

1. UNDERSTAND the problem. Reread it as many times as needed.
2. TRANSLATE into an equation. (Fill in the blanks below.)

41	increased by	8	is	what number
↓	↓	↓	↓	↓
41	___	8	___	what number

Finish with:

3. SOLVE
4. INTERPRET

2. What is 12 multiplied by 9?

Start the solution:

1. UNDERSTAND the problem. Reread it as many times as needed.
2. TRANSLATE into an equation. (Fill in the blanks below.)

what	is	12	multiplied by	9
↓	↓	↓	↓	↓
what number	___	12	___	9

Finish with:

3. SOLVE
4. INTERPRET

▶ **3.** What is the quotient of 1185 and 5?

4. 78 decreased by 12 is what number?

▶ **5.** What is the total of 35 and 7?

6. What is the difference of 48 and 8?

7. 60 times 10 is what number?

8. 60 divided by 10 is what number?

△ **9.** A vacant lot in the shape of a rectangle measures 120 feet by 80 feet.

   a. What is the perimeter of the lot?

   b. What is the area of the lot?

80 feet

120 feet

△ **10.** A parking lot in the shape of a rectangle measures 100 feet by 150 feet.

   a. What is the perimeter of the lot?

   b. What is the area of the parking lot?

100 feet

150 feet

11. A family bought a house for $185,700 and later sold the house for $201,200. How much money did they make by selling the house?

Start the solution:

1. UNDERSTAND the problem. Reread it as many times as needed.
2. TRANSLATE into an equation. (Fill in the blanks below.)

money made	is	selling price	minus	purchase price
↓	↓	↓	↓	↓

money
made   = ____ − ____

Finish with:
3. SOLVE
4. INTERPRET

12. Three people dream of equally sharing a $147 million lottery. How much would each person receive if they have the winning ticket?

Start the solution:

1. UNDERSTAND the problem. Reread it as many times as needed.
2. TRANSLATE into an equation. (Fill in the blanks below.)

each person's share	is	lottery amount	divided by	number by persons
↓	↓	↓	↓	↓

each person's
share   = ____ ÷ ____

Finish with:
3. SOLVE
4. INTERPRET

13. There are 24 hours in a day. How many hours are in a week?

14. There are 60 minutes in an hour. How many minutes are in a day?

15. The Verrazano Narrows Bridge is the longest bridge in New York, measuring 4260 feet. The George Washington Bridge, also in New York, is 760 feet shorter than the Verrazano Narrows Bridge. Find the length of the George Washington Bridge.

16. In 2013, the Goodyear Tire & Rubber Company began replacing its fleet of nonrigid GZ-20 blimps with new Goodyear NT semi-rigid airships. The new Goodyear NT airship can hold 297,527 cubic feet of helium. Its GZ-20 predecessor held 94,827 fewer cubic feet of helium. How much helium did a GZ-20 blimp hold? (*Source:* Goodyear Tire & Rubber Company)

17. Yellowstone National Park in Wyoming was the first national park in the United States. It was created in 1872. One of the more recent additions to the National Park System is First State National Monument. It was established in 2013. How much older is Yellowstone than First State? (*Source:* National Park Service)

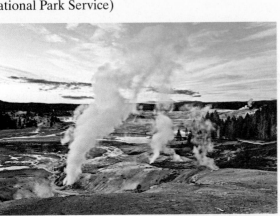

18. Razor scooters were introduced in 2000. Radio Flyer Wagons were first introduced 83 years earlier. In what year were Radio Flyer Wagons introduced? (*Source:* Toy Industry Association, Inc.)

19. Since their introduction, the number of LEGO building bricks that have been sold is equivalent to the world's current population of approximately 6 billion people owning 62 LEGO bricks each. About how many LEGO bricks have been sold since their introduction? (*Source:* LEGO Company)

20. In 2012, the average weekly pay for a home health aide in the United States was about $420. At this rate, how much will a home health aide earn working a 52-week year? (*Source:* Bureau of Labor Statistics)

21. The three most common city names in the United States are Fairview, Midway, and Riverside. There are 287 towns named Fairview, 252 named Midway, and 180 named Riverside. Find the total number of towns named Fairview, Midway, or Riverside.

22. In the game of Monopoly, a player must own all properties in a color group before building houses. The yellow color-group properties are Atlantic Avenue, Ventnor Avenue, and Marvin Gardens. These cost $260, $260, and $280, respectively, when purchased from the bank. What total amount must a player pay to the bank before houses can be built on the yellow properties? (*Source:* Hasbro, Inc.)

23. In 2012, the average weekly pay for a correctional officer in the United States was $840. If such an officer works 40 hours in one week, what is his or her hourly pay? (*Source:* Bureau of Labor Statistics)

24. In 2012, the average weekly pay for a loan officer was $1360. If a loan officer works 40 hours in one week, what is his or her hourly pay? (*Source:* Bureau of Labor Statistics)

25. Three ounces of canned tuna in oil has 165 calories. How many calories does 1 ounce have? (*Source: Home and Garden Bulletin No. 72,* U.S. Department of Agriculture)

26. A whole cheesecake has 3360 calories. If the cheesecake is cut into 12 equal pieces, how many calories will each piece have? (*Source: Home and Garden Bulletin No. 72,* U.S. Department of Agriculture)

27. The average estimated 2012 U.S. population was 313,900,000. Between Memorial Day and Labor Day, 7 billion hot dogs are consumed. Approximately how many hot dogs were consumed per person between Memorial Day and Labor Day in 2012? Divide, but do not give the remainder part of the quotient. (*Source:* U.S. Census Bureau, National Hot Dog and Sausage Council)

28. Adrian Peterson, a running back with the NFL's Minnesota Vikings, scored an average of 5 points per game during the 2012 regular season. He played in a total of 16 games during the season. What was the total number of points he scored during the 2012 football season? (*Source:* National Football League)

29. Macy's, formerly the Federated Department Stores Company, operates a total of 843 Macy's and Bloomingdale's department stores around the country. In 2012, Macy's had sales of approximately $27,685,999,890. What is the average amount of sales made by each of the 843 stores? Divide, but do not give the remainder part of the quotient. (*Source:* Macy's)

30. In 2012, PetSmart employed approximately 52,000 associates and operated roughly 1300 stores. What is the average number of associates employed at each of its stores? (*Source:* PetSmart, Inc.)

31. In 2012, the Museum of Modern Art in New York welcomed 2,805,659 visitors. The J. Paul Getty Museum in Los Angeles received 1,590,608 visitors. How many more people visited the Museum of Modern Art than the Getty Museum? (*Source: The Art Newspaper*)

32. In 2012, Target Corporation operated 1778 stores in the United States. Of these, 149 were in Texas. How many Target Stores were located in states other than Texas? (*Source:* Target Corporation)

33. The length of the southern boundary of the conterminous United States is 1933 miles. The length of the northern boundary of the conterminous United States is 2054 miles longer than this. What is the length of the northern boundary? (*Source:* U.S. Geological Survey)

34. In humans, 14 muscles are required to smile. It takes 29 more muscles to frown. How many muscles does it take to frown?

2054 miles longer

1933 miles

35. An instructor at the University of New Orleans receives a paycheck every four weeks. Find how many paychecks he receives in a year. (A year has 52 weeks.)

36. A loan of $6240 is to be paid in 48 equal payments. How much is each payment?

Objective B *Solve. See Example 5.*

37. Find the total cost of 3 sweaters at $38 each and 5 shirts at $25 each.

38. Find the total cost of 10 computers at $2100 each and 7 boxes of diskettes at $12 each.

39. A college student has $950 in an account. She spends $205 from the account on books and then deposits $300 in the account. How much money is now in the account?

40. The temperature outside was 57°F (degrees Fahrenheit). During the next few hours, it decreased by 18 degrees and then increased by 23 degrees. Find the new temperature.

The table shows the menu from a concession stand at the county fair. Use this menu to answer Exercises 41 and 42.

41. A hungry college student is debating between the following two orders:
a. a hamburger, an order of onion rings, a candy bar, and a soda.
b. a hot dog, an apple, an order of french fries, and a soda.
Which order will be cheaper? By how much?

Corky's Concession Stand Menu	
Item	**Price**
Hot dog	$3
Hamburger	$4
Soda	$1
Onion rings	$3
French fries	$2
Apple	$1
Candy bar	$2

42. A family of four is debating between the following two orders:
a. 6 hot dogs, 4 orders of onion rings, and 4 sodas.
b. 4 hamburgers, 4 orders of french fries, 2 apples, and 4 sodas.
Will the family save any money by ordering (b) instead of (a)? If so, how much?

Objectives A B Mixed Practice *Use the bar graph to answer Exercises 43 through 50. (Source: Internet World Stats)*

43. Which region of the world listed had the greatest number of Internet users in 2012?

44. Which region of the world listed had the least number of Internet users in 2012?

45. How many more Internet users (in millions) did the world region with the most Internet users have than the world region with the fewest Internet users?

46. How many more Internet users did Africa have than the Middle East in 2012?

47. How many more Internet users did North America have than Latin America/Caribbean?

48. Which region of the world had more Internet users, Europe or North America? How many more Internet users did it have?

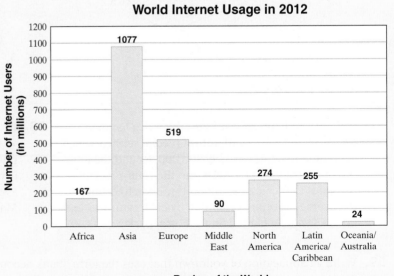

Find the average number of Internet users for the world regions listed in the graph.

49. The two world regions with the greatest number of Internet users.

50. The four world regions with the least number of Internet users.

Solve.

51. The learning lab at a local university is receiving new equipment. Twenty-two computers are purchased for $615 each and three printers for $408 each. Find the total cost for this equipment.

52. The washateria near the local community college is receiving new equipment. Thirty-six washers are purchased for $585 each and ten dryers are purchased for $388 each. Find the total cost for this equipment.

53. The American Heart Association recommends consuming no more than 2400 milligrams of salt per day. (This is about the amount in 1 teaspoon of salt.) How many milligrams of sodium is this in a week?

54. This semester a particular student pays $1750 for room and board, $709 for a meal ticket plan, and $2168 for tuition. What is her total bill?

△ **55.** The Meishs' yard is in the shape of a rectangle and measures 50 feet by 75 feet. In their yard, they have a rectangular swimming pool that measures 15 feet by 25 feet.
  a. Find the area of the entire yard.
  b. Find the area of the swimming pool.
  c. Find the area of the yard that is not part of the swimming pool.

56. The community is planning to construct a rectangular-shaped playground within the local park. The park is in the shape of a square and measures 100 yards on each side. The playground is to measure 15 yards by 25 yards.
  a. Find the area of the entire park.
  b. Find the area of the playground.
  c. Find the area of the park that is not part of the playground.

Concept Extensions

57. In 2012, the United States Postal Service handled 170,000,000 articles with return receipt service, which generated revenues of $399,000,000. Round the number of return receipt articles and the return receipt revenues to the nearest hundred-million to estimate the average revenue generated by each return receipt article. (*Source:* United States Postal Service)

58. In 2012, the United States Postal Service handled 40,000,000 pieces of Express Mail, which generated revenues of $802,000,000. Round the Express Mail revenues to the nearest ten-million to estimate the average revenue generated by each piece of Express Mail. (*Source:* United States Postal Service)

59. Write an application of your own that uses the term "bank account" and the numbers 1036 and 524.

1.9 Exponents, Square Roots, and Order of Operations ▶

Objective A Using Exponential Notation ▶

In the product $3 \cdot 3 \cdot 3 \cdot 3 \cdot 3$, notice that 3 is a factor several times. When this happens, we can use a shorthand notation, called an **exponent,** to write the repeated multiplication.

$\underbrace{3 \cdot 3 \cdot 3 \cdot 3 \cdot 3}_{\text{3 is a factor 5 times}}$ can be written as

3^5 Read as "three to the fifth power."

exponent

base

This is called **exponential notation.** The **exponent,** 5, indicates how many times the **base,** 3, is a factor.

The table below shows examples of reading exponential notation in words.

Expression	In Words
5^2	"five to the second power" or "five squared"
5^3	"five to the third power" or "five cubed"
5^4	"five to the fourth power"

Usually, an exponent of 1 is not written, so when no exponent appears, we assume that the exponent is 1. For example, $2 = 2^1$ and $7 = 7^1$.

Examples Write using exponential notation.

1. $7 \cdot 7 \cdot 7 = 7^3$
2. $3 \cdot 3 = 3^2$
3. $6 \cdot 6 \cdot 6 \cdot 6 \cdot 6 = 6^5$
4. $3 \cdot 3 \cdot 3 \cdot 3 \cdot 17 \cdot 17 \cdot 17 = 3^4 \cdot 17^3$

 Work Practice 1–4

Objective B Evaluating Exponential Expressions ▶

To **evaluate** an exponential expression, we write the expression as a product and then find the value of the product.

Examples Evaluate.

5. $9^2 = 9 \cdot 9 = 81$
6. $6^1 = 6$
7. $3^4 = 3 \cdot 3 \cdot 3 \cdot 3 = 81$
8. $5 \cdot 6^2 = 5 \cdot 6 \cdot 6 = 180$

◾ Work Practice 5–8

Objectives

A Write Repeated Factors Using Exponential Notation. ▶

B Evaluate Expressions Containing Exponents. ▶

C Evaluate the Square Root of a Perfect Square. ▶

D Use the Order of Operations. ▶

E Find the Area of a Square. ▶

Practice 1–4

Write using exponential notation.

1. $8 \cdot 8 \cdot 8 \cdot 8$
2. $3 \cdot 3 \cdot 3$
3. $10 \cdot 10 \cdot 10 \cdot 10 \cdot 10$
4. $5 \cdot 5 \cdot 4 \cdot 4 \cdot 4 \cdot 4 \cdot 4 \cdot 4$

Practice 5–8

Evaluate.

5. 4^2 6. 7^3

7. 11^1 8. $2 \cdot 3^2$

Answers

1. 8^4 2. 3^3 3. 10^5 4. $5^2 \cdot 4^6$

5. 16 6. 343 7. 11 8. 18

Example 8 illustrates an important property: An exponent applies only to its base. The exponent 2, in $5 \cdot 6^2$, applies only to its base, 6.

Helpful Hint

An exponent applies only to its base. For example, $4 \cdot 2^3$ means $4 \cdot 2 \cdot 2 \cdot 2$.

Helpful Hint

Don't forget that 2^4, for example, is *not* $2 \cdot 4$. The expression 2^4 means repeated multiplication of the same factor.

$$2^4 = 2 \cdot 2 \cdot 2 \cdot 2 = 16, \quad \text{whereas } 2 \cdot 4 = 8$$

✓**Concept Check** Which of the following statements is correct?
a. 3^6 is the same as $6 \cdot 6 \cdot 6$.
b. "Eight to the fourth power" is the same as 8^4.
c. "Ten squared" is the same as 10^3.
d. 11^2 is the same as $11 \cdot 2$.

Objective C Evaluating Square Roots ▶

A **square root** of a number is one of two identical factors of the number. For example,

$7 \cdot 7 = 49$, so a square root of 49 is 7.

We use this symbol $\sqrt{\ }$ (called a radical sign) for finding square roots. Since

$7 \cdot 7 = 49$, then $\sqrt{49} = 7$.

Practice 9–11

Find each square root.
9. $\sqrt{100}$
10. $\sqrt{4}$
11. $\sqrt{1}$

Examples   Find each square root.

9. $\sqrt{25} = 5$ because $5 \cdot 5 = 25$
10. $\sqrt{81} = 9$ because $9 \cdot 9 = 81$
11. $\sqrt{0} = 0$ because $0 \cdot 0 = 0$

▪ Work Practice 9–11

Helpful Hint

Make sure you understand the difference between squaring a number and finding the square root of a number.

$9^2 = 9 \cdot 9 = 81 \quad \sqrt{9} = 3$ because $3 \cdot 3 = 9$

Answers
9. 10 **10.** 2 **11.** 1

✓**Concept Check Answer**
b

Not every square root simplifies to a whole number. We will study this more in a later chapter. In this section, we will find square roots of perfect squares only.

Objective D Using the Order of Operations ▶

Suppose that you are in charge of taking inventory at a local cell phone store. An employee has given you the number of a certain cell phone in stock as the expression

$6 + 2 \cdot 30$

To calculate the value of this expression, do you add first or multiply first? If you add first, the answer is 240. If you multiply first, the answer is 66.

Mathematical symbols wouldn't be very useful if two values were possible for one expression. Thus, mathematicians have agreed that, given a choice, we multiply first.

$6 + 2 \cdot 30 = 6 + 60$ Multiply.
$\qquad\qquad = 66$ Add.

This agreement is one of several **order of operations** agreements.

Order of Operations

1. Perform all operations within parentheses ( ), brackets [ ], or other grouping symbols such as fraction bars or square roots, starting with the innermost set.
2. Evaluate any expressions with exponents.
3. Multiply or divide in order from left to right.
4. Add or subtract in order from left to right.

Below we practice using order of operations to simplify expressions.

Example 12 Simplify: $2 \cdot 4 - 3 \div 3$

Solution: There are no parentheses and no exponents, so we start by multiplying and dividing, from left to right.

$2 \cdot 4 - 3 \div 3 = 8 - 3 \div 3$ Multiply.
$\qquad\qquad\qquad = 8 - 1$ Divide.
$\qquad\qquad\qquad = 7$ Subtract.

■ Work Practice 12

Example 13 Simplify: $4^2 \div 2 \cdot 4$

Solution: We start by evaluating 4^2.

$4^2 \div 2 \cdot 4 = 16 \div 2 \cdot 4$ Write 4^2 as 16.

Next we multiply or divide *in order* from left to right. Since division appears before multiplication from left to right, we divide first, then multiply.

$16 \div 2 \cdot 4 = 8 \cdot 4$ Divide.
$\qquad\qquad\quad = 32$ Multiply.

■ Work Practice 13

Practice 12
Simplify: $9 \cdot 3 - 8 \div 4$

Practice 13
Simplify: $48 \div 3 \cdot 2^2$

Answers
12. 25 **13.** 64

Practice 14

Simplify: $(10 - 7)^4 + 2 \cdot 3^2$

Example 14 Simplify: $(8 - 6)^2 + 2^3 \cdot 3$

Solution: $(8 - 6)^2 + 2^3 \cdot 3 = 2^2 + 2^3 \cdot 3$ Simplify inside parentheses.

$$= 4 + 8 \cdot 3$$ Write 2^2 as 4 and 2^3 as 8.

$$= 4 + 24$$ Multiply.

$$= 28$$ Add.

■ Work Practice 14

Practice 15

Simplify:

$36 \div [20 - (4 \cdot 2)] + 4^3 - 6$

Example 15 Simplify: $4^3 + [3^2 - (10 \div 2)] - 7 \cdot 3$

Solution: Here we begin with the innermost set of parentheses.

$$4^3 + [3^2 - (10 \div 2)] - 7 \cdot 3 = 4^3 + [3^2 - 5] - 7 \cdot 3$$ Simplify inside parentheses.

$$= 4^3 + [9 - 5] - 7 \cdot 3$$ Write 3^2 as 9.

$$= 4^3 + 4 - 7 \cdot 3$$ Simplify inside brackets.

$$= 64 + 4 - 7 \cdot 3$$ Write 4^3 as 64.

$$= 64 + 4 - 21$$ Multiply.

$$= 47$$ Add and subtract from left to right.

■ Work Practice 15

Practice 16

Simplify: $\dfrac{25 + 8 \cdot 2 - 3^3}{2(3 - 2)}$

Example 16 Simplify: $\dfrac{7 - 2 \cdot 3 + 3^2}{5(2 - 1)}$

Solution: Here, the fraction bar is like a grouping symbol. We simplify above and below the fraction bar separately.

$$\frac{7 - 2 \cdot 3 + 3^2}{5(2 - 1)} = \frac{7 - 2 \cdot 3 + 9}{5(1)}$$ Evaluate 3^2 and $(2 - 1)$.

$$= \frac{7 - 6 + 9}{5}$$ Multiply $2 \cdot 3$ in the numerator and multiply 5 and 1 in the denominator.

$$= \frac{10}{5}$$ Add and subtract from left to right.

$$= 2$$ Divide.

■ Work Practice 16

Practice 17

Simplify: $81 \div \sqrt{81} \cdot 5 + 7$

Example 17 Simplify: $64 \div \sqrt{64} \cdot 2 + 4$

Solution: $64 \div \sqrt{64} \cdot 2 + 4 = 64 \div 8 \cdot 2 + 4$ Find the square root.

$$= 8 \cdot 2 + 4$$ Divide.

$$= 16 + 4$$ Multiply.

$$= 20$$ Add.

■ Work Practice 17

Answers

14. 99   **15.** 61   **16.** 7   **17.** 52

Objective E Finding the Area of a Square

Since a square is a special rectangle, we can find its area by finding the product of its length and its width.

Area of a rectangle = length · width

By recalling that each side of a square has the same measurement, we can use the following procedure to find its area:

$$\begin{aligned} \text{Area of a square} &= \text{length} \cdot \text{width} \\ &= \text{side} \cdot \text{side} \\ &= (\text{side})^2 \end{aligned}$$

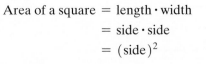
Square | Side
Side

Helpful Hint

Recall from Section 1.6 that area is measured in **square** units while perimeter is measured in units.

Example 18 Find the area of a square whose side measures 4 inches.

Solution:
$$\begin{aligned} \text{Area of a square} &= (\text{side})^2 \\ &= (4 \text{ inches})^2 \\ &= 16 \text{ square inches} \end{aligned}$$

4 inches

The area of the square is 16 square inches.

■ Work Practice 18

Practice 18

Find the area of a square whose side measures 12 centimeters.

Calculator Explorations Exponents

To evaluate an exponent such as 4^7 on a calculator, find the keys marked $\boxed{y^x}$ or $\boxed{\wedge}$ and $\boxed{=}$ or $\boxed{\text{ENTER}}$. To evaluate 4^7, press the keys $\boxed{4}$ $\boxed{y^x}$ (or $\boxed{\wedge}$) $\boxed{7}$ then $\boxed{=}$ or $\boxed{\text{ENTER}}$. The display will read $\boxed{16384}$. Thus, $4^7 = 16{,}384$.

Use a calculator to evaluate.

1. 4^6 **2.** 5^6 **3.** 5^5

4. 7^6 **5.** 2^{11} **6.** 6^8

Order of Operations

To see whether your calculator has the order of operations built in, evaluate $5 + 2 \cdot 3$ by pressing the keys $\boxed{5}$ $\boxed{+}$ $\boxed{2}$ $\boxed{\times}$ $\boxed{3}$ then $\boxed{=}$ or $\boxed{\text{ENTER}}$. If the display reads $\boxed{11}$, your calculator does have the order of operations built in. This means that most of the time,

you can key in a problem exactly as it is written and the calculator will perform operations in the proper order. When evaluating an expression containing parentheses, key in the parentheses. (If an expression contains brackets, key in parentheses.) For example, to evaluate $2[25 - (8 + 4)] - 11$, press the keys $\boxed{2}$ $\boxed{\times}$ $\boxed{(}$ $\boxed{25}$ $\boxed{-}$ $\boxed{(}$ $\boxed{8}$ $\boxed{+}$ $\boxed{4}$ $\boxed{)}$ $\boxed{)}$ $\boxed{-}$ $\boxed{11}$ then $\boxed{=}$ or $\boxed{\text{ENTER}}$.

The display will read $\boxed{15}$.

Use a calculator to evaluate.

7. $7^4 + 5^3$

8. $12^4 - 8^4$

9. $63 \cdot 75 - 43 \cdot 10$

10. $8 \cdot 22 + 7 \cdot 16$

11. $4(15 \div 3 + 2) - 10 \cdot 2$

12. $155 - 2(17 + 3) + 185$

Answer

18. 144 sq cm

Vocabulary, Readiness & Video Check

Use the choices below to fill in each blank.

addition   multiplication   exponent   base

subtraction   division   square root

1. In $2^5 = 32$, the 2 is called the _____ and the 5 is called the _____.

2. To simplify $8 + 2 \cdot 6$, which operation should be performed first? _____

3. To simplify $(8 + 2) \cdot 6$, which operation should be performed first? _____

4. To simplify $9(3 - 2) \div 3 + 6$, which operation should be performed first? _____

5. To simplify $8 \div 2 \cdot 6$, which operation should be performed first? _____

6. The _____ of a whole number is one of two identical factors of the number.

Martin-Gay Interactive Videos   Watch the section lecture video and answer the following questions.

See Video 1.9

Objective A 7. In the ⊞ Example 1 expression, what is the 3 called and what is the 12 called? ▶

Objective B 8. As mentioned in ⊞ Example 4, what "understood exponent" does any number we've worked with before have? ▶

Objective C 9. From ⊞ Example 7, how do we know that $\sqrt{64} = 8$? ▶

Objective D 10. List the three operations needed to evaluate ⊞ Example 9 in the order they should be performed. ▶

Objective E 11. As explained in the lecture before ⊞ Example 12, why does the area of a square involve an exponent whereas the area of a rectangle usually does not? ▶

1.9 Exercise Set MyMathLab® ▶

Objective A *Write using exponential notation. See Examples 1 through 4.*

1. $4 \cdot 4 \cdot 4$ 2. $5 \cdot 5 \cdot 5 \cdot 5$ 3. $7 \cdot 7 \cdot 7 \cdot 7 \cdot 7 \cdot 7$ 4. $6 \cdot 6 \cdot 6 \cdot 6 \cdot 6 \cdot 6 \cdot 6$

▶ 5. $12 \cdot 12 \cdot 12$ 6. $10 \cdot 10 \cdot 10$ ▶ 7. $6 \cdot 6 \cdot 5 \cdot 5 \cdot 5$ 8. $4 \cdot 4 \cdot 3 \cdot 3 \cdot 3$

9. $9 \cdot 8 \cdot 8$ 10. $7 \cdot 4 \cdot 4 \cdot 4$ 11. $3 \cdot 2 \cdot 2 \cdot 2 \cdot 2$ 12. $4 \cdot 6 \cdot 6 \cdot 6 \cdot 6$

13. $3 \cdot 2 \cdot 2 \cdot 2 \cdot 2 \cdot 5 \cdot 5 \cdot 5 \cdot 5 \cdot 5$ 14. $6 \cdot 6 \cdot 2 \cdot 9 \cdot 9 \cdot 9 \cdot 9$

Objective B *Evaluate. See Examples 5 through 8.*

15. 8^2 **16.** 6^2 ▶ **17.** 5^3 **18.** 6^3 **19.** 2^5 **20.** 3^5

21. 1^{10} **22.** 1^{12} ▶ **23.** 7^1 **24.** 8^1 **25.** 2^7 **26.** 5^4

27. 2^8 **28.** 3^3 **29.** 4^4 **30.** 4^3 **31.** 9^3 **32.** 8^3

33. 12^2 **34.** 11^2 ▶ **35.** 10^2 **36.** 10^3 **37.** 20^1 **38.** 14^1

39. 3^6 **40.** 4^5 **41.** $3 \cdot 2^6$ **42.** $5 \cdot 3^2$ **43.** $2 \cdot 3^4$ **44.** $2 \cdot 7^2$

Objective C *Find each square root. See Examples 9 through 11.*

▶ **45.** $\sqrt{9}$ **46.** $\sqrt{36}$ ▶ **47.** $\sqrt{64}$ **48.** $\sqrt{121}$

49. $\sqrt{144}$ **50.** $\sqrt{0}$ **51.** $\sqrt{16}$ **52.** $\sqrt{169}$

Objective D *Simplify. See Examples 12 through 16. (This section does not contain square roots.)*

▶ **53.** $15 + 3 \cdot 2$ **54.** $24 + 6 \cdot 3$ ▶ **55.** $14 \div 7 \cdot 2 + 3$ **56.** $100 \div 10 \cdot 5 + 4$

57. $32 \div 4 - 3$ **58.** $42 \div 7 - 6$ **59.** $13 + \dfrac{24}{8}$ **60.** $32 + \dfrac{8}{2}$

61. $6 \cdot 5 + 8 \cdot 2$ **62.** $3 \cdot 4 + 9 \cdot 1$ **63.** $\dfrac{5 + 12 \div 4}{1^7}$ **64.** $\dfrac{6 + 9 \div 3}{3^2}$

65. $(7 + 5^2) \div 4 \cdot 2^3$ **66.** $6^2 \cdot (10 - 8)$ **67.** $5^2 \cdot (10 - 8) + 2^3 + 5^2$

68. $5^3 \div (10 + 15) + 9^2 + 3^3$ **69.** $\dfrac{18 + 6}{2^4 - 2^2}$ **70.** $\dfrac{40 + 8}{5^2 - 3^2}$

71. $(3 + 5) \cdot (9 - 3)$ **72.** $(9 - 7) \cdot (12 + 18)$ ▶ **73.** $\dfrac{7(9 - 6) + 3}{3^2 - 3}$

74. $\dfrac{5(12-7)-4}{5^2-18}$

75. $8 \div 0 + 37$

76. $18 - 7 \div 0$

77. $2^4 \cdot 4 - (25 \div 5)$

78. $2^3 \cdot 3 - (100 \div 10)$

79. $3^4 - [35 - (12 - 6)]$

80. $[40 - (8 - 2)] - 2^5$

▶ **81.** $(7 \cdot 5) + [9 \div (3 \div 3)]$

82. $(18 \div 6) + [(3 + 5) \cdot 2]$

83. $8 \cdot [2^2 + (6 - 1) \cdot 2] - 50 \cdot 2$

84. $35 \div [3^2 + (9 - 7) - 2^2] + 10 \cdot 3$

85. $\dfrac{9^2 + 2^2 - 1^2}{8 \div 2 \cdot 3 \cdot 1 \div 3}$

86. $\dfrac{5^2 - 2^3 + 1^4}{10 \div 5 \cdot 4 \cdot 1 \div 4}$

Simplify. See Examples 12 through 17. (This section does contain square roots.)

87. $6 \cdot \sqrt{9} + 3 \cdot \sqrt{4}$

88. $3 \cdot \sqrt{25} + 2 \cdot \sqrt{81}$

89. $4 \cdot \sqrt{49} - 0 \div \sqrt{100}$

90. $7 \cdot \sqrt{36} - 0 \div \sqrt{64}$

91. $\dfrac{\sqrt{4} + 4^2}{5(20 - 16) - 3^2 - 5}$

92. $\dfrac{\sqrt{9} + 9^2}{3(10 - 6) - 2^2 - 1}$

93. $\sqrt{81} \div \sqrt{9} + 4^2 \cdot 2 - 10$

94. $\sqrt{100} \div \sqrt{4} + 3^3 \cdot 2 - 20$

95. $[\sqrt{225} \div (11 - 6) + 2^2] + (\sqrt{25} - \sqrt{1})^2$

96. $[\sqrt{169} \div (20 - 7) + 2^5] - (\sqrt{4} + \sqrt{9})^2$

97. $7^2 - \{18 - [40 \div (4 \cdot 2) + \sqrt{4}] + 5^2\}$

98. $29 - \{5 + 3[8 \cdot (10 - \sqrt{64})] - 50\}$

Objective E Mixed Practice (*Sections 1.3, 1.6*) *Find the area and perimeter of each square. See Example 18.*

▶ **99.**

7 meters

△ **100.**

9 centimeters

△ **101.**

23 miles

△ **102.**

41 feet

Concept Extensions

Answer the following true or false. See the Concept Check in this section.

103. "Six to the fifth power" is the same as 6^5.

104. "Seven squared" is the same as 7^2.

105. 2^5 is the same as $5 \cdot 5$.

106. 4^9 is the same as $4 \cdot 9$.

Insert grouping symbols (parentheses) so that each given expression evaluates to the given number.

107. $2 + 3 \cdot 6 - 2$; evaluates to 28

108. $2 + 3 \cdot 6 - 2$; evaluates to 20

109. $24 \div 3 \cdot 2 + 2 \cdot 5$; evaluates to 14

110. $24 \div 3 \cdot 2 + 2 \cdot 5$; evaluates to 15

△ **111.** A building contractor is bidding on a contract to install gutters on seven homes in a retirement community, all in the shape shown. To estimate the cost of materials, she needs to know the total perimeter of all seven homes. Find the total perimeter.

112. The building contractor from Exercise 111 plans to charge $4 per foot for installing vinyl gutters. Find the total charge for the seven homes given the total perimeter answer to Exercise 111.

Simplify.

113. $(7 + 2^4)^5 - (3^5 - 2^4)^2$

114. $25^3 \cdot (45 - 7 \cdot 5) \cdot 5$

115. Write an expression that simplifies to 5. Use multiplication, division, addition, subtraction, and at least one set of parentheses. Explain the process you would use to simplify the expression.

116. Explain why $2 \cdot 3^2$ is not the same as $(2 \cdot 3)^2$.

Chapter 1 Group Activity

Modeling Subtraction of Whole Numbers

A mathematical concept can be represented or modeled in many different ways. For instance, subtraction can be represented by the following symbolic model:

$$11 - 4$$

The following verbal models can also represent subtraction of these same quantities:

"Four subtracted from eleven" or
"Eleven take away four"

Physical models can also represent mathematical concepts. In these models, a number is represented by that many objects. For example, the number 5 can be represented by five pennies, squares, paper clips, tiles, or bottle caps.

A physical representation of the number 5

Take-Away Model for Subtraction: 11 − 4

- Start with 11 objects.
- Take 4 objects away.
- How many objects remain?

Comparison Model for Subtraction: 11 − 4

- Start with a set of 11 of one type of object and a set of 4 of another type of object.

- Make as many pairs that include one object of each type as possible.

- How many more objects left are in the larger set?

Missing Addend Model for Subtraction: 11 − 4

- Start with 4 objects.
- Continue adding objects until a total of 11 is reached.
- How many more objects were needed to give a total of 11?

Group Activity

Use an appropriate physical model for subtraction to solve each of the following problems. Explain your reasoning for choosing each model.

1. Sneha has assembled 12 computer components so far this shift. If her quota is 20 components, how many more components must she assemble to reach her quota?

2. Yuko has 14 daffodil bulbs to plant in her yard. She planted 5 bulbs in the front yard. How many bulbs does she have left for planting in the backyard?

3. Todd is 19 years old and his sister Tanya is 13 years old. How much older is Todd than Tanya?

Chapter 1 Vocabulary Check

Fill in each blank with one of the words or phrases listed below.

difference	area	square root
place value	factor	quotient
sum	whole numbers	perimeter

addend	divisor	minuend
subtrahend	exponent	digits
dividend	average	product

1. The _____ are 0, 1, 2, 3, ...

2. The _____ of a polygon is its distance around or the sum of the lengths of its sides.

3. The position of each digit in a number determines its _____.

4. A(n) _____ is a shorthand notation for repeated multiplication of the same factor.

5. To find the _____ of a rectangle, multiply length times width.

6. A(n) _____ of a number is one of two identical factors of the number.

7. The _____ used to write numbers are 0, 1, 2, 3, 4, 5, 6, 7, 8, and 9.

8. The _____ of a list of numbers is their sum divided by the number of numbers.

Use the facts below for Exercises 9 through 18.

$$2 \cdot 3 = 6 \quad 4 + 17 = 21 \quad 20 - 9 = 11 \quad 5)\overline{35}^{\,7}$$

9. The 5 above is called the _____.

10. The 35 above is called the _____.

11. The 7 above is called the _____.

12. The 3 above is called a(n) _____.

13. The 6 above is called the _____.

14. The 20 above is called the _____.

15. The 9 above is called the _____.

16. The 11 above is called the _____.

17. The 4 above is called a(n) _____.

18. The 21 above is called the _____.

Helpful Hint

▶ Are you preparing for your test? Don't forget to take the Chapter 1 Test on page 108. Then check your answers at the back of the text and use the Chapter Test Prep Videos to see the fully worked-out solutions to any of the exercises you want to review.

1 Chapter Highlights

Definitions and Concepts	Examples
Section 1.2 Place Value, Names for Numbers, and Reading Tables	

The **whole numbers** are 0, 1, 2, 3, 4, 5, ...	0, 14, 968, 5,268,619
The position of each digit in a number determines its **place value.** A place-value chart is shown next with the names of the periods given.	

(continued)

Definitions and Concepts	Examples

Section 1.2   Place Value, Names for Numbers, and Reading Tables (*continued*)

To write a whole number in words, write the number in each period followed by the name of the period. (The name of the ones period is not included.)

9,078,651,002 is written as nine billion, seventy-eight million, six hundred fifty-one thousand, two.

To write a whole number in standard form, write the number in each period, followed by a comma.

Four million, seven hundred six thousand, twenty-eight is written as 4,706,028.

Section 1.3   Adding Whole Numbers and Perimeter

To add whole numbers, add the digits in the ones place, then the tens place, then the hundreds place, and so on, carrying when necessary.

Find the sum:

$$
\begin{array}{r}
{}^{2\,1\,1} \\
2689 \leftarrow \text{addend} \\
1735 \leftarrow \text{addend} \\
+\ 662 \leftarrow \text{addend} \\
\hline
5086 \leftarrow \text{sum}
\end{array}
$$

The **perimeter** of a polygon is its distance around or the sum of the lengths of its sides.

△ Find the perimeter of the polygon shown.

5 feet
2 feet   3 feet
9 feet

The perimeter is
5 feet + 3 feet + 9 feet + 2 feet = 19 feet.

Section 1.4   Subtracting Whole Numbers

To subtract whole numbers, subtract the digits in the ones place, then the tens place, then the hundreds place, and so on, borrowing when necessary.

Subtract:

$$
\begin{array}{r}
{}^{8\,15} \\
79\cancel{5}4 \leftarrow \text{minuend} \\
-5673 \leftarrow \text{subtrahend} \\
\hline
2281 \leftarrow \text{difference}
\end{array}
$$

Section 1.5   Rounding and Estimating

Rounding Whole Numbers to a Given Place Value

Step 1: Locate the digit to the right of the given place value.

Step 2: If this digit is 5 or greater, add 1 to the digit in the given place value and replace each digit to its right with 0.

Step 3: If this digit is less than 5, replace it and each digit to its right with 0.

Round 15,721 to the nearest thousand.

15,⑦21

Add 1 — Replace with zeros.

Since the circled digit is 5 or greater, add 1 to the given place value and replace digits to its right with zeros.

15,721 rounded to the nearest thousand is 16,000.

Definitions and Concepts	Examples

Section 1.6   Multiplying Whole Numbers and Area

To multiply 73 and 58, for example, multiply 73 and 8, then 73 and 50. The sum of these partial products is the product of 73 and 58. Use the notation to the right.

$$
\begin{array}{r}
73 \leftarrow \text{factor} \\
\times\ 58 \leftarrow \text{factor} \\
\hline
584 \leftarrow 73 \times 8 \\
3650 \leftarrow 73 \times 50 \\
\hline
4234 \leftarrow \text{product}
\end{array}
$$

To find the **area** of a rectangle, multiply length times width.

△ Find the area of the rectangle shown.

11 meters

7 meters

$$
\begin{aligned}
\text{area of rectangle} &= \text{length} \cdot \text{width} \\
&= (11 \text{ meters})(7 \text{ meters}) \\
&= 77 \text{ square meters}
\end{aligned}
$$

Section 1.7   Dividing Whole Numbers

Division Properties of 0

The quotient of 0 and any number (except 0) is 0.

The quotient of any number and 0 is not a number. We say that this quotient is undefined.

To divide larger whole numbers, use the process called **long division** as shown to the right.

$\dfrac{0}{5} = 0$

$\dfrac{7}{0}$ is undefined

```
              507  R 2   ← quotient and remainder
divisor → 14)7100    ←——— dividend
          −70↓          5(14) = 70
           10           Subtract and bring down the 0.
           −0↓          0(14) = 0
           100          Subtract and bring down the 0.
           −98          7(14) = 98
             2          Subtract. The remainder is 2.
```

To check, see that $507 \cdot 14 + 2 = 7100$.

The **average** of a list of numbers is

$$
\text{average} = \frac{\text{sum of numbers}}{\textit{number} \text{ of numbers}}
$$

Find the average of 23, 35, and 38.

$$
\text{average} = \frac{23 + 35 + 38}{3} = \frac{96}{3} = 32
$$

Definitions and Concepts	Examples

Section 1.8   An Introduction to Problem Solving

Problem-Solving Steps

Suppose that 225 tickets are sold for each performance of a play. How many tickets are sold for 5 performances?

1. UNDERSTAND the problem.

1. UNDERSTAND. Read and reread the problem. Since we want the number of tickets for 5 performances, we multiply.

2. TRANSLATE the problem.

2. TRANSLATE.

number of tickets	is	number of performances	times	tickets per performance
↓	↓	↓		↓
Number of tickets	=	5	·	225

3. SOLVE the problem.

3. SOLVE: See if the answer is reasonable by also estimating.

$$\begin{array}{r} \overset{12}{225} \\ \times\ \ 5 \\ \hline 1125 \end{array} \text{ exact} \qquad \text{rounds to} \qquad \begin{array}{r} 200 \\ \times\ \ 5 \\ \hline 1000 \end{array} \text{ estimate}$$

4. INTERPRET the results.

4. INTERPRET. **Check** your work. The product is reasonable since 1125 is close to our estimated answer of 1000, and **state** your conclusion: There are 1125 tickets sold for 5 performances.

Section 1.9   Exponents, Square Roots, and Order of Operations

An **exponent** is a shorthand notation for repeated multiplication of the same factor.

A **square root** of a number is one of two identical factors of the number.

$$3^4 = \underbrace{3 \cdot 3 \cdot 3 \cdot 3}_{\text{4 factors of 3}} = 81$$

base

$$\sqrt{36} = 6 \quad \text{because} \quad 6 \cdot 6 = 36$$
$$\sqrt{121} = 11 \quad \text{because} \quad 11 \cdot 11 = 121$$
$$\sqrt{0} = 0 \quad \text{because} \quad 0 \cdot 0 = 0$$

Order of Operations

1. Perform all operations within parentheses ( ), brackets [ ], or other grouping symbols such as square roots or fraction bars, starting with the innermost set.

2. Evaluate any expressions with exponents.

3. Multiply or divide in order from left to right.

4. Add or subtract in order from left to right.

The **area of a square** is (side)^2.

Simplify: $\dfrac{5 + 3^2}{2(7 - 6)}$

Simplify above and below the fraction bar separately.

$$\frac{5 + 3^2}{2(7 - 6)} = \frac{5 + 9}{2(1)} \quad \begin{array}{l} \text{Evaluate } 3^2 \text{ above the fraction bar.} \\ \text{Subtract: } 7 - 6 \text{ below the fraction bar.} \end{array}$$

$$= \frac{14}{2} \quad \begin{array}{l} \text{Add.} \\ \text{Multiply.} \end{array}$$

$$= 7 \quad \text{Divide.}$$

Find the area of a square with side length 9 inches.

$$\begin{aligned} \text{Area of the square} &= (\text{side})^2 \\ &= (9 \text{ inches})^2 \\ &= 81 \text{ square inches} \end{aligned}$$

(1.2) *Determine the place value of the digit 4 in each whole number.*

1. 7640

2. 46,200,120

Write each whole number in words.

3. 7640

4. 46,200,120

Write each whole number in expanded form.

5. 3158

6. 403,225,000

Write each whole number in standard form.

7. Eighty-one thousand, nine hundred

8. Six billion, three hundred four million

The following table shows the Internet and Facebook use of world regions as of June 2012. Use this table to answer Exercises 9 through 12.

World Region	Internet Users	Facebook Users
Africa	167,335,676	51,612,460
Asia	1,076,681,059	254,336,520
Europe	518,512,109	250,934,000
Middle East	90,000,455	23,811,620
North America	273,785,413	184,177,220
Latin America/ Caribbean	254,915,745	188,339,620
Oceania/Australia	24,287,919	14,614,780
(*Source:* Internet World Stats)		

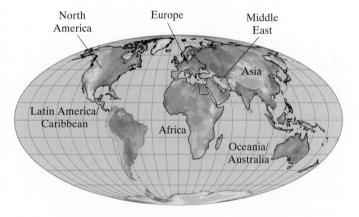

9. Find the number of Internet users in Europe.

10. Find the number of Facebook users in North America.

11. Which world region had the largest number of Facebook users?

12. Which world region had the smallest number of Internet users?

(1.3) *Add.*

13. 17 + 46

14. 28 + 39

15. 25 + 8 + 15

16. 27 + 9 + 41

17. 932 + 24

18. 819 + 21

19. 567 + 7383

20. 463 + 6787

21. 91 + 3623 + 497

22. 82 + 1647 + 238

Solve.

23. Find the sum of 86, 331, and 909.

24. Find the sum of 49, 529, and 308.

25. What is 26,481 increased by 865?

26. What is 38,556 increased by 744?

27. The distance from Chicago to New York City is 714 miles. The distance from New York City to New Delhi, India, is 7318 miles. Find the total distance from Chicago to New Delhi if traveling by air through New York City.

28. Susan Summerline earned salaries of $62,589, $65,340, and $69,770 during the years 2002, 2003, and 2004, respectively. Find her total earnings during those three years.

Find the perimeter of each figure.

△ **29.**

72 feet

50 feet

72 feet

82 feet

△ **30.**

11 kilometers   20 kilometers

35 kilometers

(1.4) *Subtract and then check.*

31. $93 - 79$ **32.** $61 - 27$ **33.** $462 - 397$ **34.** $583 - 279$ **35.** $4000 - 86$ **36.** $8000 - 92$

Solve.

37. Subtract 7965 from 25,862.

38. Subtract 4349 from 39,007.

39. Find the increase in population for San Antonio, Texas, from 2000 (population: 1,144,646) to 2012 (population: 1,382,951). (*Source:* U.S. Census Bureau)

40. Find the decrease in population for Detroit, Michigan, from 2000 (population: 951,270) to 2012 (population: 701,475). (*Source:* U.S. Census Bureau)

41. Bob Roma is proofreading the Yellow Pages for his county. If he has finished 315 pages of the total 712 pages, how many pages does he have left to proofread?

42. Shelly Winters bought a new car listed at $28,425. She received a discount of $1599 and a factory rebate of $1200. Find how much she paid for the car.

The following bar graph shows the monthly savings account balance for a freshman attending a local community college. Use this graph to answer Exercises 43 through 46.

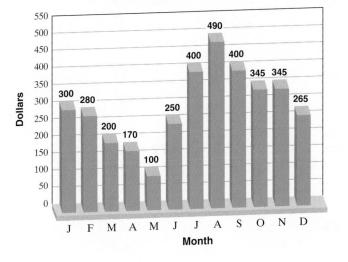

43. During what month was the balance the least?

44. During what month was the balance the greatest?

45. By how much did his balance decrease from February to April?

46. By how much did his balance increase from June to August?

(1.5) *Round to the given place.*

47. 93 to the nearest ten

48. 45 to the nearest ten

49. 467 to the nearest ten

50. 493 to the nearest hundred

51. 4832 to the nearest hundred

52. 57,534 to the nearest thousand

53. 49,683,712 to the nearest million

54. 768,542 to the nearest hundred-thousand

55. In 2012, 126,226,713 Americans cast a ballot in the presidential election. Round this number to the nearest million. (*Source:* CNN)

56. In 2011, there were 98,817 public elementary and secondary schools in the United States. Round this number to the nearest thousand. (*Source:* National Center for Education Statistics)

Estimate the sum or difference by rounding each number to the nearest hundred.

57. 4892 + 647 + 1876

58. 5925 − 1787

59. A group of students took a week-long driving trip and traveled 628, 290, 172, 58, 508, 445, and 383 miles on seven consecutive days. Round each distance to the nearest hundred to estimate the distance they traveled.

60. The estimated 2012 population of Houston, Texas, was 2,160,821, and for San Diego, California, it was 1,338,348. Round each number to the nearest hundred-thousand and estimate how much larger Houston is than San Diego. (*Source:* U.S. Census Bureau)

(1.6) *Multiply.*

61. 273
× 7

62. 349
× 4

63. 47
× 30

64. 69
× 42

65. 20(8)(5)

66. 25(9)(4)

67. 48
× 77

68. 77
× 22

69. 49 · 49 · 0

70. 62 · 88 · 0

71. 586
× 29

72. 242
× 37

73. 642
× 177

74. 347
× 129

75. 1026
× 401

76. 2107
× 302

77. 375 · 1000

78. 108 · 1000

79. 30 · 400

80. 50 · 700

81. 1700 · 3000

82. 1900 · 4000

Solve.

83. Find the product of 5 and 230.

84. Find the product of 6 and 820.

85. Multiply 9 and 12.

86. Multiply 8 and 14.

87. One ounce of Swiss cheese contains 8 grams of fat. How many grams of fat are in 3 ounces of Swiss cheese? (*Source: Home and Garden Bulletin No. 72,* U.S. Department of Agriculture)

88. The cost for a South Dakota resident to attend Black Hills State University full-time is $7617 per semester. Determine the cost for 20 students to attend full-time. (*Source:* Black Hills State University)

Find the area of each rectangle.

△**89.**

12 miles

5 miles

△**90.**

20 centimeters

25 centimeters

(1.7) *Divide and then check.*

91. $\dfrac{18}{6}$ **92.** $\dfrac{36}{9}$ **93.** $42 \div 7$ **94.** $35 \div 5$ **95.** $27 \div 5$

96. $18 \div 4$ **97.** $16 \div 0$ **98.** $0 \div 8$ **99.** $9 \div 9$ **100.** $10 \div 1$

101. $0 \div 668$ **102.** $918 \div 0$ **103.** $5\overline{)167}$ **104.** $8\overline{)159}$ **105.** $26\overline{)626}$

106. $19\overline{)680}$ **107.** $47\overline{)23,792}$ **108.** $53\overline{)48,111}$ **109.** $207\overline{)578,291}$ **110.** $306\overline{)615,732}$

Solve.

111. Find the quotient of 92 and 5.

112. Find the quotient of 86 and 4.

113. One foot is 12 inches. Find how many feet there are in 5496 inches.

114. One mile is 1760 yards. Find how many miles there are in 22,880 yards.

115. Find the average of the numbers 76, 49, 32, and 47.

116. Find the average of the numbers 23, 85, 62, and 66.

(1.8) *Solve.*

117. A box can hold 24 cans of corn. How many boxes can be filled with 648 cans of corn?

118. If a ticket to a movie costs $6, how much do 32 tickets cost?

119. In 2012, U.S. companies spent $74 billion on television advertising. By comparison, U.S. companies spent only $69 billion on television advertising in 2011. How much more did U.S. companies spend on television advertising in 2012? (*Source:* Kantar Media)

120. The cost to banks when a person uses an ATM (Automatic Teller Machine) is 27¢. The cost to banks when a person deposits a check with a teller is 48¢ more. How much is this cost?

121. A golf pro orders shirts for the company sponsoring a local charity golfing event. Shirts size large cost $32 while shirts size extra-large cost $38. If 15 large shirts and 11 extra-large shirts are ordered, find the cost.

122. Two rectangular pieces of land are purchased: one that measures 65 feet by 110 feet and one that measures 80 feet by 200 feet. Find the total area of land purchased. (*Hint:* Find the area of each rectangle, then add.)

(1.9) *Simplify.*

123. 7^2 **124.** 5^3 **125.** $5 \cdot 3^2$ **126.** $4 \cdot 10^2$

127. $18 \div 3 + 7$ **128.** $12 - 8 \div 4$ **129.** $\dfrac{5(6^2 - 3)}{3^2 + 2}$ **130.** $\dfrac{7(16 - 8)}{2^3}$

131. $48 \div 8 \cdot 2$ **132.** $27 \div 9 \cdot 3$

133. $2 + 3[1^5 + (20 - 17) \cdot 3] + 5 \cdot 2$ **134.** $21 - [2^4 - (7 - 5) - 10] + 8 \cdot 2$

Simplify. (These exercises contain square roots.)

135. $\sqrt{81}$ **136.** $\sqrt{4}$ **137.** $\sqrt{1}$ **138.** $\sqrt{0}$

139. $4 \cdot \sqrt{25} - 2 \cdot 7$ **140.** $8 \cdot \sqrt{49} - 3 \cdot 9$

141. $(\sqrt{36} - \sqrt{16})^3 \cdot [10^2 \div (3 + 17)]$ **142.** $(\sqrt{49} - \sqrt{25})^3 \cdot [9^2 \div (2 + 7)]$

143. $\dfrac{5 \cdot 7 - 3 \cdot \sqrt{25}}{2(\sqrt{121} - 3^2)}$ **144.** $\dfrac{4 \cdot 8 - 1 \cdot \sqrt{121}}{3(\sqrt{81} - 2^3)}$

Find the area of each square.

△ **145.** A square with side length of 7 meters.   △ **146.**

3 inches

Mixed Review

Perform the indicated operations.

147. $375 - 68$ **148.** $729 - 47$ **149.** 723×3 **150.** 629×4

151. $264 + 39 + 598$ **152.** $593 + 52 + 766$ **153.** $13\overline{)5962}$ **154.** $18\overline{)4267}$

155. 1968×36 **156.** 5324×18 **157.** $2000 - 356$ **158.** $9000 - 519$

Round to the given place.

159. 736 to the nearest ten

160. 258,371 to the nearest thousand

161. 1999 to the nearest hundred

162. 44,499 to the nearest ten thousand

Write each whole number in words.

163. 36,911

164. 154,863

Write each whole number in standard form.

165. Seventy thousand, nine hundred forty-three

166. Forty-three thousand, four hundred one

Simplify.

167. 4^3

168. 5^3

169. $\sqrt{144}$

170. $\sqrt{100}$

171. $24 \div 4 \cdot 2$

172. $\sqrt{256} - 3 \cdot 5$

173. $\dfrac{8(7-4)-10}{4^2-3^2}$

174. $\dfrac{(15+\sqrt{9})\cdot(8-5)}{2^3+1}$

Solve.

175. 36 divided by 9 is what number?

176. What is the product of 2 and 12?

177. 16 increased by 8 is what number?

178. 7 subtracted from 21 is what number?

The following table shows the 2012 and 2013 average Major League Baseball salaries (rounded to the nearest thousand) for the five teams with the largest payrolls for 2013. Use this table to answer Exercises 179 and 180. (Source: CBSSports.com, Associated Press)

Team	2013 Average Salary	2012 Average Salary
Los Angeles Dodgers	$7,469,000	$3,264,000
New York Yankees	$7,151,000	$6,256,000
Philadelphia Phillies	$6,125,000	$5,798,000
Detroit Tigers	$5,708,000	$4,561,000
Boston Red Sox	$5,022,000	$5,094,000

179. How much more was the average salary for a Los Angeles Dodgers player in 2013 than in 2012?

180. How much less was the average Boston Red Sox salary than the average New York Yankee salary in 2013?

181. A manufacturer of drinking glasses ships his delicate stock in special boxes that can hold 32 glasses. If 1714 glasses are manufactured, how many full boxes are filled? Are there any glasses left over?

182. A teacher orders 2 small white boards for $27 each and 8 boxes of dry erase pens for $4 each. What is her total bill before taxes?

Answers

Simplify.

1. Write 82,426 in words.

2. Write "four hundred two thousand, five hundred fifty" in standard form.

1. _____

2. _____

3. $59 + 82$

4. $600 - 487$

5. $\begin{array}{r} 496 \\ \times\ \ 30 \\ \hline \end{array}$

3. _____

4. _____

5. _____

6. $52{,}896 \div 69$

7. $2^3 \cdot 5^2$

8. $\sqrt{4} \cdot \sqrt{25}$

6. _____

7. _____

8. _____

9. $0 \div 49$

10. $62 \div 0$

11. $\left(2^4 - 5\right) \cdot 3$

9. _____

10. _____

11. _____

12. $16 + 9 \div 3 \cdot 4 - 7$

13. $\dfrac{64 \div 8 \cdot 2}{\left(\sqrt{9} - \sqrt{4}\right)^2 + 1}$

12. _____

13. _____

14. $2\left[(6 - 4)^2 + (22 - 19)^2\right] + 10$

15. $5698 \cdot 1000$

14. _____

15. _____

16. $8000 \cdot 1400$

17. Round 52,369 to the nearest thousand.

16. _____

17. _____

18. _____

Estimate each sum or difference by rounding each number to the nearest hundred.

18. $6289 + 5403 + 1957$

19. $4267 - 2738$

19. _____

Solve.

20. Subtract 15 from 107.

21. Find the sum of 15 and 107.

22. Find the product of 15 and 107.

23. Find the quotient of 107 and 15.

24. Twenty-nine cans of Sherwin-Williams paint cost $493. How much was each can?

25. Jo McElory is looking at two new refrigerators for her apartment. One costs $599 and the other costs $725. How much more expensive is the higher-priced one?

26. One tablespoon of white granulated sugar contains 45 calories. How many calories are in 8 tablespoons of white granulated sugar? (*Source: Home and Garden Bulletin No. 72,* U.S. Department of Agriculture)

27. A small business owner recently ordered 16 digital cameras that cost $430 each and 5 printers that cost $205 each. Find the total cost for these items.

Find the perimeter and the area of each figure.

△ **28.**

| Square | 5 centimeters |

△ **29.**

20 yards

| Rectangle | 10 yards |

20. _____

21. _____

22. _____

23. _____

24. _____

25. _____

26. _____

27. _____

28. _____

29. _____

Multiplying and Dividing Fractions

Fractions are numbers, and like whole numbers, they can be added, subtracted, multiplied, and divided. Fractions are very useful and appear frequently in everyday language, in common phrases like "half an hour," "quarter of a pound," and "third of a cup." This chapter introduces the concept of fractions, presents some basic vocabulary, and demonstrates how to multiply and divide fractions.

| **Plateau** | **Fin:** A thin wall of rock | **Window:** A natural hole | **Hoodoo:** A tall, thin tower of rock |

What Are Hoodoos, and How Are They Formed?

A hoodoo is a tall thin tower of sedimentary rock that is topped by a piece of harder stone. The diagrams above show how a hoodoo is formed, with the two main eroding processes being rain and also freeze/thaw cycles. In Bryce Canyon, there are over 200 freeze/thaw cycles each year that slowly pry open cracks and create many hoodoos.

Hoodoos can be anywhere from 5 to 150 feet tall. Unfortunately, hoodoos don't last long because the erosion that formed them continues. The average rate of erosion is 2 to 4 feet every 100 years.

The National Park Service (NPS) is charged with the enormous task of managing and protecting our national resources such as Bryce Canyon. In Section 2.3 and throughout Chapter 3, we will explore fractions relating to the various types of national parks and monuments that are protected by the NPS.

Overnight Stays at National Parks

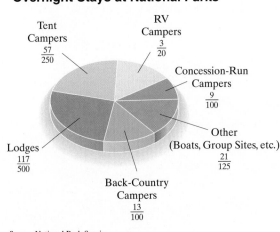

Tent Campers $\frac{57}{250}$

RV Campers $\frac{3}{20}$

Concession-Run Campers $\frac{9}{100}$

Other (Boats, Group Sites, etc.) $\frac{21}{125}$

Lodges $\frac{117}{500}$

Back-Country Campers $\frac{13}{100}$

Source: National Park Service

2.1 Introduction to Fractions and Mixed Numbers

Objective A Identifying Numerators and Denominators and Reviewing Division Properties of 0 and 1

Whole numbers are used to count whole things or units, such as cars, horses, dollars, and people. To refer to a part of a whole, fractions can be used. Here are some examples of **fractions.** Study these examples for a moment.

a cup

1 part considered

$\frac{1}{2}$ of a cup

a foot

2 parts considered

3 equal parts

$\frac{2}{3}$ of a foot

5 parts considered

6 equal parts

$\frac{5}{6}$ of a pizza

In a fraction, the top number is called the **numerator** and the bottom number is called the **denominator.** The bar between the numbers is called the **fraction bar.**

Names	Fraction	Meaning
numerator ⟶	5	⟵ number of parts being considered
denominator ⟶	6	⟵ number of equal parts in the whole

Examples Identify the numerator and the denominator of each fraction.

1. $\dfrac{3}{7}$ ← numerator
← denominator

2. $\dfrac{13}{5}$ ← numerator
← denominator

> **Helpful Hint**
> Notice the fraction
> $\dfrac{11}{1} = 11$, or also $11 = \dfrac{11}{1}$.

▶ Work Practice 1–2

Before we continue further, don't forget from Section 1.7 that the fraction bar indicates division. Let's review some division properties of 1 and 0.

$\dfrac{9}{9} = 1$ because $1 \cdot 9 = 9$ $\dfrac{11}{1} = 11$ because $11 \cdot 1 = 11$

$\dfrac{0}{6} = 0$ because $0 \cdot 6 = 0$ $\dfrac{6}{0}$ *is undefined* because there is no number that when multiplied by 0 gives 6.

In general, we can say the following.

Let *n* be any whole number except 0.

$\dfrac{n}{n} = 1$ $\dfrac{0}{n} = 0$

$\dfrac{n}{1} = n$ $\dfrac{n}{0}$ is undefined.

Objectives

A Identify the Numerator and the Denominator of a Fraction and Review Division Properties of 0 and 1.

B Write a Fraction to Represent Parts of Figures or Real-Life Data.

C Identify Proper Fractions, Improper Fractions, and Mixed Numbers.

D Write Mixed Numbers as Improper Fractions.

E Write Improper Fractions as Mixed Numbers or Whole Numbers.

Practice 1–2

Identify the numerator and the denominator of each fraction.

1. $\dfrac{9}{2}$ **2.** $\dfrac{10}{17}$

Practice 3–6

Simplify.

3. $\dfrac{0}{2}$ **4.** $\dfrac{8}{8}$

5. $\dfrac{4}{0}$ **6.** $\dfrac{20}{1}$

Practice 7–8

Write a fraction to represent the shaded part of each figure.

7.

8.

Practice 9–10

Write a fraction to represent the part of the whole shown.

9. Just consider this part of the syringe

10.

Answers

3. 0   **4.** 1   **5.** undefined   **6.** 20

7. $\dfrac{3}{8}$ **8.** $\dfrac{1}{6}$ **9.** $\dfrac{7}{10}$ **10.** $\dfrac{9}{16}$

Examples   Simplify.

3. $\dfrac{5}{5} = 1$ **4.** $\dfrac{0}{7} = 0$ **5.** $\dfrac{10}{1} = 10$ **6.** $\dfrac{3}{0}$ is undefined

■ Work Practice 3–6

Objective B   Writing Fractions to Represent Parts of Figures or Real-Life Data

One way to become familiar with the concept of fractions is to visualize fractions with shaded figures. We can then write a fraction to represent the shaded area of the figure.

Examples   Write a fraction to represent the shaded part of each figure.

7. In this figure, 2 of the 5 equal parts are shaded. Thus, the fraction is $\dfrac{2}{5}$.

$\dfrac{2}{5}$ ← number of parts shaded / ← number of equal parts

8. In this figure, 3 of the 10 rectangles are shaded. Thus, the fraction is $\dfrac{3}{10}$.

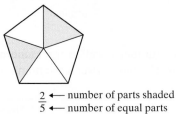

$\dfrac{3}{10}$ ← number of parts shaded / ← number of equal parts

■ Work Practice 7–8

Examples   Write a fraction to represent the shaded part of the diagram.

9.

The fraction is $\dfrac{3}{10}$.

10.

The fraction is $\dfrac{1}{3}$.

■ Work Practice 9–10

Draw a figure and then shade a part of it to represent each fraction.

11. $\frac{5}{6}$ of a figure

We will use a geometric figure such as a rectangle. Since the denominator is 6, we divide it into 6 equal parts. Then we shade 5 of the equal parts.

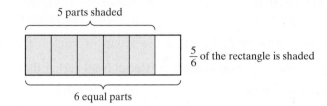

12. $\frac{3}{8}$ of a figure

If you'd like, our figure can consist of 8 triangles of the same size. We will shade 3 of the triangles.

■ Work Practice 11–12

✓**Concept Check** If represents $\frac{6}{7}$ of a whole diagram, sketch the whole diagram.

Example 13 Writing Fractions from Real-Life Data

Of the eight planets in our solar system (Pluto is now a dwarf planet), three are closer to the Sun than Mars. What fraction of the planets are closer to the Sun than Mars?

Solution: The fraction of planets closer to the Sun than Mars is:

$\frac{3}{8}$ ← number of planets closer
← number of planets in our solar system

Thus, $\frac{3}{8}$ of the planets in our solar system are closer to the Sun than Mars.

■ Work Practice 13

Practice 11–12

Draw and shade a part of a figure to represent each fraction.

11. $\frac{2}{3}$ of a figure

12. $\frac{7}{11}$ of a figure

Practice 13

Of the eight planets in our solar system, five are farther from the Sun than Earth is. What fraction of the planets are farther from the Sun than Earth is?

Answers

11. answers may vary; for example,

◯◯◯

12. answers may vary; for example,

13. $\frac{5}{8}$

✓**Concept Check Answer**

Objective C Identifying Proper Fractions, Improper Fractions, and Mixed Numbers

A **proper fraction** is a fraction whose numerator is less than its denominator. Proper fractions are less than 1. For example, the shaded portion of the triangle's area is represented by $\frac{2}{3}$.

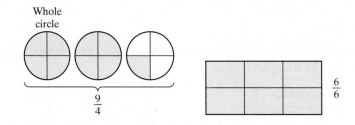

$\frac{2}{3}$

An **improper fraction** is a fraction whose numerator is greater than or equal to its denominator. Improper fractions are greater than or equal to 1. The shaded part of the group of circles' area below is $\frac{9}{4}$. The shaded part of the rectangle's area is $\frac{6}{6}$. (Recall from earlier that $\frac{6}{6}$ simplifies to 1 and notice that 1 whole figure or rectangle is shaded below.)

Whole circle

$\frac{9}{4}$

$\frac{6}{6}$

A **mixed number** contains a whole number and a fraction. Mixed numbers are greater than 1. Above, we wrote the shaded part of the group of circles below as the improper fraction $\frac{9}{4}$. Now let's write the shaded part as a mixed number. The shaded part of the group of circles' area is $2\frac{1}{4}$. (Read "two and one-fourth.")

2 whole circles $\frac{1}{4}$ circle

$2\frac{1}{4}$

Helpful Hint

The mixed number $2\frac{1}{4}$ represents $2 + \frac{1}{4}$.

Example 14 Identify each number as a proper fraction, improper fraction, or mixed number.

a. $\frac{6}{7}$ is a proper fraction

b. $\frac{13}{12}$ is an improper fraction

c. $\frac{2}{2}$ is an improper fraction

d. $\frac{99}{101}$ is a proper fraction

e. $1\frac{7}{8}$ is a mixed number

f. $\frac{93}{74}$ is an improper fraction

■ Work Practice 14

Practice 14

Identify each number as a proper fraction, improper fraction, or mixed number.

a. $\frac{5}{8}$ b. $\frac{7}{7}$

c. $\frac{14}{13}$ d. $\frac{13}{14}$

e. $5\frac{1}{4}$ f. $\frac{100}{49}$

Answers

14. a. proper fraction **b.** improper fraction **c.** improper fraction **d.** proper fraction **e.** mixed number **f.** improper fraction

Examples Represent the shaded part of each figure group's area as both an improper fraction and a mixed number.

15.

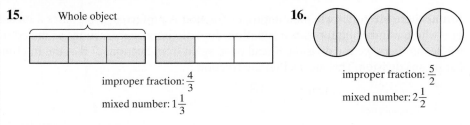

Whole object

improper fraction: $\frac{4}{3}$

mixed number: $1\frac{1}{3}$

16.

improper fraction: $\frac{5}{2}$

mixed number: $2\frac{1}{2}$

■ Work Practice 15–16

✓**Concept Check** If you were to estimate $2\frac{1}{8}$ by a whole number, would you choose 2 or 3? Why?

Objective D Writing Mixed Numbers as Improper Fractions ▶

Notice from Examples 15 and 16 that mixed numbers and improper fractions were both used to represent the shaded area of the figure groups. For example,

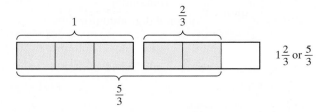

$1\frac{2}{3}$ or $\frac{5}{3}$

The following steps may be used to write a mixed number as an improper fraction:

Writing a Mixed Number as an Improper Fraction

To write a mixed number as an improper fraction:

Step 1: Multiply the denominator of the fraction by the whole number.

Step 2: Add the numerator of the fraction to the product from Step 1.

Step 3: Write the sum from Step 2 as the numerator of the improper fraction over the original denominator.

For example,

Step 1   Step 2

$$1\frac{2}{3} = \frac{\overbrace{3 \cdot 1 + 2}}{3} = \frac{3 + 2}{3} = \frac{5}{3}$$

Step 3

Example 17 Write each as an improper fraction.

a. $4\frac{2}{9} = \frac{9 \cdot 4 + 2}{9} = \frac{36 + 2}{9} = \frac{38}{9}$

b. $1\frac{8}{11} = \frac{11 \cdot 1 + 8}{11} = \frac{11 + 8}{11} = \frac{19}{11}$

■ Work Practice 17

Objective E Writing Improper Fractions as Mixed Numbers or Whole Numbers ▶

Just as there are times when an improper fraction is preferred, sometimes a mixed or a whole number better suits a situation. To write improper fractions as mixed or whole numbers, we use division. Recall once again from Section 1.7 that the fraction bar means division. This means that the fraction

$$\frac{5}{3} \begin{array}{l} \text{numerator} \\ \text{denominator} \end{array} \quad \text{means} \quad 3)\overline{5}$$

numerator
denominator

Writing an Improper Fraction as a Mixed Number or a Whole Number

To write an improper fraction as a mixed number or a whole number:

Step 1: Divide the denominator into the numerator.

Step 2: The whole number part of the mixed number is the quotient. The fraction part of the mixed number is the remainder over the original denominator.

$$\text{quotient}\,\frac{\text{remainder}}{\text{original denominator}}$$

For example,

Step 1

$$\frac{5}{3}: \begin{array}{r} 1 \\ 3)\overline{5} \\ -3 \\ \hline 2 \end{array}$$

Step 2

$$\frac{5}{3} = 1\frac{2}{3} \begin{array}{l} \leftarrow \text{remainder} \\ \leftarrow \text{original denominator} \end{array}$$

quotient

Practice 18

Write each as a mixed number or a whole number.

a. $\frac{9}{5}$ **b.** $\frac{23}{9}$ **c.** $\frac{48}{4}$

d. $\frac{62}{13}$ **e.** $\frac{51}{7}$ **f.** $\frac{21}{20}$

Example 18 Write each as a mixed number or a whole number.

a. $\frac{30}{7}$ **b.** $\frac{16}{15}$ **c.** $\frac{84}{6}$

Solution:

a. $\frac{30}{7}:$ $\begin{array}{r} 4 \\ 7)\overline{30} \\ -28 \\ \hline 2 \end{array}$ $\frac{30}{7} = 4\frac{2}{7}$

b. $\frac{16}{15}:$ $\begin{array}{r} 1 \\ 15)\overline{16} \\ -15 \\ \hline 1 \end{array}$ $\frac{16}{15} = 1\frac{1}{15}$

c. $\frac{84}{6}:$ $\begin{array}{r} 14 \\ 6)\overline{84} \\ -6 \\ \hline 24 \\ -24 \\ \hline 0 \end{array}$ $\frac{84}{6} = 14$ Since the remainder is 0, the result is the whole number 14.

Helpful Hint

When the remainder is 0, the improper fraction is a whole number. For example, $\frac{92}{4} = 23$.

$$\begin{array}{r} 23 \\ 4)\overline{92} \\ -8 \\ \hline 12 \\ -12 \\ \hline 0 \end{array}$$

■ Work Practice 18

Answers

18. **a.** $1\frac{4}{5}$ **b.** $2\frac{5}{9}$ **c.** 12   **d.** $4\frac{10}{13}$

e. $7\frac{2}{7}$ **f.** $1\frac{1}{20}$

Vocabulary, Readiness & Video Check

Use the choices below to fill in each blank.

improper	fraction	proper	is undefined	mixed number = 0
greater than or equal to 1	denominator	= 1	less than 1	numerator

1. The number $\frac{17}{31}$ is called a(n) _____. The number 31 is called its _____ and 17 is called its _____.

2. If we simplify each fraction, $\frac{9}{9}$ _____, $\frac{0}{4}$ _____, and we say $\frac{4}{0}$ _____.

3. The fraction $\frac{8}{3}$ is called a(n) _____ fraction, the fraction $\frac{3}{8}$ is called a(n) _____ fraction, and $10\frac{3}{8}$ is called a(n) _____ .

4. The value of an improper fraction is always _____, and the value of a proper fraction is always _____.

Martin-Gay Interactive Videos   Watch the section lecture video and answer the following questions.

See Video 2.1

Objective A 5. From ▥ Example 3, what can you conclude about any fraction where the numerator and denominator are the same nonzero number? ▸

Objective B 6. In ▥ Example 8, what does the denominator 50 represent? ▸

Objective C 7. In ▥ Example 11, there are two shapes in the diagram, so why do the representative fractions have a denominator 3? ▸

Objective D 8. Complete this statement based on the lecture before ▥ Example 12: The operation of _____ is understood in a mixed number notation; for example, $1\frac{1}{3}$ means 1 _____ $\frac{1}{3}$. ▸

Objective E 9. From the lecture before ▥ Example 15, what operation is used to write an improper fraction as a mixed number? ▸

2.1 Exercise Set MyMathLab® ▸

Objectives A C **Mixed Practice** *Identify the numerator and the denominator of each fraction and identify each fraction as proper or improper. See Examples 1, 2, and 14.*

▸ **1.** $\frac{1}{2}$

2. $\frac{1}{4}$

▸ **3.** $\frac{10}{3}$

4. $\frac{53}{21}$

5. $\frac{15}{15}$

6. $\frac{26}{26}$

Objective A *Simplify. See Examples 3 through 6.*

▸ 7. $\dfrac{21}{21}$ **8.** $\dfrac{14}{14}$ **▸ 9.** $\dfrac{5}{0}$ **10.** $\dfrac{1}{0}$ **▸ 11.** $\dfrac{13}{1}$ **12.** $\dfrac{14}{1}$

▸ 13. $\dfrac{0}{20}$ **14.** $\dfrac{0}{17}$ **15.** $\dfrac{10}{0}$ **16.** $\dfrac{0}{18}$ **17.** $\dfrac{16}{1}$ **18.** $\dfrac{18}{18}$

Objective B *Write a fraction to represent the shaded part of each. See Examples 7 through 10.*

19.

20.

21.

22.

▸ 23.

24.

25.

26.

27.

28.

29.   1 mile

30.

Draw and shade a part of a figure to represent each fraction. See Examples 11 and 12.

31. $\dfrac{1}{5}$ of a figure   **32.** $\dfrac{1}{16}$ of a figure   **33.** $\dfrac{7}{8}$ of a figure   **34.** $\dfrac{3}{5}$ of a figure

35. $\dfrac{6}{7}$ of a figure   **36.** $\dfrac{7}{9}$ of a figure   **37.** $\dfrac{4}{4}$ of a figure   **38.** $\dfrac{6}{6}$ of a figure

Write each fraction. See Example 13.

39. Of the 131 students at a small private school, 42 are freshmen. What fraction of the students are freshmen?

40. Of the 63 employees at a new biomedical engineering firm, 22 are men. What fraction of the employees are men?

41. Use Exercise 39 to answer a and b.

   a. How many students are *not* freshmen?

   b. What fraction of the students are *not* freshmen?

42. Use Exercise 40 to answer a and b.

   a. How many of the employees are women?

   b. What fraction of the employees are women?

43. As of the beginning of 2014, the United States has had 44 different presidents. A total of seven U.S. presidents were born in the state of Ohio, second only to the state of Virginia in producing U.S. presidents. What fraction of U.S. presidents were born in Ohio? (*Source: World Almanac and Book of Facts*)

44. Of the eight planets in our solar system, four have days that are longer than the 24-hour Earth day. What fraction of the planets have longer days than Earth has? (*Source:* National Space Science Data Center)

45. Hurricane Sandy, which struck the East Coast in October 2012, was the largest Atlantic hurricane ever documented. Sandy was just one of 19 named tropical storms that formed during the 2012 Atlantic hurricane season. A total of 10 of these tropical storms turned into hurricanes. What fraction of the 2012 Atlantic tropical storms escalated into hurricanes? (*Source:* National Oceanic and Atmospheric Administration)

46. There are 12 inches in a foot. What fractional part of a foot does 5 inches represent?

47. There are 31 days in the month of March. What fraction of the month does 11 days represent?

Mon.	Tue.	Wed.	Thu.	Fri.	Sat.	Sun.
					1	2
3	4	5	6	7	8	9
10	11	12	13	14	15	16
17	18	19	20	21	22	23
24	25	26	27	28	29	30
31						

48. There are 60 minutes in an hour. What fraction of an hour does 37 minutes represent?

49. In a basic college mathematics class containing 31 students, there are 18 freshmen, 10 sophomores, and 3 juniors. What fraction of the class is sophomores?

50. In a sports team with 20 children, there are 9 boys and 11 girls. What fraction of the team is boys?

51. Thirty-three out of the fifty total states in the United States contain federal Indian reservations.

   a. What fraction of the states contain federal Indian reservations?

   b. How many states do not contain federal Indian reservations?

   c. What fraction of the states do not contain federal Indian reservations? (*Source:* Tiller Research, Inc., Albuquerque, NM)

52. Consumer fireworks are legal in 46 out of the 50 total states in the United States. (*Source:* USA.gov)

   a. In what fraction of the states are consumer fireworks legal?

   b. In how many states are consumer fireworks illegal?

   c. In what fraction of the states are consumer fireworks illegal? (*Source:* United States Fireworks Safety Council)

53. A bag contains 50 red or blue marbles. If 21 marbles are blue,

   a. What *fraction* of the marbles are blue?

   b. How many marbles are red?

   c. What *fraction* of the marbles are red?

54. An art dealer is taking inventory. His shop contains a total of 37 pieces, which are all sculptures, watercolor paintings, or oil paintings. If there are 15 watercolor paintings and 17 oil paintings, answer each question.

   a. What fraction of the inventory is watercolor paintings?

   b. What fraction of the inventory is oil paintings?

   c. How many sculptures are there?

   d. What fraction of the inventory is sculptures?

Objective **C** *Write the shaded area in each figure group as (a) an improper fraction and (b) a mixed number. See Examples 15 and 16.*

55.

56.

57.

58.

59.

60.

2, 3, 5, 7, 11, 13, 17, 19

61.

62.

Objective D *Write each mixed number as an improper fraction. See Example 17.*

63. $2\frac{1}{3}$

64. $6\frac{3}{4}$

65. $3\frac{3}{5}$

66. $2\frac{5}{9}$

67. $6\frac{5}{8}$

68. $7\frac{3}{8}$

69. $2\frac{11}{15}$

70. $1\frac{13}{17}$

71. $11\frac{6}{7}$

72. $12\frac{2}{5}$

73. $6\frac{6}{13}$

74. $8\frac{9}{10}$

75. $4\frac{13}{24}$

76. $5\frac{17}{25}$

77. $17\frac{7}{12}$

78. $12\frac{7}{15}$

79. $9\frac{7}{20}$

80. $10\frac{14}{27}$

81. $2\frac{51}{107}$

82. $3\frac{27}{125}$

83. $166\frac{2}{3}$

84. $114\frac{2}{7}$

Objective E *Write each improper fraction as a mixed number or a whole number. See Example 18.*

85. $\frac{17}{5}$

86. $\frac{13}{7}$

87. $\frac{37}{8}$

88. $\frac{64}{9}$

89. $\frac{47}{15}$

90. $\frac{65}{12}$

91. $\frac{46}{21}$

92. $\frac{67}{17}$

93. $\frac{198}{6}$

94. $\frac{112}{7}$

95. $\frac{225}{15}$

96. $\frac{196}{14}$

97. $\frac{200}{3}$

98. $\frac{300}{7}$

99. $\frac{247}{23}$

100. $\frac{437}{53}$

101. $\frac{319}{18}$

102. $\frac{404}{21}$

103. $\frac{182}{175}$

104. $\frac{149}{143}$

105. $\frac{737}{112}$

106. $\frac{901}{123}$

Review

Simplify. See Section 1.9.

107. 3^2

108. 4^3

109. 5^3

110. 3^4

Write each using exponents.

111. $7 \cdot 7 \cdot 7 \cdot 7 \cdot 7$

112. $5 \cdot 5 \cdot 5 \cdot 5$

113. $2 \cdot 2 \cdot 2 \cdot 3$

114. $4 \cdot 4 \cdot 10 \cdot 10 \cdot 10$

Concept Extensions

Write each fraction.

115. In your own words, explain how to write an improper fraction as a mixed number.

116. In your own words, explain how to write a mixed number as an improper fraction.

Identify the larger fraction for each pair.

117. $\frac{1}{2}$ or $\frac{2}{3}$ (*Hint:* Represent each fraction by the shaded part of equivalent figures. Then compare the shaded areas.)

118. $\frac{7}{4}$ or $\frac{3}{5}$ (*Hint:* Identify each as a proper fraction or an improper fraction.)

Solve. See the first Concept Check in this section.

119. If ◯◯◯◯ represents $\frac{4}{9}$ of a whole diagram, sketch the whole diagram.

120. If △△ represents $\frac{1}{3}$ of a whole diagram, sketch the whole diagram.

121. IKEA Group employs workers in four different regions worldwide, as shown on the bar graph. What fraction of IKEA employees work in the North American region? (*Source:* IKEA Group)

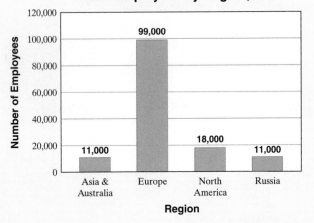

122. The Public Broadcasting Service (PBS) provides programming to the noncommercial public TV stations of the United States. The bar graph shows a breakdown of the public television licensees by type. Each licensee operates one or more PBS member TV stations. What fraction of the public television licensees are universities or colleges? (*Source:* The Public Broadcasting Service)

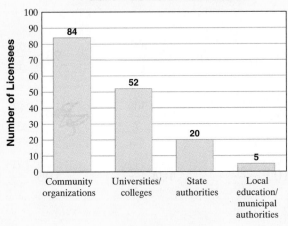

123. Heifer International is a nonprofit world hunger relief organization that focuses on sustainable agriculture programs. Currently Heifer International is working in 6 North American countries, 5 South American countries, 4 Central/Eastern European countries, 14 African countries, and 7 Asian/South Pacific countries. What fraction of the total countries in which Heifer International works is located in North America? (*Hint:* First find the total number of countries) (*Source:* Heifer International)

124. The United States Mint operates six facilities. One facility is headquarters, one facility is a depository, and four facilities mint coins. What fraction of the United States Mint facilities produces coins? (*Hint:* First find the total number of facilities.) (*Source:* United States Mint)

2.2 Factors and Prime Factorization

To perform many operations with fractions, it is necessary to be able to factor a number. In this section, only the **natural numbers**—1, 2, 3, 4, 5, and so on—will be considered.

✓**Concept Check** How are the natural numbers and the whole numbers alike? How are they different?

Objective A Finding Factors of Numbers

Recall that when numbers are multiplied to form a product, each number is called a factor. Since $5 \cdot 9 = 45$, both 5 and 9 are **factors** of 45, and $5 \cdot 9$ is called a **factorization** of 45.

The two-number factorizations of 45 are

$$1 \cdot 45 \quad 3 \cdot 15 \quad 5 \cdot 9$$

Thus, we say that the factors of 45 are $1, 3, 5, 9, 15,$ and 45.

> **Helpful Hint**
>
> From our definition of factor above, notice that a **factor** of a number divides the number evenly (with a remainder of 0). For example,
>
> $$\frac{45}{1)45} \quad \frac{15}{3)45} \quad \frac{9}{5)45} \quad \frac{5}{9)45} \quad \frac{3}{15)45} \quad \frac{1}{45)45}$$

Example 1 Find all the factors of 20.

Solution: First we write all the two-number factorizations of 20.

$$1 \cdot 20 = 20$$
$$2 \cdot 10 = 20$$
$$4 \cdot 5 = 20$$

The factors of 20 are 1, 2, 4, 5, 10, and 20.

▧ Work Practice 1

Objective B Identifying Prime and Composite Numbers

Of all the ways to factor a number, one special way is called the **prime factorization.** To help us write prime factorizations, we first review prime and composite numbers.

Prime Numbers

A **prime number** is a natural number that has exactly two different factors, 1 and itself.

The first several prime numbers are

$$2, 3, 5, 7, 11, 13, 17$$

It would be helpful to memorize these.

If a natural number other than 1 is not a prime number, it is called a **composite number.**

Objectives

A Find the Factors of a Number. ▶

B Identify Prime and Composite Numbers. ▶

C Find the Prime Factorization of a Number. ▶

Practice 1

Find all the factors of each number.
a. 15 **b.** 7 **c.** 24

Answers
1. a. $1, 3, 5, 15$ **b.** $1, 7$
c. $1, 2, 3, 4, 6, 8, 12, 24$

✓**Concept Check Answer**
answers may vary

123

Composite Numbers

A **composite number** is any natural number, other than 1, that is not prime.

> **Helpful Hint**
>
> The natural number 1 is neither prime nor composite.

Practice 2

Determine whether each number is prime or composite. Explain your answers.

21, 13, 18, 29, 39

Example 2 Determine whether each number is prime or composite. Explain your answers.

3,   9,   11,   17,   26

Solution: The number 3 is prime. Its only factors are 1 and 3 (itself).

The number 9 is composite. It has more than two factors: 1, 3, and 9.

The number 11 is prime. Its only factors are 1 and 11.

The number 17 is prime. Its only factors are 1 and 17.

The number 26 is composite. Its factors are 1, 2, 13, and 26.

▓ Work Practice 2

Objective C Finding Prime Factorizations ▶

Now we are ready to find **prime factorizations** of numbers.

Prime Factorization

The **prime factorization** of a number is the factorization in which all the factors are prime numbers.

For example, the prime factorization of 12 is $2 \cdot 2 \cdot 3$ because

$$12 = \underbrace{2 \cdot 2 \cdot 3}$$

This product is 12 and each number is a prime number.

Every whole number greater than 1 has exactly one prime factorization.

> **Helpful Hint**
>
> Don't forget that multiplication is commutative, so $2 \cdot 2 \cdot 3$ can also be written as $2 \cdot 3 \cdot 2$ or $3 \cdot 2 \cdot 2$ or $2^2 \cdot 3$. Any one of these can be called *the prime factorization of* 12.

Practice 3

Find the prime factorization of 28.

Example 3 Find the prime factorization of 45.

Solution: The first prime number, 2, does not divide 45 evenly (with a remainder of 0). The second prime number, 3, does, so we divide 45 by 3.

$$3\overline{)45}^{\,15}$$

Because 15 is not prime and 3 also divides 15 evenly, we divide by 3 again.

$$3\overline{)15}^{\,5}$$
$$3\overline{)45}$$

Answers

2. 13 and 29 are prime. 21, 18, and 39 are composite.   **3.** $2 \cdot 2 \cdot 7$ or $2^2 \cdot 7$

The quotient, 5, is a prime number, so we are finished. The prime factorization of 45 is

$$45 = 3 \cdot 3 \cdot 5 \quad \text{or} \quad 45 = 3^2 \cdot 5,$$

using exponents.

▥ Work Practice 3

There are a few quick **divisibility tests** to determine whether a number is divisible by the primes 2, 3, or 5. (A number is divisible by 2, for example, if 2 divides it evenly.)

Divisibility Tests

A whole number is divisible by:

• 2 if the last digit is 0, 2, 4, 6, or 8.

     ↓

   13**2** is divisible by 2 since the last digit is a 2.

• 3 if the sum of the digits is divisible by 3.

   144 is divisible by 3 since $1 + 4 + 4 = 9$ is divisible by 3.

• 5 if the last digit is 0 or 5.

     ↓

   111**5** is divisible by 5 since the last digit is a 5.

Helpful Hint

Here are a few other divisibility tests you may find interesting. A whole number is divisible by:

• 4 if its last two digits are divisible by 4.

   17**12** is divisible by 4.

• 6 if it's divisible by 2 and 3.

   9858 is divisible by 6.

• 9 if the sum of its digits is divisible by 9.

   5238 is divisible by 9 since $5 + 2 + 3 + 8 = 18$ is divisible by 9.

We will usually begin the division process with the smallest prime number factor of the given number. Since multiplication is commutative, this is not necessary. As long as the divisor is any prime number factor, this process works.

Example 4   Find the prime factorization of 180.

Solution: We divide 180 by 2 and continue dividing until the quotient is no longer divisible by 2. We then divide by the next largest prime number, 3, until the quotient is no longer divisible by 3. We continue this process until the quotient is a prime number.

$$
\begin{array}{r}
5 \\
3)\overline{\;15\;} \\
3)\overline{\;45\;} \\
2)\overline{\;90\;} \\
2)\overline{180}
\end{array}
$$

(Continued on next page)

Practice 4

Find the prime factorization of 120.

Answer

4. $2 \cdot 2 \cdot 2 \cdot 3 \cdot 5$ or $2^3 \cdot 3 \cdot 5$

Thus, the prime factorization of 180 is

$$180 = 2 \cdot 2 \cdot 3 \cdot 3 \cdot 5 \quad \text{or} \quad 180 = 2^2 \cdot 3^2 \cdot 5,$$

using exponents.

▪ Work Practice 4

Practice 5

Find the prime factorization of 756.

Example 5   Find the prime factorization of 945.

Solution: This number is not divisible by 2 but is divisible by 3. We will begin by dividing 945 by 3.

$$\begin{array}{r} 7 \\ 5\overline{)35} \\ 3\overline{)105} \\ 3\overline{)315} \\ 3\overline{)945} \end{array}$$

Thus, the prime factorization of 945 is

$$945 = 3 \cdot 3 \cdot 3 \cdot 5 \cdot 7 \quad \text{or} \quad 945 = 3^3 \cdot 5 \cdot 7$$

▪ Work Practice 5

Another way to find the prime factorization is to use a factor tree, as shown in the next example.

Practice 6

Use a factor tree to find the prime factorization of 45.

Example 6   Use a factor tree to find the prime factorization of 18.

Solution: We begin by writing 18 as a product of two natural numbers greater than 1, say $2 \cdot 9$.

$$\begin{array}{c} 18 \\ \swarrow \searrow \\ 2 \cdot 9. \end{array}$$

The number 2 is prime, but 9 is not. So we write 9 as $3 \cdot 3$.

$$\begin{array}{c} 18 \\ \swarrow \searrow \\ 2 \cdot 9 \\ \downarrow \quad \downarrow\searrow \\ 2 \cdot 3 \cdot 3 \end{array}$$

Each factor is now prime, so the prime factorization is

$$18 = 2 \cdot 3 \cdot 3 \quad \text{or} \quad 18 = 2 \cdot 3^2,$$

using exponents.

▪ Work Practice 6

In this text, we will write the factorization of a number from the smallest factor to the largest factor.

Answers

5. $2 \cdot 2 \cdot 3 \cdot 3 \cdot 3 \cdot 7$ or $2^2 \cdot 3^3 \cdot 7$

6. $3 \cdot 3 \cdot 5$ or $3^2 \cdot 5$

Example 7 Use a factor tree to find the prime factorization of 80.

Solution: Write 80 as a product of two numbers. Continue this process until all factors are prime.

All factors are now prime, so the prime factorization of 80 is

$$2 \cdot 2 \cdot 2 \cdot 2 \cdot 5 \quad \text{or} \quad 2^4 \cdot 5.$$

Work Practice 7

Helpful Hint

It makes no difference which factors you start with. The prime factorization of a number will be the same.

Same factors as in Example 7

✓**Concept Check** True or false? Two different numbers can have exactly the same prime factorization. Explain your answer.

Example 8 Use a factor tree to find the prime factorization of 175.

Solution: We begin by writing 175 as a product of two numbers greater than 1, say $7 \cdot 25$.

$$175$$
$$\swarrow \searrow$$
$$7 \cdot 25$$
$$\downarrow \quad \downarrow \searrow$$
$$7 \cdot 5 \cdot 5$$

The prime factorization of 175 is

$$175 = 5 \cdot 5 \cdot 7 \quad \text{or} \quad 175 = 5^2 \cdot 7$$

Work Practice 8

Practice 7

Use a factor tree to find the prime factorization of each number.

a. 30 **b.** 56 **c.** 72

Practice 8

Use a factor tree to find the prime factorization of 117.

Answers

7. a. $2 \cdot 3 \cdot 5$ **b.** $2 \cdot 2 \cdot 2 \cdot 7$ or $2^3 \cdot 7$
c. $2 \cdot 2 \cdot 2 \cdot 3 \cdot 3$ or $2^3 \cdot 3^2$
8. $3 \cdot 3 \cdot 13$ or $3^2 \cdot 13$

✓**Concept Check Answer**
false; answers may vary

Vocabulary, Readiness & Video Check

Use the choices below to fill in each blank.

factor(s)   prime factorization   prime
natural   composite

1. The number 40 equals $2 \cdot 2 \cdot 2 \cdot 5$. Since each factor is prime, we call $2 \cdot 2 \cdot 2 \cdot 5$ the _____ of 40.
2. A natural number, other than 1, that is not prime is called a(n) _____ number.
3. A natural number that has exactly two different factors, 1 and itself, is called a(n) _____ number.
4. The numbers $1, 2, 3, 4, 5, \ldots$ are called the _____ numbers.
5. Since $30 = 5 \cdot 6$, the numbers 5 and 6 are _____ of 30.
6. Answer true or false: $5 \cdot 6$ is the prime factorization of 30. _____

Martin-Gay Interactive Videos   *Watch the section lecture video and answer the following questions.*

Objective A 7. From ▢ Example 2, what aren't $3 \cdot 4$ and $4 \cdot 3$ considered different two-number factorizations of 12? ▶

Objective B 8. From the lecture before ▢ Example 3, are all natural numbers either prime or composite? ▶

Objective C 9. Complete this statement based on ▢ Example 7: You may write factors in different _____, but every natural number has only _____ prime factorization. ▶

See Video 2.2

2.2 Exercise Set MyMathLab® ▶

Objective A *List all the factors of each number. See Example 1.*

1. 8   2. 6   ▶ 3. 25   4. 30   5. 4   6. 9

7. 18   8. 48   9. 29   10. 37   11. 80   12. 100

▶ 13. 12   14. 28   15. 34   16. 26

Objective B *Identify each number as prime or composite. See Example 2.*

17. 7   18. 5   ▶ 19. 4   ▶ 20. 10   21. 23   22. 13

23. 49   24. 45   ▶ 25. 67   26. 89   27. 39   28. 21

29. 31   30. 27   31. 63   32. 51   33. 119   34. 147

Objective C   *Find the prime factorization of each number. Write any repeated factors using exponents. See Examples 3 through 8.*

35. 32   **36.** 64   ▸ **37.** 15   **38.** 21   **39.** 40   **40.** 63

▸ **41.** 36   **42.** 80   **43.** 39   **44.** 56   **45.** 60   **46.** 84

47. 110   **48.** 130   **49.** 85   **50.** 93   **51.** 128   **52.** 81

53. 154   **54.** 198   **55.** 300   **56.** 360   **57.** 240   **58.** 836

59. 828   **60.** 504   **61.** 882   **62.** 405   **63.** 637   **64.** 539

Objectives B C   Mixed Practice   *Find the prime factorization of each composite number. Write any repeated factors using exponents. Write prime if the number is prime.*

65. 33   **66.** 48   **67.** 98   **68.** 54   **69.** 67   **70.** 59

71. 459   **72.** 208   **73.** 97   **74.** 103   **75.** 700   **76.** 1000

Review

Round each whole number to the indicated place value. See Section 1.5.

77. 4267 hundreds   **78.** 32,465 thousands   **79.** 7,658,240 ten-thousands

80. 4,286,340 tens   **81.** 19,764 thousands   **82.** 10,292,876 millions

The bar graph below shows the number of patents that Apple Inc. has been granted over a three-year period. Use this bar graph to answer the questions below. See Section 2.1. (Source: IFI CLAIMS Patent Services)

Apple Inc. Patents

83. Find the total number of patents received by Apple for the years shown.

84. How many fewer patents were granted in 2010 than in 2012?

85. What fraction of the patents were granted in 2012?

86. What fraction of the patents were granted in 2011?

Concept Extensions

Find the prime factorization of each number.

87. 34,020

88. 131,625

89. In your own words, define a prime number.

90. The number 2 is a prime number. All other even natural numbers are composite numbers. Explain why.

91. Why are we interested in the prime factorizations of nonzero whole numbers only?

92. Two students have different prime factorizations for the same number. Is this possible? Explain.

2.3 Simplest Form of a Fraction

Objectives

A Write a Fraction in Simplest Form or Lowest Terms.

B Determine Whether Two Fractions Are Equivalent.

C Solve Problems by Writing Fractions in Simplest Form.

Objective A Writing Fractions in Simplest Form

Fractions that represent the same portion of a whole are called **equivalent fractions.**

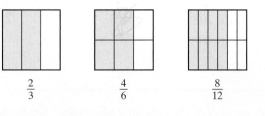

$$\frac{2}{3} \qquad \frac{4}{6} \qquad \frac{8}{12}$$

For example, $\frac{2}{3}, \frac{4}{6},$ and $\frac{8}{12}$ all represent the same shaded portion of the rectangle's area, so they are equivalent fractions.

$$\frac{2}{3} = \frac{4}{6} = \frac{8}{12}$$

There are many equivalent forms of a fraction. A special form of a fraction is called **simplest form.**

> **Simplest Form of a Fraction**
>
> A fraction is written in **simplest form** or **lowest terms** when the numerator and the denominator have no common factors other than 1.

For example, the fraction $\frac{2}{3}$ is in simplest form because 2 and 3 have no common factor other than 1. The fraction $\frac{4}{6}$ is *not* in simplest form because 4 and 6 both have a factor of 2. That is, 2 is a common factor of 4 and 6. The process of writing a fraction in simplest form is called **simplifying** the fraction.

To simplify $\frac{4}{6}$ and write it as $\frac{2}{3}$, let's first study a few properties. Recall from Section 2.1 that any nonzero whole number n divided by itself is 1.

Any nonzero number n divided by itself is 1.

$$\frac{5}{5} = 1, \ \frac{17}{17} = 1, \ \frac{24}{24} = 1, \text{ or, in general, } \frac{n}{n} = 1$$

Also, in general, if $\frac{a}{b}$ and $\frac{c}{d}$ are fractions (with b and d not 0), the following is true.

$$\frac{a \cdot c}{b \cdot d} = \frac{a}{b} \cdot \frac{c}{d}^*$$

These properties allow us to do the following:

$$\frac{4}{6} = \frac{2 \cdot 2}{2 \cdot 3} = \frac{2}{2} \cdot \frac{2}{3} = 1 \cdot \frac{2}{3} = \frac{2}{3}$$

When 1 is multiplied by a number, the result is the same number.

$\llcorner$ This is 1

Note: We will study this concept further in the next section.

Example 1   Write in simplest form: $\dfrac{12}{20}$

Solution: Notice that 12 and 20 have a common factor of 4.

$$\frac{12}{20} = \frac{4 \cdot 3}{4 \cdot 5} = \frac{4}{4} \cdot \frac{3}{5} = 1 \cdot \frac{3}{5} = \frac{3}{5}$$

Since 3 and 5 have no common factors (other than 1), $\dfrac{3}{5}$ is in simplest form.

🔳 Work Practice 1

If you have trouble finding common factors, write the prime factorizations of the numerator and the denominator.

Example 2   Write in simplest form: $\dfrac{42}{66}$

Solution: Let's write the prime factorizations of 42 and 66.

$$\frac{42}{66} = \frac{2 \cdot 3 \cdot 7}{2 \cdot 3 \cdot 11} = \frac{2}{2} \cdot \frac{3}{3} \cdot \frac{7}{11} = 1 \cdot 1 \cdot \frac{7}{11} = \frac{7}{11}$$

🔳 Work Practice 2

In the example above, you may have saved time by noticing that 42 and 66 have a common factor of 6.

$$\frac{42}{66} = \frac{6 \cdot 7}{6 \cdot 11} = \frac{6}{6} \cdot \frac{7}{11} = 1 \cdot \frac{7}{11} = \frac{7}{11}$$

Helpful Hint

Writing the prime factorizations of the numerator and the denominator is helpful in finding any common factors.

Practice 1

Write in simplest form: $\dfrac{30}{45}$

Practice 2

Write in simplest form: $\dfrac{39}{51}$

Answers

1. $\dfrac{2}{3}$ **2.** $\dfrac{13}{17}$

Practice 3

Write in simplest form: $\dfrac{9}{50}$

Example 3 Write in simplest form: $\dfrac{10}{27}$

Solution:

$$\dfrac{10}{27} = \dfrac{2 \cdot 5}{3 \cdot 3 \cdot 3} \qquad \text{Prime factorizations of 10 and 27.}$$

Since 10 and 27 have no common factors, $\dfrac{10}{27}$ is already in simplest form.

▪ Work Practice 3

Practice 4

Write in simplest form: $\dfrac{49}{112}$

Example 4 Write in simplest form: $\dfrac{30}{108}$

Solution:

$$\dfrac{30}{108} = \dfrac{2 \cdot 3 \cdot 5}{2 \cdot 2 \cdot 3 \cdot 3 \cdot 3} = \dfrac{2}{2} \cdot \dfrac{3}{3} \cdot \dfrac{5}{2 \cdot 3 \cdot 3} = 1 \cdot 1 \cdot \dfrac{5}{18} = \dfrac{5}{18}$$

▪ Work Practice 4

We can use a shortcut procedure with common factors when simplifying.

$$\dfrac{4}{6} = \dfrac{\overset{1}{\cancel{2}} \cdot 2}{\underset{1}{\cancel{2}} \cdot 3} = \dfrac{1 \cdot 2}{1 \cdot 3} = \dfrac{2}{3} \qquad \begin{array}{l}\text{Divide out the common factor of 2 in the}\\ \text{numerator and denominator.}\end{array}$$

This procedure is possible because dividing out a common factor in the numerator and denominator is the same as removing a factor of 1 in the product.

Writing a Fraction in Simplest Form

To write a fraction in simplest form, write the prime factorizations of the numerator and the denominator and then divide both by all common factors.

Practice 5

Write in simplest form: $\dfrac{64}{20}$

Example 5 Write in simplest form: $\dfrac{72}{26}$

Solution:

$$\dfrac{72}{26} = \dfrac{\overset{1}{\cancel{2}} \cdot 2 \cdot 2 \cdot 3 \cdot 3}{\underset{1}{\cancel{2}} \cdot 13} = \dfrac{1 \cdot 2 \cdot 2 \cdot 3 \cdot 3}{1 \cdot 13} = \dfrac{36}{13},$$

which can also be written as

$$2\dfrac{10}{13}$$

▪ Work Practice 5

✓**Concept Check** Which is the correct way to simplify the fraction $\dfrac{15}{25}$? Or are both correct? Explain.

a. $\dfrac{15}{25} = \dfrac{3 \cdot \overset{1}{\cancel{5}}}{5 \cdot \underset{1}{\cancel{5}}} = \dfrac{3}{5}$ 　　　　b. $\dfrac{1\overset{1}{\cancel{5}}}{2\underset{1}{\cancel{5}}} = \dfrac{11}{21}$

Answers

3. $\dfrac{9}{50}$ **4.** $\dfrac{7}{16}$ **5.** $\dfrac{16}{5}$ or $3\dfrac{1}{5}$

✓**Concept Check Answers**
a. correct **b.** incorrect

Example 6   Write in simplest form: $\dfrac{6}{60}$

Solution:

$$\frac{6}{60} = \frac{\overset{1}{\cancel{2}} \cdot \overset{1}{\cancel{3}}}{\underset{1}{\cancel{2}} \cdot 2 \cdot \underset{1}{\cancel{3}} \cdot 5} = \frac{1 \cdot 1}{1 \cdot 2 \cdot 1 \cdot 5} = \frac{1}{10}$$

■ Work Practice 6

Helpful Hint

Be careful when all factors of the numerator or denominator are divided out. In Example 6, the numerator was $1 \cdot 1 = 1$, so the final result was $\dfrac{1}{10}$.

In the fraction of Example 6, $\dfrac{6}{60}$, you may have immediately noticed that the largest common factor of 6 and 60 is 6. If so, you may simply divide out that largest common factor.

$$\frac{6}{60} = \frac{\overset{1}{\cancel{6}}}{\underset{1}{\cancel{6}} \cdot 10} = \frac{1}{1 \cdot 10} = \frac{1}{10} \qquad \text{Divide out the common factor of 6.}$$

Notice that the result, $\dfrac{1}{10}$, is in simplest form. If it were not, we would repeat the same procedure until the result was in simplest form.

Example 7   Write in simplest form: $\dfrac{45}{75}$

Solution: You may write the prime factorizations of 45 and 75 or you may notice that these two numbers have a common factor of 15.

$$\frac{45}{75} = \frac{3 \cdot \overset{1}{\cancel{15}}}{5 \cdot \underset{1}{\cancel{15}}} = \frac{3 \cdot 1}{5 \cdot 1} = \frac{3}{5}$$

The numerator and denominator of $\dfrac{3}{5}$ have no common factors other than 1, so $\dfrac{3}{5}$ is in simplest form.

■ Work Practice 7

Objective B Determining Whether Two Fractions Are Equivalent

Recall that two fractions are equivalent if they represent the same part of a whole. One way to determine whether two fractions are equivalent is to see whether they simplify to the same fraction.

Practice 6
Write in simplest form: $\dfrac{8}{56}$

Practice 7
Write in simplest form: $\dfrac{42}{48}$

Answers
6. $\dfrac{1}{7}$ **7.** $\dfrac{7}{8}$

Determine whether $\dfrac{7}{9}$ and $\dfrac{21}{27}$ are equivalent.

Example 8 Determine whether $\dfrac{16}{40}$ and $\dfrac{10}{25}$ are equivalent.

Solution: Simplify each fraction.

$$\dfrac{16}{40} = \dfrac{\overset{1}{\cancel{8}} \cdot 2}{\underset{1}{\cancel{8}} \cdot 5} = \dfrac{1 \cdot 2}{1 \cdot 5} = \dfrac{2}{5} \Big\}$$

$$\dfrac{10}{25} = \dfrac{2 \cdot \overset{1}{\cancel{5}}}{5 \cdot \underset{1}{\cancel{5}}} = \dfrac{2 \cdot 1}{5 \cdot 1} = \dfrac{2}{5} \Big\}$$

Since these fractions are the same, $\dfrac{16}{40} = \dfrac{10}{25}$.

🔲 Work Practice 8

There is a shortcut method you may use to check or test whether two fractions are equivalent. In the example above, we learned that the fractions are equivalent, or

$$\dfrac{16}{40} = \dfrac{10}{25}$$

In this example above, we call $25 \cdot 16$ and $40 \cdot 10$ **cross products** because they are the products one obtains by multiplying across.

Notice that these cross products are equal

$$25 \cdot 16 = 400, \quad 40 \cdot 10 = 400$$

In general, this is true for equivalent fractions.

Equivalent Fractions

Since the cross products ($8 \cdot 6 = 48$ and $24 \cdot 2 = 48$) are equal, the fractions are equivalent.
   Note: If the cross products are not equal, the fractions are not equivalent.

Determine whether $\dfrac{4}{13}$ and $\dfrac{5}{18}$ are equivalent.

Example 9 Determine whether $\dfrac{8}{11}$ and $\dfrac{19}{26}$ are equivalent.

Solution: Let's check cross products.

Since $208 \neq 209$, then $\dfrac{8}{11} \neq \dfrac{19}{26}$.

🔲 Work Practice 9

Helpful Hint "Not equal to" symbol.

Answer
8. equivalent   **9.** not equivalent

Objective C Solving Problems by Writing Fractions in Simplest Form ▶

Many real-life problems can be solved by writing fractions. To make the answers clearer, these fractions should be written in simplest form.

Example 10 Calculating Fraction of Parks in Pennsylvania

There are currently 46 national historical parks in the United States. Two of these historical parks are located in the state of Pennsylvania. What fraction of the United States' national historical parks can be found in Pennsylvania? Write the fraction in simplest form. (*Source:* National Park Service)

Solution: First we determine the fraction of parks found in Pennsylvania.

$$\frac{2}{46} \quad \begin{array}{l} \leftarrow \text{ national historical parks in Pennsylvania} \\ \leftarrow \text{ total national historical parks} \end{array}$$

Next we simplify the fraction.

$$\frac{2}{46} = \frac{\overset{1}{\cancel{2}}}{\underset{1}{\cancel{2}} \cdot 23} = \frac{1}{1 \cdot 23} = \frac{1}{23}$$

Thus, $\frac{1}{23}$ of the United States' national parks are in the state of Pennsylvania.

🔲 **Work Practice 10**

Practice 10

There are four national historical parks in the state of Virginia. See Example 10 and determine what fraction of the United States' national historical parks can be found in Virginia. Write the fraction in simplest form.

Answer

10. $\frac{2}{23}$

🖩 **Calculator Explorations** **Simplifying Fractions**

Scientific Calculator

Many calculators have a fraction key, such as $\boxed{a\ b/c}$, that allows you to simplify a fraction on the calculator. For example, to simplify $\frac{324}{612}$, enter

$$\boxed{3}\boxed{2}\boxed{4}\boxed{a\ b/c}\boxed{6}\boxed{1}\boxed{2}\boxed{=}$$

The display will read

$$\boxed{9 \mid 17}$$

which represents $\frac{9}{17}$, the original fraction simplified.

Helpful Hint The Calculator Explorations boxes in this chapter provide only an introduction to fraction keys on calculators. Any time you use a calculator, there are both advantages and limitations to its use. Never rely solely on your calculator. It is very important that you understand how to perform all operations on fractions by hand in order to progress through later topics. For further information, talk to your instructor.

Use your calculator to simplify each fraction.

1. $\frac{128}{224}$ 2. $\frac{231}{396}$ 3. $\frac{340}{459}$ 4. $\frac{999}{1350}$

5. $\frac{810}{432}$ 6. $\frac{315}{225}$ 7. $\frac{243}{54}$ 8. $\frac{689}{455}$

Vocabulary, Readiness & Video Check

Use the choices below to fill in each blank.

0   cross products   equivalent

1   simplest form n

1. In $\dfrac{11}{48}$, since 11 and 48 have no common factors other than 1, $\dfrac{11}{48}$ is in _____.

2. Fractions that represent the same portion of a whole are called _____ fractions.

3. In the statement $\dfrac{5}{12} = \dfrac{15}{36}$, $5 \cdot 36$ and $12 \cdot 15$ are called _____.

4. The fraction $\dfrac{7}{7}$ simplifies to _____.

5. The fraction $\dfrac{0}{7}$ simplifies to _____.

6. The fraction $\dfrac{n}{1}$ simplifies to _____.

Martin-Gay Interactive Videos   *Watch the section lecture video and answer the following questions.*

Objective A 7. Complete this statement based on the lecture before Example 1: A special form of a(n) _____ form of a fraction is called simplest form.

Objective B 8. The fractions in Example 5 are shown to be equivalent by using cross products. Describe another way to tell whether the two fractions are equivalent.

Objective C 9. Why isn't $\dfrac{10}{24}$ the final answer to Example 7? What is the final answer?

See Video 2.3

2.3 Exercise Set MyMathLab®

Objective A *Write each fraction in simplest form. See Examples 1 through 7.*

1. $\dfrac{3}{12}$

2. $\dfrac{5}{30}$

3. $\dfrac{4}{42}$

4. $\dfrac{9}{48}$

5. $\dfrac{14}{16}$

6. $\dfrac{22}{34}$

7. $\dfrac{20}{30}$

8. $\dfrac{70}{80}$

9. $\dfrac{35}{50}$

10. $\dfrac{25}{55}$

11. $\dfrac{63}{81}$

12. $\dfrac{21}{49}$

13. $\dfrac{24}{40}$

14. $\dfrac{36}{54}$

15. $\dfrac{27}{64}$

16. $\dfrac{32}{63}$

17. $\dfrac{25}{40}$

18. $\dfrac{36}{42}$

19. $\dfrac{40}{64}$

20. $\dfrac{28}{60}$

21. $\dfrac{56}{68}$

22. $\dfrac{39}{42}$

23. $\dfrac{36}{24}$

24. $\dfrac{60}{36}$

25. $\dfrac{90}{120}$

26. $\dfrac{60}{150}$

27. $\dfrac{70}{196}$

28. $\dfrac{98}{126}$

29. $\dfrac{66}{308}$ **30.** $\dfrac{65}{234}$ **31.** $\dfrac{55}{85}$ **32.** $\dfrac{78}{90}$

33. $\dfrac{75}{350}$ **34.** $\dfrac{72}{420}$ **35.** $\dfrac{189}{216}$ **36.** $\dfrac{144}{162}$

37. $\dfrac{288}{480}$ **38.** $\dfrac{135}{585}$ **39.** $\dfrac{224}{16}$ **40.** $\dfrac{270}{15}$

Objective B   *Determine whether each pair of fractions is equivalent. See Examples 8 and 9.*

41. $\dfrac{3}{6}$ and $\dfrac{4}{8}$ **42.** $\dfrac{3}{9}$ and $\dfrac{2}{6}$ ▷ **43.** $\dfrac{7}{11}$ and $\dfrac{5}{8}$

44. $\dfrac{2}{5}$ and $\dfrac{4}{11}$ **45.** $\dfrac{10}{15}$ and $\dfrac{6}{9}$ **46.** $\dfrac{4}{10}$ and $\dfrac{6}{15}$

▷ **47.** $\dfrac{3}{9}$ and $\dfrac{6}{18}$ **48.** $\dfrac{2}{8}$ and $\dfrac{7}{28}$ **49.** $\dfrac{10}{13}$ and $\dfrac{12}{15}$

50. $\dfrac{16}{20}$ and $\dfrac{9}{12}$ **51.** $\dfrac{8}{18}$ and $\dfrac{12}{24}$ **52.** $\dfrac{6}{21}$ and $\dfrac{14}{35}$

Objective C   *Solve. Write each fraction in simplest form. See Example 10.*

53. A work shift for an employee at McDonald's consists of 8 hours. What fraction of the employee's work shift is represented by 2 hours?

54. Two thousand baseball caps were sold one year at the U.S. Open Golf Tournament. What fractional part of this total does 200 caps represent?

▷ **55.** There are 5280 feet in a mile. What fraction of a mile is represented by 2640 feet?

56. There are 100 centimeters in 1 meter. What fraction of a meter is 20 centimeters?

57. Sixteen states in the United States have Ritz-Carlton hotels. (*Source:* Ritz-Carlton Hotel Company, LLC)

     a. What fraction of states can claim at least one Ritz-Carlton hotel?

     b. How many states do not have a Ritz-Carlton hotel?

     c. Write the fraction of states without a Ritz-Carlton hotel.

58. There are 78 national monuments in the United States. Ten of these monuments are located in New Mexico. (*Source:* National Park Service)

     a. What fraction of the national monuments in the United States can be found in New Mexico?

     b. How many of the national monuments in the United States are found outside New Mexico?

     c. Write the fraction of national monuments found in states other than New Mexico.

▷ **59.** The outer wall of the Pentagon is 24 inches wide. Ten inches is concrete, 8 inches is brick, and 6 inches is limestone. What fraction of the wall is concrete? (*Source: USA Today*)

60. There are 35 students in a biology class. If 10 students made an A on the first test, what fraction of the students made an A?

Limestone (6 in.)

Brick (8 in.)

Concrete (10 in.)

61. Safeway Inc. operates grocery stores under multiple banners in 20 states in the United States. (*Source:* Safeway Inc.)
   a. How many states do not have a Safeway-owned store?
   b. What fraction of states do not have a Safeway-owned store?

62. Katy Biagini just bought a brand-new 2013 Toyota Camry hybrid for $28,000. Her old car was traded in for $12,000.
   a. How much of her purchase price was not covered by her trade-in?
   b. What fraction of the purchase price was not covered by the trade-in?

63. Of the 20 top-selling video games of 2011, eight had ESRB ratings of "Everyone." What fraction of 2011's top-selling video games had "Everyone" ratings? (*Source:* The NPD Group/Retail Tracking Service)

64. Worldwide, Hallmark employs about 12,000 full-time employees. About 3200 employees work at the Hallmark headquarters in Kansas City, Missouri. What fraction of Hallmark employees work in Kansas City? (*Source:* Hallmark Cards, Inc.)

Review

Multiply. See Section 1.6.

65.   91
         $\times$ 4

66.   73
         $\times$ 8

67.   387
         $\times$ 6

68.   562
         $\times$ 9

69.   72
         $\times$ 35

70.   238
         $\times$ 26

Concept Extensions

71. In your own words, define equivalent fractions.

72. Given a fraction, say $\frac{3}{8}$, how many fractions are there that are equivalent to it? Explain your answer.

Write each fraction in simplest form.

73. $\dfrac{3975}{6625}$

74. $\dfrac{9506}{12,222}$

There are generally considered to be eight basic blood types. The table shows the number of people with the various blood types in a typical group of 100 blood donors. Use the table to answer Exercises 75 through 78. Write each answer in simplest form.

Distribution of Blood Types in Blood Donors	
Blood Type	**Number of People**
O Rh-positive	37
O Rh-negative	7
A Rh-positive	36
A Rh-negative	6
B Rh-positive	9
B Rh-negative	1
AB Rh-positive	3
AB Rh-negative	1
(*Source:* American Red Cross Biomedical Services)	

75. What fraction of blood donors have blood type A Rh-positive?

76. What fraction of blood donors have an O blood type?

77. What fraction of blood donors have an AB blood type?

78. What fraction of blood donors have a B blood type?

*The following graph is called a **circle graph** or **pie chart**. Each sector (shaped like a piece of pie) shows the fraction of entering college freshmen who expect to major in each discipline shown. The whole circle represents the entire class of college freshmen. Use this graph to answer Exercises 79 through 82.*

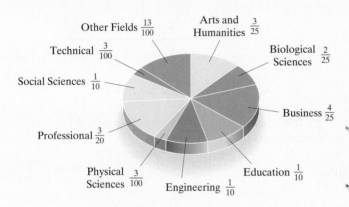

Other Fields $\frac{13}{100}$ Arts and Humanities $\frac{3}{25}$ Technical $\frac{3}{100}$ Biological Sciences $\frac{2}{25}$ Social Sciences $\frac{1}{10}$ Business $\frac{4}{25}$ Professional $\frac{3}{20}$ Physical Sciences $\frac{3}{100}$ Engineering $\frac{1}{10}$ Education $\frac{1}{10}$

Source: The Higher Education Research Institute

79. What fraction of entering college freshmen plan to major in education?

80. What fraction of entering college freshmen plan to major in biological sciences?

81. Why is the Social Sciences sector the same size as the Engineering sector?

82. Why is the Physical Sciences sector smaller than the Business sector?

Use this circle graph to answer Exercises 83 through 86.

Areas Maintained by the National Park Service

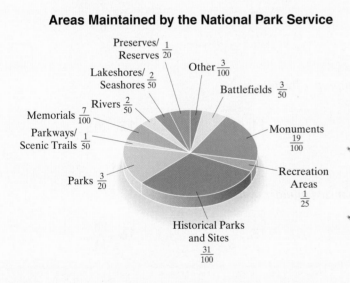

Preserves/ Reserves $\frac{1}{20}$ Other $\frac{3}{100}$ Lakeshores/ Seashores $\frac{2}{50}$ Battlefields $\frac{3}{50}$ Rivers $\frac{2}{50}$ Memorials $\frac{7}{100}$ Parkways/ Scenic Trails $\frac{1}{50}$ Monuments $\frac{19}{100}$ Parks $\frac{3}{20}$ Recreation Areas $\frac{1}{25}$ Historical Parks and Sites $\frac{31}{100}$

83. What fraction of National Park Service areas are National Memorials?

84. What fraction of National Park Service areas are National Parks?

85. Why is the National Battlefields sector smaller than the National Monuments sector?

86. Why is the National Lakeshores/National Seashores sector the same size as the National Rivers sector?

Use the following numbers for Exercises 87 through 90.

| 8691 | 786 | 1235 | 2235 | 85 | 105 | 22 | 222 | 900 | 1470 |

87. List the numbers divisible by both 2 and 3.

88. List the numbers that are divisible by both 3 and 5.

89. The answers to Exercise 87 are also divisible by what number? Tell why.

90. The answers to Exercise 88 are also divisible by what number? Tell why.

Summary on Fractions, Mixed Numbers, and Factors

Answers

1. _____

2. _____

3. _____

4. _____

5. _____

6. _____

7. _____

8. _____

9. _____

10. _____

11. _____

12. _____

13. _____

14. _____

15. _____

16. _____

17. _____

18. _____

19. _____

20. _____

Use a fraction to represent the shaded area of each figure. If the fraction is improper, also write the fraction as a mixed number.

1.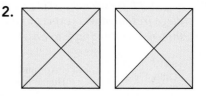

2.

Solve.

3. In a survey, 73 people out of 85 get fewer than 8 hours of sleep each night. What fraction of people in the survey get fewer than 8 hours of sleep?

4. Sketch a diagram to represent $\frac{9}{13}$.

Simplify.

5. $\frac{11}{11}$ **6.** $\frac{17}{1}$ **7.** $\frac{0}{3}$ **8.** $\frac{7}{0}$

Write each mixed number as an improper fraction.

9. $3\frac{1}{8}$ **10.** $5\frac{3}{5}$ **11.** $9\frac{6}{7}$ **12.** $20\frac{1}{7}$

Write each improper fraction as a mixed number or a whole number.

13. $\frac{20}{7}$ **14.** $\frac{55}{11}$ **15.** $\frac{39}{8}$ **16.** $\frac{98}{11}$

List the factors of each number.

17. 35   **18.** 40

Determine whether each number is prime or composite.

19. 72   **20.** 13

140

Write the prime factorization of each composite number. Write prime if the number is prime.
Write any repeated factors using exponents.

21. 65

22. 70

23. 96

24. 132

25. 252

26. 31

27. 315

28. 441

29. 286

30. 41

Write each fraction in simplest form.

31. $\dfrac{2}{14}$

32. $\dfrac{24}{20}$

33. $\dfrac{18}{38}$

34. $\dfrac{42}{110}$

35. $\dfrac{56}{60}$

36. $\dfrac{72}{80}$

37. $\dfrac{54}{135}$

38. $\dfrac{90}{240}$

39. $\dfrac{165}{210}$

40. $\dfrac{245}{385}$

Determine whether each pair of fractions is equivalent.

41. $\dfrac{7}{8}$ and $\dfrac{9}{10}$

42. $\dfrac{10}{12}$ and $\dfrac{15}{18}$

Solve. Write fraction answers in simplest form.

43. Of the 50 states, 2 states are not adjacent to any other states.

    a. What fraction of the states are not adjacent to other states?

    b. How many states are adjacent to other states?

    c. What fraction of the states are adjacent to other states?

44. In a recent year, 540 films were released and rated. Of these, 145 were rated PG-13. (*Source:* Nash Information, LLC.)

    a. What fraction were rated PG-13?

    b. How many films were rated other than PG-13?

    c. What fraction of films were rated other than PG-13?

21. _____
22. _____
23. _____
24. _____
25. _____
26. _____
27. _____
28. _____
29. _____
30. _____
31. _____
32. _____
33. _____
34. _____
35. _____
36. _____
37. _____
38. _____
39. _____
40. _____
41. _____
42. _____
43. _____
44. _____

2.4 Multiplying Fractions and Mixed Numbers

Objectives

A Multiply Fractions.

B Multiply Fractions and Mixed Numbers or Whole Numbers.

C Solve Problems by Multiplying Fractions.

Objective A Multiplying Fractions

Let's use a diagram to discover how fractions are multiplied. For example, to multiply $\frac{1}{2}$ and $\frac{3}{4}$, we find $\frac{1}{2}$ of $\frac{3}{4}$. To do this, we begin with a diagram showing $\frac{3}{4}$ of a rectangle's area shaded.

$\frac{3}{4}$ of the rectangle's area is shaded.

To find $\frac{1}{2}$ of $\frac{3}{4}$, we heavily shade $\frac{1}{2}$ of the part that is already shaded.

By counting smaller rectangles, we see that $\frac{3}{8}$ of the larger rectangle is now heavily shaded, so that

$$\frac{1}{2} \text{ of } \frac{3}{4} \text{ is } \frac{3}{8}, \text{ or } \frac{1}{2} \cdot \frac{3}{4} = \frac{3}{8}$$ Notice that $\frac{1}{2} \cdot \frac{3}{4} = \frac{1 \cdot 3}{2 \cdot 4} = \frac{3}{8}$.

Multiplying Fractions

To multiply two fractions, multiply the numerators and multiply the denominators. If $a, b, c,$ and d represent positive whole numbers, we have

$$\frac{a}{b} \cdot \frac{c}{d} = \frac{a \cdot c}{b \cdot d}$$

Practice 1–2

Multiply.

1. $\frac{3}{8} \cdot \frac{5}{7}$ **2.** $\frac{1}{3} \cdot \frac{1}{6}$

Examples Multiply.

1. $\frac{2}{3} \cdot \frac{5}{11} = \frac{2 \cdot 5}{3 \cdot 11} = \frac{10}{33}$ Multiply numerators.
Multiply denominators.

This fraction is in simplest form since 10 and 33 have no common factors other than 1.

2. $\frac{1}{4} \cdot \frac{1}{2} = \frac{1 \cdot 1}{4 \cdot 2} = \frac{1}{8}$ This fraction is in simplest form.

■ Work Practice 1–2

Practice 3

Multiply and simplify: $\frac{6}{55} \cdot \frac{5}{8}$

Answers

1. $\frac{15}{56}$ **2.** $\frac{1}{18}$ **3.** $\frac{3}{44}$

Example 3 Multiply and simplify: $\frac{6}{7} \cdot \frac{14}{27}$

Solution:

$$\frac{6}{7} \cdot \frac{14}{27} = \frac{6 \cdot 14}{7 \cdot 27}$$

We can simplify by finding the prime factorizations and using our shortcut procedure of dividing out common factors in the numerator and denominator.

$$\frac{6 \cdot 14}{7 \cdot 27} = \frac{2 \cdot \overset{1}{\cancel{3}} \cdot 2 \cdot \overset{1}{\cancel{7}}}{\underset{1}{\cancel{7}} \cdot \underset{1}{\cancel{3}} \cdot 3 \cdot 3} = \frac{2 \cdot 2}{3 \cdot 3} = \frac{4}{9}$$

🔲 **Work Practice 3**

Helpful Hint

Remember that the shortcut procedure above is the same as removing factors of 1 in the product.

$$\frac{6 \cdot 14}{7 \cdot 27} = \frac{2 \cdot 3 \cdot 2 \cdot 7}{7 \cdot 3 \cdot 3 \cdot 3} = \frac{7}{7} \cdot \frac{3}{3} \cdot \frac{2 \cdot 2}{3 \cdot 3} = 1 \cdot 1 \cdot \frac{4}{9} = \frac{4}{9}$$

Helpful Hint

In simplifying a product, don't forget that it may be possible to identify common factors without actually writing the prime factorizations. For example,

$$\frac{10}{11} \cdot \frac{1}{20} = \frac{10 \cdot 1}{11 \cdot 20} = \frac{\overset{1}{\cancel{10}} \cdot 1}{11 \cdot \underset{2}{\cancel{20}}} = \frac{1}{11 \cdot 2} = \frac{1}{22}$$

Example 4   Multiply and simplify: $\dfrac{23}{32} \cdot \dfrac{4}{7}$

Solution: Notice that 4 and 32 have a common factor of 4.

$$\frac{23}{32} \cdot \frac{4}{7} = \frac{23 \cdot 4}{32 \cdot 7} = \frac{23 \cdot \overset{1}{\cancel{4}}}{\underset{}{\cancel{4}} \cdot 8 \cdot 7} = \frac{23}{8 \cdot 7} = \frac{23}{56}$$

🔲 **Work Practice 4**

Practice 4
Multiply and simplify: $\dfrac{4}{15} \cdot \dfrac{3}{8}$

After multiplying two fractions, always check to see whether the product can be simplified.

Examples   Multiply.

5. $\dfrac{3}{4} \cdot \dfrac{8}{5} = \dfrac{3 \cdot 8}{4 \cdot 5} = \dfrac{3 \cdot \overset{1}{\cancel{4}} \cdot 2}{\underset{1}{\cancel{4}} \cdot 5} = \dfrac{6}{5}$

6. $\dfrac{6}{13} \cdot \dfrac{26}{30} = \dfrac{6 \cdot 26}{13 \cdot 30} = \dfrac{\overset{1}{\cancel{6}} \cdot \overset{1}{\cancel{13}} \cdot 2}{\underset{1}{\cancel{13}} \cdot \underset{1}{\cancel{6}} \cdot 5} = \dfrac{2}{5}$

7. $\dfrac{1}{3} \cdot \dfrac{2}{5} \cdot \dfrac{9}{16} = \dfrac{1 \cdot 2 \cdot 9}{3 \cdot 5 \cdot 16} = \dfrac{1 \cdot \overset{1}{\cancel{2}} \cdot \overset{1}{\cancel{3}} \cdot 3}{\underset{1}{\cancel{3}} \cdot 5 \cdot \underset{1}{\cancel{2}} \cdot 8} = \dfrac{3}{40}$

🔲 **Work Practice 5–7**

Practice 5–7
Multiply.

5. $\dfrac{2}{5} \cdot \dfrac{20}{7}$

6. $\dfrac{4}{11} \cdot \dfrac{33}{16}$

7. $\dfrac{1}{6} \cdot \dfrac{3}{10} \cdot \dfrac{25}{16}$

Answers

4. $\dfrac{1}{10}$ **5.** $\dfrac{8}{7}$ **6.** $\dfrac{3}{4}$ **7.** $\dfrac{5}{64}$

Objective B Multiplying Fractions and Mixed Numbers or Whole Numbers ▶

When multiplying a fraction and a mixed or a whole number, remember that mixed and whole numbers can be written as fractions.

> ### Multiplying Fractions and Mixed Numbers or Whole Numbers
>
> To multiply with mixed numbers or whole numbers, first write any mixed or whole numbers as fractions and then multiply as usual.

Practice 8

Multiply and simplify: $2\frac{1}{2} \cdot \frac{8}{15}$

Example 8 Multiply: $3\frac{1}{3} \cdot \frac{7}{8}$

Solution: The mixed number $3\frac{1}{3}$ can be written as the fraction $\frac{10}{3}$. Then,

$$3\frac{1}{3} \cdot \frac{7}{8} = \frac{10}{3} \cdot \frac{7}{8} = \frac{\overset{1}{2} \cdot 5 \cdot 7}{3 \cdot \underset{1}{2} \cdot 4} = \frac{35}{12} \quad \text{or} \quad 2\frac{11}{12}$$

■ Work Practice 8

Don't forget that a whole number can be written as a fraction by writing the whole number over 1. For example,

$$20 = \frac{20}{1} \quad \text{and} \quad 7 = \frac{7}{1}$$

Practice 9

Multiply.

$\frac{2}{3} \cdot 18$

Example 9 Multiply.

$$\frac{3}{4} \cdot 20 = \frac{3}{4} \cdot \frac{20}{1} = \frac{3 \cdot 20}{4 \cdot 1} = \frac{3 \cdot \overset{1}{4} \cdot 5}{\underset{1}{4} \cdot 1} = \frac{15}{1} \quad \text{or} \quad 15$$

■ Work Practice 9

When both numbers to be multiplied are mixed or whole numbers, it is a good idea to estimate the product to see if your answer is reasonable. To do this, we first practice rounding mixed numbers to the nearest whole. If the fraction part of the mixed number is $\frac{1}{2}$ or greater, we round the whole number part up. If the fraction part of the mixed number is less than $\frac{1}{2}$, then we do not round the whole number part up. Study the table below for examples.

Mixed Number	Rounding
$5\frac{1}{4}$ $\frac{1}{4}$ is less than $\frac{1}{2}$	Thus, $5\frac{1}{4}$ rounds to 5.
$3\frac{9}{16}$ ← 9 is greater than 8. → Half of 16 is 8.	Thus, $3\frac{9}{16}$ rounds to 4.
$1\frac{3}{7}$ ← 3 is less than $3\frac{1}{2}$. → Half of 7 is $3\frac{1}{2}$.	Thus, $1\frac{3}{7}$ rounds to 1.

Answers

8. $\frac{4}{3}$ or $1\frac{1}{3}$ 9. 12

Examples Multiply. Check by estimating.

10. $1\frac{2}{3} \cdot 2\frac{1}{4} = \frac{5}{3} \cdot \frac{9}{4} = \frac{5 \cdot 9}{3 \cdot 4} = \frac{5 \cdot \overset{1}{\cancel{3}} \cdot 3}{\cancel{3} \cdot 4} = \frac{15}{4}$ or $3\frac{3}{4}$ Exact

Let's check by estimating.

$1\frac{2}{3}$ rounds to 2, $2\frac{1}{4}$ rounds to 2, and $2 \cdot 2 = 4$ Estimate

The estimate is close to the exact value, so our answer is reasonable.

11. $7 \cdot 2\frac{11}{14} = \frac{7}{1} \cdot \frac{39}{14} = \frac{7 \cdot 39}{1 \cdot 14} = \frac{\overset{1}{\cancel{7}} \cdot 39}{1 \cdot 2 \cdot \underset{1}{\cancel{7}}} = \frac{39}{2}$ or $19\frac{1}{2}$ Exact

To estimate,

$2\frac{11}{14}$ rounds to 3 and $7 \cdot 3 = 21$. Estimate

The estimate is close to the exact value, so our answer is reasonable

Work Practice 10–11

Recall from Section 1.6 that 0 multiplied by any number is 0. This is true of fractions and mixed numbers also.

Examples Multiply.

12. $0 \cdot \frac{3}{5} = 0$

13. $2\frac{3}{8} \cdot 0 = 0$

Work Practice 12–13

✓Concept Check Find the error.

$2\frac{1}{4} \cdot \frac{1}{2} = 2\frac{1 \cdot 1}{4 \cdot 2} = 2\frac{1}{8}$

Objective C Solving Problems by Multiplying Fractions ▶

To solve real-life problems that involve multiplying fractions, we use our four problem-solving steps from Chapter 1. In Example 14, a key word that implies multiplication is used. That key word is "**of.**"

Helpful Hint

"of" usually translates to multiplication.

Practice 10–11

Multiply.

10. $3\frac{1}{5} \cdot 2\frac{3}{4}$ **11.** $5 \cdot 3\frac{11}{15}$

Practice 12–13

Multiply.

12. $\frac{9}{11} \cdot 0$

13. $0 \cdot 4\frac{1}{8}$

Answers

10. $\frac{44}{5}$ or $8\frac{4}{5}$ **11.** $\frac{56}{3}$ or $18\frac{2}{3}$

12. 0 **13.** 0

✓Concept Check Answer

forgot to change mixed number to fraction

Practice 14

Kings Dominion is an amusement park in Doswell, Virginia. Of its 48 rides, $\frac{5}{16}$ of them are roller coasters. How many roller coasters are in Kings Dominion? (*Source:* Cedar Fair Parks)

 Example 14 Finding the Number of Roller Coasters in an Amusement Park

Cedar Point is an amusement park located in Sandusky, Ohio. Its collection of 72 rides is the largest in the world. Of the rides, $\frac{2}{9}$ are roller coasters. How many roller coasters are in Cedar Point's collection of rides? (*Source:* Cedar Fair Parks)

Solution:

1. UNDERSTAND the problem. To do so, read and reread the problem. We are told that $\frac{2}{9}$ of Cedar Point's rides are roller coasters. The word "of" here means multiplication.

2. TRANSLATE.

In words:	number of roller coasters	is	$\frac{2}{9}$	of	total rides at Cedar Point
	↓	↓	↓	↓	↓
Translate:	number of roller coasters	=	$\frac{2}{9}$	·	72

3. SOLVE: Before we solve, let's estimate a reasonable answer. The fraction $\frac{2}{9}$ is less than $\frac{1}{3}$ (draw a diagram, if needed), and $\frac{1}{3}$ of 72 rides is 24 rides, so the number of roller coasters should be less than 24.

$$\frac{2}{9} \cdot 72 = \frac{2}{9} \cdot \frac{72}{1} = \frac{2 \cdot 72}{9 \cdot 1} = \frac{2 \cdot \overset{1}{\cancel{9}} \cdot 8}{\underset{1}{\cancel{9}} \cdot 1} = \frac{16}{1} \quad \text{or} \quad 16$$

4. INTERPRET. *Check* your work. From our estimate, our answer is reasonable. *State* your conclusion: The number of roller coasters at Cedar Point is 16.

■ Work Practice 14

Helpful Hint

To help visualize a fractional part of a whole number, look at the diagram below.

$$\frac{1}{5} \text{ of } 60 = ?$$

$\frac{1}{5}$ of 60 is 12.

Answer

14. 15 roller coasters

Vocabulary, Readiness & Video Check

Use the choices below to fill in each blank. Not all choices will be used.

multiplication $\dfrac{a \cdot d}{b \cdot c}$ $\dfrac{a \cdot c}{b \cdot d}$ $\dfrac{2 \cdot 2 \cdot 2}{7}$ $\dfrac{2}{7} \cdot \dfrac{2}{7} \cdot \dfrac{2}{7}$

division   0

1. To multiply two fractions, we write $\dfrac{a}{b} \cdot \dfrac{c}{d} = $ _____ .

2. Using the definition of an exponent, the expression $\dfrac{2^3}{7} = $ _____ while $\left(\dfrac{2}{7}\right)^3 = $ _____ .

3. The word "of" indicates _____ .

4. $\dfrac{1}{5} \cdot 0 = $ _____ .

Martin-Gay Interactive Videos   *Watch the section lecture video and answer the following questions.*

Objective A 5. In ▣ Example 2, why isn't the multiplication of fractions done immediately? ▶

Objective B 6. Why do we need to know how to multiply fractions to perform the operation in ▣ Example 5? ▶

Objective C 7. What relationship between radius and diameter is used to solve ▣ Example 7? ▶

See Video 2.4

2.4 Exercise Set MyMathLab® ▶

Objective A *Multiply. Write each answer in simplest form. See Examples 1 through 7 and 12.*

1. $\dfrac{1}{3} \cdot \dfrac{2}{5}$

2. $\dfrac{2}{3} \cdot \dfrac{4}{7}$

▶ **3.** $\dfrac{6}{5} \cdot \dfrac{1}{7}$

4. $\dfrac{7}{3} \cdot \dfrac{1}{4}$

5. $\dfrac{3}{10} \cdot \dfrac{3}{8}$

6. $\dfrac{2}{5} \cdot \dfrac{7}{11}$

▶ **7.** $\dfrac{2}{7} \cdot \dfrac{5}{8}$

8. $\dfrac{7}{8} \cdot \dfrac{2}{3}$

9. $\dfrac{16}{5} \cdot \dfrac{3}{4}$

10. $\dfrac{8}{3} \cdot \dfrac{5}{12}$

11. $\dfrac{5}{28} \cdot \dfrac{2}{25}$

12. $\dfrac{4}{35} \cdot \dfrac{5}{24}$

13. $0 \cdot \dfrac{8}{9}$

14. $\dfrac{11}{12} \cdot 0$

15. $\dfrac{1}{10} \cdot \dfrac{1}{11}$

16. $\dfrac{1}{9} \cdot \dfrac{1}{13}$

17. $\dfrac{18}{20} \cdot \dfrac{36}{99}$

18. $\dfrac{5}{32} \cdot \dfrac{64}{100}$

19. $\dfrac{3}{8} \cdot \dfrac{9}{10}$

20. $\dfrac{4}{5} \cdot \dfrac{8}{25}$

▶ **21.** $\dfrac{11}{20} \cdot \dfrac{1}{7} \cdot \dfrac{5}{22}$

22. $\dfrac{27}{32} \cdot \dfrac{10}{13} \cdot \dfrac{16}{30}$

23. $\dfrac{1}{3} \cdot \dfrac{2}{7} \cdot \dfrac{1}{5}$

24. $\dfrac{3}{5} \cdot \dfrac{1}{2} \cdot \dfrac{3}{7}$

25. $\dfrac{9}{20} \cdot 0 \cdot \dfrac{4}{19}$

26. $\dfrac{8}{11} \cdot \dfrac{4}{7} \cdot 0$

27. $\dfrac{3}{14} \cdot \dfrac{6}{25} \cdot \dfrac{5}{27} \cdot \dfrac{7}{6}$

28. $\dfrac{7}{8} \cdot \dfrac{9}{20} \cdot \dfrac{12}{22} \cdot \dfrac{11}{14}$

Objective **B** *Round each mixed number to the nearest whole number. See the table at the bottom of page 144.*

29. $7\dfrac{7}{8}$ **30.** $11\dfrac{3}{4}$ **31.** $6\dfrac{1}{5}$ **32.** $4\dfrac{1}{9}$ **33.** $19\dfrac{11}{20}$ **34.** $18\dfrac{12}{22}$

Multiply. Write each answer in simplest form. For those exercises marked, find both an exact product and an estimated product. See Examples 8 through 13.

35. $12 \cdot \dfrac{1}{4}$ **36.** $\dfrac{2}{3} \cdot 6$ ▶ **37.** $\dfrac{5}{8} \cdot 4$ **38.** $10 \cdot \dfrac{7}{8}$ **39.** $1\dfrac{1}{4} \cdot \dfrac{4}{25}$

40. $\dfrac{3}{22} \cdot 3\dfrac{2}{3}$ **41.** $\dfrac{2}{5} \cdot 4\dfrac{1}{6}$ **42.** $2\dfrac{1}{9} \cdot \dfrac{6}{7}$ **43.** $\dfrac{2}{3} \cdot 1$ **44.** $1 \cdot \dfrac{5}{9}$

▶ **45.** $2\dfrac{1}{5} \cdot 3\dfrac{1}{2}$ **46.** $2\dfrac{1}{4} \cdot 7\dfrac{1}{8}$ **47.** $3\dfrac{4}{5} \cdot 6\dfrac{2}{7}$ **48.** $5\dfrac{5}{6} \cdot 7\dfrac{3}{5}$ **49.** $5 \cdot 2\dfrac{1}{2}$

    Exact: Exact: Exact: Exact:

    Estimate: Estimate: Estimate: Estimate:

50. $6 \cdot 3\dfrac{1}{3}$ **51.** $1\dfrac{1}{5} \cdot 12\dfrac{1}{2}$ **52.** $1\dfrac{1}{6} \cdot 7\dfrac{1}{5}$ **53.** $\dfrac{3}{4} \cdot 16 \cdot \dfrac{1}{2}$ **54.** $\dfrac{7}{8} \cdot 24 \cdot \dfrac{1}{3}$

55. $\dfrac{3}{10} \cdot 15 \cdot 2\dfrac{1}{2}$ **56.** $\dfrac{11}{20} \cdot 12 \cdot 3\dfrac{1}{3}$ **57.** $3\dfrac{1}{2} \cdot 1\dfrac{3}{4} \cdot 2\dfrac{2}{3}$ **58.** $4\dfrac{1}{2} \cdot 2\dfrac{1}{9} \cdot 1\dfrac{1}{5}$

Objectives **A** **B** **Mixed Practice** *Multiply and simplify. See Examples 1 through 13.*

59. $\dfrac{1}{4} \cdot \dfrac{2}{15}$ **60.** $\dfrac{3}{8} \cdot \dfrac{5}{12}$ **61.** $\dfrac{19}{37} \cdot 0$ **62.** $0 \cdot \dfrac{3}{31}$ **63.** $2\dfrac{4}{5} \cdot 1\dfrac{1}{7}$

64. $3\dfrac{1}{5} \cdot 2\dfrac{11}{32}$ **65.** $\dfrac{3}{2} \cdot \dfrac{7}{3}$ **66.** $\dfrac{15}{2} \cdot \dfrac{3}{5}$ **67.** $\dfrac{6}{15} \cdot \dfrac{5}{16}$ **68.** $\dfrac{9}{20} \cdot \dfrac{10}{90}$

69. $\dfrac{7}{72} \cdot \dfrac{9}{49}$ **70.** $\dfrac{3}{80} \cdot \dfrac{2}{27}$ **71.** $20 \cdot \dfrac{11}{12}$ **72.** $30 \cdot \dfrac{8}{9}$ **73.** $9\dfrac{5}{7} \cdot 8\dfrac{1}{5} \cdot 0$

74. $4\dfrac{11}{13} \cdot 0 \cdot 12\dfrac{1}{13}$ **75.** $12\dfrac{4}{5} \cdot 6\dfrac{7}{8} \cdot \dfrac{26}{77}$ **76.** $14\dfrac{2}{5} \cdot 8\dfrac{1}{3} \cdot \dfrac{11}{16}$

Objective C *Solve. Write each answer in simplest form. For Exercises 77 through 80, recall that "of" translates to multiplication. See Example 14.*

77. Find $\frac{1}{4}$ of 200.   **78.** Find $\frac{1}{5}$ of 200.   **79.** Find $\frac{5}{6}$ of 24.   **80.** Find $\frac{5}{8}$ of 24.

Solve. For Exercises 81 and 82, the solutions have been started for you. See Example 14.

81. In the United States, $\frac{4}{25}$ of college freshmen major in business. A community college in Pennsylvania has a freshman enrollment of approximately 800 students. How many of these freshmen might we expect to major in business?

Start the solution:

1. UNDERSTAND the problem. Reread it as many times as needed.
2. TRANSLATE into an equation. (Fill in the blank below.)

freshmen majoring in business	is	$\frac{4}{25}$	of	community college freshmen enrollment
↓	↓	↓	↓	↓

$$\text{freshmen majoring in business} = \frac{4}{25} \cdot \underline{\hspace{2cm}}$$

Finish with:
3. SOLVE
4. INTERPRET

82. A patient was told that, at most, $\frac{1}{5}$ of his calories should come from fat. If his diet consists of 3000 calories a day, find the maximum number of calories that can come from fat.

Start the solution:

1. UNDERSTAND the problem. Reread it as many times as needed.
2. TRANSLATE into an equation. (Fill in the blank below.)

patient's fat calories	is	$\frac{1}{5}$	of	his daily calories
↓	↓	↓	↓	↓

$$\text{patient's fat calories} = \frac{1}{5} \cdot \underline{\hspace{2cm}}$$

Finish with:
3. SOLVE
4. INTERPRET

83. In 2012, there were approximately 225 million moviegoers in the United States and Canada. Of these, about $\frac{13}{25}$ were female. Find the approximate number of females who attended the movies in that year. (*Source:* Motion Picture Association of America)

84. In 2012, cinemas in the United States and Canada sold about 1400 million movie tickets. About $\frac{57}{100}$ of these tickets were purchased by frequent moviegoers who go to the cinema once or more per month. Find the number of tickets purchased by frequent moviegoers in 2012. (*Source:* Motion Picture Association of America)

85. The Oregon National Historic Trail is 2170 miles long. It begins in Independence, Missouri, and ends in Oregon City, Oregon. Manfred Coulon has hiked $\frac{2}{5}$ of the trail before. How many miles has he hiked?

(*Source:* National Park Service)

86. Each turn of a screw sinks it $\frac{3}{16}$ of an inch deeper into a piece of wood. Find how deep the screw is after 8 turns.

87. The radius of a circle is one-half of its diameter, as shown. If the diameter of a circle is $\frac{3}{8}$ of an inch, what is its radius?

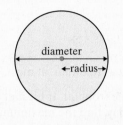

88. The diameter of a circle is twice its radius, as shown in the Exercise 87 illustration. If the radius of a circle is $\frac{7}{20}$ of a foot, what is its diameter?

89. A veterinarian's dipping vat holds 36 gallons of liquid. She normally fills it $\frac{5}{6}$ full of a medicated flea dip solution. Find how many gallons of solution are normally in the vat.

90. The plans for a deck call for $\frac{2}{5}$ of a 4-foot post to be underground. Find the length of the post that is to be buried.

91. An estimate for the measure of an adult's wrist is $\frac{1}{4}$ of the waist size. If Jorge has a 34-inch waist, estimate the size of his wrist.

92. An estimate for an adult's waist measurement is found by multiplying the neck size (in inches) by 2. Jock's neck measures $17\frac{1}{2}$ inches. Estimate his waist measurement.

93. A sidewalk is built 6 bricks wide by laying each brick side by side. How many inches wide is the sidewalk if each brick measures $3\frac{1}{4}$ inches wide?

94. A recipe calls for $\frac{1}{3}$ of a cup of flour. How much flour should be used if only $\frac{1}{2}$ of the recipe is being made?

95. A Japanese company called Che-ez! manufactures a small digital camera, the SPYZ camera. The face of the camera measures $2\frac{9}{25}$ inches by $1\frac{13}{25}$ inches and is slightly bigger than a Zippo lighter. Find the area of the face of this camera. (Area = length · width)

$1\frac{13}{25}$ in.

$2\frac{9}{25}$ in.

96. As part of his research, famous tornado expert Dr. T. Fujita studied approximately 31,050 tornadoes that occurred in the United States between 1916 and 1985. He found that roughly $\frac{7}{10}$ of these tornadoes occurred during April, May, June, and July. How many of these tornadoes occurred during these four months? (*Source: U.S. Tornadoes Part 1*, T. Fujita, University of Chicago)

Find the area of each rectangle. Recall that area = length · width.

97.

$\frac{1}{5}$ foot

$\frac{5}{14}$ foot

98. $\frac{1}{2}$ mile

$\frac{3}{8}$ mile

99.

$1\frac{3}{4}$ yards

2 yards

100.

5 inches

$3\frac{1}{2}$ inches

*Recall that the following graph is called a **circle graph** or **pie chart**. Each sector (shaped like a piece of pie) shows the fractional part of a car's total mileage that falls into a particular category. The whole circle represents a car's total mileage.*

Shopping $\frac{3}{25}$

Vacation/other $\frac{3}{50}$

Work $\frac{8}{25}$

Social/ recreational $\frac{13}{100}$

Medical $\frac{1}{100}$

Family business $\frac{1}{5}$

Visit friends $\frac{3}{25}$

School/church $\frac{2}{50}$

Source: The American Automobile Manufacturers Association and The National Automobile Dealers Association

In one year, a family drove 12,000 miles in the family car. Use the circle graph to determine how many of these miles might be expected to fall in the categories shown in Exercises 101 through 104.

101. Work

102. Shopping

103. Family business

104. Medical

Review

Divide. See Section 1.7.

105. $8\overline{)1648}$ **106.** $7\overline{)3920}$ **107.** $23\overline{)1300}$ **108.** $31\overline{)2500}$

Concept Extensions

109. In your own words, explain how to multiply
  a. fractions
  b. mixed numbers

110. In your own words, explain how to round a mixed number to the nearest whole number.

Find the error in each calculation. See the Concept Check in this section.

111. $3\dfrac{2}{3} \cdot 1\dfrac{1}{7} = 3\dfrac{2}{21}$

112. $5 \cdot 2\dfrac{1}{4} = 10\dfrac{1}{4}$

Choose the best estimate for each product.

113. $3\dfrac{1}{5} \cdot 4\dfrac{5}{8}$
  a. 7
  b. 15
  c. 8
  d. $12\dfrac{1}{8}$

114. $\dfrac{11}{12} \cdot 4\dfrac{1}{16}$
  a. 16
  b. 1
  c. 4
  d. 8

115. $9 \cdot \dfrac{10}{11}$
  a. 9
  b. 90
  c. 99
  d. 0

116. $7\dfrac{1}{4} \cdot 4\dfrac{1}{5}$
  a. 40
  b. $\dfrac{7}{5}$
  c. 35
  d. 28

117. If $\dfrac{3}{4}$ of 36 students on a first bus are girls and $\dfrac{2}{3}$ of the 30 students on a second bus are *boys*, how many students on the two buses are girls?

118. In 2013, a survey found that about $\dfrac{11}{20}$ of all adults in the United States owned a smartphone. There were roughly 240 million U.S. adults at that time. How many U.S. adults owned a smartphone in 2013? (*Source:* Pew Research Center, U.S. Census Bureau)

119. The estimated population of New Zealand was 4,433,000 in 2012. About $\dfrac{3}{20}$ of New Zealand's population is of Māori descent. How many Māori lived in New Zealand in 2012? (*Source:* Statistics New Zealand)

120. Approximately $\dfrac{1}{8}$ of the U.S. population lives in the state of California. If the U.S. population is approximately 313,914,000, find the approximate population of California. (*Source:* U.S. Census Bureau)

2.5 Ddividing Fractions and Mixed Numbers

Objective A Finding Reciprocals of Fractions

Before we can divide fractions, we need to know how to find the **reciprocal** of a fraction or whole number.

> **Reciprocal of a Fraction**
>
> Two numbers are **reciprocals** of each other if their product is 1. The reciprocal of the fraction $\frac{a}{b}$ is $\frac{b}{a}$ because $\frac{a}{b} \cdot \frac{b}{a} = \frac{a \cdot b}{b \cdot a} = 1$.

> **Finding the Reciprocal of a Fraction**
>
> To find the reciprocal of a fraction, interchange its numerator and denominator.

For example,

The reciprocal of $\frac{2}{5}$ is $\frac{5}{2}$ because $\frac{2}{5} \cdot \frac{5}{2} = \frac{10}{10} = 1$.

The reciprocal of 7, or $\frac{7}{1}$, is $\frac{1}{7}$ because $7 \cdot \frac{1}{7} = \frac{7}{1} \cdot \frac{1}{7} = \frac{7}{7} = 1$.

Examples Find the reciprocal of each number.

1. The reciprocal of $\frac{5}{6}$ is $\frac{6}{5}$. $\qquad \frac{5}{6} \cdot \frac{6}{5} = \frac{5 \cdot 6}{6 \cdot 5} = \frac{30}{30} = 1$

2. The reciprocal of $\frac{11}{8}$ is $\frac{8}{11}$. $\qquad \frac{11}{8} \cdot \frac{8}{11} = \frac{11 \cdot 8}{8 \cdot 11} = \frac{88}{88} = 1$

3. The reciprocal of $\frac{1}{3}$ is $\frac{3}{1}$ or 3. $\qquad \frac{1}{3} \cdot \frac{3}{1} = \frac{1 \cdot 3}{3 \cdot 1} = \frac{3}{3} = 1$

4. The reciprocal of 5, or $\frac{5}{1}$, is $\frac{1}{5}$. $\qquad \frac{5}{1} \cdot \frac{1}{5} = \frac{5 \cdot 1}{1 \cdot 5} = \frac{5}{5} = 1$

■ Work Practice 1–4

> **Helpful Hint**
>
> Every number except 0 has a reciprocal. The number 0 has no reciprocal because there is no number that when multiplied by 0 gives a result of 1.

Objective B Dividing Fractions

Division of fractions has the same meaning as division of whole numbers. For example,

$10 \div 5$ means: How many 5s are there in 10?

There are two 5s in 10, so $10 \div 5 = 2$.

Objectives

A Find the Reciprocal of a Fraction.

B Divide Fractions.

C Divide Fractions and Mixed Numbers or Whole Numbers.

D Solve Problems by Dividing Fractions.

Practice 1–4

Find the reciprocal of each number.

1. $\frac{4}{9}$ 2. $\frac{15}{7}$

3. 9 4. $\frac{1}{8}$

Answers

1. $\frac{9}{4}$ 2. $\frac{7}{15}$ 3. $\frac{1}{9}$ 4. 8

$\dfrac{3}{4} \div \dfrac{1}{8}$ means: How many $\dfrac{1}{8}$s are there in $\dfrac{3}{4}$?

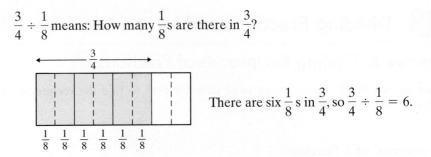

There are six $\dfrac{1}{8}$ s in $\dfrac{3}{4}$, so $\dfrac{3}{4} \div \dfrac{1}{8} = 6$.

We use reciprocals to divide fractions.

Dividing Fractions

To divide two fractions, multiply the first fraction by the reciprocal of the second fraction.

If $a, b, c,$ and d represent numbers, and $b, c,$ and d are not 0, then

$$\underbrace{\dfrac{a}{b} \div \dfrac{c}{d} = \dfrac{a}{b} \cdot \dfrac{d}{c}}_{\text{reciprocal}} = \dfrac{a \cdot d}{b \cdot c}$$

For example,

$$\dfrac{3}{4} \div \dfrac{1}{8} = \dfrac{3}{4} \cdot \dfrac{8}{1} = \dfrac{3 \cdot 8}{4 \cdot 1} = \dfrac{3 \cdot 2 \cdot \cancel{4}^{1}}{\cancel{4}_{1} \cdot 1} = \dfrac{6}{1} \text{ or } 6$$

Just as when you are multiplying fractions, always check to see whether your answer can be simplified when you divide fractions.

Practice 5–7

Divide and simplify.

5. $\dfrac{3}{2} \div \dfrac{14}{5}$ **6.** $\dfrac{8}{7} \div \dfrac{2}{9}$

7. $\dfrac{4}{9} \div \dfrac{1}{2}$

Examples   Divide and simplify.

5. $\dfrac{7}{8} \div \dfrac{2}{9} = \dfrac{7}{8} \cdot \dfrac{9}{2} = \dfrac{7 \cdot 9}{8 \cdot 2} = \dfrac{63}{16}$

6. $\dfrac{5}{16} \div \dfrac{3}{4} = \dfrac{5}{16} \cdot \dfrac{4}{3} = \dfrac{5 \cdot 4}{16 \cdot 3} = \dfrac{5 \cdot \cancel{4}^{1}}{\cancel{4}_{1} \cdot 4 \cdot 3} = \dfrac{5}{12}$

7. $\dfrac{2}{5} \div \dfrac{1}{2} = \dfrac{2}{5} \cdot \dfrac{2}{1} = \dfrac{2 \cdot 2}{5 \cdot 1} = \dfrac{4}{5}$

■ Work Practice 5–7

Helpful Hint

When dividing fractions, do *not* look for common factors to divide out until you rewrite the division as multiplication.

Do not try to divide out these two 2s.

$$\dfrac{1}{2} \div \dfrac{2}{3} = \dfrac{1}{2} \cdot \dfrac{3}{2} = \dfrac{3}{4}$$

Answers

5. $\dfrac{15}{28}$ **6.** $\dfrac{36}{7}$ **7.** $\dfrac{8}{9}$

Objective C Dividing Fractions and Mixed Numbers or Whole Numbers ▶

Just as with multiplying, mixed or whole numbers should be written as fractions before you divide them.

> ### Dividing Fractions and Mixed Numbers or Whole Numbers
>
> To divide with a mixed number or a whole number, first write the mixed or whole number as a fraction and then divide as usual.

Examples Divide.

8. $\dfrac{3}{4} \div 5 = \dfrac{3}{4} \div \dfrac{5}{1} = \dfrac{3}{4} \cdot \dfrac{1}{5} = \dfrac{3 \cdot 1}{4 \cdot 5} = \dfrac{3}{20}$

9. $\dfrac{11}{18} \div 2\dfrac{5}{6} = \dfrac{11}{18} \div \dfrac{17}{6} = \dfrac{11}{18} \cdot \dfrac{6}{17} = \dfrac{11 \cdot 6}{18 \cdot 17} = \dfrac{11 \cdot \overset{1}{\cancel{6}}}{\cancel{6} \cdot 3 \cdot 17} = \dfrac{11}{51}$

10. $5\dfrac{2}{3} \div 2\dfrac{5}{9} = \dfrac{17}{3} \div \dfrac{23}{9} = \dfrac{17}{3} \cdot \dfrac{9}{23} = \dfrac{17 \cdot 9}{3 \cdot 23} = \dfrac{17 \cdot \overset{1}{\cancel{3}} \cdot 3}{\cancel{3} \cdot 23} = \dfrac{51}{23} \text{ or } 2\dfrac{5}{23}$

■ Work Practice 8–10

Recall from Section 1.7 that the quotient of 0 and any number (except 0) is 0. This is true of fractions and mixed numbers also. For example,

$0 \div \dfrac{7}{8} = 0 \cdot \dfrac{8}{7} = 0$ Recall that 0 multiplied by any number is 0.

Also recall from Section 1.7 that the quotient of any number and 0 is undefined. This is also true of fractions and mixed numbers. For example, to find $\dfrac{7}{8} \div 0$, or $\dfrac{7}{8} \div \dfrac{0}{1}$, we would need to find the reciprocal of 0 $\left(\text{or } \dfrac{0}{1} \right)$. As we mentioned in the helpful hint at the beginning of this section, 0 has no reciprocal because there is no number that when multiplied by 0 gives a result of 1. Thus,

$\dfrac{7}{8} \div 0$ is undefined.

Examples Divide.

11. $0 \div \dfrac{2}{21} = 0 \cdot \dfrac{21}{2} = 0$ **12.** $1\dfrac{3}{4} \div 0$ is undefined.

■ Work Practice 11–12

✓ **Concept Check** Which of the following is the correct way to divide $\dfrac{2}{5}$ by $\dfrac{3}{4}$?

Or are both correct? Explain.

a. $\dfrac{5}{2} \cdot \dfrac{3}{4}$ **b.** $\dfrac{2}{5} \cdot \dfrac{4}{3}$

Practice 8–10

Divide.

8. $\dfrac{4}{9} \div 7$ **9.** $\dfrac{8}{15} \div 3\dfrac{4}{5}$

10. $3\dfrac{2}{7} \div 2\dfrac{3}{14}$

Practice 11–12

Divide.

11. $\dfrac{14}{17} \div 0$ **12.** $0 \div 2\dfrac{1}{8}$

Answers

8. $\dfrac{4}{63}$ **9.** $\dfrac{8}{57}$ **10.** $\dfrac{46}{31}$ or $1\dfrac{15}{31}$

11. undefined   **12.** 0

✓ **Concept Check Answers**

a. incorrect   **b.** correct

Objective D Solving Problems by Dividing Fractions

To solve real-life problems that involve dividing fractions, we continue to use our four problem-solving steps.

Practice 13

A designer of clothing designs an outfit that requires $2\frac{1}{7}$ yards of material. How many outfits can be made from a 30-yard bolt of material?

Example 13   Calculating Manufacturing Materials Needed

In a manufacturing process, a metal-cutting machine cuts strips $1\frac{3}{5}$ inches wide from a piece of metal stock. How many such strips can be cut from a 48-inch piece of stock?

Solution:

1. UNDERSTAND the problem. To do so, read and reread the problem. Then draw a diagram:

```
|←———— 48 inches ————→|
[ metal stock diagram ]
↑↑↑↑↑↑ ···
 1 3/5 inches
```

We want to know how many $1\frac{3}{5}$s there are in 48.

2. TRANSLATE.

In words:

Number of strips	is	48	divided by	$1\frac{3}{5}$
↓	↓	↓	↓	↓

Translate:

$$\text{Number of strips} = 48 \div 1\frac{3}{5}$$

3. SOLVE: Let's estimate a reasonable answer. The mixed number $1\frac{3}{5}$ rounds to 2 and $48 \div 2 = 24$.

$$48 \div 1\frac{3}{5} = 48 \div \frac{8}{5} = \frac{48}{1} \cdot \frac{5}{8} = \frac{48 \cdot 5}{1 \cdot 8} = \frac{\cancel{8} \cdot 6 \cdot 5}{1 \cdot \cancel{8}} = \frac{30}{1} \text{ or } 30$$

4. INTERPRET. *Check* your work. Since the exact answer of 30 is close to our estimate of 24, our answer is reasonable. *State* your conclusion: Thirty strips can be cut from the 48-inch piece of stock.

Answer

13. 14 outfits

■ Work Practice 13

Vocabulary, Readiness & Video Check

Use the choices below to fill in each blank. Not all choices will be used.

multiplication $\dfrac{a \cdot d}{b \cdot c}$ $\dfrac{a \cdot c}{b \cdot d}$

division   `0   reciprocals

1. Two numbers are _____ of each other if their product is 1.

2. Every number has a reciprocal except _____.

3. To divide two fractions, we write $\dfrac{a}{b} \div \dfrac{c}{d} =$ _____ .

4. The word "per" usually indicates _____ .

Martin-Gay Interactive Videos

Watch the section lecture video and answer the following questions.

Objective A **5.** From ▯ Example 2, what can we conclude is the reciprocal of any nonzero number *n*? ▶

Objective B **6.** From ▯ Example 6, what number has no reciprocal? ▶

Objective C **7.** In ▯ Example 8, why can't we divide out common factors once we've written the mixed numbers as fractions? ▶

Objective C **8.** In ▯ Example 9, what phrase tells us that we have a division problem? ▶

See Video 2.5

2.5 Exercise Set MyMathLab® ▶

Objective A *Find the reciprocal of each number. See Examples 1 through 4.*

▶ **1.** $\dfrac{4}{7}$ **2.** $\dfrac{9}{10}$ **3.** $\dfrac{1}{11}$ **4.** $\dfrac{1}{20}$

▶ **5.** 15 **6.** 13 **7.** $\dfrac{12}{7}$ **8.** $\dfrac{10}{3}$

Objective B *Divide. Write each answer in simplest form. See Examples 5 through 7 and 11 and 12.*

▶ **9.** $\dfrac{2}{3} \div \dfrac{5}{6}$ **10.** $\dfrac{5}{8} \div \dfrac{2}{3}$ **11.** $\dfrac{8}{9} \div \dfrac{1}{2}$ **12.** $\dfrac{10}{11} \div \dfrac{4}{5}$

13. $\dfrac{3}{7} \div \dfrac{5}{6}$ **14.** $\dfrac{16}{27} \div \dfrac{8}{15}$ **15.** $\dfrac{3}{5} \div \dfrac{4}{5}$ **16.** $\dfrac{11}{16} \div \dfrac{13}{16}$

▶ **17.** $\dfrac{1}{10} \div \dfrac{10}{1}$ **18.** $\dfrac{3}{13} \div \dfrac{13}{3}$ **19.** $\dfrac{7}{9} \div \dfrac{7}{3}$ **20.** $\dfrac{6}{11} \div \dfrac{6}{5}$

21. $\dfrac{5}{8} \div \dfrac{3}{8}$ **22.** $\dfrac{7}{8} \div \dfrac{5}{6}$ **23.** $\dfrac{7}{45} \div \dfrac{4}{25}$ **24.** $\dfrac{14}{52} \div \dfrac{1}{13}$

25. $\dfrac{2}{37} \div \dfrac{1}{7}$ **26.** $\dfrac{100}{158} \div \dfrac{10}{79}$ ▶ **27.** $\dfrac{3}{25} \div \dfrac{27}{40}$ **28.** $\dfrac{6}{15} \div \dfrac{7}{10}$

29. $\dfrac{11}{12} \div \dfrac{11}{12}$ **30.** $\dfrac{7}{13} \div \dfrac{7}{13}$ ▶ **31.** $\dfrac{8}{13} \div 0$ **32.** $0 \div \dfrac{4}{11}$

33. $0 \div \dfrac{7}{8}$ **34.** $\dfrac{2}{3} \div 0$ **35.** $\dfrac{25}{126} \div \dfrac{125}{441}$ **36.** $\dfrac{65}{495} \div \dfrac{26}{231}$

Objective C *Divide. Write each answer in simplest form. See Examples 8 through 12.*

37. $\dfrac{2}{3} \div 4$ **38.** $\dfrac{5}{6} \div 10$ **39.** $8 \div \dfrac{3}{5}$ **40.** $7 \div \dfrac{2}{11}$

41. $2\dfrac{1}{2} \div \dfrac{1}{2}$ **42.** $4\dfrac{2}{3} \div \dfrac{2}{5}$ **43.** $\dfrac{5}{12} \div 2\dfrac{1}{3}$ **44.** $\dfrac{4}{15} \div 2\dfrac{1}{2}$

45. $3\dfrac{3}{7} \div 3\dfrac{1}{3}$ **46.** $2\dfrac{5}{6} \div 4\dfrac{6}{7}$ **47.** $1\dfrac{4}{9} \div 2\dfrac{5}{6}$ **48.** $3\dfrac{1}{10} \div 2\dfrac{1}{5}$

49. $0 \div 15\dfrac{4}{7}$ **50.** $\dfrac{33}{50} \div 1$ **51.** $1 \div \dfrac{13}{17}$ **52.** $0 \div 7\dfrac{9}{10}$

53. $1 \div \dfrac{18}{35}$ **54.** $\dfrac{17}{75} \div 1$ **55.** $10\dfrac{5}{9} \div 16\dfrac{2}{3}$ **56.** $20\dfrac{5}{6} \div 137\dfrac{1}{2}$

Objectives B C Mixed Practice *Divide. Write each answer in simplest form. See Examples 5 through 12.*

57. $\dfrac{6}{15} \div \dfrac{12}{5}$ **58.** $\dfrac{4}{15} \div \dfrac{8}{3}$ **59.** $\dfrac{11}{20} \div \dfrac{3}{11}$ **60.** $\dfrac{9}{20} \div \dfrac{2}{9}$

61. $12 \div \dfrac{1}{8}$ **62.** $9 \div \dfrac{1}{6}$ **63.** $\dfrac{3}{7} \div \dfrac{4}{7}$ **64.** $\dfrac{3}{8} \div \dfrac{5}{8}$

65. $2\dfrac{3}{8} \div 0$ **66.** $20\dfrac{1}{5} \div 0$ **67.** $\dfrac{11}{85} \div \dfrac{7}{5}$ **68.** $\dfrac{13}{84} \div \dfrac{3}{16}$

69. $4\dfrac{5}{11} \div 1\dfrac{2}{5}$ **70.** $8\dfrac{2}{7} \div 3\dfrac{1}{7}$ **71.** $\dfrac{27}{100} \div \dfrac{3}{20}$ **72.** $\dfrac{25}{128} \div \dfrac{5}{32}$

Objective D *Solve. For Exercises 73 and 74, the solutions have been started for you. Write each answer in simplest form. See Example 13.*

73. A heart attack patient in rehabilitation walked on a treadmill $12\frac{3}{4}$ miles over 4 days. How many miles is this per day?

Start the solution:

1. UNDERSTAND the problem. Reread it as many times as needed.
2. TRANSLATE into an equation. (Fill in the blanks.)

miles per day	is	total miles	divided by	number of days
↓	↓	↓	↓	↓

$$\text{miles per day} = \underline{\hspace{1cm}} \div \underline{\hspace{1cm}}$$

Finish with:

3. SOLVE and
4. INTERPRET

74. A local restaurant is selling hamburgers from a booth on Memorial Day. A total of $27\frac{3}{4}$ pounds of hamburger have been ordered. How many quarter-pound hamburgers can this make?

Start the solution:

1. UNDERSTAND the problem. Reread it as many times as needed.
2. TRANSLATE into an equation. (Fill in the blanks.)

how many quarter-pound hamburgers	is	total pounds of hamburger	divided by	a quarter-pound
↓	↓	↓	↓	↓

$$\text{how many quarter-pound hamburgers} = \underline{\hspace{1cm}} \div \underline{\hspace{1cm}}$$

Finish with:

3. SOLVE and
4. INTERPRET

75. A patient is to take $3\frac{1}{3}$ tablespoons of medicine per day in 4 equally divided doses. How much medicine is to be taken in each dose?

76. If there are $13\frac{1}{3}$ grams of fat in 4 ounces of lean hamburger meat, how many grams of fat are in an ounce?

77. The record for rainfall during a 24-hour period in Alaska is $15\frac{1}{5}$ inches. This record was set in Angoon, Alaska, in October 1982. How much rain fell per hour on average? (*Source:* National Climatic Data Center)

78. An order for 125 custom-made candle stands was placed with Mr. Levi, the manager of Just For You, Inc. The worker assigned to the job can produce $2\frac{3}{5}$ candle stands per hour. Using this worker, how many work hours will be required to complete the order?

79. In September 2013, the average price of aluminum was $82\frac{1}{10}$ ¢ per pound. During that time, a family received 1642¢ for aluminum cans that they sold for recycling at a scrap metal center. Assuming that they received the average price, how many pounds of aluminum cans did they recycle? (*Source:* London Metal Exchange)

80. Yoko's Fine Jewelry paid $450 for a $\frac{3}{4}$-carat gem. At this price, what is the cost of one carat?

△ **81.** The area of the rectangle below is 12 square meters. If its width is $2\frac{4}{7}$ meters, find its length.

Rectangle $2\frac{4}{7}$ meters

△ **82.** The perimeter of the square below is $23\frac{1}{2}$ feet. Find the length of each side.

Square

Mixed Practice (*Sections 2.4, 2.5*)   *Perform the indicated operation.*

83. $\dfrac{2}{5} \cdot \dfrac{4}{7}$

84. $\dfrac{2}{5} \div \dfrac{4}{7}$

85. $2\dfrac{2}{3} \div 1\dfrac{1}{16}$

86. $2\dfrac{2}{3} \cdot 1\dfrac{1}{16}$

87. $5\dfrac{1}{7} \cdot \dfrac{2}{9} \cdot \dfrac{14}{15}$

88. $8\dfrac{1}{6} \cdot \dfrac{3}{7} \cdot \dfrac{18}{25}$

89. $\dfrac{11}{20} \div \dfrac{20}{11}$

90. $2\dfrac{1}{5} \div 1\dfrac{7}{10}$

Review

Perform each indicated operation. See Sections 1.3 and 1.4.

91.
$$\begin{array}{r} 27 \\ 76 \\ + 98 \\ \hline \end{array}$$

92.
$$\begin{array}{r} 811 \\ 42 \\ + 69 \\ \hline \end{array}$$

93.
$$\begin{array}{r} 968 \\ -772 \\ \hline \end{array}$$

94.
$$\begin{array}{r} 882 \\ -773 \\ \hline \end{array}$$

95.
$$\begin{array}{r} 2000 \\ - 431 \\ \hline \end{array}$$

96.
$$\begin{array}{r} 500 \\ - 92 \\ \hline \end{array}$$

Concept Extensions

A student asked you to find the error in the work below. Find the error and correct it. See the Concept Check in this section.

97. $20\dfrac{2}{3} \div 10\dfrac{1}{2} = 2\dfrac{1}{3}$

98. $6\dfrac{1}{4} \div \dfrac{1}{2} = 3\dfrac{1}{8}$

Choose the best estimate for each quotient.

99. $20\dfrac{1}{4} \div \dfrac{5}{6}$

    a. 5 **b.** $5\dfrac{1}{8}$ **c.** 20 **d.** 10

100. $\dfrac{11}{12} \div 16\dfrac{1}{5}$

    a. $\dfrac{1}{16}$ **b.** 4 **c.** 8 **d.** 16

101. $12\frac{2}{13} \div 3\frac{7}{8}$

    a. 4   **b.** 9   **c.** 36   **d.** 3

102. $10\frac{1}{4} \div 2\frac{1}{16}$

    a. 8   **b.** 5   **c.** 20   **d.** 12

Simplify.

103. $\dfrac{42}{25} \cdot \dfrac{125}{36} \div \dfrac{7}{6}$

104. $\left(\dfrac{8}{13} \cdot \dfrac{39}{16} \cdot \dfrac{8}{9}\right)^2 \div \dfrac{1}{2}$

105. The FedEx Express air fleet includes 64 Boeing MD11s. These Boeing MD11s make up $\dfrac{32}{317}$ of the FedEx Express fleet. How many aircraft make up the entire FedEx Express air fleet? (*Source:* FedEx Corporation)

106. One-third of all native flowering plant species in the United States are at risk of becoming extinct. That translates into 5144 at-risk flowering plant species. Based on this data, how many flowering plant species are native to the United States overall? (*Source:* The Nature Conservancy)

(*Hint:* How many $\dfrac{1}{3}$s are in 5144?)

107. In your own words, describe how to find the reciprocal of a number.

108. In your own words, describe how to divide fractions.

Chapter 2 Group Activity

Blood and Blood Donation (Sections 2.1, 2.2, 2.3)

Blood is the workhorse of the body. It carries to the body's tissues everything they need, from nutrients to antibodies to heat. Blood also carries away waste products like carbon dioxide. Blood contains three types of cells—red blood cells, white blood cells, and platelets—suspended in clear, watery fluid called plasma. Blood is $\dfrac{11}{20}$ plasma, and plasma itself is $\dfrac{9}{10}$ water. In the average healthy adult human, blood accounts for $\dfrac{1}{11}$ of a person's body weight.

Roughly every 2 seconds someone in the United States needs blood. Although only $\dfrac{1}{20}$ of eligible donors donate blood, the American Red Cross is still able to collect nearly 6 million volunteer donations of blood each year. This volume makes Red Cross Biomedical Services the largest blood supplier for blood transfusions in the United States.

Group Activity

Contact your local Red Cross Blood Service office. Find out how many people donated blood in your area in the past two months. Ask whether it is possible to get a breakdown of the blood donations by blood type. (For more on blood types, see Exercises 75 through 78 in Section 2.3.)

1. Research the population of the area served by your local Red Cross Blood Service office. Write the fraction of the local population who gave blood in the past two months.

2. Use the breakdown by blood type to write the fraction of donors giving each type of blood.

Chapter 2 Vocabulary Check

Fill in each blank with one of the words or phrases listed below.

mixed number	equivalent	0	undefined
composite number	improper fraction	simplest form	prime factorization
prime number	proper fraction	numerator	denominator
reciprocals	cross products		

1. Two numbers are _____ of each other if their product is 1.

2. A(n) _____ is a natural number greater than 1 that is not prime.

3. Fractions that represent the same portion of a whole are called _____ fractions.

4. A(n) _____ is a fraction whose numerator is greater than or equal to its denominator.

5. A(n) _____ is a natural number greater than 1 whose only factors are 1 and itself.

6. A fraction is in _____ when the numerator and the denominator have no factors in common other than 1.

7. A(n) _____ is one whose numerator is less than its denominator.

8. A(n) _____ contains a whole number part and a fraction part.

9. In the fraction $\frac{7}{9}$, the 7 is called the _____ and the 9 is called the _____.

10. The _____ of a number is the factorization in which all the factors are prime numbers.

11. The fraction $\frac{3}{0}$ is _____.

12. The fraction $\frac{0}{5} = $ _____.

13. In $\frac{a}{b} = \frac{c}{d}$, $a \cdot d$ and $b \cdot c$ are called _____.

> **Helpful Hint** ▶ Are you preparing for your test? Don't forget to take the Chapter 2 Test on page 168. Then check your answers at the back of the text and use the Chapter Test Prep Videos to see the fully worked-out solutions to any of the exercises you want to review.

2 Chapter Highlights

Definitions and Concepts	Examples
Section 2.1 Introduction to Fractions and Mixed Numbers	
A **fraction** is of the form: $\frac{\text{numerator}}{\text{denominator}}$ ⟵ number of parts being considered ⟵ number of equal parts in the whole	Write a fraction to represent the shaded part of the figure. $\dfrac{3}{8}$ ⟵ number of parts shaded ⟵ number of equal parts
A fraction is called a **proper fraction** if its numerator is less than its denominator.	$\dfrac{1}{3}, \dfrac{2}{5}, \dfrac{7}{8}, \dfrac{100}{101}$
A fraction is called an **improper fraction** if its numerator is greater than or equal to its denominator.	$\dfrac{5}{4}, \dfrac{2}{2}, \dfrac{9}{7}, \dfrac{101}{100}$
A **mixed number** contains a whole number and a fraction.	$1\dfrac{1}{2}, 5\dfrac{7}{8}, 25\dfrac{9}{10}$

Definitions and Concepts	Examples

Section 2.1   Introduction to Fractions and Mixed Numbers (*continued*)

To Write a Mixed Number as an Improper Fraction

1. Multiply the denominator of the fraction by the whole number.

2. Add the numerator of the fraction to the product from step 1.

3. Write the sum from step 2 as the numerator of the improper fraction over the original denominator.

$$5\frac{2}{7} = \frac{7 \cdot 5 + 2}{7} = \frac{35 + 2}{7} = \frac{37}{7}$$

To Write an Improper Fraction as a Mixed Number or a Whole Number

1. Divide the denominator into the numerator.

2. The whole number part of the mixed number is the quotient. The fraction is the remainder over the original denominator.

$$\text{quotient}\frac{\text{remainder}}{\text{original denominator}}$$

$$\frac{17}{3} = 5\frac{2}{3}$$

$$\begin{array}{r} 5 \\ 3\overline{)17} \\ -15 \\ \hline 2 \end{array}$$

Section 2.2   Factors and Prime Factorization

A **prime number** is a natural number that has exactly two different factors, 1 and itself.

$2, 3, 5, 7, 11, 13, 17, \ldots$

A **composite number** is any natural number other than 1 that is not prime.

$4, 6, 8, 9, 10, 12, 14, 15, 16, \ldots$

The prime factorization of a number is the factorization in which all the factors are prime numbers.

Write the prime factorization of 60.
$$60 = 6 \cdot 10$$
$$= 2 \cdot 3 \cdot 2 \cdot 5 \quad \text{or} \quad 2^2 \cdot 3 \cdot 5$$

Section 2.3   Simplest Form of a Fraction

Fractions that represent the same portion of a whole are called **equivalent fractions.**

$$\frac{3}{4} = \frac{12}{16}$$

A fraction is in **simplest form** or **lowest terms** when the numerator and the denominator have no common factors other than 1.

The fraction $\frac{2}{3}$ is in simplest form.

To write a fraction in simplest form, write the prime factorizations of the numerator and the denominator and then divide both by all common factors.

Write in simplest form: $\frac{30}{36}$

$$\frac{30}{36} = \frac{2 \cdot 3 \cdot 5}{2 \cdot 2 \cdot 3 \cdot 3} = \frac{2}{2} \cdot \frac{3}{3} \cdot \frac{5}{2 \cdot 3} = 1 \cdot 1 \cdot \frac{5}{6} = \frac{5}{6}$$

$$\text{or} \quad \frac{30}{36} = \frac{2 \cdot \overset{1}{3} \cdot 5}{\underset{1}{2} \cdot 2 \cdot \underset{1}{3} \cdot 3} = \frac{5}{6}$$

(continued)

Definitions and Concepts	Examples

Section 2.3   Simplest Form of a Fraction *(continued)*

Definitions and Concepts	Examples
Two fractions are equivalent if	Determine whether $\dfrac{7}{8}$ and $\dfrac{21}{24}$ are equivalent.
Method 1. They simplify to the same fraction.	**Method 1.** $\dfrac{7}{8}$ is in simplest form; $\dfrac{21}{24} = \dfrac{\overset{1}{\cancel{3}} \cdot 7}{\underset{1}{\cancel{3}} \cdot 8} = \dfrac{7}{8}$
	Since both simplify to $\dfrac{7}{8}$, then $\dfrac{7}{8} = \dfrac{21}{24}$.
Method 2. Their cross products are equal.	**Method 2.**
	$\begin{array}{ccc} 24 \cdot 7 & \dfrac{7}{8} = \dfrac{21}{24} & 8 \cdot 21 \\ = 168 & & = 168 \end{array}$
	Since $168 = 168$, $\dfrac{7}{8} = \dfrac{21}{24}$

Section 2.4   Multiplying Fractions and Mixed Numbers

Definitions and Concepts	Examples
To multiply two fractions, multiply the numerators and multiply the denominators.	Multiply.
	$\dfrac{7}{8} \cdot \dfrac{3}{5} = \dfrac{7 \cdot 3}{8 \cdot 5} = \dfrac{21}{40}$
	$\dfrac{3}{4} \cdot \dfrac{1}{6} = \dfrac{3 \cdot 1}{4 \cdot 6} = \dfrac{\overset{1}{\cancel{3}} \cdot 1}{4 \cdot \underset{1}{\cancel{3}} \cdot 2} = \dfrac{1}{8}$
To multiply with mixed numbers or whole numbers, first write any mixed or whole numbers as fractions and then multiply as usual.	$2\dfrac{1}{3} \cdot \dfrac{1}{9} = \dfrac{7}{3} \cdot \dfrac{1}{9} = \dfrac{7 \cdot 1}{3 \cdot 9} = \dfrac{7}{27}$

Section 2.5   Dividing Fractions and Mixed Numbers

Definitions and Concepts	Examples
To find the **reciprocal** of a fraction, interchange its numerator and denominator.	The reciprocal of $\dfrac{3}{5}$ is $\dfrac{5}{3}$.
To divide two fractions, multiply the first fraction by the reciprocal of the second fraction.	Divide.
	$\dfrac{3}{10} \div \dfrac{7}{9} = \dfrac{3}{10} \cdot \dfrac{9}{7} = \dfrac{3 \cdot 9}{10 \cdot 7} = \dfrac{27}{70}$
To divide with mixed numbers or whole numbers, first write any mixed or whole numbers as fractions and then divide as usual.	$2\dfrac{5}{8} \div 3\dfrac{7}{16} = \dfrac{21}{8} \div \dfrac{55}{16} = \dfrac{21}{8} \cdot \dfrac{16}{55} = \dfrac{21 \cdot 16}{8 \cdot 55}$
	$= \dfrac{21 \cdot 2 \cdot \overset{1}{\cancel{8}}}{\underset{1}{\cancel{8}} \cdot 55} = \dfrac{42}{55}$

(2.1) *Determine whether each number is an improper fraction, a proper fraction, or a mixed number.*

1. $\dfrac{11}{23}$

2. $\dfrac{9}{8}$

3. $\dfrac{1}{2}$

4. $2\dfrac{1}{4}$

Write a fraction to represent the shaded area.

5.

6.

7.

8.

9. A basketball player made 11 free throws out of 12 during a game. What fraction of free throws did the player make?

10. A new car lot contained 23 blue cars out of a total of 131 cars.
   a. How many cars on the lot are not blue?
   b. What fraction of cars on the lot are not blue?

Write each improper fraction as a mixed number or a whole number.

11. $\dfrac{15}{4}$

12. $\dfrac{275}{6}$

13. $\dfrac{39}{13}$

14. $\dfrac{60}{12}$

Write each mixed number as an improper fraction.

15. $1\dfrac{1}{5}$

16. $1\dfrac{1}{21}$

17. $2\dfrac{8}{9}$

18. $3\dfrac{11}{12}$

(2.2) *Identify each number as prime or composite.*

19. 51

20. 17

List all factors of each number.

21. 42

22. 20

Find the prime factorization of each number.

23. 68

24. 90

25. 785

26. 255

(2.3) *Write each fraction in simplest form.*

27. $\dfrac{12}{28}$

28. $\dfrac{15}{27}$

29. $\dfrac{25}{75}$

30. $\dfrac{36}{72}$

31. $\dfrac{29}{32}$

32. $\dfrac{18}{23}$

33. $\dfrac{48}{6}$

34. $\dfrac{54}{9}$

Solve.

35. There are 12 inches in a foot. What fractional part of a foot does 8 inches represent?

36. Six out of 15 cars are white. What fraction of the cars are *not* white?

Determine whether each two fractions are equivalent.

37. $\frac{10}{34}$ and $\frac{4}{14}$

38. $\frac{30}{50}$ and $\frac{9}{15}$

(2.4) *Multiply. Write each answer in simplest form. Estimate where noted.*

39. $\frac{3}{5} \cdot \frac{1}{2}$

40. $\frac{6}{7} \cdot \frac{5}{12}$

41. $\frac{24}{5} \cdot \frac{15}{8}$

42. $\frac{27}{21} \cdot \frac{7}{18}$

43. $5 \cdot \frac{7}{8}$

44. $6 \cdot \frac{5}{12}$

45. $\frac{39}{3} \cdot \frac{7}{13} \cdot \frac{5}{21}$

46. $\frac{42}{5} \cdot \frac{15}{6} \cdot \frac{7}{9}$

47. $1\frac{5}{8} \cdot 3\frac{1}{5}$

Exact:

Estimate:

48. $3\frac{6}{11} \cdot 1\frac{7}{13}$

Exact:

Estimate:

49. $\frac{3}{4} \cdot 8 \cdot 4\frac{1}{8}$

50. $2\frac{1}{9} \cdot 3 \cdot \frac{1}{38}$

51. There are $7\frac{1}{3}$ grams of fat in each ounce of hamburger. How many grams of fat are in a 5-ounce hamburger patty?

52. An art teacher needs 45 pieces of PVC piping for an art project. If each piece needs to be $\frac{3}{4}$ inch long, find the total length of piping she needs.

△ **53.** Find the area of each rectangle.

$\frac{7}{10}$ inch

$2\frac{1}{8}$ inches

△ **54.**

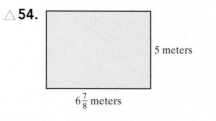

5 meters

$6\frac{7}{8}$ meters

(2.5) *Find the reciprocal of each number.*

55. 7

56. $\frac{1}{8}$

57. $\frac{14}{23}$

58. $\frac{17}{5}$

Divide. Write each answer in simplest form.

59. $\frac{3}{4} \div \frac{3}{8}$

60. $\frac{21}{4} \div \frac{7}{5}$

61. $\frac{5}{3} \div 2$

62. $5 \div \frac{15}{8}$

63. $6\frac{3}{4} \div 1\frac{2}{7}$

64. $5\frac{1}{2} \div 2\frac{1}{11}$

65. A truck traveled 341 miles on $15\frac{1}{2}$ gallons of gas. How many miles might we expect the truck to travel on 1 gallon of gas?

66. Herman Heltznutt walks 5 days a week for a total distance of $5\frac{1}{4}$ miles per week. If he walks the same distance each day, find the distance he walks each day.

Mixed Review

Determine whether each number is an improper fraction, a proper fraction, or a mixed number.

67. $\frac{0}{3}$ **68.** $\frac{12}{12}$ **69.** $5\frac{6}{7}$ **70.** $\frac{13}{9}$

Write each improper fraction as a mixed number or a whole number. Write each mixed number as an improper fraction.

71. $\frac{125}{4}$ **72.** $\frac{54}{9}$ **73.** $5\frac{10}{17}$ **74.** $7\frac{5}{6}$

Identify each number as prime or composite.

75. 27   **76.** 23

Find the prime factorization of each number.

77. 180   **78.** 98

Write each fraction in simplest form.

79. $\frac{45}{50}$ **80.** $\frac{30}{42}$ **81.** $\frac{140}{150}$ **82.** $\frac{84}{140}$

Multiply or divide as indicated. Write each answer in simplest form. Estimate where noted.

83. $\frac{7}{8}\cdot\frac{2}{3}$ **84.** $\frac{6}{15}\cdot\frac{5}{8}$ **85.** $\frac{18}{5}\div\frac{2}{5}$ **86.** $\frac{9}{2}\div\frac{1}{3}$

87. $4\frac{1}{6}\cdot2\frac{2}{5}$ **88.** $5\frac{2}{3}\cdot2\frac{1}{4}$ **89.** $\frac{7}{2}\div1\frac{1}{2}$ **90.** $1\frac{3}{5}\div\frac{1}{4}$
  Exact: Exact:
  Estimate: Estimate:

△**91.** A slab of natural granite is purchased and a rectangle with length $7\frac{4}{11}$ feet and width $5\frac{1}{2}$ feet is cut from it. Find the area of the rectangle.

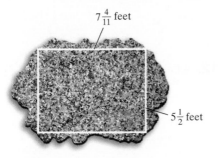

$7\frac{4}{11}$ feet

$5\frac{1}{2}$ feet

92. An area of Mississippi received $23\frac{1}{2}$ inches of rain in $30\frac{1}{2}$ hours. How many inches per 1 hour is this?

Chapter 2

Test

Step-by-step test solutions are found on the Chapter Test Prep Videos. Where available:

MyMathLab® or You Tube™

Answers

Write a fraction to represent the shaded area.

1.

2.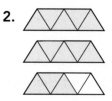

1. _____

Write each mixed number as an improper fraction.

2. _____

3. $7\frac{2}{3}$

4. $3\frac{6}{11}$

3. _____

Write each improper fraction as a mixed number or a whole number.

4. _____

5. $\frac{23}{5}$

6. $\frac{75}{4}$

5. _____

6. _____

Write each fraction in simplest form.

7. _____

7. $\frac{24}{210}$

8. $\frac{42}{70}$

8. _____

9. _____

Determine whether these fractions are equivalent.

9. $\frac{5}{7}$ and $\frac{8}{11}$

10. $\frac{6}{27}$ and $\frac{14}{63}$

10. _____

11. _____

Find the prime factorization of each number.

12. _____

11. 84

12. 495

Perform each indicated operation. Write each answer in simplest form.

13. $\dfrac{4}{4} \div \dfrac{3}{4}$ **14.** $\dfrac{4}{3} \cdot \dfrac{4}{4}$ **15.** $2 \cdot \dfrac{1}{8}$ **16.** $\dfrac{2}{3} \cdot \dfrac{8}{15}$

17. $8 \div \dfrac{1}{2}$ **18.** $13\dfrac{1}{2} \div 3$ **19.** $\dfrac{3}{8} \cdot \dfrac{16}{6} \cdot \dfrac{4}{11}$ **20.** $5\dfrac{1}{4} \div \dfrac{7}{12}$

21. $\dfrac{16}{3} \div \dfrac{3}{12}$ **22.** $3\dfrac{1}{3} \cdot 6\dfrac{3}{4}$ **23.** $12 \div 3\dfrac{1}{3}$ **24.** $\dfrac{14}{5} \cdot \dfrac{25}{21} \cdot 2$

△**25.** Find the area of the figure.

$\dfrac{2}{3}$ mile

$1\dfrac{8}{9}$ miles

26. During a 258-mile trip, a car used $10\dfrac{3}{4}$ gallons of gas. How many miles would we expect the car to travel on 1 gallon of gas?

27. How many square yards of artificial turf are necessary to cover a football field, *not* including the end zones and the sidelines? (*Hint:* A football field measures $100 \times 53\dfrac{1}{3}$ yards.)

$53\dfrac{1}{3}$ yards

100 yards

28. Prior to an oil spill, the stock in an oil company sold for $120 per share. As a result of the liability that the company incurred from the spill, the price per share fell to $\dfrac{3}{4}$ of the price before the spill. What did the stock sell for after the spill?

13. _____

14. _____

15. _____

16. _____

17. _____

18. _____

19. _____

20. _____

21. _____

22. _____

23. _____

24. _____

25. _____

26. _____

27. _____

28. _____

Answers

1. _____

2. _____

3. _____

4. _____

5. _____

6. _____

7. _____

8. _____

9. _____

10. _____

11. _____

12. _____

13. a. _____

   b. _____

1. Find the place value of the digit 3 in the whole number 396,418.

2. Write 2036 in words.

3. Write the number, eight hundred five, in standard form.

4. Add: $7 + 6 + 10 + 3 + 5$

5. Add: $34,285 + 149,761$

6. Find the average of 56, 18, and 43.

△ 7. Find the perimeter of the polygon shown.

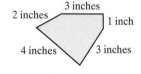

8. Subtract 8 from 25.

9. In 2011, a total of 12,734,424 passenger vehicles were sold in the United States. In 2012, total passenger vehicle sales in the United States had increased by 1,705,636 vehicles. Find the total number of passenger vehicles sold in the United States in 2012. (*Source:* Alliance of Automobile Manufacturers)

10. Find $\sqrt{25}$.

11. Subtract: $7826 - 505$
    Check by adding.

12. Find 8^2.

13. In the following graph, each bar represents a country and the height of each bar represents the number of threatened mammal species identified in that country.

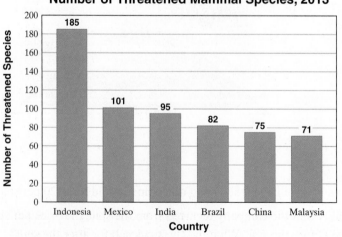

Source: International Union for Conservation of Nature

   a. Which country shown has the greatest number of threatened mammal species?
   b. Find the total number of threatened mammal species for Malaysia, China, and Indonesia.

14. Find $205 \div 8$.

15. Round 568 to the nearest ten.

16. Round 2366 to the nearest hundred.

17. Round each number to the nearest hundred to find an estimated difference.

$$\begin{array}{r} 4725 \\ -2879 \\ \hline \end{array}$$

18. Round each number to the nearest ten to find an estimated sum.
$38 + 43 + 126 + 92$

19. Multiply.
  a. 6×1 **b.** $0(18)$
  c. $1 \cdot 45$ **d.** $(75)(0)$

20. Simplify: $30 \div 3 \cdot 2$

21. Rewrite each using the distributive property.
  a. $3(4 + 5)$ **b.** $10(6 + 8)$
  c. $2(7 + 3)$

22. Multiply: 12×15

23. Find each quotient. Check by multiplying.
  a. $9\overline{)0}$ **b.** $0 \div 12$
  c. $\dfrac{0}{5}$ **d.** $\dfrac{3}{0}$

24. Find the area.

7 miles	Rectangle
	22 miles

25. Divide and check: $1872 \div 9$

26. Subtract: $5000 - 986$

27. As part of a promotion, an executive receives 238 cards, each good for one free song download. If she wants to share them evenly with 19 friends, how many download cards will each friend receive? How many will be left over?

28. Find the product of 9 and 7.

29. A gardener bought enough plants to fill a rectangular garden with length 30 feet and width 20 feet. Because of shading problems from a nearby tree, the gardener changed the width of the garden to 15 feet. If the area is to remain the same, what is the new length of the garden?

30. Find the sum of 9 and 7.

14. _____
15. _____
16. _____
17. _____
18. _____
19. a. _____
    b. _____
    c. _____
    d. _____
20. _____
21. a. _____
    b. _____
    c. _____
22. _____
23. a. _____
    b. _____
    c. _____
    d. _____
24. _____
25. _____
26. _____
27. _____
28. _____
29. _____
30. _____

31. _____

32. _____

33. _____

34. _____

35. _____

36. _____

37. _____

38. _____

39. a. _____

     b. _____

40. _____

41. _____

42. _____

43. _____

44. _____

45. _____

46. _____

47. _____

48. _____

49. _____

50. _____

Write using exponential notation.

31. $7 \cdot 7 \cdot 7$

32. $7 \cdot 7 \cdot 7 \cdot 7$

33. $3 \cdot 3 \cdot 3 \cdot 3 \cdot 17 \cdot 17 \cdot 17$

34. $2 \cdot 2 \cdot 3 \cdot 3 \cdot 3 \cdot 3$

35. Simplify: $2 \cdot 4 - 3 \div 3$

36. Simplify: $8 \cdot \sqrt{100} - 4^2 \cdot 5$

37. Write a fraction to represent the shaded part of the figure.

38. Write the prime factorization of 156.

39. Write each as an improper fraction.

   a. $4\frac{2}{9}$ **b.** $1\frac{8}{11}$

40. Write $7\frac{4}{5}$ as an improper fraction.

41. Find all the factors of 20.

42. Determine whether $\frac{8}{20}$ and $\frac{14}{35}$ are equivalent.

43. Write in simplest form: $\frac{42}{66}$

44. Write in simplest form: $\frac{70}{105}$

45. Multiply: $3\frac{1}{3} \cdot \frac{7}{8}$

46. Multiply: $\frac{2}{3} \cdot 4$

47. Find the reciprocal of $\frac{1}{3}$.

48. Find the reciprocal of 9.

49. Divide and simplify: $\frac{5}{16} \div \frac{3}{4}$

50. Divide: $1\frac{1}{10} \div 5\frac{3}{5}$

Adding and Subtracting Fractions

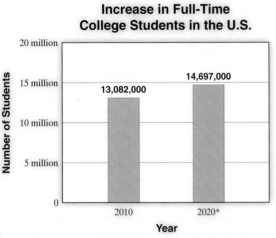

Increase in Full-Time College Students in the U.S.

Source: U.S. Department of Education's National Center for Education Statistics

* prediction

Having learned what fractions are and how to multiply and divide them in Chapter 2, we are ready to continue our study of fractions. In this chapter, we learn how to add and subtract fractions and mixed numbers. We then conclude this chapter with solving problems using fractions.

About $\frac{3}{4}$ of College Students Have a Job. What Else Do They Do?

The circle graph below shows how an average full-time college student spends his or her weekdays and the portion of each weekday that is devoted to each type of activity. For example, on average, a full-time U.S. college student spends $\frac{36}{100}$ of each weekday sleeping, and so on. In Section 3.1, Exercises 49–52, we use fractions to help us understand the time use illustrated in this circle graph.

Full-Time College Students' Average Weekday Time Use

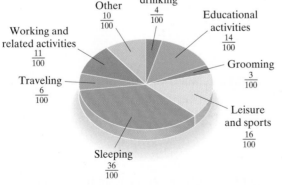

Source: Bureau of Labor Statistics

3.1 Adding and Subtracting Like Fractions

Objectives

A Add Like Fractions.

B Subtract Like Fractions.

C Solve Problems by Adding or Subtracting Like Fractions.

Fractions with the same denominator are called **like fractions.** Fractions that have different denominators are called **unlike fractions.**

Like Fractions	Unlike Fractions
$\frac{2}{5}$ and $\frac{3}{5}$ ↑ ↑ ——— same denominator	$\frac{2}{5}$ and $\frac{3}{4}$ ↑ ↑ ——— different denominators
$\frac{5}{21}, \frac{16}{21},$ and $\frac{7}{21}$ ↑ ↑ ↑ ——— same denominator	$\frac{5}{7}$ and $\frac{5}{9}$ ↑ ↑ ——— different denominators

Objective A Adding Like Fractions

To see how we add like fractions (fractions with the same denominator), study the figures below:

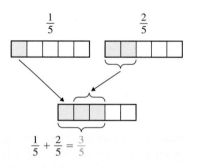

$$\frac{1}{5} + \frac{2}{5} = \frac{3}{5}$$

Adding Like Fractions (Fractions with the Same Denominator)

To add like fractions, add the numerators and write the sum over the common denominator.

If $a, b,$ and c represent nonzero whole numbers, we have

$$\frac{a}{c} + \frac{b}{c} = \frac{a + b}{c}$$

For example,

$$\frac{1}{4} + \frac{2}{4} = \frac{1 + 2}{4} = \frac{3}{4}$$ ← Add the numerators.
← Keep the denominator.

Helpful Hint

As usual, don't forget to write all answers in simplest form.

Examples Add and simplify.

1. $\dfrac{2}{7} + \dfrac{3}{7} = \dfrac{2+3}{7} = \dfrac{5}{7}$ ←— Add the numerators.
←— Keep the common denominator.

2. $\dfrac{3}{16} + \dfrac{7}{16} = \dfrac{3+7}{16} = \dfrac{10}{16} = \dfrac{\overset{1}{\cancel{2}} \cdot 5}{\cancel{2} \cdot 8} = \dfrac{5}{8}$

3. $\dfrac{7}{13} + \dfrac{6}{13} + \dfrac{3}{13} = \dfrac{7+6+3}{13} = \dfrac{16}{13} \text{ or } 1\dfrac{3}{13}$

■ Work Practice 1–3

✓**Concept Check** Find and correct the error in the following:

$\dfrac{1}{5} + \dfrac{1}{5} \;\cancel{=}\; \dfrac{2}{10}$

Objective B Subtracting Like Fractions

To see how we subtract like fractions (fractions with the same denominator), study the following figure:

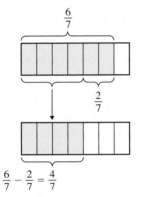

$\dfrac{6}{7} - \dfrac{2}{7} = \dfrac{4}{7}$

Subtracting Like Fractions (Fractions with the Same Denominator)

To subtract like fractions, subtract the numerators and write the difference over the common denominator.

If a, b, and c represent nonzero whole numbers, then

$$\dfrac{a}{c} - \dfrac{b}{c} = \dfrac{a-b}{c}$$

For example,

$\dfrac{4}{5} - \dfrac{2}{5} = \dfrac{4-2}{5} = \dfrac{2}{5}$ ←— Subtract the numerators.
←— Keep the denominator.

Examples Subtract and simplify.

4. $\dfrac{8}{9} - \dfrac{1}{9} = \dfrac{8-1}{9} = \dfrac{7}{9}$ ←— Subtract the numerators.
←— Keep the common denominator.

5. $\dfrac{7}{8} - \dfrac{5}{8} = \dfrac{7-5}{8} = \dfrac{2}{8} = \dfrac{\overset{1}{\cancel{2}}}{\cancel{2} \cdot 4} = \dfrac{1}{4}$

■ Work Practice 4–5

Objective C Solving Problems by Adding or Subtracting Like Fractions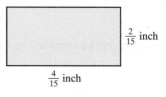

Many real-life problems involve finding the perimeters of square or rectangular areas such as pastures, swimming pools, and so on. We can use our knowledge of adding fractions to find perimeters.

Practice 6

Find the perimeter of the square.

$\frac{3}{20}$ mile

Example 6   Find the perimeter of the rectangle.

$\frac{2}{15}$ inch

$\frac{4}{15}$ inch

Solution: Recall that perimeter means distance around and that opposite sides of a rectangle are the same length.

$\frac{4}{15}$ inch

$\frac{2}{15}$ inch $\frac{2}{15}$ inch

$\frac{4}{15}$ inch

$$\text{Perimeter} = \frac{2}{15} + \frac{4}{15} + \frac{2}{15} + \frac{4}{15} = \frac{2 + 4 + 2 + 4}{15}$$

$$= \frac{12}{15} = \frac{\overset{1}{\cancel{3}} \cdot 4}{\underset{1}{\cancel{3}} \cdot 5} = \frac{4}{5}$$

The perimeter of the rectangle is $\frac{4}{5}$ inch.

■ Work Practice 6

We can combine our skills in adding and subtracting fractions with our four problem-solving steps from Chapter 1 to solve many kinds of real-life problems.

Practice 7

If a piano student practices the piano $\frac{3}{8}$ of an hour in the morning and $\frac{1}{8}$ of an hour in the evening, how long did she practice that day?

Example 7   Total Amount of an Ingredient in a Recipe

A recipe calls for $\frac{1}{3}$ of a cup of honey at the beginning and $\frac{2}{3}$ of a cup of honey later. How much total honey is needed to make the recipe?

$\frac{1}{3}$ cup $\frac{2}{3}$ cup

Solution:

1. UNDERSTAND the problem. To do so, read and reread the problem. Since we are finding total honey, we add.

Answers

6. $\frac{3}{5}$ mi   7. $\frac{1}{2}$ hr

2. TRANSLATE.

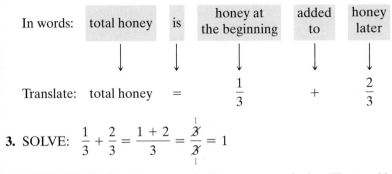

In words:	total honey	is	honey at the beginning	added to	honey later

Translate:	total honey	=	$\dfrac{1}{3}$	+	$\dfrac{2}{3}$

3. SOLVE: $\dfrac{1}{3} + \dfrac{2}{3} = \dfrac{1+2}{3} = \dfrac{\cancel{3}^{1}}{\cancel{3}_{1}} = 1$

4. INTERPRET. *Check* your work. *State* your conclusion: The total honey needed for the recipe is 1 cup.

■ Work Practice 7

Example 8 Calculating Distance

The distance from home to the World Gym is $\dfrac{7}{8}$ of a mile and from home to the post office is $\dfrac{3}{8}$ of a mile. How much farther is it from home to the World Gym than from home to the post office?

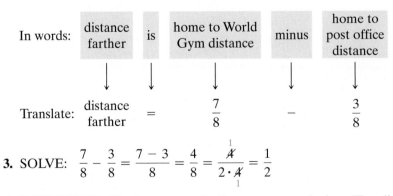

Solution:

1. UNDERSTAND. Read and reread the problem. The phrase "How much farther" tells us to subtract distances.

2. TRANSLATE.

In words:	distance farther	is	home to World Gym distance	minus	home to post office distance

Translate:	distance farther	=	$\dfrac{7}{8}$	−	$\dfrac{3}{8}$

3. SOLVE: $\dfrac{7}{8} - \dfrac{3}{8} = \dfrac{7-3}{8} = \dfrac{4}{8} = \dfrac{\cancel{4}^{1}}{2 \cdot \cancel{4}_{1}} = \dfrac{1}{2}$

4. INTERPRET. *Check* your work. *State* your conclusion: The distance from home to the World Gym is $\dfrac{1}{2}$ mile farther than from home to the post office.

■ Work Practice 8

Practice 8

A jogger ran $\dfrac{13}{4}$ miles on Monday and $\dfrac{7}{4}$ miles on Wednesday. How much farther did he run on Monday than on Wednesday?

Answer

8. $\dfrac{3}{2}$ or $1\dfrac{1}{2}$ mi

Vocabulary, Readiness & Video Check

Use the choices below to fill in each blank. Not all choices will be used.

perimeter like $\dfrac{a-c}{b}$ $\dfrac{a+c}{b}$

equivalent unlike

1. The fractions $\dfrac{9}{11}$ and $\dfrac{13}{11}$ are called _____ fractions while $\dfrac{3}{4}$ and $\dfrac{1}{3}$ are called _____ fractions.

2. $\dfrac{a}{b} + \dfrac{c}{b} =$ _____ .

3. $\dfrac{a}{b} - \dfrac{c}{b} =$ _____ .

4. The distance around a figure is called its _____ .

State whether the fractions in each list are like or unlike fractions.

5. $\dfrac{7}{8}, \dfrac{7}{10}$

6. $\dfrac{2}{3}, \dfrac{4}{9}$

7. $\dfrac{9}{10}, \dfrac{1}{10}$

8. $\dfrac{8}{11}, \dfrac{2}{11}$

9. $\dfrac{2}{31}, \dfrac{30}{31}, \dfrac{19}{31}$

10. $\dfrac{3}{10}, \dfrac{3}{11}, \dfrac{3}{13}$

11. $\dfrac{5}{12}, \dfrac{7}{12}, \dfrac{12}{11}$

12. $\dfrac{1}{5}, \dfrac{2}{5}, \dfrac{4}{5}$

Martin-Gay Interactive Videos Watch the section lecture video and answer the following questions.

Objective A 13. In ▤ Example 2, why is $\dfrac{6}{9}$ not the final answer? ▶

Objective B 14. What two questions are asked during the solving of ▤ Example 5? What are the answers to these questions? ▶

Objective C 15. What is the perimeter equation used to solve ▤ Example 6? What is the final answer? ▶

See Video 3.1 🍎

3.1 Exercise Set MyMathLab® ▶

Objective A *Add and simplify. See Examples 1 through 3.*

1. $\dfrac{1}{7} + \dfrac{2}{7}$

2. $\dfrac{9}{17} + \dfrac{2}{17}$

3. $\dfrac{1}{10} + \dfrac{1}{10}$

4. $\dfrac{1}{4} + \dfrac{1}{4}$

5. $\dfrac{2}{9} + \dfrac{4}{9}$

6. $\dfrac{3}{10} + \dfrac{2}{10}$

7. $\dfrac{6}{20} + \dfrac{1}{20}$

8. $\dfrac{2}{8} + \dfrac{3}{8}$

9. $\dfrac{3}{14} + \dfrac{4}{14}$

10. $\dfrac{5}{24} + \dfrac{7}{24}$

11. $\dfrac{10}{11} + \dfrac{3}{11}$

12. $\dfrac{13}{17} + \dfrac{9}{17}$

13. $\dfrac{4}{13} + \dfrac{2}{13} + \dfrac{1}{13}$ **14.** $\dfrac{5}{11} + \dfrac{1}{11} + \dfrac{2}{11}$ **15.** $\dfrac{7}{18} + \dfrac{3}{18} + \dfrac{2}{18}$ **16.** $\dfrac{7}{15} + \dfrac{4}{15} + \dfrac{1}{15}$

Objective B *Subtract and simplify. See Examples 4 and 5.*

17. $\dfrac{10}{11} - \dfrac{4}{11}$ **18.** $\dfrac{9}{13} - \dfrac{5}{13}$ **19.** $\dfrac{4}{5} - \dfrac{1}{5}$ **20.** $\dfrac{7}{8} - \dfrac{4}{8}$

21. $\dfrac{7}{4} - \dfrac{3}{4}$ **22.** $\dfrac{18}{5} - \dfrac{3}{5}$ **23.** $\dfrac{7}{8} - \dfrac{1}{8}$ **24.** $\dfrac{5}{6} - \dfrac{1}{6}$

25. $\dfrac{25}{12} - \dfrac{15}{12}$ **26.** $\dfrac{30}{20} - \dfrac{15}{20}$ **27.** $\dfrac{11}{10} - \dfrac{3}{10}$ **28.** $\dfrac{14}{15} - \dfrac{4}{15}$

29. $\dfrac{86}{90} - \dfrac{85}{90}$ **30.** $\dfrac{74}{80} - \dfrac{73}{80}$ **31.** $\dfrac{27}{33} - \dfrac{8}{33}$ **32.** $\dfrac{37}{45} - \dfrac{18}{45}$

Objectives A B Mixed Practice *Perform the indicated operation. See Examples 1 through 5.*

33. $\dfrac{8}{21} + \dfrac{5}{21}$ **34.** $\dfrac{7}{37} + \dfrac{9}{37}$ **35.** $\dfrac{99}{100} - \dfrac{9}{100}$ **36.** $\dfrac{85}{200} - \dfrac{15}{200}$

37. $\dfrac{13}{28} - \dfrac{13}{28}$ **38.** $\dfrac{15}{26} - \dfrac{15}{26}$ **39.** $\dfrac{3}{16} + \dfrac{7}{16} + \dfrac{2}{16}$ **40.** $\dfrac{5}{18} + \dfrac{1}{18} + \dfrac{6}{18}$

Objective C *Find the perimeter of each figure. (Hint: Recall that perimeter means distance around.) See Example 6.*

41.
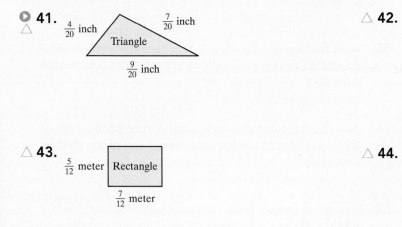

$\dfrac{4}{20}$ inch $\dfrac{7}{20}$ inch

Triangle

$\dfrac{9}{20}$ inch

42.
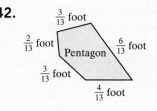

$\dfrac{3}{13}$ foot

$\dfrac{2}{13}$ foot   Pentagon $\dfrac{6}{13}$ foot

$\dfrac{3}{13}$ foot

$\dfrac{4}{13}$ foot

43.

$\dfrac{5}{12}$ meter | Rectangle

$\dfrac{7}{12}$ meter

44.

Square | $\dfrac{1}{6}$ centimeter

Solve. For Exercises 45 and 46, the solutions have been started for you. Write each answer in simplest form.
See Examples 7 and 8.

45. A railroad inspector must inspect $\frac{19}{20}$ of a mile of railroad track. If she has already inspected $\frac{5}{20}$ of a mile, how much more does she need to inspect?

Start the solution:
1. UNDERSTAND the problem. Reread it as many times as needed.
2. TRANSLATE into an equation. (Fill in the blanks.)

distance left to inspect	is	distance needed to inspect	minus	distance already inspected
↓	↓	↓	↓	↓

distance left to inspect = _____ − _____

Finish with:
3. SOLVE. and
4. INTERPRET.

46. Scott Davis has run $\frac{11}{8}$ miles already and plans to complete $\frac{16}{8}$ miles. To do this, how much farther must he run?

Start the solution:
1. UNDERSTAND the problem. Reread it as many times as needed.
2. TRANSLATE into an equation. (Fill in the blanks.)

distance left to run	is	distance planned to run	minus	distance already run
↓	↓	↓	↓	↓

distance left to run = _____ − _____

Finish with:
3. SOLVE. and
4. INTERPRET.

47. Emil Vasquez, a bodybuilder, worked out $\frac{7}{8}$ of an hour one morning before school and $\frac{5}{8}$ of an hour that evening. How long did he work out that day?

48. A recipe for Heavenly Hash cake calls for $\frac{3}{4}$ cup of sugar and later $\frac{1}{4}$ cup of sugar. How much sugar is needed to make the recipe?

The circle graph below shows full-time U.S. college students' time use on an average weekday. Use this graph for Exercises 49–52. Write your answers in simplest form.

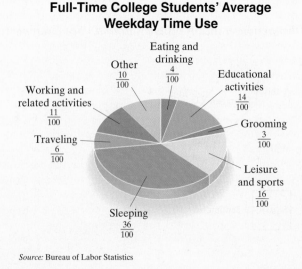

Full-Time College Students' Average Weekday Time Use

Other $\frac{10}{100}$

Eating and drinking $\frac{4}{100}$

Educational activities $\frac{14}{100}$

Working and related activities $\frac{11}{100}$

Grooming $\frac{3}{100}$

Traveling $\frac{6}{100}$

Leisure and sports $\frac{16}{100}$

Sleeping $\frac{36}{100}$

Source: Bureau of Labor Statistics

49. What fraction of a full-time U.S. college student's weekday is spent on eating and drinking, grooming, and sleeping?

50. What fraction of a full-time U.S. college student's weekday is spent on working and related activities and educational activities?

51. How much greater is the fractional part of a college student's weekday that is spent on leisure and sports than on traveling?

52. How much greater is the fractional part of a college student's weekday that is spent on sleeping than on educational activities?

Solve. Write your answers in simplest form.

53. In 2012, $\frac{5}{20}$ of Target's total retail sales were in the health, beauty, and household essentials category, and $\frac{4}{20}$ of Target's total retail sales were in the food and pet supplies category. What fraction of Target's total retail sales were made in these two categories combined? (*Source:* Target Corporation)

54. Road congestion can be caused by a variety of problems. Approximately $\frac{4}{10}$ of all road congestion in the United States is caused by bottlenecks, while $\frac{1}{10}$ of all road congestion in the United States is caused by road construction zones. What fraction of U.S. road congestion is caused by bottlenecks or construction zones? (*Source:* Nationwide Insurance)

55. Effective January 2014, the fraction of states in the United States with maximum interstate highway speed limits up to and including 70 mph was $\frac{34}{50}$. The fraction of states with 70 mph speed limits was $\frac{21}{50}$. What fraction of states had speed limits that were less than 70 mph? (*Source:* Insurance Institute for Highway Safety)

56. When people take aspirin, $\frac{31}{50}$ of the time it is used to treat some type of pain. Approximately $\frac{7}{50}$ of all aspirin use is for treating headaches. What fraction of aspirin use is for treating pain other than headaches? (*Source:* Bayer Market Research)

The map of the world below shows the fraction of the world's surface land area taken up by each continent. In other words, the continent of Africa makes up $\frac{20}{100}$ of the land in the world. Use this map for Exercises 57 through 60. Write your answers in simplest form.

57. Find the fractional part of the world's land area within the continents of North America and South America.

58. Find the fractional part of the world's land area within the continents of Asia and Africa.

59. How much greater is the fractional part of the continent of Antarctica than the fractional part of the continent of Europe?

60. How much greater is the fractional part of the continent of Asia than the continent of Australia?

The theater industry is shifting away from analog movie screens toward digital and digital 3-D movie screens. Use the circle graph to answer Exercises 61 and 62. Write your answers in simplest form.

U.S. Theater Screens by Type, 2012

Analog $\frac{8}{50}$

Digital 3-D $\frac{17}{50}$

Digital non-3-D $\frac{25}{50}$

Source: IHS Screen Digest

61. What fraction of U.S. theater screens are digital, either 3-D or non-3-D?

62. Home much greater is the fraction of screens that are digital 3-D than the fraction of screens that are analog?

Review

Write the prime factorization of each number. See Section 2.2.

63. 10

64. 12

65. 8

66. 20

67. 55

68. 28

Concept Extensions

Perform each indicated operation.

69. $\dfrac{3}{8} + \dfrac{7}{8} - \dfrac{5}{8}$

70. $\dfrac{12}{20} - \dfrac{1}{20} - \dfrac{3}{20}$

71. $\dfrac{4}{11} + \dfrac{5}{11} - \dfrac{3}{11} + \dfrac{2}{11}$

72. $\dfrac{9}{12} + \dfrac{1}{12} - \dfrac{3}{12} - \dfrac{5}{12}$

Find and correct the error. See the Concept Check in this section.

73. $\dfrac{2}{7} + \dfrac{9}{7} = \dfrac{11}{14}$

74. $\dfrac{3}{4} - \dfrac{1}{4} = \dfrac{2}{8} = \dfrac{1}{4}$

Solve. For Exercises 77 through 80, write each answer in simplest form.

75. In your own words, explain how to add like fractions.

76. In your own words, explain how to subtract like fractions.

77. Use the circle graph for Exercises 49 through 52 and find the sum of all the daily time-use fractions. Explain your answer.

78. Use the map of the world for Exercises 57 through 60 and find the sum of all the continents' fractions. Explain your answer.

79. Mike Cannon jogged $\frac{3}{8}$ of a mile from home and then rested. Then he continued jogging farther from home for another $\frac{3}{8}$ of a mile until he discovered his watch had fallen off. He walked back along the same path for $\frac{4}{8}$ of a mile until he found his watch. Find how far he was from his home.

80. A trim carpenter needs the following lengths of boards: $\frac{5}{4}$ feet, $\frac{15}{4}$ feet, $\frac{9}{4}$ feet, and $\frac{13}{4}$ feet. Is a 10-foot board long enough for the carpenter to cut these lengths? If not, how much more length is needed?

3.2 Least Common Multiple

Objective A Finding the Least Common Multiple Using Multiples

A multiple of a number is the product of that number and a natural number. For example, multiples of 5 are

$$5 \cdot 1 \quad 5 \cdot 2 \quad 5 \cdot 3 \quad 5 \cdot 4 \quad 5 \cdot 5 \quad 5 \cdot 6 \quad 5 \cdot 7 \quad 5 \cdot 8$$
$$\downarrow \quad \downarrow \quad \downarrow \quad \downarrow \quad \downarrow \quad \downarrow \quad \downarrow \quad \downarrow$$
$$5, \quad 10, \quad 15, \quad 20, \quad 25, \quad 30, \quad 35, \quad 40, \dots$$

Multiples of 4 are

4, 8, 12, 16, 20, 24, 28, 32, 36, 40, 44, …

Common multiples of both 4 and 5 are numbers that are found in both lists above. If we study the lists of multiples and extend them we have

Common multiples of 4 and 5: 20, 40, 60, 80, …

We call the smallest number in the list of common multiples the **least common multiple (LCM)**. From the list of common multiples of 4 and 5, we see that the LCM of 4 and 5 is 20.

Objectives

A Find the Least Common Multiple (LCM) Using Multiples.

B Find the LCM Using Prime Factorization.

C Write Equivalent Fractions.

Example 1 Find the LCM of 6 and 8.

Solution: Multiples of 6: 6, 12, 18, ㉔, 30, 36, 42, ㊽, …

Multiples of 8: 8, 16, ㉔, 32, 40, ㊽, 56, …

The common multiples are 24, 48, …. The least common multiple (LCM) is 24.

Work Practice 1

Practice 1

Find the LCM of 15 and 50.

Answer

1. 150

Listing all the multiples of every number in a list can be cumbersome and tedious. We can condense the procedure shown in Example 1 with the following steps:

> **Method 1: Finding the LCM of a List of Numbers Using Multiples of the Largest Number**
>
> **Step 1:** Write the multiples of the largest number (starting with the number itself) until a multiple common to all numbers in the list is found.
>
> **Step 2:** The multiple found in Step 1 is the LCM.

Example 2   Find the LCM of 9 and 12.

Solution: We write the multiples of 12 until we find a number that is also a multiple of 9.

$12 \cdot 1 = 12$ Not a multiple of 9.

$12 \cdot 2 = 24$ Not a multiple of 9.

$12 \cdot 3 = 36$ A multiple of 9.

The LCM of 9 and 12 is 36.

■ Work Practice 2

Practice 2

Find the LCM of 8 and 10.

Example 3   Find the LCM of 7 and 14.

Solution: We write the multiples of 14 until we find one that is also a multiple of 7.

$14 \cdot 1 = 14$ A multiple of 7

The LCM of 7 and 14 is 14.

■ Work Practice 3

Practice 3

Find the LCM of 8 and 16.

Example 4   Find the LCM of 12 and 20.

Solution: We write the multiples of 20 until we find one that is also a multiple of 12.

$20 \cdot 1 = 20$ Not a multiple of 12

$20 \cdot 2 = 40$ Not a multiple of 12

$20 \cdot 3 = 60$ A multiple of 12

The LCM of 12 and 20 is 60.

■ Work Practice 4

Practice 4

Find the LCM of 25 and 30.

Objective B Finding the LCM Using Prime Factorization

Method 1 for finding multiples works fine for smaller numbers, but may get tedious for larger numbers. A second method that uses prime factorization may be easier to use for larger numbers.

For example, to find the LCM of 270 and 84, let's look at the prime factorization of each.

$270 = 2 \cdot 3 \cdot 3 \cdot 3 \cdot 5$

$84 = 2 \cdot 2 \cdot 3 \cdot 7$

Answers

2. 40   **3.** 16   **4.** 150

Recall that the LCM must be a multiple of both 270 and 84. Thus, to build the LCM, we will circle the greatest number of factors for each different prime number. The LCM is the product of the circled factors.

Prime Number Factors

$270 =$ 2· $(3 \cdot 3 \cdot 3)$· (5) Circle the greatest number of factors for each
$84 =$ $(2 \cdot 2)$· 3· (7) different prime number.

$\text{LCM} = 2 \cdot 2 \cdot 3 \cdot 3 \cdot 3 \cdot 5 \cdot 7 = 3780$

The number 3780 is the smallest number that both 270 and 84 divide into evenly.
This Method 2 is summarized below:

Method 2: Finding the LCM of a List of Numbers Using Prime Factorization

Step 1: Write the prime factorization of each number.

Step 2: For each different prime factor in step 1, circle the greatest number of times that factor occurs in any one factorization.

Step 3: The LCM is the product of the circled factors.

Example 5   Find the LCM of 72 and 60.

Solution: First we write the prime factorization of each number.

$72 = 2 \cdot 2 \cdot 2 \cdot 3 \cdot 3$
$60 = 2 \cdot 2 \cdot 3 \cdot 5$

For the prime factors shown, we circle the greatest number of prime factors found in either factorization.

$72 = (2 \cdot 2 \cdot 2) \cdot (3 \cdot 3)$
$60 = 2 \cdot 2 \cdot 3 \cdot (5)$

The LCM is the product of the circled factors.

$\text{LCM} = 2 \cdot 2 \cdot 2 \cdot 3 \cdot 3 \cdot 5 = 360$

The LCM is 360.

▇ Work Practice 5

Practice 5

Find the LCM of 40 and 108.

Helpful Hint

If you prefer working with exponents, circle the factor with the greatest exponent.
   Example 5:

$72 = (2^3) \cdot (3^2)$
$60 = 2^2 \cdot 3 \cdot (5)$
$\text{LCD} = 2^3 \cdot 3^2 \cdot 5 = 360$

Answer

5. 1080

Helpful Hint

If the number of factors of a prime number are equal, circle either one, but not both. For example,

$$12 = \boxed{2 \cdot 2} \cdot \boxed{3}$$
$$15 = 3 \cdot \boxed{5} \qquad \text{Circle either 3 but not both.}$$

The LCM is $2 \cdot 2 \cdot 3 \cdot 5 = 60$.

Practice 6

Find the LCM of 20, 24, and 45.

Example 6 Find the LCM of 15, 18, and 54.

Solution:
$$15 = 3 \cdot \boxed{5}$$
$$18 = \boxed{2} \cdot 3 \cdot 3$$
$$54 = 2 \cdot \boxed{3 \cdot 3 \cdot 3}$$

The LCM is $2 \cdot 3 \cdot 3 \cdot 3 \cdot 5$ or 270.

■ Work Practice 6

Practice 7

Find the LCM of 7 and 21.

Example 7 Find the LCM of 11 and 33.

Solution:
$$11 = \boxed{11}$$
$$33 = 3 \cdot \boxed{11} \qquad \begin{array}{l}\text{It makes no difference}\\\text{which 11 is circled.}\end{array}$$

The LCM is $3 \cdot 11$ or 33.

■ Work Practice 7

Objective C Writing Equivalent Fractions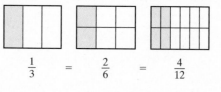

To add or subtract unlike fractions in the next section, we first write equivalent fractions with the LCM as the denominator. Recall from Section 2.3 that fractions that represent the same portion of a whole are called "equivalent fractions."

$$\frac{1}{3} \quad = \quad \frac{2}{6} \quad = \quad \frac{4}{12}$$

To write $\dfrac{1}{3}$ as an equivalent fraction with a denominator of 12, we multiply by 1 in the form of $\dfrac{4}{4}$.

$$\frac{1}{3} = \frac{1}{3} \cdot 1 = \frac{1}{3} \cdot \frac{4}{4} = \frac{1 \cdot 4}{3 \cdot 4} = \frac{4}{12}$$

$$\frac{4}{4} = 1$$

So $\dfrac{1}{3} = \dfrac{4}{12}$.

To write an equivalent fraction,

$$\frac{a}{b} = \frac{a}{b} \cdot \frac{c}{c} = \frac{a \cdot c}{b \cdot c}$$

where a, b, and c are nonzero numbers.

✓ **Concept Check** Which of the following is not equivalent to $\frac{3}{4}$?

a. $\frac{6}{8}$ **b.** $\frac{18}{24}$ **c.** $\frac{9}{14}$ **d.** $\frac{30}{40}$

Example 8 Write an equivalent fraction with the indicated denominator.

$$\frac{3}{4} = \frac{}{20}$$

Solution: In the denominators, since $4 \cdot 5 = 20$, we will multiply by 1 in the form of $\frac{5}{5}$.

$$\frac{3}{4} = \frac{3}{4} \cdot \frac{5}{5} = \frac{3 \cdot 5}{4 \cdot 5} = \frac{15}{20}$$

Thus, $\frac{3}{4} = \frac{15}{20}$.

■ **Work Practice 8**

Practice 8

Write an equivalent fraction with the indicated denominator:
$$\frac{7}{8} = \frac{}{56}$$

Helpful Hint

To check Example 8, write $\frac{15}{20}$ in simplest form.

$$\frac{15}{20} = \frac{3 \cdot \cancel{5}^{1}}{4 \cdot \cancel{5}_{1}} = \frac{3}{4},$$ the original fraction.

If the original fraction is in lowest terms, we can check our work by writing the new equivalent fraction in simplest form. This form should be the original fraction.

✓ **Concept Check** True or false? When the fraction $\frac{2}{9}$ is rewritten as an equivalent fraction with 27 as the denominator, the result is $\frac{2}{27}$.

Example 9 Write an equivalent fraction with the indicated denominator.

$$\frac{1}{2} = \frac{}{24}$$

Solution: Since $2 \cdot 12 = 24$, we multiply by 1 in the form of $\frac{12}{12}$.

$$\frac{1}{2} = \frac{1}{2} \cdot \frac{12}{12} = \frac{1 \cdot 12}{2 \cdot 12} = \frac{12}{24}$$

Thus, $\frac{1}{2} = \frac{12}{24}$.

■ **Work Practice 9**

Practice 9

Write an equivalent fraction with the indicated denominator.
$$\frac{3}{5} = \frac{}{15}$$

Answers

8. $\frac{49}{56}$ **9.** $\frac{9}{15}$

✓ **Concept Check Answers**

c

false; the correct result would be $\frac{6}{27}$

Practice 10

Write an equivalent fraction with the given denominator.

$$4 = \frac{}{6}$$

Answer

10. $\frac{24}{6}$

Example 10   Write an equivalent fraction with the given denominator.

$$3 = \frac{}{7}$$

Solution: Recall that $3 = \frac{3}{1}$. Since $1 \cdot 7 = 7$, multiply by 1 in the form of $\frac{7}{7}$.

$$\frac{3}{1} = \frac{3}{1} \cdot \frac{7}{7} = \frac{3 \cdot 7}{1 \cdot 7} = \frac{21}{7}$$

■ Work Practice 10

Vocabulary, Readiness & Video Check

Use the choices below to fill in each blank.

    least common multiple (LCM)   multiple   equivalent

1. Fractions that represent the same portion of a whole are called _____ fractions.
2. The smallest positive number that is a multiple of all numbers in a list is called the _____.
3. A(n) _____ of a number is the product of that number and a natural number.

Martin-Gay Interactive Videos

Watch the section lecture video and answer the following questions.

Objective A 4. From the lecture before ⊞ Example 1, why do the multiples of a number continue on indefinitely? ▶

Objective B 5. In ⊞ Example 2, why does it make sense that the LCM of 8 and 24 is 24? ▶

Objective C 6. Why isn't the answer to ⊞ Example 5, $\frac{20}{35}$, simplified? ▶

See Video 3.2 🍎

3.2   Exercise Set   MyMathLab® ▶

Objective A B Mixed Practice *Find the LCM of each list of numbers. See Examples 1 through 7.*

1. $3, 4$
2. $4, 6$
▶ 3. $9, 15$
4. $15, 20$
5. $12, 18$
6. $10, 15$

7. $24, 36$
8. $42, 70$
9. $18, 21$
10. $24, 45$
11. $15, 25$
12. $21, 14$

▶ 13. $8, 24$
14. $15, 90$
15. $6, 7$
16. $13, 8$
17. $8, 6, 27$
18. $6, 25, 10$

▶ 19. $25, 15, 6$
20. $4, 14, 20$
21. $34, 68$
22. $25, 175$
23. $84, 294$
24. $48, 54$

25. 30, 36, 50   **26.** 21, 28, 42   **27.** 50, 72, 120   **28.** 70, 98, 100   **29.** 11, 33, 121   **30.** 10, 15, 100

31. 4, 6, 10, 15   **32.** 25, 3, 15, 10

Objective C Write each fraction or whole number as an equivalent fraction with the given denominator. See Examples 8 through 10.

33. $\dfrac{4}{7} = \dfrac{}{35}$

34. $\dfrac{3}{5} = \dfrac{}{20}$

35. $\dfrac{2}{3} = \dfrac{}{21}$

36. $6 = \dfrac{}{10}$

37. $5 = \dfrac{}{3}$

38. $\dfrac{9}{10} = \dfrac{}{70}$

39. $\dfrac{1}{2} = \dfrac{}{30}$

40. $\dfrac{1}{3} = \dfrac{}{30}$

41. $\dfrac{10}{7} = \dfrac{}{21}$

42. $\dfrac{5}{3} = \dfrac{}{21}$

43. $\dfrac{3}{4} = \dfrac{}{28}$

44. $\dfrac{4}{5} = \dfrac{}{45}$

45. $\dfrac{2}{3} = \dfrac{}{45}$

46. $\dfrac{2}{3} = \dfrac{}{75}$

47. $\dfrac{4}{9} = \dfrac{}{81}$

48. $\dfrac{5}{11} = \dfrac{}{88}$

49. $\dfrac{15}{13} = \dfrac{}{78}$

50. $\dfrac{9}{7} = \dfrac{}{84}$

51. $\dfrac{14}{17} = \dfrac{}{68}$

52. $\dfrac{19}{21} = \dfrac{}{126}$

A non-store retailer is a mail-order business that sells goods via catalogs, toll-free telephone numbers, or online media. The table shows the fraction of non-store retailers' goods that were sold online in 2011 by type of goods. Use this table to answer Exercises 53 through 56.

Type of Goods Sold by Non-Store Retailers	Fraction of Goods That Were Sold Online	Equivalent Fraction with a Denominator of 100
Books and magazines	$\dfrac{43}{50}$	
Clothing and accessories	$\dfrac{4}{5}$	
Computer hardware	$\dfrac{29}{50}$	
Computer software	$\dfrac{17}{25}$	
Drugs, health and beauty aids	$\dfrac{3}{25}$	
Electronics and appliances	$\dfrac{21}{25}$	
Food, beer, and wine	$\dfrac{17}{25}$	
Home furnishings	$\dfrac{81}{100}$	
Music and videos	$\dfrac{9}{10}$	
Office equipment and supplies	$\dfrac{79}{100}$	
Sporting goods	$\dfrac{37}{50}$	
Toys, hobbies, and games	$\dfrac{39}{50}$	

(*Source:* U.S. Census Bureau)

53. Complete the table by writing each fraction as an equivalent fraction with a denominator of 100.

54. Which of these types of goods has the largest fraction sold online?

55. Which of these types of goods has the smallest fraction sold online?

56. Which of the types of goods has **more than** $\dfrac{4}{5}$ of the goods sold online? (*Hint:* Write $\dfrac{4}{5}$ as an equivalent fraction with a denominator of 100.)

Review

Add or subtract as indicated. See Section 3.1.

57. $\dfrac{7}{10} - \dfrac{2}{10}$ **58.** $\dfrac{8}{13} - \dfrac{3}{13}$ **59.** $\dfrac{1}{5} + \dfrac{1}{5}$ **60.** $\dfrac{1}{8} + \dfrac{3}{8}$

61. $\dfrac{23}{18} - \dfrac{15}{18}$ **62.** $\dfrac{36}{30} - \dfrac{12}{30}$ **63.** $\dfrac{2}{9} + \dfrac{1}{9} + \dfrac{6}{9}$ **64.** $\dfrac{2}{12} + \dfrac{7}{12} + \dfrac{3}{12}$

Concept Extensions

Write each fraction as an equivalent fraction with the indicated denominator.

65. $\dfrac{37}{165} = \dfrac{}{3630}$

66. $\dfrac{108}{215} = \dfrac{}{4085}$

67. In your own words, explain how to find the LCM of two numbers.

68. In your own words, explain how to write a fraction as an equivalent fraction with a given denominator.

Solve. See the Concept Checks in this section.

69. Which of the following are equivalent to $\dfrac{2}{3}$?

   a. $\dfrac{10}{15}$ **b.** $\dfrac{40}{60}$

   c. $\dfrac{16}{20}$ **d.** $\dfrac{200}{300}$

70. True or False? When the fraction $\dfrac{7}{12}$ is rewritten with a denominator of 48, the result is $\dfrac{11}{48}$. If false, give the correct fraction.

3.3 Adding and Subtracting Unlike Fractions

Objectives

A Add Unlike Fractions.

B Subtract Unlike Fractions.

C Solve Problems by Adding or Subtracting Unlike Fractions.

Objective A Adding Unlike Fractions

In this section we add and subtract fractions with unlike denominators. To add or subtract these unlike fractions, we first write the fractions as equivalent fractions with a common denominator and then add or subtract the like fractions. The common denominator that we use is the least common multiple (LCM) of the denominators. This denominator is called the **least common denominator (LCD).**

To begin, let's add the unlike fractions $\dfrac{3}{4} + \dfrac{1}{6}$. The LCM of denominators 4 and 6 is 12. This means that the number 12 is also the LCD. So we write each fraction as an equivalent fraction with a denominator of 12, then add as usual. This addition process is shown next and also illustrated by figures.

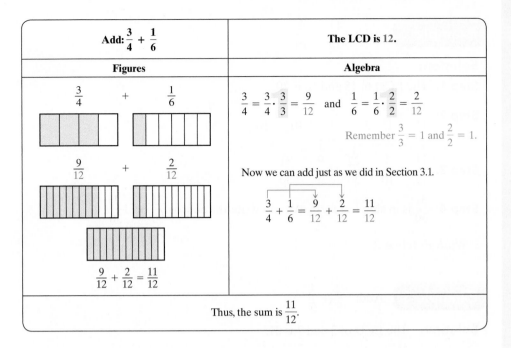

Add: $\dfrac{3}{4} + \dfrac{1}{6}$	The LCD is 12.
Figures	**Algebra**

$$\frac{3}{4} = \frac{3}{4} \cdot \frac{3}{3} = \frac{9}{12} \quad \text{and} \quad \frac{1}{6} = \frac{1}{6} \cdot \frac{2}{2} = \frac{2}{12}$$

Remember $\dfrac{3}{3} = 1$ and $\dfrac{2}{2} = 1$.

Now we can add just as we did in Section 3.1.

$$\frac{3}{4} + \frac{1}{6} = \frac{9}{12} + \frac{2}{12} = \frac{11}{12}$$

Thus, the sum is $\dfrac{11}{12}$.

Adding or Subtracting Unlike Fractions

Step 1: Find the LCM of the denominators of the fractions. This number is the least common denominator (LCD).

Step 2: Write each fraction as an equivalent fraction whose denominator is the LCD.

Step 3: Add or subtract the like fractions.

Step 4: Write the sum or difference in simplest form.

Example 1 Add: $\dfrac{2}{5} + \dfrac{4}{15}$

Solution:

Step 1: The LCM of the denominators 5 and 15 is 15. Thus, the LCD is 15. In later examples, we shall simply say, for example, that the LCD of 5 and 15 is 15.

Step 2: $\dfrac{2}{5} = \dfrac{2}{5} \cdot \dfrac{3}{3} = \dfrac{6}{15}, \dfrac{4}{15} = \dfrac{4}{15}$ ← This fraction already has a denominator of 15.

└ Multiply by 1 in the form $\dfrac{3}{3}$

Step 3: $\dfrac{2}{5} + \dfrac{4}{15} = \dfrac{6}{15} + \dfrac{4}{15} = \dfrac{10}{15}$

Step 4: Write in simplest form.

$$\frac{10}{15} = \frac{2 \cdot \cancel{5}^{1}}{3 \cdot \cancel{5}_{1}} = \frac{2}{3}$$

■ Work Practice 1

Practice 1

Add: $\dfrac{1}{6} + \dfrac{5}{18}$

Answer

1. $\dfrac{4}{9}$

Practice 2

Add: $\dfrac{5}{6} + \dfrac{2}{9}$

Example 2 Add: $\dfrac{11}{15} + \dfrac{3}{10}$

Solution:

Step 1: The LCD of 15 and 10 is 30.

Step 2: $\dfrac{11}{15} = \dfrac{11}{15} \cdot \dfrac{2}{2} = \dfrac{22}{30}$ $\qquad \dfrac{3}{10} = \dfrac{3}{10} \cdot \dfrac{3}{3} = \dfrac{9}{30}$

Step 3: $\dfrac{11}{15} + \dfrac{3}{10} = \dfrac{22}{30} + \dfrac{9}{30} = \dfrac{31}{30}$

Step 4: $\dfrac{31}{30}$ is in simplest form. We can write the sum as $\dfrac{31}{30}$ or $1\dfrac{1}{30}$.

■ Work Practice 2

Practice 3

Add: $\dfrac{2}{5} + \dfrac{4}{9}$

Example 3 Add: $\dfrac{2}{3} + \dfrac{1}{7}$

Solution: The LCD of 3 and 7 is 21.

$$\dfrac{2}{3} + \dfrac{1}{7} = \dfrac{2}{3} \cdot \dfrac{7}{7} + \dfrac{1}{7} \cdot \dfrac{3}{3}$$

$$= \dfrac{14}{21} + \dfrac{3}{21}$$

$$= \dfrac{17}{21} \quad \text{Simplest form.}$$

■ Work Practice 3

Practice 4

Add: $\dfrac{1}{2} + \dfrac{4}{5} + \dfrac{7}{10}$

Example 4 Add: $\dfrac{1}{2} + \dfrac{2}{3} + \dfrac{5}{6}$

Solution: The LCD of 2, 3, and 6 is 6.

$$\dfrac{1}{2} + \dfrac{2}{3} + \dfrac{5}{6} = \dfrac{1}{2} \cdot \dfrac{3}{3} + \dfrac{2}{3} \cdot \dfrac{2}{2} + \dfrac{5}{6}$$

$$= \dfrac{3}{6} + \dfrac{4}{6} + \dfrac{5}{6}$$

$$= \dfrac{12}{6} = 2$$

■ Work Practice 4

Answers

2. $\dfrac{19}{18}$ or $1\dfrac{1}{18}$ 3. $\dfrac{38}{45}$ 4. 2

✓**Concept Check Answer**

When adding unlike fractions, we don't add the denominators.
Correct solution:

$\dfrac{2}{9} + \dfrac{4}{11} = \dfrac{22}{99} + \dfrac{36}{99} = \dfrac{58}{99}$

✓**Concept Check** Find and correct the error in the following:

$\dfrac{2}{9} + \dfrac{4}{11} = \dfrac{6}{20} = \dfrac{3}{10}$

Objective B Subtracting Unlike Fractions

As indicated in the box on page 191, we follow the same steps when subtracting unlike fractions as when adding them.

Example 5   Subtract: $\dfrac{2}{5} - \dfrac{3}{20}$

Solution:

Step 1: The LCD of 5 and 20 is 20.

Step 2: $\dfrac{2}{5} = \dfrac{2}{5} \cdot \dfrac{4}{4} = \dfrac{8}{20}$ $\dfrac{3}{20} = \dfrac{3}{20}$ $\longleftarrow$ The fraction already has a denominator of 20.

Step 3: $\dfrac{2}{5} - \dfrac{3}{20} = \dfrac{8}{20} - \dfrac{3}{20} = \dfrac{5}{20}$

Step 4: Write in simplest form.

$$\dfrac{5}{20} = \dfrac{\overset{1}{\cancel{5}}}{\underset{1}{\cancel{5}} \cdot 4} = \dfrac{1}{4}$$

■ **Work Practice 5**

Practice 5

Subtract: $\dfrac{7}{12} - \dfrac{5}{24}$

Example 6   Subtract: $\dfrac{10}{11} - \dfrac{2}{3}$

Solution:

Step 1: The LCD of 11 and 3 is 33.

Step 2: $\dfrac{10}{11} = \dfrac{10}{11} \cdot \dfrac{3}{3} = \dfrac{30}{33}$ $\dfrac{2}{3} = \dfrac{2}{3} \cdot \dfrac{11}{11} = \dfrac{22}{33}$

Step 3: $\dfrac{10}{11} - \dfrac{2}{3} = \dfrac{30}{33} - \dfrac{22}{33} = \dfrac{8}{33}$

Step 4: $\dfrac{8}{33}$ is in simplest form.

■ **Work Practice 6**

Practice 6

Subtract: $\dfrac{9}{10} - \dfrac{3}{7}$

Example 7   Subtract: $\dfrac{11}{12} - \dfrac{2}{9}$

Solution: The LCD of 12 and 9 is 36.

$$\dfrac{11}{12} - \dfrac{2}{9} = \dfrac{11}{12} \cdot \dfrac{3}{3} - \dfrac{2}{9} \cdot \dfrac{4}{4}$$

$$= \dfrac{33}{36} - \dfrac{8}{36}$$

$$= \dfrac{25}{36}$$

■ **Work Practice 7**

Practice 7

Subtract: $\dfrac{7}{8} - \dfrac{5}{6}$

Answers

5. $\dfrac{3}{8}$ **6.** $\dfrac{33}{70}$ **7.** $\dfrac{1}{24}$

✓**Concept Check Answer**

When subtracting unlike fractions, we don't subtract the denominators. Correct solution:

$$\dfrac{11}{12} - \dfrac{3}{4} = \dfrac{11}{12} - \dfrac{9}{12} = \dfrac{2}{12} = \dfrac{1}{6}$$

✓**Concept Check** Find and correct the error in the following:

$$\dfrac{11}{12} - \dfrac{3}{4} = \dfrac{8}{8} = 1$$

Objective C Solving Problems by Adding or Subtracting Unlike Fractions

Very often, real-world problems involve adding or subtracting unlike fractions.

Practice 8

To repair her sidewalk, a homeowner must pour small amounts of cement in three different locations. She needs $\frac{3}{5}$ of a cubic yard, $\frac{2}{10}$ of a cubic yard, and $\frac{2}{15}$ of a cubic yard for these locations. Find the total amount of cement the homeowner needs.

Example 8 Finding Total Weight

A freight truck has $\frac{1}{4}$ ton of computers, $\frac{1}{3}$ ton of televisions, and $\frac{3}{8}$ ton of small appliances. Find the total weight of its load.

Solution:

1. **UNDERSTAND.** Read and reread the problem. The phrase "total weight" tells us to add.

2. **TRANSLATE.**

In words:	total weight	is	weight of computers	plus	weight of televisions	plus	weight of appliances
	↓	↓	↓	↓	↓	↓	↓
Translate:	total weight	=	$\frac{1}{4}$	+	$\frac{1}{3}$	+	$\frac{3}{8}$

3. **SOLVE:** The LCD is 24.

$$\frac{1}{4} + \frac{1}{3} + \frac{3}{8} = \frac{1}{4} \cdot \frac{6}{6} + \frac{1}{3} \cdot \frac{8}{8} + \frac{3}{8} \cdot \frac{3}{3}$$

$$= \frac{6}{24} + \frac{8}{24} + \frac{9}{24}$$

$$= \frac{23}{24}$$

4. **INTERPRET.** *Check* the solution. *State* your conclusion: The total weight of the truck's load is $\frac{23}{24}$ ton.

■ Work Practice 8

Answer

8. $\frac{14}{15}$ cu yd

Example 9 Calculating Flight Time

A flight from Tucson to Phoenix, Arizona, requires $\frac{5}{12}$ of an hour. If the plane has been flying $\frac{1}{4}$ of an hour, find how much time remains before landing.

Solution:

1. **UNDERSTAND.** Read and reread the problem. The phrase "how much time remains" tells us to subtract.

2. **TRANSLATE.**

In words:	time remaining	is	flight time from Tucson of Phoenix	minus	flight time already passed
	↓	↓	↓	↓	↓
Translate:	time remaining	$=$	$\frac{5}{12}$	$-$	$\frac{1}{4}$

3. **SOLVE:** The LCD is 12.

$$\frac{5}{12} - \frac{1}{4} = \frac{5}{12} - \frac{1}{4} \cdot \frac{3}{3}$$

$$= \frac{5}{12} - \frac{3}{12}$$

$$= \frac{2}{12}$$

$$= \frac{\overset{1}{\cancel{2}}}{\underset{1}{\cancel{2} \cdot 6}}$$

$$= \frac{1}{6}$$

4. **INTERPRET.** *Check* the solution. *State* your conclusion: The flight time remaining is $\frac{1}{6}$ of an hour.

▨ Work Practice 9

Calculator Explorations Performing Operations on Fractions

Scientific Calculator

Many calculators have a fraction key, such as $\boxed{a_{b/c}}$, that allows you to enter fractions and perform operations on them, and then it gives the result as a fraction. If your calculator has a fraction key, use it to calculate

$$\frac{3}{5} + \frac{4}{7}$$

Enter the keystrokes

$$\boxed{3}\ \boxed{a_{b/c}}\ \boxed{5}\ \boxed{+}\ \boxed{4}\ \boxed{a_{b/c}}\ \boxed{7}\ \boxed{=}$$

The display should read $\boxed{1\_6\ |\ 35}$, which represents the mixed number $1\frac{6}{35}$. Let's write the result as a fraction. To convert from mixed number notation to fractional notation, press

$$\boxed{2^{\text{nd}}}\ \boxed{d/c}$$

The display now reads $\boxed{41\ |\ 35}$, which represents $\frac{41}{35}$, the sum in fractional notation.

Graphing Calculator

Graphing calculators also allow you to perform operations on fractions and will give exact fractional results. The fraction option on a graphing calculator may be found under the $\boxed{\text{MATH}}$ menu. To perform the addition to the left, try the keystrokes.

$$\boxed{3}\ \boxed{\div}\ \boxed{5}\ \boxed{+}\ \boxed{4}\ \boxed{\div}\ \boxed{7}\ \boxed{\text{MATH}}\ \boxed{\text{ENTER}}$$
$$\boxed{\text{ENTER}}$$

The display should read

$$\boxed{3/5 + 4/7 \blacktriangleright \text{Frac } 41/35}$$

Use a calculator to add the following fractions. Give each sum as a fraction.

1. $\dfrac{1}{16} + \dfrac{2}{5}$ 2. $\dfrac{3}{20} + \dfrac{2}{25}$ 3. $\dfrac{4}{9} + \dfrac{7}{8}$

4. $\dfrac{9}{11} + \dfrac{5}{12}$ 5. $\dfrac{10}{17} + \dfrac{12}{19}$ 6. $\dfrac{14}{31} + \dfrac{15}{21}$

Vocabulary, Readiness & Video Check

Use the choices below to fill in each blank. Any numerical answers are not listed.

least common denominator   equivalent

1. To add or subtract unlike fractions, we first write the fractions as _____ fractions with a common denominator. The common denominator we use is called the _____.

2. The LCD for $\dfrac{5}{8}$ and $\dfrac{1}{6}$ is _____.

3. $\dfrac{5}{8} + \dfrac{1}{6} = \dfrac{5}{8} \cdot \dfrac{3}{3} + \dfrac{1}{6} \cdot \dfrac{4}{4} = \underline{} + \underline{} = \underline{}$.

4. $\dfrac{5}{8} - \dfrac{1}{6} = \dfrac{5}{8} \cdot \dfrac{3}{3} - \dfrac{1}{6} \cdot \dfrac{4}{4} = \underline{} - \underline{} = \underline{}$.

Martin-Gay Interactive Videos

See Video 3.3

Watch the section lecture video and answer the following questions.

Objective A 5. In Example 1, why does multiplying $\dfrac{2}{11}$ by $\dfrac{3}{3}$ not change the value of the fraction?

Objective B 6. In Example 3, how did we know we needed to multiply the two denominators to get the LCD?

Objective C 7. What are the two forms of the answer to Example 5?

3.3 Exercise Set MyMathLab®

Objective A *Add and simplify. See Examples 1 through 4.*

1. $\frac{2}{3} + \frac{1}{6}$

2. $\frac{5}{6} + \frac{1}{12}$

3. $\frac{1}{2} + \frac{1}{3}$

4. $\frac{2}{3} + \frac{1}{4}$

5. $\frac{2}{11} + \frac{2}{33}$

6. $\frac{5}{9} + \frac{1}{3}$

7. $\frac{3}{14} + \frac{3}{7}$

8. $\frac{2}{5} + \frac{2}{15}$

9. $\frac{11}{35} + \frac{2}{7}$

10. $\frac{4}{5} + \frac{3}{40}$

11. $\frac{8}{25} + \frac{7}{35}$

12. $\frac{5}{14} + \frac{10}{21}$

13. $\frac{7}{15} + \frac{5}{12}$

14. $\frac{5}{8} + \frac{3}{20}$

15. $\frac{2}{28} + \frac{2}{21}$

16. $\frac{6}{25} + \frac{7}{35}$

17. $\frac{9}{44} + \frac{17}{36}$

18. $\frac{2}{33} + \frac{2}{21}$

19. $\frac{5}{11} + \frac{3}{13}$

20. $\frac{3}{7} + \frac{9}{17}$

21. $\frac{1}{3} + \frac{1}{9} + \frac{1}{27}$

22. $\frac{1}{4} + \frac{1}{16} + \frac{1}{64}$

23. $\frac{5}{7} + \frac{1}{8} + \frac{1}{2}$

24. $\frac{10}{13} + \frac{7}{10} + \frac{1}{5}$

25. $\frac{5}{36} + \frac{3}{4} + \frac{1}{6}$

26. $\frac{7}{18} + \frac{2}{9} + \frac{5}{6}$

27. $\frac{13}{20} + \frac{3}{5} + \frac{1}{3}$

28. $\frac{2}{7} + \frac{13}{28} + \frac{2}{5}$

Objective B *Subtract and simplify. See Examples 5 through 7.*

29. $\frac{7}{8} - \frac{3}{16}$

30. $\frac{5}{13} - \frac{3}{26}$

31. $\frac{5}{6} - \frac{3}{7}$

32. $\frac{3}{4} - \frac{1}{7}$

33. $\frac{5}{7} - \frac{1}{8}$

34. $\frac{10}{13} - \frac{7}{10}$

35. $\frac{9}{11} - \frac{4}{9}$

36. $\frac{7}{18} - \frac{2}{9}$

37. $\frac{11}{35} - \frac{2}{7}$

38. $\frac{2}{5} - \frac{3}{25}$

39. $\frac{5}{12} - \frac{1}{9}$

40. $\frac{7}{12} - \frac{5}{18}$

41. $\frac{7}{15} - \frac{5}{12}$

42. $\frac{5}{8} - \frac{3}{20}$

43. $\frac{3}{28} - \frac{2}{21}$

44. $\frac{6}{25} - \frac{7}{35}$

45. $\frac{1}{100} - \frac{1}{1000}$

46. $\frac{1}{50} - \frac{1}{500}$

47. $\frac{21}{44} - \frac{11}{36}$

48. $\frac{7}{18} - \frac{2}{45}$

Objectives A B Mixed Practice *Perform the indicated operation. See Examples 1 through 7.*

49. $\dfrac{5}{12} + \dfrac{1}{9}$

50. $\dfrac{7}{12} + \dfrac{5}{18}$

51. $\dfrac{17}{35} - \dfrac{2}{7}$

52. $\dfrac{13}{24} - \dfrac{1}{6}$

53. $\dfrac{9}{28} - \dfrac{3}{40}$

54. $\dfrac{10}{26} - \dfrac{3}{8}$

55. $\dfrac{2}{3} + \dfrac{4}{45} + \dfrac{4}{5}$

56. $\dfrac{3}{16} + \dfrac{1}{4} + \dfrac{1}{16}$

Objective C *Find the perimeter of each geometric figure. (Hint: Recall that perimeter means distance around.)*

57.
Parallelogram
$\frac{4}{5}$ cm, $\frac{1}{3}$ cm, $\frac{1}{3}$ cm, $\frac{4}{5}$ cm

△**58.**
Triangle
$\frac{5}{8}$ mile, $\frac{3}{8}$ mile, $\frac{1}{2}$ mile

△**59.**
Quadrilateral
$\frac{1}{4}$ meter, $\frac{1}{5}$ meter, $\frac{1}{2}$ meter, $\frac{3}{4}$ meter

△**60.**
Rectangle
$\frac{1}{7}$ yard, $\frac{10}{21}$ yard

Solve. For Exercises 61 and 62, the solutions have been started for you. See Examples 8 and 9.

61. The slowest mammal is the three-toed sloth from South America. The sloth has an average ground speed of $\dfrac{1}{10}$ mph. In the trees, it can accelerate to $\dfrac{17}{100}$ mph. How much faster can a sloth travel in the trees? (*Source: Guinness World Records*)

Start the solution:

1. UNDERSTAND the problem. Reread it as many times as needed.

2. TRANSLATE into an equation. (Fill in the blanks.)

how much faster sloth travels in trees	is	sloth speed in trees	minus	sloth speed on ground
↓	↓	↓	↓	↓

how much faster sloth travels in trees $= \underline{\hspace{1cm}} - \underline{\hspace{1cm}}$

Finish with:

3. SOLVE. and
4. INTERPRET.

62. Killer bees have been known to chase people for up to $\dfrac{1}{4}$ of a mile, while domestic European honeybees will normally chase a person for no more than 100 feet, or $\dfrac{5}{264}$ of a mile. How much farther will a killer bee chase a person than a domestic honeybee? (*Source: Coachella Valley Mosquito & Vector Control District*)

Start the solution:

1. UNDERSTAND the problem. Reread it as many times as needed.

2. TRANSLATE into an equation. (Fill in the blanks.)

how much farther killer bee will chase than honeybee	is	distance killer bee chases	minus	distance honeybee chases
↓	↓	↓	↓	↓

how much farther killer bee will chase than honeybee $= \underline{\hspace{1cm}} - \underline{\hspace{1cm}}$

Finish with:

3. SOLVE. and
4. INTERPRET.

63. Find the inner diameter of the washer. (*Hint:* Use the outer diameter and subtract the washer widths.)

Inner
diameter

$\frac{3}{16}$ inch $\frac{3}{16}$ inch

1 inch

64. Find the inner diameter of the tubing. (See the hint for Exercise 63.)

Inner
diameter

$\frac{1}{8}$ inch $\frac{1}{8}$ inch

$\frac{3}{4}$ inch

65. Given the following diagram, find its total length. (*Hint:* Find the sum of the partial lengths.)

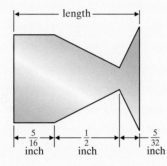

length

$\frac{5}{16}$ inch $\frac{1}{2}$ inch $\frac{5}{32}$ inch

66. Given the following diagram, find its total width. (*Hint:* Find the sum of the partial widths.)

width

$\frac{7}{32}$ inch

$\frac{3}{16}$ inch

$\frac{7}{32}$ inch

67. Together, Thin Mints and Samoas account for $\frac{11}{25}$ of the Girl Scout cookies sold each year. Thin Mints alone account for $\frac{1}{4}$ of all Girl Scout cookie sales. What fraction of Girl Scout cookies sold are Samoas? (*Source:* Girl Scouts of the United States of America)

68. About $\frac{13}{20}$ of American students ages 10 to 17 name math, science, or art as their favorite subject in school. Art is the favorite subject for about $\frac{4}{25}$ of the American students ages 10 to 17. For what fraction of students this age is math or science their favorite subject? (*Source:* Peter D. Hart Research Associates for the National Science Foundation)

The table below shows the fraction of the Earth's water area taken up by each ocean. Use this table for Exercises 69 and 70.

Fraction of Earth's Water Area per Ocean	
Ocean	**Fraction**
Arctic	$\frac{1}{25}$
Atlantic	$\frac{13}{50}$
Pacific	$\frac{1}{2}$
Indian	$\frac{1}{5}$

69. What fraction of the world's water surface area is accounted for by the Pacific and Atlantic Oceans?

70. What fraction of the world's water surface area is accounted for by the Arctic and Indian Oceans?

We first viewed this circle graph in Section 2.3. In this section we study it further. Use it to answer Exercises 71 through 74.

Areas Maintained by the National Park Service

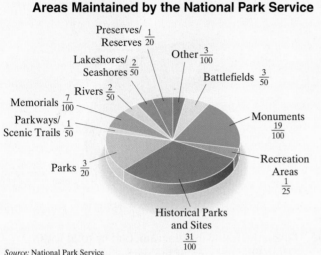

Source: National Park Service

71. What fraction of areas maintained by the National Park Service is designated as either National Parks or National Monuments?

72. What fraction of areas maintained by the National Park Service is designated as either National Memorials or National Battlefields?

73. What fraction of areas maintained by the National Park Service is NOT designated as National Monuments?

74. What fraction of areas maintained by the National Park Service is NOT designated as National Preserves or National Reserves?

Review

Multiply or divide as indicated. See Sections 2.4 and 2.5.

75. $1\frac{1}{2} \cdot 3\frac{1}{3}$

76. $2\frac{5}{6} \div 5$

77. $4 \div 7\frac{1}{4}$

78. $4\frac{3}{4} \cdot 5\frac{1}{5}$

79. $3 \cdot 2\frac{1}{9}$

80. $6\frac{2}{7} \cdot 14$

Concept Extensions

For Exercises 81 and 82 below, do the following:

a. *Draw three rectangles of the same size and represent each fraction in the sum or difference, one fraction per rectangle, by shading.*

b. *Using these rectangles as estimates, determine whether there is an error in the sum or difference.*

c. *If there is an error, correctly calculate the sum or difference.*

See the Concept Checks in this section.

81. $\frac{3}{5} + \frac{4}{5} \stackrel{?}{=} \frac{7}{10}$

82. $\frac{3}{4} - \frac{5}{8} \stackrel{?}{=} \frac{2}{4}$

Subtract from left to right.

83. $\frac{2}{3} - \frac{1}{4} - \frac{2}{540}$

84. $\frac{9}{10} - \frac{7}{200} - \frac{1}{3}$

Perform each indicated operation.

85. $\frac{30}{55} + \frac{1000}{1760}$

86. $\frac{19}{26} - \frac{968}{1352}$

87. In your own words, describe how to add or subtract two fractions with different denominators.

88. Find the sum of the fractions in the circle graph above. Did the sum surprise you? Why or why not?

Operations on Fractions and Mixed Numbers

Answers

Find the LCM of each list of numbers.

1. 5, 6 **2.** 3, 7 **3.** 2, 14

4. 5, 25 **5.** 4, 20, 25 **6.** 6, 18, 30

Write each fraction as an equivalent fraction with the indicated denominator.

7. $\dfrac{3}{8} = \dfrac{}{24}$ **8.** $\dfrac{7}{9} = \dfrac{}{36}$ **9.** $\dfrac{1}{4} = \dfrac{}{40}$

10. $\dfrac{2}{5} = \dfrac{}{30}$ **11.** $\dfrac{11}{15} = \dfrac{}{75}$ **12.** $\dfrac{5}{6} = \dfrac{}{48}$

Add or subtract as indicated. Simplify if necessary.

13. $\dfrac{3}{8} + \dfrac{1}{8}$ **14.** $\dfrac{7}{10} - \dfrac{3}{10}$ **15.** $\dfrac{17}{24} - \dfrac{3}{24}$

16. $\dfrac{4}{15} + \dfrac{9}{15}$ **17.** $\dfrac{1}{4} + \dfrac{1}{2}$ **18.** $\dfrac{1}{3} - \dfrac{1}{5}$

19. $\dfrac{7}{9} - \dfrac{2}{5}$ **20.** $\dfrac{3}{10} + \dfrac{2}{25}$ **21.** $\dfrac{7}{8} + \dfrac{1}{20}$

1. _____
2. _____
3. _____
4. _____
5. _____
6. _____
7. _____
8. _____
9. _____
10. _____
11. _____
12. _____
13. _____
14. _____
15. _____
16. _____
17. _____
18. _____
19. _____
20. _____
21. _____

201

22. _____

23. _____

24. _____

25. _____

26. _____

27. _____

28. _____

29. _____

30. _____

31. _____

32. _____

33. _____

34. _____

35. _____

36. _____

37. _____

38. _____

39. _____

40. _____

41. _____

42. _____

43. _____

44. _____

45. _____

46. _____

22. $\dfrac{5}{12} - \dfrac{2}{18}$

23. $\dfrac{1}{11} - \dfrac{1}{11}$

24. $\dfrac{3}{17} - \dfrac{2}{17}$

25. $\dfrac{9}{11} - \dfrac{2}{3}$

26. $\dfrac{1}{6} - \dfrac{1}{7}$

27. $\dfrac{2}{9} + \dfrac{1}{18}$

28. $\dfrac{4}{13} + \dfrac{2}{26}$

29. $\dfrac{2}{9} + \dfrac{1}{18} + \dfrac{1}{3}$

30. $\dfrac{3}{10} + \dfrac{1}{5} + \dfrac{6}{25}$

Mixed Practice (*Sections 2.4, 2.5, 3.1, 3.2, 3.3*) *Perform the indicated operation.*

31. $\dfrac{9}{10} + \dfrac{2}{3}$

32. $\dfrac{9}{10} - \dfrac{2}{3}$

33. $\dfrac{9}{10} \cdot \dfrac{2}{3}$

34. $\dfrac{9}{10} \div \dfrac{2}{3}$

35. $\dfrac{21}{25} - \dfrac{3}{70}$

36. $\dfrac{21}{25} + \dfrac{3}{70}$

37. $\dfrac{21}{25} \div \dfrac{3}{70}$

38. $\dfrac{21}{25} \cdot \dfrac{3}{70}$

39. $3\dfrac{7}{8} \cdot 2\dfrac{2}{3}$

40. $3\dfrac{7}{8} \div 2\dfrac{2}{3}$

41. $\dfrac{2}{9} + \dfrac{5}{27} + \dfrac{1}{2}$

42. $\dfrac{3}{8} + \dfrac{11}{16} + \dfrac{2}{3}$

43. $11\dfrac{7}{10} \div 3\dfrac{3}{100}$

44. $7\dfrac{1}{4} \cdot 3\dfrac{1}{5}$

45. $\dfrac{14}{15} - \dfrac{4}{27}$

46. $\dfrac{9}{14} - \dfrac{11}{32}$

3.4 Adding and Subtracting Mixed Numbers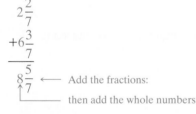

Objective A Adding Mixed Numbers

Recall that a mixed number has a whole number part and a fraction part.

$$2\frac{3}{8} \text{ means } 2 + \frac{3}{8}$$

whole number ↓ ↓
fraction

✓**Concept Check** Which of the following are equivalent to 7?

a. $6\frac{5}{5}$ **b.** $6\frac{7}{7}$ **c.** $5\frac{8}{4}$

d. $6\frac{17}{17}$ **e.** all of these

Adding or Subtracting Mixed Numbers

To add or subtract mixed numbers, add or subtract the fraction parts and then add or subtract the whole number parts.

For example,

$$
\begin{array}{r}
2\frac{2}{7} \\
+6\frac{3}{7} \\
\hline
8\frac{5}{7}
\end{array}
$$
← Add the fractions:
then add the whole numbers

Example 1 Add: $2\frac{1}{3} + 5\frac{3}{8}$. Check by estimating.

Solution: The LCD of 3 and 8 is 24.

$$
\begin{array}{r}
2\frac{1\cdot 8}{3\cdot 8} = 2\frac{8}{24} \\
+5\frac{3\cdot 3}{8\cdot 3} = +5\frac{9}{24} \\
\hline
7\frac{17}{24}
\end{array}
$$
← Add the fractions
Add the whole numbers

To check by estimating, we round as usual. The fraction $2\frac{1}{3}$ rounds to 2, $5\frac{3}{8}$ rounds to 5, and $2 + 5 = 7$, our estimate.

Our exact answer is close to 7, so our answer is reasonable.

■ Work Practice 1

Objectives

A Add Mixed Numbers.

B Subtract Mixed Numbers.

C Solve Problems by Adding or Subtracting Mixed Numbers.

Practice 1

Add: $4\frac{2}{5} + 5\frac{1}{6}$

Answer

1. $9\frac{17}{30}$

✓**Concept Check Answer**

e

Helpful Hint

When adding or subtracting mixed numbers and whole numbers, it is a good idea to estimate to see if your answer is reasonable.

For the rest of this section, we leave most of the checking by estimating to you.

Practice 2

Add: $2\frac{5}{14} + 5\frac{6}{7}$

 Example 2 Add: $3\frac{4}{5} + 1\frac{4}{15}$

Solution: The LCD of 5 and 15 is 15.

$$
\begin{array}{rcl}
3\frac{4}{5} &=& 3\frac{12}{15} \\
+1\frac{4}{15} &=& +1\frac{4}{15} \\
\hline
&& 4\frac{16}{15}
\end{array}
$$

Add the fractions; then add the whole numbers.

Notice that the fraction part is improper.

Since $\frac{16}{15}$ is $1\frac{1}{15}$ we can write the sum as

$$
4\frac{16}{15} = 4 + 1\frac{1}{15} = 5\frac{1}{15}
$$

Work Practice 2

✔**Concept Check** Explain how you could estimate the following sum:

$$5\frac{1}{9} + 14\frac{10}{11}.$$

Practice 3

Add: $10 + 2\frac{6}{7} + 3\frac{1}{5}$

Example 3 Add: $1\frac{4}{5} + 4 + 2\frac{1}{2}$

Solution: The LCD of 5 and 2 is 10.

$$
\begin{array}{rcl}
1\frac{4}{5} &=& 1\frac{8}{10} \\
4 &=& 4 \\
+2\frac{1}{2} &=& +2\frac{5}{10} \\
\hline
&& 7\frac{13}{10} = 7 + 1\frac{3}{10} = 8\frac{3}{10}
\end{array}
$$

Work Practice 3

Answers

2. $8\frac{3}{14}$ **3.** $16\frac{2}{35}$

✔**Concept Check Answer**

Round each mixed number to the nearest whole number and add. $5\frac{1}{9}$ rounds to 5 and $14\frac{10}{11}$ rounds to 15, and the estimated sum is $5 + 15 = 20$.

Objective B Subtracting Mixed Numbers

Example 4 Subtract: $9\dfrac{3}{7} - 5\dfrac{2}{21}$. Check by estimating.

Solution: The LCD of 7 and 21 is 21.

$$
\begin{aligned}
9\tfrac{3}{7} &= 9\tfrac{9}{21} \quad \leftarrow \text{The LCD of 7 and 21 is 21.}\\
-5\tfrac{2}{21} &= -5\tfrac{2}{21}\\
\hline
&\;\;\;4\tfrac{7}{21} \quad \leftarrow \text{Subtract the fractions.}
\end{aligned}
$$

Subtract the whole numbers.

Then $4\dfrac{7}{21}$ simplifies to $4\dfrac{1}{3}$. The difference is $4\dfrac{1}{3}$.

To check, $9\dfrac{3}{7}$ rounds to 9, $5\dfrac{2}{21}$ rounds to 5, and $9 - 5 = 4$, our estimate.

Our exact answer is close to 4, so our answer is reasonable.

■ Work Practice 4

When subtracting mixed numbers, borrowing may be needed, as shown in the next example.

Example 5 Subtract: $7\dfrac{3}{14} - 3\dfrac{6}{7}$

Solution: The LCD of 7 and 14 is 14.

$$
\begin{aligned}
7\tfrac{3}{14} &= 7\tfrac{3}{14}\\
-3\tfrac{6}{7} &= -3\tfrac{12}{14}\\
\hline
\end{aligned}
$$

Notice that we cannot subtract $\dfrac{12}{14}$ from $\dfrac{3}{14}$, so we borrow from the whole number 7.

borrow 1 from 7

$$7\tfrac{3}{14} = 6 + 1\tfrac{3}{14} = 6 + \tfrac{17}{14} \text{ or } 6\tfrac{17}{14}$$

Now subtract.

$$
\begin{aligned}
7\tfrac{3}{14} &= 7\tfrac{3}{14} = 6\tfrac{17}{14}\\
-3\tfrac{6}{7} &= -3\tfrac{12}{14} = -3\tfrac{12}{14}\\
\hline
&\qquad\qquad\qquad 3\tfrac{5}{14} \quad \leftarrow \text{Subtract the fractions.}
\end{aligned}
$$

Subtract the whole numbers.

■ Work Practice 5

✓**Concept Check** In the subtraction problem $5\dfrac{1}{4} - 3\dfrac{3}{4}$, $5\dfrac{1}{4}$ must be rewritten because $\dfrac{3}{4}$ cannot be subtracted from $\dfrac{1}{4}$. Why is it incorrect to rewrite $5\dfrac{1}{4}$ as $5\dfrac{5}{4}$?

Practice 4

Subtract: $29\dfrac{7}{9} - 13\dfrac{5}{18}$

Practice 5

Subtract: $9\dfrac{7}{15} - 5\dfrac{3}{5}$

Answers

4. $16\dfrac{1}{2}$ **5.** $3\dfrac{13}{15}$

✓**Concept Check Answer**

Rewrite $5\dfrac{1}{4}$ as $4\dfrac{5}{4}$ by borrowing from the 5.

Practice 6

Subtract: $25 - 10\frac{2}{9}$

Example 6   Subtract: $12 - 8\frac{3}{7}$

Solution:

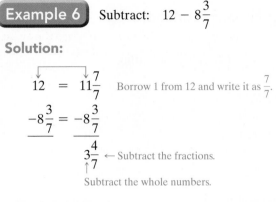

$$12 = 11\frac{7}{7} \qquad \text{Borrow 1 from 12 and write it as } \frac{7}{7}.$$

$$-8\frac{3}{7} = -8\frac{3}{7}$$

$$3\frac{4}{7} \leftarrow \text{Subtract the fractions.}$$
$$\uparrow \text{Subtract the whole numbers.}$$

■ **Work Practice 6**

Objective C   Solving Problems by Adding or Subtracting Mixed Numbers ▶

Now that we know how to add and subtract mixed numbers, we can solve real-life problems.

Example 7   Calculating Total Weight

Two packages of ground round are purchased. One package weighs $2\frac{3}{8}$ pounds and the other $1\frac{4}{5}$ pounds. What is the combined weight of the ground round?

Solution:

1. **UNDERSTAND.** Read and reread the problem. The phrase "combined weight" tells us to add.

2. **TRANSLATE.**

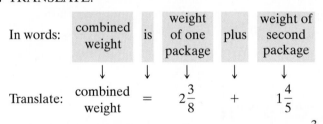

In words:	combined weight	is	weight of one package	plus	weight of second package
	↓	↓	↓	↓	↓
Translate:	combined weight	=	$2\frac{3}{8}$	+	$1\frac{4}{5}$

Practice 7

Two rainbow trout weigh $2\frac{1}{2}$ pounds and $3\frac{2}{3}$ pounds. What is the total weight of the two trout?

3. **SOLVE:** Before we solve, let's estimate. The fraction $2\frac{3}{8}$ rounds to 2, $1\frac{4}{5}$ rounds to 2, and $2 + 2 = 4$. The combined weight should be close to 4.

$$2\frac{3}{8} = 2\frac{15}{40}$$
$$+1\frac{4}{5} = +1\frac{32}{40}$$
$$\overline{\qquad\qquad}$$
$$3\frac{47}{40} = 4\frac{7}{40}$$

4. **INTERPRET.** *Check* your work. Our estimate of 4 tells us that the exact answer of $4\frac{7}{40}$ is reasonable. *State* your conclusion: The combined weight of the ground round is $4\frac{7}{40}$ pounds.

■ **Work Practice 7**

Answers

6. $14\frac{7}{9}$ **7.** $6\frac{1}{6}$ lb

Example 8 Finding Legal Lobster Size

Lobster fishermen must measure the upper body shells of the lobsters they catch. Lobsters that are too small are thrown back into the ocean. Each state has its own size standard for lobsters to help control the breeding stock. Massachusetts divided its waters into four Lobster Conservation Management Areas, with a different minimum lobster size permitted in each area. In the off-shore area, the legal lobster size increased from $3\frac{13}{32}$ inches in 2006 to $3\frac{1}{2}$ inches in 2008. How much of an increase was this? (*Source:* Massachusetts Division of Marine Fisheries)

Solution:

1. **UNDERSTAND.** Read and reread the problem carefully. The word "increase" found in the problem might make you think that we add to solve the problem. But the phrase "how much of an increase" tells us to subtract to find the increase.

2. **TRANSLATE.**

In words:	increase	is	new lobster size	minus	old lobster size
	↓	↓	↓	↓	↓
Translate:	increase	$=$	$3\frac{1}{2}$	$-$	$3\frac{13}{32}$

3. **SOLVE.** Before we solve, let's estimate. The fraction $3\frac{1}{2}$ can be rounded to 4, $3\frac{13}{32}$ can be rounded to 3, and $4 - 3 = 1$. The increase is not 1, but will be smaller since we rounded $3\frac{1}{2}$ up more than we rounded $3\frac{13}{32}$ down.

$$3\frac{1}{2} = 3\frac{16}{32}$$
$$-3\frac{13}{32} = -3\frac{13}{32}$$
$$\overline{\frac{3}{32}}$$

4. **INTERPRET.** *Check* your work. Our estimate tells us that the exact increase of $\frac{3}{32}$ is reasonable. *State* your conclusion: The increase in lobster size is $\frac{3}{32}$ of an inch.

▪ Work Practice 8

Practice 8

The measurement around the trunk of a tree just below shoulder height is called its girth. The largest known American beech tree in the United States has a girth of $24\frac{1}{6}$ feet. The largest known sugar maple tree in the United States has a girth of $19\frac{5}{12}$ feet. How much larger is the girth of the largest known American beech tree than the girth of the largest known sugar maple tree? (*Source: American Forests*)

Girth

Vocabulary, Readiness & Video Check

Use the choices below to fill in each blank.

round   fraction   whole number

improper   mixed number

1. The number $5\frac{3}{4}$ is called a(n) _____.

2. For $5\frac{3}{4}$, the 5 is called the _____ part and $\frac{3}{4}$ is called the _____ part.

3. To estimate operations on mixed numbers, we _____ mixed numbers to the nearest whole number.

4. The mixed number $2\frac{5}{8}$ written as a(n) _____ fraction is $\frac{21}{8}$.

Choose the best estimate for each sum or difference.

5. $3\frac{7}{8} + 2\frac{1}{5}$

   a. 6 **b.** 5 **c.** 1 **d.** 2

6. $3\frac{7}{8} - 2\frac{1}{5}$

   a. 6 **b.** 5 **c.** 1 **d.** 2

7. $8\frac{1}{3} - 1\frac{1}{2}$

   a. 4 **b.** 10 **c.** 6 **d.** 16

8. $8\frac{1}{3} + 1\frac{1}{2}$

   a. 4 **b.** 10 **c.** 6 **d.** 16

Martin-Gay Interactive Videos   *Watch the section lecture video and answer the following questions.*

Objective A 9. In ▤ Example 2, why is the first form of the answer not in a good format? ▶

Objective B 10. In ▤ Example 3, how is 6 rewritten in the subtraction problem? ▶

Objective C 11. In ▤ Example 4, why can't we subtract immediately once we rewrite the fraction parts of the mixed numbers with the LCD? ▶

See Video 3.4 🍎

3.4 Exercise Set MyMathLab® ▶

Objective A *Add. For those exercises marked, find an exact sum and an estimated sum. See Examples 1 through 3.*

1. $4\frac{7}{10}$
  $+2\frac{1}{10}$

  Exact:

  Estimate:

2. $7\frac{4}{9}$
  $+3\frac{2}{9}$

  Exact:

  Estimate:

▶3. $10\frac{3}{14}$
  $+3\frac{4}{7}$

  Exact:

  Estimate:

4. $12\frac{5}{12}$
  $+4\frac{1}{6}$

  Exact:

  Estimate:

5. $9\frac{1}{5}$
  $+8\frac{2}{25}$

6. $6\frac{2}{13}$

$+8\frac{7}{26}$

7. $3\frac{1}{2}$

$+4\frac{1}{8}$

8. $9\frac{3}{4}$

$+2\frac{1}{8}$

9. $1\frac{5}{6}$

$+5\frac{3}{8}$

10. $2\frac{5}{12}$

$+1\frac{5}{8}$

11. $8\frac{2}{5}$

$+11\frac{2}{3}$

12. $7\frac{3}{7}$

$+3\frac{3}{5}$

13. $11\frac{3}{5}$

$+7\frac{2}{5}$

14. $19\frac{7}{9}$

$+8\frac{2}{9}$

15. $40\frac{9}{10}$

$+15\frac{8}{27}$

16. $102\frac{5}{8}$

$+96\frac{21}{25}$

17. $3\frac{5}{8}$

$2\frac{1}{6}$

$+7\frac{3}{4}$

18. $4\frac{1}{3}$

$9\frac{2}{5}$

$+3\frac{1}{6}$

19. $12\frac{3}{14}$

10

$+25\frac{5}{12}$

20. $8\frac{2}{9}$

32

$+9\frac{10}{21}$

Objectives B *Subtract. For those exercises marked, find an exact difference and an estimated difference. See Examples 4 through 6.*

21. $4\frac{7}{10}$

$-2\frac{1}{10}$

Exact:

Estimate:

22. $7\frac{4}{9}$

$-3\frac{2}{9}$

Exact:

Estimate:

23. $10\frac{13}{14}$

$-3\frac{4}{7}$

Exact:

Estimate:

24. $12\frac{5}{12}$

$-4\frac{1}{6}$

Exact:

Estimate:

25. $9\frac{1}{5}$

$-8\frac{6}{25}$

26. $5\frac{2}{13}$

$-4\frac{7}{26}$

27. $5\frac{2}{3} - 3\frac{1}{5}$

28. $23\frac{3}{5}$

$-8\frac{8}{15}$

29. $15\frac{4}{7}$

$-9\frac{11}{14}$

30. $5\frac{3}{8} - 2\frac{13}{20}$

31. $47\frac{4}{18} - 23\frac{19}{24}$

32. $6\frac{1}{6} - 5\frac{11}{14}$

33. 10

$-8\frac{1}{5}$

34. 23

$-17\frac{3}{4}$

35. $11\frac{3}{5}$

$-9\frac{11}{15}$

36. $9\frac{1}{10}$
$-7\frac{2}{5}$
_____

◑ 37. 6
$-2\frac{4}{9}$
_____

38. 8
$-1\frac{7}{10}$
_____

39. $63\frac{1}{6}$
$-47\frac{5}{12}$
_____

40. $86\frac{2}{15}$
$-27\frac{3}{10}$
_____

Objectives A B Mixed Practice *Perform the indicated operation. See Examples 1 through 6.*

41. $15\frac{1}{6}$
$+13\frac{5}{12}$
_____

42. $21\frac{3}{10}$
$+11\frac{3}{5}$
_____

43. $22\frac{7}{8}$
$-\ 7$
_____

44. $27\frac{3}{21}$
$-\ 9$
_____

45. $5\frac{8}{9} + 2\frac{1}{9}$

46. $12\frac{13}{16} + 7\frac{3}{16}$

47. $33\frac{11}{20} - 15\frac{19}{30}$

48. $54\frac{7}{30} - 38\frac{29}{50}$

Objective C *Solve. For Exercises 49 and 50, the solutions have been started for you. Write each answer in simplest form. See Examples 7 and 8.*

49. To prevent intruding birds, birdhouses built for Eastern Bluebirds should have an entrance hole measuring $1\frac{1}{2}$ inches in diameter. Entrance holes in birdhouses for Mountain Bluebirds should measure $1\frac{9}{16}$ inches in diameter. How much wider should entrance holes for Mountain Bluebirds be than for Eastern Bluebirds? (*Source:* North American Bluebird Society)

Start the solution:

1. UNDERSTAND the problem. Reread it as many times as needed.

2. TRANSLATE into an equation. (Fill in the blanks.)

Finish with:

3. SOLVE and

4. INTERPRET

50. If the total weight allowable without overweight charges is 50 pounds and the traveler's luggage weighs $60\frac{5}{8}$ pounds, on how many pounds will the traveler's overweight charges be based?

Start the solution:

1. UNDERSTAND the problem. Reread it as many times as needed.

2. TRANSLATE into an equation. (Fill in the blanks.)

Finish with:

3. SOLVE and

4. INTERPRET

51. Charlotte Dowlin has $15\frac{2}{3}$ feet of plastic pipe. She cuts off a $2\frac{1}{2}$-foot length and then a $3\frac{1}{4}$-foot length. If she now needs a 10-foot piece of pipe, will the remaining piece do? If not, by how much will the piece be short?

$3\frac{1}{4}$ feet   ? $2\frac{1}{2}$ feet

$15\frac{2}{3}$ feet

52. A trim carpenter cuts a board $3\frac{3}{8}$ feet long from one 6 feet long. How long is the remaining piece?

$3\frac{3}{8}$ feet   ?

6 feet

53. If Tucson's average annual rainfall is $11\frac{1}{4}$ inches and Yuma's is $3\frac{3}{5}$ inches, how much more rain, on average, does Tucson get than Yuma?

54. A pair of crutches needs adjustment. One crutch is 43 inches and the other is $41\frac{5}{8}$ inches. Find how much the short crutch should be lengthened to make both crutches the same length.

55. On four consecutive days, a concert pianist practiced for $2\frac{1}{2}$ hours, $1\frac{2}{3}$ hours, $2\frac{1}{4}$ hours, and $3\frac{5}{6}$ hours. Find his total practice time.

56. A tennis coach was preparing her team for a tennis tournament and enforced this practice schedule: Monday, $2\frac{1}{2}$ hours; Tuesday, $2\frac{2}{3}$ hours; Wednesday, $1\frac{3}{4}$ hours; and Thursday, $1\frac{9}{16}$ hours. How long did the team practice that week before Friday's tournament?

57. Jerald Divis, a tax consultant, takes $3\frac{1}{2}$ hours to prepare a personal tax return and $5\frac{7}{8}$ hours to prepare a small business return. How much longer does it take him to prepare the small business return?

58. Jessica Callac takes $2\frac{3}{4}$ hours to clean her room. Her brother Matthew takes $1\frac{1}{3}$ hours to clean his room. If they start at the same time, how long does Matthew have to wait for Jessica to finish?

59. Located on an island in New York City's harbor, the Statue of Liberty is one of the largest statues in the world. The copper figure is $46\frac{1}{20}$ meters tall from feet to tip of torch. The figure stands on a pedestal that is $46\frac{47}{50}$ meters tall. What is the overall height of the Statue of Liberty from the base of the pedestal to the tip of the torch? (*Source:* National Park Service)

60. The record for largest rainbow trout ever caught is 48 pounds and was set in Saskatchewan in 2009. The record for largest brown trout ever caught is $42\frac{1}{16}$ pounds and was set in New Zealand in 2013. How much more did the record-setting rainbow trout weigh than the record-setting brown trout? (*Source:* International Game Fish Association)

61. The longest floating pontoon bridge in the United States is the Evergreen Point Bridge in Seattle, Washington. It is 2526 yards long. The second-longest pontoon bridge in the United States is the Lacey V. Murrow Memorial Bridge, also in Seattle. It is $2206\frac{2}{3}$ yards long. How much longer is the Evergreen Point Bridge than the Lacey V. Murrow Memorial Bridge? (*Source:* Federal Highway Administration)

62. What is the difference between interest rates of $11\frac{1}{2}$ percent and $9\frac{3}{4}$ percent?

The following table lists some upcoming total eclipses of the Sun that will be visible in North America. The duration of each eclipse is listed in the table. Use the table to answer Exercises 63 through 66.

Total Solar Eclipses Visible from North America	
Date of Eclipse	**Duration (in Minutes)**
August 21, 2017	$2\frac{2}{3}$
April 8, 2024	$4\frac{7}{15}$
August 12, 2026	$2\frac{3}{10}$
(*Source:* NASA/Goddard Space Flight Center)	

63. What is the total duration for the three eclipses?

64. What is the total duration for the two eclipses occurring in even-numbered years?

65. How much longer will the April 8, 2024, eclipse be than the August 21, 2017, eclipse?

66. How much longer will the August 21, 2017, eclipse be than the August 12, 2026, eclipse?

Find the perimeter of each figure.

△ **67.**

$2\frac{1}{3}$ miles $2\frac{1}{3}$ miles

$2\frac{1}{3}$ miles

△ **68.**

7 inches $11\frac{1}{5}$ inches

$12\frac{1}{3}$ inches

△ **69.**

$5\frac{1}{3}$ meters

3 meters   5 meters

$7\frac{7}{8}$ meters

△ **70.**

$3\frac{1}{4}$ yards $3\frac{1}{4}$ yards

$3\frac{1}{4}$ yards $3\frac{1}{4}$ yards

$3\frac{1}{4}$ yards

Review

Evaluate each expression. See Section 1.9.

71. 2^3

72. 3^2

73. 5^2

74. 2^5

75. $20 \div 10 \cdot 2$

76. $36 - 5 \cdot 6 + 10$

77. $2 + 3(8 \cdot 7 - 1)$

78. $2(10 - 2 \cdot 5) + 13$

Simplify. Write any mixed number whose fraction part is not a proper fraction in simplest form. See Sections 2.1 and 2.3.

79. $3\dfrac{5}{5}$

80. $10\dfrac{8}{7}$

81. $9\dfrac{10}{16}$

82. $6\dfrac{7}{14}$

Concept Extensions

Solve. See the Concept Checks in this section.

83. Which of the following are equivalent to 10?

   a. $9\dfrac{5}{5}$ **b.** $9\dfrac{100}{100}$ **c.** $6\dfrac{44}{11}$ **d.** $8\dfrac{13}{13}$

84. Which of the following are equivalent to $7\dfrac{3}{4}$?

   a. $6\dfrac{7}{4}$ **b.** $5\dfrac{11}{4}$ **c.** $7\dfrac{12}{16}$ **d.** all of them

85. Explain in your own words why $9\dfrac{13}{9}$ is equal to $10\dfrac{4}{9}$.

86. In your own words, explain

   a. when to borrow when subtracting mixed numbers, and

   b. how to borrow when subtracting mixed numbers.

Solve.

87. Carmen's Candy Clutch is famous for its "Nutstuff," a special blend of nuts and candy. A Supreme box of Nutstuff has $2\dfrac{1}{4}$ pounds of nuts and $3\dfrac{1}{2}$ pounds of candy. A Deluxe box has $1\dfrac{3}{8}$ pounds of nuts and $4\dfrac{1}{4}$ pounds of candy. Which box is heavier and by how much?

88. Willie Cassidie purchased three Supreme boxes and two Deluxe boxes of Nutstuff from Carmen's Candy Clutch. (See Exercise 87.) What is the total weight of his purchase?

3.5 Order, Exponents, and the Order of Operations

Objectives

A Compare Fractions.

B Evaluate Fractions Raised to Powers.

C Review Operations on Fractions.

D Use the Order of Operations.

Objective A Comparing Fractions

Recall that whole numbers can be shown on a number line using equally spaced distances.

$$\begin{array}{c} \mid \quad \mid \quad \mid \quad \bullet \quad \mid \quad \bullet \quad \mid \quad \mid \\ 0 \quad 1 \quad 2 \quad 3 \quad 4 \quad 5 \quad 6 \quad 7 \end{array}$$

From the number line, we can see the order of numbers. For example, we can see that 3 is less than 5 because 3 is to the left of 5.

For any two numbers on a number line, the number to the left is always the smaller number, and the number to the right is always the larger number.

We use the **inequality symbols** $<$ or $>$ to write the order of numbers.

> ### Inequality Symbols
>
> $<$ means *is less than*.
>
> $>$ means *is greater than*.

For example,

$$\underbrace{3 \text{ is less than } 5}_{3 < 5} \quad \text{or} \quad \underbrace{5 \text{ is greater than } 3}_{5 > 3}$$

We can compare fractions the same way. To see fractions on a number line, divide the spaces between whole numbers into equal parts.

For example, let's compare $\dfrac{2}{5}$ and $\dfrac{4}{5}$.

$$\begin{array}{c} \mid \quad \mid \quad \bullet \quad \mid \quad \bullet \quad \mid \quad \mid \quad \mid \quad \mid \quad \mid \\ 0 \quad \frac{1}{5} \quad \frac{2}{5} \quad \frac{3}{5} \quad \frac{4}{5} \quad 1 \qquad\qquad 2 \end{array}$$
$$\frac{5}{5} = 1$$

Since $\dfrac{4}{5}$ is to the right of $\dfrac{2}{5}$,

$$\frac{2}{5} < \frac{4}{5} \quad \text{\small Notice that } 2 < 4 \text{ also.}$$

Helpful Hint Notice that to compare like fractions, we compare the numerators. The order of the like fractions is the same as the order of the numerators.

> ### Comparing Fractions
>
> To determine which of two fractions is greater,
>
> **Step 1:** Write the fractions as like fractions.
>
> **Step 2:** The fraction with the greater numerator is the greater fraction.

Practice 1

Insert $<$ or $>$ to form a true statement.

$$\frac{8}{9} \qquad \frac{10}{11}$$

Answer

1. $<$

Example 1 Insert $<$ or $>$ to form a true statement.

$$\frac{3}{10} \qquad \frac{2}{7}$$

Solution:

Step 1: The LCD of 10 and 7 is 70.

$$\frac{3}{10} = \frac{3}{10} \cdot \frac{7}{7} = \frac{21}{70}; \quad \frac{2}{7} = \frac{2}{7} \cdot \frac{10}{10} = \frac{20}{70}$$

214

Copyright 2015 Pearson Education, Inc.

Step 2: Since $21 > 20$, then $\dfrac{21}{70} > \dfrac{20}{70}$ or

$$\frac{3}{10} > \frac{2}{7}$$

■ Work Practice 1

| Example 2 | Insert $<$ or $>$ to form a true statement.

$$\frac{9}{10} \quad \frac{11}{12}$$

Solution:

Step 1: The LCD of 10 and 12 is 60.

$$\frac{9}{10} = \frac{9}{10} \cdot \frac{6}{6} = \frac{54}{60} \qquad \frac{11}{12} = \frac{11}{12} \cdot \frac{5}{5} = \frac{55}{60}$$

Step 2: Since $54 < 55$, then $\dfrac{54}{60} < \dfrac{55}{60}$ or

$$\frac{9}{10} < \frac{11}{12}$$

■ Work Practice 2

Helpful Hint

If we think of $<$ and $>$ as arrowheads, a true statement is always formed when the arrow points to the smaller number.

$$\frac{2}{3} > \frac{1}{3} \qquad\qquad \frac{5}{6} < \frac{7}{6}$$

$\uparrow$ $\uparrow$
points to smaller number   points to smaller number

Objective B Evaluating Fractions Raised to Powers ▶

Recall from Section 1.9 that exponents indicate repeated multiplication.

exponent
↓
$$5^{3} = \underbrace{5 \cdot 5 \cdot 5}_{} = 125$$
↑
base   3 factors of 5

Exponents mean the same when the base is a fraction. For example,

$$\left(\frac{1}{3}\right)^{4} = \underbrace{\frac{1}{3} \cdot \frac{1}{3} \cdot \frac{1}{3} \cdot \frac{1}{3}}_{} = \frac{1}{81}$$
↑
base   4 factors of $\frac{1}{3}$

| Examples | Evaluate each expression.

3. $\left(\dfrac{1}{4}\right)^{2} = \dfrac{1}{4} \cdot \dfrac{1}{4} = \dfrac{1}{16}$

4. $\left(\dfrac{3}{5}\right)^{3} = \dfrac{3}{5} \cdot \dfrac{3}{5} \cdot \dfrac{3}{5} = \dfrac{27}{125}$

5. $\left(\dfrac{1}{6}\right)^{2} \cdot \left(\dfrac{3}{4}\right)^{3} = \left(\dfrac{1}{6} \cdot \dfrac{1}{6}\right) \cdot \left(\dfrac{3}{4} \cdot \dfrac{3}{4} \cdot \dfrac{3}{4}\right) = \dfrac{1 \cdot 1 \cdot \overset{1}{\cancel{3}} \cdot \overset{1}{\cancel{3}} \cdot 3}{2 \cdot \underset{1}{\cancel{3}} \cdot 2 \cdot \underset{1}{\cancel{3}} \cdot 4 \cdot 4 \cdot 4} = \dfrac{3}{256}$

■ Work Practice 3–5

Practice 2

Insert $<$ or $>$ to form a true statement.

$$\frac{3}{5} \quad \frac{2}{9}$$

Practice 3–5

Evaluate each expression.

3. $\left(\dfrac{1}{5}\right)^{2}$ **4.** $\left(\dfrac{2}{3}\right)^{3}$

5. $\left(\dfrac{1}{4}\right)^{2}\left(\dfrac{2}{3}\right)^{3}$

Answers

2. $>$ **3.** $\dfrac{1}{25}$ **4.** $\dfrac{8}{27}$ **5.** $\dfrac{1}{54}$

Objective C Reviewing Operations on Fractions ▶

To get ready to use the order of operations with fractions, let's first review the operations on fractions that we have learned.

Review of Operations on Fractions

Operation	Procedure	Example
Multiply	Multiply the numerators and multiply the denominators.	$\dfrac{5}{9} \cdot \dfrac{1}{2} = \dfrac{5 \cdot 1}{9 \cdot 2} = \dfrac{5}{18}$
Divide	Multiply the first fraction by the reciprocal of the second fraction.	$\dfrac{2}{3} \div \dfrac{11}{13} = \dfrac{2}{3} \cdot \dfrac{13}{11} = \dfrac{2 \cdot 13}{3 \cdot 11} = \dfrac{26}{33}$
Add or Subtract	1. Write each fraction as an equivalent fraction whose denominator is the LCD 2. Add or subtract numerators and write the result over the common denominator.	$\dfrac{3}{4} + \dfrac{1}{8} = \dfrac{3}{4} \cdot \dfrac{2}{2} + \dfrac{1}{8} = \dfrac{6}{8} + \dfrac{1}{8} = \dfrac{7}{8}$

Practice 6–9

Perform each indicated operation.

6. $\dfrac{3}{7} \div \dfrac{10}{11}$ **7.** $\dfrac{4}{15} + \dfrac{2}{5}$

8. $\dfrac{2}{3} \cdot \dfrac{9}{10}$ **9.** $\dfrac{11}{12} - \dfrac{2}{5}$

Examples Perform each indicated operation.

6. $\dfrac{1}{2} \div \dfrac{8}{7} = \dfrac{1}{2} \cdot \dfrac{7}{8} = \dfrac{1 \cdot 7}{2 \cdot 8} = \dfrac{7}{16}$ To divide: multiply by the reciprocal.

7. $\dfrac{6}{35} + \dfrac{3}{7} = \dfrac{6}{35} + \dfrac{3}{7} \cdot \dfrac{5}{5} = \dfrac{6}{35} + \dfrac{15}{35} = \dfrac{21}{35}$ To add: need the LCD. The LCD is 35.

$$= \dfrac{\cancel{7} \cdot 3}{\cancel{7} \cdot 5} = \dfrac{3}{5}$$

8. $\dfrac{2}{9} \cdot \dfrac{3}{11} = \dfrac{2 \cdot 3}{9 \cdot 11} = \dfrac{2 \cdot \overset{1}{\cancel{3}}}{\cancel{3} \cdot 3 \cdot 11} = \dfrac{2}{33}$ To multiply: multiply numerators and multiply denominators.

9. $\dfrac{6}{7} - \dfrac{1}{3} = \dfrac{6}{7} \cdot \dfrac{3}{3} - \dfrac{1}{3} \cdot \dfrac{7}{7} = \dfrac{18}{21} - \dfrac{7}{21} = \dfrac{11}{21}$ To subtract: need the LCD. The LCD is 21.

■ Work Practice 6–9

Objective D Using the Order of Operations ▶

The order of operations that we use on whole numbers applies to expressions containing fractions and mixed numbers also.

Order of Operations

1. Perform all operations within parentheses ( ), brackets [ ], or other grouping symbols such as square roots or fraction bars, starting with the innermost set.
2. Evaluate any expressions with exponents.
3. Multiply or divide in order from left to right.
4. Add or subtract in order from left to right.

Answers

6. $\dfrac{33}{70}$ **7.** $\dfrac{2}{3}$ **8.** $\dfrac{3}{5}$ **9.** $\dfrac{31}{60}$

Example 10   Simplify: $\dfrac{1}{5} \div \dfrac{2}{3} \cdot \dfrac{4}{5}$

Solution: Multiply or divide *in order* from left to right. We divide first.

$$\dfrac{1}{5} \div \dfrac{2}{3} \cdot \dfrac{4}{5} = \dfrac{1}{5} \cdot \dfrac{3}{2} \cdot \dfrac{4}{5}$$

↑   To divide, multiply by the reciprocal.

$$= \dfrac{3}{10} \cdot \dfrac{4}{5}$$

$$= \dfrac{3 \cdot 4}{10 \cdot 5} \qquad \text{Multiply.}$$

$$= \dfrac{3 \cdot 2 \cdot \overset{1}{\cancel{2}}}{\underset{1}{\cancel{2}} \cdot 5 \cdot 5} \qquad \text{Simplify.}$$

$$= \dfrac{6}{25} \qquad \text{Simplify.}$$

■ Work Practice 10

Practice 10

Simplify: $\dfrac{2}{9} \div \dfrac{4}{7} \cdot \dfrac{3}{10}$

Example 11   Simplify: $\left(\dfrac{2}{3}\right)^2 \div \left(\dfrac{8}{27} + \dfrac{2}{3}\right)$

Solution: Start within the right set of parentheses. We add.

$$\left(\dfrac{2}{3}\right)^2 \div \left(\dfrac{8}{27} + \dfrac{2}{3}\right) = \left(\dfrac{2}{3}\right)^2 \div \left(\dfrac{8}{27} + \dfrac{18}{27}\right) \quad \text{The LCD is 27. Write } \dfrac{2}{3} \text{ as } \dfrac{18}{27}.$$

$$= \left(\dfrac{2}{3}\right)^2 \div \dfrac{26}{27} \qquad \text{Simplify inside the parentheses.}$$

$$= \dfrac{4}{9} \div \dfrac{26}{27} \qquad \text{Write } \left(\dfrac{2}{3}\right)^2 \text{ as } \dfrac{4}{9}.$$

$$= \dfrac{4}{9} \cdot \dfrac{27}{26}$$

$$= \dfrac{\overset{1}{\cancel{2}} \cdot 2 \cdot 3 \cdot \overset{1}{\cancel{9}}}{\underset{1}{\cancel{9}} \cdot \underset{1}{\cancel{2}} \cdot 13}$$

$$= \dfrac{6}{13}$$

■ Work Practice 11

Practice 11

Simplify: $\left(\dfrac{2}{5}\right)^2 \div \left(\dfrac{3}{5} - \dfrac{11}{25}\right)$

✔ **Concept Check** What should be done first to simplify $3\left[\left(\dfrac{1}{4}\right)^2 + \dfrac{3}{2}\left(\dfrac{6}{7} - \dfrac{1}{3}\right)\right]$?

Recall from Section 1.7 that the average of a list of numbers is their sum divided by the number of numbers in the list.

Example 12   Find the average of $\dfrac{1}{3}, \dfrac{2}{5},$ and $\dfrac{2}{9}$.

Solution: The average is their sum, divided by 3.

$$\left(\dfrac{1}{3} + \dfrac{2}{5} + \dfrac{2}{9}\right) \div 3 = \left(\dfrac{15}{45} + \dfrac{18}{45} + \dfrac{10}{45}\right) \div 3 \quad \text{The LCD is 45.}$$

$$= \dfrac{43}{45} \div 3 \qquad \text{Add.}$$

$$= \dfrac{43}{45} \cdot \dfrac{1}{3}$$

$$= \dfrac{43}{135} \qquad \text{Multiply.}$$

■ Work Practice 12

Practice 12

Find the average of $\dfrac{1}{2}, \dfrac{3}{8},$ and $\dfrac{7}{24}$.

Answers

10. $\dfrac{7}{60}$ **11.** 1   **12.** $\dfrac{7}{18}$

✔ **Concept Check Answer**

$\dfrac{6}{7} - \dfrac{1}{3}$

Vocabulary, Readiness & Video Check

Use the choices below to fill in each blank. Not all choices will be used.

addition   multiplication   evaluate the exponential expression
subtraction   division

1. To simplify $\frac{1}{2} + \frac{2}{3} \cdot \frac{7}{8}$, which operation do we perform first? _____

2. To simplify $\frac{1}{2} \div \frac{2}{3} \cdot \frac{7}{8}$, which operation do we perform first? _____

3. To simplify $\frac{7}{8} \cdot \left(\frac{1}{2} - \frac{2}{3} \right)$, which operation do we perform first? _____

4. To simplify $9 - \left(\frac{3}{4} \right)^2$, which operation do we perform first? _____

Martin-Gay Interactive Videos   Watch the section lecture video and answer the following questions.

See Video 3.5

Objective A 5. Complete this statement based on ▤ Example 1: When comparing fractions, as long as the _____ are the same, we can just compare _____. ▶

Objective B 6. Complete this statement based on ▤ Example 3: The meaning of an exponent is the same whether the base is a _____ or the base is a _____. ▶

Objective C 7. Fraction operations are reviewed in the lecture before ▤ Example 4. How are denominators treated differently when adding and subtracting fractions than when multiplying and dividing? ▶

Objective D 8. In ▤ Example 6, why did we subtract before applying the exponent? ▶

3.5 Exercise Set MyMathLab® ▶

Objective A *Insert < or > to form a true statement. See Examples 1 and 2.*

1. $\frac{7}{9}$ $\frac{6}{9}$

2. $\frac{12}{17}$ $\frac{13}{17}$

▶ 3. $\frac{3}{3}$ $\frac{5}{3}$

4. $\frac{3}{23}$ $\frac{4}{23}$

5. $\frac{9}{42}$ $\frac{5}{21}$

6. $\frac{17}{32}$ $\frac{5}{16}$

7. $\frac{9}{8}$ $\frac{17}{16}$

8. $\frac{3}{8}$ $\frac{14}{40}$

9. $\frac{3}{4}$ $\frac{2}{3}$

10. $\frac{2}{5}$ $\frac{1}{3}$

▶ 11. $\frac{3}{5}$ $\frac{9}{14}$

12. $\frac{3}{10}$ $\frac{7}{25}$

13. $\frac{1}{10}$ $\frac{1}{11}$

14. $\frac{1}{13}$ $\frac{1}{14}$

15. $\frac{27}{100}$ $\frac{7}{25}$

16. $\frac{37}{120}$ $\frac{9}{30}$

Objective B *Evaluate each expression. See Examples 3 through 5.*

17. $\left(\dfrac{1}{2}\right)^4$

18. $\left(\dfrac{1}{7}\right)^2$

19. $\left(\dfrac{2}{5}\right)^3$

20. $\left(\dfrac{3}{4}\right)^3$

21. $\left(\dfrac{4}{7}\right)^3$

22. $\left(\dfrac{2}{3}\right)^4$

23. $\left(\dfrac{2}{9}\right)^2$

24. $\left(\dfrac{7}{11}\right)^2$

25. $\left(\dfrac{3}{4}\right)^2 \cdot \left(\dfrac{2}{3}\right)^3$

26. $\left(\dfrac{1}{6}\right)^2 \cdot \left(\dfrac{9}{10}\right)^2$

27. $\dfrac{9}{10}\left(\dfrac{2}{5}\right)^2$

28. $\dfrac{7}{11}\left(\dfrac{3}{10}\right)^2$

Objective C *Perform each indicated operation. See Examples 6 through 9.*

29. $\dfrac{2}{15} + \dfrac{3}{5}$

30. $\dfrac{5}{12} + \dfrac{5}{6}$

31. $\dfrac{3}{7} \cdot \dfrac{1}{5}$

32. $\dfrac{9}{10} \div \dfrac{2}{3}$

33. $1 - \dfrac{4}{9}$

34. $5 - \dfrac{2}{3}$

35. $4\dfrac{2}{9} + 5\dfrac{9}{11}$

36. $7\dfrac{3}{7} + 6\dfrac{3}{5}$

37. $\dfrac{5}{6} - \dfrac{3}{4}$

38. $\dfrac{7}{10} - \dfrac{3}{25}$

39. $\dfrac{6}{11} \div \dfrac{2}{3}$

40. $\dfrac{3}{8} \cdot \dfrac{1}{11}$

41. $0 \cdot \dfrac{9}{10}$

42. $\dfrac{5}{6} \cdot 0$

43. $0 \div \dfrac{9}{10}$

44. $\dfrac{5}{6} \div 0$

45. $\dfrac{20}{35} \cdot \dfrac{7}{10}$

46. $\dfrac{18}{25} \div \dfrac{3}{5}$

47. $\dfrac{4}{7} - \dfrac{6}{11}$

48. $\dfrac{11}{20} + \dfrac{7}{15}$

Objective D *Use the order of operations to simplify each expression. See Examples 10 and 11.*

49. $\dfrac{1}{5} + \dfrac{1}{3} \cdot \dfrac{1}{4}$

50. $\dfrac{1}{2} + \dfrac{1}{6} \cdot \dfrac{1}{3}$

51. $\dfrac{5}{6} \div \dfrac{1}{3} \cdot \dfrac{1}{4}$

52. $\dfrac{7}{8} \div \dfrac{1}{4} \cdot \dfrac{1}{7}$

53. $\dfrac{1}{5} \cdot \left(2\dfrac{5}{6} - \dfrac{1}{3}\right)$

54. $\dfrac{4}{7} \cdot \left(6 - 2\dfrac{1}{2}\right)$

55. $2 \cdot \left(\dfrac{1}{4} + \dfrac{1}{5}\right) + 2$

56. $\dfrac{2}{5} \cdot \left(5 - \dfrac{1}{2}\right) - 1$

57. $\left(\dfrac{3}{4}\right)^2 \div \left(\dfrac{3}{4} - \dfrac{1}{12}\right)$

58. $\left(\dfrac{8}{9}\right)^2 \div \left(2 - \dfrac{2}{3}\right)$

59. $\left(\dfrac{2}{3} - \dfrac{5}{9}\right)^2$

60. $\left(1 - \dfrac{2}{5}\right)^3$

61. $\dfrac{5}{9} \cdot \dfrac{1}{2} + \dfrac{2}{3} \cdot \dfrac{5}{6}$

62. $\dfrac{7}{10} \cdot \dfrac{1}{2} + \dfrac{3}{4} \cdot \dfrac{3}{5}$

63. $\dfrac{27}{16} \cdot \left(\dfrac{2}{3}\right)^2 - \dfrac{3}{20}$

64. $\dfrac{64}{27} \cdot \left(\dfrac{3}{4}\right)^2 - \dfrac{7}{10}$

65. $\dfrac{3}{13} \div \dfrac{9}{26} - \dfrac{7}{24} \cdot \dfrac{8}{14}$ **66.** $\dfrac{5}{11} \div \dfrac{15}{77} - \dfrac{7}{10} \cdot \dfrac{5}{14}$ **67.** $\dfrac{3}{14} + \dfrac{10}{21} \div \left(\dfrac{3}{7}\right)\left(\dfrac{9}{4}\right)$ **68.** $\dfrac{11}{15} + \dfrac{7}{9} \div \left(\dfrac{14}{3}\right)\left(\dfrac{2}{3}\right)$

69. $\left(\dfrac{3}{4} + \dfrac{1}{8}\right)^2 - \left(\dfrac{1}{2} + \dfrac{1}{8}\right)$ **70.** $\left(\dfrac{1}{6} + \dfrac{1}{3}\right)^3 + \left(\dfrac{2}{5} \cdot \dfrac{3}{4}\right)^2$

Find the average of each list of numbers. See Example 12.

71. $\dfrac{5}{6}$ and $\dfrac{2}{3}$ **72.** $\dfrac{1}{2}$ and $\dfrac{4}{7}$ **73.** $\dfrac{1}{5}, \dfrac{3}{10},$ and $\dfrac{3}{20}$ **74.** $\dfrac{1}{3}, \dfrac{1}{4},$ and $\dfrac{1}{6}$

Objective C D Mixed Practice

75. The average fraction of online sales of computer hardware is $\dfrac{23}{50}$, of computer software is $\dfrac{1}{2}$, and of movies and movies is $\dfrac{3}{5}$. Find the average of these fractions.

76. The average fraction of online sales of sporting goods is $\dfrac{12}{25}$, of toys and hobbies and games is $\dfrac{1}{2}$, and of computer hardware is $\dfrac{23}{50}$. Find the average of these fractions.

Review

Identify each key word with the operation it most likely translates to. After each word, write A for addition, S for subtraction, M for multiplication, and D for division. See Sections 1.3, 1.4, 1.6, and 1.7.

77. increased by   **78.** sum   **79.** triple   **80.** product

81. subtracted from   **82.** decreased by   **83.** quotient   **84.** divided by

85. times   **86.** difference   **87.** total   **88.** more than

Concept Extensions

Solve.

89. Calculate $\dfrac{2^3}{3}$ and $\left(\dfrac{2}{3}\right)^3$. Do both of these expressions simplify to the same number? Explain why or why not.

90. Calculate $\left(\dfrac{1}{2}\right)^2 \cdot \left(\dfrac{3}{4}\right)^2$ and $\left(\dfrac{1}{2} \cdot \dfrac{3}{4}\right)^2$. Do both of these expressions simplify to the same number? Explain why or why not.

Each expression contains one addition, one subtraction, one multiplication, and one division. Write the operations in the order that they should be performed. Do not actually simplify. See the Concept Check in this section.

91. $[9 + 3(4 - 2)] \div \dfrac{10}{21}$ **92.** $[30 - 4(3 + 2)] \div \dfrac{5}{2}$

93. $\dfrac{1}{3} \div \left(\dfrac{2}{3}\right)\left(\dfrac{4}{5}\right) - \dfrac{1}{4} + \dfrac{1}{2}$

94. $\left(\dfrac{5}{6} - \dfrac{1}{3}\right) \cdot \dfrac{1}{3} + \dfrac{1}{2} \div \dfrac{9}{8}$

Solve.

95. In 2012, about $\dfrac{4}{25}$ of the total weight of mail delivered by the United States Postal Service was first-class mail. That same year, about $\dfrac{81}{200}$ of the total weight of mail delivered by the United States Postal Service was standard mail. Which of these two categories accounts for a greater portion of the mail handled by weight? (*Source:* U.S. Postal Service)

96. The National Park System (NPS) in the United States includes a wide variety of park types. National military parks account for $\dfrac{11}{500}$ of all NPS parks, and $\dfrac{9}{200}$ of NPS parks are classified as national preserves. Which category, national military park or national preserve, is bigger? (*Source:* National Park Service)

97. A recent survey reported that $\dfrac{2}{5}$ of the average college student's spending is for discretionary purchases, such as clothing, entertainment, and technology. About $\dfrac{13}{50}$ of the average college student's spending is for room and board. In which category, discretionary purchases or room and board, does the average college student spend more? (*Source:* Nationwide Bank)

98. On a normal work day in 2012, the average working parent in the U.S. spent $\dfrac{1}{20}$ of his or her day caring for others and $\dfrac{77}{240}$ of his or her day sleeping. Which did the average working parent spend more time doing, caring for others or sleeping? (*Source:* U.S. Bureau of Labor Statistics)

3.6 Fractions and Problem Solving

Objective A Solving Problems Containing Fractions or Mixed Numbers

Now that we know how to add, subtract, multiply, and divide fractions and mixed numbers, we can solve problems containing these numbers.

Don't forget the key words and phrases listed below that help indicate which operation to use. Also included are several words and phrases that translate to the symbol "=".

Objective

A Solve Problems by Performing Operations on Fractions or Mixed Numbers.

Addition (+)	Subtraction (−)	Multiplication (·)	Division (÷)	Equality (=)
sum	difference	product	quotient	equals
plus	minus	times	divide	is equal to
added to	subtract	multiply	shared equally	is/was
more than	less than	multiply by	among	yields
increased by	decreased by	of	divided by	
total	less	double/triple	divided into	

Recall the following problem-solving steps introduced in Section 1.8. They may be helpful to you:

Problem-Solving Steps

1. **UNDERSTAND** the problem. Some ways of doing this are to read and reread the problem, construct a drawing, and look for key words to identify an operation.
2. **TRANSLATE** the problem. That is, write the problem in short form using words, and then translate to numbers and symbols.
3. **SOLVE** the problem. It is helpful to estimate the solution by rounding. Then carry out the indicated operation from step 2.
4. **INTERPRET** the results. *Check* the proposed solution in the stated problem and *state* your conclusions. Write your results with the correct units attached.

In the first example, we find the volume of a box. Volume measures the space enclosed by a region and is measured in cubic units. We study volume further in a later chapter.

$$\text{Volume of a box} = \text{length} \cdot \text{width} \cdot \text{height}$$

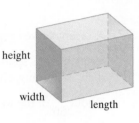

Helpful Hint

Remember:

Perimeter measures the distance around a figure. It is measured in **units.**

☐ Perimeter

Area measures the amount of surface of a figure. It is measured in **square units.**

▭ Area

Volume measures the amount of space enclosed by a region. It is measured in **cubic units.**

▱ Volume

Practice 1

Find the volume of a box that measures $4\frac{1}{3}$ feet by $1\frac{1}{2}$ feet by $3\frac{1}{3}$ feet.

$3\frac{1}{3}$ feet

$1\frac{1}{2}$ feet $4\frac{1}{3}$ feet

Answer

1. $21\frac{2}{3}$ cu ft

Example 1   Finding Volume of a Camcorder Box

Toshiba recently produced a small camcorder. It measures approximately $1\frac{1}{2}$ inches by $1\frac{3}{5}$ inches by $1\frac{7}{25}$ inches. Find the volume of a box with these dimensions. (*Note:* The camcorder weighs only 65 grams—about the weight of 65 standard paper clips.) (*Source: Guinness World Records*)

Solution:

1. **UNDERSTAND.** Read and reread the problem. The phrase "volume of a box" tells us what to do. The volume of a box is the product of its length, width, and height. Since we are multiplying, it makes no difference which measurement we call length, width, or height.

2. TRANSLATE.

In words:

volume of a box	is	length	·	width	·	height
↓	↓	↓		↓		↓

Translate: volume of a box $= 1\frac{1}{2}$ in. $\cdot$ $1\frac{3}{5}$ in. $\cdot$ $1\frac{7}{25}$ in.

3. SOLVE: Before we multiply, let's estimate by rounding each dimension to a whole number. The number $1\frac{1}{2}$ rounds to 2, $1\frac{3}{5}$ rounds to 2, and $1\frac{7}{25}$ rounds to 1, so our estimate is $2 \cdot 2 \cdot 1$ or 4 cubic inches.

$$1\frac{1}{2} \text{ in.} \cdot 1\frac{3}{5} \text{ in.} \cdot 1\frac{7}{25} \text{ in.} = \frac{3}{2} \cdot \frac{8}{5} \cdot \frac{32}{25} \quad \text{cubic inches}$$

$$= \frac{3 \cdot \overset{4}{\cancel{8}} \cdot 32}{\underset{1}{\cancel{2}} \cdot 5 \cdot 25} \quad \text{cubic inches}$$

$$= \frac{384}{125} \text{ or } 3\frac{9}{125} \quad \text{cubic inches}$$

Helpful Hint
Notice a short-cut taken when simplifying the fraction. Here, a common factor of 2 is recognized in 8 and 2. Then $8 \div 2$ is 4 in the numerator and $2 \div 2$ is 1 in the denominator.

4. INTERPRET. *Check* your work. The exact answer is close to our estimate, so it is reasonable. *State* your conclusion: The volume of a box that measures $1\frac{1}{2}$ inches by $1\frac{3}{5}$ inches by $1\frac{7}{25}$ inches is $3\frac{9}{125}$ cubic inches.

■ Work Practice 1

Example 2 Finding Unknown Length

Given the following diagram, find its total length.

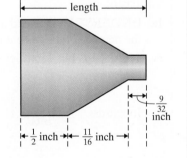

Practice 2

Given the following diagram, find its total width.

Solution:

1. UNDERSTAND. Read and reread the problem. Then study the diagram. The phrase "total length" tells us to add.

2. TRANSLATE. It makes no difference which length we call first, second, or third length.

In words:

total length	is	first length	+	second length	+	third length
↓	↓	↓		↓		↓

Translate: total length $= \frac{1}{2}$ in. $+ \frac{11}{16}$ in. $+ \frac{9}{32}$ in.

(Continued on next page)

Answer

2. 2 in.

3. SOLVE:

$$\frac{1}{2} + \frac{11}{16} + \frac{9}{32} = \frac{1 \cdot 16}{2 \cdot 16} + \frac{11 \cdot 2}{16 \cdot 2} + \frac{9}{32}$$

$$= \frac{16}{32} + \frac{22}{32} + \frac{9}{32}$$

$$= \frac{47}{32} \text{ or } 1\frac{15}{32}$$

4. INTERPRET. *Check* your work. *State* your conclusion: The total length is $1\frac{15}{32}$ inches.

■ Work Practice 2

Many problems require more than one operation to solve, as shown in the next application.

Practice 3

Suppose that 25 acres of land are purchased, but because of roads and wetlands concerns, $6\frac{2}{3}$ acres cannot be developed into lots. How many $\frac{5}{6}$-acre lots can the rest of the land be divided into?

Example 3 Acreage for Single-Family Home Lots

A contractor is considering buying land to develop a subdivision for single-family homes. Suppose she buys 44 acres and calculates that $4\frac{1}{4}$ acres of this land will be used for roads and a retention pond. How many $\frac{3}{4}$-acre lots can she sell using the rest of the acreage?

Solution:

1a. UNDERSTAND. Read and reread the problem. The phrase "using the rest of the acreage" tells us that initially we are to subtract.

2a. TRANSLATE. First, let's calculate the amount of acreage that can be used for lots.

In words:	acreage for lots	is	total acreage	minus	acreage for roads and a pond
	↓	↓	↓	↓	↓
Translate:	acreage for lots	=	44	−	$4\frac{1}{4}$

3a. SOLVE:

$$\begin{array}{r} 44 = 43\frac{4}{4} \\ -4\frac{1}{4} = -4\frac{1}{4} \\ \hline 39\frac{3}{4} \end{array}$$

1b. UNDERSTAND. Now that we know $39\frac{3}{4}$ acres can be used for lots, we calculate how many $\frac{3}{4}$ acres are in $39\frac{3}{4}$. This means that we divide.

Answer

3. 22 lots

2b. TRANSLATE.

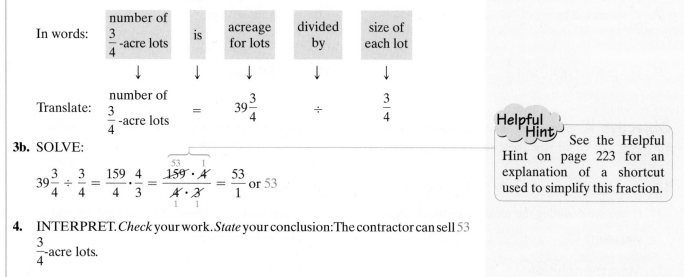

In words:

number of $\frac{3}{4}$-acre lots	is	acreage for lots	divided by	size of each lot
↓	↓	↓	↓	↓

Translate:

$$\text{number of } \frac{3}{4}\text{-acre lots} = 39\frac{3}{4} \div \frac{3}{4}$$

3b. SOLVE:

$$39\frac{3}{4} \div \frac{3}{4} = \frac{159}{4} \cdot \frac{4}{3} = \frac{\overset{53}{\cancel{159}} \cdot \overset{1}{\cancel{4}}}{\underset{1}{\cancel{4}} \cdot \underset{1}{\cancel{3}}} = \frac{53}{1} \text{ or } 53$$

> **Helpful Hint**
>
> See the Helpful Hint on page 223 for an explanation of a shortcut used to simplify this fraction.

4. INTERPRET. *Check* your work. *State* your conclusion: The contractor can sell 53 $\frac{3}{4}$-acre lots.

■ Work Practice 3

Vocabulary, Readiness & Video Check

Martin-Gay Interactive Videos   Watch the section lecture video and answer the following question.

Objective A 1. From the lecture before ⊞ Example 1, what's the purpose of interpreting the results when problem solving? ▶

See Video 3.6

3.6 **Exercise Set** MyMathLab® ▶

To prepare for problem solving, translate each phrase to an expression. Do not simplify the expression.

1. The sum of $\frac{1}{2}$ and $\frac{1}{3}$.

2. The product of $\frac{1}{2}$ and $\frac{1}{3}$.

3. The quotient of 20 and $6\frac{2}{5}$.

4. The difference of 20 and $6\frac{2}{5}$.

5. Subtract $\frac{5}{8}$ from $\frac{15}{16}$.

6. The total of $\frac{15}{36}$ and $\frac{18}{30}$.

7. $\frac{21}{68}$ increased by $\frac{7}{34}$.

8. $\frac{21}{68}$ decreased by $\frac{7}{34}$.

9. The product of $8\frac{1}{3}$ and $\frac{7}{9}$.

10. $37\frac{1}{2}$ divided by $9\frac{1}{2}$.

Objective A *Solve. Write any improper-fraction answers as mixed numbers. For Exercises 11 and 12, the solutions have been started for you. Write each answer in simplest form. See Examples 1 through 3.*

11. A recipe for brownies calls for $1\frac{2}{3}$ cups of sugar. If you are doubling the recipe, how much sugar do you need?

Start the solution:

1. UNDERSTAND the problem. Reread it as many times as needed.
2. TRANSLATE into an equation. (Fill in the blanks below.)

sugar needed	is	double	recipe amount of sugar
↓	↓	↓	↓
sugar needed	=	$2 \cdot$ *	_____

Finish with:

3. SOLVE
4. INTERPRET

*Note: Another way to double a number is to add the number to the same number.

12. A nacho recipe calls for $\frac{1}{3}$ cup cheddar cheese and $\frac{1}{2}$ cup jalapeño cheese. Find the total amount of cheese in the recipe.

Start the solution:

1. UNDERSTAND the problem. Reread it as many times as needed.
2. TRANSLATE into an equation. (Fill in the blanks below.)

total cheese	is	how much cheddar	added to	how much jalapeño cheese
↓	↓	↓	↓	↓
total cheese	=	_____	+	_____

Finish with:

3. SOLVE
4. INTERPRET

13. A decorative wall in a garden is to be built using bricks that are $2\frac{3}{4}$ inches wide and mortar joints that are $\frac{1}{2}$ inch wide. Use the diagram to find the height of the wall.

14. Suppose that the contractor building the wall in Exercise 13 decides that he wants one more layer of bricks with a mortar joint below and above that layer. Find the new height of the wall.

height   Mortar joint

15. Doug and Claudia Scaggs recently drove $290\frac{1}{4}$ miles on $13\frac{1}{2}$ gallons of gas. Calculate how many miles per gallon they get in their vehicle.

16. A contractor is using 18 acres of his land to sell $\frac{3}{4}$-acre lots. How many lots can he sell?

▶ **17.** The life expectancy of a circulating coin is 30 years. The life expectancy of a circulating dollar bill is only $\frac{1}{20}$ as long. Find the life expectancy of circulating paper money. (*Source:* The U.S. Mint)

18. The Indian head one-cent coin of 1859–1864 was made of copper and nickel only. If $\frac{3}{25}$ of the coin was nickel, what part of the whole coin was copper? (*Source:* The U.S. Mint)

19. The Gauge Act of 1846 set the standard gauge for U.S. railroads at $56\frac{1}{2}$ inches. (See figure.) If the standard gauge in Spain is $65\frac{9}{10}$ inches, how much wider is Spain's standard gauge than the U.S. standard gauge? (*Source:* San Diego Railroad Museum)

20. The standard railroad track gauge (see figure) in Spain is $65\frac{9}{10}$ inches, while in neighboring Portugal it is $65\frac{11}{20}$ inches. Which gauge is wider and by how much? (*Source:* San Diego Railroad Museum)

Track gauge (U.S. $56\frac{1}{2}$ inches)

$\frac{5}{8}$ inch

Point of measurement of gauge

21. Mark Nguyen is a tailor making costumes for a play. He needs enough material for 1 large shirt that requires $1\frac{1}{2}$ yards of material and 5 small shirts that each require $\frac{3}{4}$ yard of material. He finds a 5-yard remnant of material on sale. Is 5 yards of material enough to make all 6 shirts? If not, how much more material does he need?

22. A beanbag manufacturer makes a large beanbag requiring $4\frac{1}{3}$ yards of vinyl fabric and a smaller size requiring $3\frac{1}{4}$ yards. A 100-yard roll of fabric is to be used to make 12 large beanbags. How many smaller beanbags can be made from the remaining piece?

23. A plumber has a 10-foot piece of PVC pipe. How many $\frac{9}{5}$-foot pieces can be cut from the 10-foot piece?

24. A carpenter has a 12-foot board to be used to make windowsills. If each sill requires $2\frac{5}{16}$ feet, how many sills can be made from the 12-foot board?

25. Suppose that the cross section of a piece of pipe looks like the diagram shown. Find the total outer diameter.

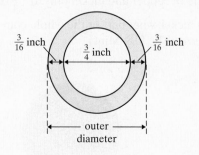

$\frac{3}{16}$ inch $\frac{3}{4}$ inch $\frac{3}{16}$ inch

outer diameter

26. Suppose that the cross section of a piece of pipe looks like the diagram shown. Find the inner diameter.

$\frac{7}{8}$ foot

$\frac{1}{16}$ foot $\frac{1}{16}$ foot

inner diameter

27. A recipe for chocolate chip cookies calls for $2\frac{1}{2}$ cups of flour. If you are making $1\frac{1}{2}$ recipes, how many cups of flour are needed?

28. A recipe for a homemade cleaning solution calls for $1\frac{3}{4}$ cups of vinegar. If you are tripling the recipe, how much vinegar is needed?

△ **29.** The Polaroid Pop Shot, the world's first disposable instant camera, can take color photographs measuring $4\frac{1}{2}$ inches by $2\frac{1}{2}$ inches. Find the area of a photograph. (*Source: Guinness World Records*)

△ **30.** A model for a proposed computer chip measures $\frac{3}{4}$ inch by $1\frac{1}{4}$ inches. Find its area.

31. A total solar eclipse on March 9, 2016, will last $4\frac{3}{20}$ minutes and can be viewed from Sumatra, Borneo, and Sulawesi. The next total solar eclipse, on August 21, 2017, will last $2\frac{2}{3}$ minutes and can be viewed in the northern Pacific Ocean, United States, and southern Atlantic Ocean. How much longer is the 2016 eclipse? (*Source:* NASA/Goddard Space Flight Center)

32. The pole vault record for the 1908 Summer Olympics was $12\frac{1}{6}$ feet. The record for the 2012 Summer Olympics was a little over $19\frac{9}{16}$ feet. Find the difference in the heights. (*Source:* International Olympic Committee)

△ **33.** The world's lightest handheld mobile phone measures approximately $2\frac{4}{5}$ inches by $1\frac{1}{2}$ inches by $\frac{3}{10}$ inch. Find the volume of a box with those dimensions. (*Source: Guinness World Records*)

△ **34.** Early cell phones were large and heavy. One early model measured approximately 8 inches by $2\frac{1}{2}$ inches by $2\frac{1}{2}$ inches. Find the volume of a box with those dimensions.

35. A stack of $\frac{5}{8}$-inch-wide sheetrock has a height of $41\frac{7}{8}$ inches. How many sheets of sheetrock are in the stack?

36. A stack of $\frac{5}{4}$-inch-thick books has a height of $28\frac{3}{4}$ inches. How many books are in the stack?

37. William Arcencio is remodeling his home. In order to save money, he is upgrading the plumbing himself. He needs 12 pieces of copper tubing, each $\frac{3}{4}$ of a foot long.

  a. If he has a 10-foot piece of tubing, will that be enough?

  b. How much more does he need or how much tubing will he have left over?

38. Trishelle Dallam is building a bookcase. Each shelf will be $2\frac{3}{8}$ feet long, and she needs wood for 7 shelves.

  a. How many shelves can she cut from an 8-foot board?

  b. Based on your answer for part a, how many 8-foot boards will she need?

Recall that the average of a list of numbers is their sum divided by the number of numbers in the list. Use this procedure for Exercises 39 and 40.

39. A female lion had 4 cubs. They weighed $2\frac{1}{8}$, $2\frac{7}{8}$, $3\frac{1}{4}$, and $3\frac{1}{2}$ pounds. What is the average cub weight?

40. Three brook trout were caught, tagged, and then released. They weighed $1\frac{1}{2}$, $1\frac{3}{8}$, and $1\frac{7}{8}$ pounds. Find their average weight.

Find the area and perimeter of each figure.

△ **41.**

| Rectangle | $\frac{3}{16}$ inch |

$\frac{3}{8}$ inch

△ **42.**

| Square | $1\frac{7}{10}$ mile |

△ **43.**

| Square | $\frac{5}{9}$ meter |

△ **44.**

| Rectangle | 5 inches |

$3\frac{1}{2}$ inches

For Exercises 45 through 48, see the diagram. (Source: www.usflag.org)

45. The length of the U.S. flag is $1\frac{9}{10}$ its width. If a flag is being designed with a width of $2\frac{1}{2}$ feet, find its length.

46. The width of the Union portion of the U.S. flag is $\frac{7}{13}$ of the width of the flag. If a flag is being designed with a width of $2\frac{1}{2}$ feet, find the width of the Union portion.

$\frac{19}{25}$ of total width

$\frac{7}{13}$ of width

width = $2\frac{1}{2}$ feet

$1\frac{9}{10}$ of width

47. There are 13 stripes of equal width in the flag. If the width of a flag is $2\frac{1}{2}$ feet, find the width of each stripe.

48. The length of the Union portion of the flag is $\frac{19}{25}$ of the total width. If the width of a flag is $2\frac{1}{2}$ feet, find the length of the Union portion.

Review

Simplify. See Section 1.9.

49. $\sqrt{9}$

50. $\sqrt{4}$

51. 9^2

52. 4^2

53. $8 \div 4 \cdot 2$

54. $20 \div 5 \cdot 2$

55. $3^2 - 2^2 + 5^2$

56. $8^2 - 6^2 + 7^2$

57. $5 + 3[14 - (12 \div 3)]$

58. $7 + 2[20 - (35 \div 5)]$

Concept Extensions

59. Suppose you are finding the average of $7\frac{1}{9}$ and $12\frac{19}{20}$. Can the average be $1\frac{1}{2}$? Can the average be $15\frac{1}{2}$? Why or why not?

60. Suppose that you are finding the average of $1\frac{3}{4}$, $1\frac{1}{8}$, and $1\frac{9}{10}$. Can the average be $2\frac{1}{4}$? Can the average be $\frac{15}{16}$? Why or why not?

The figure shown is for Exercises 61 and 62.

△ **61.** Find the area of the figure. (*Hint:* The area of the figure can be found by finding the sum of the areas of the rectangles shown in the figure.)

△ **62.** Find the perimeter of the figure.

Rectangle $3\frac{8}{9}$ feet

$6\frac{2}{3}$ feet   Rectangle $4\frac{4}{9}$ feet

$6\frac{7}{9}$ feet

63. On a particular day, 240 customers ate lunch at a local restaurant. If $\frac{3}{10}$ of them ordered a $7 lunch, $\frac{5}{12}$ of them ordered a $5 lunch, and the remaining customers ordered a $9 lunch, how many customers ordered a $9 lunch?

64. Scott purchased a case of 24 apples. He used $\frac{1}{3}$ of them to make an apple pie, $\frac{1}{4}$ of them to make apple crisp, and kept the rest for after-school snacks for his children. How many apples did Scott keep for snacks?

65. Coins were practically made by hand in the late 1700s. Back then, it took 3 years to produce our nation's first million coins. Today, it takes only $\frac{11}{13,140}$ as long to produce the same amount. Calculate how long it takes today in hours to produce one million coins. (*Hint:* First convert 3 years to equivalent hours. Use 365 days for each of the 3 years.) (*Source:* The U.S. Mint)

66. The largest suitcase measures $13\frac{1}{3}$ feet by $8\frac{3}{4}$ feet by $4\frac{4}{25}$ feet. Find its volume. (*Source: Guinness World Records*)

Chapter 3 Group Activity

Sections 3.1–3.6

This activity may be completed by working in groups or individually.

Lobsters are normally classified by weight. Use the weight classification table to answer the questions in this activity.

Classification of Lobsters	
Class	**Weight (in Pounds)**
Chicken	1 to $1\frac{1}{8}$
Quarter	$1\frac{1}{4}$
Half	$1\frac{1}{2}$ to $1\frac{3}{4}$
Select	$1\frac{3}{4}$ to $2\frac{1}{2}$
Large select	$2\frac{1}{2}$ to $3\frac{1}{2}$
Jumbo	Over $3\frac{1}{2}$

(*Source:* The Maine Lobster Promotion Council)

1. A lobster fisher has kept four lobsters from a lobster trap. Classify each lobster if they have the following weights:

    a. $1\frac{7}{8}$ pounds   **b.** $1\frac{9}{16}$ pounds

    c. $2\frac{3}{4}$ pounds   **d.** $2\frac{3}{8}$ pounds

2. A recipe requires 5 pounds of lobster. Using the minimum weight for each class, decide whether a chicken, half, and select lobster will be enough for the recipe, and explain your reasoning. If not, suggest a better choice of lobsters to meet the recipe requirements.

3. A lobster market customer has selected two chickens, a select, and a large select. What is the most that these four lobsters could weigh? What is the least that these four lobsters could weigh?

4. A lobster market customer wishes to buy four quarters. If lobsters sell for $7 per pound, how much will the customer owe for her purchase?

5. Why do you think there is no classification for lobsters weighing under 1 pound?

Chapter 3 Vocabulary Check

Fill in each blank with one of the words or phrases listed below.

equivalent	least common multiple	exponent	unlike
mixed number	<	>	least common denominator	like

1. Fractions that have the same denominator are called _____ fractions.

2. The _____ is the smallest number that is a multiple of all numbers in a list of numbers.

3. _____ fractions represent the same portion of a whole.

4. A(n)_____ has a whole number part and a fraction part.

5. The symbol _____ means is greater than.

6. The symbol _____ means is less than.

7. The LCM of the denominators in a list of fractions is called the _____.

8. Fractions that have different denominators are called _____ fractions.

9. A shorthand notation for repeated multiplication of the same factor is a(n) _____.

Helpful Hint

▶ Are you preparing for your test? Don't forget to take the Chapter 3 Test on page 240. Then check your answers at the back of the text and use the Chapter Test Prep Videos to see the fully worked-out solutions to any of the exercises you want to review.

3 Chapter Highlights

Definitions and Concepts	Examples
Section 3.1 Adding and Subtracting Like Fractions	
Fractions that have the same denominator are called **like fractions.**	$\frac{1}{3}$ and $\frac{2}{3}$; $\frac{5}{7}$ and $\frac{6}{7}$

To add or subtract like fractions, combine the numerators and place the sum or difference over the common denominator.	$\frac{2}{7} + \frac{3}{7} = \frac{5}{7}$ ← Add the numerators.
	    ← Keep the common denominator.
	
	$\frac{7}{8} - \frac{4}{8} = \frac{3}{8}$ ← Subtract the numerators.
	    ← Keep the common denominator.

Section 3.2 Least Common Multiple	
The **least common multiple (LCM)** is the smallest number that is a multiple of all numbers in a list of numbers.	The LCM of 2 and 6 is 6 because 6 is the smallest number that is a multiple of both 2 and 6.
Method 1 for Finding the LCM of a List of Numbers Using Multiples	Find the LCM of 4 and 6 using Method 1.
Step 1: Write the multiples of the largest number (starting with the number itself) until a multiple common to all numbers in the list is found.	$6 \cdot 1 = 6$ Not a multiple of 4 $6 \cdot 2 = 12$ A multiple of 4 The LCM is 12.
Step 2: The multiple found in step 1 is the LCM.	

Definitions and Concepts	Examples

Method 2 for Finding the LCM of a List of Numbers Using Prime Factorization	Find the LCM of 6 and 20 using Method 2.
Step 1: Write the prime factorization of each number. **Step 2:** For each different prime factor in step 1, circle the greatest number of times that factor occurs in any one factorization. **Step 3:** The LCM is the product of the circled factors.	$6 = 2 \cdot \boxed{3}$ $20 = \boxed{2 \cdot 2} \cdot \boxed{5}$ The LCM is $2 \cdot 2 \cdot 3 \cdot 5 = 60$
Equivalent fractions represent the same portion of a whole.	Write an equivalent fraction with the indicated denominator. $$\frac{2}{8} = \frac{}{16}$$ $$\frac{2 \cdot 2}{8 \cdot 2} = \frac{4}{16}$$

Section 3.3   Adding and Subtracting Unlike Fractions

To Add or Subtract Fractions with Unlike Denominators	Add: $\dfrac{3}{20} + \dfrac{2}{5}$
Step 1: Find the LCD.	**Step 1:** The LCD of 20 and 5 is 20.
Step 2: Write each fraction as an equivalent fraction whose denominator is the LCD.	**Step 2:** $\dfrac{3}{20} = \dfrac{3}{20}; \dfrac{2}{5} = \dfrac{2}{5} \cdot \dfrac{4}{4} = \dfrac{8}{20}$
Step 3: Add or subtract the like fractions.	**Step 3:** $\dfrac{3}{20} + \dfrac{2}{5} = \dfrac{3}{20} + \dfrac{8}{20} = \dfrac{11}{20}$
Step 4: Write the sum or difference in simplest form.	**Step 4:** $\dfrac{11}{20}$ is in simplest form.

Section 3.4   Adding and Subtracting Mixed Numbers

To add or subtract with mixed numbers, add or subtract the fractions and then add or subtract the whole numbers.	Add: $2\dfrac{1}{2} + 5\dfrac{7}{8}$ $2\dfrac{1}{2} = 2\dfrac{4}{8}$ $+5\dfrac{7}{8} = +5\dfrac{7}{8}$ $\rule{4cm}{0.4pt}$ $7\dfrac{11}{8} = 7 + 1\dfrac{3}{8} = 8\dfrac{3}{8}$

Section 3.5   Order, Exponents, and the Order of Operations

To compare like fractions, compare the numerators. The order of the fractions is the same as the order of the numerators.	Compare $\dfrac{3}{10}$ and $\dfrac{4}{10}$. $\dfrac{3}{10} < \dfrac{4}{10}$ since $3 < 4$

(continued)

Definitions and Concepts	Examples

Section 3.5   Order, Exponents, and the Order of Operations (*continued*)

To compare unlike fractions, first write the fractions as like fractions. Then the fraction with the greater numerator is the greater fraction.

Compare $\frac{2}{5}$ and $\frac{3}{7}$.

$$\frac{2}{5} = \frac{2}{5} \cdot \frac{7}{7} = \frac{14}{35} \qquad \frac{3}{7} = \frac{3}{7} \cdot \frac{5}{5} = \frac{15}{35}$$

Since $14 < 15$, then

$$\frac{14}{35} < \frac{15}{35} \quad \text{or} \quad \frac{2}{5} < \frac{3}{7}$$

Exponents mean repeated multiplication whether the base is a whole number or a fraction.

$$\left(\frac{1}{2}\right)^3 = \frac{1}{2} \cdot \frac{1}{2} \cdot \frac{1}{2} = \frac{1}{8}$$

Order of Operations

Perform each indicated operation.

1. Perform all operations within parentheses ( ), brackets [ ], or other grouping symbols such as square roots or fraction bars.

2. Evaluate any expressions with exponents.

3. Multiply or divide in order from left to right.

4. Add or subtract in order from left to right.

$$\frac{1}{2} + \frac{2}{3} \cdot \frac{1}{5} = \frac{1}{2} + \frac{2}{15} \qquad \text{Multiply.}$$

$$= \frac{1}{2} \cdot \frac{15}{15} + \frac{2}{15} \cdot \frac{2}{2} \qquad \text{The LCD is 30.}$$

$$= \frac{15}{30} + \frac{4}{30}$$

$$= \frac{19}{30} \qquad \text{Add.}$$

Section 3.6   Fractions and Problem Solving

Problem-Solving Steps

A stack of $\frac{3}{4}$-inch plywood has a height of $50\frac{1}{4}$ inches. How many sheets of plywood are in the stack?

1. UNDERSTAND the problem.

1. UNDERSTAND. Read and reread the problem. We want to know how many $\frac{3}{4}$'s are in $50\frac{1}{4}$, so we divide.

2. TRANSLATE the problem.

2. TRANSLATE.

number of sheets in stack	is	height of stack	÷	height of a sheet

$$\text{number of sheets in stack} = 50\frac{1}{4} \div \frac{3}{4}$$

3. SOLVE the problem.

3. SOLVE. $50\frac{1}{4} \div \frac{3}{4} = \frac{201}{4} \cdot \frac{4}{3}$

$$= \frac{\overset{67}{201} \cdot \overset{1}{4}}{\underset{1}{4} \cdot \underset{1}{3}}$$

$$= 67$$

4. INTERPRET the results.

4. INTERPRET. *Check* your work and *state* your conclusion: There are 67 sheets of plywood in the stack.

(3.1) *Add or subtract as indicated. Simplify your answers.*

1. $\dfrac{7}{11} + \dfrac{3}{11}$

2. $\dfrac{4}{50} + \dfrac{2}{50}$

3. $\dfrac{11}{15} - \dfrac{1}{15}$

4. $\dfrac{4}{21} - \dfrac{1}{21}$

5. $\dfrac{4}{15} + \dfrac{3}{15} + \dfrac{2}{15}$

6. $\dfrac{3}{20} + \dfrac{7}{20} + \dfrac{2}{20}$

7. $\dfrac{1}{12} + \dfrac{11}{12}$

8. $\dfrac{3}{4} + \dfrac{1}{4}$

9. $\dfrac{11}{25} + \dfrac{6}{25} + \dfrac{2}{25}$

10. $\dfrac{4}{21} + \dfrac{1}{21} + \dfrac{11}{21}$

Solve.

11. One evening Mark Alorenzo did $\dfrac{3}{8}$ of his homework before supper, another $\dfrac{2}{8}$ of it while his children did their homework, and $\dfrac{1}{8}$ after his children went to bed. What part of his homework did he do that evening?

△ 12. The Simpsons will be fencing in their land, which is in the shape of a rectangle. In order to do this, they need to find its perimeter. Find the perimeter of their land.

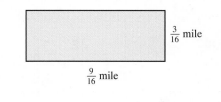

$\frac{3}{16}$ mile

$\frac{9}{16}$ mile

(3.2) *Find the LCM of each list of numbers.*

13. $5, 11$

14. $20, 30$

15. $20, 24$

16. $16, 5$

17. $12, 21, 63$

18. $6, 8, 18$

Write each fraction as an equivalent fraction with the given denominator.

19. $\dfrac{7}{8} = \dfrac{}{64}$

20. $\dfrac{2}{3} = \dfrac{}{30}$

21. $\dfrac{7}{11} = \dfrac{}{33}$

22. $\dfrac{10}{13} = \dfrac{}{26}$

23. $\dfrac{4}{15} = \dfrac{}{60}$

24. $\dfrac{5}{12} = \dfrac{}{60}$

(3.3) *Add or subtract as indicated. Simplify your answers.*

25. $\dfrac{7}{18} + \dfrac{2}{9}$

26. $\dfrac{4}{15} + \dfrac{1}{5}$

27. $\dfrac{4}{13} - \dfrac{1}{26}$

28. $\dfrac{7}{12} - \dfrac{1}{9}$

29. $\dfrac{1}{3} + \dfrac{9}{14}$

30. $\dfrac{7}{18} + \dfrac{5}{24}$

31. $\dfrac{11}{15} - \dfrac{4}{9}$

32. $\dfrac{9}{14} - \dfrac{3}{35}$

Find the perimeter of each figure.

△ **33.**

$\frac{2}{9}$ meter   Rectangle

$\frac{5}{6}$ meter

△ **34.** $\frac{1}{5}$ foot $\frac{3}{5}$ foot

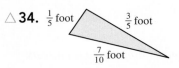

$\frac{7}{10}$ foot

35. Find the difference in length of two scarves if one scarf is $\frac{5}{12}$ of a yard long and the other is $\frac{2}{3}$ of a yard long.

36. Truman Kalzote cleaned $\frac{3}{5}$ of his house yesterday and $\frac{1}{10}$ of it today. How much of the house has been cleaned?

(3.4) *Add or subtract as indicated. Simplify your answers.*

37. $31\frac{2}{7} + 14\frac{10}{21}$

38. $24\frac{4}{5} + 35\frac{1}{5}$

39. $69\frac{5}{22} - 36\frac{7}{11}$

40. $36\frac{3}{20} - 32\frac{5}{6}$

41.
$\begin{aligned} & 29\frac{2}{9} \\ & 27\frac{7}{18} \\ +\, & 54\frac{2}{3} \\ \hline \end{aligned}$

42.
$\begin{aligned} & 7\frac{3}{8} \\ & 9\frac{5}{6} \\ +\, & 3\frac{1}{12} \\ \hline \end{aligned}$

43.
$\begin{aligned} & 9\frac{3}{5} \\ -\, & 4\frac{1}{7} \\ \hline \end{aligned}$

44.
$\begin{aligned} & 8\frac{3}{11} \\ -\, & 5\frac{1}{5} \\ \hline \end{aligned}$

Solve.

45. The average annual snowfall at a certain ski resort is $62\frac{3}{10}$ inches. Last year it had $54\frac{1}{2}$ inches. How many inches below average was last year's snowfall?

46. Dinah's homemade canned peaches contain $15\frac{3}{5}$ ounces per can. A can of Amy's brand contains $15\frac{5}{8}$ ounces per can. Amy's brand weighs how much more than Dinah's?

△ **47.** Find the perimeter of a sheet of shelf paper needed to fit exactly a square drawer $1\frac{1}{4}$ feet long on each side.

$1\frac{1}{4}$ feet

△ **48.** Find the perimeter of a rectangular sheet of gift wrap that is $2\frac{1}{4}$ feet by $3\frac{1}{3}$ feet.

$2\frac{1}{4}$ feet

$3\frac{1}{3}$ feet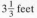

(3.5) *Insert* $<$ *or* $>$ *to form a true statement.*

49. $\dfrac{5}{11}$ $\quad$ $\dfrac{6}{11}$

50. $\dfrac{4}{35}$ $\quad$ $\dfrac{3}{35}$

51. $\dfrac{5}{14}$ $\quad$ $\dfrac{16}{42}$

52. $\dfrac{6}{35}$ $\quad$ $\dfrac{17}{105}$

53. $\dfrac{7}{8}$ $\quad$ $\dfrac{6}{7}$

54. $\dfrac{7}{10}$ $\quad$ $\dfrac{2}{3}$

Evaluate each expression. Use the order of operations to simplify.

55. $\left(\dfrac{3}{7}\right)^2$

56. $\left(\dfrac{4}{5}\right)^3$

57. $\left(\dfrac{1}{2}\right)^4 \cdot \left(\dfrac{3}{5}\right)^2$

58. $\left(\dfrac{1}{3}\right)^2 \cdot \left(\dfrac{9}{10}\right)^2$

59. $\dfrac{5}{13} \div \dfrac{1}{2} \cdot \dfrac{4}{5}$

60. $\dfrac{8}{11} \div \dfrac{1}{3} \cdot \dfrac{11}{12}$

61. $\left(\dfrac{6}{7} - \dfrac{3}{14}\right)^2$

62. $\left(\dfrac{1}{3}\right)^2 - \dfrac{2}{27}$

63. $\dfrac{8}{9} - \dfrac{1}{8} \div \dfrac{3}{4}$

64. $\dfrac{9}{10} - \dfrac{1}{9} \div \dfrac{2}{3}$

65. $\dfrac{2}{7} \cdot \left(\dfrac{1}{5} + \dfrac{3}{10}\right)$

66. $\dfrac{9}{10} \div \left(\dfrac{1}{5} + \dfrac{1}{20}\right)$

67. $\left(\dfrac{3}{4} + \dfrac{1}{2}\right) \div \left(\dfrac{4}{9} + \dfrac{1}{3}\right)$

68. $\left(\dfrac{3}{8} - \dfrac{1}{16}\right) \div \left(\dfrac{1}{2} - \dfrac{1}{8}\right)$

69. $\dfrac{6}{7} \cdot \dfrac{5}{2} - \dfrac{3}{4} \cdot \dfrac{1}{2}$

70. $\dfrac{9}{10} \cdot \dfrac{1}{3} - \dfrac{2}{5} \cdot \dfrac{1}{11}$

Find the average of each list of fractions.

71. $\dfrac{2}{3}, \dfrac{5}{6}, \dfrac{1}{9}$

72. $\dfrac{4}{5}, \dfrac{9}{10}, \dfrac{3}{20}$

(3.6)

73. Our solar system has 146 known and officially confirmed moons. The planet Jupiter can claim $\frac{25}{73}$ of these moons. How many moons does Jupiter have? (*Source:* NASA)

74. James Hardaway just bought $5\frac{7}{8}$ acres of land adjacent to the $9\frac{3}{4}$ acres he already owned. How much land does he now own?

Find the unknown measurements.

△ **75.**

△ **76.**

Find the perimeter and area of each rectangle. Attach the proper units to each. Remember that perimeter is measured in units and area is measured in square units.

△ **77.**

△ **78.**

Mixed Review

Find the LCM of each list of numbers.

79. $15, 30, 45$

80. $6, 15, 20$

Write each fraction as an equivalent fraction with the given denominator.

81. $\dfrac{5}{6} = \dfrac{}{48}$

82. $\dfrac{7}{8} = \dfrac{}{72}$

Add or subtract as indicated. Simplify your answers.

83. $\dfrac{5}{12} - \dfrac{3}{12}$

84. $\dfrac{3}{10} - \dfrac{1}{10}$

85. $\dfrac{2}{3} + \dfrac{1}{4}$

86. $\dfrac{5}{11} + \dfrac{2}{55}$

87. $7\dfrac{3}{4}$
$+5\dfrac{2}{3}$
___

88. $2\dfrac{7}{8}$
$+9\dfrac{1}{2}$
___

89. $12\dfrac{3}{5}$
$-9\dfrac{1}{7}$
___

90. $32\dfrac{10}{21}$
$-24\dfrac{3}{7}$
___

Evaluate each expression. Use the order of operations to simplify.

91. $\dfrac{2}{5} + \left(\dfrac{2}{5}\right)^2 - \dfrac{3}{25}$

92. $\dfrac{1}{4} + \left(\dfrac{1}{2}\right)^2 - \dfrac{3}{8}$

93. $\left(\dfrac{5}{6} - \dfrac{3}{4}\right)^2$

94. $\left(2 - \dfrac{2}{3}\right)^3$

95. $\dfrac{2}{3} \div \left(\dfrac{3}{5} + \dfrac{5}{3}\right)$

96. $\dfrac{3}{8} \cdot \left(\dfrac{2}{3} - \dfrac{4}{9}\right)$

Insert $<$ or $>$ to form a true statement.

97. $\dfrac{3}{14}$ __ $\dfrac{2}{3}$

98. $\dfrac{7}{23}$ __ $\dfrac{3}{16}$

Solve.

99. Gregor Krowsky studied math for $\dfrac{3}{8}$ of an hour and geography for $\dfrac{1}{8}$ of an hour. How long did he study?

100. Two packages to be mailed weigh $3\dfrac{3}{4}$ pounds and $2\dfrac{3}{5}$ pounds. Find their combined weight.

101. A ribbon $5\dfrac{1}{2}$ yards long is cut from a reel of ribbon with 50 yards on it. Find the length of the piece remaining on the reel.

102. Linda Taneff has a board that is $10\dfrac{2}{3}$ feet in length. She plans to cut it into 5 equal lengths to use for a bookshelf. Find the length of each piece.

103. A recipe for pico de gallo calls for $1\dfrac{1}{2}$ tablespoons of cilantro. Five recipes will be made for a charity event. How much cilantro is needed?

104. Beryl Goldstein mixed $\dfrac{5}{8}$ of a gallon of water with $\dfrac{1}{8}$ of a gallon of punch concentrate. Then she and her friends drank $\dfrac{3}{8}$ of a gallon of the punch. How much of the punch was left?

Chapter 3 | Test

Step-by-step test solutions are found on the Chapter Test Prep Videos. Where available: MyMathLab® or You Tube™

Answers

1. Find the LCM of 4 and 15.

2. Find the LCM of 8, 9, and 12.

Insert < or > to form a true statement.

3. $\dfrac{5}{6}$ ___ $\dfrac{26}{30}$

4. $\dfrac{7}{8}$ ___ $\dfrac{8}{9}$

1. _____

2. _____

3. _____

Perform each indicated operation. Simplify your answers.

4. _____

5. $\dfrac{7}{9} + \dfrac{1}{9}$

6. $\dfrac{8}{15} - \dfrac{2}{15}$

7. $\dfrac{9}{10} + \dfrac{2}{5}$

5. _____

6. _____

7. _____

8. $\dfrac{1}{6} + \dfrac{3}{14}$

9. $\dfrac{7}{8} - \dfrac{1}{3}$

10. $\dfrac{17}{21} - \dfrac{1}{7}$

8. _____

9. _____

10. _____

11. $\dfrac{9}{20} + \dfrac{2}{3}$

12. $\dfrac{16}{25} - \dfrac{1}{2}$

13. $\dfrac{11}{12} + \dfrac{3}{8} + \dfrac{5}{24}$

11. _____

12. _____

13. _____

14. $3\dfrac{7}{8}$
$7\dfrac{2}{5}$
$+2\dfrac{3}{4}$

15. $8\dfrac{2}{9}$
12
$+10\dfrac{1}{15}$

16. $5\dfrac{1}{6}$
$-3\dfrac{7}{8}$

14. _____

15. _____

16. _____

17. _____

18. _____

17. 19
$-2\dfrac{3}{11}$

18. $\dfrac{2}{7} \cdot \left(6 - \dfrac{1}{6}\right)$

19. $\left(\dfrac{2}{3}\right)^4$

19. _____

20. $\dfrac{1}{2} \div \dfrac{2}{3} \cdot \dfrac{3}{4}$

21. $\left(\dfrac{4}{5}\right)^2 + \left(\dfrac{1}{2}\right)^3$

22. $\left(\dfrac{3}{4}\right)^2 \div \left(\dfrac{2}{3} + \dfrac{5}{6}\right)$

23. Find the average of $\dfrac{5}{6}, \dfrac{4}{3}$, and $\dfrac{7}{12}$.

Solve.

24. A carpenter cuts a piece $2\dfrac{3}{4}$ feet long from a cedar plank that is $6\dfrac{1}{2}$ feet long. How long is the remaining piece?

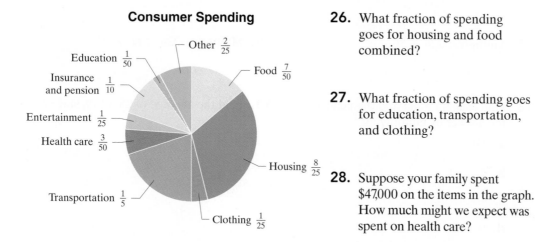

$6\frac{1}{2}$ feet

$2\frac{3}{4}$ feet

25. A small airplane used $58\dfrac{3}{4}$ gallons of fuel on a $7\dfrac{1}{2}$-hour trip. How many gallons of fuel were used for each hour?

The circle graph below shows us how the average consumer spends money. For example, $\dfrac{7}{50}$ of your spending goes for food. Use this information for Exercises 26 through 28.

Consumer Spending

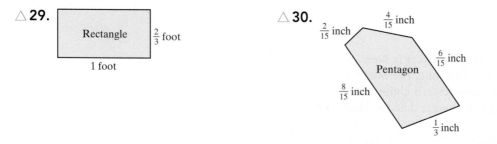

Other $\frac{2}{25}$

Education $\frac{1}{50}$

Insurance and pension $\frac{1}{10}$

Entertainment $\frac{1}{25}$

Health care $\frac{3}{50}$

Transportation $\frac{1}{5}$

Food $\frac{7}{50}$

Housing $\frac{8}{25}$

Clothing $\frac{1}{25}$

26. What fraction of spending goes for housing and food combined?

27. What fraction of spending goes for education, transportation, and clothing?

28. Suppose your family spent $47,000 on the items in the graph. How much might we expect was spent on health care?

Find the perimeter of each figure. For Exercise 29, find the area also.

△ **29.**

Rectangle $\frac{2}{3}$ foot

1 foot

△ **30.**

$\frac{4}{15}$ inch

$\frac{2}{15}$ inch

$\frac{6}{15}$ inch

Pentagon

$\frac{8}{15}$ inch

$\frac{1}{3}$ inch

20. _____

21. _____

22. _____

23. _____

24. _____

25. _____

26. _____

27. _____

28. _____

29. _____

30. _____

Answers

1. _____

2. _____

3. _____

4. _____

5. _____

6. _____

7. _____

8. _____

9. _____

10. _____

11. _____

12. _____

13. a. _____

b. _____

c. _____

d. _____

e. _____

f. _____

14. _____

15. _____

16. _____

Write each number in words.

1. 85

2. 107

3. 126

4. 5026

5. Add: 23 + 136

6. Find the perimeter.

3 in.   7 in.   9 in.

7. Subtract: 543 − 29. Then check by adding.

8. Divide: 3268 ÷ 27

9. Round 278,362 to the nearest thousand.

10. Find all the factors of 30.

11. Multiply: 236 × 86

12. Multiply: 236 × 86 × 0

13. Find each quotient and then check the answer by multiplying.

a. $1\overline{)7}$

b. 12 ÷ 1

c. $\dfrac{6}{6}$

d. 9 ÷ 9

e. $\dfrac{20}{1}$

f. $18\overline{)18}$

14. Find the average of 25, 17, 19, and 39.

15. The Hudson River in New York State is 306 miles long. The Snake River, in the northwestern United States, is 732 miles longer than the Hudson River. How long is the Snake River? (*Source:* U.S. Department of the Interior)

16. Evaluate: $\sqrt{121}$

Evaluate.

17. 9^2 **18.** 5^3 **19.** 3^4 **20.** 10^3

Write the shaded part of each diagram as an improper fraction and a mixed number.

21.   **22.**

23.   **24.**

25. Of the numbers 3, 9, 11, 17, 26, which are prime and which are composite?

26. Simplify: $\dfrac{6^2 + 4 \cdot 4 + 2^3}{37 - 5^2}$

27. Find the prime factorization of 180.

28. Find the difference of 87 and 25.

29. Write $\dfrac{72}{26}$ in simplest form.

30. Write $9\dfrac{7}{8}$ as an improper fraction.

31. Determine whether $\dfrac{16}{40}$ and $\dfrac{10}{25}$ are equivalent.

32. Insert $<$ or $>$ to form a true statement. $\dfrac{4}{7}$ $\dfrac{5}{9}$

Multiply.

33. $\dfrac{2}{3} \cdot \dfrac{5}{11}$ **34.** $2\dfrac{5}{8} \cdot \dfrac{4}{7}$

35. $\dfrac{1}{4} \cdot \dfrac{1}{2}$ **36.** $7 \cdot 5\dfrac{2}{7}$

17. _____

18. _____

19. _____

20. _____

21. _____

22. _____

23. _____

24. _____

25. _____

26. _____

27. _____

28. _____

29. _____

30. _____

31. _____

32. _____

33. _____

34. _____

35. _____

36. _____

37. _____

38. _____

39. _____

40. _____

41. _____

42. _____

43. _____

44. _____

45. _____

46. _____

47. _____

48. _____

49. _____

50. _____

Divide.

37. $\dfrac{11}{18} \div 2\dfrac{5}{6}$

38. $\dfrac{15}{19} \div \dfrac{3}{5}$

39. $5\dfrac{2}{3} \div 2\dfrac{5}{9}$

40. $\dfrac{8}{11} \div \dfrac{1}{22}$

41. Add and simplify: $\dfrac{3}{16} + \dfrac{7}{16}$

42. Subtract and simplify: $\dfrac{11}{20} - \dfrac{7}{20}$

43. Find the LCM of 6 and 8.

44. Find the LCM of 7 and 5.

45. Add: $\dfrac{1}{2} + \dfrac{2}{3} + \dfrac{5}{6}$

46. Evaluate: $\left(\dfrac{5}{9}\right)^2$

47. Subtract: $9\dfrac{3}{7} - 5\dfrac{2}{21}$

48. Subtract: $\dfrac{31}{100} - \dfrac{5}{25}$

49. Simplify: $\left(\dfrac{2}{3}\right)^2 \div \left(\dfrac{8}{27} + \dfrac{2}{3}\right)$

50. $\dfrac{1}{10} \div \dfrac{7}{8} \cdot \dfrac{2}{5}$

Decimals

Layout of the 2014
Circuit de la Sarthe

What Sports Car Endurance Race Is the Oldest?

The 24 Hours of Le Mans is the oldest active sports car endurance race in the world. First held in 1923, the race is run each year at Circuit de la Sarthe near the village of Le Mans, France. On race day, each qualifying car is driven in roughly two-hour shifts by a team of three drivers. The team that completes the most laps of the circuit after 24 hours is declared the winner. Some of the world's most-recognized car manufacturers, such as Porsche and Audi, have taken top honors at Le Mans.

In Section 4.2, Exercises 69–72, we will explore the average race speeds of recent 24 Hours of Le Mans winners. We will revisit Le Mans in Exercise 69 of Section 4.5 and Exercise 36 of Section 4.6.

Decimal numbers represent parts of a whole, just like fractions. In this chapter, we learn to perform arithmetic operations using decimals and to analyze the relationship between fractions and decimals. We also learn how decimals are used in the real world.

Average Speed of 24 Hours of Le Mans Winners

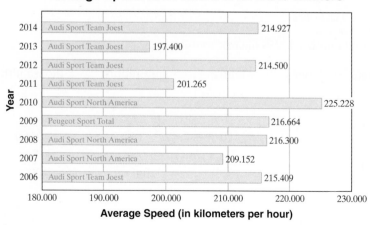

Year	Team	Average Speed (in kilometers per hour)
2014	Audi Sport Team Joest	214.927
2013	Audi Sport Team Joest	197.400
2012	Audi Sport Team Joest	214.500
2011	Audi Sport Team Joest	201.265
2010	Audi Sport North America	225.228
2009	Peugeot Sport Total	216.664
2008	Audi Sport North America	216.300
2007	Audi Sport North America	209.152
2006	Audi Sport Team Joest	215.409

Source: lemans-history.com

Objectives

A Know the Meaning of Place Value for a Decimal Number, and Write Decimals in Words. ▶

B Write Decimals in Standard Form. ▶

C Write Decimals as Fractions. ▶

D Write Fractions as Decimals. ▶

Helpful Hint   Notice that place values to the left of the decimal point end in "s." Place values to the right of the decimal point end in "ths."

Objective A Decimal Notation and Writing Decimals in Words ▶

Like fractional notation, decimal notation is used to denote a part of a whole. Numbers written in decimal notation are called **decimal numbers**, or simply **decimals.** The decimal 17.758 has three parts.

$$1\ 7\ .\ 7\ 5\ 8$$

Whole number part ← Decimal point → Decimal part

In Section 1.2, we introduced place value for whole numbers. Place names and place values for the whole number part of a decimal number are exactly the same, as shown next. Place names and place values for the decimal part are also shown.

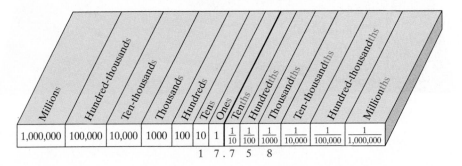

Notice that the value of each place is $\dfrac{1}{10}$ of the value of the place to its left. For example,

$$1 \cdot \frac{1}{10} = \frac{1}{10}$$

↑ ones   ↑ tenths

$$\frac{1}{10} \cdot \frac{1}{10} = \frac{1}{100}$$

↑ tenths   ↑ hundredths

The decimal number 17.758 means

1 ten	+	7 ones	+	7 tenths	+	5 hundredths	+	8 thousandths

or $\quad 1 \cdot 10 \quad + \quad 7 \cdot 1 \quad + \quad 7 \cdot \dfrac{1}{10} \quad + \quad 5 \cdot \dfrac{1}{100} \quad + \quad 8 \cdot \dfrac{1}{1000}$

or $\quad\ \ 10 \quad\ \ + \quad\ \ 7 \quad\ \ + \quad\ \ \dfrac{7}{10} \quad\ + \quad\ \ \dfrac{5}{100} \quad\ \ + \quad\ \dfrac{8}{1000}$

Writing (or Reading) a Decimal in Words

Step 1: Write the whole number part in words.

Step 2: Write "and" for the decimal point.

Step 3: Write the decimal part in words as though it were a whole number, followed by the place value of the last digit.

Example 1 Write the decimal 1.3 in words.

Solution: one and three tenths

 Work Practice 1

Practice 1

Write the decimal 8.7 in words.

Example 2

Write the decimal in the following sentence in words: The Golden Jubilee Diamond is a 545.67-carat cut diamond. (*Source: Guinness World Records*)

Solution: five hundred forty-five and sixty-seven hundredths

Work Practice 2

Practice 2

Write the decimal 97.28 in words.

Example 3 Write the decimal 19.5023 in words.

Solution: nineteen and five thousand twenty-three ten-thousandths

Work Practice 3

Practice 3

Write the decimal 302.105 in words.

Example 4

Write the decimal in the following sentence in words: The oldest known fragments of the Earth's crust are zircon crystals; they were discovered in Australia and are thought to be 4.276 billion years old. (*Source: Guinness World Records*)

Solution: four and two hundred seventy-six thousandths

Work Practice 4

Practice 4

Write the decimal 72.1085 in words.

Suppose that you are paying $368.42 for an automotive repair job at Jake's Body Shop by writing a check. Checks are usually written using the following format.

Answers

1. eight and seven tenths
2. ninety-seven and twenty-eight hundredths   **3.** three hundred two and one hundred five thousandths
4. seventy-two and one thousand eighty-five ten-thousandths

Practice 5

Fill in the check to CLECO (Central Louisiana Electric Company) to pay for your monthly electric bill of $207.40.

Example 5 Fill in the check to Camelot Music to pay for your purchase of $92.98.

Solution:

■ Work Practice 5

Objective B Writing Decimals in Standard Form ▶

A decimal written in words can be written in standard form by reversing the preceding procedure.

Practice 6–7

Write each decimal in standard form.

6. Three hundred and ninety-six hundredths

7. Thirty-nine and forty-two thousandths

Examples Write each decimal in standard form.

6. Forty-eight and twenty-six hundredths is

48.26

└── hundredths place

7. Six and ninety-five thousandths is

6.095

└── thousandths place

■ Work Practice 6–7

Helpful Hint

When converting a decimal from words to decimal notation, make sure the last digit is in the correct place by inserting 0s if necessary. For example,

Two and thirty-eight thousandths is 2.038

thousandths place

Objective C Writing Decimals as Fractions ▶

Once you master reading and writing decimals, writing a decimal as a fraction follows naturally.

Decimal	In Words	Fraction
0.7	seven tenths	$\frac{7}{10}$
0.51	fifty-one hundredths	$\frac{51}{100}$
0.009	nine thousandths	$\frac{9}{1000}$
0.05	five hundredths	$\frac{5}{100} = \frac{1}{20}$

Answers

5. CLECO; 207.40; Two hundred seven and $\frac{40}{100}$ 6. 300.96   7. 39.042

Notice that the number of decimal places in a decimal number is the same as the number of zeros in the denominator of the equivalent fraction. We can use this fact to write decimals as fractions.

$$0.\underbrace{51}_{} = \frac{51}{100} \qquad\qquad 0.\underbrace{009}_{} = \frac{9}{1000}$$

2 decimal   2 zeros   3 decimal   3 zeros
places   places

Example 8   Write 0.43 as a fraction.

Solution: $0.43 = \dfrac{43}{100}$

$\uparrow$ $\uparrow$
2   2
decimal places   zeros

▪ **Work Practice 8**

Example 9   Write 5.7 as a mixed number.

Solution: $5.7 = 5\dfrac{7}{10}$

$\uparrow$ $\uparrow$
1   1
decimal place   zero

▪ **Work Practice 9**

Examples   Write each decimal as a fraction or a mixed number. Write your answer in simplest form.

10. $0.125 = \dfrac{125}{1000} = \dfrac{\overset{1}{\cancel{125}}}{8 \cdot \cancel{125}} = \dfrac{1}{8}$

11. $23.5 = 23\dfrac{5}{10} = 23\dfrac{\overset{1}{\cancel{5}}}{2 \cdot \cancel{5}} = 23\dfrac{1}{2 \cdot 1} = 23\dfrac{1}{2}$

12. $105.083 = 105\dfrac{83}{1000}$

▪ **Work Practice 10–12**

Objective D   Writing Fractions as Decimals ▶

If the denominator of a fraction is a power of 10, we can write it as a decimal by reversing the procedure above.

Examples   Write each fraction as a decimal.

13. $\dfrac{8}{10} = 0.8$

$\uparrow$ $\uparrow$
1 zero   1 decimal place

14. $\dfrac{87}{10} = 8.7$

$\uparrow$ $\uparrow$
1 zero   1 decimal place

15. $\dfrac{18}{1000} = 0.018$

$\uparrow$ $\uparrow$
3 zeros   3 decimal places

16. $\dfrac{507}{100} = 5.07$

$\uparrow$ $\uparrow$
2 zeros   2 decimal places

▪ **Work Practice 13–16**

Practice 8
Write 0.037 as a fraction.

Practice 9
Write 14.97 as a mixed number.

Practice 10–12
Write each decimal as a fraction or mixed number. Write your answer in simplest form.
10. 0.12
11. 57.8
12. 209.986

Practice 13–16
Write each fraction as a decimal.

13. $\dfrac{58}{100}$ **14.** $\dfrac{59}{100}$

15. $\dfrac{6}{1000}$ **16.** $\dfrac{172}{10}$

Answers
8. $\dfrac{37}{1000}$ **9.** $14\dfrac{97}{100}$ **10.** $\dfrac{3}{25}$

11. $57\dfrac{4}{5}$ **12.** $209\dfrac{493}{500}$ **13.** 0.58

14. 0.59   **15.** 0.006   **16.** 17.2

Vocabulary, Readiness & Video Check

Use the choices below to fill in each blank.

words	decimals	and
tens	tenths	standard form

1. The number "twenty and eight hundredths" is written in _____ and "20.08" is written in _____.
2. Like fractions, _____ are used to denote parts of a whole.
3. When writing a decimal number in words, the decimal point is written as _____.
4. The place value _____ is to the right of the decimal point while _____ is to the left of the decimal point.

Determine the place value for the digit 7 in each number.

5. 70   6. 700   7. 0.7   8. 0.07

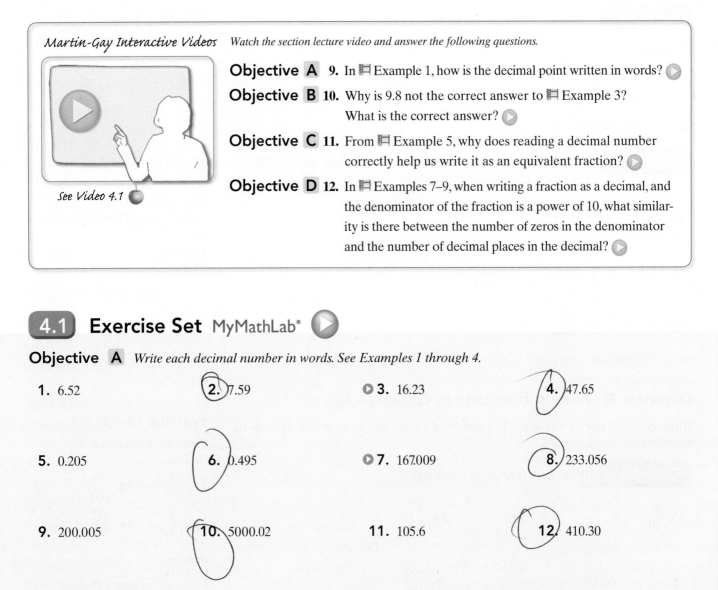

Martin-Gay Interactive Videos   Watch the section lecture video and answer the following questions.

Objective A 9. In ▦ Example 1, how is the decimal point written in words? ▶

Objective B 10. Why is 9.8 not the correct answer to ▦ Example 3? What is the correct answer? ▶

Objective C 11. From ▦ Example 5, why does reading a decimal number correctly help us write it as an equivalent fraction? ▶

Objective D 12. In ▦ Examples 7–9, when writing a fraction as a decimal, and the denominator of the fraction is a power of 10, what similarity is there between the number of zeros in the denominator and the number of decimal places in the decimal? ▶

See Video 4.1

4.1 Exercise Set MyMathLab® ▶

Objective A *Write each decimal number in words. See Examples 1 through 4.*

1. 6.52   2. 7.59   ▶ 3. 16.23   4. 47.65

5. 0.205   6. 0.495   ▶ 7. 167.009   8. 233.056

9. 200.005   10. 5000.02   11. 105.6   12. 410.30

13. The Akashi Kaikyo Bridge, between Kobe and Awaji-Shima, Japan, is approximately 2.43 miles long.

14. The English Channel Tunnel is 31.04 miles long. (*Source: Railway Directory & Year Book*)

15. Mercury makes a complete orbit of the Sun every 87.97 days. (*Source:* National Space Science Data Center)

16. Saturn makes a complete orbit of the Sun every 29.48 years. (*Source:* National Space Science Data Center)

17. The total number of television households within the United States for the 2013–2014 season was estimated at 115.6 million. (*Source:* Nielsen Media Research)

18. In 2012, the average retail customer's wait time in line at a U.S. post office was 2.6 minutes. (*Source:* USPS)

Fill in each check for the described purchase. See Example 5.

19. Your monthly car loan of $321.42 to R. W. Financial.

Your Preprinted Name	60–8124/7233	1407
Your Preprinted Address	1000613331	
	DATE	
PAY TO THE ORDER OF		$
		DOLLARS
FIRST STATE BANK		
OF FARTHINGTON		
FARTHINGTON, IL 64422		
MEMO		
⑆621497260⑆ 1000613331⑈ 1407		

20. Your part of the monthly apartment rent, which is $213.70. You pay this to Amanda Dupre.

Your Preprinted Name	60–8124/7233	1408
Your Preprinted Address	1000613331	
	DATE	
PAY TO THE ORDER OF		$
		DOLLARS
FIRST STATE BANK		
OF FARTHINGTON		
FARTHINGTON, IL 64422		
MEMO		
⑆621497260⑆ 1000613331⑈ 1408		

21. Your cell phone bill of $59.68 to Bell South.

Your Preprinted Name	60–8124/7233	1409
Your Preprinted Address	1000613331	
	DATE	
PAY TO THE ORDER OF		$
		DOLLARS
FIRST STATE BANK		
OF FARTHINGTON		
FARTHINGTON, IL 64422		
MEMO		
⑆621497260⑆ 1000613331⑈ 1409		

22. Your grocery bill of $87.49 to Albertsons.

Your Preprinted Name	60–8124/7233	1410
Your Preprinted Address	1000613331	
	DATE	
PAY TO THE ORDER OF		$
		DOLLARS
FIRST STATE BANK		
OF FARTHINGTON		
FARTHINGTON, IL 64422		
MEMO		
⑆621497260⑆ 1000613331⑈ 1410		

Objective B *Write each decimal number in standard form. See Examples 6 and 7.*

23. Six and five tenths

24. Three and nine tenths

25. Nine and eight hundredths

26. Twelve and six hundredths

27. Seven hundred five and six hundred twenty-five thousandths

28. Eight hundred four and three hundred ninety-nine thousandths

▶ **29.** Forty-six ten-thousandths

30. Thirty-eight ten-thousandths

31. The record rainfall amount for a 24-hour period in Alabama is thirty-two and fifty-two hundredths inches. This record was set at Dauphin Island Sea Lab in 1997. (*Source:* National Climatic Data Center)

32. For model year 2013, the average fuel economy of all imported passenger cars sold in the United States was thirty-seven and one tenth miles per gallon. (*Source:* U.S. Department of Transporation)

33. The average IndyCar burns one and three-tenths gallons of fuel per lap at the Indianapolis Motor Speedway. (*Source:* Indianapolis Motor Speedway)

34. The IZOD IndyCar series races at the Mid-Ohio Sports Car Course each season. The track length there is two and two hundred fifty-eight thousandths miles. (*Source:* IndyCar.com)

Objective C *Write each decimal as a fraction or a mixed number. Write your answer in simplest form. See Examples 8 through 12.*

35. 0.3

36. 0.9

▶ **37.** 0.27

38. 0.39

39. 0.8

40. 0.4

41. 0.15

42. 0.64

43. 5.47

44. 6.3

45. 0.048

46. 0.082

▶ **47.** 7.008

48. 9.005

49. 15.802

50. 11.406

51. 0.3005

52. 0.2006

53. 487.32

54. 298.62

Objective D *Write each fraction as a decimal. See Examples 13 through 16.*

▶ **55.** $\dfrac{6}{10}$

56. $\dfrac{3}{10}$

▶ **57.** $\dfrac{45}{100}$

58. $\dfrac{75}{100}$

59. $\dfrac{37}{10}$

60. $\dfrac{28}{10}$

61. $\dfrac{268}{1000}$

62. $\dfrac{709}{1000}$

63. $\dfrac{9}{100}$

64. $\dfrac{7}{100}$

65. $\dfrac{4026}{1000}$

66. $\dfrac{3601}{1000}$

▶ **67.** $\dfrac{28}{1000}$

68. $\dfrac{63}{1000}$

69. $\dfrac{563}{10}$

70. $\dfrac{206}{10}$

Objectives Ⓐ Ⓑ Ⓒ Ⓓ **Mixed Practice** *Fill in the chart. The first row is completed for you. See Examples 1 through 16.*

	Decimal Number in Standard Form	In Words	Fraction
	0.37	thirty-seven hundredths	$\frac{37}{100}$
71.			$\frac{43}{100}$
72.			$\frac{89}{100}$
73.		eight tenths	
74.		five tenths	
75.	0.077		
76.	0.019		

Review

Round 47,261 to the indicated place value. See Section 1.5.

77. tens

78. hundreds

79. thousands

80. ten-thousands

Concept Extensions

81. In your own words, describe how to write a decimal as a fraction or a mixed number.

82. In your own words, describe how to write a fraction as a decimal.

83. Write 0.00026849576 in words.

84. Write 0.00026849576 as a fraction. Do not simplify the resulting fraction.

85. Write $17\frac{268}{1000}$ as a decimal.

86. Write $7\frac{12}{100}$ as a decimal.

4.2 **Order and Rounding** ▶

Objective Ⓐ Comparing Decimals ▶

One way to compare decimals is to compare their graphs on a number line. Recall from Section 3.5 that for any two numbers on a number line, the number to the left is smaller and the number to the right is larger. The decimals 0.5 and 0.8 are graphed as follows:

Objectives

Ⓐ Compare Decimals. ▶

Ⓑ Round a Decimal Number to a Given Place Value. ▶

Comparing decimals by comparing their graphs on a number line can be time consuming. Another way to compare the size of decimals is to compare digits in corresponding places.

Comparing Two Decimals

Compare digits in the same places from left to right. When two digits are not equal, the number with the larger digit is the larger decimal. If necessary, insert 0s after the last digit to the right of the decimal point to continue comparing.

Compare hundredths-place digits

28.253   28.263
↑   ↑

5   <   6

so 28.253   <   28.263

Before we continue, let's take a moment and convince ourselves that inserting a zero after the last digit to the right of a decimal point does not change the value of the number.

For example, let's show that

$$0.7 = 0.70$$

If we write 0.7 as a fraction, we have

$$0.7 = \frac{7}{10}$$

Let's now multiply by 1. Recall that multiplying a number by 1 does not change the value of the number.

$$0.7 = \frac{7}{10} = \frac{7}{10} \cdot 1 = \frac{7}{10} \cdot \frac{10}{10} = \frac{7 \cdot 10}{10 \cdot 10} = \frac{70}{100} = 0.70$$

Thus $0.7 = 0.70$ and so on.

Helpful Hint

For any decimal, inserting 0s after the last digit to the right of the decimal point does not change the value of the number.

$$7.6 = 7.60 = 7.600, \text{ and so on}$$

When a whole number is written as a decimal, the decimal point is placed to the right of the ones digit.

$$25 = 25.0 = 25.00, \text{ and so on}$$

Practice 1

Insert $<$, $>$, or $=$ to form a true statement.

13.208   13.281

Example 1   Insert $<$, $>$, or $=$ to form a true statement.

0.378   0.368

Solution:

0.3 7 8   0.3 6 8   The tenths places are the same.

0.3 7 8   0.3 6 8   The hundredths places are different.

Since $7 > 6$, then $0.378 > 0.368$.

■ Work Practice 1

Answer

1. $<$

Example 2 Insert <, >, or = to form a true statement.

0.052   0.236

Solution:

0.**0**52 < 0.**2**36   0 is smaller than 2 in the tenths place.
　↑　　　↑

■ Work Practice 2

Example 3 Insert <, >, or = to form a true statement.

0.52   0.063

Solution:

0.**5**2 > 0.**0**63   5 is larger than 0 in the tenths place.
　↑　　　↑

■ Work Practice 3

Example 4 Write the decimals in order from smallest to largest.

7.035, 8.12, 7.03, 7.1

Solution: By comparing the ones digits, the decimal 8.12 is the largest number. To write the rest of the decimals in order, we compare digits to the right of the decimal point. We will insert zeros to help us compare.

7.035   7.030   7.100

Helpful Hint

You may also immediately notice that 7.1 is larger than both 7.035 and 7.03.

By comparing digits to the right of the decimal point, we can now arrange the decimals from smallest to largest.

7.030,   7.035,   7.100,   8.12   or

7.03,   7.035,   7.1,   8.12

■ Work Practice 4

Objective B Rounding Decimals ▷

We **round the decimal part** of a decimal number in nearly the same way as we round whole numbers. The only difference is that we delete digits to the right of the rounding place, instead of replacing these digits by 0s. For example,

24.954   rounded to the nearest hundredth is   24.95
　↑
hundredths place

Rounding Decimals to a Place Value to the Right of the Decimal Point

Step 1: Locate the digit to the right of the given place value.

Step 2: If this digit is 5 or greater, add 1 to the digit in the given place value and delete all digits to its right. If this digit is less than 5, delete all digits to the right of the given place value.

Practice 2

Insert <, >, or = to form a true statement.

0.124   0.086

Practice 3

Insert <, >, or = to form a true statement.

0.61   0.076

Practice 4

Write the decimals in order from smallest to largest.

14.605, 14.65, 13.9, 14.006

Answers
2. > **3.** >
4. 13.9, 14.006, 14.605, 14.65

Practice 5

Round 123.7814 to the nearest thousandth.

Example 5 Round 736.2359 to the nearest tenth.

Solution:

Step 1: We locate the digit to the right of the tenths place.

736.2③59

Step 2: Since the digit to the right is less than 5, we delete it and all digits to its right.

Thus, 736.2359 rounded to the nearest tenth is 736.2.

■ Work Practice 5

Practice 6

Round 123.7817 to the nearest tenth.

Example 6 Round 736.2359 to the nearest hundredth.

Solution:

Step 1: We locate the digit to the right of the hundredths place.

hundredths place

736.23⑤9

digit to the right

Step 2: Since the digit to the right is 5, we add 1 to the digit in the hundredths place and delete all digits to the right of the hundredths place.

736.23⑤9

Add 1. ⌐ Delete these digits.

Thus, 736.2359 rounded to the nearest hundredth is 736.24.

■ Work Practice 6

Rounding often occurs with money amounts. Since there are 100 cents in a dollar, each cent is $\frac{1}{100}$ of a dollar. This means that if we want to round to the nearest cent, we round to the nearest hundredth of a dollar.

Practice 7

In Sandersville, the price of a gallon of premium gasoline is $3.1589. Round this to the nearest cent.

Example 7 The price of a gallon of premium gasoline in Cross City is currently $3.1779. Round this to the nearest cent.

Solution:

hundredths place ─────┐ ┌─ 7 is greater than 5

$3.17⑦9

Add 1. ⌐ Delete these digits.

Since the digit to the right is greater than 5, we add 1 to the hundredths digit and delete all digits to the right of the hundredths digit.

Thus, $3.1779 rounded to the nearest cent is $3.18.

■ Work Practice 7

Answers

5. 123.781   **6.** 123.8   **7.** $3.16

Example 8   Round $0.098 to the nearest cent.

Solution:

hundredths place ————┐ ┌——— 8 is greater than 5
$0.098
        ↗ ↑——— Delete this digit.
     Add 1.

          1 0
        $0.09̸
          ↗
       Add 1.

$9 + 1 = 10$, so replace the digit 9 by 0 and carry the 1 to the place value to the left. Thus, $0.098 rounded to the nearest cent is $0.10.

▥ **Work Practice 8**

✓**Concept Check**   1756.0894 rounded to the nearest ten is
a. 1756.1   **b.** 1760.0894   **c.** 1760   **d.** 1750

Example 9   Determining State Taxable Income

A high school teacher's taxable income is $41,567.72. The tax tables in the teacher's state use amounts to the nearest dollar. Round the teacher's income to the nearest dollar.

Solution: Rounding to the nearest dollar means rounding to the ones place.

ones place ————┐ ┌——— 7 is greater than 5
$41,567.72
        ↑ ↑——— Delete these digits.
   Add 1.

Thus, the teacher's income rounded to the nearest dollar is $41,568.

▥ **Work Practice 9**

In Section 4.4, we will introduce a formula for the distance around a circle. The distance around a circle is given the special name **circumference.**

The symbol π is the Greek letter pi, pronounced "pie." We use π to denote the following constant:

$$\pi = \frac{\text{circumference of a circle}}{\text{diameter of a circle}}$$

The value π is an **irrational number.** This means if we try to write it as a decimal, it neither ends nor repeats in a pattern.

Example 10 $\pi \approx 3.14159265$. Round π to the nearest hundredth.

Solution:

hundredths place ———┐ ┌——— 1 is less than 5.
3.14159265
        ↑——— Delete these digits.

Thus, 3.14159265 rounded to the nearest hundredth is 3.14. In other words, $\pi \approx 3.14$.

▥ **Work Practice 10**

Practice 8
Round $1.095 to the nearest cent.

Practice 9
Water bills in Gotham City are always rounded to the nearest dollar. Round a water bill of $24.62 to the nearest dollar.

Practice 10
$\pi \approx 3.14159265$. Round π to the nearest ten-thousandth.

Answers
8. $1.10   **9.** $25   **10.** $\pi \approx 3.1416$

✓**Concept Check Answer**
c

Vocabulary, Readiness & Video Check

Use the choices below to fill in each blank. Some choices may be used more than once or not at all.

before	7.0	diameter
after	0.7	circumference

1. Another name for the distance around a circle is its _____.

2. $\pi = \dfrac{\text{_____ of a circle}}{\text{_____ of a circle}}$

3. The decimal point in a whole number is _____ the last digit.

4. The whole number 7 = _____.

Martin-Gay Interactive Videos

Watch the section lecture video and answer the following questions.

Objective A 5. In ▯ Example 2, we compare place value by place value in which direction? ▶

Objective B 6. The decimal number in ▯ Example 5 is being rounded to the nearest thousandth. Why is the digit 5, which is not in the thousandth place, looked at? ▶

See Video 4.2

4.2 Exercise Set MyMathLab® ▶

Objective A *Insert* $<$, $>$, *or* $=$ *to form a true statement. See Examples 1 through 3.*

1. 0.15 0.16

2. 0.12 0.15

▶ **3.** 0.57 0.54

4. 0.59 0.52

5. 0.098 0.1

6. 0.0756 0.2

7. 0.54900 0.549

8. 0.98400 0.984

▶ **9.** 167.908 167.980

10. 519.3405 519.3054

11. 420,000 0.000042

12. 0.000987 987,000

Write the decimals in order from smallest to largest. See Example 4.

13. 0.006, 0.06, 0.0061

14. 0.082, 0.008, 0.080

15. 0.042, 0.36, 0.03

16. 0.21, 0.056, 0.065

17. 1.1, 1.16, 1.01, 1.09

18. 3.6, 3.069, 3.09, 3.06

19. 21.001, 20.905, 21.03, 21.12

20. 36.050, 35.72, 35.702, 35.072

Objective B *Round each decimal to the given place value. See Examples 5 through 10.*

21. 0.57, to the nearest tenth

22. 0.54, to the nearest tenth

▶ **23.** 0.234, to the nearest hundredth

24. 0.452, to the nearest hundredth

25. 0.5942, to the nearest thousandth

26. 63.4523, to the nearest thousandth

27. 98,207.23, to the nearest ten

28. 68,934.543, to the nearest ten

29. 12.342, to the nearest tenth

30. 42.9878, to the nearest thousandth

31. 17.667, to the nearest hundredth

32. 0.766, to the nearest hundredth

33. 0.501, to the nearest tenth

34. 0.602, to the nearest tenth

35. 0.1295, to the nearest thousandth

36. 0.8295, to the nearest thousandth

37. 3829.34, to the nearest ten

38. 4520.876, to the nearest hundred

Round each monetary amount to the nearest cent or dollar as indicated. See Examples 7 through 9.

39. $0.067, to the nearest cent

40. $0.025, to the nearest cent

41. $42,650.14, to the nearest dollar

42. $768.45, to the nearest dollar

43. $26.95, to the nearest dollar

44. $14,769.52, to the nearest dollar

45. $0.1992, to the nearest cent

46. $0.7633, to the nearest cent

Round each number to the given place value. See Examples 5 through 10.

47. The latest generation Apple MacBook Air, at its thinnest point, measures 0.2794 cm. Round this number to the nearest tenth. (*Source:* Apple, Inc.)

48. A large tropical cockroach of the family Dictyoptera is the fastest-moving insect. This insect was clocked at a speed of 3.36 miles per hour. Round this number to the nearest tenth. (*Source:* University of California, Berkeley)

49. During the 2013 Boston Marathon, Hiroyuki Yamamoto of Japan was the first wheelchair competitor to cross the finish line. His time was 1.4256 hours. Round this time to the nearest hundredth. (*Source:* Boston Athletic Association)

50. Lindsey Vonn of the U.S. ski team took first place in the women's giant slalom in the 2013 International Ski Federation World Cup. Her winning time was 2.372 minutes. Round this time to the nearest hundredth of a minute. (*Source:* International Ski Federation)

51. A used biology textbook is priced at $47.89. Round this price to the nearest dollar.

52. A used office desk is advertised at $49.95 by Drawley's Office Furniture. Round this price to the nearest dollar.

53. The 2012 estimated population density of the state of Louisiana is 106.5157 people per square mile. Round this population density to the nearest tenth. (*Source:* U.S. Census Bureau)

54. The 2012 estimated population density of the state of Arkansas is 56.675 people per square mile. Round this population density to the nearest tenth. (*Source:* U.S. Census Bureau)

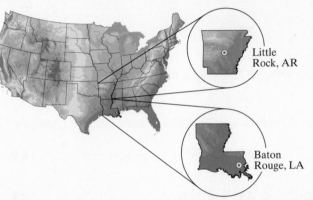

Little Rock, AR

Baton Rouge, LA

55. The length of a day on Mars is 24.6229 hours. Round this figure to the nearest thousandth. (*Source:* National Space Science Data Center)

56. Venus makes a complete orbit around the Sun every 224.695 days. Round this figure to the nearest whole day. (*Source:* National Space Science Data Center)

Sun

Mars

24.6229 hours

Sun

Venus

224.695 days

57. Kingda Ka is a hydraulic-launch roller coaster at Six Flags Great Adventure, an amusement park in Jackson, New Jersey. Currently it is the world's tallest roller coaster. A ride on the Kingda Ka lasts about 0.4667 minute. Round this figure to the nearest tenth. (*Source:* Roller Coaster DataBase)

58. During the 2012 NFL season, the average length of an Indianapolis Colts' punt was 47.6 yards. Round this figure to the nearest whole yard. (*Source:* National Football League)

Review

Perform each indicated operation. See Sections 1.3 and 1.4.

59. $3452 + 2314$

60. $8945 + 4536$

61. $94 - 23$

62. $82 - 47$

63. $482 - 239$

64. $4002 - 3897$

Concept Extensions

Solve. See the Concept Check in this section.

65. 2849.1738 rounded to the nearest hundred is
   a. 2849.17   **b.** 2800   **c.** 2850   **d.** 2849.174

66. 146.059 rounded to the nearest ten is
   a. 146.0   **b.** 146.1   **c.** 140   **d.** 150

67. 2849.1738 rounded to the nearest hundredth is
  a. 2849.17   **b.** 2800   **c.** 2850   **d.** 2849.174

68. 146.059 rounded to the nearest tenth is
  a. 146.0   **b.** 146.1   **c.** 140   **d.** 150

Mixed Practice (Sections 4.1, 4.2) *The table gives the average speed, in kilometers per hour, for the winners of the 24 Hours of Le Mans for each of the years listed. Use the table to answer Exercises 69 through 72. (Source: lemans-history.com)*

Year	Team	Average Speed (in kph)
2006	Audi Sport Team Joest	215.409
2007	Audi Sport North America	209.152
2008	Audi Sport North America	216.300
2009	Peugeot Sport Total	216.664
2010	Audi Sport North America	225.228
2011	Audi Sport Team Joest	201.265
2012	Audi Sport Team Joest	214.500
2013	Audi Sport Team Joest	197.400
2014	Audi Sport Team Joest	214.927

69. What is the fastest average speed on the list? Write this speed as a mixed number. Which team achieved this average speed?

70. What is the slowest average speed on the list? Write this speed as a mixed number. Which team achieved this average speed?

71. Make a list of the average winning speeds in order from fastest to slowest for the years 2010 through 2013.

72. Make a list of the average winning speeds in order from fastest to slowest for the years 2006 through 2009.

73. Write a 5-digit number that rounds to 1.7.

74. Write a 4-digit number that rounds to 26.3.

75. Write a decimal number that is greater than 8 but less than 9.

76. Write a decimal number that is greater than 48.1, but less than 48.2.

77. Which number(s) rounds to 0.26?
  0.26559   0.26499   0.25786   0.25186

78. Which number(s) rounds to 0.06?
  0.0612   0.066   0.0586   0.0506

Write these numbers from smallest to largest.

79. 0.9
  0.1038
  0.10299
  0.1037

80. 0.01
  0.0839
  0.09
  0.1

81. The all-time top six movies* (those that earned the most money in the United States) along with the approximate amount of money they have earned are listed in the table. Estimate the total amount of money that these movies have earned by first rounding each earning to the nearest hundred million. (*Source:* The Internet Movie Database)

Top All-Time American Movies	
Movie	**Gross Domestic Earnings**
Avatar (2009)	$760.5 million
Titanic (1997)	$658.7 million
Marvel's The Avengers (2012)	$623.4 million
The Dark Knight (2008)	$534.9 million
Star Wars: Episode I—The Phantom Menace (1999)	$474.5 million
Star Wars (1977)	$461.0 million
*Note: Many of these movies are still earning substantial amounts of money.	

82. In 2012, American manufacturers shipped approximately 6.2 million music videos to retailers. The value of these shipments was approximately $118.2 million. Estimate the value of an individual music video by rounding 118.2 and 6.2 to the nearest ten, then dividing. (*Source:* Recording Industry Association of America)

4.3 Adding and Subtracting Decimals

Objectives

A Add Decimals.

B Subtract Decimals.

C Estimate When Adding or Subtracting Decimals.

D Solve Problems That Involve Adding or Subtracting Decimals.

Objective A Adding Decimals

Adding decimals is similar to adding whole numbers. We add digits in corresponding place values from right to left, carrying if necessary. To make sure that digits in corresponding place values are added, we line up the decimal points vertically.

> **Adding or Subtracting Decimals**
>
> **Step 1:** Write the decimals so that the decimal points line up vertically.
>
> **Step 2:** Add or subtract as with whole numbers.
>
> **Step 3:** Place the decimal point in the sum or difference so that it lines up vertically with the decimal points in the problem.

In this section, we will insert zeros in decimal numbers so that place-value digits line up neatly. For instance, see Example 1.

Practice 1

Add.

a. $15.52 + 2.371$

b. $20.06 + 17.612$

c. $0.125 + 122.8$

Example 1 Add: $23.85 + 1.604$

Solution: First we line up the decimal points vertically.

$$
\begin{array}{r}
23.85\underline{0} \\
+\ 1.604 \\
\hline
\end{array}
$$
Insert one 0 so that digits line up neatly.

↑
Line up decimal points.

Then we add the digits from right to left as for whole numbers.

$$
\begin{array}{r}
\overset{1}{2}3.850 \\
+\ 1.604 \\
\hline
25.454
\end{array}
$$

└── Place the decimal point in the sum so that all decimal points line up.

■ **Work Practice 1**

Helpful Hint

Recall that 0's may be placed after the last digit to the right of the decimal point without changing the value of the decimal. This may be used to help line up place values when adding decimals.

$$
\begin{array}{rcl}
3.2 & \text{becomes} & 3.2\underline{00} \quad \text{Insert two 0s.} \\
15.567 & & 15.567 \\
+\ 0.11 & & +\ 0.11\underline{0} \quad \text{Insert one 0.} \\
\hline
& & 18.877 \quad \text{Add.}
\end{array}
$$

Answers

1. **a.** 17.891 **b.** 37.672 **c.** 122.925

Example 2 Add: 763.7651 + 22.001 + 43.89

Solution: First we line up the decimal points.

$$
\begin{array}{r}
\overset{1}{7}\overset{1}{6}\overset{1}{3}.7651 \\
22.0010 \quad \text{Insert one 0.} \\
+\ 43.8900 \quad \text{Insert two 0s.} \\
\hline
829.6561 \quad \text{Add.}
\end{array}
$$

■ Work Practice 2

Practice 2
Add.
a. 34.567 + 129.43 + 2.8903
b. 11.21 + 46.013 + 362.526

Helpful Hint

Don't forget that the decimal point in a whole number is after the last digit.

Example 3 Add: 45 + 2.06

Solution:
$$
\begin{array}{r}
45.00 \quad \text{Insert a decimal point and two 0s.} \\
+\ 2.06 \quad \text{Line up decimal points.} \\
\hline
47.06 \quad \text{Add.}
\end{array}
$$

■ Work Practice 3

Practice 3
Add: 26.072 + 119

✓**Concept Check** What is wrong with the following calculation of the sum of 7.03, 2.008, 19.16, and 3.1415?

$$
\begin{array}{r}
7.03 \\
2.008 \\
19.16 \\
+\ 3.1415 \\
\hline
3.6042
\end{array}
$$

Objective B Subtracting Decimals

Subtracting decimals is similar to subtracting whole numbers. We line up digits and subtract from right to left, borrowing when needed.

Example 4 Subtract: 35.218 − 23.65. Check your answer.

Solution: First we line up the decimal points.

$$
\begin{array}{r}
3\overset{4}{\cancel{5}}.\overset{11}{\cancel{2}}\overset{11}{1}8 \\
-23.650 \quad \text{Insert one 0.} \\
\hline
11.568 \quad \text{Subtract.}
\end{array}
$$

Recall that we can check a subtraction problem by adding.

$$
\begin{array}{r}
\overset{1}{1}1.\overset{1}{5}68 \quad \text{Difference} \\
+23.650 \quad \text{Subtrahend} \\
\hline
35.218 \quad \text{Minuend}
\end{array}
$$

■ Work Practice 4

Practice 4
Subtract. Check your answers.
a. 82.75 − 15.9
b. 126.032 − 95.71

Answers
2. a. 166.8873 **b.** 419.749 **3.** 145.072
4. a. 66.85 **b.** 30.322

✓**Concept Check Answer**
The decimal points and places are not lined up properly.

Practice 5

Subtract. Check your answers.
a. 5.8 − 3.92
b. 9.72 − 4.068

Example 5   Subtract: 3.5 − 0.068. Check your answer.

Solution:

$$
\begin{array}{r}
\overset{4\;\overset{9}{\cancel{10}}\;10}{3.\cancel{5}\,\cancel{0}\,\cancel{0}} \\
-\,0.0\,6\,8 \\
\hline
3.4\,3\,2
\end{array}
$$

Insert two 0s.
Line up decimal points.
Subtract.

Check:

$$
\begin{array}{r}
\overset{1\;1}{} \\
3.432 \\
+\,0.068 \\
\hline
3.500
\end{array}
$$

Difference
Subtrahend
Minuend

■ Work Practice 5

Practice 6

Subtract. Check your answers.
a. 53 − 29.31
b. 120 − 68.22

Example 6   Subtract: 85 − 17.31. Check your answer.

Solution:

$$
\begin{array}{r}
\overset{7\;14\;\overset{9}{\cancel{10}}\;10}{8\,\cancel{5}.\cancel{0}\,\cancel{0}} \\
-\,1\,7.3\,1 \\
\hline
6\,7.6\,9
\end{array}
$$

Check:

$$
\begin{array}{r}
\overset{1\;1\;\;1}{} \\
67.69 \\
+\,17.31 \\
\hline
85.00
\end{array}
$$

Difference
Subtrahend
Minuend

■ Work Practice 6

Objective C Estimating When Adding or Subtracting Decimals ▶

To help avoid errors, we can also estimate to see if our answer is reasonable when adding or subtracting decimals. Although only one estimate is needed per operation, we show two to show variety.

Practice 7

Add or subtract as indicated. Then estimate to see if the answer is reasonable by rounding the given numbers and adding or subtracting the rounded numbers.
a. 48.1 + 326.97
b. 18.09 − 0.746

Example 7   Add or subtract as indicated. Then estimate to see if the answer is reasonable by rounding the given numbers and adding or subtracting the rounded numbers.

a. 27.6 + 519.25

Solution:

Exact		Estimate 1		Estimate 2
$\overset{1}{}27.60$	rounds to	30		30
$+\,519.25$	rounds to	$+\,500$	or	$+\,520$
546.85		530		550

Since the exact answer is close to either estimate, it is reasonable. (In the first estimate, each number is rounded to the place value of the leftmost digit. In the second estimate, each number is rounded to the nearest ten.)

b. 11.01 − 0.862

Solution:

Exact		Estimate 1		Estimate 2
$\overset{0\;\;9\;10\,10}{1\cancel{1}.\cancel{0}\cancel{1}\cancel{0}}$	rounds to	10		11
$-\,0.862$	rounds to	$-\,1$	or	$-\,1$
10.148		9		10

In the first estimate, we rounded the first number to the nearest ten and the second number to the nearest one. In the second estimate, we rounded both numbers to the nearest one. Both estimates show us that our answer is reasonable.

■ Work Practice 7

> **Helpful Hint**
> Remember that estimates are for our convenience to quickly check the reasonableness of an answer.

Answers
5. a. 1.88 b. 5.652
6. a. 23.69 b. 51.78
7. a. 375.07 b. 17.344

✓**Concept Check** Why shouldn't the sum 21.98 + 42.36 be estimated as 30 + 50 = 80?

Objective D Solving Problems by Adding or Subtracting Decimals ▶

Decimals are very common in real-life problems.

Example 8 Calculating the Cost of Owning an Automobile

Find the total monthly cost of owning and operating a certain automobile given the expenses shown.

Monthly car payment:	$256.63
Monthly insurance cost:	$47.52
Average gasoline bill per month:	$95.33

Solution:

1. UNDERSTAND. Read and reread the problem. The phrase "total monthly cost" tells us to add.

2. TRANSLATE.

In words:	total monthly cost	is	car payment	plus	insurance cost	plus	gasoline bill
	↓	↓	↓	↓	↓	↓	↓
Translate:	total monthly cost	=	$256.63	+	$47.52	+	$95.33

3. SOLVE: Let's also estimate by rounding each number to the nearest ten.

$$
\begin{array}{rll}
\overset{111}{256.63} & \text{rounds to} & 260 \\
47.52 & \text{rounds to} & 50 \\
+\ \ 95.33 & \text{rounds to} & 100 \\
\hline
399.48 & \text{Exact} \quad 410 & \text{Estimate}
\end{array}
$$

4. INTERPRET. *Check* your work. Since our estimate is close to our exact answer, our answer is reasonable. *State* your conclusion: The total monthly cost is $399.48.

◼ Work Practice 8

The bar graph in Example 9 has horizontal bars. To visualize the value represented by a bar, see how far it extends to the right. The value of each bar is labeled and we will study bar graphs further in a later chapter.

Example 9 Comparing Average Heights

The bar graph on the next page shows the current average heights for male adults in various countries. How much greater is the average male height in Denmark than the average male height in the United States?

(Continued on next page)

Practice 8

Find the total monthly cost of owning and operating a certain automobile given the expenses shown.

Monthly car payment:	$536.52
Monthly insurance cost:	$52.68
Average gasoline bill per month:	$87.50

Answer
8. $676.70

✓**Concept Check Answer**
Each number is rounded incorrectly. The estimate is too high.

Practice 9

Use the bar graph in Example 9. How much greater is the average height in the Netherlands than the average height in Israel?

Average Male Adult Height

Netherlands — 72.8 inches
Denmark — 71.1 inches
Sweden — 70.9 inches
Norway — 70.8 inches
United States — 70.2 inches
Australia — 70.2 inches
Israel — 69.2 inches

0 * 68 69 70 71 72 73 74 75

Source: Disabled World

* The ⌁ means that some numbers are purposefully missing on the axis.

Solution:

1. **UNDERSTAND.** Read and reread the problem. Since we want to know "how much greater," we subtract.

2. **TRANSLATE.**

In words:	How much greater	is	Denmark's average male height	minus	United States's average male height
	↓	↓	↓	↓	↓
Translate:	How much greater	=	71.1	−	70.2

3. **SOLVE:** We also estimate by rounding each number to the nearest whole.

$$
\begin{array}{r}
7\overset{0}{\cancel{1}}.\overset{11}{\cancel{1}} \\
-\ 70.2 \\
\hline
0.9
\end{array}
\quad
\begin{array}{l}
\text{rounds to} \quad 71 \\
\text{rounds to} \quad -70 \\
\hline
\text{Exact} \qquad 1 \quad \text{Estimate}
\end{array}
$$

4. **INTERPRET.** *Check* your work. Since our estimate is close to our exact answer, 0.9 inch is reasonable. *State* your conclusion: The average male height in Denmark is 0.9 inch greater than the U.S. average male height.

◼ Work Practice 9

🖩 Calculator Explorations

Entering Decimal Numbers

To enter a decimal number, find the key marked ⟨ · ⟩. To enter the number 2.56, for example, press the keys ⟨2⟩⟨·⟩⟨5⟩⟨6⟩.

The display will read ⟨ 2.56 ⟩.

Operations on Decimal Numbers

Operations on decimal numbers are performed in the same way as operations on whole or signed numbers. For example, to find 8.625 − 4.29, press the keys ⟨8.625⟩⟨−⟩⟨4.29⟩⟨=⟩ or ⟨ENTER⟩.

The display will read ⟨ 4.335 ⟩. (Although entering 8.625, for example, requires pressing more than one key, we group numbers together here for easier reading.)

Use a calculator to perform each indicated operation.

1. 315.782 + 12.96

2. 29.68 + 85.902

3. 6.249 − 1.0076

4. 5.238 − 0.682

5.
$$
\begin{array}{r}
12.555 \\
224.987 \\
5.2 \\
+\ 622.65 \\
\hline
\end{array}
$$

6.
$$
\begin{array}{r}
47.006 \\
0.17 \\
313.259 \\
+\ 139.088 \\
\hline
\end{array}
$$

Answer

9. 3.6 in.

Vocabulary, Readiness & Video Check

Use the choices below to fill in each blank. Not all choices will be used.

minuend	vertically	first	true	37.0	horizontally
difference	subtrahend	last	false	0.37	

1. The number 37 equals _____.
2. The decimal point in a whole number is positioned after the _____ digit.
3. In $89.2 - 14.9 = 74.3$, the number 74.3 is called the _____, 89.2 is the _____, and 14.9 is the _____.
4. To add or subtract decimals, we line up the decimal points _____.
5. True or false: The number 5.6 is closer to 5 than 6 on a number line. _____.
6. True or false: The number 10.48 is closer to 10 than 11 on a number line. _____.

Martin-Gay Interactive Videos

See Video 4.3

Watch the section lecture video and answer the following questions.

Objective A 7. From ▦ Example 2, what does lining up the decimal points also line up? Why is this important? ▶

Objective B 8. From ▦ Example 3, where is the decimal point in a whole number? ▶

Objective C 9. In ▦ Example 5, estimating is used to check whether the answer to the subtraction problem is reasonable, but what is the best way to fully check? ▶

Objective D 10. Complete this statement based on ▦ Example 6: To calculate the amount of border material needed, we are actually calculating the _____ of the triangle. ▶

4.3 Exercise Set MyMathLab® ▶

Objectives A C **Mixed Practice** *Add. See Examples 1 through 3 and 7. For those exercises marked, also estimate to see if the answer is reasonable.*

▶ **1.** $1.3 + 2.2$ **2.** $2.5 + 4.1$ **3.** $5.7 + 1.13$ **4.** $2.31 + 6.4$ **5.** $0.003 + 0.091$

6. $0.004 + 0.085$ **7.** $19.23 + 602.782$ **8.** $47.14 + 409.567$ **9.** $490 + 93.09$ **10.** $600 + 83.0062$

11.
$$\begin{array}{r} 234.89 \\ + 230.67 \\ \hline \end{array}$$
Exact: Estimate:

12.
$$\begin{array}{r} 734.89 \\ + 640.56 \\ \hline \end{array}$$
Exact: Estimate:

13.
$$\begin{array}{r} 100.009 \\ 6.08 \\ + 9.034 \\ \hline \end{array}$$
Exact: Estimate:

14.
$$\begin{array}{r} 200.89 \\ 7.49 \\ + 62.83 \\ \hline \end{array}$$
Exact: Estimate:

▶ **15.** $24.6 + 2.39 + 0.0678$ **16.** $32.4 + 1.58 + 0.0934$

17. Find the sum of 45.023, 3.006, and 8.403 **18.** Find the sum of 65.0028, 5.0903, and 6.9003

Objectives B C Mixed Practice *Subtract and check. See Examples 4 through 7. For those exercises marked, also estimate to see if the answer is reasonable.*

19. 8.8 − 2.3

20. 7.6 − 2.1

21. 18 − 2.7

22. 28 − 3.3

23.   654.9
        − 56.67

24.   863.23
        − 39.453

25. 5.9 − 4.07
Exact:
Estimate:

26. 6.4 − 3.04
Exact:
Estimate:

27. 923.5 − 61.9

28. 845.93 − 45.8

29. 500.34 − 123.45

30. 600.74 − 463.98

31.   1000
        − 123.4
Exact:

Estimate:

32.   2000
        − 327.47
Exact:

Estimate:

33. 200 − 5.6

34. 800 − 8.9

35. 3 − 0.0012

36. 7 − 0.097

37. Subtract 6.7 from 23.

38. Subtract 9.2 from 45.

Objectives A B Mixed Practice *Perform the indicated operation. See Examples 1 through 6.*

39. 86.05 + 1.978

40. 95.07 + 4.216

41. 86.05 − 1.978

42. 95.07 − 4.216

43. Add 150 and 93.17.

44. Add 250 and 86.07.

45. 150 − 93.17

46. 250 − 86.07

47. Subtract 8.94 from 12.1.

48. Subtract 6.73 from 20.2.

Objective D *Solve. For Exercises 49 and 50, the solutions have been started for you. See Examples 8 and 9.*

49. Ann-Margaret Tober bought a book for $32.48. If she paid with two $20 bills, what was her change?

Start the solution:

1. UNDERSTAND the problem. Reread it as many times as needed.
2. TRANSLATE into an equation. (Fill in the blank.)

change	is	two $20 bills	minus	cost of book
↓	↓	↓	↓	↓
change	=	40	−	_____

Finish with
3. SOLVE and 4. INTERPRET

50. Phillip Guillot bought a car part for $18.26. If he paid with two $10 bills, what was his change?

Start the solution:

1. UNDERSTAND the problem. Reread it as many times as needed.
2. TRANSLATE into an equation. (Fill in the blank.)

change	is	two $10 bills	minus	cost of car part
↓	↓	↓	↓	↓
change	=	20	−	_____

Finish with
3. SOLVE and 4. INTERPRET

51. Find the total monthly cost of owning and maintaining a car given the information shown.

Monthly car payment:	$275.36
Monthly insurance cost:	$ 83.00
Average cost of gasoline per month:	$ 81.60
Average maintenance cost per month:	$ 14.75

52. Find the total monthly cost of owning and maintaining a car given the information shown.

Monthly car payment:	$306.42
Monthly insurance cost:	$ 53.50
Average cost of gasoline per month:	$123.00
Average maintenance cost per month:	$ 23.50

53. Gasoline was $2.839 per gallon one week and $2.979 per gallon the next. By how much did the price change?

54. A pair of eyeglasses costs a total of $347.89. The frames of the glasses are $97.23. How much do the lenses of the eyeglasses cost?

55. Find the perimeter.

Square   7.14 meters

56. Find the perimeter.

4.2 in.   5.78 in.   7.8 in.

57. The top face of the latest iPod nano measures 3.01 inches by 1.56 inches. Find the perimeter of the rectangular face. (*Source:* Apple Inc.)

58. The top face of the latest iPod touch measures 4.86 inches by 2.31 inches. Find the perimeter of this rectangular face. (*Source:* Apple Inc.)

59. The normal monthly average wind speed for April at the weather station on Mt. Washington in New Hampshire is 34.7 miles per hour. The highest speed ever recorded at the station, in April 1934, was 231.0 miles per hour. How much faster was the highest speed than the average April wind speed? (*Source:* Mount Washington Observatory)

60. In 2012, the average adult American spent 9.49 hours per day on personal care activities (including sleeping) and 5.37 hours per day on leisure and sports. How much more time on average do Americans spend on personal care activities than on leisure and sports? (*Source:* U.S. Bureau of Labor Statistics)

61. The average temperature for the contiguous United States during September 2013 was 67.3 degrees Fahrenheit. This is 2.5 degrees Fahrenheit above the 20th-century average temperature for September. What is the United States' 20th-century average temperature for September? (*Source:* National Climatic Data Center)

62. Historically, the average rainfall for the month of May in Omaha, Nebraska, is 4.79 inches. In May 2013, Omaha received 6.30 inches of rain. By how much was Omaha's May rainfall above average? (*Source:* National Oceanic and Atmospheric Administration)

63. Andy Green still holds the record for one-mile land speed. This record was 129.567 miles per hour faster than a previous record of 633.468 set in 1983. What was Green's record-setting speed? (*Source:* United States Auto Club; this record was made in October 1997)

64. It costs $5.80 to send a 2-pound package locally via Priority Mail at a U.S. Post Office. To send the same package as Express Mail, it costs $18.10. How much more does it cost to send a package as Express Mail? (*Source:* USPS)

65. The face of the Apple iPhone 5c is a rectangle measuring 4.90 inches by 2.33 inches. Find the perimeter of this rectangular phone. (*Source:* Apple Inc.)

66. The Samsung Galaxy S4, an Android-based smartphone and a leading iPhone competitor, measures 5.38 inches by 2.75 inches. Find the perimeter of this rectangular phone. (*Source:* based on data from Samsung Electronics Co.)

67. The average U.S. movie theater ticket price in 2012 was $7.96. In 2011, it was $7.93. Find the increase in average movie theater ticket price from 2011 to 2012. (*Source:* MPAA)

68. The average U.S. movie theater ticket price in 2003 was $6.03. For 2013, it was estimated to be $8.05. Find the increase in average movie theater ticket price for this 10-year period. (*Source:* MPAA and Box Office Mojo)

This bar graphs shows the average number of text messages sent each month in the United States for the years shown. Use this graph to answer Exercises 69 and 70. (Source: CTIA — The Wireless Association)

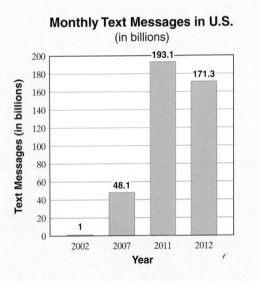

Monthly Text Messages in U.S.
(in billions)

69. Find the decrease in monthly text messages from 2011 to 2012.

70. Find the increase in monthly text messages from 2002 to 2007.

71. The snowiest city in the United States is Valdez, Alaska, which receives an average of 85.7 more inches of snow than the second snowiest city. The second snowiest city in the United States is Blue Canyon, California. Blue Canyon receives an average of 240.3 inches of snow annually. How much snow does Valdez receive on average each year? (*Source:* National Climatic Data Center)

72. The driest city in the world is Aswan, Egypt, which receives an average of only 0.02 inch of rain per year. Yuma, Arizona, is the driest city in the United States. Yuma receives an average of 2.63 more inches of rain each year than Aswan. What is the average annual rainfall in Yuma? (*Source:* National Climatic Data Center)

73. A landscape architect is planning a border for a flower garden shaped like a triangle. The sides of the garden measure 12.4 feet, 29.34 feet, and 25.7 feet. Find the amount of border material needed.

74. A contractor purchased enough railing to completely enclose the newly built deck shown below. Find the amount of railing purchased.

15.7 feet
10.6 feet

The table shows the average retail price of a gallon of gasoline (all grades and formulations) in the United States in each of the years shown. Use this table to answer Exercises 75 and 76. (Source: Energy Information Administration)

Year	Gasoline Price (dollars per gallon)
2008	3.317
2009	2.401
2010	2.836
2011	3.577
2012	3.695

75. How much more was the average cost of a gallon of gasoline in 2012 than in 2009?

76. How much less was the average cost of a gallon of gasoline in 2010 than in 2008?

The following table shows spaceflight information for astronaut James A. Lovell. Use this table to answer Exercises 77 and 78.

Spaceflights of James A. Lovell		
Year	Mission	Duration (in hours)
1965	Gemini 6	330.583
1966	Gemini 12	94.567
1968	Apollo 8	147.0
1970	Apollo 13	142.9
(Source: NASA)		

77. Find the total time spent in spaceflight by astronaut James A. Lovell.

78. Find the total time James A. Lovell spent in space-flight on all Apollo missions.

The bar graph shows the top five chocolate-consuming nations in the world in 2012. Use this table to answer Exercises 79 through 84.

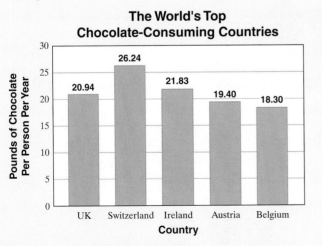

The World's Top Chocolate-Consuming Countries

UK 20.94, Switzerland 26.24, Ireland 21.83, Austria 19.40, Belgium 18.30

Pounds of Chocolate Per Person Per Year

Country

Source: Based on data from confectionarynews.com

79. Which country in the bar graph has the greatest chocolate consumption per person?

80. Which country in the bar graph has the least chocolate consumption per person?

81. How much more is the greatest chocolate consumption than the least chocolate consumption shown in the bar graph?

82. How much more chocolate does the average Irish citizen consume per year than the average Austrian?

83. Make a table listing the countries and their corresponding chocolate consumptions in order from greatest to least.

84. Find the sum of the five bar heights shown in the graph. What type of company might be interested in this sum?

Review

Multiply. See Sections 1.6 and 3.5.

85. $23 \cdot 2$ **86.** $46 \cdot 3$ **87.** $43 \cdot 90$ **88.** $30 \cdot 32$ **89.** $\left(\dfrac{2}{3}\right)^2$ **90.** $\left(\dfrac{1}{5}\right)^3$

Concept Extensions

A friend asks you to check his calculations for Exercises 91 and 92. Are they correct? If not, explain your friend's errors and correct the calculations. See the first Concept Check in this section.

91.

$$\begin{array}{r} \overset{1}{9}.2 \\ \overset{1}{8}.63 \\ +\ 4.005 \\ \hline 4.960 \end{array}$$

92.
$$\begin{array}{r} \overset{8\ 9\ 9\ 9}{9\cancel{0}\cancel{0}.\cancel{0}} \\ -\ 96.4 \\ \hline 8\ 03.5 \end{array}$$

Find the unknown length in each figure.

△**93.**

2.3 inches   ?   2.3 inches

10.68 inches

△**94.**

5.26 meters | 7.82 meters | ? meters

17.67 meters

Let's review the values of these common U.S. coins in order to answer the following exercises.

Penny   Nickel   Dime   Quarter

$0.01 $0.05 $0.10 $0.25

For Exercises 95 and 96, write the value of each group of coins. To do so, it is usually easiest to start with the coin(s) of greatest value and end with the coin(s) of least value.

95.

96.

97. Name the different ways that coins can have a value of $0.17 given that you may use no more than 10 coins.

98. Name the different ways that coin(s) can have a value of $0.25 given that there are no pennies.

99. Why shouldn't the sum
$$82.95 + 51.26$$
be estimated as $90 + 60 = 150$?
See the second Concept Check in this section.

100. Laser beams can be used to measure the distance to the moon. One measurement showed the distance to the moon to be 256,435.235 miles. A later measurement showed that the distance is 256,436.012 miles. Find how much farther away the moon is in the second measurement as compared to the first.

101. Explain how adding or subtracting decimals is similar to adding or subtracting whole numbers.

102. Explain how adding or subtracting decimals is different from adding or subtracting whole numbers.

4.4 Multiplying Decimals and Circumference of a Circle ▷

Objective A Multiplying Decimals ▷

Multiplying decimals is similar to multiplying whole numbers. The only difference is that we place a decimal point in the product. To discover where a decimal point is placed in the product, let's multiply 0.6×0.03. We first write each decimal as an equivalent fraction and then multiply.

$$\underset{\substack{\uparrow \\ \text{1 decimal} \\ \text{place}}}{0.6} \times \underset{\substack{\uparrow \\ \text{2 decimal} \\ \text{places}}}{0.03} = \frac{6}{10} \times \frac{3}{100} = \frac{18}{1000} = \underset{\substack{\uparrow \\ \text{3 decimal} \\ \text{places}}}{0.018}$$

Notice that $1 + 2 = 3$, the number of decimal places in the product. Now let's multiply 0.03×0.002.

$$\underset{\substack{\uparrow \\ \text{2 decimal} \\ \text{places}}}{0.03} \times \underset{\substack{\uparrow \\ \text{3 decimal} \\ \text{places}}}{0.002} = \frac{3}{100} \times \frac{2}{1000} = \frac{6}{100,000} = \underset{\substack{\uparrow \\ \text{5 decimal} \\ \text{places}}}{0.00006}$$

Again, we see that $2 + 3 = 5$, the number of decimal places in the product.

Instead of writing decimals as fractions each time we want to multiply, we notice a pattern from these examples and state a rule that we can use:

Objectives

A Multiply Decimals. ▷

B Estimate When Multiplying Decimals. ▷

C Multiply by Powers of 10. ▷

D Find the Circumference of a Circle. ▷

E Solve Problems by Multiplying Decimals. ▷

Multiplying Decimals

Step 1: Multiply the decimals as though they are whole numbers.

Step 2: The decimal point in the product is placed so that the number of decimal places in the product is equal to the *sum* of the number of decimal places in the factors.

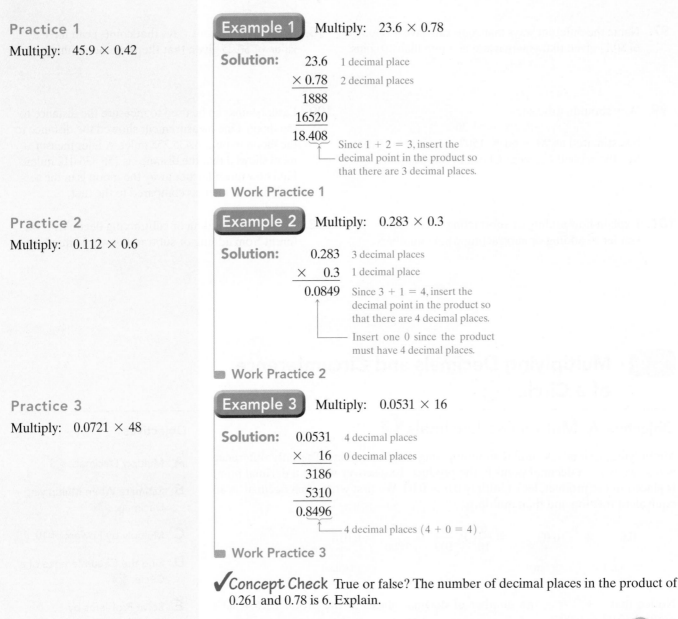

Practice 1

Multiply: 45.9×0.42

Example 1 Multiply: 23.6×0.78

Solution:

$$\begin{array}{r} 23.6 \\ \times\ 0.78 \\ \hline 1888 \\ 16520 \\ \hline 18.408 \end{array}$$

23.6 1 decimal place
$\times\ 0.78$ 2 decimal places

Since $1 + 2 = 3$, insert the decimal point in the product so that there are 3 decimal places.

▪ Work Practice 1

Practice 2

Multiply: 0.112×0.6

Example 2 Multiply: 0.283×0.3

Solution:

$$\begin{array}{r} 0.283 \\ \times\ \ \ \ 0.3 \\ \hline 0.0849 \end{array}$$

0.283 3 decimal places
$\times\ \ \ 0.3$ 1 decimal place

Since $3 + 1 = 4$, insert the decimal point in the product so that there are 4 decimal places.

Insert one 0 since the product must have 4 decimal places.

▪ Work Practice 2

Practice 3

Multiply: 0.0721×48

Example 3 Multiply: 0.0531×16

Solution:

$$\begin{array}{r} 0.0531 \\ \times\ \ \ \ 16 \\ \hline 3186 \\ 5310 \\ \hline 0.8496 \end{array}$$

0.0531 4 decimal places
$\times\ \ \ \ 16$ 0 decimal places

4 decimal places $(4 + 0 = 4)$

▪ Work Practice 3

✓**Concept Check** True or false? The number of decimal places in the product of 0.261 and 0.78 is 6. Explain.

Objective B Estimating When Multiplying Decimals ▶

Just as for addition and subtraction, we can estimate when multiplying decimals to check the reasonableness of our answer.

Practice 4

Multiply: 30.26×2.98. Then estimate to see whether the answer is reasonable.

Example 4 Multiply: 28.06×1.95. Then estimate to see whether the answer is reasonable by rounding each factor, then multiplying the rounded numbers.

Solution:

Exact:	**Estimate 1**	**Estimate 2**
28.06	28 Rounded to ones	30 Rounded to tens
$\times\ \ 1.95$	$\times\ 2$	$\times\ 2$
14030	56	60
252540		
280600		
54.7170		

The answer 54.7170 (or 54.717) is reasonable.

▪ Work Practice 4

Answers

1. 19.278 **2.** 0.0672 **3.** 3.4608
4. 90.1748

✓**Concept Check Answer**

false: 3 decimal places and 2 decimal places means 5 decimal places in the product

As shown in Example 4, estimated results will vary depending on what estimates are used. Notice that estimating results is a good way to see whether the decimal point has been correctly placed.

Objective C Multiplying by Powers of 10 ▷

There are some patterns that occur when we multiply a number by a power of 10 such as 10, 100, 1000, 10,000, and so on.

$23.6951 \times 10 = 236.951$ Move the decimal point *1 place* to the *right*.
1 zero

$23.6951 \times 100 = 2369.51$ Move the decimal point *2 places* to the *right*.
2 zeros

$23.6951 \times 100,000 = 2,369,510.$ Move the decimal point *5 places* to the *right* (insert a 0).
5 zeros

Notice that we move the decimal point the same number of places as there are zeros in the power of 10.

> **Multiplying Decimals by Powers of 10 such as 10, 100, 1000, 10,000...**
>
> Move the decimal point to the *right* the same number of places as there are *zeros* in the power of 10.

Examples Multiply.

5. $7.68 \times 10 = 76.8$ 7.68
6. $23.702 \times 100 = 2370.2$ 23.702
7. $76.3 \times 1000 = 76,300$ 76.300

▨ Work Practice 5–7

Practice 5–7

Multiply.
5. 23.7×10
6. 203.004×100
7. 1.15×1000

There are also powers of 10 that are less than 1. The decimals 0.1, 0.01, 0.001, 0.0001, and so on are examples of powers of 10 less than 1. Notice the pattern when we multiply by these powers of 10:

$569.2 \times 0.1 = 56.92$ Move the decimal point *1 place* to the *left*.
1 decimal place

$569.2 \times 0.01 = 5.692$ Move the decimal point *2 places* to the *left*.
2 decimal places

$569.2 \times 0.0001 = 0.05692$ Move the decimal point *4 places* to the *left* (insert one 0).
4 decimal places

> **Multiplying Decimals by Powers of 10 such as 0.1, 0.01, 0.001, 0.0001...**
>
> Move the decimal point to the *left* the same number of places as there are *decimal places* in the power of 10.

Answers
5. 237 **6.** 20,300.4 **7.** 1150

Practice 8–10

Multiply.
8. 7.62 × 0.1
9. 1.9 × 0.01
10. 7682 × 0.001

Examples Multiply.

8. 42.1 × 0.1 = 4.21   42.1
9. 76,805 × 0.01 = 768.05   76,805.
10. 9.2 × 0.001 = 0.0092   0009.2

■ **Work Practice 8–10**

Many times we see large numbers written, for example, in the form 451.8 million rather than in the longer standard notation. The next example shows us how to interpret these numbers.

Practice 11

In 2020, the population of the United States is projected to be 333.9 million. Write this number in standard notation. (*Source:* U.S. Census Bureau)

Example 11 In 2050, the population of the United States is projected to be 399.8 million. Write this number in standard notation. (*Source:* U.S. Census Bureau)

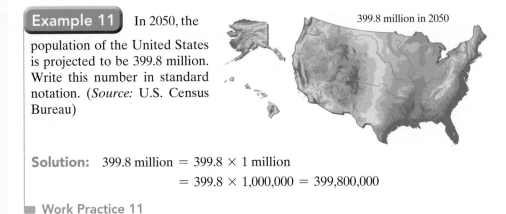

399.8 million in 2050

Solution: 399.8 million = 399.8 × 1 million
                       = 399.8 × 1,000,000 = 399,800,000

■ **Work Practice 11**

Objective D Finding the Circumference of a Circle

Recall from Section 1.3 that the distance around a polygon is called its perimeter. The distance around a circle is given the special name **circumference,** and this distance depends on the radius or the diameter of the circle.

Circumference of a Circle

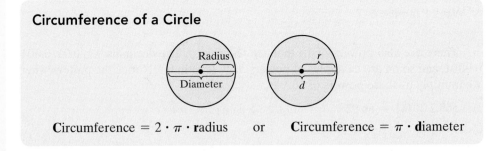

Circumference = 2 · π · **r**adius   or   Circumference = π · **d**iameter

In Section 4.2, we learned about the symbol π as the Greek letter pi, pronounced "pie." It is a constant between 3 and 4.

Approximations for π

Two common approximations for π are:

$$\pi \approx 3.14 \qquad \text{or} \qquad \pi \approx \frac{22}{7}$$

a decimal approximation   a fraction approximation

Answers

8. 0.762 **9.** 0.019 **10.** 7.682
11. 333,900,000

△ **Example 12**   Circumference of a Circle

Find the circumference of a circle whose radius is 5 inches. Then use the approximation 3.14 for π to approximate the circumference.

Solution: Circumference $= 2 \cdot \pi \cdot$ radius

$$= 2 \cdot \pi \cdot 5 \text{ inches}$$

$$= 10\pi \text{ inches}$$

5 inches

Next, we replace π with the approximation 3.14.

Circumference $= 10\pi$ inches

("is approximately") $\rightarrow$ $\approx 10(3.14)$ inches

$$= 31.4 \text{ inches}$$

The *exact* circumference or distance around the circle is 10π inches, which is *approximately* 31.4 inches.

■ **Work Practice 12**

Practice 12

Find the circumference of a circle whose radius is 11 meters. Then use the approximation 3.14 for π to approximate this circumference.

Objective E Solving Problems by Multiplying Decimals ▶

The solutions to many real-life problems are found by multiplying decimals. We continue using our four problem-solving steps to solve such problems.

Example 13   Finding the Total Cost of Materials for a Job

A college student is hired to paint a billboard with paint costing $2.49 per quart. If the job requires 3 quarts of paint, what is the total cost of the paint?

Solution:

1. **UNDERSTAND.** Read and reread the problem. The phrase "total cost" might make us think addition, but since this problem requires repeated addition, let's multiply.
2. **TRANSLATE.**

In words:	Total cost	is	cost per quart of paint	times	number of quarts
	↓	↓	↓	↓	↓
Translate:	Total cost	=	2.49	×	3

3. **SOLVE.** We can estimate to check our calculations. The number 2.49 rounds to 2 and $2 \times 3 = 6$.

$$\begin{array}{r} \overset{1\ 2}{2.49} \\ \times \quad 3 \\ \hline 7.47 \end{array}$$

4. **INTERPRET.** *Check* your work. Since 7.47 is close to our estimate of 6, our answer is reasonable. *State* your conclusion: The total cost of the paint is $7.47.

■ **Work Practice 13**

Practice 13

A biology major is fertilizing her garden. She uses 5.6 ounces of fertilizer per square yard. The garden measures 60.5 square yards. How much fertilizer does she need?

Answers

12. 22π m; 69.08 m   **13.** 338.8 oz

Vocabulary, Readiness & Video Check

Use the choices below to fill in each blank.

circumference	left	sum	zeros
decimal places	right	product	factor

1. When multiplying decimals, the number of decimal places in the product is equal to the _____ of the number of decimal places in the factors.

2. In $8.6 \times 5 = 43$, the number 43 is called the _____, while 8.6 and 5 are each called a(n) _____.

3. When multiplying a decimal number by powers of 10, such as 10, 100, 1000, and so on, we move the decimal point in the number to the _____ the same number of places as there are _____ in the power of 10.

4. When multiplying a decimal number by powers of 10, such as 0.1, 0.01, and so on, we move the decimal point in the number to the _____ the same number of places as there are _____ in the power of 10.

5. The distance around a circle is called its _____.

Do not multiply. Just give the number of decimal places in the product. See the Concept Check in this section.

6.	7.	8.	9.	10.	11.
0.46	57.9	0.428	0.0073	0.028	5.1296
× 0.81	× 0.36	× 0.2	× 21	× 1.36	× 7.3987

Martin-Gay Interactive Videos   *Watch the section lecture video and answer the following questions.*

See Video 4.4

Objective A   12. From ▣ Example 1, explain the difference between multiplying whole numbers and multiplying decimal numbers. ▶

Objective B   13. From ▣ Example 2, what does estimating especially help us with? ▶

Objective C   14. In ▣ Example 3, why don't we multiply as we did in ▣ Example 2? ▶

Objective D   15. Why is 25.12 meters not the exact answer to ▣ Example 5? ▶

Objective E   16. In ▣ Example 6, why is 24.8 not the complete answer? What is the complete answer? ▶

4.4   Exercise Set MyMathLab® ▶

Objectives **A** **B** **Mixed Practice** *Multiply. See Examples 1 through 4. For those exercises marked, also estimate to see if the answer is reasonable.*

1.	2.	▶ 3.	4.
0.2	0.7	1.2	6.8
× 0.6	× 0.9	× 0.5	× 0.3

5. 0.26×5

6. 0.19×6

7. 5.3×4.2
   Exact:
   Estimate:

8. 6.2×3.8
   Exact:
   Estimate:

9.	10.	▶ 11.	12.
0.576	0.971	1.0047	2.0005
× 0.7	× 0.5	× 8.2	× 5.5
		Exact: Estimate:	Exact: Estimate:

13.   490.2
    × 0.023

14.   300.9
    × 0.032

15. Multiply 16.003 and 5.31

16. Multiply 31.006 and 3.71

Objective C   *Multiply. See Examples 5 through 10.*

17. 6.5 × 10

18. 7.2 × 100

19. 6.5 × 0.1

20. 4.7 × 0.1

21. 7.2 × 0.01

22. 0.06 × 0.01

▶ **23.** 7.093 × 100

24. 0.5 × 100

25. 6.046 × 1000

26. 9.1 × 1000

▶ **27.** 37.62 × 0.001

28. 14.3 × 0.001

Objectives A B C   **Mixed Practice** *Multiply. See Examples 1 through 10.*

29. 0.123 × 0.4

30. 0.216 × 0.3

31. 0.123 × 100

32. 0.216 × 100

33. 8.6 × 0.15

34. 0.42 × 5.7

35. 9.6 × 0.01

36. 5.7 × 0.01

37. 562.3 × 0.001

38. 993.5 × 0.001

39.   5.62
    × 7.7

40.   8.03
    × 5.5

Write each number in standard notation. See Example 11.

41. The storage silos at the main Hershey chocolate factory in Hershey, Pennsylvania, can hold enough cocoa beans to make 5.5 billion Hershey's milk chocolate bars. (*Source:* Hershey Foods Corporation)

42. The total domestic revenue collected by Netflix in 2012 was $3.322 billion. (*Source:* Netflix, Inc.)

43. The Racer is the most-ridden roller coaster at King's Island, an amusement park near Cincinnati, Ohio. Since 1972, it has given more than 97.8 million rides. (*Source:* Cedar Fair, L.P.)

44. About 45.6 million American households own at least one dog. (*Source:* American Pet Products Association)

45. The most-visited national park in the United States in 2012 was the Blue Ridge Parkway in Virginia and North Carolina. An estimated 292 thousand people visited the park each week that year. (*Source:* National Park Service)

46. In 2012, approximately 17.2 thousand vessels passed through the Suez Canal. (*Source:* suezcanal. gov)

Objective **D** *Find the circumference of each circle. Then use the approximation 3.14 for π and approximate each circumference. See Example 12.*

47.   4 meters

48.   8 feet

49.   10 centimeters

50.   22 inches

51.   9.1 yards

52.   5.9 kilometers

Objectives **D** **E** **Mixed Practice** *Solve. For Exercises 53 and 54, the solutions have been started for you. See Examples 12 and 13. For circumference applications, find the exact circumference and then use 3.14 for π to approximate the circumference.*

53. An electrician for Central Power and Light worked 40 hours last week. Calculate his pay before taxes for last week if his hourly wage is $17.88.

Start the solution:

1. UNDERSTAND the problem. Reread it as many times as needed.
2. TRANSLATE into an equation. (Fill in the blanks.)

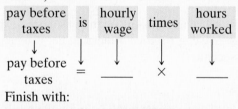

| pay before taxes | is | hourly wage | times | hours worked |

$$\text{pay before taxes} \quad = \quad \underline{\hspace{1cm}} \quad \times \quad \underline{\hspace{1cm}}$$

Finish with:
3. SOLVE and 4. INTERPRET.

54. An assembly line worker worked 20 hours last week. Her hourly rate is $19.52 per hour. Calculate her pay before taxes.

Start the solution:

1. UNDERSTAND the problem. Reread it as many times as needed.
2. TRANSLATE into an equation. (Fill in the blanks.)

| pay before taxes | is | hourly rate | times | hours worked |

$$\text{pay before taxes} \quad = \quad \underline{\hspace{1cm}} \quad \times \quad \underline{\hspace{1cm}}$$

Finish with:
3. SOLVE and 4. INTERPRET.

55. Under certain conditions, the average cost of driving a medium sedan in 2012 was $0.61 per mile. How much would it have cost to drive such a car 15,000 miles in 2012? (*Source:* American Automobile Association)

56. In September 2013, a U.S. airline passenger paid an average of $0.1643, disregarding taxes and fees, to fly 1 mile. Use this number to calculate the cost before taxes and fees to fly from Atlanta, Georgia, to Minneapolis, Minnesota, a distance of 905 miles. Round to the nearest cent. (*Source:* Airlines for America)

57. A 1-ounce serving of cream cheese contains 6.2 grams of saturated fat. How much saturated fat is in 4 ounces of cream cheese? (*Source: Home and Garden Bulletin No. 72;* U.S. Department of Agriculture)

58. A 3.5-ounce serving of lobster meat contains 0.1 gram of saturated fat. How much saturated fat do 3 servings of lobster meat contain? (*Source:* The National Institutes of Health)

△ **59.** Recall that the face of the Apple iPhone 5c (see Section 4.3) is a rectangle measuring 4.90 inches by 2.33 inches. Find the area of the face of the Apple iPhone 5c. (*Source:* Apple Inc.)

△ **60.** Recall that the rectangular face of the Samsung Galaxy S4 smartphone (see Section 4.3) measures 5.38 inches by 2.75 inches. Find the area of the face of the Samsung Galaxy S4. (*Source:* Based on data from Samsung Electronics Co.)

△ **61.** In 1893, the first ride called a Ferris wheel was constructed by Washington Gale Ferris. Its diameter was 250 feet. Find its circumference. Give an exact answer and an approximation using 3.14 for π. (*Source: The Handy Science Answer Book,* Visible Ink Press, 1994)

△ **62.** The radius of Earth is approximately 3950 miles. Find the distance around Earth at the equator. Give an exact answer and an approximation using 3.14 for π. (*Hint:* Find the circumference of a circle with radius 3950 miles.)

△ **63.** The London Eye, built for the Millennium celebration in London, resembles a gigantic Ferris wheel with a diameter of 135 meters. If Adam Hawn rides the Eye for one revolution, find how far he travels. Give an exact answer and an approximation using 3.14 for π. (*Source:* Londoneye.com)

△ **64.** The world's longest suspension bridge is the Akashi Kaikyo Bridge in Japan. This bridge has two circular caissons, which are underwater foundations. If the diameter of a caisson is 80 meters, find its circumference. Give an exact answer and an approximation using 3.14 for π. (*Source: Scientific American; How Things Work Today*)

80 meters

Caisson

65. A meter is a unit of length in the metric system that is approximately equal to 39.37 inches. Sophia Wagner is 1.65 meters tall. Find her approximate height in inches.

66. The doorway to a room is 2.15 meters tall. Approximate this height in inches. (*Hint:* See Exercise 65.)

△ **67. a.** Approximate the circumference of each circle.

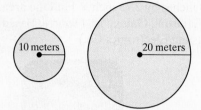

b. If the radius of a circle is doubled, is its corresponding circumference doubled?

69. Recall that the top face of the latest iPod nano (see Section 4.3) measures 3.01 inches by 1.56 inches. Find the area of the face of the iPod nano. Round to the nearest hundredth. (*Source:* Apple Inc.)

△ **68. a.** Approximate the circumference of each circle.

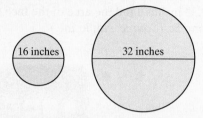

b. If the diameter of a circle is doubled, is its corresponding circumference doubled?

70. Recall that the top face of the latest iPod touch (see Section 4.3) measures 4.86 inches by 2.31 inches. Find the area of the face of the iPod touch. Round to the nearest hundredth. (*Source:* Apple Inc.)

Review

Divide. See Sections 1.7 and 2.5.

71. $130 \div 5$ **72.** $486 \div 27$ **73.** $2016 \div 56$ **74.** $1863 \div 69$

75. $2920 \div 365$ **76.** $2916 \div 6$ **77.** $\dfrac{24}{7} \div \dfrac{8}{21}$ **78.** $\dfrac{162}{25} \div \dfrac{9}{75}$

Concept Extensions

Mixed Practice (*Sections 4.3, 4.4*) *Perform the indicated operations.*

79. $3.6 + 0.04$ **80.** $7.2 + 0.14 + 98.6$ **81.** $3.6 - 0.04$

82. $100 - 48.6$ **83.** 0.221×0.5 **84.** 3.6×0.04

85. Find how far radio waves travel in 20.6 seconds. (Radio waves travel at a speed of $1.86 \times 100{,}000$ miles per second.)

86. If it takes radio waves approximately 8.3 minutes to travel from the Sun to the Earth, find approximately how far it is from the Sun to the Earth. (*Hint:* See Exercise 85.)

87. In your own words, explain how to find the number of decimal places in a product of decimal numbers.

88. In your own words, explain how to multiply by a power of 10.

89. Write down two decimal numbers whose product will contain 5 decimal places. Without multiplying, explain how you know your answer is correct.

90. Explain the process for multiplying a decimal number by a power of 10.

Operations on Decimals

Perform the indicated operations.

1. $1.6 + 0.97$ **2.** $3.2 + 0.85$ **3.** $9.8 - 0.9$ **4.** $10.2 - 6.7$

5. $\begin{array}{r} 0.8 \\ \times\, 0.2 \\ \hline \end{array}$ **6.** $\begin{array}{r} 0.6 \\ \times\, 0.4 \\ \hline \end{array}$ **7.** $8 + 2.16 + 0.9$ **8.** $6 + 3.12 + 0.6$

9. $\begin{array}{r} 9.6 \\ \times\, 0.5 \\ \hline \end{array}$ **10.** $\begin{array}{r} 8.7 \\ \times\, 0.7 \\ \hline \end{array}$ **11.** $\begin{array}{r} 123.6 \\ -\ 48.04 \\ \hline \end{array}$ **12.** $\begin{array}{r} 325.2 \\ -\ 36.08 \\ \hline \end{array}$

13. $25 + 0.026$ **14.** $0.125 + 44$ **15.** $100 - 17.3$ **16.** $300 - 26.1$

17. 2.8×100 **18.** 1.6×1000 **19.** $\begin{array}{r} 96.21 \\ 7.028 \\ +\ 121.7 \\ \hline \end{array}$ **20.** $\begin{array}{r} 0.268 \\ 1.93 \\ +\ 142.881 \\ \hline \end{array}$

21. Find the product of 1.2 and 5.   **22.** Find the sum of 1.2 and 5.

23. $\begin{array}{r} 12.004 \\ \times\quad 2.3 \\ \hline \end{array}$ **24.** $\begin{array}{r} 28.006 \\ \times\quad 5.2 \\ \hline \end{array}$

25. Subtract 4.6 from 10.   **26.** Subtract 0.26 from 18.

Answers

1. _____
2. _____
3. _____
4. _____
5. _____
6. _____
7. _____
8. _____
9. _____
10. _____
11. _____
12. _____
13. _____
14. _____
15. _____
16. _____
17. _____
18. _____
19. _____
20. _____
21. _____
22. _____
23. _____
24. _____
25. _____
26. _____

27. _____

28. _____

29. _____

30. _____

31. _____

32. _____

33. _____

34. _____

35. _____

36. _____

37. _____

38. _____

39. _____

40. _____

41. _____

42. _____

43. _____

27. 268.19
$+\,146.25$

28. 860.18
$+\,434.85$

29. $160 - 43.19$

30. $120 - 101.21$

31. 15.62×10

32. $15.62 + 10$

33. $15.62 - 10$

34. 117.26×2.6

35. $117.26 - 2.6$

36. $117.26 + 2.6$

37. 0.0072×0.06

38. 0.0025×0.03

39. $0.0072 + 0.06$

40. $0.03 - 0.0025$

41. 0.862×1000

42. 2.93×0.01

43. Estimate the distance in miles between Garden City, Kansas, and Wichita, Kansas, by rounding each given distance to the nearest ten.

4.5 Dividing Decimals and Order of Operations ▶

Objective A Dividing Decimals ▶

Dividing decimal numbers is similar to dividing whole numbers. The only difference is that we place a decimal point in the quotient. If the divisor is a whole number, we place the decimal point in the quotient directly above the decimal point in the dividend, and then divide as with whole numbers. Recall that division can be checked by multiplication.

Dividing by a Whole Number

Step 1: Place the decimal point in the quotient directly above the decimal point in the dividend.

Step 2: Divide as with whole numbers.

Example 1   Divide: $270.2 \div 7$. Check your answer.

Solution: We divide as usual. The decimal point in the quotient is directly above the decimal point in the dividend.

```
                    ┌─── Write the decimal point.
                   38.6  ← quotient
divisor →    7)270.2  ← dividend
            −21
              60
            −56
              4 2
            −4 2
                0
```

```
                      6 4
Check:    38.6   ← quotient
        ×    7   ← divisor
        270.2   ← dividend
```

The quotient is 38.6.

■ Work Practice 1

Example 2   Divide: $32\overline{)8.32}$

Solution: We divide as usual. The decimal point in the quotient is directly above the decimal point in the dividend.

```
                 0.26  ← quotient
divisor →  32)8.32  ← dividend
            −64
            192
           −192
              0
```

```
Check:    0.26   quotient
        × 32   divisor
          52
         7 80
         8.32   dividend
```

■ Work Practice 2

Sometimes to continue dividing we need to insert zeros after the last digit in the dividend.

Objectives

A Divide Decimals. ▶

B Estimate When Dividing Decimals. ▶

C Divide Decimals by Powers of 10. ▶

D Solve Problems by Dividing Decimals. ▶

E Review Order of Operations by Simplifying Expressions Containing Decimals. ▶

Practice 1

Divide: $370.4 \div 8$. Check your answer.

Practice 2

Divide: $48\overline{)34.08}$. Check your answer.

Answers
1. 46.3   **2.** 0.71

285

Practice 3

Divide and check.

a. $0.4 \div 8$

b. $13.62 \div 12$

Example 3 Divide and check: $0.5 \div 4$.

Solution:

$$
\begin{array}{r}
0.125 \\
4\overline{)0.500} \\
-4 \\
\hline
10 \\
-8 \\
\hline
20 \\
-20 \\
\hline
0
\end{array}
$$

Insert two 0s to continue dividing.

Check:

$$
\begin{array}{r}
\overset{12}{0.1\overset{}{2}5} \\
\times \quad 4 \\
\hline
0.500
\end{array}
$$

■ Work Practice 3

If the divisor is not a whole number, before we divide we need to move the decimal point to the right until the divisor is a whole number.

$$1.5\overline{)64.85}$$

divisor ⌐   ⌐ dividend

To understand how this works, let's rewrite

$$1.5\overline{)64.85} \quad \text{as} \quad \frac{64.85}{1.5}$$

and then multiply by 1 in the form of $\dfrac{10}{10}$. We use the form $\dfrac{10}{10}$ so that the denominator (divisor) becomes a whole number.

$$\frac{64.85}{1.5} = \frac{64.85}{1.5} \cdot 1 = \frac{64.85}{1.5} \cdot \frac{10}{10} = \frac{64.85 \cdot 10}{1.5 \cdot 10} = \frac{648.5}{15},$$

which can be written as $15\overline{)648.5}$. Notice that

$$1.5\overline{)64.85} \quad \text{is equivalent to} \quad 15\overline{)648.5}$$

The decimal points in the dividend and the divisor were both moved one place to the right, and the divisor is now a whole number. This procedure is summarized next:

Dividing by a Decimal

Step 1: Move the decimal point in the divisor to the right until the divisor is a whole number.

Step 2: Move the decimal point in the dividend to the right the *same number of places* as the decimal point was moved in Step 1.

Step 3: Divide. Place the decimal point in the quotient directly over the moved decimal point in the dividend.

Answers

3. a. 0.05 **b.** 1.135

Example 4 Divide: $10.764 \div 2.3$

Solution: We move the decimal points in the divisor and the dividend one place to the right so that the divisor is a whole number.

$2.3\overline{)10.764}$ becomes
$$
\begin{array}{r}
4.68 \\
23\overline{)107.64} \\
-92 \\
\hline
15\,6 \\
-13\,8 \\
\hline
1\,84 \\
-1\,84 \\
\hline
0
\end{array}
$$

■ Work Practice 4

Example 5 Divide: $5.264 \div 0.32$

Solution:

$0.32\overline{)5.264}$ becomes
$$
\begin{array}{r}
16.45 \\
32\overline{)526.40} \quad \text{Insert one 0.}\\
-32 \\
\hline
206 \\
-192 \\
\hline
14\,4 \\
-12\,8 \\
\hline
1\,60 \\
-1\,60 \\
\hline
0
\end{array}
$$

■ Work Practice 5

✓**Concept Check** Is it always true that the number of decimal places in a quotient equals the sum of the decimal places in the dividend and divisor?

Example 6 Divide: $17.5 \div 0.48$. Round the quotient to the nearest hundredth.

Solution: First we move the decimal points in the divisor and the dividend two places. Then we divide and round the quotient to the nearest hundredth.

hundredths place
$$
\begin{array}{r}
36.458 \approx 36.46 \\
48\overline{)1750.000} \\
-144 \\
\hline
310 \\
-288 \\
\hline
22\,0 \\
-19\,2 \\
\hline
2\,80 \\
-2\,40 \\
\hline
400 \\
-384 \\
\hline
16
\end{array}
$$

"is approximately"

When rounding to the nearest hundredth, carry the division process out to one more decimal place, the thousandths place.

■ Work Practice 6

Practice 4
Divide: $166.88 \div 5.6$

Practice 5
Divide: $1.976 \div 0.16$

Practice 6
Divide $23.4 \div 0.57$. Round the quotient to the nearest hundredth.

Answers
4. 29.8 **5.** 12.35 **6.** 41.05

✓Concept Check Answer
no

Objective B Estimating When Dividing Decimals

Just as for addition, subtraction, and multiplication of decimals, we can estimate when dividing decimals to check the reasonableness of our answer.

Example 7 Divide: $272.356 \div 28.4$. Then estimate to see whether the proposed result is reasonable.

Solution:

Exact:	Estimate 1		Estimate 2
9.59	9		10
$284.\overline{)2723.56}$	$30\overline{)270}$	or	$30\overline{)300}$
-2556			
$\overline{167\,5}$			
$-142\,0$			
$\overline{25\,56}$			
$-25\,56$			
$\overline{0}$			

The estimate is 9 or 10, so 9.59 is reasonable.

■ Work Practice 7

✓**Concept Check** If a quotient is to be rounded to the nearest thousandth, to what place should the division be carried out? (Assume that the division carries out to your answer.)

Objective C Dividing Decimals by Powers of 10

As with multiplication, there are patterns that occur when we divide decimals by powers of 10 such as 10, 100, 1000, and so on.

$$\frac{569.2}{10} = 56.92 \qquad \text{Move the decimal point } \textit{1 place} \text{ to the } \textit{left}.$$
— 1 zero

$$\frac{569.2}{10,000} = 0.05692 \qquad \text{Move the decimal point } \textit{4 places} \text{ to the } \textit{left}.$$
— 4 zeros

This pattern suggests the following rule:

> **Dividing Decimals by Powers of 10 such as 10, 100, or 1000**
>
> Move the decimal point of the dividend to the *left* the same number of places as there are *zeros* in the power of 10.

Examples Divide.

8. $\dfrac{786.1}{1000} = 0.7861$ Move the decimal point *3 places* to the *left*.
— 3 zeros

9. $\dfrac{0.12}{10} = 0.012$ Move the decimal point *1 place* to the *left*.
— 1 zero

■ Work Practice 8–9

Objective D Solving Problems by Dividing Decimals ▶

Many real-life problems involve dividing decimals.

Example 10 Calculating Materials Needed for a Job

A gallon of paint covers a 250-square-foot area. How many gallons of paint are needed to cover a wall that measures 1450 square feet? If only gallon containers of paint are available, how many gallon containers are needed?

Solution:

1. UNDERSTAND. Read and reread the problem. We need to know how many 250s are in 1450, so we divide.

2. TRANSLATE.

In words:
number of gallons	is	square feet	divided by	square feet per gallon
↓	↓	↓	↓	↓

Translate: number of gallons = 1450 ÷ 250

3. SOLVE. Let's see if our answer is reasonable by estimating. The dividend 1450 rounds to 1500 and divisor 250 rounds to 300. Then $1500 \div 300 = 5$.

$$
\begin{array}{r}
5.8 \\
250\overline{)1450.0} \\
-1250 \\
\hline
200\ 0 \\
-200\ 0 \\
\hline
0
\end{array}
$$

4. INTERPRET. *Check* your work. Since our estimate is close to our answer of 5, our answer is reasonable. *State* your conclusion: To paint the wall, 5.8 gallons of paint are needed. If only gallon containers of paint are available, then 6 gallon containers of paint are needed to complete the job.

■ Work Practice 10

Objective E Simplifying Expressions with Decimals ▶

In the remaining examples, we will review the order of operations by simplifying expressions that contain decimals.

Order of Operations

1. Perform all operations within parentheses ( ), brackets [ ], or other grouping symbols such as square roots or fraction bars, starting with the innermost set.
2. Evaluate any expressions with exponents.
3. Multiply or divide in order from left to right.
4. Add or subtract in order from left to right.

Example 11 Simplify: $723.6 \div 1000 \times 10$

Solution: Multiply or divide in order from left to right.

$$723.6 \div 1000 \times 10 = 0.7236 \times 10 \quad \text{Divide.}$$
$$= 7.236 \quad \text{Multiply.}$$

■ Work Practice 11

Practice 11
Simplify: $897.8 \div 100 \times 10$

Answers
10. 11.84 bags; 12 bags **11.** 89.78

Practice 12

Simplify: $8.69(3.2 - 1.8)$

Example 12   Simplify: $0.5(8.6 - 1.2)$

Solution: According to the order of operations, we simplify inside the parentheses first.

$$0.5(8.6 - 1.2) = 0.5(7.4) \quad \text{Subtract.}$$
$$= 3.7 \quad \text{Multiply.}$$

■ Work Practice 12

Practice 13

Simplify: $\dfrac{20.06 - (1.2)^2 \div 10}{0.02}$

Example 13   Simplify: $\dfrac{5.68 + (0.9)^2 \div 100}{0.2}$

Solution: First we simplify the numerator of the fraction. Then we divide.

$$\frac{5.68 + (0.9)^2 \div 100}{0.2} = \frac{5.68 + 0.81 \div 100}{0.2} \quad \text{Simplify } (0.9)^2.$$
$$= \frac{5.68 + 0.0081}{0.2} \quad \text{Divide.}$$
$$= \frac{5.6881}{0.2} \quad \text{Add.}$$
$$= 28.4405 \quad \text{Divide.}$$

■ Work Practice 13

Answers

12. 12.166 **13.** 995.8

Calculator Explorations

Calculator errors can easily be made by pressing an incorrect key or by not pressing a correct key hard enough. Estimation is a valuable tool that can be used to check calculator results.

Example   Use estimation to determine whether the calculator result is reasonable or not. (For example, a result that is not reasonable can occur if proper keys are not pressed.)

Simplify: $82.064 \div 23$

Calculator display: $\boxed{35.68}$

Solution: Round each number to the nearest 10. Since $80 \div 20 = 4$, the calculator display 35.68 is not reasonable.

Use estimation to determine whether each result is reasonable or not.

1. 102.62×41.8 Result: 428.9516

2. $174.835 \div 47.9$ Result: 3.65

3. $1025.68 - 125.42$ Result: 900.26

4. $562.781 + 2.96$ Result: 858.781

Vocabulary, Readiness & Video Check

Use the choices below to fill in each blank. Some choices may be used more than once, and some not used at all.

dividend	divisor	quotient	true
zeros	left	right	false

1. In $6.5 \div 5 = 1.3$, the number 1.3 is called the _____, 5 is the _____, and 6.5 is the _____.

2. To check a division exercise, we can perform the following multiplication: quotient · _____ = _____.

3. To divide a decimal number by a power of 10, such as 10, 100, 1000, and so on, we move the decimal point in the number to the _____ the same number of places as there are _____ in the power of 10.

4. True or false: If $1.058 \div 0.46 = 2.3$, then $2.3 \times 0.46 = 1.058$. _____

Recall properties of division and simplify.

5. $\dfrac{5.9}{1}$

6. $\dfrac{0.7}{0.7}$

7. $\dfrac{0}{9.86}$

8. $\dfrac{2.36}{0}$

9. $\dfrac{7.261}{7.261}$

10. $\dfrac{8.25}{1}$

11. $\dfrac{11.1}{0}$

12. $\dfrac{0}{89.96}$

Martin-Gay Interactive Videos

Watch the section lecture video and answer the following questions.

See Video 4.5

Objective A 13. From the lecture before ⊞ Example 2, what must we make sure the divisor is before dividing decimals? ▶

Objective B 14. In ⊞ Example 4, what is the estimated answer and what is the exact answer? ▶

Objective C 15. In ⊞ Example 6, why don't we divide as we did in ⊞ Examples 1–3? ▶

Objective D 16. In ⊞ Example 7, why is the division carried to the hundredths place? ▶

Objective E 17. In ⊞ Example 9, besides meaning division, what other purpose does the fraction bar serve? ▶

4.5 **Exercise Set** MyMathLab® ▶

Objectives A B Mixed Practice *Divide. See Examples 1 through 5 and 7. For those exercises marked, also estimate to see if the answer is reasonable.*

1. $3\overline{)13.8}$

2. $2\overline{)11.8}$

▶ 3. $5\overline{)0.47}$

4. $6\overline{)0.51}$

5. $0.06\overline{)18}$

6. $0.04\overline{)20}$

▶ 7. $0.82\overline{)4.756}$

8. $0.92\overline{)3.312}$

▶ 9. $5.5\overline{)36.3}$
Exact:
Estimate:

10. $2.2\overline{)21.78}$
Exact:
Estimate:

11. $6.195 \div 15$

12. $8.823 \div 17$

13. $0.54 \div 12$

14. $1.35 \div 18$

15. Divide 4.2 by 0.6.

16. Divide 3.6 by 0.9.

17. $0.27\overline{)1.296}$

18. $0.34\overline{)2.176}$

19. $0.02\overline{)42}$

20. $0.03\overline{)24}$

21. $0.6\overline{)18}$

22. $0.4\overline{)20}$

23. $0.005\overline{)35}$

24. $0.0007\overline{)35}$

25. $7.2\overline{)70.56}$
Exact:
Estimate:

26. $6.3\overline{)54.18}$ **27.** $5.4\overline{)51.84}$ **28.** $7.7\overline{)33.88}$ **29.** $\dfrac{1.215}{0.027}$ **30.** $\dfrac{3.213}{0.051}$
Exact:
Estimate:

31. $0.25\overline{)13.648}$ **32.** $0.75\overline{)49.866}$ **33.** $3.78\overline{)0.02079}$ **34.** $2.96\overline{)0.01332}$

Divide. Round the quotients as indicated. See Example 6.

35. Divide 429.34 by 2.4 and round the quotient to the nearest whole number.

36. Divide 54.8 by 2.6 and round the quotient to the nearest whole number.

▶ **37.** Divide 0.549 by 0.023 and round the quotient to the nearest hundredth.

38. Divide 0.0453 by 0.98 and round the quotient to the nearest thousandth.

39. Divide 45.23 by 0.4 and round the quotient to the nearest tenth.

40. Divide 83.32 by 0.6 and round the quotient to the nearest tenth.

Objective C *Divide. See Examples 8 and 9.*

▶ **41.** $\dfrac{54.982}{100}$ **42.** $\dfrac{342.54}{100}$ **43.** $\dfrac{26.87}{10}$ **44.** $\dfrac{13.49}{10}$ ▶ **45.** $\dfrac{12.9}{1000}$ **46.** $\dfrac{0.27}{1000}$

Objectives A C Mixed Practice *Divide. See Examples 1–5, 8, and 9.*

47. $7\overline{)88.2}$ **48.** $9\overline{)130.5}$ **49.** $\dfrac{13.1}{10}$ **50.** $\dfrac{17.7}{10}$

51. $6.8\overline{)83.13}$ **52.** $4.8\overline{)123.72}$ **53.** $\dfrac{456.25}{10,000}$ **54.** $\dfrac{986.11}{10,000}$

Objective D *Solve. For Exercises 55 and 56, the solutions have been started for you. See Example 10.*

55. Josef Jones is painting the walls of a room. The walls have a total area of 546 square feet. A quart of paint covers 52 square feet. If he must buy paint in whole quarts, how many quarts does he need?

Start the solution:
1. UNDERSTAND the problem. Reread it as many times as needed.
2. TRANSLATE into an equation. (Fill in the blanks.)

number of quarts	is	square feet	divided by	square feet per quart
↓	↓	↓	↓	↓

$$\text{number of quarts} = \underline{\quad\quad} \div \underline{\quad\quad}$$

3. SOLVE. Don't forget to round up your quotient.
4. INTERPRET.

56. A shipping box can hold 36 books. If 486 books must be shipped, how many boxes are needed?

Start the solution:
1. UNDERSTAND the problem. Reread it as many times as needed.
2. TRANSLATE into an equation. (Fill in the blanks.)

number of boxes	is	number of books	divided by	books per box
↓	↓	↓	↓	↓

$$\text{number of boxes} = \underline{\quad\quad} \div \underline{\quad\quad}$$

3. SOLVE. Don't forget to round up your quotient.
4. INTERPRET.

△ **57.** A pound of fertilizer covers 39 square feet of lawn. Vivian Bulgakov's lawn measures 7883.5 square feet. How much fertilizer, to the nearest tenth of a pound, does she need to buy?

58. A page of a book contains about 1.5 kilobytes of information. If a computer disk can hold 740 kilobytes of information, how many pages of a book can be stored on one computer disk? Round to the nearest tenth of a page.

59. There are approximately 39.37 inches in 1 meter. How many meters, to the nearest tenth of a meter, are there in 200 inches?

←——— 1 meter ———→
←—— ≈39.37 inches ——→

60. There are 2.54 centimeters in 1 inch. How many inches are there in 50 centimeters? Round to the nearest tenth.

←——— 1 inch ———→
←—— 2.54 cm ——→

▶ **61.** In the United States, an average child will wear down 730 crayons by his or her tenth birthday. Find the number of boxes of 64 crayons this is equivalent to. Round to the nearest tenth. (*Source:* Binney & Smith Inc.)

62. During a recent year, American farmers received an average of $71.90 per hundred pounds of turkey. What was the average price per pound for turkeys? Round to the nearest cent. (*Source:* National Agricultural Statistics Service)

A child is to receive a dose of 0.5 teaspoon of cough medicine every 4 hours. If the bottle contains 4 fluid ounces, answer Exercises 63 through 66.

63. A fluid ounce equals 6 teaspoons. How many teaspoons are in 4 fluid ounces?

64. The bottle of medicine contains how many doses for the child? (*Hint:* See Exercise 63.)

65. If the child takes a dose every four hours, how many days will the medicine last?

66. If the child takes a dose every six hours, how many days will the medicine last?

Solve.

67. Americans ages 20–34 drive, on average, 15,098 miles per year. About how many miles each week is that? Round to the nearest tenth. (*Note:* There are 52 weeks in a year.) (*Source:* U.S. Office of Highway Policy Information)

68. Drake Saucier was interested in the gas mileage on his "new" used car. He filled the tank, drove 423.8 miles, and filled the tank again. When he refilled the tank, it took 19.35 gallons of gas. Calculate the miles per gallon for Drake's car. Round to the nearest tenth.

69. During the 24 Hours of the Le Mans endurance auto race in 2013, the winning team of Tom Kristensen, Allan McNish, and Loïc Duval drove a total of 2947.1 miles in 24 hours. What was their average speed in miles per hour? Round to the nearest tenth. (*Source:* Based on data from lemans-history.com)

70. During the 2012 Summer Olympics, Ethiopian runner Meseret Defar took the gold medal in the women's 5000-meter event. Her time for the event was 904.25 seconds. What was her average speed in meters per second? Round to the nearest tenth. (*Source:* International Olympic Committee)

71. Elena Delle Donne of the Chicago Sky was the WNBA's Rookie of the Year for 2013. She scored a total of 543 points in the 30 games she played in the 2013 regular season. What was the average number of points she scored per game? (*Source:* Women's National Basketball Association)

72. During the 2012 National Football League regular season, the top-scoring team was the New England Patriots with a total of 557 points throughout the season. The Patriots played 16 games. What was the average number of points the team scored per game? Round to the nearest hundredth. (*Source:* National Football League)

Objective E *Simplify each expression. See Examples 11 through 13.*

▶ 73. $0.7(6 - 2.5)$

74. $1.4(2 - 1.8)$

75. $\dfrac{0.29 + 1.69}{3}$

76. $\dfrac{1.697 - 0.29}{0.7}$

77. $30.03 + 5.1 \times 9.9$

78. $60 - 6.02 \times 8.97$

79. $7.8 - 4.83 \div 2.1 + 9.2$

80. $90 - 62.1 \div 2.7 + 8.6$

81. $93.07 \div 10 \times 100$

82. $35.04 \div 100 \times 10$

▶ 83. $\dfrac{7.8 + 1.1 \times 100 - 3.6}{0.2}$

84. $\dfrac{9.6 - 7.8 \div 10 + 1.2}{0.02}$

85. $5(20.6 - 2.06) - (0.8)^2$

86. $(10.6 - 9.8)^2 \div 0.01 + 8.6$

87. $6 \div 0.1 + 8.9 \times 10 - 4.6$

88. $8 \div 10 + 7.6 \times 0.1 - (0.1)^2$

Review

Write each decimal as a fraction. See Section 4.1.

89. 0.9

90. 0.7

91. 0.05

92. 0.08

Concept Extensions

Mixed Practice (Sections 4.3, 4.4, 4.5) *Perform the indicated operation.*

93. $1.278 \div 0.3$

94. 1.278×0.3

95. $1.278 + 0.3$

96. $1.278 - 0.3$

97. $\begin{array}{r} 8.6 \\ \times\ 3.1 \\ \hline \end{array}$

98. $7.2 + 0.05 + 49.1$

99. $\begin{array}{r} 1000 \\ -\ 95.71 \\ \hline \end{array}$

100. $\dfrac{87.2}{10{,}000}$

Choose the best estimate.

101. 8.62×41.7
    a. 36
    b. 32
    c. 360
    d. 3.6

102. $1.437 + 20.69$
    a. 34
    b. 22
    c. 3.4
    d. 2.2

103. $78.6 \div 97$
    a. 7.86
    b. 0.786
    c. 786
    d. 7860

104. $302.729 - 28.697$
    a. 270
    b. 20
    c. 27
    d. 300

Recall from Section 1.7 that the average of a list of numbers is their total divided by how many numbers there are in the list. Use this procedure to find the average of the test scores listed in Exercises 105 and 106. If necessary, round to the nearest tenth.

105. 86, 78, 91, 87

106. 56, 75, 80

△**107.** The area of a rectangle is 38.7 square feet. If its width is 4.5 feet, find its length.

38.7 square feet   4.5 feet

?

△**108.** The perimeter of a square is 180.8 centimeters. Find the length of a side.

Perimeter is 180.8 centimeters

?

109. When dividing decimals, describe the process you use to place the decimal point in the quotient.

110. In your own words, describe how to quickly divide a number by a power of 10 such as 10, 100, 1000, etc.

To convert wind speeds in miles per hour to knots, divide by 1.15. Use this information and the Saffir-Simpson Hurricane Intensity chart below to answer Exercises 111 and 112. Round to the nearest tenth.

Saffir-Simpson Hurricane Intensity Scale				
Category	Wind Speed	Barometric Pressure [inches of mercury (Hg)]	Storm Surge	Damage Potential
1 (Weak)	75–95 mph	≥ 28.94 in.	4–5 ft	Minimal damage to vegetation
2 (Moderate)	96–110 mph	28.50–28.93 in.	6–8 ft	Moderate damage to houses
3 (Strong)	111–130 mph	27.91–28.49 in.	9–12 ft	Extensive damage to small buildings
4 (Very Strong)	131–155 mph	27.17–27.90 in.	13–18 ft	Extreme structural damage
5 (Devastating)	> 155 mph	< 27.17 in.	>18 ft	Catastrophic building failures possible

111. The chart gives wind speeds in miles per hour. What is the range of wind speeds for a Category 1 hurricane in knots?

112. What is the range of wind speeds for a Category 4 hurricane in knots?

113. A rancher is building a horse corral that's shaped like a rectangle with dimensions of 24.28 meters by 15.675 meters. He plans to make a four-wire fence; that is, he will string four wires around the corral. How much wire will he need?

114. A college student signed up for a new credit card that guarantees her no interest charges on transferred balances for a year. She transferred over a $2523.86 balance from her old credit card. Her minimum payment is $185.35 per month. If she only pays the minimum, will she pay off her balance before interest charges start again?

Objective A Writing Fractions as Decimals ▶

To write a fraction as a decimal, we interpret the fraction bar to mean division and find the quotient.

> **Writing Fractions as Decimals**
>
> To write a fraction as a decimal, divide the numerator by the denominator.

Practice 1

a. Write $\dfrac{2}{5}$ as a decimal.

b. Write $\dfrac{9}{40}$ as a decimal.

Example 1 Write $\dfrac{1}{4}$ as a decimal.

Solution: $\dfrac{1}{4} = 1 \div 4$

$$
\begin{array}{r}
0.25 \\
4\overline{)1.00} \\
-8 \\
\hline
20 \\
-20 \\
\hline
0
\end{array}
$$

Thus, $\dfrac{1}{4}$ written as a decimal is 0.25.

▬ Work Practice 1

Practice 2

a. Write $\dfrac{5}{6}$ as a decimal.

b. Write $\dfrac{2}{9}$ as a decimal.

Example 2 Write $\dfrac{2}{3}$ as a decimal.

Solution:

$$
\begin{array}{r}
0.666\ldots \\
3\overline{)2.000} \\
-18 \\
\hline
20 \\
-18 \\
\hline
20 \\
-18 \\
\hline
2
\end{array}
$$

This pattern will continue because $\dfrac{2}{3} = 0.6666\ldots$

Remainder is 2, then 0 is brought down.

Remainder is 2, then 0 is brought down.

Remainder is 2.

Notice the digit 2 keeps occurring as the remainder. This will continue so that the digit 6 will keep repeating in the quotient. We place a bar over the digit 6 to indicate that it repeats.

$$\frac{2}{3} = 0.666\ldots = 0.\overline{6}$$

We can also write a decimal approximation for $\dfrac{2}{3}$. For example, $\dfrac{2}{3}$ rounded to the nearest hundredth is 0.67. This can be written as $\dfrac{2}{3} \approx 0.67$.

▬ Work Practice 2

Answers

1. a. 0.4 **b.** 0.225
2. a. 0.8$\overline{3}$ **b.** 0.$\overline{2}$

Example 3 Write $\frac{22}{7}$ as a decimal. (The fraction $\frac{22}{7}$ is an approximation for π.) Round to the nearest hundredth.

Practice 3

Write $\frac{28}{13}$ as a decimal. Round to the nearest thousandth.

Solution:

$$
\begin{array}{r}
3.142 \approx 3.14 \quad \text{\small Carry the division out to the thousandths place.} \\
7\overline{)22.000} \\
-21 \quad\;\;\;\; \\
\hline
1\,0 \quad\;\; \\
-\;7 \quad\;\; \\
\hline
30 \quad \\
-28 \quad \\
\hline
20 \\
-14 \\
\hline
6
\end{array}
$$

The fraction $\frac{22}{7}$ in decimal form is approximately 3.14. Thus, $\pi \approx \frac{22}{7}$ (a fraction approximation for π) and $\pi \approx 3.14$ (a decimal approximation for π).

■ Work Practice 3

Example 4 Write $2\frac{3}{16}$ as a decimal.

Solution:

Option 1. Write the fractional part only as a decimal.

$$
\frac{3}{16} \longrightarrow
\begin{array}{r}
0.1875 \\
16\overline{)3.0000} \\
-1\,6\;\;\;\;\; \\
\hline
1\,40\;\;\; \\
-1\,28\;\;\; \\
\hline
120\; \\
-112\; \\
\hline
80 \\
-80 \\
\hline
0
\end{array}
$$

Thus $2\frac{3}{16} = 2.1875$

Option 2. Write $2\frac{3}{16}$ as an improper fraction, and divide.

$$
2\frac{3}{16} = \frac{35}{16} \longrightarrow
\begin{array}{r}
2.1875 \\
16\overline{)35.0000} \\
-32\;\;\;\;\; \\
\hline
3\,0\;\;\; \\
-1\,6\;\;\; \\
\hline
1\,40\; \\
-1\,28\; \\
\hline
120 \\
-112 \\
\hline
80 \\
-80 \\
\hline
0
\end{array}
$$

Thus $2\frac{3}{16} = 2.1875$

■ Work Practice 4

Some fractions may be written as decimals using our knowledge of decimals. From Section 4.1, we know that if the denominator of a fraction is 10, 100, 1000, or so on, we can immediately write the fraction as a decimal. For example,

$$\frac{4}{10} = 0.4, \qquad \frac{12}{100} = 0.12, \text{ and so on.}$$

Practice 5

Write $\frac{3}{5}$ as a decimal.

Example 5 Write $\frac{4}{5}$ as a decimal.

Solution: Let's write $\frac{4}{5}$ as an equivalent fraction with a denominator of 10.

$$\frac{4}{5} = \frac{4}{5} \cdot \frac{2}{2} = \frac{8}{10} = 0.8$$

◼ Work Practice 5

Practice 6

Write $\frac{3}{50}$ as a decimal.

Example 6 Write $\frac{1}{25}$ as a decimal.

Solution: $\frac{1}{25} = \frac{1}{25} \cdot \frac{4}{4} = \frac{4}{100} = 0.04$

◼ Work Practice 6

✓**Concept Check** Suppose you are writing the fraction $\frac{9}{16}$ as a decimal. How do you know you have made a mistake if your answer is 1.735?

Objective B Comparing Fractions and Decimals

Now we can compare decimals and fractions by writing fractions as equivalent decimals.

Practice 7

Insert $<$, $>$, or $=$ to form a true statement.

$$\frac{1}{5} \qquad 0.25$$

Example 7 Insert $<$, $>$, or $=$ to form a true statement.

$$\frac{1}{8} \qquad 0.12$$

Solution: First we write $\frac{1}{8}$ as an equivalent decimal. Then we compare decimal places.

$$\begin{array}{r} 0.125 \\ 8)\overline{1.000} \\ -8 \\ \hline 20 \\ -16 \\ \hline 40 \\ -40 \\ \hline 0 \end{array}$$

Original numbers	$\frac{1}{8}$	0.12
Decimals	0.125	0.120
Compare	0.125 > 0.12	

Thus, $\frac{1}{8} > 0.12$

◼ Work Practice 7

Practice 8

Insert $<$, $>$, or $=$ to form a true statement.

a. $\frac{1}{2}$ 0.54 **b.** $0.\overline{4}$ $\frac{4}{9}$

c. $\frac{5}{7}$ 0.72

Answers

5. 0.6 6. 0.06 7. <
8. a. < b. = c. <

✓**Concept Check Answer**

$\frac{9}{16}$ is less than 1 while 1.735 is greater than 1.

Example 8 Insert $<$, $>$, or $=$ to form a true statement.

$$0.\overline{7} \qquad \frac{7}{9}$$

Solution: We write $\frac{7}{9}$ as a decimal and then compare.

$$\begin{array}{r} 0.77\ldots = 0.\overline{7} \\ 9)\overline{7.00} \\ -6\,3 \\ \hline 70 \\ -63 \\ \hline 7 \end{array}$$

Original numbers	$0.\overline{7}$	$\frac{7}{9}$
Decimals	$0.\overline{7}$	$0.\overline{7}$
Compare	$0.\overline{7} = 0.\overline{7}$	

Thus, $0.\overline{7} = \frac{7}{9}$

◼ Work Practice 8

Example 9 Write the numbers in order from smallest to largest.

$$\frac{9}{20}, \frac{4}{9}, 0.456$$

Solution:

Original numbers	$\frac{9}{20}$	$\frac{4}{9}$	0.456
Decimals	0.450	0.444 . . .	0.456
Compare in order	2nd	1st	3rd

Written in order, we have

1st 2nd 3rd
↓   ↓   ↓
$$\frac{4}{9}, \frac{9}{20}, 0.456$$

■ Work Practice 9

Practice 9

Write the numbers in order from smallest to largest.

a. $\frac{1}{3}, 0.302, \frac{3}{8}$ **b.** $1.26, 1\frac{1}{4}, 1\frac{2}{5}$

c. $0.4, 0.41, \frac{5}{7}$

Objective C Solving Area Problems Containing Fractions and Decimals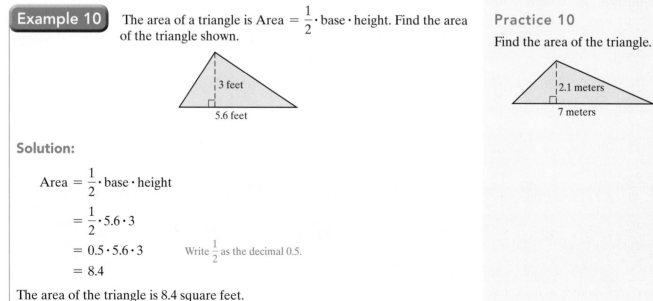

Sometimes real-life problems contain both fractions and decimals. In this section, we solve such problems concerning area. In the next example, we review the area of a triangle. This concept will be studied more in depth in a later chapter.

Example 10 The area of a triangle is Area $= \frac{1}{2} \cdot$ base $\cdot$ height. Find the area of the triangle shown.

3 feet

5.6 feet

Solution:

$$\text{Area} = \frac{1}{2} \cdot \text{base} \cdot \text{height}$$

$$= \frac{1}{2} \cdot 5.6 \cdot 3$$

$$= 0.5 \cdot 5.6 \cdot 3 \qquad \text{Write } \frac{1}{2} \text{ as the decimal } 0.5.$$

$$= 8.4$$

The area of the triangle is 8.4 square feet.

■ Work Practice 10

Practice 10

Find the area of the triangle.

2.1 meters

7 meters

Answers

9. a. $0.302, \frac{1}{3}, \frac{3}{8}$ **b.** $1\frac{1}{4}, 1.26, 1\frac{2}{5}$

c. $0.4, 0.41, \frac{5}{7}$ **10.** 7.35 sq m

Vocabulary, Readiness & Video Check

Answer each exercise "true" or "false."

1. The number $0.\overline{5}$ means 0.555.

2. To write $\dfrac{9}{19}$ as a decimal, perform the division $9\overline{)19}$.

3. $(1.2)^2$ means $(1.2)(1.2)$ or 1.44.

4. The area of a figure is written in *square* units.

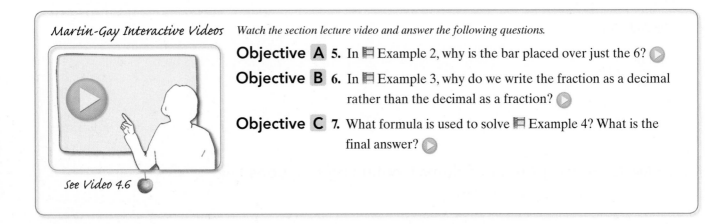

Martin-Gay Interactive Videos   Watch the section lecture video and answer the following questions.

Objective A **5.** In Example 2, why is the bar placed over just the 6?

Objective B **6.** In Example 3, why do we write the fraction as a decimal rather than the decimal as a fraction?

Objective C **7.** What formula is used to solve Example 4? What is the final answer?

See Video 4.6

4.6 Exercise Set MyMathLab®

Objective A *Write each number as a decimal. See Examples 1 through 6.*

1. $\dfrac{1}{5}$ **2.** $\dfrac{1}{20}$ **3.** $\dfrac{17}{25}$ **4.** $\dfrac{13}{25}$ **5.** $\dfrac{3}{4}$

6. $\dfrac{3}{8}$ **7.** $\dfrac{2}{25}$ **8.** $\dfrac{3}{25}$ **9.** $\dfrac{6}{5}$ **10.** $\dfrac{5}{4}$

11. $\dfrac{11}{12}$ **12.** $\dfrac{5}{12}$ **13.** $\dfrac{17}{40}$ **14.** $\dfrac{19}{25}$ **15.** $\dfrac{9}{20}$

16. $\dfrac{31}{40}$ **17.** $\dfrac{1}{3}$ **18.** $\dfrac{7}{9}$ **19.** $\dfrac{7}{16}$ **20.** $\dfrac{9}{16}$

21. $\dfrac{7}{11}$ **22.** $\dfrac{9}{11}$ **23.** $5\dfrac{17}{20}$ **24.** $4\dfrac{7}{8}$ **25.** $\dfrac{78}{125}$ **26.** $\dfrac{159}{375}$

Round each number as indicated.

27. Round your decimal answer to Exercise 17 to the nearest hundredth.

28. Round your decimal answer to Exercise 18 to the nearest hundredth.

29. Round your decimal answer to Exercise 19 to the nearest hundredth.

30. Round your decimal answer to Exercise 20 to the nearest hundredth.

31. Round your decimal answer to Exercise 21 to the nearest tenth.

32. Round your decimal answer to Exercise 22 to the nearest tenth.

Write each fraction as a decimal. If necessary, round to the nearest hundredth. See Examples 1 through 6.

33. Of the U.S. mountains that are over 14,000 feet in elevation, $\frac{56}{91}$ are located in Colorado. (*Source:* U.S. Geological Survey)

34. About $\frac{21}{50}$ of all blood donors have type A blood. (*Source:* American Red Cross Biomedical Services)

35. As of the end of 2012, $\frac{191}{500}$ of all U.S. households were wireless-only households, meaning they no longer subscribe to landline telephone services. (*Source:* National Center for Health Statistics)

36. Porsche is the auto manufacturer with the most wins at the 24 Hours of Le Mans endurance race. By 2013, $\frac{16}{81}$ of Le Mans races had been won in Porsche vehicles. (*Source:* lemans-history.com)

37. When first launched, the Hubble Space Telescope's primary mirror was out of shape on the edges by $\frac{1}{50}$ of a human hair. This very small defect made it difficult to focus on faint objects being viewed. Because the HST was in low Earth orbit, it was serviced by a shuttle and the defect was corrected.

38. The two mirrors currently in use in the Hubble Space Telescope were ground so that they do not deviate from a perfect curve by more than $\frac{1}{800,000}$ of an inch. Do not round this number.

Objective B *Insert $<$, $>$, or $=$ to form a true statement. See Examples 7 and 8.*

39. 0.562 ___ 0.569

40. 0.983 ___ 0.988

41. 0.215 ___ $\frac{43}{200}$

42. $\frac{29}{40}$ ___ 0.725

43. $\frac{9}{100}$ ___ 0.0932

44. $\frac{1}{200}$ ___ 0.00563

45. $0.\overline{6}$ ___ $\frac{5}{6}$

46. $0.\overline{1}$ ___ $\frac{2}{17}$

47. $\frac{51}{91}$ ___ $0.56\overline{4}$

48. $0.58\overline{3}$ ___ $\frac{6}{11}$

49. $\frac{1}{9}$ ___ 0.1

50. 0.6 ___ $\frac{2}{3}$

51. 1.38 ___ $\frac{18}{13}$

52. 0.372 ___ $\frac{22}{59}$

53. 7.123 ___ $\frac{456}{64}$

54. 12.713 ___ $\frac{89}{7}$

Write the numbers in order from smallest to largest. See Example 9.

55. 0.34, 0.35, 0.32

56. 0.47, 0.42, 0.40

57. 0.49, 0.491, 0.498

58. 0.72, 0.727, 0.728

59. $\frac{3}{4}$, 0.78, 0.73

60. $\frac{2}{5}$, 0.49, 0.42

61. $\frac{4}{7}$, 0.453, 0.412

62. $\frac{6}{9}$, 0.663, 0.668

63. 5.23, $\frac{42}{8}$, 5.34

64. 7.56, $\frac{67}{9}$, 7.562

65. $\frac{12}{5}$, 2.37, $\frac{17}{8}$

66. $\frac{29}{16}$, 1.75, $\frac{59}{32}$

Objective C *Find the area of each triangle or rectangle. See Example 10.*

△ **67.** 9 inches   5.7 inches

△ **68.** 4.4 feet   17 feet

△ **69.** 3.6 centimeters   5.2 centimeters

△ **70.** 10 meters   25.6 meters

▶ **71.** 0.62 yard $\frac{2}{5}$ yard
△

△ **72.** 1.2 miles $\frac{7}{8}$ mile

Review

Simplify. See Sections 1.9 and 3.5.

73. 2^3

74. 5^4

75. $6^2 \cdot 2$

76. $4 \cdot 3^4$

77. $\left(\dfrac{1}{3}\right)^4$

78. $\left(\dfrac{4}{5}\right)^3$

79. $\left(\dfrac{3}{5}\right)^2$

80. $\left(\dfrac{7}{2}\right)^2$

81. $\left(\dfrac{2}{5}\right)\left(\dfrac{5}{2}\right)^2$

82. $\left(\dfrac{2}{3}\right)^2\left(\dfrac{3}{2}\right)^3$

Concept Extensions

Without calculating, describe each number as < 1, $= 1$, or > 1. See the Concept Check in this section.

83. 1.0

84. 1.0000

85. 1.00001

86. $\dfrac{101}{99}$

87. $\dfrac{99}{100}$

88. $\dfrac{99}{99}$

In 2012, there were 15,012 broadcast radio stations in the United States. The most popular formats are shown in the graph along with their counts. Use this graph to answer Exercises 89–92. (Source: Arbitron Radio Station Information Database)

Top Broadcast Radio Station Formats in 2012

Format (Total stations: 15,012)

89. Write the fraction of radio stations with a news/talk format as a decimal. Round to the nearest hundredth.

90. Write the fraction of radio stations with a country music format as a decimal. Round to the nearest thousandth.

91. Estimate, by rounding each number in the table to the nearest hundred, the total number of stations with the top eight formats in 2012.

92. Use your estimate from Exercise 91 to write the fraction of radio stations accounted for by the top eight formats as a decimal. Round to the nearest hundredth.

93. Describe two ways to determine the larger of two fractions.

94. Describe two ways to write fractions as decimals.

95. Describe two ways to write mixed numbers as decimals.

96. Do you prefer performing operations on decimals or fractions? Why?

Find the value of each expression. Give the result as a decimal.

97. $(9.6)(5) - \dfrac{3}{4}$

98. $2(7.8) - \dfrac{1}{5}$

99. $\left(\dfrac{1}{10}\right)^2 + (1.6)(2.1)$

100. $8.25 - \left(\dfrac{1}{2}\right)^2$

101. $\dfrac{3}{8}(5.9 - 4.7)$

102. $\dfrac{1}{4}(9.6 + 5.2)$

Chapter 4 Group Activity

Maintaining a Checking Account

(Sections 4.1, 4.2, 4.3, 4.4)

This activity may be completed by working in groups or individually.

A checking account is a convenient way of handling money and paying bills. To open a checking account, the bank or savings and loan association requires a customer to make a deposit. Then the customer receives a checkbook that contains checks, deposit slips, and a register for recording checks written and deposits made. It is important to record all payments and deposits that affect the account. It is also important to keep the checkbook balance current by subtracting checks written and adding deposits made.

About once a month checking customers receive a statement from the bank listing all activity that the account has had in the last month. The statement lists a beginning balance, all checks and deposits, any service charges made against the account, and an ending balance. Because it may take several days for checks that a customer has written to clear the banking system, the check register may list checks that do not appear on the monthly bank statement. These checks are called **outstanding checks.** Deposits that are recorded in the check register but do not appear on the statement are called **deposits**

in transit. Because of these differences, it is important to balance, or reconcile, the checkbook against the monthly statement. The steps for doing so are listed below.

Balancing or Reconciling a Checkbook

Step 1: Place a check mark in the checkbook register next to each check and deposit listed on the monthly bank statement. Any entries in the register without a check mark are outstanding checks or deposits in transit.

Step 2: Find the ending checkbook register balance and add to it any outstanding checks and any interest paid on the account.

Step 3: From the total in Step 2, subtract any deposits in transit and any service charges.

Step 4: Compare the amount found in Step 3 with the ending balance listed on the bank statement. If they are the same, the checkbook balances with the bank statement. Be sure to update the check register with service charges and interest.

Step 5: If the checkbook does not balance, recheck the balancing process. Next, make sure that the running checkbook register balance was calculated correctly. Finally, compare the checkbook register with the statement to make sure that each check was recorded for the correct amount.

For the checkbook register and monthly bank statement given:

a. *update the checkbook register*   **b.** *list the outstanding checks and deposits in transit*
c. *balance the checkbook—be sure to update the register with any interest or service fees*

#	Date	Description	Payment	✓	Deposit	Balance 425.86
114	4/1	Market Basket	30.27			
115	4/3	May's Texaco	8.50			
	4/4	Cash at ATM	50.00			
116	4/6	UNO Bookstore	121.38			
	4/7	Deposit			100.00	
117	4/9	MasterCard	84.16			
118	4/10	Blockbuster	6.12			
119	4/12	Kroger	18.72			
120	4/14	Parking sticker	18.50			
	4/15	Direct deposit			294.36	
121	4/20	Rent	395.00			
122	4/25	Student fees	20.00			
	4/28	Deposit			75.00	

Checkbook Register

First National Bank Monthly Statement 4/30

BEGINNING BALANCE:		425.86
Date	Number	Amount

CHECKS AND ATM WITHDRAWALS

Date	Number	Amount
4/3	114	30.27
4/4	ATM	50.00
4/11	117	84.16
4/13	115	8.50
4/15	119	18.72
4/22	121	395.00

DEPOSITS

4/7		100.00
4/15	Direct deposit	294.36

SERVICE CHARGES

Low balance fee	7.50

INTEREST

Credited 4/30	1.15
ENDING BALANCE:	227.22

Chapter 4　Vocabulary Check

Fill in each blank with one of the choices listed below. Some choices may be used more than once or not at all.

vertically	decimal	and	right triangle	diameter
standard form	product	quotient	circumference	difference
sum	denominator	numerator		

1. Like fractional notation, _____ notation is used to denote a part of a whole.

2. To write fractions as decimals, divide the _____ by the _____.

3. To add or subtract decimals, write the decimals so that the decimal points line up _____.

4. When writing decimals in words, write " _____ " for the decimal point.

5. When multiplying decimals, the decimal point in the product is placed so that the number of decimal places in the product is equal to the _____ of the number of decimal places in the factors.

6. The distance around a circle is called the _____.

7. When 2 million is written as 2,000,000, we say it is written in _____.

8. $\pi = \dfrac{\underline{\hspace{2cm}} \text{ of a circle}}{\underline{\hspace{2cm}} \text{ of the same circle}}$

9. In $3.4 - 2 = 1.4$, the number 1.4 is called the _____.

10. In $3.4 \div 2 = 1.7$, the 1.7 is called the _____.

11. In $3.4 \times 2 = 6.8$, the 6.8 is called the _____.

12. In $3.4 + 2 = 5.4$, the 5.4 is called the _____.

Helpful Hint

▶ Are you preparing for your test? Don't forget to take the Chapter 4 Test on page 312. Then check your answers at the back of the text and use the Chapter Test Prep Videos to see the fully worked-out solutions to any of the exercises you want to review.

4　Chapter Highlights

Definitions and Concepts	Examples
Section 4.1　Introduction to Decimals	

Place-Value Chart

hundreds	tens	ones	.	tenths	hundredths	thousandths	ten-thousandths	hundred-thousandths
		4	decimal point	2	6	5		
100	10	1		$\dfrac{1}{10}$	$\dfrac{1}{100}$	$\dfrac{1}{1000}$	$\dfrac{1}{10,000}$	$\dfrac{1}{100,000}$

4.265 means

$$4 \cdot 1 + 2 \cdot \frac{1}{10} + 6 \cdot \frac{1}{100} + 5 \cdot \frac{1}{1000}$$

or

$$4 + \frac{2}{10} + \frac{6}{100} + \frac{5}{1000}$$

(continued)

Definitions and Concepts	Examples

Section 4.1   Introduction to Decimals (continued)

Writing (or Reading) a Decimal in Words

Step 1: Write the whole number part in words.

Step 2: Write "and" for the decimal point.

Step 3: Write the decimal part in words as though it were a whole number, followed by the place value of the last digit.

A decimal written in words can be written in standard form by reversing the above procedure.

Write 3.08 in words.
Three and eight hundredths

Write "four and twenty-one thousandths" in standard form.
　4.021

Section 4.2   Order and Rounding

To **compare decimals,** compare digits in the same place from left to right. When two digits are not equal, the number with the larger digit is the larger decimal.

To Round Decimals to a Place Value to the Right of the Decimal Point

Step 1: Locate the digit to the right of the given place value.

Step 2: If this digit is 5 or greater, add 1 to the digit in the given place value and delete all digits to its right. If this digit is less than 5, delete all digits to the right of the given place value.

3.0261 > 3.0186 because

$\uparrow$　　　$\uparrow$

2　>　1

Round 86.1256 to the nearest hundredth.

　　　　　　　　hundredths place
Step 1: 86.12⑤6
　　　　　　　　digit to the right

Step 2: Since the digit to the right is 5 or greater, we add 1 to the digit in the hundredths place and delete all digits to its right.

86.1256 rounded to the nearest hundredth is 86.13.

Section 4.3   Adding and Subtracting Decimals

To Add or Subtract Decimals

Step 1: Write the decimals so that the decimal points line up vertically.

Step 2: Add or subtract as with whole numbers.

Step 3: Place the decimal point in the sum or difference so that it lines up vertically with the decimal points in the problem.

Add: 4.6 + 0.28　　　　Subtract: 2.8 − 1.04

$$\begin{array}{r} 4.60 \\ +0.28 \\ \hline 4.88 \end{array}$$
$$\begin{array}{r} \overset{7\ 10}{2.8\,\cancel{0}} \\ -1.04 \\ \hline 1.76 \end{array}$$

Section 4.4   Multiplying Decimals and Circumference of a Circle

To Multiply Decimals

Step 1: Multiply the decimals as though they are whole numbers.

Step 2: The decimal point in the product is placed so that the number of decimal places in the product is equal to the *sum* of the number of decimal places in the factors.

The **circumference** of a circle is the distance around the circle.

radius
diameter

$C = 2 \cdot \pi \cdot \text{radius}$ or
$C = \pi \cdot \text{diameter},$

where $\pi \approx 3.14$ or $\dfrac{22}{7}$.

Multiply: 1.48×5.9

$$\begin{array}{r} 1.48 \quad \leftarrow 2 \text{ decimal places} \\ \times\ 5.9 \quad \leftarrow 1 \text{ decimal places} \\ \hline 1332 \\ 7400 \\ \hline 8.732 \quad \leftarrow 3 \text{ decimal places} \end{array}$$

Find the exact circumference of a circle with radius 5 miles and an approximation by using 3.14 for π.

$$\begin{aligned} C &= 2 \cdot \pi \cdot \text{radius} \\ &= 2 \cdot \pi \cdot 5 \\ &= 10\pi \\ &\approx 10(3.14) \\ &= 31.4 \end{aligned}$$

The circumference is exactly 10π miles and *approximately* 31.4 miles.

Definitions and Concepts	Examples

Section 4.5   Dividing Decimals and Order of Operations

To Divide Decimals

Step 1: If the divisor is not a whole number, move the decimal point in the divisor to the right until the divisor is a whole number.

Step 2: Move the decimal point in the dividend to the right the *same number of places* as the decimal point was moved in step 1.

Step 3: Divide. The decimal point in the quotient is directly over the moved decimal point in the dividend.

Divide: $1.118 \div 2.6$

$$
\begin{array}{r}
0.43 \\
2.6)\overline{1.118} \\
-1\,04 \\
\hline
78 \\
-78 \\
\hline
0
\end{array}
$$

Order of Operations

1. Perform all operations within parentheses ( ), brackets [ ], or grouping symbols such as square roots or fraction bars.

2. Evaluate any expressions with exponents.

3. Multiply or divide in order from left to right.

4. Add or subtract in order from left to right.

Simplify.

$$
\begin{aligned}
1.9(12.8 - 4.1) &= 1.9(8.7) \quad \text{Subtract.} \\
&= 16.53 \quad \text{Multiply.}
\end{aligned}
$$

Section 4.6   Fractions and Decimals

To **write fractions as decimals,** divide the numerator by the denominator.

Write $\dfrac{3}{8}$ as a decimal.

$$
\begin{array}{r}
0.375 \\
8)\overline{3.000} \\
-2\,4 \\
\hline
60 \\
-56 \\
\hline
40 \\
-40 \\
\hline
0
\end{array}
$$

Chapter 4   Review

(4.1) *Determine the place value of the digit 4 in each decimal.*

1. 23.45

2. 0.000345

Write each decimal in words.

3. 0.45   **4.** 0.00345   **5.** 109.23   **6.** 46.007

Write each decimal in standard form.

7. Two and fifteen hundredths

8. Five hundred three and one hundred two thousandths

Write the decimal as a fraction or a mixed number. Write your answer in simplest form.

9. 0.16

10. 12.023

11. 1.0045

12. 25.25

Write each fraction as a decimal.

13. $\dfrac{9}{10}$

14. $\dfrac{25}{100}$

15. $\dfrac{45}{1000}$

16. $\dfrac{261}{10}$

(4.2) *Insert* $<, >,$ *or* $=$ *to make a true statement.*

17. 0.49   0.43

18. 0.973   0.9730

Write the decimals in order from smallest to largest.

19. 8.6, 8.09, 0.92

20. 0.09, 0.1, 0.091

Round each decimal to the given place value.

21. 0.623, nearest tenth

22. 0.9384, nearest hundredth

Round each money amount to the nearest cent.

23. $0.259

24. $12.461

Solve.

25. In a recent year, engaged couples in the United States spent an average of $31,304.35 on their wedding. Round this number to the nearest dollar.

26. A certain kind of chocolate candy bar contains 10.75 teaspoons of sugar. Write this number as a mixed number.

(4.3) *Add or subtract as indicated.*

27. 2.4 + 7.12

28. 3.9 − 1.2

29. 6.4 + 0.88

30. 19.02 + 6.98 + 0.007

31. 892.1 − 432.4   **32.** 100.342 − 0.064

33. Subtract 34.98 from 100.   **34.** Subtract 10.02 from 200.

35. Find the total distance between Grove City and Jerome.

36. The price of oil was $49.02 per barrel on October 23. It was $51.46 on October 24. Find by how much the price of oil increased from the 23rd to the 24th.

△ **37.** Find the perimeter.

△ **38.** Find the perimeter.

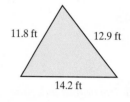

(4.4) *Multiply.*

39.   3.7
      × 5

40.   9.1
      × 6

41. 7.2 × 10

42. 9.345 × 1000

43.   4.02
      × 2.3

44.   39.02
      × 87.3

Solve.

△ **45.** Find the exact circumference of the circle. Then use the approximation 3.14 for π and approximate the circumference.

46. A kilometer is approximately 0.625 mile. It is 102 kilometers from Hays to Colby. Write 102 kilometers in miles to the nearest tenth of a mile.

Write each number in standard notation.

47. Saturn is a distance of about 887 million miles from the Sun.

48. The tail of a comet can be over 600 thousand miles long.

(4.5) *Divide. Round the quotient to the nearest thousandth if necessary.*

49. 3)0.261   **50.** 20)316.5   **51.** 21 ÷ 0.3   **52.** 0.0063 ÷ 0.03

53. $0.34\overline{)2.74}$ **54.** $19.8\overline{)601.92}$ **55.** $\dfrac{2.67}{100}$ **56.** $\dfrac{93}{10}$

57. There are approximately 3.28 feet in 1 meter. Find how many meters are in 24 feet to the nearest tenth of a meter.

←——— 1 meter ———→
←—— ≈3.28 feet ——→

58. George Strait pays $69.71 per month to pay back a loan of $3136.95. In how many months will the loan be paid off?

Simplify each expression.

59. $7.6 \times 1.9 + 2.5$ **60.** $(2.3)^2 - 1.4$ **61.** $\dfrac{7 + 0.74}{0.06}$

62. $\dfrac{(1.5)^2 + 0.5}{0.05}$ **63.** $0.9(6.5 - 5.6)$ **64.** $0.0726 \div 10 \times 1000$

(4.6) *Write each fraction as a decimal. Round to the nearest thousandth if necessary.*

65. $\dfrac{4}{5}$ **66.** $\dfrac{12}{13}$ **67.** $2\dfrac{1}{3}$ **68.** $\dfrac{13}{60}$

Insert <, >, or = to make a true statement.

69. 0.392 0.3920 **70.** $0.\overline{4}$ $\dfrac{4}{9}$ **71.** 0.293 $\dfrac{5}{17}$ **72.** $\dfrac{4}{7}$ 0.625

Write the numbers in order from smallest to largest.

73. $0.839, \dfrac{17}{20}, 0.837$ **74.** $\dfrac{18}{11}, 1.63, \dfrac{19}{12}$

Find each area.

△ **75.**

3 feet
←——— 4.6 feet ———→

△ **76.**

2.1 inches
←—— 5.2 inches ——→

Mixed Review

77. Write 200.0032 in words.

78. Write sixteen thousand twenty-five and fourteen thousandths in standard form.

79. Write 0.00231 as a fraction or a mixed number.

80. Write the numbers $\dfrac{6}{7}, \dfrac{8}{9}, 0.75$ in order from smallest to largest.

Write each fraction as a decimal. Round to the nearest thousandth, if necessary.

81. $\dfrac{7}{100}$

82. $\dfrac{9}{80}$ (Do not round.)

83. $\dfrac{8935}{175}$

Insert <, >, or = to make a true statement.

84. 402.00032   402.000032

85. 0.230505   0.23505

86. $\dfrac{6}{11}$ 0.55

Round each decimal to the given place value.

87. 42.895, nearest hundredth

88. 16.34925, nearest thousandth

Round each money amount to the nearest dollar.

89. $123.46

90. $3645.52

Add or subtract as indicated.

91. $4.9 - 3.2$

92. $5.23 - 2.74$

93. $200.49 + 16.82 + 103.002$

94. $0.00236 + 100.45 + 48.29$

Multiply or divide as indicated. Round to the nearest thousandth, if necessary.

95. $\begin{array}{r} 2.54 \\ \times\ 3.2 \\ \hline \end{array}$

96. $\begin{array}{r} 3.45 \\ \times\ 2.1 \\ \hline \end{array}$

97. $0.005\overline{)24.5}$

98. $2.3\overline{)54.98}$

Solve.

△ **99.** Tomaso is going to fertilize his lawn, a rectangle that measures 77.3 feet by 115.9 feet. Approximate the area of the lawn by rounding each measurement to the nearest ten feet.

77.3 feet

115.9 feet

100. Estimate the cost of the items to see whether the groceries can be purchased with a $5 bill.

$1.89

$1.07

3 cans for $0.99

Simplify each expression.

101. $\dfrac{(3.2)^2}{100}$

102. $(2.6 + 1.4)(4.5 - 3.6)$

Answers

Write the decimal as indicated.

1. 45.092, in words

2. Three thousand and fifty-nine thousandths, in standard form

Round the decimal to the indicated place value.

1. _____

3. 34.8923, nearest tenth

4. 0.8623, nearest thousandth

2. _____

3. _____

5. Insert <, >, or = to make a true statement. 25.0909   25.9090

6. Write the numbers in order from smallest to largest. $\dfrac{4}{9}$ 0.454   0.445

4. _____

5. _____

6. _____

Write the decimal as a fraction or a mixed number in simplest form.

7. _____

7. 0.345

8. 24.73

8. _____

9. _____

Write the fraction or mixed number as a decimal. If necessary, round to the nearest thousandth.

10. _____

9. $\dfrac{13}{20}$

10. $5\dfrac{8}{9}$

11. $\dfrac{16}{17}$

11. _____

12. _____

13. _____

Perform the indicated operations. Round the result to the nearest thousandth if necessary.

14. _____

12. 2.893 + 4.2 + 10.49

13. Subtract 8.6 from 20.

14.   10.2
     ×   4.3

15. _____

16. _____

17. _____

312

15. $0.23\overline{)12.88}$

16.   0.165
   × 0.47

17. $7\overline{)46.71}$

18. 126.9×100

19. $\dfrac{47.3}{10}$

20. $0.3[1.57 - (0.6)^2]$

21. $\dfrac{0.23 + 1.63}{0.3}$

22. At its farthest, Pluto is 4583 million miles from the Sun. Write this number using standard notation.

△ **23.** Find the area.

△ **24.** Find the exact circumference of the circle. Then use the approximation 3.14 for π and approximate the circumference.

25. Vivian Thomas is going to put insecticide on her lawn to control grubworms. The lawn is a rectangle that measures 123.8 feet by 80 feet. The amount of insecticide required is 0.02 ounce per square foot.

a. Find the area of her lawn.

b. Find how much insecticide Vivian needs to purchase.

26. Find the total distance from Bayette to Center City.

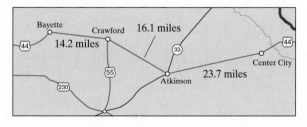

18. _____

19. _____

20. _____

21. _____

22. _____

23. _____

24. _____

25. a. _____

b. _____

26. _____

Answers

1. _____

2. _____

3. _____

4. _____

5. _____

6. _____

7. _____

8. _____

9. _____

10. _____

11. _____

12. _____

13. _____

14. _____

15. _____

16. _____

17. _____

18. _____

1. Write 106,052,447 in words.

2. Write two hundred seventy-six thousand, four in standard form.

3. In 2011, a total of 12,734,424 passenger vehicles were sold in the United States. In 2012, total passenger vehicle sales in the United States had increased by 1,705,636 vehicles. Find the total number of passenger vehicles sold in the United States in 2012. (*Source:* Alliance of Automobile Manufacturers)

4. There are 12 fluid ounces of soda in a can. How many fluid ounces of soda are in a case (24 cans) of soda?

5. Subtract: $900 - 174$. Then check by adding.

6. Simplify: $5^2 \cdot 2^3$

7. Round each number to the nearest hundred to find an estimated sum.

$$\begin{array}{r} 294 \\ 625 \\ 1071 \\ + \ 349 \end{array}$$

8. Simplify: $7 \cdot \sqrt{144}$

9. A digital video disc (DVD) can hold about 4800 megabytes (MB) of information. How many megabytes can 12 DVDs hold?

10. Find the perimeter and area of the square.

7 feet

11. Divide: $6819 \div 17$

12. Write $2\dfrac{5}{8}$ as an improper fraction.

13. Simplify: $4^3 + [3^2 - (10 \div 2)] - 7 \cdot 3$

14. Write $\dfrac{64}{5}$ as a mixed number.

15. Identify the numerator and the denominator: $\dfrac{3}{7}$

16. Simplify: $24 \div 8 \cdot 3$

17. Write $\dfrac{6}{60}$ in simplest form.

18. Simplify: $(8 - 5)^2 + (10 - 8)^3$

19. Multiply: $\dfrac{3}{4} \cdot 20$

20. Simplify: $1 + 2[30 \div (7 - 2)]$

21. Divide: $\dfrac{7}{8} \div \dfrac{2}{9}$

22. Find the average of 117, 125, and 142.

23. Multiply: $1\dfrac{2}{3} \cdot 2\dfrac{1}{4}$

24. A total of \$324 is paid for 36 tickets to the Audubon Zoo. How much did each ticket cost?

25. Divide: $\dfrac{3}{4} \div 5$

26. Simplify: $\left(\dfrac{3}{4} \div \dfrac{1}{2}\right) \cdot \dfrac{9}{10}$

Simplify.

27. $\dfrac{8}{9} - \dfrac{1}{9}$

28. $\dfrac{4}{15} + \dfrac{2}{15}$

29. $\dfrac{7}{8} - \dfrac{5}{8}$

30. $\dfrac{1}{20} + \dfrac{3}{20} + \dfrac{4}{20}$

Write an equivalent fraction with the indicated denominator.

31. $\dfrac{3}{4} = \dfrac{}{20}$

32. $\dfrac{7}{9} = \dfrac{}{45}$

Perform the indicated operations.

33. $\dfrac{11}{15} + \dfrac{3}{10}$

34. $\dfrac{7}{30} - \dfrac{2}{9}$

35. Two packages of ground round are purchased. One package weighs $2\dfrac{3}{8}$ pounds and the other $1\dfrac{4}{5}$ pounds. What is the combined weight of the ground round?

36. A color cartridge for a business printer weighs $2\dfrac{5}{16}$ pounds. How much do 12 cartridges weigh?

19. _____

20. _____

21. _____

22. _____

23. _____

24. _____

25. _____

26. _____

27. _____

28. _____

29. _____

30. _____

31. _____

32. _____

33. _____

34. _____

35. _____

36. _____

37. _____

38. _____

39. _____

40. _____

41. _____

42. _____

43. _____

44. _____

45. _____

46. _____

47. _____

48. _____

49. _____

50. _____

Evaluate each expression.

37. $\left(\dfrac{1}{4}\right)^2$

38. $\left(\dfrac{7}{11}\right)^2$

39. $\left(\dfrac{1}{6}\right)^2 \cdot \left(\dfrac{3}{4}\right)^3$

40. $\left(\dfrac{1}{2}\right)^3 \cdot \left(\dfrac{4}{9}\right)^2$

41. Write 0.43 as a fraction.

42. Write $\dfrac{3}{4}$ as a decimal.

43. Insert $<$, $>$, or $=$ to form a true statement.
0.378   0.368

44. Write "five and six hundredths" in standard form.

45. Subtract: $35.218 - 23.65$ Check your answer.

46. Add: $75.1 + 0.229$

Multiply.

47. 23.702×100

48. 1.7×0.07

49. $76,805 \times 0.01$

50. Divide: $0.1157 \div 0.013$

Ratio and Proportion

One of many 3-D cameras

Note: To date, there continues to be much research and controversy over 3-D films. The main controversy is focused on inferior conversions of 2-D filmed movies to 3-D. The main research is focused on watching 3-D movies with adequate light and glasses or without glasses at all.

How Popular Are 3-D Films Now?

A 3-D (three-dimensional) film is a film that enhances the illusion of depth perception. Believe it or not, 3-D films have existed in some form since 1890, but because of high cost and lack of a standardized format, these films are only now starting to be widely shown and produced. The graph shows the trends in the releases of both 2-D and 3-D films in recent years.

In Section 5.1, Exercise 47, we calculate the ratio of 3-D to total films.

Films Released by U.S. Production Companies

Source: Motion Picture Association of America

Having studied fractions in Chapters 2 and 3, we are ready to explore the useful notations of ratio and proportion. Ratio is another name for quotient and is usually written in fraction form. A proportion is an equation with two equal ratios.

Objectives

A Write Ratios as Fractions.

B Write Ratios in Simplest Form.

Objective A Writing Ratios as Fractions

A **ratio** is the quotient of two quantities. A ratio, in fact, is no different from a fraction, except that a ratio is sometimes written using notation other than fractional notation. For example, the ratio of 1 to 2 can be written as

$$1 \text{ to } 2 \quad \text{or} \quad \frac{1}{2} \quad \text{or} \quad 1 : 2$$

fractional notation   colon notation

These ratios are all read as, "the ratio of 1 to 2."

✓**Concept Check** How should each ratio be read aloud?

a. $\frac{8}{5}$ **b.** $\frac{5}{8}$

In this section, we write ratios using fractional notation. If the fraction happens to be an improper fraction, do not write the fraction as a mixed number. Why? The mixed number form is not a ratio or quotient of two quantities.

Writing a Ratio as a Fraction

The order of the quantities is important when writing ratios. To write a ratio as a fraction, write the *first number* of the ratio as the *numerator* of the fraction and the *second number* as the *denominator*.

Helpful Hint

The ratio of 6 to 11 is $\frac{6}{11}$, *not* $\frac{11}{6}$.

Practice 1

Write the ratio of 20 to 23 using fractional notation.

Example 1 Write the ratio of 12 to 17 using fractional notation.

Solution: The ratio is $\frac{12}{17}$.

Helpful Hint Don't forget that order is important when writing ratios. The ratio $\frac{17}{12}$ is *not* the same as the ratio $\frac{12}{17}$.

■ Work Practice 1

Practice 2–3

Write each ratio using fractional notation.

2. The ratio of 10.3 to 15.1

3. The ratio of $3\frac{1}{3}$ to $12\frac{1}{5}$

Examples Write each ratio using fractional notation.

2. The ratio of 2.6 to 3.1 is $\frac{2.6}{3.1}$.

3. The ratio of $1\frac{1}{2}$ to $7\frac{3}{4}$ is $\dfrac{1\frac{1}{2}}{7\frac{3}{4}}$.

■ Work Practice 2–3

Answers

1. $\frac{20}{23}$ **2.** $\frac{10.3}{15.1}$ **3.** $\dfrac{3\frac{1}{3}}{12\frac{1}{5}}$

✓**Concept Check Answers**

a. "the ratio of eight to five"
b. "the ratio of five to eight"

Objective B Writing Ratios in Simplest Form ▶

To simplify a ratio, we just write the fraction in simplest form. Common factors as well as common units can be divided out.

Example 4　Write the ratio of $15 to $10 as a fraction in simplest form.

Solution:

$$\frac{\$15}{\$10} = \frac{15}{10} = \frac{3 \cdot \cancel{5}^{1}}{2 \cdot \cancel{5}_{1}} = \frac{3}{2}$$

■ Work Practice 4

Practice 4

Write the ratio of $8 to $6 as a fraction in simplest form.

Helpful Hint

In the example above, although $\frac{3}{2} = 1\frac{1}{2}$, a ratio is a quotient of *two* quantities. For that reason, ratios are not written as mixed numbers.

If a ratio contains decimal numbers or mixed numbers, we simplify by writing the ratio as a ratio of whole numbers.

Example 5　Write the ratio of 2.6 to 3.1 as a fraction in simplest form.

Solution:　The ratio in fraction form is

$$\frac{2.6}{3.1}$$

Now let's clear the ratio of decimals.

$$\frac{2.6}{3.1} = \frac{2.6}{3.1} \cdot 1 = \frac{2.6}{3.1} \cdot \frac{10}{10} = \frac{2.6 \cdot 10}{3.1 \cdot 10} = \frac{26}{31} \quad \text{Simplest form}$$

■ Work Practice 5

Practice 5

Write the ratio of 3.9 to 8.8 as a fraction in simplest form.

Example 6　Write the ratio of $1\frac{1}{5}$ to $2\frac{7}{10}$ as a fraction in simplest form.

Solution:　The ratio in fraction form is $\dfrac{1\frac{1}{5}}{2\frac{7}{10}}$.

To simplify, remember that the fraction bar means division.

$$\frac{1\frac{1}{5}}{2\frac{7}{10}} = 1\frac{1}{5} \div 2\frac{7}{10} = \frac{6}{5} \div \frac{27}{10} = \frac{6}{5} \cdot \frac{10}{27} = \frac{6 \cdot 10}{5 \cdot 27} = \frac{2 \cdot \cancel{3}^{1} \cdot 2 \cdot \cancel{5}^{1}}{\cancel{5}_{1} \cdot \cancel{3}_{1} \cdot 3 \cdot 3} = \frac{4}{9} \quad \text{Simplest form.}$$

■ Work Practice 6

Practice 6

Write the ratio of $2\frac{2}{3}$ to $1\frac{13}{15}$ as a fraction in simplest form.

Answers

4. $\dfrac{4}{3}$　**5.** $\dfrac{39}{88}$　**6.** $\dfrac{10}{7}$

Practice 7

Use the circle graph for Example 7 to write the ratio of work miles to total miles as a fraction in simplest form.

 Example 7 Writing a Ratio from a Circle Graph

The circle graph at the right shows the part of a car's total mileage that falls into a particular category. Write the ratio of family business miles to total miles as a fraction in simplest form.

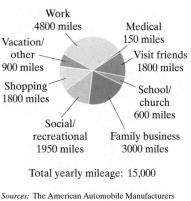

Solution:

$$\frac{\text{family business miles}}{\text{total miles}} = \frac{3000 \text{ miles}}{15{,}000 \text{ miles}}$$

$$= \frac{3000}{15{,}000}$$

$$= \frac{\overset{1}{\cancel{3000}}}{5 \cdot \underset{1}{\cancel{3000}}}$$

$$= \frac{1}{5}$$

Total yearly mileage: 15,000

Sources: The American Automobile Manufacturers Association and The National Automobile Dealers Association.

■ Work Practice 7

 Practice 8

Given the triangle shown:

6 meters, 10 meters, 8 meters

a. Find the ratio of the length of the shortest side to the length of the longest side.

b. Find the ratio of the length of the longest side to the perimeter of the triangle.

 Example 8 Given the rectangle shown:

a. Find the ratio of its width to its length.
b. Find the ratio of its length to its perimeter.

8 feet, 5 feet

Solution:

a. The ratio of its width to its length is

$$\frac{\text{width}}{\text{length}} = \frac{5 \text{ feet}}{8 \text{ feet}} = \frac{5}{8}$$

b. Recall that the perimeter of the rectangle is the distance around the rectangle: $8 + 5 + 8 + 5 = 26$ feet. The ratio of its length to its perimeter is

$$\frac{\text{length}}{\text{perimeter}} = \frac{8 \text{ feet}}{26 \text{ feet}} = \frac{8}{26} = \frac{\overset{1}{\cancel{2}} \cdot 2 \cdot 2}{\underset{1}{\cancel{2}} \cdot 13} = \frac{4}{13}$$

■ Work Practice 8

✓**Concept Check** Explain why the answer $\frac{8}{5}$ would be incorrect for part **a** of Example 8.

Answers

7. $\frac{8}{25}$ **8. a.** $\frac{3}{5}$ **b.** $\frac{5}{12}$

✓**Concept Check Answer**

$\frac{8}{5}$ would be the ratio of the rectangle's length to its width.

Vocabulary, Readiness & Video Check

Answer each statement true or false.

1. The quotient of two quantities is called a ratio. _____

2. The ratio $\frac{7}{5}$ means the same as the ratio $\frac{5}{7}$. _____

3. The ratio $\frac{7.2}{8.1}$ is in simplest form. _____

4. The ratio $\frac{10 \text{ feet}}{30 \text{ feet}}$ is in simplest form. _____

5. The ratio $\frac{9}{10}$ is in simplest form. _____

6. The ratio 2 to 5 equals $\frac{5}{2}$ in fractional notation. _____

7. The ratio 30 : 41 equals $\frac{30}{41}$ in fractional notation. _____

8. The ratio 15 to 45 equals $\frac{3}{1}$ in fractional notation. _____

Martin-Gay Interactive Videos   *Watch the section lecture video and answer the following questions.*

Objective A **9.** Based on the lecture before ⊞ Example 1, what three notations can we use for a ratio? For your answer, use the ratio example given in the lecture. ▶

Objective B **10.** Why is the ratio in ⊞ Example 3 not in simplest form? ▶

See Video 5.1 🍎

5.1 Exercise Set MyMathLab® ▶

Objective A *Write each ratio using fractional notation. Do not simplify. See Examples 1 through 3.*

1. 11 to 14   **2.** 7 to 12   ▶**3.** 23 to 10   **4.** 14 to 5   **5.** 151 to 201   **6.** 673 to 1000

7. 2.8 to 7.6   **8.** 3.9 to 4.2   **9.** 5 to $7\frac{1}{2}$ **10.** $5\frac{3}{4}$ to 3   **11.** $3\frac{3}{4}$ to $1\frac{2}{3}$ **12.** $2\frac{2}{5}$ to $6\frac{1}{2}$

Objectives A B Mixed Practice *Write each ratio as a ratio of whole numbers using fractional notation. Write the fraction in simplest form. See Examples 1 through 6.*

▶**13.** 16 to 24   **14.** 25 to 150   ▶**15.** 7.7 to 10   **16.** 8.1 to 10

17. 4.63 to 8.21

18. 9.61 to 7.62

19. 9 inches to 12 inches

20. 14 centimeters to 20 centimeters

21. 10 hours to 24 hours

22. 18 quarts to 30 quarts

23. $32 to $100

24. $46 to $102

25. 24 days to 14 days

26. 80 miles to 120 miles

27. 32,000 bytes to 46,000 bytes

28. 600 copies to 150 copies

29. 8 inches to 20 inches

30. 9 yards to 2 yards

31. $3\frac{1}{2}$ to $12\frac{1}{4}$

32. $3\frac{1}{3}$ to $4\frac{1}{6}$

33. $7\frac{3}{5}$ hours to $1\frac{9}{10}$ hours

34. $25\frac{1}{2}$ days to $2\frac{5}{6}$ days

Write the ratio described in each exercise as a fraction in simplest form. See Examples 7 and 8.

35.

Average Weight of Mature Whales	
Blue Whale	**Fin Whale**
145 tons	50 tons

Use the table to find the ratio of the weight of an average mature Fin Whale to the weight of an average mature Blue Whale.

36.

Countries with Small Land Areas	
Tuvalu	**San Marino**
10 sq mi	24 sq mi

(Source: World Almanac)

Use the table to find the ratio of the land area of Tuvalu to the land area of San Marino.

37. The mural *Independence and the Opening of the West,* painted by Thomas Hart Benton on the wall of the lobby of the Harry Truman Presidential Library in Independence, Missouri, occupies a space that measures 32 feet wide by 19 feet high. Find the ratio of the width to the perimeter of this mural.

38. More than one billion people per year see Wyland's Whaling Wall murals throughout the United States, Canada, Japan, Australia, New Zealand, and Europe. The mural on the Marketplace Design Center in Philadelphia, *East Coast Humpbacks,* measures 125 feet long by 130 feet high. Find the ratio of the height to the perimeter of this mural.

39. Find the ratio of the width of a regulation size basketball court to its perimeter.

40. Find the ratio of the width to the perimeter shown of the swimming pool.

50 feet (width)

94 feet (length)

30 feet (width)

45 feet (length)

At the Hidalgo County School Board meeting one night, there were 125 women and 100 men present.

41. Find the ratio of women to men.

42. Find the ratio of men to the total number of people present.

Blood contains three types of cells: red blood cells, white blood cells, and platelets. For approximately every 600 red blood cells in healthy humans, there are 40 platelets and 1 white blood cell. Use this information for Exercises 43 and 44. (Source: American Red Cross Biomedical Services)

43. Write the ratio of red blood cells to platelet cells.

44. Write the ratio of white blood cells to red blood cells.

△ **45.** Find the ratio of the longest side to the perimeter of the right-triangular-shaped billboard.

△ **46.** Find the ratio of the base to the perimeter of the triangular mainsail.

8 feet   15 feet   17 feet

13 feet   12 feet (height)   5 feet (base)

In 2013, 659 films by U.S. production companies were released. Use this information for Exercises 47 and 48.

47. In 2013, 45 digital 3-D films were released by U.S. production companies. Find the ratio of digital films to total films for 2013. (*Source:* Motion Picture Association of America)

48. In 2013, 545 independent films were released by U.S. production companies. Find the ratio of independent films to total films for 2013. (*Source:* Motion Picture Association of America)

49. Of the U.S. mountains that are over 14,000 feet in elevation, 57 are located in Colorado and 19 are located in Alaska. Find the ratio of the number of mountains over 14,000 feet found in Alaska to the number of mountains over 14,000 feet found in Colorado. (*Source:* U.S. Geological Survey)

50. Citizens of the United States eat an average of 25 pints of ice cream per year. Residents of the New England states eat an average of 39 pints of ice cream per year. Find the ratio of the amount of ice cream eaten by New Englanders to the amount eaten by the average U.S. citizen. (*Source:* International Dairy Foods Association)

51. At the Summer Olympics in London, England, a total of 302 gold medals were awarded, and Russian athletes won a total of 24 gold medals. Find the ratio of gold medals won by the Russian athletes to the total gold medals awarded. (*Source:* espn.go.com)

52. For the 2013 London Marathon, 22,045 males and 12,233 females completed the race. Find the ratio of female finishers to male finishers. (*Source:* World Marathon Majors)

53. As of 2012, Target stores operate in 49 states. Find the ratio of states without Target stores to states with Target stores. (*Source:* Target Corporation)

54. Four states in the United States officially designate themselves as "commonwealths": Kentucky, Massachusetts, Pennsylvania, and Virginia. Find the ratio of states that are designated as commonwealths to the states that are not designed as commonwealths.

Review

Divide. See Section 4.5.

55. $9\overline{)20.7}$

56. $7\overline{)60.2}$

57. $3.7\overline{)0.555}$

58. $4.6\overline{)1.15}$

Concept Extensions

Solve. See the Concept Checks in this section. Write how each should be read as a ratio.

59. $\dfrac{7}{9}$

60. $\dfrac{12}{5}$

61. $30:1$

62. 5 to 4

63. Is the ratio $\dfrac{11}{15}$ the same as the ratio of $\dfrac{15}{11}$? Explain your answer.

64. Explain why the ratio $\dfrac{40}{17}$ is incorrect for Exercise 45.

Decide whether each value is a ratio written as a fraction in simplest form. If not, write it as a fraction in simplest form.

65. $\dfrac{\$3}{\$2}$

66. $\dfrac{7.1}{4.3}$

67. $\dfrac{1 \text{ foot}}{30 \text{ inches}}$

68. $\dfrac{2\frac{1}{10}}{3\frac{3}{14}}$

69. $4\frac{1}{2}$

70. $\dfrac{12 \text{ inches}}{2 \text{ feet}}$

71. A grocer will refuse a shipment of tomatoes if the ratio of bruised tomatoes to the total batch is at least 1 to 10. A sample is found to contain 3 bruised tomatoes and 33 good tomatoes. Determine whether the shipment should be refused.

72. A panty hose manufacturing machine will be repaired if the ratio of defective panty hose to good panty hose is at least 1 to 20. A quality control engineer found 10 defective panty hose in a batch of 200. Determine whether the machine should be repaired.

73. In 2013, 12 states had primary laws prohibiting all drivers from using hand-held cell phones while driving. These laws allow law enforcement officers to ticket a driver for using a hand-held cell phone, even if no other traffic offense has occurred. (*Source:* Governors Highway Safety Association)

   a. Find the ratio of states with primary hand-held cell phone laws to total U.S. states.

   b. Find the number of states with no primary law prohibiting hand-held cell phone use while driving.

   c. Find the ratio of states with primary hand-held cell phone laws to states without such laws.

5.2 Rates

Objective A Writing Rates as Fractions

A special type of ratio is a rate. **Rates** are used to compare *different* kinds of quantities. For example, suppose that a recreational runner can run 3 miles in 33 minutes. If we write this rate as a fraction, we have

$$\frac{3 \text{ miles}}{33 \text{ minutes}} = \frac{1 \text{ mile}}{11 \text{ minutes}} \quad \text{In simplest form}$$

> **Helpful Hint**
>
> When comparing quantities with different units, write the units as part of the comparison. They do not divide out.
>
> **Same Units:** $\dfrac{3 \text{ inches}}{12 \text{ inches}} = \dfrac{1}{4}$
>
> **Different Units:** $\dfrac{2 \text{ miles}}{20 \text{ minutes}} = \dfrac{1 \text{ mile}}{10 \text{ minutes}}$ Units are still written.

Example 1 Write the rate as a fraction in simplest form: 10 nails every 6 feet

Solution:

$$\frac{10 \text{ nails}}{6 \text{ feet}} = \frac{5 \text{ nails}}{3 \text{ feet}}$$

■ Work Practice 1

Examples Write each rate as a fraction in simplest form.

2. $2160 for 12 weeks is $\dfrac{2160 \text{ dollars}}{12 \text{ weeks}} = \dfrac{180 \text{ dollars}}{1 \text{ week}}$

3. 360 miles on 16 gallons of gasoline is $\dfrac{360 \text{ miles}}{16 \text{ gallons}} = \dfrac{45 \text{ miles}}{2 \text{ gallons}}$

■ Work Practice 2–3

✓ **Concept Check** True or false? $\dfrac{16 \text{ gallons}}{4 \text{ gallons}}$ is a rate. Explain.

Objective B Finding Unit Rates

A **unit rate** is a rate with a denominator of 1. A familiar example of a unit rate is 55 mph (or 55 mi/hr), read as "55 **miles per hour.**" This means 55 miles per 1 hour or

$$\frac{55 \text{ miles}}{1 \text{ hour}} \quad \text{Denominator of 1}$$

> **Helpful Hint** In this context, the word "per" translates to division.

Objectives

A Write Rates as Fractions.

B Find Unit Rates.

C Find Unit Prices.

Practice 1

Write the rate as a fraction in simplest form: 12 commercials every 45 minutes

Practice 2–3

Write each rate as a fraction in simplest form.

2. $1680 for 8 weeks

3. 236 miles on 12 gallons of gasoline

Answers

1. $\dfrac{4 \text{ commercials}}{15 \text{ min}}$ 2. $\dfrac{\$210}{1 \text{ wk}}$ 3. $\dfrac{59 \text{ mi}}{3 \text{ gal}}$

✓ **Concept Check Answer**

false; a rate compares different kinds of quantities

> **Writing a Rate as a Unit Rate**
>
> To write a rate as a unit rate, divide the numerator of the rate by the denominator.

Practice 4

Write as a unit rate: 3200 feet every 8 seconds

Example 4   Write as a unit rate: $31,500 every 7 months

Solution:

$$\frac{31,500 \text{ dollars}}{7 \text{ months}} \qquad 7\overline{)31,500}^{\;4\,500}$$

The unit rate is

$$\frac{4500 \text{ dollars}}{1 \text{ month}} \text{ or } 4500 \text{ dollars/month} \quad \text{Read as, "4500 dollars per month."}$$

■ Work Practice 4

Practice 5

Write as a unit rate: 78 bushels of fruit from 12 trees

Example 5   Write as a unit rate: 337.5 miles every 15 gallons of gas

Solution:

$$\frac{337.5 \text{ miles}}{15 \text{ gallons}} \qquad 15\overline{)337.5}^{\;22.5}$$

The unit rate is

$$\frac{22.5 \text{ miles}}{1 \text{ gallon}} \text{ or } 22.5 \text{ miles/gallon} \quad \text{Read as, "22.5 miles per gallon."}$$

■ Work Practice 5

Objective C Finding Unit Prices

Rates are used extensively in sports, business, medicine, and science. One of the most common uses of rates is in consumer economics. When a unit rate is "money per item," it is also called a **unit price**.

$$\text{unit price} = \frac{\text{price}}{\text{number of units}}$$

Practice 6

An automobile rental agency charges $170 for 5 days for a certain model car. What is the unit price in dollars per day?

Example 6   Finding Unit Price

A store charges $3.36 for a 16-ounce jar of picante sauce. What is the unit price in dollars per ounce?

Solution:

$$\frac{\text{unit}}{\text{price}} = \frac{\text{price}}{\text{number of units}} = \frac{\$3.36}{16 \text{ ounces}} = \frac{\$0.21}{1 \text{ ounce}} \text{ or } \$0.21 \text{ per ounce}$$

■ Work Practice 6

Answers

4. $\dfrac{400 \text{ ft}}{1 \text{ sec}}$ or 400 ft/sec

5. $\dfrac{6.5 \text{ bushels}}{1 \text{ tree}}$ or 6.5 bushels/tree

6. $34 per day

| **Example 7** | Finding the Best Buy |

Approximate each unit price to decide which is the better buy: 4 bars of soap for $0.99 or 5 bars of soap for $1.19.

Solution:

$$\frac{\text{unit}}{\text{price}} = \frac{\text{price}}{\text{no. of units}} = \frac{\$0.99}{4 \text{ bars}} \approx \begin{array}{c}\$0.25 \text{ per bar}\\ \text{of soap}\end{array} \qquad \begin{array}{r} 0.247 \\ 4\overline{)0.990} \end{array} \approx 0.25$$

("is approximately")

$$\frac{\text{unit}}{\text{price}} = \frac{\text{price}}{\text{no. of units}} = \frac{\$1.19}{5 \text{ bars}} \approx \begin{array}{c}\$0.24 \text{ per bar}\\ \text{of soap}\end{array} \qquad \begin{array}{r} 0.238 \\ 5\overline{)1.190} \end{array} \approx 0.24$$

Thus, the 5-bar package is the better buy.

■ **Work Practice 7**

Practice 7

Approximate each unit price to decide which is the better buy for a bag of nacho chips: 11 ounces for $2.32 or 16 ounces for $3.59.

Answer

7. 11-oz bag

Vocabulary, Readiness & Video Check

Use the choices below to fill in each blank. Not all choices will be used.

| rate | division | unit price | unit |
| numerator | different | denominator | |

1. A rate with a denominator of 1 is called a(n) _____ rate.
2. When a rate is written as money per item, a unit rate is called a(n) _____.
3. The word *per* translates to "_____."
4. Rates are used to compare _____ types of quantities.
5. To write a rate as a unit rate, divide the _____ of the rate by the _____.

Martin-Gay Interactive Videos   Watch the section lecture video and answer the following questions.

Objective A 6. Why can't we divide out the units in ▤ Example 2? ▶

Objective B 7. Why did we divide the first quantity of the rate in ▤ Example 4 by the second quantity? ▶

Objective C 8. From ▤ Example 5, unit prices can be especially helpful when? ▶

See Video 5.2 🍒

5.2 Exercise Set   MyMathLab® ▶

Objective A *Write each rate as a fraction in simplest form. See Examples 1 through 3.*

1. 5 shrubs every 15 feet

2. 14 lab tables for 28 students

3. 15 returns for 100 sales

4. 8 phone lines for 36 employees

5. 6 laser printers for 28 computers

6. 4 inches of rain in 18 hours

7. 18 gallons of pesticide for 4 acres of crops

8. 150 graduate students for 8 advisors

9. 6 flight attendants for 200 passengers

10. 240 pounds of grass seed for 9 lawns

11. 355 calories in a 10-fluid-ounce chocolate milkshake (*Source: Home and Garden Bulletin No. 72*, U.S. Department of Agriculture)

12. 160 calories in an 8-fluid-ounce serving of cream of tomato soup (*Source: Home and Garden Bulletin No. 72*, U.S. Department of Agriculture)

Objective B *Write each rate as a unit rate. See Examples 4 and 5.*

13. 330 calories in a 3-ounce serving

14. 275 miles in 11 hours

15. 375 riders in 5 subway cars

16. 18 signs in 6 blocks

17. A hummingbird moves its wings at a rate of 5400 wingbeats a minute. Write this rate in wingbeats per second.

18. A bat moves its wings at a rate of 1200 wingbeats a minute. Write this rate in wingbeats per second.

19. $1,000,000 lottery winnings paid over 20 years

20. 400,000 library books for 8000 students

21. The state of Delaware has 638,736 registered voters for two senators. (*Source:* Delaware.gov)

22. The 2020 projected population of Louisiana is approximately 4,758,720 residents for 64 parishes. (*Note:* Louisiana is the only U.S. state with parishes instead of counties.) (*Source:* Louisiana.gov)

23. 12,000 good assembly-line products to 40 defective products

24. 5,000,000 lottery tickets for 4 lottery winners

25. The combined salary for the 20 highest-paid players of the 2013 World Series Champion Boston Red Sox was approximately $137,492,000. (*Source:* espn.go.com)

26. The top-grossing concert tour of all time was the 2009–2011 US 360° Tour, which grossed approximately $737 million for 110 shows worldwide. (*Source:* Billboard)

27. Doug Mehrens uses more crayons annually than anyone else in the world. The Phoenix-based artist goes through about 24,000 a year, many of them melted, to complete his contemporary art works. If he works on his art every day, determine how many crayons Doug Mehrens uses each day. Round your answer to the nearest whole crayon. (*Note:* use 1 year = 365 days) (*Source:* Crayola, LLC)

28. 12,000,000 tons of dust and dirt are trapped by the 25,000,000 acres of lawns in the United States each year. (*Source:* Professional Lawn Care Association of America)

29. For fiscal year 2013, the National Zoo in Washington, D.C., requested an annual budget of roughly $24,300,000 for its 400 different species. (*Source:* Smithsonian Institution)

30. On average in 2013, it cost each passenger about $750 to travel 5000 miles internationally by plane. (*Source:* Airlines for America)

31. On average, it cost $1,680,000 to build 12 Habitat for Humanity houses in the Philadelphia area in 2012. (*Source:* Habitat for Humanity Philadelphia)

32. In 2013, the number of movie admissions (tickets) sold was about 1340 million for 228 million moviegoers in the U.S. and Canada. Find the number of tickets sold per moviegoer. Round to the nearest whole. (*Source:* MPAA)

Find each unit rate.

33. The record number of tornadoes in the U.S. for a single calendar day is 209. Write this as a unit rate of tornadoes per hour rounded to the nearest tenth. (*Source:* Storm Prediction Center)

34. Sammy Joe Wingfield from Arlington, Texas, is still the fastest bricklayer on record. On May 20, 1994, he laid 1048 bricks in 60 minutes. Find his unit rate of bricks per minute rounded to the nearest tenth. (*Source: The Guinness Book of Records*)

35. Charlie Catlett can assemble 250 computer boards in an 8-hour shift while Suellen Catlett can assemble 402 computer boards in a 12-hour shift.
  a. Find the unit rate of Charlie.
  b. Find the unit rate of Suellen.
  c. Who can assemble computer boards faster, Charlie or Suellen?

36. Jerry Stein laid 713 bricks in 46 minutes while his associate, Bobby Burns, laid 396 bricks in 30 minutes.
  a. Find the unit rate of Jerry.
  b. Find the unit rate of Bobby.
  c. Who is the faster bricklayer?

For Exercises 37 and 38, round the rates to the nearest tenth.

37. One student drove 400 miles in his car on 14.5 gallons of gasoline. His sister drove 270 miles in her truck on 9.25 gallons of gasoline.
  a. Find the unit rate of the car.
  b. Find the unit rate of the truck.
  c. Which vehicle gets better gas mileage?

38. Charlotte Leal is a grocery scanner who can scan an average of 100 items in 3.5 minutes while her cousin Leo can scan 148 items in 5.5 minutes.
  a. Find the unit rate of Charlotte.
  b. Find the unit rate of Leo.
  c. Who is the faster scanner?

Objective C *Find each unit price. See Example 6.*

39. $57.50 for 5 DVDs

40. $0.87 for 3 apples

41. $1.19 for 7 bananas

42. $73.50 for 6 lawn chairs

Find each unit price and decide which is the better buy. Round to three decimal places. Assume that we are comparing different sizes of the same brand. See Examples 6 and 7.

43. Crackers:
  $1.19 for 8 ounces
  $1.59 for 12 ounces

44. Pickles:
  $1.89 for 32 ounces
  $0.89 for 18 ounces

45. Frozen orange juice:
  $1.69 for 16 ounces
  $0.69 for 6 ounces

46. Eggs:
  $0.69 for a dozen
  $2.10 for a flat $\left(2\frac{1}{2}\text{ dozen}\right)$

47. Soy sauce:
12 ounces for $2.29
8 ounces for $1.49

48. Shampoo:
20 ounces for $1.89
32 ounces for $3.19

49. Napkins:
100 for $0.59
180 for $0.93

50. Crackers:
20 ounces for $2.39
8 ounces for $0.99

Review

Multiply or divide as indicated. See Sections 4.4 and 4.5.

51.   1.7
$\times\ 6$

52.   2.3
$\times\ 9$

53.   3.7
$\times 1.2$

54.   6.6
$\times 2.5$

55. $2.3\overline{)4.37}$

56. $3.5\overline{)22.75}$

Concept Extensions

Fill in the table to calculate miles per gallon.

	Beginning Odometer Reading	Ending Odometer Reading	Miles Driven	Gallons of Gas Used	Miles per Gallon (round to the nearest tenth)
57.	29,286	29,543		13.4	
58.	16,543	16,895		15.8	
59.	79,895	80,242		16.1	
60.	31,623	32,056		11.9	

Find each unit rate.

61. The longest stairway is the service stairway for the Niesenbahn Cable railway near Spiez, Switzerland. It has 11,674 steps and rises to a height of 7759 feet. Find the unit rate of steps per foot rounded to the nearest tenth of a step. (*Source: Guinness World Records*)

62. In the United States, the total number of students enrolled in public schools is 49,800,000. There are about 98,800 public schools. Write a unit rate in students per school. Round to the nearest whole. (*Source: National Center for Education Statistics*)

63. In your own words, define the phrase "unit rate."

64. In your own words, define the phrase "unit price."

65. Should the rate $\dfrac{3 \text{ lights}}{2 \text{ feet}}$ be written as $\dfrac{3}{2}$? Explain why or why not.

66. Find an item in the grocery store and calculate its unit price.

Ratio and Rate

Answers

Write each ratio as a ratio of whole numbers using fractional notation. Write the fraction in simplest form.

1. 18 to 20

2. 36 to 100

3. 8.6 to 10

4. 1.6 to 4.6

5. $8.65 to $6.95

6. 7.2 ounces to 8.4 ounces

7. $3\frac{1}{2}$ to 13

8. $1\frac{2}{3}$ to $2\frac{3}{4}$

9. 8 inches to 12 inches

10. 3 hours to 24 hours

Find the ratio described in each problem.

11. During the 2012–13 academic year, a full college professor in Columbia University's doctoral program earned $212.3 thousand. By contrast, a Columbia doctoral associate professor earned only $132.4 thousand. Find the ratio of full professor salary to associate professor salary at the doctoral level at Columbia University. (*Source: American Association of University Professors*)

12. The New York Yankees are a dynastic powerhouse. They have won 27 out of the 108 Major League Baseball World Series played through 2012. (*Source: Major League Baseball*)

13. The circle graph below shows how the top 25 movies of 2013 were rated. Use this graph to answer the questions.

  a. How many top 25 movies were rated R?

  b. Find the ratio of top 25 PG-rated movies to PG-13-rated movies for 2013.

14. Find the ratio of the width to the length of the sign below.

Top 25 Movies of 2013

R
4 films

G
1 film

PG
5 films

PG-13
15 films

Source: MPAA

12 inches

RESERVED PARKING

18 inches

Answers	
1.	
2.	
3.	
4.	
5.	
6.	
7.	
8.	
9.	
10.	
11.	
12.	
13. a.	
b.	
14.	

332

Write each rate as a fraction in simplest form.

15. 5 offices for every 20 graduate
assistants

16. 6 lights every 15 feet

17. 100 U.S. senators for 50 states

18. 5 teachers for every 140 students

19. 64 computers every 100 households

20. 45 students for every 10 computers

Write each rate as a unit rate.

21. 165 miles in 3 hours

22. 560 feet in 4 seconds

23. 63 employees for 3 fax lines

24. 85 phone calls for 5 teenagers

25. 115 miles every 5 gallons

26. 112 teachers for 7 computers

27. 7524 books for 1254 college students

28. 2002 pounds for 13 adults

Write each unit price, rounded to the nearest hundredth, and decide which is the better buy.

29. Dog food:
    $2.16 for 8 pounds
    $4.99 for 18 pounds

30. Paper plates:
    $1.98 for 100
    $8.99 for 500
    (Round to the nearest thousandth.)

31. Microwave popcorn:
    3 packs for $2.39
    8 packs for $5.99

32. AA batteries:
    4 for $3.69
    10 for $9.89

15. _____

16. _____

17. _____

18. _____

19. _____

20. _____

21. _____

22. _____

23. _____

24. _____

25. _____

26. _____

27. _____

28. _____

29. _____

30. _____

31. _____

32. _____

Objectives

A Write Sentences as Proportions.

B Determine Whether Proportions Are True.

C Find an Unknown Number in a Proportion.

Objective A Writing Proportions

A **proportion** is a statement that two ratios or rates are equal. For example,

$$\frac{5}{6} = \frac{10}{12}$$

is a proportion. We can read this as, "5 is to 6 as 10 is to 12."

Practice 1

Write each sentence as a proportion.

a. 24 right is to 6 wrong as 4 right is to 1 wrong.

b. 32 Cubs fans is to 18 Mets fans as 16 Cubs fans is to 9 Mets fans.

Example 1 Write each sentence as a proportion.

a. 12 diamonds is to 15 rubies as 4 diamonds is to 5 rubies.

b. 5 hits is to 9 at bats as 20 hits is to 36 at bats.

Solution:

a. diamonds $\rightarrow$ $\dfrac{12}{15} = \dfrac{4}{5}$ $\leftarrow$ diamonds
  rubies $\rightarrow$ $\leftarrow$ rubies

b. hits $\rightarrow$ $\dfrac{5}{9} = \dfrac{20}{36}$ $\leftarrow$ hits
  at bats $\rightarrow$ $\leftarrow$ at bats

■ Work Practice 1

Helpful Hint

Notice in the above examples of proportions that the numerators contain the same units and the denominators contain the same units. In this text, proportions will be written so that this is the case.

Objective B Determining Whether Proportions Are True

Like other mathematical statements, a proportion may be either true or false. A proportion is true if its ratios are equal. Since ratios are fractions, one way to determine whether a proportion is true is to write both fractions in simplest form and compare them.

Another way is to compare cross products as we did in Section 2.3.

Note: In the box below, we are using letters to represent numbers. We later call these letters *variables*.

Using Cross Products to Determine Whether Proportions Are True or False

Cross products
$a \cdot d$ $b \cdot c$

$$\frac{a}{b} = \frac{c}{d}$$

If cross products are *equal*, the proportion is *true*.
If cross products are *not equal*, the proportion is *false*.

Answers

1. a. $\dfrac{24}{6} = \dfrac{4}{1}$ **b.** $\dfrac{32}{18} = \dfrac{16}{9}$

Example 2 Is $\dfrac{2}{3} = \dfrac{4}{6}$ a true proportion?

Solution:

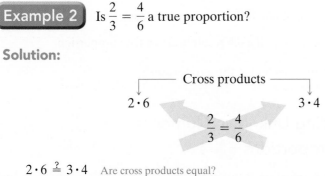

Cross products

$2 \cdot 6$ $3 \cdot 4$

$\dfrac{2}{3} = \dfrac{4}{6}$

$2 \cdot 6 \overset{?}{=} 3 \cdot 4$ Are cross products equal?

$12 = 12$ Equal, so proportion is true.

Since the cross products are equal, the proportion is true.

■ **Work Practice 2**

Practice 2

Is $\dfrac{3}{6} = \dfrac{4}{8}$ a true proportion?

Example 3 Is $\dfrac{4.1}{7} = \dfrac{2.9}{5}$ a true proportion?

Solution:

Cross products

$4.1 \cdot 5$ $7 \cdot 2.9$

$\dfrac{4.1}{7} = \dfrac{2.9}{5}$

$4.1 \cdot 5 \overset{?}{=} 7 \cdot 2.9$ Are cross products equal?

$20.5 \neq 20.3$ Not equal, so proportion is false.

Since the cross products are not equal, $\dfrac{4.1}{7} \neq \dfrac{2.9}{5}$. The proportion is false.

■ **Work Practice 3**

Practice 3

Is $\dfrac{3.6}{6} = \dfrac{5.4}{8}$ a true proportion?

Example 4 Is $\dfrac{1\frac{1}{6}}{10\frac{1}{2}} = \dfrac{\frac{1}{2}}{4\frac{1}{2}}$ a true proportion?

Solution:

$\dfrac{1\frac{1}{6}}{10\frac{1}{2}} = \dfrac{\frac{1}{2}}{4\frac{1}{2}}$

$1\dfrac{1}{6} \cdot 4\dfrac{1}{2} \overset{?}{=} 10\dfrac{1}{2} \cdot \dfrac{1}{2}$ Are cross products equal?

$\dfrac{7}{6} \cdot \dfrac{9}{2} \overset{?}{=} \dfrac{21}{2} \cdot \dfrac{1}{2}$ Write mixed numbers as improper fractions.

$\dfrac{21}{4} = \dfrac{21}{4}$ Equal, so proportion is true.

Since the cross products are equal, the proportion is true.

■ **Work Practice 4**

Practice 4

Is $\dfrac{4\frac{1}{5}}{2\frac{1}{3}} = \dfrac{3\frac{3}{10}}{1\frac{5}{6}}$ a true proportion?

Answers

2. yes   **3.** no   **4.** yes

✓**Concept Check** Think about cross products and write the true proportion $\frac{5}{8} = \frac{10}{16}$ in two other ways so that each result is also a true proportion.

(*Note:* There are no units attached in this proportion.)

Objective C Finding Unknown Numbers in Proportions

When one number of a proportion is unknown, we can use cross products to find the unknown number. For example, to find the unknown number n in the proportion $\frac{n}{30} = \frac{2}{3}$, we first find the cross products.

$n \cdot 3$ $30 \cdot 2$ Find the cross products.

$$\frac{n}{30} = \frac{2}{3}$$

If the proportion is true, then cross products are equal.

$n \cdot 3 = 30 \cdot 2$ Set the cross products equal to each other.

$n \cdot 3 = 60$ Write $2 \cdot 30$ as 60.

To find the unknown number n, we ask ourselves, "What number times 3 is 60?" The number is 20 and can be found by dividing 60 by 3.

$n = \dfrac{60}{3}$ Divide 60 by the number multiplied by n.

$n = 20$ Simplify.

Thus, the unknown number is 20.

Check: To *check*, let's replace n with this value, 20, and verify that a true proportion results.

$$\frac{20}{30} \stackrel{?}{=} \frac{2}{3}$$ ← Replace n with 20.

$$\frac{20}{30} \stackrel{?}{=} \frac{2}{3}$$

$3 \cdot 20 \stackrel{?}{=} 2 \cdot 30$

$60 = 60$ Cross products are equal.

Finding an Unknown Value *n* in a Proportion

Step 1: Set the cross products equal to each other.

Step 2: Divide the number not multiplied by n by the number multiplied by n.

✓**Concept Check Answer**

possible answers: $\dfrac{8}{5} = \dfrac{16}{10}$ and $\dfrac{5}{10} = \dfrac{8}{16}$

Example 5   Find the value of the unknown number n.

$$\frac{51}{34} = \frac{3}{n}$$

Solution:

Step 1:

$51 \cdot n = 34 \cdot 3$ Set cross products equal.

$51 \cdot n = 102$ Multiply.

Step 2:

$n = \dfrac{102}{51}$ Divide 102 by 51, the number multiplied by n.

$n = 2$ Simplify.

Check: $\dfrac{51}{34} \overset{?}{=} \dfrac{3}{2}$ Replace n with its value, 2.

$$\frac{51}{34} \overset{?}{=} \frac{3}{2}$$

$51 \cdot 2 \overset{?}{=} 34 \cdot 3$ Cross products are equal,

$102 = 102$ so the proportion is true.

■ **Work Practice 5**

Example 6   Find the unknown number n.

$$\frac{7}{n} = \frac{6}{5}$$

Solution:

Step 1:

$$\frac{7}{n} = \frac{6}{5}$$

$7 \cdot 5 = n \cdot 6$ Set the cross products equal to each other.

$35 = n \cdot 6$ Multiply.

Step 2:

$\dfrac{35}{6} = n$ Divide 35 by 6, the number multiplied by n.

$5\dfrac{5}{6} = n$

Check: Check to see that $5\dfrac{5}{6}$ is the unknown number.

■ **Work Practice 6**

Practice 5

Find the value of the unknown number n.

$$\frac{15}{2} = \frac{60}{n}$$

Practice 6

Find the unknown number n.

$$\frac{8}{n} = \frac{5}{9}$$

Answers

5. $n = 8$

6. $n = 14\dfrac{2}{5}$

Practice 7

Find the unknown number n.

$$\frac{n}{6} = \frac{0.7}{1.2}$$

Example 7 Find the unknown number n.

$$\frac{n}{3} = \frac{0.8}{1.5}$$

Solution:

Step 1:

$$\frac{n}{3} = \frac{0.8}{1.5}$$

$n \cdot 1.5 = 3 \cdot 0.8$ Set the cross products equal to each other.

$n \cdot 1.5 = 2.4$ Multiply.

Step 2:

$n = \dfrac{2.4}{1.5}$ Divide 2.4 by 1.5, the number multiplied by n.

$n = 1.6$ Simplify.

Check: Check to see that 1.6 is the unknown number.

■ Work Practice 7

Practice 8

Find the unknown number n.

$$\frac{n}{4\frac{1}{3}} = \frac{4\frac{1}{2}}{1\frac{3}{4}}$$

Example 8 Find the unknown number n.

$$\frac{1\frac{2}{3}}{3\frac{1}{4}} = \frac{n}{2\frac{3}{5}}$$

Solution:

Step 1:

$$\frac{1\frac{2}{3}}{3\frac{1}{4}} = \frac{n}{2\frac{3}{5}}$$

$1\frac{2}{3} \cdot 2\frac{3}{5} = 3\frac{1}{4} \cdot n$ Set the cross products equal to each other.

$\dfrac{13}{3} = 3\frac{1}{4} \cdot n$ Multiply. $1\frac{2}{3} \cdot 2\frac{3}{5} = \dfrac{5}{3} \cdot \dfrac{13}{5} = \dfrac{\overset{1}{\cancel{5}} \cdot 13}{3 \cdot \underset{1}{\cancel{5}}} = \dfrac{13}{3}$

$\dfrac{13}{3} = \dfrac{13}{4} \cdot n$ Write $3\frac{1}{4}$ as $\dfrac{13}{4}$.

Step 2:

$\dfrac{13}{3} \div \dfrac{13}{4} = n$ Divide $\dfrac{13}{3}$ by $\dfrac{13}{4}$, the number multiplied by n.

or

$n = \dfrac{13}{3} \cdot \dfrac{4}{13} = \dfrac{4}{3}$ or $1\frac{1}{3}$ Divide by multiplying by the reciprocal.

Check: Check to see that $1\frac{1}{3}$ is the unknown number.

■ Work Practice 8

Answers

7. $n = 3.5$

8. $n = 11\frac{1}{7}$

Vocabulary, Readiness & Video Check

Use the words and phrases below to fill in each blank.

ratio	cross products	true
false	proportion	

1. $\frac{4.2}{8.4} = \frac{1}{2}$ is called a _____ while $\frac{7}{8}$ is called a(n) _____.

2. In $\frac{a}{b} = \frac{c}{d}$, $a \cdot d$ and $b \cdot c$ are called in _____.

3. In a proportion, if cross products are equal, the proportion is _____.

4. In a proportion, if cross products are not equal, the proportion is _____.

Use cross products and mentally determine whether each proportion is true or false.

5. $\frac{2}{1} = \frac{6}{3}$ **6.** $\frac{3}{1} = \frac{15}{5}$ **7.** $\frac{1}{2} = \frac{3}{5}$ **8.** $\frac{2}{11} = \frac{1}{5}$ **9.** $\frac{2}{3} = \frac{40}{60}$ **10.** $\frac{3}{4} = \frac{6}{8}$

Martin-Gay Interactive Videos

Watch the section lecture video and answer the following questions.

Objective A **11.** From Example 1, what does "as" translate to in a proportion statement?

Objective B **12.** In Example 2, what are the cross products of the proportion? Is the proportion true or false?

Objective C **13.** As briefly mentioned in Example 4, what's another word for the unknown value n?

See Video 5.3

5.3 Exercise Set MyMathLab®

Objective A *Write each sentence as a proportion. See Example 1.*

1. 10 diamonds is to 6 opals as 5 diamonds is to 3 opals.

2. 8 books is to 6 courses as 4 books is to 3 courses.

3. 3 printers is to 12 computers as 1 printer is to 4 computers.

4. 4 hit songs is to 16 releases as 1 hit song is to 4 releases.

5. 6 eagles is to 58 sparrows as 3 eagles is to 29 sparrows.

6. 12 errors is to 8 pages as 1.5 errors is to 1 page.

7. $2\frac{1}{4}$ cups of flour is to 24 cookies as $6\frac{3}{4}$ cups of flour is to 72 cookies.

8. $1\frac{1}{2}$ cups milk is to 10 bagels as $\frac{3}{4}$ cup milk is to 5 bagels.

9. 22 vanilla wafers is to 1 cup of cookie crumbs as 55 vanilla wafers is to 2.5 cups of cookie crumbs. (*Source:* Based on data from *Family Circle* magazine)

10. 1 cup of instant rice is to 1.5 cups cooked rice as 1.5 cups of instant rice is to 2.25 cups of cooked rice. (*Source:* Based on data from *Family Circle* magazine)

Objective B *Determine whether each proportion is a true proportion. See Examples 2 through 4.*

11. $\dfrac{15}{9} = \dfrac{5}{3}$

12. $\dfrac{8}{6} = \dfrac{20}{15}$

▶13. $\dfrac{8}{6} = \dfrac{9}{7}$

14. $\dfrac{7}{12} = \dfrac{4}{7}$

▶15. $\dfrac{9}{36} = \dfrac{2}{8}$

16. $\dfrac{8}{24} = \dfrac{3}{9}$

17. $\dfrac{5}{8} = \dfrac{625}{1000}$

18. $\dfrac{30}{50} = \dfrac{600}{1000}$

19. $\dfrac{0.8}{0.3} = \dfrac{0.2}{0.6}$

20. $\dfrac{0.7}{0.4} = \dfrac{0.3}{0.1}$

21. $\dfrac{8}{10} = \dfrac{5.6}{0.7}$

22. $\dfrac{4.2}{8.4} = \dfrac{5}{10}$

23. $\dfrac{\frac{3}{4}}{\frac{4}{3}} = \dfrac{\frac{1}{2}}{\frac{8}{9}}$

24. $\dfrac{\frac{2}{5}}{\frac{2}{7}} = \dfrac{\frac{1}{10}}{\frac{1}{3}}$

25. $\dfrac{2\frac{2}{5}}{\frac{2}{3}} = \dfrac{1\frac{1}{9}}{\frac{1}{4}}$

26. $\dfrac{5\frac{5}{8}}{\frac{5}{3}} = \dfrac{4\frac{1}{2}}{1\frac{1}{5}}$

27. $\dfrac{\frac{4}{5}}{\frac{6}{6}} = \dfrac{\frac{6}{5}}{9}$

28. $\dfrac{\frac{6}{7}}{3} = \dfrac{\frac{10}{7}}{5}$

Objectives A B Mixed Practice—Translating *Write each sentence as a proportion. Then determine whether the proportion is a true proportion. See Examples 1 through 4.*

29. eight is to twelve as four is to six

30. six is to eight as nine is to twelve

31. five is to two as thirteen is to five

32. four is to three as seven is to five

33. one and eight tenths is to two as four and five tenths is to five

34. fifteen hundredths is to three as thirty-five hundredths is to seven

35. two thirds is to one fifth as two fifths is to one ninth

36. ten elevenths is to three fourths as one fourth is to one half

Objective C *For each proportion, find the unknown number n. See Examples 5 through 8.*

37. $\dfrac{n}{5} = \dfrac{6}{10}$

38. $\dfrac{n}{3} = \dfrac{12}{9}$

39. $\dfrac{18}{54} = \dfrac{3}{n}$

40. $\dfrac{25}{100} = \dfrac{7}{n}$

▶41. $\dfrac{n}{8} = \dfrac{50}{100}$

42. $\dfrac{n}{21} = \dfrac{12}{18}$

43. $\dfrac{8}{15} = \dfrac{n}{6}$

44. $\dfrac{12}{10} = \dfrac{n}{16}$

45. $\dfrac{24}{n} = \dfrac{60}{96}$

46. $\dfrac{26}{n} = \dfrac{28}{49}$

47. $\dfrac{3.5}{12.5} = \dfrac{7}{n}$

48. $\dfrac{0.2}{0.7} = \dfrac{8}{n}$

49. $\dfrac{0.05}{12} = \dfrac{n}{0.6}$

50. $\dfrac{7.8}{13} = \dfrac{n}{2.6}$

51. $\dfrac{8}{1\frac{1}{3}} = \dfrac{24}{n}$

52. $\dfrac{12}{3\frac{3}{4}} = \dfrac{48}{n}$

53. $\dfrac{\frac{1}{3}}{\frac{3}{8}} = \dfrac{\frac{2}{5}}{n}$

54. $\dfrac{\frac{7}{9}}{\frac{8}{27}} = \dfrac{\frac{1}{4}}{n}$

55. $\dfrac{12}{n} = \dfrac{\frac{2}{3}}{\frac{6}{9}}$

56. $\dfrac{24}{n} = \dfrac{\frac{8}{15}}{\frac{5}{9}}$

57. $\dfrac{n}{1\frac{1}{5}} = \dfrac{4\frac{1}{6}}{6\frac{2}{3}}$

58. $\dfrac{n}{3\frac{1}{8}} = \dfrac{7\frac{3}{5}}{2\frac{3}{8}}$

59. $\dfrac{25}{n} = \dfrac{3}{7\frac{7}{30}}$

60. $\dfrac{9}{n} = \dfrac{5}{11\frac{1}{15}}$

Review

Insert < or > to form a true statement. See Sections 3.5 and 4.2.

61. 8.01   8.1

62. 7.26   7.026

63. $2\frac{1}{2}$ $2\frac{1}{3}$

64. $9\frac{1}{5}$ $9\frac{1}{4}$

65. $5\frac{1}{3}$ $6\frac{2}{3}$

66. $1\frac{1}{2}$ $2\frac{1}{2}$

Concept Extensions

Use the numbers in each proportion to write two other true proportions. See the Concept Check in this section.

67. $\dfrac{9}{15} = \dfrac{3}{5}$

68. $\dfrac{1}{4} = \dfrac{5}{20}$

69. $\dfrac{6}{18} = \dfrac{1}{3}$

70. $\dfrac{2}{7} = \dfrac{4}{14}$

71. If the proportion $\dfrac{a}{b} = \dfrac{c}{d}$ is a true proportion, write two other true proportions using the same letters.

72. Write a true proportion.

73. Explain the difference between a ratio and a proportion.

74. Explain how to find the unknown number in a proportion such as $\dfrac{n}{18} = \dfrac{12}{8}$.

For each proportion, find the unknown number n. For Exercises 75 through 80, round your answer to the given place value.

75. $\dfrac{3.2}{0.3} = \dfrac{n}{1.4}$
Round to the nearest tenth.

76. $\dfrac{1.8}{n} = \dfrac{2.5}{8.4}$
Round to the nearest tenth.

77. $\dfrac{n}{5.2} = \dfrac{0.08}{6}$
Round to the nearest hundredth.

78. $\dfrac{4.25}{6.03} = \dfrac{5}{n}$
Round to the nearest hundredth.

79. $\dfrac{43}{17} = \dfrac{8}{n}$
Round to the nearest thousandth.

80. $\dfrac{n}{12} = \dfrac{18}{7}$
Round to the nearest hundredth.

81. $\dfrac{n}{7} = \dfrac{0}{8}$

82. $\dfrac{0}{2} = \dfrac{n}{3.5}$

83. $\dfrac{n}{1150} = \dfrac{588}{483}$

84. $\dfrac{585}{n} = \dfrac{117}{474}$

85. $\dfrac{222}{1515} = \dfrac{37}{n}$

86. $\dfrac{1425}{1062} = \dfrac{n}{177}$

5.4 Proportions and Problem Solving

Objective A Solving Problems by Writing Proportions

Objective

A Solve Problems by Writing Proportions.

Writing proportions is a powerful tool for solving problems in almost every field, including business, chemistry, biology, health sciences, and engineering, as well as in daily life. Given a specified ratio (or rate) of two quantities, a proportion can be used to determine an unknown quantity.

In this section, we use the same problem-solving steps that we have used earlier in this text.

Practice 1

On an architect's blueprint, 1 inch corresponds to 4 feet. How long is a wall represented by a $4\dfrac{1}{4}$-inch line on the blueprint?

Example 1 Determining Distances from a Map

On a chamber of commerce map of Abita Springs, 5 miles corresponds to 2 inches. How many miles correspond to 7 inches?

Answer
1. 17 ft

Solution:

1. UNDERSTAND. Read and reread the problem. You may want to draw a diagram.

From the diagram we can see that a reasonable solution should be between 15 and 20 miles.

2. TRANSLATE. We will let n represent our unknown number. Since 5 miles corresponds to 2 inches as n miles corresponds to 7 inches, we have the proportion

$$\begin{array}{lll} \text{miles} & \rightarrow & \dfrac{5}{2} = \dfrac{n}{7} \quad \leftarrow \quad \text{miles} \\ \text{inches} & \rightarrow & \phantom{\dfrac{5}{2} = \dfrac{n}{7}} \leftarrow \quad \text{inches} \end{array}$$

3. SOLVE: In earlier sections, we estimated to obtain a reasonable answer. Notice we did this in Step 1 above.

$$\dfrac{5}{2} = \dfrac{n}{7}$$

$5 \cdot 7 = 2 \cdot n$ 　　　Set the cross products equal to each other.

$35 = 2 \cdot n$ 　　　Multiply.

$\dfrac{35}{2} = n$ 　　　Divide 35 by 2, the number multiplied by n.

$n = 17\dfrac{1}{2}$ or 17.5 　　Simplify.

4. INTERPRET. *Check* your work. This result is reasonable since it is between 15 and 20 miles. *State* your conclusion: 7 inches corresponds to 17.5 miles.

■ Work Practice 1

Helpful Hint

We can also solve Example 1 by writing the proportion

$$\dfrac{2 \text{ inches}}{5 \text{ miles}} = \dfrac{7 \text{ inches}}{n \text{ miles}}$$

Although other proportions may be used to solve Example 1, we will solve by writing proportions so that the numerators have the same unit measures and the denominators have the same unit measures.

Example 2 Finding Medicine Dosage

The standard dose of an antibiotic is 4 cc (cubic centimeters) for every 25 pounds (lb) of body weight. At this rate, find the standard dose for a 140-lb woman.

Solution:

1. UNDERSTAND. Read and reread the problem. You may want to draw a diagram to estimate a reasonable solution.

(Continued on next page)

Practice 2

An auto mechanic recommends that 3 ounces of isopropyl alcohol be mixed with a tankful of gas (14 gallons) to increase the octane of the gasoline for better engine performance. At this rate, how many gallons of gas can be treated with a 16-ounce bottle of alcohol?

Answer

2. $74\dfrac{2}{3}$ or $74.\overline{6}$ gal

140–pound woman

25 pounds ⟶ 4 cc
25 pounds ⟶ 4 cc
25 pounds ⟶ 4 cc
25 pounds ⟶ 4 cc
25 pounds ⟶ 4 cc
15 pounds ⟶ ?
_____
140 pounds   over   20 cc

From the diagram, we can see that a reasonable solution is a little over 20 cc.

2. **TRANSLATE.** We will let n represent the unknown number. From the problem, we know that 4 cc is to 25 pounds as n cc is to 140 pounds, or

$$\begin{array}{ccccc} \text{cubic centimeters} & \rightarrow & \dfrac{4}{25} = \dfrac{n}{140} & \leftarrow & \text{cubic centimeters} \\ \text{pounds} & \rightarrow & & \leftarrow & \text{pounds} \end{array}$$

3. **SOLVE:**

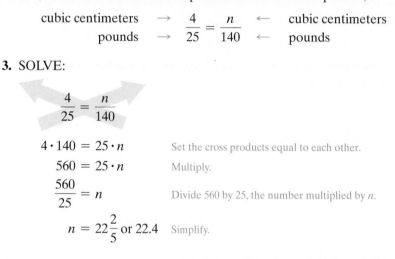

$$\frac{4}{25} = \frac{n}{140}$$

$4 \cdot 140 = 25 \cdot n$ Set the cross products equal to each other.

$560 = 25 \cdot n$ Multiply.

$\dfrac{560}{25} = n$ Divide 560 by 25, the number multiplied by n.

$n = 22\dfrac{2}{5}$ or 22.4   Simplify.

4. **INTERPRET.** *Check* your work. This result is reasonable since it is a little over 20 cc. *State* your conclusion: The standard dose for a 140-lb woman is 22.4 cc.

▬ **Work Practice 2**

Practice 3

If a gallon of paint covers 400 square feet, how many gallons are needed to paint a retaining wall that is 260 feet long and 4 feet high? Round the answer up to the nearest whole gallon.

△ **Example 3**   Calculating Supplies Needed to Fertilize a Lawn

A 50-pound bag of fertilizer covers 2400 square feet of lawn. How many bags of fertilizer are needed to cover a town square containing 15,360 square feet of lawn? Round the answer up to the nearest whole bag.

Answer

3. 3 gal

Solution:

1. UNDERSTAND. Read and reread the problem. Draw a picture.

1 bag covers

How many bags cover ... ?

Since one bag covers 2400 square feet, let's see how many 2400's there are in 15,360. We will estimate. The number 15,360 rounded to the nearest thousand is 15,000 and 2400 rounded to the nearest thousand is 2000. Then

$$15,000 \div 2000 = 7\frac{1}{2} \text{ or } 7.5.$$

2. TRANSLATE. We'll let n represent the unknown number. From the problem, we know that 1 bag is to 2400 square feet as n bags is to 15,360 square feet.

$$\begin{array}{ccccc} \text{bags} & \rightarrow & \dfrac{1}{2400} = \dfrac{n}{15{,}360} & \leftarrow & \text{bags} \\ \text{square feet} & \rightarrow & & \leftarrow & \text{square feet} \end{array}$$

3. SOLVE:

$$\dfrac{1}{2400} = \dfrac{n}{15{,}360}$$

$1 \cdot 15{,}360 = 2400 \cdot n$ Set the cross products equal to each other.

$15{,}360 = 2400 \cdot n$ Multiply.

$\dfrac{15{,}360}{2400} = n$ Divide 15,360 by 2400, the number multiplied by n.

$n = 6.4$ Simplify.

4. INTERPRET. *Check* that replacing n with 6.4 makes the proportion true. Is the answer reasonable? Yes, because it's close to $7\frac{1}{2}$ or 7.5. Because we must buy whole bags of fertilizer, 7 bags are needed. *State* your conclusion: To cover 15,360 square feet of lawn, 7 bags are needed.

▧ **Work Practice 3**

✓**Concept Check** You are told that 12 ounces of ground coffee will brew enough coffee to serve 20 people. How could you estimate how much ground coffee will be needed to serve 95 people?

✓**Concept Check Answer**

Find how much will be needed for 100 people (20×5) by multiplying 12 ounces by 5, which is 60 ounces.

Vocabulary, Readiness & Video Check

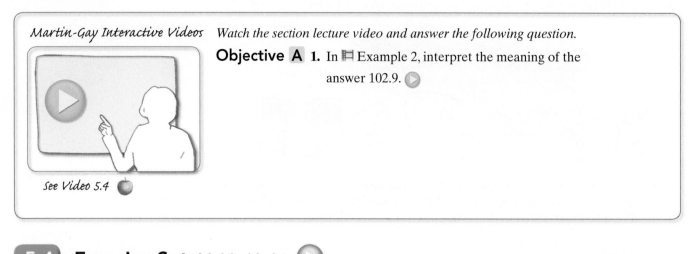

Martin-Gay Interactive Videos   Watch the section lecture video and answer the following question.

Objective A 1. In ⊟ Example 2, interpret the meaning of the answer 102.9. ▶

See Video 5.4 🍎

5.4 **Exercise Set** MyMathLab® ▶

Objective A *Solve. For Exercises 1 and 2, the solutions have been started for you. See Examples 1 through 3.*

An NBA basketball player averages 45 baskets for every 100 attempts.

1. If he attempted 800 field goals, how many field goals did he make?

Start the solution:

1. UNDERSTAND the problem. Reread it as many times as needed. Let's let
   n = how many field goals he made
2. TRANSLATE into an equation.

baskets (field goals) $\rightarrow \dfrac{45}{100} = \dfrac{n}{800}$ $\leftarrow$ baskets (field goals)
attempts $\rightarrow$ $\qquad\qquad\quad$ $\leftarrow$ attempts

3. SOLVE the equation. Set cross products equal to each other and solve.

$$\dfrac{45}{100} \diagdown \dfrac{n}{800}$$

   After SOLVING, then
4. INTERPRET.

2. If he made 225 baskets, how many did he attempt?

Start the solution:

1. UNDERSTAND the problem. Reread it as many times as needed. Let's let
   n = how many baskets attempted
2. TRANSLATE into an equation.

baskets $\rightarrow \dfrac{45}{100} = \dfrac{225}{n}$ $\leftarrow$ baskets
attempts $\rightarrow$ $\qquad\qquad\quad$ $\leftarrow$ attempts

3. SOLVE the equation. Set cross products equal to each other and solve.

$$\dfrac{45}{100} \diagup\diagdown \dfrac{225}{n}$$

   After SOLVING, then
4. INTERPRET.

It takes a word processor 30 minutes to word process and spell check 4 pages.

3. Find how long it takes her to word process and spell check 22 pages.

4. Find how many pages she can word process and spell check in 4.5 hours.

University Law School accepts 2 out of every 7 applicants.

5. If the school accepted 180 students, find how many applications they received.

6. If the school accepted 150 students, find how many applications they received.

On an architect's blueprint, 1 inch corresponds to 8 feet.

7. Find the length of a wall represented by a line $2\frac{7}{8}$ inches long on the blueprint.

8. Find the length of a wall represented by a line $5\frac{1}{4}$ inches long on the blueprint.

A human-factors expert recommends that there be at least 9 square feet of floor space in a college classroom for every student in the class.

△ **9.** Find the minimum floor space that 30 students require.

△ **10.** Due to a lack of space, a university converts a 21-by-15-foot conference room into a classroom. Find the maximum number of students the room can accommodate.

A Honda Civic Hybrid car averages 627 miles on a 12.3-gallon tank of gas.

11. Manuel Lopez is planning a 1250-mile vacation trip in his Honda Civic Hybrid. Find how many gallons of gas he can expect to burn. Round to the nearest gallon.

12. Ramona Hatch has enough money to put 6.9 gallons of gas in her Honda Civic Hybrid. She is planning on driving home from college for the weekend. If her home is 290 miles away, should she make it home before she runs out of gas?

The scale on an Italian map states that 1 centimeter corresponds to 30 kilometers.

13. Find how far apart Milan and Rome are if their corresponding points on the map are 15 centimeters apart.

14. On the map, a small Italian village is located 0.4 centimeter from the Mediterranean Sea. Find the actual distance.

A bag of Scotts fertilizer covers 3000 square feet of lawn.

△ **15.** Find how many bags of fertilizer should be purchased to cover a rectangular lawn 260 feet by 180 feet.

△ **16.** Find how many bags of fertilizer should be purchased to cover a square lawn measuring 160 feet on each side.

A Cubs baseball player gets 3 hits every 8 times at bat.

17. If this Cubs player comes up to bat 40 times in the World Series, find how many hits he would be expected to get.

18. At this rate, if he got 12 hits, find how many times he batted.

A survey reveals that 2 out of 3 people prefer Coke to Pepsi.

19. In a room of 40 people, how many people are likely to prefer Coke? Round the answer to the nearest person.

20. In a college class of 36 students, find how many students are likely to prefer Pepsi.

A self-tanning lotion advertises that a 3-oz bottle will provide four applications.

21. Jen Haddad found a great deal on a 14-oz bottle of the self-tanning lotion she had been using. Based on the advertising claims, how many applications of the self-tanner should Jen expect? Round down to the smaller whole number.

22. The Community College thespians need fake tans for a play they are doing. If the play has a cast of 35, how many ounces of self-tanning lotion should the cast purchase? Round up to the next whole number of ounces.

The school's computer lab goes through 5 reams of printer paper every 3 weeks.

23. Find out how long a case of printer paper is likely to last (a case of paper holds 8 reams of paper). Round to the nearest week.

24. How many cases of printer paper should be purchased to last the entire semester of 15 weeks? Round up to the next case.

A recipe for pancakes calls for 2 cups flour and $1\frac{1}{2}$ cups milk to make a serving for four people.

25. Ming has plenty of flour, but only 4 cups milk. How many servings can he make?

26. The swim team has a weekly breakfast after early practice. How much flour will it take to make pancakes for 18 swimmers?

Solve.

27. In the Seattle Space Needle, the elevators whisk you to the revolving restaurant at a speed of 800 feet in 60 seconds. If the revolving restaurant is 500 feet up, how long will it take you to reach the restaurant by elevator? (*Source:* Seattle Space Needle)

28. A 16-oz grande Tazo Black Iced Tea at Starbucks has 80 calories. How many calories are there in a 24-oz venti Tazo Black Iced Tea? (*Source:* Starbucks Coffee Company)

29. Mosquitos are annoying insects. To eliminate mosquito larvae, a certain granular substance can be applied to standing water in a ratio of 1 tsp per 25 sq ft of standing water.

   a. At this rate, find how many teaspoons of granules must be used for 450 square feet.

   b. If 3 tsp = 1 tbsp, how many tablespoons of granules must be used?

31. The daily supply of oxygen for one person is provided by 625 square feet of lawn. A total of 3750 square feet of lawn would provide the daily supply of oxygen for how many people? (*Source:* Professional Lawn Care Association of America)

30. Another type of mosquito control is liquid, where 3 oz of pesticide is mixed with 100 oz of water. This mixture is sprayed on roadsides to control mosquito breeding grounds hidden by tall grass.

   a. If one mixture of water with this pesticide can treat 150 feet of roadway, how many ounces of pesticide are needed to treat one mile? (*Hint:* 1 mile = 5280 feet)

   b. If 8 liquid ounces equals one cup, write your answer to part **a** in cups. Round to the nearest cup.

32. In 2012, approximately $21 billion of the $53 billion Americans spent on their pets was spent on pet food. Petsmart had $6,758,237,000 in net sales that year. How much of Petsmart's net sales would you expect to have been spent on pet food? Round to the nearest thousand. (*Source:* American Pet Products Manufacturers Association and Petsmart)

33. A student would like to estimate the height of the Statue of Liberty in New York City's harbor. The length of the Statue of Liberty's right arm is 42 feet. The student's right arm is 2 feet long and her height is $5\frac{1}{3}$ feet. Use this information to estimate the height of the Statue of Liberty. How close is your estimate to the statue's actual height of 111 feet, 1 inch from heel to top of head? (*Source:* National Park Service)

34. The length of the Statue of Liberty's index finger is 8 feet while the height to the top of the head is about 111 feet. Suppose your measurements are proportionally the same as this statue and your height is 5 feet.

   a. Use this information to find the proposed length of your index finger. Give an exact measurement and then a decimal rounded to the nearest hundredth.

   b. Measure your index finger and write it as a decimal in feet rounded to the nearest hundredth. How close is the length of your index finger to the answer to part **a**? Explain why.

35. There are 72 milligrams of cholesterol in a 3.5-ounce serving of lobster. How much cholesterol is in 5 ounces of lobster? Round to the nearest tenth of a milligram. (*Source:* The National Institutes of Health)

36. There are 76 milligrams of cholesterol in a 3-ounce serving of skinless chicken. How much cholesterol is in 8 ounces of chicken? (*Source:* USDA)

37. Trump World Tower in New York City is 881 feet tall and contains 72 stories. The Empire State Building contains 102 stories. If the Empire State Building has the same number of feet per floor as the Trump World Tower, approximate its height rounded to the nearest foot. (*Source:* skyscrapers.com)

38. In 2013, approximately 96 million of the 136 million U.S. employees worked in service industries. In a town of 6800 workers, how many would be expected to work in service-industry jobs? (*Source:* U.S. Bureau of Labor Statistics)

39. Medication is prescribed in 7 out of every 10 hospital emergency room visits that involve an injury. If a large urban hospital had 620 emergency room visits involving an injury in the past month, how many of these visits would you expect included a prescription for medication? (*Source:* National Center for Health Statistics)

40. One pound of firmly packed brown sugar yields $2\frac{1}{4}$ cups. How many pounds of brown sugar will be required in a recipe that calls for 6 cups of firmly packed brown sugar? (*Source:* Based on data from *Family Circle* magazine)

41. In 2012, two out of every three wireless phones sold in the U.S. were smartphones. Approximately 170 million wireless phones were sold in the U.S. in 2012. How many of them were smartphones? Round to the nearest million. (*Source:* Consumer Electronics Association)

42. In 2012, nearly one out of every four autos sold in the U.S. was a crossover utility vehicle (CUV). If an auto dealership sold 4500 vehicles in total that year, how many of these sales would you expect involved a CUV? (*Source:* Alliance of Automobile Manufacturers)

When making homemade ice cream in a hand-cranked freezer, the tub containing the ice cream mix is surrounded by a brine (water/salt) solution. To freeze the ice cream mix rapidly so that smooth and creamy ice cream results, the brine solution should combine crushed ice and rock salt in a ratio of 5 to 1. Use this information for Exercises 43 and 44. (Source: White Mountain Freezers, The Rival Company)

43. A small ice cream freezer requires 12 cups of crushed ice. How much rock salt should be mixed with the ice to create the necessary brine solution?

44. A large ice cream freezer requires $18\frac{3}{4}$ cups of crushed ice. How much rock salt will be needed to create the necessary brine solution?

45. The gas/oil ratio for a certain chainsaw is 50 to 1.
  a. How much oil (in gallons) should be mixed with 5 gallons of gasoline?
  b. If 1 gallon equals 128 fluid ounces, write the answer to part **a** in fluid ounces. Round to the nearest whole ounce.

46. The gas/oil ratio for a certain tractor mower is 20 to 1.
  a. How much oil (in gallons) should be mixed with 10 gallons of gas?
  b. If 1 gallon equals 4 quarts, write the answer to part **a** in quarts.

47. The adult daily dosage for a certain medicine is 150 mg (milligrams) of medicine for every 20 pounds of body weight.
  a. At this rate, find the daily dose for a man who weighs 275 pounds.
  b. If the man is to receive 500 mg of this medicine every 8 hours, is he receiving the proper dosage?

48. The adult daily dosage for a certain medicine is 80 mg (milligrams) for every 25 pounds of body weight.
  a. At this rate, find the daily dose for a woman who weighs 190 pounds.
  b. If she is to receive this medicine every 6 hours, find the amount to be given every 6 hours.

Review

Find the prime factorization of each number. See Section 2.2.

49. 15   **50.** 21   **51.** 20   **52.** 24   **53.** 200   **54.** 300   **55.** 32   **56.** 81

Concept Extensions

As we have seen earlier, proportions are often used in medicine dosage calculations. The exercises below have to do with liquid drug preparations, where the weight of the drug is contained in a volume of solution. The description of mg and ml below will help. We will study metric units further in Chapter 7.

mg means milligrams (A paper clip weighs about a gram. A milligram is about the weight of $\frac{1}{1000}$ of a paper clip.)

ml means milliliter (A liter is about a quart. A milliliter is about the amount of liquid in $\frac{1}{1000}$ of a quart.)

One way to solve the applications below is to set up the proportion $\frac{mg}{ml} = \frac{mg}{ml}$.
A solution strength of 15 mg of medicine in 1 ml of solution is available.

57. If a patient needs 12 mg of medicine, how many ml do you administer?

58. If a patient needs 33 mg of medicine, how many ml do you administer?

A solution strength of 8 mg of medicine in 1 ml of solution is available.

59. If a patient needs 10 mg of medicine, how many ml do you administer?

60. If a patient needs 6 mg of medicine, how many ml do you administer?

Estimate the following. See the Concept Check in this section.

61. It takes 1.5 cups of milk to make 11 muffins. Estimate the amount of milk needed to make 8 dozen muffins. Explain your calculation.

62. A favorite chocolate chip recipe calls for $2\frac{1}{2}$ cups of flour to make 2 dozen cookies. Estimate the amount of flour needed to make 50 cookies. Explain your calculation.

A board such as the one pictured below will balance if the following proportion is true:

$$\frac{\text{first weight}}{\text{second distance}} = \frac{\text{second weight}}{\text{first distance}}$$

Use this proportion to solve Exercises 63 and 64.

63. Find the distance *n* that will allow the board to balance.

40 pounds   60 pounds

64. Find the length *n* needed to lift the weight below.

125 pounds   3 feet 400 pounds

65. Describe a situation in which writing a proportion might solve a problem related to driving a car.

Chapter 5 Group Activity

Consumer Price Index

Sections 5.1–5.4

Do you remember when the regular price of a candy bar was 5¢, 10¢, or 25¢? It is certainly difficult to find a candy bar for that price these days. The reason is inflation: the tendency for the price of a given product to increase over time. Businesses and government agencies use the Consumer Price Index (CPI) to track inflation. The CPI measures the change in prices over time of basic consumer goods and services.

The CPI is very useful for comparing the prices of fixed items in various years. For instance, suppose an insurance company customer submits a claim for the theft of a fishing boat purchased in 1975. Because the customer's policy includes replacement cost coverage, the insurance company must calculate how much it would cost to replace the boat at the time of the theft. (Let's assume the theft took place in August 2013.) The customer has a receipt for the boat showing that it cost $598 in 1975. The insurance company can use the following proportion to calculate the replacement cost:

$$\frac{\text{price in earlier year}}{\text{price in later year}} = \frac{\text{CPI value in earlier year}}{\text{CPI value in later year}}$$

The CPI value is 53.8 for 1975. In August 2013, the CPI value is 233.877. The insurance company would then use the following proportion for this situation. (We will let n represent the unknown price in August 2013.)

$$\frac{\text{price in 1975}}{\text{price in 2013}} = \frac{\text{CPI value in 1975}}{\text{CPI value in Aug. 2013}}$$

$$\frac{598}{n} = \frac{53.8}{233.877}$$

$$53.8 \cdot n = 598(233.877)$$

$$\frac{53.8 \cdot n}{53.8} = \frac{598(233.877)}{53.8}$$

$$n \approx 2600$$

The replacement cost of the fishing boat at August 2013 prices is $2600.

Critical Thinking

1. What trends do you see in the CPI values in the table? Do you think these trends make sense? Explain.

2. A piece of jewelry cost $800 in 1985. What is its 2012 replacement value?

3. In 2000, the cost of a loaf of bread was about $1.89. What would an equivalent loaf of bread cost in 1950?

4. Suppose a couple purchased a house for $12,000 in 1940. At what price could they have been expected to sell the house in 1990?

5. An original Ford Model T cost about $850 in 1915. What is the equivalent cost of a Model T in 2010 dollars?

Consumer Price Index	
Year	CPI
1915	10.1
1920	20.0
1925	17.5
1930	16.7
1935	13.7
1940	14.0
1945	18.0
1950	24.1
1955	26.8
1960	29.6
1965	31.5
1970	38.8
1975	53.8
1980	82.4
1985	107.6
1990	130.7
1995	152.4
2000	172.2
2005	195.3
2010	218.1
2011	224.9
2012	229.6
2013	233.0

(*Source:* Bureau of Labor Statistics, U.S. Department of Labor)

Chapter 5 Vocabulary Check

Fill in each blank with one of the words or phrases listed below.

not equal	equal	cross products	rate
unit rate	ratio	unit price	proportion

1. A(n) _____ is the quotient of two numbers. It can be written as a fraction, using a colon, or using the word *to*.

2. $\dfrac{x}{2} = \dfrac{7}{16}$ is an example of a(n) _____.

3. A(n) _____ is a rate with a denominator of 1.

4. A(n) _____ is a "money per item" unit rate.

5. A(n) _____ is used to compare different kinds of quantities.

6. In the proportion $\dfrac{x}{2} = \dfrac{7}{16}$, $x \cdot 16$ and $2 \cdot 7$ are called _____.

7. If cross products are _____, the proportion is true.

8. If cross products are _____, the proportion is false.

Helpful Hint ▶ Are you preparing for your test? Don't forget to take the Chapter 5 Test on page 359. Then check your answers at the back of the text and use the Chapter Test Prep Videos to see the fully worked-out solutions to any of the exercises you want to review.

5 Chapter Highlights

Definitions and Concepts	Examples
Section 5.1 Ratios	
A **ratio** is the quotient of two quantities.	The ratio of 3 to 4 can be written as $$\frac{3}{4} \quad \text{or} \quad 3 : 4$$ fraction notation colon notation
Section 5.2 Rates	
Rates are used to compare different kinds of quantities.	Write the rate 12 spikes every 8 inches as a fraction in simplest form. $$\frac{12 \text{ spikes}}{8 \text{ inches}} = \frac{3 \text{ spikes}}{2 \text{ inches}}$$
A **unit rate** is a rate with a denominator of 1.	Write as a unit rate: 117 miles on 5 gallons of gas $$\frac{117 \text{ miles}}{5 \text{ gallons}} = \frac{23.4 \text{ miles}}{1 \text{ gallon}} \quad \text{or 23.4 miles per gallon}$$ or 23.4 miles/gallon
A **unit price** is a "money per item" unit rate.	Write as a unit price: \$5.88 for 42 ounces of detergent $$\frac{\$5.88}{42 \text{ ounces}} = \frac{\$0.14}{1 \text{ ounce}} = \$0.14 \text{ per ounce}$$

Definitions and Concepts	Examples

Section 5.3   Proportions

A **proportion** is a statement that two ratios or rates are equal.

$\dfrac{1}{2} = \dfrac{4}{8}$ is a proportion.

Using Cross Products to Determine Whether Proportions Are True or False

Is $\dfrac{6}{10} = \dfrac{9}{15}$ a true proportion?

Cross products

$a \cdot d$ $b \cdot c$

$$\dfrac{a}{b} \Join \dfrac{c}{d}$$

Cross products

$6 \cdot 15$ $10 \cdot 9$

$$\dfrac{6}{10} \Join \dfrac{9}{15}$$

If cross products are equal, the proportion is true.
If $ad = bc$, then the proportion is true.
If cross products are not equal, the proportion is false.
If $ad \neq bc$, then the proportion is false.

$6 \cdot 15 \overset{?}{=} 10 \cdot 9$ Are cross products equal?

$90 = 90$

Since cross products are equal, the proportion is a true proportion.

Finding an Unknown Value *n* in a Proportion

Find n: $\dfrac{n}{7} = \dfrac{5}{8}$

Step 1:   Set the cross products equal to each other.

Step 1:

$$\dfrac{n}{7} \Join \dfrac{5}{8}$$

$n \cdot 8 = 7 \cdot 5$ Set the cross products equal to each other.
$n \cdot 8 = 35$ Multiply.

Step 2:   Divide the number not multiplied by n by the number multiplied by n.

Step 2:

$n = \dfrac{35}{8}$ Divide 35 by 8, the number multiplied by n.

$n = 4\dfrac{3}{8}$

Section 5.4   Proportions and Problem Solving

Given a specified ratio (or rate) of two quantities, a proportion can be used to determine an unknown quantity.

On a map, 50 miles corresponds to 3 inches. How many miles correspond to 10 inches?

1. UNDERSTAND. Read and reread the problem.

2. TRANSLATE. We let n represent the unknown number. We are given that 50 miles is to 3 inches as n miles is to 10 inches.

$$\begin{array}{ccc} \text{miles} \rightarrow & \dfrac{50}{3} = & \dfrac{n}{10} \leftarrow \text{miles} \\ \text{inches} \rightarrow & & \quad\leftarrow \text{inches} \end{array}$$

(continued)

Definitions and Concepts	Examples

Section 5.4 Proportions and Problem Solving (*continued*)

3. SOLVE:

$$\frac{50}{3} = \frac{n}{10}$$

$50 \cdot 10 = 3 \cdot n$ Set the cross products equal to each other.

$500 = 3 \cdot n$ Multiply.

$\dfrac{500}{3} = n$ Divide 500 by 3, the number multiplied by n.

$n = 166\dfrac{2}{3}$

4. INTERPRET. *Check* your work. *State* your conclusion:

On the map, $166\dfrac{2}{3}$ miles corresponds to 10 inches.

Chapter 5   Review

(5.1) *Write each ratio as a fraction in simplest form.*

1. 23 to 37

2. 14 to 51

3. 6000 people to 4800 people

4. $121 to $143

5. 3.5 centimeters to 7.5 centimeters

6. 4.25 yards to 8.75 yards

7. $2\dfrac{1}{4}$ to $4\dfrac{3}{8}$

8. $3\dfrac{1}{2}$ to $2\dfrac{7}{10}$

The circle graph below shows how the top 25 movies (or films) of 2013 were rated. Use this graph to answer the questions.

Top 25 Movies of 2013

R — 4 films
G — 1 film
PG — 5 films
PG-13 — 15 films

Source: MPAA

9. a. How many top 25 movies were rated G?
   b. Find the ratio of top 25 G-rated movies to total movies for that year.

10. a. How many top 25 movies were rated PG-13?
   b. Find the ratio of top 25 PG-13-rated movies to total movies for that year.

(5.2) *Write each rate as a fraction in simplest form.*

11. 8 stillborn births to 1000 live births

12. 6 professors for 20 graduate research assistants

13. 15 word processing pages printed in 6 minutes

14. 8 computers assembled in 6 hours

Write each rate as a unit rate.

15. 468 miles in 9 hours

16. 180 feet in 12 seconds

17. $27.84 for 4 CDs

18. 8 gallons of pesticide for 6 acres of crops

19. $234 for books for 5 college courses

20. 104 bushels of fruit from 8 trees

Find each unit price and decide which is the better buy. Round to 3 decimal places. Assume that we are comparing different sizes of the same brand.

21. Taco sauce: 8 ounces for $0.99 or 12 ounces for $1.69

22. Peanut butter: 18 ounces for $1.49 or 28 ounces for $2.39

23. 2% milk: $0.59 for 16 ounces, $1.69 for 64 ounces, or $2.29 for 1 gallon (1 gallon = 128 fluid ounces)

24. Coca-Cola: $0.59 for 12 ounces, $0.79 for 16 ounces, or $1.19 for 32 ounces

(5.3) *Write each sentence as a proportion.*

25. 20 men is to 14 women as 10 men is to 7 women.

26. 50 tries is to 4 successes as 25 tries is to 2 successes.

27. 16 sandwiches is to 8 players as 2 sandwiches is to 1 player.

28. 12 tires is to 3 cars as 4 tires is to 1 car.

Determine whether each proportion is true.

29. $\dfrac{21}{8} = \dfrac{14}{6}$

30. $\dfrac{3}{5} = \dfrac{60}{100}$

31. $\dfrac{3.75}{3} = \dfrac{7.5}{6}$

32. $\dfrac{3.1}{6.2} = \dfrac{0.8}{0.16}$

Find the unknown number n in each proportion.

33. $\dfrac{n}{6} = \dfrac{15}{18}$ 　　　**34.** $\dfrac{n}{9} = \dfrac{5}{3}$ 　　　**35.** $\dfrac{4}{13} = \dfrac{10}{n}$ 　　　**36.** $\dfrac{8}{5} = \dfrac{9}{n}$

37. $\dfrac{8}{\frac{3}{2}} = \dfrac{n}{6}$ 　　　**38.** $\dfrac{9}{2} = \dfrac{n}{\frac{3}{2}}$ 　　　**39.** $\dfrac{27}{\frac{9}{4}} = \dfrac{n}{5}$ 　　　**40.** $\dfrac{6}{\frac{5}{2}} = \dfrac{n}{3}$

41. $\dfrac{0.4}{n} = \dfrac{2}{4.7}$ 　　　**42.** $\dfrac{7.2}{n} = \dfrac{6}{0.3}$ 　　　**43.** $\dfrac{n}{4\frac{1}{2}} = \dfrac{2\frac{1}{10}}{8\frac{2}{5}}$ 　　　**44.** $\dfrac{n}{4\frac{2}{7}} = \dfrac{3\frac{1}{9}}{9\frac{1}{3}}$

(5.4) *Solve.*

The ratio of a quarterback's completed passes to attempted passes is 3 to 7.

45. If he attempted 32 passes, find how many passes he completed. Round to the nearest pass.

46. If he completed 15 passes, find how many passes he attempted.

One bag of pesticide covers 4000 square feet of garden.

△ **47.** Find how many bags of pesticide should be purchased to cover a rectangular garden that is 180 feet by 175 feet.

△ **48.** Find how many bags of pesticide should be purchased to cover a square garden that is 250 feet on each side.

An owner of a Ford Escort can drive 420 miles on 11 gallons of gas.

49. If Tom Aloiso runs out of gas in an Escort and AAA comes to his rescue with $1\frac{1}{2}$ gallons of gas, determine whether Tom can then drive to a gas station 65 miles away.

50. Find how many gallons of gas Tom can expect to burn on a 3000-mile trip. Round to the nearest gallon.

Yearly homeowner property taxes are figured at a rate of $1.15 tax for every $100 of house value.

51. If a homeowner pays $627.90 in property taxes, find the value of his home.

52. Find the property taxes on a townhouse valued at $89,000.

On an architect's blueprint, 1 inch = 12 feet.

53. Find the length of a wall represented by a $3\frac{3}{8}$-inch line on the blueprint.

54. If an exterior wall is 99 feet long, find how long the blueprint measurement should be.

Mixed Review

Write each ratio as a fraction in simplest form.

55. 15 to 25

56. 16 to 36

57. 14 feet to 28 feet

58. 25 feet to 60 feet

59. 3 pints to 81 pints

60. 6 pints to 48 pints

Write each rate as a fraction in simplest form.

61. 2 teachers for 18 students

62. 6 nurses for 24 patients

Write each rate as a unit rate.

63. 24 cups for 6 people

64. 18 toys for 3 children

65. 136 miles in 4 hours

66. 12 gallons of milk from 6 cows

Find each unit price and decide which is the better buy. Round to 3 decimal places. Assume that we are comparing different sizes of the same brand.

67. cold medicine:
$4.94 for 4 oz,
$9.98 for 8 oz

68. juice:
12 oz for $0.65,
64 oz for $2.98

Write each sentence as a proportion.

69. 2 cups of cookie dough is to 30 cookies as 4 cups of cookie dough is to 60 cookies

70. 5 nickels is to 3 dollars as 20 nickels is to 12 dollars

Determine whether each proportion is a true proportion.

71. $\dfrac{3}{4} = \dfrac{87}{116}$

72. $\dfrac{2}{3} = \dfrac{4}{9}$

Find the unknown number n in each proportion.

73. $\dfrac{3}{n} = \dfrac{15}{8}$

74. $\dfrac{6}{n} = \dfrac{30}{24}$

75. $\dfrac{42}{5} = \dfrac{n}{10}$

76. $\dfrac{5}{4} = \dfrac{n}{20}$

Solve. The monthly loan payment for a car is $39.75 for each $1500 borrowed.

77. Find the monthly payment for a $23,000 car loan.

78. Find the monthly payment for an $18,000 car loan.

An investment of $1200 yields $152 each year.

79. At the same rate, how much will an investment of $1350 yield in one year?

80. At the same rate, how much will an investment of $750 yield in one year?

Write each ratio or rate as a fraction in simplest form.

1. $75 to $10

2. 4500 trees to 6500 trees

3. 28 men to every 4 women

4. 9 inches of rain in 30 days

5. 8.6 to 10

6. $5\frac{7}{8}$ to $9\frac{3}{4}$

7. The world's largest yacht, the *Azzam*, measures in at 591 feet long. A Boeing 787-8 Dreamliner measures 186 feet long. Find the ratio of the length of the *Azzam* to the length of a 787-8. (*Source:* Superyachts.com, Boeing)

186 ft

591 ft

Find each unit rate.

8. 650 kilometers in 8 hours

9. 8 inches of rain in 12 hours

10. 140 students for 5 teachers

11. QRIO (Quest for Curiosity) is the world's first bipedal robot capable of running (moving with both legs off the ground at the same time) at a rate of 108 inches each 12 seconds. (*Source*: *Guinness World Records*)

Find each unit price and use them to decide which is the better buy.

12. Steak sauce:
8 ounces for $1.19
12 ounces for $1.89

13. Jelly:
$1.49 for 16 ounces
$2.39 for 24 ounces

Determine whether each proportion is a true proportion.

14. $\frac{28}{16} = \frac{14}{8}$

15. $\frac{3.6}{2.2} = \frac{1.9}{1.2}$

16. Write the sentence as a proportion.
25 computers is to 600 students as 1 computer is to 24 students.

Answers

1. _____
2. _____
3. _____
4. _____
5. _____
6. _____
7. _____
8. _____
9. _____
10. _____
11. _____
12. _____
13. _____
14. _____
15. _____
16. _____

17. _____

18. _____

19. _____

20. _____

21. _____

22. _____

23. _____

24. _____

25. _____

26. _____

Find the unknown number n in each proportion.

17. $\dfrac{n}{3} = \dfrac{15}{9}$

18. $\dfrac{8}{n} = \dfrac{11}{6}$

19. $\dfrac{\frac{15}{12}}{\frac{3}{7}} = \dfrac{n}{\frac{4}{5}}$

20. $\dfrac{1.5}{5} = \dfrac{2.4}{n}$

21. $\dfrac{n}{2\frac{5}{8}} = \dfrac{1\frac{1}{6}}{3\frac{1}{2}}$

Solve.

22. On an architect's drawing, 2 inches corresponds to 9 feet. Find the length of a home represented by a line that is 11 inches long.

23. If a car can be driven 80 miles in 3 hours, how long will it take to travel 100 miles?

24. The standard dose of medicine for a dog is 10 grams for every 15 pounds of body weight. What is the standard dose for a dog that weighs 80 pounds?

25. Jerome Grant worked 6 hours and packed 86 cartons of books. At this rate, how many cartons can he pack in 8 hours?

26. Currently 3 out of every 4 American adults drink coffee at least once per week. In a town with a population of 7900 adults, how many of these adults would you expect to drink coffee at least once per week? (*Source:* National Coffee Association)

Answers

1. Subtract. Check each answer by adding.
   a. $12 - 9$
   b. $22 - 7$
   c. $35 - 35$
   d. $70 - 0$

2. Multiply
   a. $20 \cdot 0$
   b. $20 \cdot 1$
   c. $0 \cdot 20$
   d. $1 \cdot 20$

1. a. _____

   b. _____

   c. _____

   d. _____

2. a. _____

   b. _____

3. Round 248,982 to the nearest hundred.

4. Round 248,982 to the nearest thousand.

   c. _____

   d. _____

3. _____

5. Multiply:
   a. $\begin{array}{r} 25 \\ \times\ 8 \\ \hline \end{array}$
   b. $\begin{array}{r} 246 \\ \times\ 5 \\ \hline \end{array}$

6. Divide: $10,468 \div 28$

4. _____

5. a. _____

   b. _____

7. The director of a learning lab at a local community college is working on next year's budget. Thirty-three new DVD players are needed at a cost of $187 each. What is the total cost of these DVD players?

8. A study is being conducted for erecting soundproof walls along the interstate of a metropolitan area. The following lengths of walls are part of the proposal. Find their total: 4800 feet, 3270 feet, 2761 feet, 5760 feet.

6. _____

7. _____

8. _____

9. Find the prime factorization of 45.

10. Find $\sqrt{64}$.

9. _____

10. _____

11. Write $\dfrac{12}{20}$ in simplest form.

12. Find $9^2 \cdot \sqrt{9}$.

11. _____

Multiply.

13. $\dfrac{3}{4} \cdot \dfrac{8}{5}$

14. $3\dfrac{3}{8} \cdot 4\dfrac{5}{9}$

12. _____

13. _____

14. _____

15. $\dfrac{6}{13} \cdot \dfrac{26}{30}$

16. $\dfrac{2}{11} \cdot \dfrac{5}{8} \cdot \dfrac{22}{27}$

15. _____

16. _____

Perform the indicated operation and simplify.

17. $\dfrac{2}{7} + \dfrac{3}{7}$

18. $\dfrac{26}{30} - \dfrac{7}{30}$

17. _____

18. _____

19. _____

20. _____

21. _____

22. _____

23. _____

24. _____

25. _____

26. _____

27. _____

28. _____

29. _____

30. _____

31. _____

32. _____

33. _____

34. _____

35. _____

36. _____

37. _____

19. $\dfrac{7}{13} + \dfrac{6}{13} + \dfrac{3}{13}$

20. $\dfrac{7}{10} - \dfrac{3}{10} + \dfrac{4}{10}$

21. Find the LCM of 9 and 12.

22. Add: $\dfrac{17}{25} + \dfrac{3}{10}$

23. Write an equivalent fraction with the indicated denominator. $\dfrac{1}{2} = \dfrac{}{24}$

24. Determine whether these fractions are equivalent.
$\dfrac{10}{55}, \dfrac{6}{33}$

25. Subtract: $\dfrac{10}{11} - \dfrac{2}{3}$

26. Subtract: $17\dfrac{5}{24} - 9\dfrac{5}{9}$

27. A flight from Tucson to Phoenix, Arizona, requires $\dfrac{5}{12}$ of an hour. If the plane has been flying $\dfrac{1}{4}$ of an hour, find how much time remains before landing.

28. Simplify: $80 \div 8 \cdot 2 + 7$

29. Add: $2\dfrac{1}{3} + 5\dfrac{3}{8}$

30. Find the average of $\dfrac{3}{5}, \dfrac{4}{9}$, and $\dfrac{11}{15}$.

31. Insert $<$ or $>$ to form a true statement.
$\dfrac{3}{10} \quad \dfrac{2}{7}$

32. Multiply: $28{,}000 \times 500$

33. Write the decimal 1.3 in words.

34. Write "seventy-five thousandths" in standard form.

35. Round 736.2359 to the nearest tenth.

36. Round 736.2359 to the nearest thousandth.

37. Add: $23.85 + 1.604$

38. Subtract: $700 - 18.76$

39. Multiply: 0.283×0.3

40. Write $\dfrac{3}{8}$ as a decimal.

41. Divide and check: $0.5 \div 4$

42. Write 7.9 as an improper fraction.

43. Simplify: $0.5(8.6 - 1.2)$

44. Find the unknown number n.

$$\dfrac{n}{4} = \dfrac{12}{16}$$

45. Write the numbers in order from smallest to largest.

$$\dfrac{9}{20}, \dfrac{4}{9}, 0.456$$

46. Write the rate as a unit rate.
700 meters in 5 seconds

Write each ratio using fractional notation. Simplify Exercises 48 and 50 only.

47. The ratio of 2.6 to 3.1

48. The ratio of 7 to 21

49. The ratio of $1\dfrac{1}{2}$ to $7\dfrac{3}{4}$

50. The ratio of 900 to 9000

38. _____

39. _____

40. _____

41. _____

42. _____

43. _____

44. _____

45. _____

46. _____

47. _____

48. _____

49. _____

50. _____

This chapter is devoted to percent, a concept used virtually every day in ordinary and business life. Understanding percent and using it efficiently depend on understanding ratios because a percent is a ratio whose denominator is 100. We present techniques to write percents as fractions and as decimals and then solve problems relating to sales tax, commission, discounts, interest, and other real-life situations that use percents.

6 Percent

Cell-Phone-Only Households in the United States Continue to Increase

The number of U.S. households using landline telephone service continues to decrease while the number of households using cellular telephone service only continues to increase. Based on a survey conducted in 2012, the National Center for Health Statistics (NCHS) reported that 50.8% of all U.S. households have both landline and cellular telephone service. In Section 6.1, Exercises 27 and 28, we learn the percent of all U.S. households using cell phones only and landlines only, and we convert these percents to decimals.

Cell-Phone-Only Households by State, 2013

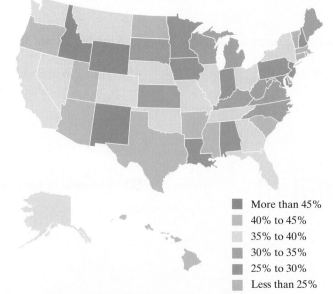

■ More than 45%
■ 40% to 45%
■ 35% to 40%
■ 30% to 35%
■ 25% to 30%
■ Less than 25%

Source: Based on estimates from Marketing Systems Group

6.1 Introduction to Percent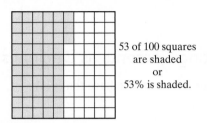

Objective A Understanding Percent

The word **percent** comes from the Latin phrase *per centum*, which means **"per 100."** For example, 53% (percent) means 53 per 100. In the square below, 53 of the 100 squares are shaded. Thus, 53% of the figure is shaded.

53 of 100 squares
are shaded
or
53% is shaded.

Since 53% means 53 per 100, 53% is the ratio of 53 to 100, or $\frac{53}{100}$.

$$53\% = \frac{53}{100}$$

Also,

$$7\% = \frac{7}{100} \quad \text{7 parts per 100 parts}$$

$$73\% = \frac{73}{100} \quad \text{73 parts per 100 parts}$$

$$109\% = \frac{109}{100} \quad \text{109 parts per 100 parts}$$

> **Percent**
>
> **Percent** means **per one hundred.** The "%" symbol is used to denote percent.

Percent is used in a variety of everyday situations. For example,

- 77.6% of the U.S. population uses the Internet.
- The store is having a 25%-off sale.
- 78% of us trust our local fire department.
- The enrollment in community colleges is predicted to increase 1.3% each year.
- The South is the home of 49% of all frequent paintball participants.

Example 1 Since 2011, white has been the world's most popular color for cars. For 2013 model cars, 25 out of every 100 were painted white. What percent of model-year 2013 cars were white? (*Source:* PPG Industries)

Solution: Since 25 out of 100 cars were painted white, the fraction is $\frac{25}{100}$. Then

$$\frac{25}{100} = 25\%$$

■ Work Practice 1

Objectives

A Understand Percent.

B Write Percents as Decimals.

C Write Decimals as Percents.

Practice 1

Of 100 students in a club, 23 are freshmen. What percent of the students are freshmen?

Answer
1. 23%

365

Practice 2
29 out of 100 executives are in their forties. What percent of executives are in their forties?

Example 2 46 out of every 100 college students live at home. What percent of students live at home? (*Source:* Independent Insurance Agents of America)

Solution:

$$\frac{46}{100} = 46\%$$

■ Work Practice 2

Objective B Writing Percents as Decimals ▶

Since percent means "per hundred," we have that

$$1\% = \frac{1}{100} = 0.01$$

In other words, the percent symbol means "per hundred" or, equivalently, "$\frac{1}{100}$" or "0.01." Thus

$$87\% = 87 \times \frac{1}{100} = \frac{87}{100}$$

or

$$87\% = 87 \times (0.01) = 0.87$$

Results are the same

Of course, we know that the end results are the same, that is,

$$\frac{87}{100} = 0.87$$

The above gives us two options for converting percents. We can replace the percent symbol, %, by $\frac{1}{100}$ or 0.01 and then multiply.

> For consistency, when we
> - convert from a percent to a *decimal*, we will drop the % symbol and multiply by 0.01 (this section).
> - convert from a percent to a *fraction*, we will drop the % symbol and multiply by $\frac{1}{100}$ (next section).

Thus, to write 53% (or 53.%) as a decimal,

$$53\% = 53(0.01) = 0.53 \quad \text{Replace the percent symbol with 0.01. Then multiply.}$$

Writing a Percent as a Decimal

Replace the percent symbol with its decimal equivalent, 0.01; then multiply.

$$43\% = 43(0.01) = 0.43$$

Helpful Hint

If it helps, think of writing a percent as a decimal by

Percent → | Remove the % symbol and move decimal point 2 places to the left | → Decimal

Answer

2. 29%

Example 3 Write 23% as a decimal.

Solution:

$$23\% = 23(0.01) \quad \text{Replace the percent symbol with 0.01.}$$
$$= 0.23 \quad \text{Multiply.}$$

 Work Practice 3

Practice 3
Write 89% as a decimal.

Examples Write each percent as a decimal.

4. $4.6\% = 4.6(0.01) = 0.046$ Replace the percent symbol with 0.01. Then multiply.

5. $190\% = 190(0.01) = 1.90$ or 1.9

6. $0.74\% = 0.74(0.01) = 0.0074$

7. $100\% = 100(0.01) = 1.00$ or 1

Helpful Hint We just learned that $100\% = 1$

Work Practice 4–7

Practice 4–7
Write each percent as a decimal.
4. 2.7%   **5.** 150%
6. 0.69%   **7.** 800%

✓**Concept Check** Why is it incorrect to write the percent 0.033% as 3.3 in decimal form?

Objective C Writing Decimals as Percents ▶

To write a decimal as a percent, we use the result of Example 7 above. In this example, we found that $1 = 100\%$.

$$0.38 = 0.38(1) = 0.38(100\%) = 38\%$$

Notice that the result is

$$0.38 = 0.38(100\%) = 38.\% \quad \text{Multiply by 1 in the form of 100\%.}$$

Writing a Decimal as a Percent

Multiply by 1 in the form of 100%.

$$0.27 = 0.27(100\%) = 27.\%$$

Helpful Hint

If it helps, think of writing a decimal as a percent by reversing the steps in the Helpful Hint on the previous page.

Percent ← Move the decimal point 2 places to the right and attach a % symbol. ← Decimal

Answers
3. 0.89 **4.** 0.027 **5.** 1.5
6. 0.0069 **7.** 8.00 or 8

✓**Concept Check Answer**
To write a percent as a decimal, the decimal point should be moved two places to the left, not to the right. So the correct answer is 0.00033.

Practice 8

Write 0.19 as a percent.

Example 8 Write 0.65 as a percent.

Solution:

$$0.65\% = 0.65(100\%) = 65.\%$$ Multiply by 100%.
$$= 65\%$$

■ Work Practice 8

Practice 9–11

Write each decimal as a percent.
9. 1.75   **10.** 0.044   **11.** 0.7

Examples Write each decimal as a percent.

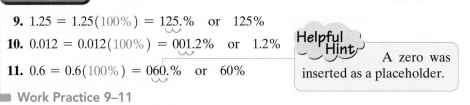

9. $1.25 = 1.25(100\%) = 125.\%$ or 125%
10. $0.012 = 0.012(100\%) = 001.2\%$ or 1.2%
11. $0.6 = 0.6(100\%) = 060.\%$ or 60%

Helpful Hint
A zero was inserted as a placeholder.

■ Work Practice 9–11

Answers
8. 19%   **9.** 175%   **10.** 4.4%
11. 70%

✓**Concept Check Answer**

To change a decimal to a percent, multiply by 100%, or move the decimal point *only* two places to the right. So the correct answer is 3.45%.

✓**Concept Check** Why is it incorrect to write the decimal 0.0345 as 34.5% in percent form?

Vocabulary, Readiness & Video Check

Use the choices below to fill in each blank. Some choices may be used more than once or not at all.

$\frac{1}{100}$ 0.01   100%   percent

1. _____ means "per hundred."
2. _____ = 1.
3. The % symbol is read as _____.
4. To write a decimal as a *percent*, multiply by 1 in the form of _____.
5. To write a percent as a *decimal*, drop the % symbol and multiply by _____.

Martin-Gay Interactive Videos

Watch the section lecture video and answer the following questions.

Objective A 6. From the lecture before ▤ Example 1, what is the most important thing to remember about percent?

Objective B 7. From the lecture before ▤ Example 2, what does 1% equal in fraction form? In decimal form?

Objective C 8. Complete this statement based on ▤ Example 7: Multiplying a decimal by _____ is the same as multiplying it by 1 and does not change the _____ of the decimal.

See Video 6.1

6.1 Exercise Set MyMathLab®

Objective A *Solve. See Examples 1 and 2.*

1. In a survey of 100 college students, 96 use the Internet. What percent use the Internet?

2. A basketball player makes 81 out of 100 attempted free throws. What percent of free throws are made?

3. Michigan leads the United States in tart cherry production, producing 75 out of every 100 tart cherries each year.
   a. What percent of tart cherries are produced in Michigan?
   b. What percent of tart cherries are *not* produced in Michigan? (*Source:* Cherry Marketing Institute)

4. The United States is the world's second-largest producer of apples. Twenty-five out of every 100 apples harvested in the United States are exported (shipped to other countries). (*Source:* U.S. Apple Association)
   a. What percent of U.S.-grown apples are exported?
   b. What percent of U.S.-grown apples are *not* exported?

One hundred adults were asked to name their favorite sport, and the results are shown in the circle graph.

5. What sport was preferred by most adults? What percent preferred this sport?

6. What sport was preferred by the least number of adults? What percent preferred this sport?

7. What percent of adults preferred football or soccer?

8. What percent of adults preferred basketball or baseball?

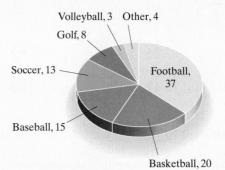

Volleyball, 3   Other, 4
Golf, 8
Soccer, 13
Football, 37
Baseball, 15
Basketball, 20

Objective B *Write each percent as a decimal. See Examples 3 through 7.*

9. 41%

10. 62%

11. 6%

12. 3%

13. 100%

14. 136%

15. 73.6%

16. 45.7%

17. 2.8%

18. 1.4%

19. 0.6%

20. 0.9%

21. 300%

22. 500%

23. 32.58%

24. 72.18%

Write each percent as a decimal. See Examples 3 through 7.

25. People take aspirin for a variety of reasons. The most common use of aspirin is to prevent heart disease, accounting for 38% of all aspirin use. (*Source:* Bayer Market Research)

26. Exports to Europe accounted for 17.7% of all of Japan's motor vehicle exports. (*Source:* Japan Automobile Manufacturers Association)

27. In 2012, 38.2% of households in the United States had no landline telephones, just cell phones. (*Source:* National Center for Health Statistics)

28. In 2012, 8.6% of households in the United States had no cell phones, just landline telephones. (*Source:* National Center for Health Statistics)

29. A $\frac{1}{2}$-cup serving of dried tart cherries delivers 45% of an adult's Daily Value of vitamin A. (*Source: USDA Nutrient Data Laboratory*)

30. In 2010, 63.5% of the paper used in the United States was recovered for recycling. (*Source:* Keep America Beautiful, Inc.)

Objective C *Write each decimal as a percent. See Examples 8 through 11.*

31. 0.98   **32.** 0.75   **33.** 3.1   **34.** 4.8   **35.** 29

36. 56   **37.** 0.003   **38.** 0.006   ◗ **39.** 0.22   **40.** 0.45

41. 5.3   **42.** 1.6   ◗ **43.** 0.056   **44.** 0.027   **45.** 0.3328

46. 0.1115   ◗ **47.** 3   **48.** 5   ◗ **49.** 0.7   **50.** 0.8

Write each decimal as a percent. See Examples 8 through 11.

51. Leisure travel accounted for 0.77 of all domestic trips in the United States. (*Source:* U.S. Travel Association)

52. According to a recent survey, 0.34 of American adults reported having used the Internet to research nutritional information about restaurant foods. (*Source:* National Restaurant Association's *2013 Restaurant Industry Forecast*)

53. In 2012, about 0.281 of full-time workers in the United States spent time volunteering in the previous year. (*Source:* Bureau of Labor Statistics)

54. In 2012, about 0.334 of part-time workers in the United States participated in volunteer activities in the previous year. (*Source:* Bureau of Labor Statistics)

55. Nearly 0.081 of the United States labor force was unemployed in 2012. (*Source:* Bureau of Labor Statistics)

56. In 2012, an estimated 0.235 of the American population was under the age of 18. (*Source:* U.S. Census Bureau)

Review

Write each fraction as a decimal. See Section 4.6.

57. $\frac{1}{4}$ **58.** $\frac{3}{5}$ **59.** $\frac{13}{20}$ **60.** $\frac{11}{40}$ **61.** $\frac{9}{10}$ **62.** $\frac{7}{10}$

Concept Extensions

Solve. See the Concept Checks in this section.

63. Which of the following are correct?

a. 6.5% = 0.65

b. 7.8% = 0.078

c. 120% = 0.12

d. 0.35% = 0.0035

64. Which of the following are correct?

a. 0.231 = 23.1%

b. 5.12 = 0.0512%

c. 3.2 = 320%

d. 0.0175 = 0.175%

Recall that 1 = 100%. This means that 1 whole is 100%. Use this for Exercises 65 and 66. (Source: Some Body by Dr. Pete Rowen)

65. The four blood types are A, B, O, and AB. (Each blood type can also be further classified as Rh-positive or Rh-negative depending upon whether your blood contains protein or not.) Given the percent blood types for the United States below, calculate the percent of U.S. population with AB blood type.

45%   40%   11%   ?%

66. The components of bone are all listed in the categories below. Find the missing percent.

1. Minerals–45%

2. Living tissue–30%

3. Water–20%

4. Other–?

The bar graph shows the predicted fastest-growing occupations. Use the graph for Exercises 67 through 70. (Source: Bureau of Labor Statistics)

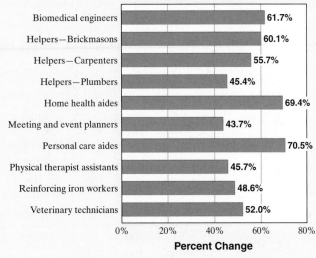

Fastest-Growing Occupations 2010–2020

Occupation	Percent Change
Biomedical engineers	61.7%
Helpers—Brickmasons	60.1%
Helpers—Carpenters	55.7%
Helpers—Plumbers	45.4%
Home health aides	69.4%
Meeting and event planners	43.7%
Personal care aides	70.5%
Physical therapist assistants	45.7%
Reinforcing iron workers	48.6%
Veterinary technicians	52.0%

Source: Bureau of Labor Statistics

67. What occupation is predicted to be the fastest growing?

68. What occupation is predicted to be the second fastest growing?

69. Write the percent change for biomedical engineers as a decimal.

70. Write the percent change for meeting and event planners as a decimal.

71. In your own words, explain how to write a percent as a decimal.

72. In your own words, explain how to write a decimal as a percent.

Objectives

A Write Percents as Fractions.

B Write Fractions as Percents.

C Convert Percents, Decimals, and Fractions.

Objective A Writing Percents as Fractions

Recall from Section 6.1 that percent means per hundred. Thus

$$1\% = \frac{1}{100} = 0.01$$

For example,

$$87\% = 87 \times \frac{1}{100} = \frac{87}{100} \quad \text{Writing 87\% as a fraction.}$$

or

$$87\% = 87 \times 0.01 = 0.87 \quad \text{Writing 87\% as a decimal.}$$

In this section we are writing percents as fractions, so we do the following.

> **Writing a Percent as a Fraction**
>
> Replace the percent symbol with its fraction equivalent, $\frac{1}{100}$; then multiply. Don't forget to simplify the fraction if possible.
>
> $$7\% = 7 \cdot \frac{1}{100} = \frac{7}{100}$$

Practice 1–5

Write each percent as a fraction or mixed number in simplest form.

1. 25%

2. 2.3%

3. 225%

4. $66\frac{2}{3}\%$

5. 8%

Examples Write each percent as a fraction or mixed number in simplest form.

1. $40\% = 40 \cdot \frac{1}{100} = \frac{40}{100} = \frac{2 \cdot \cancel{20}^{1}}{5 \cdot \cancel{20}_{1}} = \frac{2}{5}$

2. $1.9\% = 1.9 \cdot \frac{1}{100} = \frac{1.9}{100}$. We don't want the numerator of the fraction to contain a decimal, so we multiply by 1 in the form of $\frac{10}{10}$.

$$= \frac{1.9}{100} \cdot \frac{10}{10} = \frac{1.9 \cdot 10}{100 \cdot 10} = \frac{19}{1000}$$

3. $125\% = 125 \cdot \frac{1}{100} = \frac{125}{100} = \frac{5 \cdot \cancel{25}^{1}}{4 \cdot \cancel{25}_{1}} = \frac{5}{4} \text{ or } 1\frac{1}{4}$

4. $33\frac{1}{3}\% = \underbrace{33\frac{1}{3}}_{\substack{\text{Write as} \\ \text{an improper fraction.}}} \cdot \frac{1}{100} = \frac{100}{3} \cdot \frac{1}{100} = \frac{\cancel{100}^{1} \cdot 1}{3 \cdot \cancel{100}_{1}} = \frac{1}{3}$

5. $100\% = 100 \cdot \frac{1}{100} = \frac{100}{100} = 1$

> **Helpful Hint** Just as in the previous section, we confirm that 100% = 1

■ Work Practice 1–5

Objective B Writing Fractions as Percents

Recall that to write a percent as a fraction, we replace the percent symbol by its fraction equivalent, $\frac{1}{100}$. We reverse these steps to write a fraction as a percent.

Writing a Fraction as a Percent

Multiply by 1 in the form of 100%.

$$\frac{1}{8} = \frac{1}{8} \cdot 100\% = \frac{1}{8} \cdot \frac{100}{1}\% = \frac{100}{8}\% = 12\frac{1}{2}\% \quad \text{or} \quad 12.5\%$$

Helpful Hint

From Example 5, we know that

$$100\% = 1$$

Recall that when we multiply a number by 1, we are not changing the value of that number. This means that when we multiply a number by 100%, we are not changing its value but rather writing the number as an equivalent percent.

Examples Write each fraction or mixed number as a percent.

6. $\frac{9}{20} = \frac{9}{20} \cdot 100\% = \frac{9}{20} \cdot \frac{100}{1}\% = \frac{900}{20}\% = 45\%$

7. $\frac{2}{3} = \frac{2}{3} \cdot 100\% = \frac{2}{3} \cdot \frac{100}{1}\% = \frac{200}{3}\% = 66\frac{2}{3}\%$

8. $1\frac{1}{2} = \frac{3}{2} \cdot 100\% = \frac{3}{2} \cdot \frac{100}{1}\% = \frac{300}{2}\% = 150\%$

Helpful Hint

$$\frac{200}{3} = 66.\overline{6}.$$

Thus, another way to write $\frac{200}{3}\%$ is 66.$\overline{6}$%.

■ **Work Practice 6–8**

Practice 6–8

Write each fraction or mixed number as a percent.

6. $\frac{1}{2}$ **7.** $\frac{7}{40}$ **8.** $2\frac{1}{4}$

✓**Concept Check** Which digit in the percent 76.4582% represents

a. A tenth percent?
b. A thousandth percent?
c. A hundredth percent?
d. A whole percent?

Example 9 Write $\frac{1}{12}$ as a percent. Round to the nearest hundredth percent.

Solution:

$$\frac{1}{12} = \frac{1}{12} \cdot 100\% = \frac{1}{12} \cdot \frac{100}{1}\% = \frac{100}{12}\% \approx 8.33\%$$

"approximately"

$$\begin{array}{r} 8.333 \approx 8.33 \\ 12\overline{)100.000} \\ -96 \\ \hline 4\,0 \\ -3\,6 \\ \hline 40 \\ -36 \\ \hline 40 \\ -36 \\ \hline 4 \end{array}$$

Thus, $\frac{1}{12}$ is approximately 8.33%.

■ **Work Practice 9**

Practice 9

Write $\frac{3}{17}$ as a percent. Round to the nearest hundredth percent.

Answers

6. 50% **7.** $17\frac{1}{2}\%$ **8.** 225% **9.** 17.65%

✓**Concept Check Answers**
a. 4 **b.** 8 **c.** 5 **d.** 6

Objective C Converting Percents, Decimals, and Fractions ▶

Let's summarize what we have learned so far about percents, decimals, and fractions:

Summary of Converting Percents, Decimals, and Fractions

- *To write a percent as a decimal,* replace the % symbol with its decimal equivalent, 0.01; then multiply.
- *To write a percent as a fraction,* replace the % symbol with its fraction equivalent, $\frac{1}{100}$; then multiply.
- *To write a decimal or fraction as a percent,* multiply by 100%.

If we let p represent a number, below we summarize using symbols.

Write a percent as a decimal:	Write a percent as a fraction:	Write a number as a percent:
$p\% = p(0.01)$	$p\% = p \cdot \dfrac{1}{100}$	$p = p \cdot 100\%$

Practice 10

A family decides to spend no more than 22.5% of its monthly income on rent. Write 22.5% as a decimal and as a fraction.

Example 10 37.8% of automobile thefts in the continental United States occur in the South, the greatest percent. Write this percent as a decimal and as a fraction. (*Source:* Federal Bureau of Investigation's (FBI) Uniform Crime Report)

Solution:

As a decimal: $37.8\% = 37.8(0.01) = 0.378$

As a fraction: $37.8\% = 37.8 \cdot \dfrac{1}{100} = \dfrac{37.8}{100} = \dfrac{37.8}{100} \cdot \dfrac{10}{10} = \dfrac{378}{1000} = \dfrac{\cancel{2} \cdot 189}{\cancel{2} \cdot 500} = \dfrac{189}{500}.$

Thus, 37.8% written as a decimal is 0.378 and written as a fraction is $\dfrac{189}{500}$.

▪ Work Practice 10

Practice 11

Provincetown's budget for waste disposal increased by $1\frac{1}{4}$ times over the budget from last year. What percent increase is this?

Example 11 An advertisement for a stereo system reads "$\frac{1}{4}$ off." What percent off is this?

Solution: Write $\dfrac{1}{4}$ as a percent.

$$\frac{1}{4} = \frac{1}{4} \cdot 100\% = \frac{1}{4} \cdot \frac{100}{1}\% = \frac{100}{4}\% = 25\%$$

Thus, "$\dfrac{1}{4}$ off" is the same as "25% off."

▪ Work Practice 11

Note: It is helpful to know a few basic percent conversions. Appendix A.4 contains a handy reference of percent, decimal, and fraction equivalencies.

Also, Appendix A.5 shows how to find common percents of a number.

Vocabulary, Readiness & Video Check

Use the choices below to fill in each blank. Some choices may be used more than once.

$\dfrac{1}{100}$ 100%   percent

1. _____ means "per hundred."

2. _____ = 1.

3. To write a decimal or a fraction as a *percent*, multiply by 1 in the form of _____ .

4. To write a percent as a *fraction*, drop the % symbol and multiply by _____ .

Write each fraction as a percent.

5. $\dfrac{13}{100}$ **6.** $\dfrac{92}{100}$ **7.** $\dfrac{87}{100}$ **8.** $\dfrac{71}{100}$ **9.** $\dfrac{1}{100}$ **10.** $\dfrac{2}{100}$

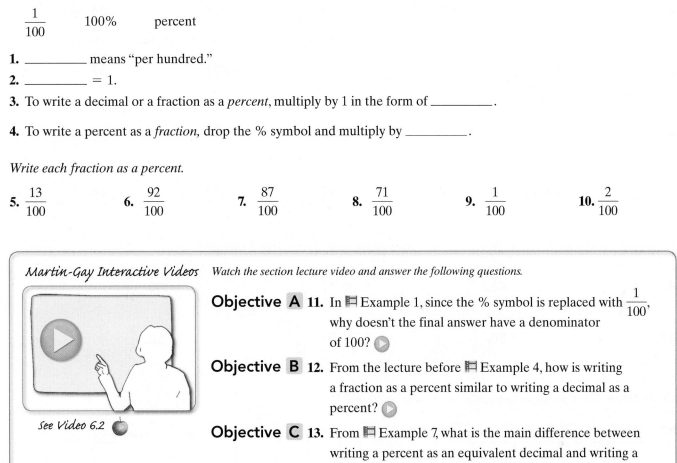

Martin-Gay Interactive Videos   Watch the section lecture video and answer the following questions.

See Video 6.2

Objective A 11. In ⊞ Example 1, since the % symbol is replaced with $\dfrac{1}{100}$, why doesn't the final answer have a denominator of 100? ▶

Objective B 12. From the lecture before ⊞ Example 4, how is writing a fraction as a percent similar to writing a decimal as a percent? ▶

Objective C 13. From ⊞ Example 7, what is the main difference between writing a percent as an equivalent decimal and writing a percent as an equivalent fraction? ▶

6.2 Exercise Set MyMathLab® ▶

Objective A *Write each percent as a fraction or mixed number in simplest form. See Examples 1 through 5.*

1. 12%   **2.** 24%   ▶**3.** 4%   **4.** 2%   **5.** 4.5%

6. 7.5%   ▶**7.** 175%   **8.** 250%   **9.** 73%   **10.** 86%

11. 12.5%   **12.** 62.5%   **13.** 6.25%   **14.** 3.75%   **15.** 6%

16. 16%   ▶**17.** $10\dfrac{1}{3}\%$ **18.** $7\dfrac{3}{4}\%$ **19.** $22\dfrac{3}{8}\%$ **20.** $15\dfrac{5}{8}\%$

Objective B   *Write each fraction or mixed number as a percent. See Examples 6 through 8.*

21. $\dfrac{3}{4}$ **22.** $\dfrac{1}{4}$ ▶ **23.** $\dfrac{7}{10}$ **24.** $\dfrac{3}{10}$ **25.** $\dfrac{2}{5}$ **26.** $\dfrac{4}{5}$

27. $\dfrac{59}{100}$ **28.** $\dfrac{83}{100}$ **29.** $\dfrac{17}{50}$ **30.** $\dfrac{47}{50}$ ▶ **31.** $\dfrac{3}{8}$ **32.** $\dfrac{5}{8}$

33. $\dfrac{5}{16}$ **34.** $\dfrac{7}{16}$ **35.** $1\dfrac{3}{5}$ **36.** $1\dfrac{3}{4}$ **37.** $\dfrac{7}{9}$ **38.** $\dfrac{1}{3}$

39. $\dfrac{13}{20}$ **40.** $\dfrac{3}{20}$ **41.** $2\dfrac{1}{2}$ **42.** $2\dfrac{1}{5}$ **43.** $1\dfrac{9}{10}$ **44.** $2\dfrac{7}{10}$

Write each fraction as a percent. Round to the nearest hundredth percent. See Example 9.

45. $\dfrac{7}{11}$ **46.** $\dfrac{5}{12}$ ▶ **47.** $\dfrac{4}{15}$ **48.** $\dfrac{10}{11}$

49. $\dfrac{1}{7}$ **50.** $\dfrac{1}{9}$ **51.** $\dfrac{11}{12}$ **52.** $\dfrac{5}{6}$

Objective C   *Complete each table. See Examples 10 and 11.*

53.

Percent	Decimal	Fraction
35%		
		$\dfrac{1}{5}$
	0.5	
70%		
		$\dfrac{3}{8}$

54.

Percent	Decimal	Fraction
60%		
		$\dfrac{2}{5}$
	0.25	
12.5%		
		$\dfrac{5}{8}$
		$\dfrac{7}{50}$

55.

Percent	Decimal	Fraction
40%		
	0.235	
		$\dfrac{4}{5}$
$33\dfrac{1}{3}\%$		
		$\dfrac{7}{8}$
7.5%		

56.

Percent	Decimal	Fraction
	0.525	
		$\dfrac{3}{4}$
$66\dfrac{2}{3}\%$		
		$\dfrac{5}{6}$
100%		

57.

Percent	Decimal	Fraction
200%		
	2.8	
705%		
		$4\frac{27}{50}$

58.

Percent	Decimal	Fraction
800%		
	3.2	
608%		
		$9\frac{13}{50}$

Solve. See Examples 10 and 11.

59. A 2012 poll revealed that 67% of American adults are in favor of keeping the penny in circulation. Write this percent as a decimal and a fraction. (*Source:* Americans for Common Cents)

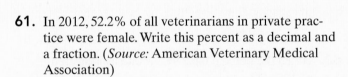

60. China was responsible for 42% of the world's smelter production of aluminum in 2012. Write this percent as a decimal and a fraction. (*Source:* U.S. Geological Survey)

61. In 2012, 52.2% of all veterinarians in private practice were female. Write this percent as a decimal and a fraction. (*Source:* American Veterinary Medical Association)

62. The U.S. penny is 97.5% zinc. Write this percent as a decimal and a fraction. (*Source:* Americans for Common Cents)

63. According to a recent survey, $\frac{4}{5}$ of American restaurant owners started out in the restaurant industry in an entry-level position. Write this fraction as a percent. (*Source:* National Restaurant Association's *2013 Restaurant Industry Forecast*)

64. Of all U.S. veterinarians in private practice in 2012, $\frac{3}{50}$ focused exclusively on horses. Write this fraction as a percent. (*Source:* American Veterinary Medical Association)

65. The sales tax in Slidell, Louisiana, is 8.75%. Write this percent as a decimal.

66. A real estate agent receives a commission of 3% of the sale price of a house. Write this percent as a decimal.

67. The average American wastes $\frac{9}{50}$ of all grain products brought into the home. Write this fraction as a percent. (*Source:* Natural Resources Defense Council)

68. Canada produces $\frac{1}{4}$ of the uranium produced in the world. Write this fraction as a percent. (*Source:* World Nuclear Association)

Canada

In Exercises 69 through 74, write the percent from the circle graph as a decimal and a fraction.

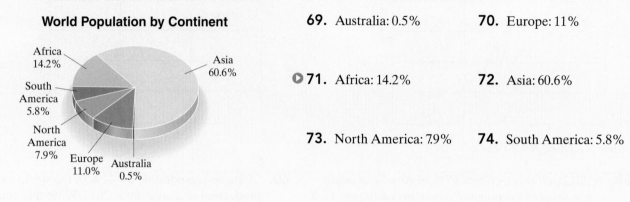

World Population by Continent

Africa 14.2%
South America 5.8%
North America 7.9%
Europe 11.0%
Australia 0.5%
Asia 60.6%

69. Australia: 0.5%

70. Europe: 11%

71. Africa: 14.2%

72. Asia: 60.6%

73. North America: 7.9%

74. South America: 5.8%

Review

Find the value of n. See Section 5.3.

75. $3 \cdot n = 45$

76. $7 \cdot n = 48$

77. $8 \cdot n = 80$

78. $2 \cdot n = 16$

79. $6 \cdot n = 72$

80. $5 \cdot n = 35$

Concept Extensions

Solve. See the Concept Check in this section.

81. Given the percent 52.8647%, round as indicated.
  a. Round to a tenth of a percent.
  b. Round to a hundredth of a percent.

82. Given the percent 0.5269%, round as indicated.
  a. Round to a tenth of a percent.
  b. Round to a hundredth of a percent.

83. Write 1.07835 as a percent rounded to the nearest tenth of a percent.

84. Write 1.25348 as a percent rounded to the nearest tenth of a percent.

85. Write 0.65794 as a percent rounded to the nearest hundredth of a percent.

86. Write 0.92571 as a percent rounded to the nearest hundredth of a percent.

87. Write 0.7682 as a percent rounded to the nearest percent.

88. Write 0.2371 as a percent rounded to the nearest percent.

What percent of the figure is shaded?

89.

90.

91.

92.

Fill in the blanks.

93. A fraction written as a percent is greater than 100% when the numerator is _____ than the denominator. (greater/less)

94. A decimal written as a percent is less than 100% when the decimal is _____ than 1. (greater/less)

95. In your own words, explain how to write a percent as a fraction.

96. In your own words, explain how to write a fraction as a decimal.

Write each fraction as a decimal and then write each decimal as a percent. Round the decimal to three decimal places (nearest thousandth) and the percent to the nearest tenth of a percent.

97. $\dfrac{21}{79}$

98. $\dfrac{56}{102}$

99. $\dfrac{850}{736}$

100. $\dfrac{506}{248}$

6.3 Solving Percent Problems Using Equations

Note: Sections 6.3 and 6.4 introduce two methods for solving percent problems. It is not necessary that you study both sections. You may want to check with your instructor for further advice.

Throughout this text, we have written mathematical statements such as $3 + 10 = 13$, or area = length · width. These statements are called "equations." An **equation** is a mathematical statement that contains an equal sign. To solve percent problems in this section, we translate the problems into such mathematical statements, or equations.

Objectives

A Write Percent Problems as Equations.

B Solve Percent Problems

Objective A Writing Percent Problems as Equations

Recognizing key words in a percent problem is helpful in writing the problem as an equation. Three key words in the statement of a percent problem and their meanings are as follows:

of means **multiplication** (·)

is means **equals** (=)

what (or some equivalent) means **the unknown number**

In our examples, we let the letter *n* stand for the unknown number.

Helpful Hint

Any letter of the alphabet can be used to represent the unknown number. In this section, we mostly use the letter *n*.

Practice 1

Translate: 6 is what percent of 24?

Example 1 Translate to an equation.

5 is what percent of 20?

Solution: 5   is   what percent of   20?
↓   ↓   ↓   ↓   ↓
5   = n ·   20

■ Work Practice 1

Helpful Hint

Remember that an equation is simply a mathematical statement that contains an equal sign (=).

$$5 = n \cdot 20$$
↑
equal sign

Practice 2

Translate: 1.8 is 20% of what number?

Example 2 Translate to an equation.

1.2 is 30% of what number?

Solution: 1.2   is   30%   of   what number?
↓   ↓   ↓   ↓   ↓
1.2   =   30%   · n

■ Work Practice 2

Practice 3

Translate: What number is 40% of 3.6?

Example 3 Translate to an equation.

What number is 25% of 0.008?

Solution: What number   is   25%   of   0.008?
↓   ↓   ↓   ↓   ↓
n =   25%   ·   0.008

■ Work Practice 3

Practice 4–6

Translate each to an equation.
4. 42% of 50 is what number?
5. 15% of what number is 9?
6. What percent of 150 is 90?

Examples Translate each of the following to an equation:

4. 38%   of   200   is   what number?
↓   ↓   ↓   ↓   ↓
38%   ·   200   = n

5. 40%   of   what number   is   80?
↓   ↓   ↓   ↓   ↓
40%   · n =   80

6. What percent   of   85   is   34?
↓   ↓   ↓   ↓   ↓
n ·   85   =   34

■ Work Practice 4–6

Answers

1. $6 = n \cdot 24$ **2.** $1.8 = 20\% \cdot n$
3. $n = 40\% \cdot 3.6$ **4.** $42\% \cdot 50 = n$
5. $15\% \cdot n = 9$ **6.** $n \cdot 150 = 90$

✓**Concept Check** In the equation $2 \cdot n = 10$, what step should be taken to solve the equation for n?

Objective B Solving Percent Problems ▶

You may have noticed by now that each percent problem has contained three numbers–in our examples, two are known and one is unknown. Each of these numbers is given a special name.

$$
\begin{array}{ccccc}
15\% & \text{of} & 60 & \text{is} & 9 \\
\downarrow & \downarrow & \downarrow & \downarrow & \downarrow \\
\underset{\text{percent}}{15\%} & \cdot & \underset{\text{base}}{60} & = & \underset{\text{amount}}{9}
\end{array}
$$

We call this equation the **percent equation.**

> ### Percent Equation
>
> percent · base = amount

> **Helpful Hint**
>
> Notice that the percent equation given above is a true statement. To see this, simplify the left side as shown:
>
> $15\% \cdot 60 = 9$
>
> $0.15 \cdot 60 = 9$ Write 15% as 0.15.
>
> $9 = 9$ Multiply.
>
> The statement $9 = 9$ is true.

After a percent problem has been written as a percent equation, we can use the equation to find the unknown number. This is called **solving** the equation.

Example 7   Solving Percent Equation for the Amount

$$
\underbrace{\text{What number}}_{\downarrow} \text{ is } \overset{\downarrow\ \downarrow\ \ \downarrow\ \downarrow}{35\% \text{ of } 40}?
$$

Solution:

$n = 35\% \cdot 40$ Translate to an equation.

$n = 0.35 \cdot 40$ Write 35% as 0.35.

$n = 14$ Multiply $0.35 \cdot 40 = 14$.

Thus, 14 is 35% of 40.

Is this reasonable? To see, round 35% to 40%. Then 40% of 40 or 0.40(40) is 16. Our result is reasonable since 16 is close to 14.

■ Work Practice 7

> **Helpful Hint**
>
> When solving a percent equation, write the percent as a decimal (or fraction).

Practice 7

What number is 20% of 85?

Answer

7. 17

✓**Concept Check Answer**

If $2 \cdot n = 10$, then $n = \dfrac{10}{2}$, or $n = 5$.

Practice 8

90% of 150 is what number?

Example 8 Solving Percent Equation for the Amount

85% of 300 is what number?

Solution:

85%	·	300	=	n	Translate to an equation.
0.85	·	300	=	n	Write 85% as 0.85.
		255	=	n	Multiply $0.85 \cdot 300 = 255$.

Thus, 85% of 300 is 255.

Is this result reasonable? To see, round 85% to 90%. Then 90% of 300 or $0.90(300) = 270$, which is close to 255.

■ Work Practice 8

Practice 9

15% of what number is 1.2?

Example 9 Solving Percent Equation for the Base

12% of what number is 0.6?

Solution:

| 12% | · | n | = 0.6 | Translate to an equation. |
| 0.12 | · | n | = 0.6 | Write 12% as 0.12. |

Recall from Section 5.3 that if "0.12 times some number is 0.6," then the number is 0.6 divided by 0.12.

$$n = \frac{0.6}{0.12} \quad \text{Divide 0.6 by 0.12, the number multiplied by } n.$$

$$n = 5$$

Thus, 12% of 5 is 0.6.

Is this reasonable? To see, round 12% to 10%. Then 10% of 5 or $0.10(5) = 0.5$, which is close to 0.6.

■ Work Practice 9

Practice 10

27 is $4\frac{1}{2}$% of what number?

Example 10 Solving Percent Equation for the Base

13 is $6\frac{1}{2}$% of what number?

Solution:

13	=	$6\frac{1}{2}$%	·	n	Translate to an equation.
13	=	0.065	·	n	$6\frac{1}{2}\% = 6.5\% = 0.065$.
$\frac{13}{0.065}$	=			n	Divide 13 by 0.065, the number multiplied by n.
200	=			n	

Thus, 13 is $6\frac{1}{2}$% of 200.

Check to see if this result is reasonable.

■ Work Practice 10

Answers

8. 135 **9.** 8 **10.** 600

Example 11 Solving Percent Equation for the Percent

What percent of 12 is 9?

Solution: $n \cdot 12 = 9$ Translate to an equation.

$n = \dfrac{9}{12}$ Divide 9 by 12, the number multiplied by n.

$n = 0.75$

Next, since we are looking for percent, we write 0.75 as a percent.

$n = 75\%$

So, 75% of 12 is 9. To check, see that $75\% \cdot 12 = 9$.

■ Work Practice 11

Helpful Hint

If your unknown in the percent equation is the percent, don't forget to convert your answer to a percent.

Example 12 Solving Percent Equation for the Percent

78 is what percent of 65?

Solution: $78 = n \cdot 65$ Translate to an equation.

$\dfrac{78}{65} = n$ Divide 78 by 65, the number multiplied by n.

$1.2 = n$

$120\% = n$ Write 1.2 as a percent.

So, 78 is 120% of 65. Check this result.

■ Work Practice 12

✓Concept Check Consider these problems.

1. 75% of 50 =
   a. 50   b. a number greater than 50   c. a number less than 50
2. 40% of a number is 10. Is the number
   a. 10?   b. less than 10?   c. greater than 10?
3. 800 is 120% of what number? Is the number
   a. 800?   b. less than 800?   c. greater than 800?

Helpful Hint

Use the following to see if your answers are reasonable.

(100%) of a number = the number

$\left(\begin{array}{c}\text{a percent}\\\text{greater than}\\100\%\end{array}\right)$ of a number = $\begin{array}{c}\text{a number greater}\\\text{than the original number}\end{array}$

$\left(\begin{array}{c}\text{a percent}\\\text{less than }100\%\end{array}\right)$ of a number = $\begin{array}{c}\text{a number less}\\\text{than the original number}\end{array}$

Practice 11
What percent of 80 is 8?

Practice 12
35 is what percent of 25?

Answers
11. 10% **12.** 140%

✓Concept Check Answers
1. c **2.** c **3.** b

Vocabulary, Readiness & Video Check

Use the choices below to fill in each blank.

percent	amount	of	less
base	the number	is	greater

1. The word _____ translates to " =".
2. The word _____ usually translates to "multiplication."
3. In the statement "10% of 90 is 9," the number 9 is called the _____, 90 is called the _____, and 10 is called the _____.
4. 100% of a number = _____.
5. Any "percent greater than 100%" of "a number" = "a number _____ than the original number."
6. Any "percent less than 100%" of "a number" = "a number _____ than the original number."

Identify the percent, the base, and the amount in each equation. Recall that percent · base = amount.

7. $42\% \cdot 50 = 21$

8. $30\% \cdot 65 = 19.5$

9. $107.5 = 125\% \cdot 86$

10. $99 = 110\% \cdot 90$

Martin-Gay Interactive Videos   *Watch the section lecture video and answer the following questions.*

Objective A **11.** From the lecture before ▣ Example 1, what are the key words and their translations that we need to remember? ▶

Objective B **12.** What is the difference between the translated equation in ▣ Example 5 and those in ▣ Examples 4 and 6? ▶

See Video 6.3 🍎

6.3 Exercise Set MyMathLab® ▶

Objective A **Translating** *Translate each to an equation. Do not solve. See Examples 1 through 6.*

▶ **1.** 18% of 81 is what number?

2. 36% of 72 is what number?

3. 20% of what number is 105?

4. 40% of what number is 6?

5. 0.6 is 40% of what number?

6. 0.7 is 20% of what number?

▶ **7.** What percent of 80 is 3.8?

8. 9.2 is what percent of 92?

9. What number is 9% of 43?

10. What number is 25% of 55?

11. What percent of 250 is 150?

12. What percent of 375 is 300?

Objective B *Solve. See Examples 7 and 8.*

13. 10% of 35 is what number?

14. 25% of 68 is what number?

15. What number is 14% of 205?

16. What number is 18% of 425?

Solve. See Examples 9 and 10.

17. 1.2 is 12% of what number?

18. 0.22 is 44% of what number?

19. $8\frac{1}{2}$% of what number is 51?

20. $4\frac{1}{2}$% of what number is 45?

Solve. See Examples 11 and 12.

21. What percent of 80 is 88?

22. What percent of 40 is 60?

23. 17 is what percent of 50?

24. 48 is what percent of 50?

Objectives A B Mixed Practice *Solve. See Examples 1 through 12.*

25. 0.1 is 10% of what number?

26. 0.5 is 5% of what number?

27. 150% of 430 is what number?

28. 300% of 56 is what number?

29. 82.5 is $16\frac{1}{2}$% of what number?

30. 7.2 is $6\frac{1}{4}$% of what number?

31. 2.58 is what percent of 50?

32. 2.64 is what percent of 25?

33. What number is 42% of 60?

34. What number is 36% of 80?

35. What percent of 184 is 64.4?

36. What percent of 120 is 76.8?

37. 120% of what number is 42?

38. 160% of what number is 40?

39. 2.4% of 26 is what number?

40. 4.8% of 32 is what number?

41. What percent of 600 is 3?

42. What percent of 500 is 2?

43. 6.67 is 4.6% of what number?

44. 9.75 is 7.5% of what number?

45. 1575 is what percent of 2500?

46. 2520 is what percent of 3500?

47. 2 is what percent of 50?

48. 2 is what percent of 40?

Review

Find the value of n in each proportion. See Section 5.3.

49. $\dfrac{27}{n} = \dfrac{9}{10}$

50. $\dfrac{35}{n} = \dfrac{7}{5}$

51. $\dfrac{n}{5} = \dfrac{8}{11}$

52. $\dfrac{n}{3} = \dfrac{6}{13}$

Write each phrase as a proportion.

53. 17 is to 12 as n is to 20   **54.** 20 is to 25 as n is to 10   **55.** 8 is to 9 as 14 is to n **56.** 5 is to 6 as 15 is to n

Concept Extensions

For each equation, determine the next step taken to find the value of n. See the first Concept Check in this section.

57. $5 \cdot n = 32$

    a. $n = 5 \cdot 32$ **b.** $n = \dfrac{5}{32}$ **c.** $n = \dfrac{32}{5}$ **d.** none of these

58. $n = 0.7 \cdot 12$

    a. $n = 8.4$ **b.** $n = \dfrac{12}{0.7}$ **c.** $n = \dfrac{0.7}{12}$ **d.** none of these

59. $0.06 = n \cdot 7$

    a. $n = 0.06 \cdot 7$ **b.** $n = \dfrac{0.06}{7}$ **c.** $n = \dfrac{7}{0.06}$ **d.** none of these

60. $0.01 = n \cdot 8$

    a. $n = 0.01 \cdot 8$ **b.** $n = \dfrac{8}{0.01}$ **c.** $n = \dfrac{0.01}{8}$ **d.** none of these

61. Write a word statement for the equation $20\% \cdot n = 18.6$. Use the phrase "some number" for "n".

62. Write a word statement for the equation $n = 33\frac{1}{3}\% \cdot 24$. Use the phrase "some number" for "n".

For each exercise, determine whether the percent, n, is (a) 100%, (b) greater than 100%, or (c) less than 100%. See the last Concept Check in this section.

63. $n\%$ of 20 is 30   **64.** $n\%$ of 98 is 98   **65.** $n\%$ of 120 is 85   **66.** $n\%$ of 35 is 50

For each exercise, determine whether the number, n, is (a) equal to 45, (b) greater than 45, or (c) less than 45.

67. 55% of 45 is n **68.** 230% of 45 is n **69.** 100% of 45 is n

70. 30% of n is 45   **71.** 100% of n is 45   **72.** 180% of n is 45

73. In your own words, explain how to solve a percent equation.

74. Write a percent problem that uses the percent 50%.

Solve.

75. 1.5% of 45,775 is what number?

76. What percent of 75,528 is 27,945.36?

77. 22,113 is 180% of what number?

6.4 Solving Percent Problems Using Proportions ▶

There is more than one method that can be used to solve percent problems. (See the note at the beginning of Section 6.3.) In the last section, we used the percent equation. In this section, we will use proportions.

Objectives

A Write Percent Problems as Proportions. ▶

B Solve Percent Problems. ▶

Objective A Writing Percent Problems as Proportions ▶

To understand the proportion method, recall that 70% means the ratio of 70 to 100, or $\frac{70}{100}$.

$$70\% = \frac{70}{100} = \frac{7}{10}$$

$\frac{7}{10}$ shaded

70% or $\frac{70}{100}$ shaded

Since the ratio $\frac{70}{100}$ is equal to the ratio $\frac{7}{10}$, we have the proportion

$$\frac{7}{10} = \frac{70}{100}.$$

We call this proportion the "percent proportion." In general, we can name the parts of this proportion as follows:

Percent Proportion

$$\frac{\text{amount}}{\text{base}} = \frac{\text{percent}}{100} \quad \leftarrow \text{always } 100$$

or

$$\text{amount} \rightarrow \quad \frac{a}{b} = \frac{p}{100} \quad \leftarrow \text{percent}$$
$$\text{base} \rightarrow$$

When we translate percent problems to proportions, the **percent,** p, can be identified by looking for the symbol % or the word *percent*. The **base,** b, usually follows the word *of*. The **amount,** a, is the part compared to the whole.

Helpful Hint

Part of Proportion	How It's Identified
Percent	% or percent
Base	Appears after *of*
Amount	Part compared to whole

Practice 1

Translate to a proportion.
15% of what number is 55?

Practice 2

Translate to a proportion.
35 is what percent of 70?

Practice 3

Translate to a proportion.
What number is 25% of 68?

Practice 4

Translate to a proportion.
520 is 65% of what number?

Answers

1. $\dfrac{55}{b} = \dfrac{15}{100}$ 2. $\dfrac{35}{70} = \dfrac{p}{100}$

3. $\dfrac{a}{68} = \dfrac{25}{100}$ 4. $\dfrac{520}{b} = \dfrac{65}{100}$

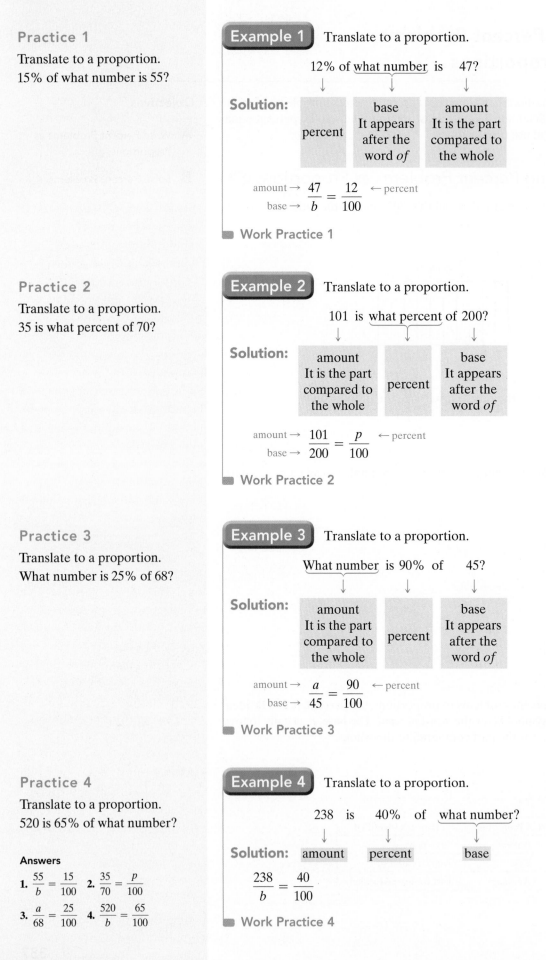

Example 1   Translate to a proportion.

12% of what number is 47?

Solution:

| percent | base It appears after the word *of* | amount It is the part compared to the whole |

amount → $\dfrac{47}{b} = \dfrac{12}{100}$ ← percent
base →

■ Work Practice 1

Example 2   Translate to a proportion.

101 is what percent of 200?

Solution:

| amount It is the part compared to the whole | percent | base It appears after the word *of* |

amount → $\dfrac{101}{200} = \dfrac{p}{100}$ ← percent
base →

■ Work Practice 2

Example 3   Translate to a proportion.

What number is 90% of 45?

Solution:

| amount It is the part compared to the whole | percent | base It appears after the word *of* |

amount → $\dfrac{a}{45} = \dfrac{90}{100}$ ← percent
base →

■ Work Practice 3

Example 4   Translate to a proportion.

238 is 40% of what number?

Solution: amount   percent   base

$$\dfrac{238}{b} = \dfrac{40}{100}$$

■ Work Practice 4

Example 5 Translate to a proportion.

What percent of 30 is 75?

Solution: ↓ percent ↓ base ↓ amount

$$\frac{75}{30} = \frac{p}{100}$$

■ Work Practice 5

Practice 5

Translate to a proportion.
What percent of 50 is 65?

Example 6 Translate to a proportion.

45% of 105 is what number?

Solution: ↓ percent ↓ base ↓ amount

$$\frac{a}{105} = \frac{45}{100}$$

■ Work Practice 6

Practice 6

Translate to a proportion.
36% of 80 is what number?

Objective B Solving Percent Problems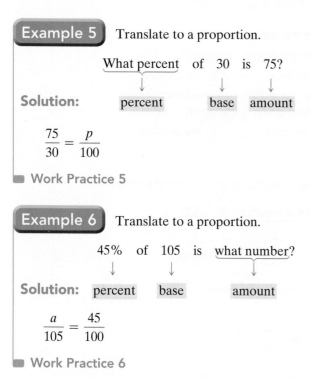

The proportions that we have written in this section contain three values that can change: the percent, the base, and the amount. If any two of these values are known, we can find the third (the unknown value). To do this, we write a percent proportion and find the unknown value as we did in Section 5.3.

Example 7 Solving Percent Proportions for the Amount

What number is 30% of 9?

Solution: ↓ amount ↓ percent ↓ base

$$\frac{a}{9} = \frac{30}{100}$$

To solve, we set cross products equal to each other.

$$\frac{a}{9} = \frac{30}{100}$$

$a \cdot 100 = 9 \cdot 30$ Set cross products equal.

$a \cdot 100 = 270$ Multiply.

Recall from Section 5.3 that if "some number times 100 is 270," then the number is 270 divided by 100.

$a = \dfrac{270}{100}$ Divide 270 by 100, the number multiplied by a.

$a = 2.7$ Simplify.

Thus, 2.7 is 30% of 9.

■ Work Practice 7

Practice 7

What number is 8% of 120?

> **Helpful Hint** The proportion in Example 7 contains the ratio $\dfrac{30}{100}$. A ratio in a proportion may be simplified before solving the proportion. The unknown number in both $\dfrac{a}{9} = \dfrac{30}{100}$ and $\dfrac{a}{9} = \dfrac{3}{10}$ is 2.7

Answers

5. $\dfrac{65}{50} = \dfrac{p}{100}$ 6. $\dfrac{a}{80} = \dfrac{36}{100}$

7. 9.6

✓**Concept Check** Consider the statement: "78 is what percent of 350?"
Which part of the percent proportion is unknown?

a. the amount   **b.** the base   **c.** the percent

Consider another statement: "14 is 10% of some number."
Which part of the percent proportion is unknown?

a. the amount   **b.** the base   **c.** the percent

Example 8   Solving Percent Proportion for the Base

$$\underset{\downarrow}{150\%} \quad \text{of} \quad \underset{\downarrow}{\underline{\text{what number}}} \quad \text{is} \quad \underset{\downarrow}{30?}$$

Solution: percent   base   amount

$$\frac{30}{b} = \frac{150}{100} \qquad \text{Write the proportion.}$$

$$\frac{30}{b} = \frac{3}{2} \qquad \text{Write } \frac{150}{100} \text{ as } \frac{3}{2}.$$

$$30 \cdot 2 = b \cdot 3 \qquad \text{Set cross products equal.}$$

$$60 = b \cdot 3 \qquad \text{Multiply.}$$

$$\frac{60}{3} = b \qquad \text{Divide 60 by 3, the number multiplied by } b.$$

$$20 = b \qquad \text{Simplify.}$$

Thus, 150% of 20 is 30.

▬ **Work Practice 8**

✓**Concept Check** When solving a percent problem by using a proportion, describe how you can check the result.

Example 9   Solving Percent Proportion for the Base

$$\underset{\downarrow}{20.8} \quad \text{is} \quad \underset{\downarrow}{40\%} \quad \text{of} \quad \underset{\downarrow}{\underline{\text{what number}}}?$$

Solution: amount   percent   base

$$\frac{20.8}{b} = \frac{40}{100} \quad \text{or} \quad \frac{20.8}{b} = \frac{2}{5} \qquad \text{Write the proportion and simplify } \frac{40}{100}.$$

$$20.8 \cdot 5 = b \cdot 2 \qquad \text{Set cross products equal.}$$

$$104 = b \cdot 2 \qquad \text{Multiply.}$$

$$\frac{104}{2} = b \qquad \text{Divide 104 by 2, the number multiplied by } b.$$

$$52 = b \qquad \text{Simplify.}$$

So, 20.8 is 40% of 52.

▬ **Work Practice 9**

Practice 8

75% of what number is 60?

Practice 9

15.2 is 5% of what number?

Answers

8. 80   **9.** 304

✓**Concept Check Answers**

c, b; by putting the result into the proportion and checking that the proportion is true

Example 10 Solving Percent Proportion for the Percent

What <u>percent</u> of 50 is 8?

    ↓   ↓   ↓

Solution: percent   base   amount

$$\frac{8}{50} = \frac{p}{100} \quad \text{or} \quad \frac{4}{25} = \frac{p}{100}$$ Write the proportion and simplify $\frac{8}{50}$.

$4 \cdot 100 = 25 \cdot p$ Set cross products equal.

$400 = 25 \cdot p$ Multiply.

$\dfrac{400}{25} = p$ Divide 400 by 25, the number multiplied by p.

$16 = p$ Simplify.

So, 16% of 50 is 8.

■ **Work Practice 10**

Helpful Hint

Recall from our percent proportion that this number already is a percent. Just keep the number as is and attach a % symbol.

Practice 10

What percent of 40 is 6?

Example 11 Solving Percent Proportion for the Percent

504 is <u>what percent</u> of 360?

  ↓   ↓   ↓

Solution: amount   percent   base

$$\frac{504}{360} = \frac{p}{100}$$

Let's choose not to simplify the ratio $\dfrac{504}{360}$.

$504 \cdot 100 = 360 \cdot p$ Set cross products equal.

$50{,}400 = 360 \cdot p$ Multiply.

$\dfrac{50{,}400}{360} = p$ Divide 50,400 by 360, the number multiplied by p.

$140 = p$ Simplify.

Notice that by choosing not to simplify $\dfrac{504}{360}$, we had larger numbers in our equation. Either way, we find that 504 is 140% of 360.

■ **Work Practice 11**

You may have noticed the following while working examples.

Practice 11

336 is what percent of 160?

Helpful Hint

Use the following to see whether your answers are reasonable.

$$100\% \text{ of a number} = \text{the number}$$

$$\left(\begin{array}{c} \text{a percent} \\ \text{greater than} \\ 100\% \end{array} \right) \text{of a number} = \begin{array}{l} \text{a number larger} \\ \text{than the original number} \end{array}$$

$$\left(\begin{array}{c} \text{a percent} \\ \text{less than } 100\% \end{array} \right) \text{of a number} = \begin{array}{l} \text{a number less} \\ \text{than the original number} \end{array}$$

Answers

10. 15% **11.** 210%

Vocabulary, Readiness & Video Check

Use the choices below to fill in each blank. These choices will be used more than once.

amount   base   percent

1. When translating the statement "20% of 15 is 3" to a proportion, the number 3 is called the _____, 15 is the _____, and 20 is the _____ .

2. In the question "50% of what number is 28?", which part of the percent proportion is unknown? _____

3. In the question "What number is 25% of 200?", which part of the percent proportion is unknown? _____

4. In the question "38 is what percent of 380?", which part of the percent proportion is unknown? _____

Identify the amount, the base, and the percent in each equation. Recall that $\dfrac{\text{amount}}{\text{base}} = \dfrac{\text{percent}}{100}$.

5. $\dfrac{12.6}{42} = \dfrac{30}{100}$ 6. $\dfrac{201}{300} = \dfrac{67}{100}$ 7. $\dfrac{20}{100} = \dfrac{102}{510}$ 8. $\dfrac{40}{100} = \dfrac{248}{620}$

Martin-Gay Interactive Videos   *Watch the section lecture video and answer the following questions.*

Objective A 9. In Example 1, how did we identify what part of the percent proportion 45 is?

Objective B 10. From Examples 4–6, what number is *always* part of the cross product equation of a percent proportion?

See Video 6.4

6.4 Exercise Set MyMathLab®

Objective A Translating *Translate each to a proportion. Do not solve. See Examples 1 through 6.*

1. 98% of 45 is what number?

2. 92% of 30 is what number?

3. What number is 4% of 150?

4. What number is 7% of 175?

5. 14.3 is 26% of what number?

6. 1.2 is 47% of what number?

7. 35% of what number is 84?

8. 85% of what number is 520?

9. What percent of 400 is 70?

10. What percent of 900 is 216?

11. 8.2 is what percent of 82?

12. 9.6 is what percent of 96?

Objective B *Solve. See Example 7.*

13. 40% of 65 is what number?

14. 25% of 84 is what number?

15. What number is 18% of 105?

16. What number is 60% of 29?

Solve. See Examples 8 and 9.

17. 15% of what number is 90?

18. 55% of what number is 55?

19. 7.8 is 78% of what number?

20. 1.1 is 44% of what number?

Solve. See Examples 10 and 11.

21. What percent of 35 is 42?

22. What percent of 98 is 147?

23. 14 is what percent of 50?

24. 24 is what percent of 50?

Objectives A B Mixed Practice *Solve. See Examples 1 through 11.*

25. 3.7 is 10% of what number?

26. 7.4 is 5% of what number?

27. 2.4% of 70 is what number?

28. 2.5% of 90 is what number?

29. 160 is 16% of what number?

30. 30 is 6% of what number?

31. 394.8 is what percent of 188?

32. 550.4 is what percent of 172?

33. What number is 89% of 62?

34. What number is 53% of 130?

35. What percent of 6 is 2.7?

36. What percent of 5 is 1.6?

37. 140% of what number is 105?

38. 170% of what number is 221?

39. 1.8% of 48 is what number?

40. 7.8% of 24 is what number?

41. What percent of 800 is 4?

42. What percent of 500 is 3?

43. 3.5 is 2.5% of what number?

44. 9.18 is 6.8% of what number?

45. 20% of 48 is what number?

46. 75% of 14 is what number?

47. 2486 is what percent of 2200?

48. 9310 is what percent of 3800?

Review

Add or subtract the fractions. See Sections 3.1, 3.3, and 3.4.

49. $\dfrac{11}{16} + \dfrac{3}{16}$

50. $\dfrac{5}{8} - \dfrac{7}{12}$

51. $3\dfrac{1}{2} - \dfrac{11}{30}$

52. $2\dfrac{2}{3} + 4\dfrac{1}{2}$

Add or subtract the decimals. See Section 4.3.

53. $\begin{array}{r} 0.41 \\ + 0.29 \\ \hline \end{array}$

54. $\begin{array}{r} 10.78 \\ 4.3 \\ + \ 0.21 \\ \hline \end{array}$

55. $\begin{array}{r} 2.38 \\ - 0.19 \\ \hline \end{array}$

56. $\begin{array}{r} 16.37 \\ - \ 2.61 \\ \hline \end{array}$

Concept Extensions

57. Write a word statement for the proportion $\dfrac{x}{28} = \dfrac{25}{100}$. Use the phrase "what number" for "x."

58. Write a percent statement that translates to $\dfrac{16}{80} = \dfrac{20}{100}$.

Suppose you have finished solving four percent problems using proportions that you set up correctly. Check each answer to see if each makes the proportion a true proportion. If any proportion is not true, solve it to find the correct solution. See the Concept Checks in this section.

59. $\dfrac{a}{64} = \dfrac{25}{100}$
Is the amount equal to 17?

60. $\dfrac{520}{b} = \dfrac{65}{100}$
Is the base equal to 800?

61. $\dfrac{p}{100} = \dfrac{13}{52}$
Is the percent equal to 25 (25%)?

62. $\dfrac{36}{12} = \dfrac{p}{100}$
Is the percent equal to 50 (50%)?

63. In your own words, describe how to identify the percent, the base, and the amount in a percent problem.

64. In your own words, explain how to use a proportion to solve a percent problem.

Solve. Round to the nearest tenth, if necessary.

65. What number is 22.3% of 53,862?

66. What percent of 110,736 is 88,542?

67. 8652 is 119% of what number?

Percent and Percent Problems

Write each number as a percent.

1. 0.12   **2.** 0.68   **3.** $\frac{1}{8}$ **4.** $\frac{5}{2}$

5. 5.2   **6.** 8   **7.** $\frac{3}{50}$ **8.** $\frac{11}{25}$

9. $7\frac{1}{2}$ **10.** $3\frac{1}{4}$ **11.** 0.03   **12.** 0.05

Write each percent as a decimal.

13. 65%   **14.** 31%   **15.** 8%   **16.** 7%

17. 142%   **18.** 400%   **19.** 2.9%   **20.** 6.6%

Write each percent as a decimal and as a fraction or mixed number in simplest form. (If necessary when writing as a decimal, round to the nearest thousandth.)

21. 3%   **22.** 5%   **23.** 5.25%   **24.** 12.75%

Answers

1. _____
2. _____
3. _____
4. _____
5. _____
6. _____
7. _____
8. _____
9. _____
10. _____
11. _____
12. _____
13. _____
14. _____
15. _____
16. _____
17. _____
18. _____
19. _____
20. _____
21. _____
22. _____
23. _____
24. _____

25. _____

26. _____

27. _____

28. _____

29. _____

30. _____

31. _____

32. _____

33. _____

34. _____

35. _____

36. _____

37. _____

38. _____

39. _____

40. _____

25. 38%   **26.** 45%   **27.** $12\frac{1}{3}\%$ **28.** $16\frac{2}{3}\%$

Solve each percent problem.

29. 12% of 70 is what number?   **30.** 36 is 36% of what number?

31. 212.5 is 85% of what number?   **32.** 66 is what percent of 55?

33. 23.8 is what percent of 85?   **34.** 38% of 200 is what number?

35. What number is 25% of 44?   **36.** What percent of 99 is 128.7?

37. What percent of 250 is 215?   **38.** What number is 45% of 84?

39. 42% of what number is 63?   **40.** 95% of what number is 58.9?

Applications of Percent

Objective A Solving Applications Involving Percent

Percent is used in a variety of everyday situations. The next examples show just a few ways that percent occurs in real-life settings. (Each of these examples shows two ways of solving these problems. If you studied Section 6.3 only, see *Method 1*. If you studied Section 6.4 only, see *Method 2*.)

The first example has to do with the Appalachian Trail, a hiking trail conceived by a forester in 1921 and diagrammed to the right.

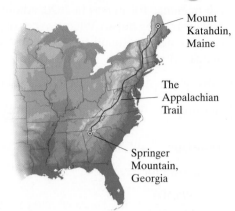

Mount Katahdin, Maine

The Appalachian Trail

Springer Mountain, Georgia

Objectives

A Solve Applications Involving Percent.

B Find Percent of Increase and Percent of Decrease.

Example 1 The circle graph in the margin shows the Appalachian Trail mileage by state. If the total mileage of the trail is 2174, use the circle graph to determine the number of miles in the state of New York. Round to the nearest whole mile.

Solution: *Method 1.* First, we state the problem in words.

In words: What number is 4% of 2174?

Translate: $n = 4\% \cdot 2174$

To solve for *n*, we find 4% · 2174.

$n = 0.04 \cdot 2174$ Write 4% as a decimal.

$n = 86.96$ Multiply.

$n \approx 87$ Round to the nearest whole.

Rounded to the nearest whole mile, we have that approximately 87 miles of the Appalachian Trail are in New York state.

Method 2. State the problem in words; then translate.

In words: What number is 4% of 2174?

       amount   percent   base

Translate: amount → $\dfrac{a}{2174} = \dfrac{4}{100}$ ← percent, base →

Next, we solve for *a*.

$a \cdot 100 = 2174 \cdot 4$ Set cross products equal.

$a \cdot 100 = 8696$ Multiply.

$\dfrac{a \cdot 100}{100} = \dfrac{8696}{100}$ Divide both sides by 100.

$a = 86.96$ Simplify.

$a \approx 87$ Round to the nearest whole.

Rounded to the nearest whole mile, we have that approximately 87 miles of the Appalachian Trail are in New York state.

■ Work Practice 1

Practice 1

If the total mileage of the Appalachian Trail is 2174, use the circle graph to determine the number of miles in the state of Virginia.

Appalachian Trail Mileage by State Percent

Georgia 4%
Maine 13%
North Carolina 4%
Tennessee 14%
New Hampshire 7%
Vermont 7%
Virginia 25%
Massachusetts 4%
Connecticut 2%
New York 4%
New Jersey 3%
Pennsylvania 11%
West Virginia 0.2%
Maryland 2%

Total miles: 2174
Note: The sum of the percents is 100.2% because of rounding.
Source: purebound.com

Answer

1. 543.5 mi

397

Practice 2

In Florida, about 34,000 new nurses were recently needed and hired. If there are now 130,000 nurses, what percent of new nurses were needed in Florida? Round to the nearest whole percent. (*Source*: *St. Petersburg Times* and *The Registered Nurse Population*)

Example 2 Finding Percent of Nursing Job Openings Due to Retirements

There is a worldwide shortage of nurses. It is expected that there will be 1,207,500 job openings for nurses in 2020. About 495,500 of these job openings will be to replace retiring nurses. What percent of job openings for nurses in 2020 will be due to retirements? Round to the nearest whole percent. (*Source:* Bureau of Labor Statistics)

Solution: *Method 1.*   First, we state the problem in words.

In words: 495,500   is   what percent   of   1,207,500?

Translate: 495,500   = n ·   1,207,500

Next, solve for n.

$$\frac{495,500}{1,207,500} = n \qquad \text{Divide 495,500 by 1,207,500, the number multiplied by } n.$$

$$0.41 \approx n \qquad \text{Divide and round to the nearest hundredth.}$$

$$41 \approx n \qquad \text{Write as a percent.}$$

In 2020, about 41% of job openings for nurses will be due to retirements.

Method 2.

In words: 495,500   is   what percent   of   1,207,500?

    amount   percent   base

Translate: amount → $\dfrac{495,500}{1,207,500} = \dfrac{p}{100}$ ← percent
             base →

Next, solve for p.

$$495,500 \cdot 100 = 1,207,500 \cdot p \qquad \text{Set cross products equal.}$$

$$49,550,000 = 1,207,500 \cdot p \qquad \text{Multiply.}$$

$$\frac{49,550,000}{1,207,500} = p \qquad \text{Divide 49,550,000 by 1,207,500, the number multiplied by } p.$$

$$41 \approx p$$

In 2020, about 41% of job openings for nurses will be due to retirements.

■ Work Practice 2

Practice 3

The freshmen class of 775 students is 31% of all students at Euclid University. How many students go to Euclid University?

Example 3 Finding the Base Number of Absences

Mr. Buccaran, the principal at Slidell High School, counted 31 freshmen absent during a particular day. If this is 4% of the total number of freshmen, how many freshmen are there at Slidell High School?

Solution: *Method 1.*   First we state the problem in words; then we translate.

In words: 31   is   4%   of   what number?

Translate: 31   =   4%   · n

Answers

2. 26%   **3.** 2500

Next, we solve for n.

$$31 = 0.04 \cdot n \quad \text{Write 4\% as a decimal.}$$

$$\frac{31}{0.04} = n \quad \text{Divide 31 by 0.04, the number multiplied by } n.$$

$$775 = n \quad \text{Simplify.}$$

There are 775 freshmen at Slidell High School.

Method 2. First we state the problem in words; then we translate.

In words: 31   is   4%   of   what number ?

amount   percent   base

Translate: amount → $\dfrac{31}{b} = \dfrac{4}{100}$ ← percent
base →

Next, we solve for b.

$$31 \cdot 100 = b \cdot 4 \quad \text{Set cross products equal.}$$

$$3100 = b \cdot 4 \quad \text{Multiply.}$$

$$\frac{3100}{4} = b \quad \text{Divide 3100 by 4, the number multiplied by } b.$$

$$775 = b \quad \text{Simplify.}$$

There are 775 freshmen at Slidell High School.

▸ **Work Practice 3**

Example 4   Finding the Base Increase in Part-Time Workers

From 2003 to 2013, the number of part-time workers in the United States increased by 12%. In 2003, there were 24.5 million part-time workers. (*Source:* U.S. Bureau of Labor Statistics)

a. Find the increase in the number of part-time workers from 2003 to 2013.
b. Find the number of part-time workers in 2013.

Solution: *Method 1.*   First we find the increase in the number of part-time workers.

In words: What number   is   12%   of   24.5?

Translate: n $=$ 12% $\cdot$ 24.5

(Continued on next page)

Practice 4

From 2010 to 2013, the number of full-time workers in the United States increased by approximately 5%. In 2010, the number of full-time workers was 110 million. (*Source:* U.S. Bureau of Labor Statistics)

a. Find the increase in the number of full-time workers from 2010 to 2013.

b. Find the total number of full-time workers in 2013.

Answers
4. a. 5.5 million   **b.** 115.5 million

Next, we solve for *n*.

$$n = 0.12 \cdot 24.5 \quad \text{Write 12\% as a decimal.}$$
$$n = 2.94 \quad \text{Multiply.}$$

a. The increase in the number of part-time workers was 2.94 million.

b. This means that the number of part-time workers in 2013 was

$$
\begin{array}{ccc}
\text{Number of} & \text{Number of} & \text{Increase} \\
\text{part-time workers} = \text{part-time workers} + & \text{in number of} \\
\text{in 2013} & \text{in 2003} & \text{part-time workers}
\end{array}
$$
$$= 24.5 \text{ million} + 2.94 \text{ million}$$
$$= 27.44 \text{ million}$$

Method 2.   First we find the increase in the number of part-time workers.

In words: What number   is   12%   of   24.5?

   ↓   ↓   ↓

   amount   percent   base

Translate: $\underset{\text{base} \to 24.5}{\overset{\text{amount} \to a}{\dfrac{a}{24.5}}} = \dfrac{12}{100} \leftarrow \text{precent}$

Next, we solve for *a*.

$$a \cdot 100 = 24.5 \cdot 12 \quad \text{Set cross products equal.}$$
$$a \cdot 100 = 294 \quad \text{Multiply.}$$
$$\frac{a \cdot 100}{100} = \frac{294}{100} \quad \text{Divide both sides by 100.}$$
$$a = 2.94 \quad \text{Simplify.}$$

a. The increase in the number of part-time workers was 2.94 million.

b. This means that the number of part-time workers in 2013 was

$$
\begin{array}{ccc}
\text{Number of} & \text{Number of} & \text{Increase} \\
\text{part-time workers} = \text{part-time workers} + & \text{in number of} \\
\text{in 2013} & \text{in 2003} & \text{part-time workers}
\end{array}
$$
$$= 24.5 \text{ million} + 2.94 \text{ million}$$
$$= 27.44 \text{ million}$$

▨ Work Practice 4

Objective B Finding Percent of Increase and Percent of Decrease

We often use percents to show how much an amount has increased or decreased.

Suppose that the population of a town is 10,000 people and then it increases by 2000 people. The **percent of increase** is

$$\underset{\text{original amount} \to}{\overset{\text{amount of increase} \to}{}} \frac{2000}{10{,}000} = 0.2 = 20\%$$

In general, we have the following.

Percent of Increase

$$\text{percent of increase} = \frac{\text{amount of increase}}{\text{original amount}}$$

Then write the quotient as a percent.

Example 5 Finding Percent of Increase

The number of applications for a mathematics scholarship at Yale increased from 34 to 45 in one year. What is the percent of increase? Round to the nearest whole percent.

Solution: First we find the amount of increase by subtracting the original number of applicants from the new number of applicants.

$$\text{amount of increase} = 45 - 34 = 11$$

The amount of increase is 11 applicants. To find the percent of increase,

$$\text{percent of increase} = \frac{\text{amount of increase}}{\text{original amount}} = \frac{11}{34} \approx 0.32 = 32\%$$

The number of applications increased by about 32%.

■ Work Practice 5

✓**Concept Check** A student is calculating the percent of increase in enrollment from 180 students one year to 200 students the next year. Explain what is wrong with the following calculations:

$$\frac{\text{Amount}}{\text{of increase}} = 200 - 180 = 20$$

$$\frac{\text{Percent of}}{\text{increase}} = \frac{20}{200} = 0.1 = 10\%$$

Suppose that your income was $300 a week and then it decreased by $30. The **percent of decrease** is

$$\text{amount of decrease} \rightarrow \frac{\$30}{\$300} = 0.1 = 10\%$$
$$\text{original amount} \rightarrow$$

> **Percent of Decrease**
>
> $$\text{percent of decrease} = \frac{\text{amount of decrease}}{\text{original amount}}$$
>
> Then write the quotient as a percent.

Example 6 Finding Percent of Decrease

In response to a decrease in sales, a company with 1500 employees reduces the number of employees to 1230. What is the percent of decrease?

Solution: First we find the amount of decrease by subtracting 1230 from 1500.

$$\text{amount of decrease} = 1500 - 1230 = 270$$

The amount of decrease is 270. To find the percent of decrease,

$$\frac{\text{percent of}}{\text{decrease}} = \frac{\text{amount of decrease}}{\text{original amount}} = \frac{270}{1500} = 0.18 = 18\%$$

The number of employees decreased by 18%.

■ Work Practice 6

✓**Concept Check** An ice cream stand sold 6000 ice cream cones last summer. This year the same stand sold 5400 cones. Was there a 10% increase, a 10% decrease, or neither? Explain.

Practice 5

The number of people attending the local play, *Peter Pan*, increased from 285 on Friday to 333 on Saturday. Find the percent of increase in attendance. Round to the nearest tenth percent.

Helpful Hint Make sure that this number is the original number and not the new number.

Practice 6

A town's population of 20,200 in 1995 decreased to 18,483 in 2005. What was the percent of decrease?

Answers

5. 16.8% **6.** 8.5%

✓**Concept Check Answers**

To find the percent of increase, you have to divide the amount of increase (20) by the original amount (180); 10% decrease.

Vocabulary, Readiness & Video Check

Martin-Gay Interactive Videos   Watch the section lecture video and answer the following questions.

Objective A **1.** How do we interpret the answer 175,000 in ▣ Example 1? ▶

Objective B **2.** In ▣ Example 3, what does the improper fraction tell us? ▶

See Video 6.5

6.5 Exercise Set MyMathLab® ▶

Objective A *Solve. For Exercises 1 and 2, the solutions have been started for you. See Examples 1 through 4. If necessary, round percents to the nearest tenth and all other answers to the nearest whole.*

1. An inspector found 24 defective bolts during an inspection. If this is 1.5% of the total number of bolts inspected, how many bolts were inspected?

Start the solution:

1. UNDERSTAND the problem. Reread it as many times as needed.

Go to *Method 1* or *Method 2*.

Method 1.

2. TRANSLATE into an equation. (Fill in the boxes.)

   24   is   1.5%   of   what number?
   ↓   ↓   ↓   ↓   ↓
   24   □   1.5%   □ n

3. SOLVE for n. (See Example 3, Method 1, for help.)

4. INTERPRET. The total number of bolts inspected was _____.

Method 2.

2. TRANSLATE into a proportion. (Fill in the blanks with "amount" or "base.")

   24   is   1.5%   of   what number?
   ↓   ↘   ↓
   _____   percent   _____

$$\text{amount} \rightarrow \frac{\rule{1cm}{0.4pt}}{\rule{1cm}{0.4pt}} = \frac{1.5}{100} \leftarrow \text{percent} \\ \text{base} \rightarrow$$

3. SOLVE the proportion. (See Example 3, Method 2, for help.)

4. INTERPRET. The total number of bolts inspected was _____.

2. A day care worker found 28 children absent one day during an epidemic of chicken pox. If this was 35% of the total number of children attending the day care center, how many children attend this day care center?

Start the solution:

1. UNDERSTAND the problem. Reread it as many times as needed.

Go to *Method 1* or *Method 2*.

Method 1.

2. TRANSLATE into an equation. (Fill in the boxes.)

   28   is   35%   of   what number?
   ↓   ↓   ↓   ↓   ↓
   28   □   35%   □ n

3. SOLVE for n. (See Example 3, Method 1, for help.)

4. INTERPRET. The total number of children attending the day care center was _____.

Method 2.

2. TRANSLATE into a proportion. (Fill in the blanks with "amount" or "base.")

   28   is   35%   of   what number?
   ↓   ↘   ↓
   _____   percent   _____

$$\text{amount} \rightarrow \frac{\rule{1cm}{0.4pt}}{\rule{1cm}{0.4pt}} = \frac{35}{100} \leftarrow \text{percent} \\ \text{base} \rightarrow$$

3. SOLVE the proportion. (See Example 3, Method 2, for help.)

4. INTERPRET. The total number of children attending the day care center was _____.

3. The Total Gym® provides weight resistance through adjustments of incline. The minimum weight resistance is 4% of the weight of the person using the Total Gym. Find the minimum weight resistance possible for a 220-pound man. (*Source:* Total Gym)

4. The maximum weight resistance for the Total Gym is 60% of the weight of the person using it. Find the maximum weight resistance possible for a 220-pound man. (See Exercise 3 if needed.)

5. A student's cost for last semester at her community college was $2700. She spent $378 of that on books. What percent of last semester's college costs was spent on books?

6. Pierre Sampeau belongs to his local food cooperative, where he receives a percentage of what he spends each year as a dividend. He spent $3850 last year at the food cooperative store and received a dividend of $154. What percent of his total spending at the food cooperative did he receive as a dividend?

7. The United States' motion picture and television industry is made up of over 108,000 businesses. About 85% of these are small businesses with fewer than 10 employees. How many motion picture and television industry businesses have fewer than 10 employees? (*Source:* Motion Picture Association of America)

8. In 2012, 13% of the population of the United States and Canada were considered frequent moviegoers, purchasing movie tickets at least once per month. If the combined population of the United States and Canada was 331,700,000 in 2012, how many people were considered frequent moviegoers? (*Source:* Motion Picture Association of America)

9. The average wedding in the United States cost $28,400 in 2013. The average cost of a wedding reception venue was approximately $12,780. Determine the percent of an average wedding budget that is devoted to the reception venue. (*Source:* Wedding Channel/TheKnot.com)

10. Of the 64,500 veterinarians in private practice in the United States in 2012, approximately 33,540 were female. Determine the percent of female veterinarians in private practice in the United States. (*Source:* American Veterinary Medical Association)

11. A furniture company currently produces 6200 chairs per month. If production decreases by 8%, find the decrease and the new number of chairs produced each month.

12. The enrollment at a local college decreased by 5% over last year's enrollment of 7640. Find the decrease in enrollment and the current enrollment.

13. From 2010 to 2020, the number of people employed as physician assistants in the United States is expected to increase by 30%. The number of people employed as physician assistants in 2010 was 83,600. Find the predicted number of physician assistants in 2020. (*Source:* Bureau of Labor Statistics)

14. From 2007 to 2012, the number of U.S. veterinarians increased by 16%. The number of U.S. veterinarians in 2007 was about 83,700. Find the number of U.S. veterinarians in 2012. (*Source:* American Veterinary Medical Association)

Let's look at the populations of two states, Michigan and Louisiana. Their locations are shown on the partial U.S. map below. Round each answer to the nearest thousand. (Source: U. S. Census Bureau)

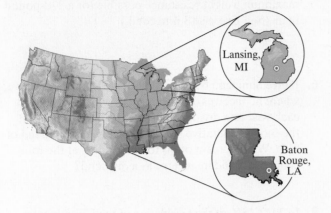

Lansing, MI

Baton Rouge, LA

15. In 2000, the population of Michigan was approximately 9940 thousand. If the population decreased by 0.6% between 2000 and 2012, find the population of Michigan in 2012.

16. In 2000, the population of Louisiana was approximately 4470 thousand. If the population increased about 3% between 2000 and 2012, find the population of Louisiana in 2012.

A popular extreme sport is snowboarding. Ski trails are marked with difficulty levels of easy ●, intermediate ■, difficult ◆, expert ◆◆, and other variations. Use this information for Exercises 17 and 18. Round each percent to the nearest whole. See Example 2.

17. At Keystone ski area in Colorado, 38 of the 131 total ski runs are rated intermediate. What percent of the runs are intermediate? (*Source:* Vail Resorts Management Company)

18. At Telluride ski area in Colorado, 29 of the 127 total ski trails are rated easy. What percent of the trails are easy? (*Source:* Telluride Ski & Golf Resort)

For each food described, find the percent of total calories from fat. If necessary, round to the nearest tenth percent. See Example 2.

19. Ranch dressing serving size of 2 tablespoons

	Calories
Total	40
From fat	20

20. Unsweetened cocoa powder serving size of 1 tablespoon

	Calories
Total	20
From fat	5

21.

Nutrition Facts

Serving Size 1 pouch (20g)
Servings Per Container 6

Amount Per Serving

Calories	80
Calories from fat	10

	% Daily Value*
Total Fat 1g	**2%**
Sodium 45mg	**2%**
Total Carbohydrate 17g	**6%**
Sugars 9g	
Protein 0g	

Vitamin C	25%

Not a significant source of saturated fat, cholesterol, dietary fiber, vitamin A, calcium and iron.

*Percent Daily Values are based on a 2,000 calorie diet.

Artificial Fruit Snacks

22.

Nutrition Facts

Serving Size ¼ cup (33g)
Servings Per Container About 9

Amount Per Serving

Calories 190	Calories from Fat 130

	% Daily Value
Total Fat 16g	**24%**
Saturated Fat 3g	**16%**
Cholesterol 0mg	**0%**
Sodium 135mg	**6%**
Total Carbohydrate 9g	**3%**
Dietary Fiber 1g	**5%**
Sugars 2g	
Protein 5g	

Vitamin A 0%	•	Vitamin C 0%
Calcium 0%	•	Iron 8%

Peanut Mixture

23.

Nutrition Facts

Serving Size 18 crackers (29g)
Servings Per Container About 9

Amount Per Serving

Calories 120 Calories from Fat 35

	% Daily Value*
Total Fat 4g	**6%**
Saturated Fat 0.5g	**3%**
Polyunsaturated Fat 0g	
Monounsaturated Fat 1.5g	
Cholesterol 0mg	**0%**
Sodium 220mg	**9%**
Total Carbohydrate 21g	**7%**
Dietary Fiber 2g	**7%**
Sugars 3g	
Protein 2g	

Vitamin A 0% • Vitamin C 0%
Calcium 2% • Iron 4%
Phosphorus 10%

Snack Crackers

24.

Nutrition Facts

Serving Size 28 crackers (31g)
Servings Per Container About 6

Amount Per Serving

Calories 130 Calories from Fat 35

	% Daily Value*
Total Fat 4g	**6%**
Saturated Fat 2g	**10%**
Polyunsaturated Fat 1g	
Monounsaturated Fat 1g	
Cholesterol 0mg	**0%**
Sodium 470mg	**20%**
Total Carbohydrate 23g	**8%**
Dietary Fiber 1g	**4%**
Sugars 4g	
Protein 2g	

Vitamin A 0% • Vitamin C 0%
Calcium 0% • Iron 2%

Snack Crackers

Solve. If necessary, round money amounts to the nearest cent and all other amounts to the nearest tenth. See Examples 1 through 4.

25. A family paid $26,250 as a down payment for a home. If this represents 15% of the price of the home, find the price of the home.

26. A banker learned that $842.40 is withheld from his monthly check for taxes and insurance. If this represents 18% of his total pay, find the total pay.

27. An owner of a repair service company estimates that for every 40 hours a repairperson is on the job, he can bill for only 78% of the hours. The remaining hours, the repairperson is idle or driving to or from a job. Determine the number of hours per 40-hour week the owner can bill for a repairperson.

28. A manufacturer of electronic components expects 1.04% of its products to be defective. Determine the number of defective components expected in a batch of 28,350 components. Round to the nearest whole component.

29. A car manufacturer announced that next year the price of a certain model of car will increase by 4.5%. This year the price is $19,286. Find the increase in price and the new price.

30. A union contract calls for a 6.5% salary increase for all employees. Determine the increase and the new salary that a worker currently making $58,500 under this contract can expect.

A popular extreme sport is artificial wall climbing. The photo shown is an artificial climbing wall. Exercises 31 and 32 are about the Footsloggers Climbing Tower in Boone, North Carolina.

31. A climber is resting at a height of 24 feet while on the Footsloggers Climbing Tower. If this is 60% of the tower's total height, find the height of the tower.

32. A group plans to climb the Footsloggers Climbing Tower at the group rate, once they save enough money. Thus far, $175 has been saved. If this is 70% of the total amount needed for the group, find the total price.

Solve.

33. Tuition for an Ohio resident at the Columbus campus of Ohio State University was $8406 in 2009. The tuition increased by 14.4% during the period from 2009 to 2013. Find the increase and the tuition for the 2013–2014 school year. Round the increase to the nearest whole dollar. (*Source:* The Ohio State University)

34. The population of Americans aged 65 and older was 43 million in 2012. That population is projected to increase by 69% by 2030. Find the increase and the projected 2030 population. (*Source:* Bureau of the Census)

35. From 2010–2011 to 2020–2021, the number of associate degrees awarded is projected to increase by 14.5%. If the number of associate degrees awarded in 2010–2011 was 888,000, find the increase and the projected number of associate degrees awarded in the 2020–2021 school year. (*Source:* National Center for Education Statistics)

36. From 2010–2011 to 2020–2021, the number of bachelor degrees awarded is projected to increase by 16%. If the number of bachelor degrees awarded in 2010–2011 was 1,703,000, find the increase and the projected number of bachelor degrees awarded in the 2020–2021 school year. (*Source:* National Center for Education Statistics)

Objective B *Find the amount of increase and the percent of increase. See Example 5.*

	Original Amount	New Amount	Amount of Increase	Percent Increase
37.	50	80		
38.	8	12		
39.	65	117		
40.	68	170		

Find the amount of decrease and the percent of decrease. See Example 6.

	Original Amount	New Amount	Amount of Decrease	Percent Decrease
41.	8	6		
42.	25	20		
43.	160	40		
44.	200	162		

Solve. Round percents to the nearest tenth, if necessary. See Examples 5 and 6.

45. There are 150 calories in a cup of whole milk and only 84 in a cup of skim milk. In switching to skim milk, find the percent of decrease in number of calories per cup.

46. In reaction to a slow economy, the number of employees at a soup company decreased from 530 to 477. What was the percent of decrease in the number of employees?

47. The number of cable TV systems recently decreased from 10,845 to 10,700. Find the percent of decrease.

48. Before taking a typing course, Geoffry Landers could type 32 words per minute. By the end of the course, he was able to type 76 words per minute. Find the percent of increase.

49. In 1940, the average size of a privately owned farm in the United States was 174 acres. In a recent year, the average size of a privately owned farm in the United States had increased to 421 acres. Find the percent of increase. (*Source:* National Agricultural Statistics Service)

50. In 2006, the average size of a privately owned farm in the United States was 443 acres. In 2012, the average size of a privately owned farm in the United States had decreased to 421 acres. Find the percent of decrease. (*Source:* National Agricultural Statistics Service)

51. Total U.S. music industry revenues fell from $7.8 billion in 2009 to $7.1 billion in 2012. Find the percent of decrease. (*Source:* Recording Industry Association of America)

52. In 2008, the number of milk cow operations in the United States was 67,000. By 2012, this number had decreased to 58,000. What was the percent of decrease? (*Source:* National Agricultural Statistics Service)

53. In 2004, 21 million people in the U.S. subscribed to high-speed internet service via their cable provider. By 2012, this number had increased to 50.3 million. What was the percent of increase? (*Source:* SNL Kagan)

54. In 2010, there were 1038 thousand high school teachers employed in the United States. This number is expected to increase to 1110 thousand teachers in 2020. What is the percent of increase? (*Source:* Bureau of Labor Statistics)

55. Between 2000 and 2012, the number of indoor cinema sites in the United States decreased from 6550 to 5317. Find the percent of decrease. (*Source:* National Association of Theater Owners)

56. In 2003, the average price of a cinema ticket was $6.03. By 2012, this price had increased to $7.96. What was the percent of increase? (*Source:* National Association of Theatre Owners)

57. In 2010, there were 64,400 dieticians employed in the United States. This number is expected to increase to 77,100 dieticians in 2020. What is the percent of increase? (*Source:* Bureau of Labor Statistics)

58. In 2010, there were 242,900 coaches and scouts employed in the United States. This number is expected to increase to 314,300 in 2020. What is the percent of increase? (*Source:* Bureau of Labor Statistics)

59. The number of cell phone tower sites in the United States was 195,613 in 2006. By 2012, the number of cell sites had increased to 301,779. Find the percent of increase. (*Source:* CTIA—The Wireless Association)

60. The population of Japan is expected to decrease from 127,050 thousand in 2014 to 100,600 thousand in 2050. Find the percent of decrease. (*Source:* Department of Population Dynamics Research)

Japan

Tokyo

Review

Perform each indicated operation. See Sections 4.3 and 4.4.

61. 0.12
       × 38

62.   42
        × 0.7

63. 9.20 + 1.98

64. 46 + 7.89

65. 78 − 19.46

66. 64.80 − 10.72

Concept Extensions

67. If a number is increased by 100%, how does the increased number compare with the original number? Explain your answer.

68. In your own words, explain what is wrong with the following statement. "Last year we had 80 students attend. This year we have a 50% increase or a total of 160 students attending."

Explain what errors were made by each student when solving percent of increase or decrease problems and then correct the errors. See the Concept Checks in this section.

The population of a certain rural town was 150 in 1980, 180 in 1990, and 150 in 2000.

69. Find the percent of increase in population from 1980 to 1990.

Miranda's solution: Percent of increase $= \dfrac{30}{180} = 0.1\overline{6} \approx 16.7\%$

70. Find the percent of decrease in population from 1990 to 2000.

Jeremy's solution: Percent of decrease $= \dfrac{30}{150} = 0.20 = 20\%$

71. The percent of increase from 1980 to 1990 is the same as the percent of decrease from 1990 to 2000. True or false.

Chris's answer: True because they had the same amount of increase as the amount of decrease.

72. Refer to Exercises 49 and 50. They are similar except one asks us to find the percent of increase in the size of U.S. privately owned farms and one asks us to find the percent of decrease. In your own words, explain how these can both be correct.

6.6 Percent and Problem Solving: Sales Tax, Commission, and Discount

Objectives

A Calculate Sales Tax and Total Price.

B Calculate Commissions.

C Calculate Discount and Sale Price.

Objective A Calculating Sales Tax and Total Price

Percents are frequently used in the retail trade. For example, most states charge a tax on certain items when purchased. This tax is called a **sales tax,** and retail stores collect it for the state. Sales tax is almost always stated as a percent of the purchase price.

A 9% sales tax rate on a purchase of a $10 calculator gives a sales tax of

sales tax = 9% of $10 = 0.09 · $10.00 = $0.90

The total price to the customer would be

$$\underbrace{\text{purchase price}}_{\$10.00} \quad \underbrace{\text{plus}}_{+} \quad \underbrace{\text{sales tax}}_{\$0.90} = \$10.90$$

This example suggests the following equations:

> ### Sales Tax and Total Price
>
> sales tax = tax rate · purchase price
>
> total price = purchase price + sales tax

In this section we round dollar amounts to the nearest cent.

Example 1   Finding Sales Tax and Purchase Price

Find the sales tax and the total price on the purchase of an $85.50 atlas in a city where the sales tax rate is 7.5%.

Solution: The purchase price is $85.50 and the tax rate is 7.5%.

$$\underbrace{\text{sales tax}}_{} = \underbrace{\text{tax rate}}_{} \cdot \underbrace{\text{purchase price}}_{}$$

$$
\begin{aligned}
\text{sales tax} &= 7.5\% \cdot \$85.50 \\
&= 0.075 \cdot \$85.50 \quad \text{Write 7.5\% as a decimal.} \\
&\approx \$6.41 \quad\quad\quad\; \text{Round to the nearest cent}
\end{aligned}
$$

Thus, the sales tax is $6.41. Next find the total price.

$$\underbrace{\text{total price}}_{} = \underbrace{\text{purchase price}}_{} + \underbrace{\text{sales tax}}_{}$$

$$
\begin{aligned}
\text{total price} &= \$85.50 + \$6.41 \\
&= \$91.91
\end{aligned}
$$

The sales tax on $85.50 is $6.41, and the total price is $91.91.

■ Work Practice 1

✓**Concept Check** The purchase price of a textbook is $50 and sales tax is 10%. If you are told by the cashier that the total price is $75, how can you tell that a mistake has been made?

Practice 1

If the sales tax rate is 8.5%, what is the sales tax and the total amount due on a $59.90 Goodgrip tire? (Round the sales tax to the nearest cent.)

Answer

1. tax: $5.09; total: $64.99

✓**Concept Check Answer**

Since $10\% = \dfrac{1}{10}$, the sales tax is $\dfrac{\$50}{10} = \$5$. The total price should have been $55.

Practice 2

The sales tax on an $18,500 automobile is $1665. Find the sales tax rate.

 Example 2 Finding a Sales Tax Rate

The sales tax on a $406 Sony flat-screen digital 27-inch television is $34.51. Find the sales tax rate.

Solution: Let r represent the unknown sales tax rate. Then

$$\text{sales tax} = \text{tax rate} \cdot \text{purchase price}$$

$$\$34.51 = r \cdot \$406$$

$$\frac{34.51}{406} = \frac{r \cdot 406}{406} \qquad \text{Divide both sides by 406.}$$

$$0.085 = r \qquad \text{Simplify.}$$

$$8.5\% = r \qquad \text{Write 0.085 as a percent.}$$

The sales tax rate is 8.5%.

■ Work Practice 2

Objective B Calculating Commissions

A **wage** is payment for performing work. Hourly wage, commissions, and salary are some of the ways wages can be paid. Many people who work in sales are paid a commission. An employee who is paid a **commission** is paid a percent of his or her total sales.

> ### Commission
>
> commission = commission rate · sales

Practice 3

A sales representative for Office Product Copiers sold $47,632 worth of copy equipment and supplies last month. What is his commission for the month if he is paid a commission rate of 6.6% of his total sales for the month?

 Example 3 Finding the Amount of Commission

Sherry Souter, a real estate broker for Wealth Investments, sold a house for $214,000 last week. If her commission rate is 1.5% of the selling price of the home, find the amount of her commission.

Solution:

commission	=	commission rate	·	sales
commission	=	1.5%	·	$214,000
	=	0.015	·	$214,000 Write 1.5% as 0.015.
	=	$3210		Multiply.

Her commission on the house is $3210.

■ Work Practice 3

Answers

2. 9% **3.** $3143.71

Example 4 Finding a Commission Rate

A salesperson earned $1560 for selling $13,000 worth of electronics equipment. Find the commission rate.

Solution: Let r stand for the unknown commission rate. Then

commission = commission rate · sales

$$\$1560 = r \cdot \$13,000$$

$$\frac{1560}{13,000} = r \quad \text{Divide 1560 by 13,000, the number multiplied by } r.$$

$$0.12 = r \quad \text{Simplify.}$$

$$12\% = r \quad \text{Write 0.12 as a percent.}$$

The commission rate is 12%.

🔲 Work Practice 4

Practice 4

A salesperson earns $645 for selling $4300 worth of appliances. Find the commission rate.

Objective C Calculating Discount and Sale Price ▶

Suppose that an item that normally sells for $40 is on sale for 25% off. This means that the **original price** of $40 is reduced, or **discounted**, by 25% of $40, or $10. The **discount rate** is 25%, the **amount of discount** is $10, and the **sale price** is $40 − $10, or $30. Study the diagram below to visualize these terms.

To calculate discounts and sale prices, we can use the following equations:

Discount and Sale Price

amount of discount = discount rate · original price

sale price = original price − amount of discount

Example 5 Finding a Discount and a Sale Price

An electric rice cooker that normally sells for $65 is on sale for 25% off. What is the amount of discount and what is the sale price?

Solution: First we find the amount of discount, or simply the discount.

amount of discount = discount rate · original price

$$\text{amount of discount} = 25\% \cdot \$65$$

$$= 0.25 \cdot \$65 \quad \text{Write 25\% as 0.25.}$$

$$= \$16.25 \quad \text{Multiply.}$$

(Continued on next page)

Practice 5

A discontinued washer and dryer combo is advertised on sale for 35% off the regular price of $700. Find the amount of discount and the sale price.

Answers
4. 15% **5.** $245; $455

The discount is $16.25. Next, find the sale price.

$$\text{sale price} = \text{original price} - \text{discount}$$

$$\text{sale price} = \quad \$65 \quad - \quad \$16.25$$

$$= \$48.75 \quad \text{Subtract.}$$

The sale price is $48.75.

■ Work Practice 5

Vocabulary, Readiness & Video Check

Use the choices below to fill in each blank. Some choices may be used more than once.

amount of discount   sale price   sales tax
commission   total price

1. _____ = tax rate · purchase price.

2. _____ = purchase price + sales tax.

3. _____ = commission rate · sales.

4. _____ = discount rate · original price.

5. _____ = original price − amount of discount.

6. sale price = original price − _____ .

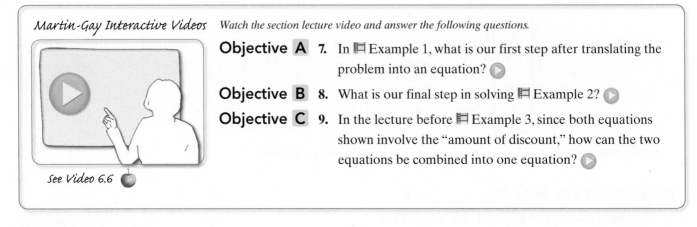

Martin-Gay Interactive Videos   *Watch the section lecture video and answer the following questions.*

See Video 6.6

Objective A 7. In ▣ Example 1, what is our first step after translating the problem into an equation? ▶

Objective B 8. What is our final step in solving ▣ Example 2? ▶

Objective C 9. In the lecture before ▣ Example 3, since both equations shown involve the "amount of discount," how can the two equations be combined into one equation? ▶

6.6 Exercise Set MyMathLab® ▶

Objective A *Solve. See Examples 1 and 2.*

1. What is the sales tax on a jacket priced at $150 if the sales tax rate is 5%?

2. If the sales tax rate is 6%, find the sales tax on a microwave oven priced at $188.

3. The purchase price of a camcorder is $799. What is the total price if the sales tax rate is 7.5%?

4. A stereo system has a purchase price of $426. What is the total price if the sales tax rate is 8%?

5. A new large-screen television has a purchase price of $4790. If the sales tax on this purchase is $335.30, find the sales tax rate.

6. The sales tax on the purchase of a $6800 used car is $374. Find the sales tax rate.

7. The sales tax on a table saw is $10.20.

   a. What is the purchase price of the table saw (before tax) if the sales tax rate is 8.5%? (*Hint:* Use the sales tax equation and insert the replacement values.)

   b. Find the total price of the table saw.

8. The sales tax on a one-half-carat diamond ring is $76.

   a. Find the purchase price of the ring (before tax) if the sales tax rate is 9.5%. (See the hint for Exercise 7a.)

   b. Find the total price of the ring.

9. A gold and diamond bracelet sells for $1800. Find the sales tax and the total price if the sales tax rate is 6.5%.

10. The purchase price of a personal computer is $1890. If the sales tax rate is 8%, what is the sales tax and the total price?

11. The sales tax on the purchase of a futon is $24.25. If the tax rate is 5%, find the purchase price of the futon.

12. The sales tax on the purchase of a TV-DVD combination is $32.85. If the tax rate is 9%, find the purchase price of the TV-DVD.

13. The sales tax is $98.70 on a stereo sound system purchase of $1645. Find the sales tax rate.

14. The sales tax is $103.50 on a necklace purchase of $1150. Find the sales tax rate.

15. A cell phone costs $210, a battery recharger costs $15, and batteries cost $5. What is the sales tax and total price for purchasing these items if the sales tax rate is 7%?

16. Ms. Warner bought a blouse for $35, a skirt for $55, and a blazer for $95. Find the sales tax and the total price she paid, given a sales tax rate of 6.5%.

Objective B *Solve. See Examples 3 and 4.*

17. A sales representative for a large furniture warehouse is paid a commission rate of 4%. Find her commission if she sold $1,329,401 worth of furniture last year.

18. Rosie Davis-Smith is a beauty consultant for a home cosmetic business. She is paid a commission rate of 12.8%. Find her commission if she sold $1638 in cosmetics last month.

19. A salesperson earned a commission of $1380.40 for selling $9860 worth of paper products. Find the commission rate.

20. A salesperson earned a commission of $3575 for selling $32,500 worth of books to various bookstores. Find the commission rate.

21. How much commission will Jack Pruet make on the sale of a $325,900 house if he receives 1.5% of the selling price?

22. Frankie Lopez sold $9638 of jewelry this week. Find her commission for the week if she receives a commission rate of 5.6%.

23. A real estate agent earned a commission of $5565 for selling a house. If his rate is 3%, find the selling price of the house. (*Hint:* Use the commission equation and insert the replacement values.)

24. A salesperson earned $1750 for selling fertilizer. If her commission rate is 7%, find the selling price of the fertilizer. (See the hint for Exercise 23.)

Objective C *Find the amount of discount and the sale price. See Example 5.*

	Original Price	Discount Rate	Amount of Discount	Sale Price
25.	$89	10%		
26.	$74	20%		
27.	$196.50	50%		
28.	$110.60	40%		
29.	$410	35%		
30.	$370	25%		
31.	$21,700	15%		
32.	$17,800	12%		

33. A $300 fax machine is on sale for 15% off. Find the amount of discount and the sale price.

34. A $4295 designer dress is on sale for 30% off. Find the amount of discount and the sale price.

Objectives A B Mixed Practice *Complete each table.*

	Purchase Price	Tax Rate	Sales Tax	Total Price
35.	$305	9%		
36.	$243	8%		
37.	$56	5.5%		
38.	$65	8.4%		

	Sale	Commission Rate	Commission
39.	$235,800	3%	
40.	$195,450	5%	
41.	$17,900		$1432
42.	$25,600		$2304

Review

Multiply. See Sections 4.4 and 4.6.

43. $2000 \cdot \dfrac{3}{10} \cdot 2$

44. $500 \cdot \dfrac{2}{25} \cdot 3$

45. $400 \cdot \dfrac{3}{100} \cdot 11$

46. $1000 \cdot \dfrac{1}{20} \cdot 5$

47. $600 \cdot 0.04 \cdot \dfrac{2}{3}$

48. $6000 \cdot 0.06 \cdot \dfrac{3}{4}$

Concept Extensions

Solve. See the Concept Check in this section.

49. Your purchase price is $68 and the sales tax rate is 9.5%. Round each amount and use the rounded amounts to estimate the total price. Choose the best estimate.
   a. $105   **b.** $58   **c.** $93   **d.** $77

50. Your purchase price is $200 and the tax rate is 10%. Choose the best estimate of the total price.
   a. $190   **b.** $210   **c.** $220   **d.** $300

Tipping

One very useful application of percent is mentally calculating a tip. Recall that to find 10% of a number, simply move the decimal point one place to the left. To find 20% of a number, just double 10% of the number. To find 15% of a number, find 10% and then add to that number half of the 10% amount. Mentally fill in the chart below. To do so, start by rounding the bill amount to the nearest dollar.

	Tipping Chart			
	Bill Amount	**10%**	**15%**	**20%**
51.	$40.21			
52.	$15.89			
53.	$72.17			
54.	$9.33			

55. Suppose that the original price of a shirt is $50. Which is better, a 60% discount or a discount of 30% followed by a discount of 35% of the reduced price? Explain your answer.

56. Which is better, a 30% discount followed by an additional 25% off or a 20% discount followed by an additional 40% off? To see, suppose an item costs $100 and calculate each discounted price. Explain your answer.

57. A diamond necklace sells for $24,966. If the tax rate is 7.5%, find the total price.

58. A house recently sold for $562,560. The commission rate on the sale is 5.5%. If the real estate agent is to receive 60% of the commission, find the amount received by the agent.

6.7 Percent and Problem Solving: Interest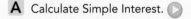

Objective A Calculating Simple Interest

Interest is money charged for using other people's money. When you borrow money, you pay interest. When you loan or invest money, you earn interest. The money borrowed, loaned, or invested is called the **principal amount,** or simply **principal.** Interest is normally stated in terms of a percent of the principal for a given period of time. The **interest rate** is the percent used in computing the interest. Unless stated otherwise, *the rate is understood to be per year.* When the interest is computed on the original principal, it is called **simple interest.** Simple interest is calculated using the following equation:

Objectives

A Calculate Simple Interest. ⊙

B Calculate Compound Interest. ⊙

C Calculate Monthly Payments. ⊙

Simple Interest

Simple Interest = Principal · Rate · Time

$$I = P \cdot R \cdot T$$

where the rate is understood to be per year and time is in years.

Practice 1

Find the simple interest after 5 years on $875 at an interest rate of 7%.

 Example 1 Finding Simple Interest

Find the simple interest after 2 years on $500 at an interest rate of 12%.

Solution: In this example, $P = \$500$, $R = 12\%$, and $T = 2$ years. Replace the variables with values in the formula $I = PRT$.

$$I = P \cdot R \cdot T$$
$$I = \$500 \cdot 12\% \cdot 2 \quad \text{Let } P = \$500, R = 12\%, \text{ and } T = 2.$$
$$= \$500 \cdot (0.12) \cdot 2 \quad \text{Write 12\% as a decimal.}$$
$$= \$120 \quad \text{Multiply.}$$

The simple interest is $120.

🔲 **Work Practice 1**

If time is not given in years, we need to convert the given time to years.

Practice 2

A student borrowed $1500 for 9 months on her credit card at a simple interest rate of 20%. How much interest did she pay?

Example 2 Finding Simple Interest

Ivan Borski borrowed $2400 at 10% simple interest for 8 months to buy a used Toyota Corolla. Find the simple interest he paid.

Solution: Since there are 12 months in a year, we first find what part of a year 8 months is.

$$8 \text{ months} = \frac{8}{12} \text{ year} = \frac{2}{3} \text{ year}$$

Now we find the simple interest.

simple interest	=	principal	·	rate	·	time
↓		↓		↓		↓
simple interest	=	$2400	·	10%	·	$\frac{2}{3}$
	=	$2400	·	0.10	·	$\frac{2}{3}$
	=	$160				

The interest on Ivan's loan is $160.

🔲 **Work Practice 2**

✓**Concept Check** Suppose in Example 2 you had obtained an answer of $16,000. How would you know that you had made a mistake in this problem?

When money is borrowed, the borrower pays the original amount borrowed, or the principal, as well as the interest. When money is invested, the investor receives the original amount invested, or the principal, as well as the interest. In either case, the **total amount** is the sum of the principal and the interest.

Finding the Total Amount of a Loan or Investment

total amount (paid or received) = principal + interest

Answers

1. $306.25 **2.** $225

✓**Concept Check Answer**

$16,000 is too much interest.

Example 3 Finding the Total Amount of an Investment

An accountant invested $2000 at a simple interest rate of 10% for 2 years. What total amount of money will she have from her investment in 2 years?

Solution: First we find her interest.

$$I = P \cdot R \cdot T$$
$$= \$2000 \cdot (0.10) \cdot 2 \quad \text{Let } P = \$2000, R = 10\% \text{ or } 0.10, \text{ and } T = 2.$$
$$= \$400$$

The interest is $400.

Next, we add the interest to the principal.

total amount = principal + interest
↓ ↓ ↓
total amount = $2000 + $400
= $2400

After 2 years, she will have a total amount of $2400.

▪ Work Practice 3

✓**Concept Check** Which investment would earn more interest: an amount of money invested at 8% interest for 2 years, or the same amount of money invested at 8% for 3 years? Explain.

Objective B Calculating Compound Interest ▶

Recall that simple interest depends on the original principal only. Another type of interest is compound interest. **Compound interest** is computed not only on the principal, but also on the interest already earned in previous compounding periods. Compound interest is used more often than simple interest.

Let's see how compound interest differs from simple interest. Suppose that $2000 is invested at 7% interest **compounded annually** for 3 years. This means that interest is added to the principal at the end of each year and that next year's interest is computed on this new amount. In this section, we round dollar amounts to the nearest cent.

	Amount at Beginning of Year	Principal	·	Rate	·	Time	= Interest	Amount at End of Year
1st year	$2000	$2000	·	0.07	·	1	= $140	$2000 + 140 = $2140
2nd year	$2140	$2140	·	0.07	·	1	= $149.80	$2140 + 149.80 = $2289.80
3rd year	$2289.80	$2289.80	·	0.07	·	1	= $160.29	$2289.80 + 160.29 = $2450.09

The compound interest earned can be found by

total amount − original principal = compound interest
↓ ↓ ↓
$2450.09 − $2000 = $450.09

The simple interest earned would have been

principal · rate · time = interest
↓ ↓ ↓ ↓
$2000 · 0.07 · 3 = $420

Practice 3

If $2100 is borrowed at a simple interest rate of 13% for 6 months, find the total amount paid.

Answer
3. $2236.50

✓**Concept Check Answer**
8% for 3 years. Since the interest rate is the same, the longer you keep the money invested, the more interest you earn.

Since compound interest earns "interest on interest," compound interest earns more than simple interest.

Computing compound interest using the method above can be tedious. We can use a calculator and the compound interest formula below to compute compound interest more quickly.

Compound Interest Formula

The total amount A in an account is given by

$$A = P\left(1 + \frac{r}{n}\right)^{n \cdot t}$$

where P is the principal, r is the interest rate written as a decimal, t is the length of time in years, and n is the number of times compounded per year.

Practice 4

$3000 is invested at 4% interest compounded annually. Find the total amount after 6 years.

Example 4 $1800 is invested at 2% interest compounded annually. Find the total amount after 3 years.

Solution: "Compounded annually" means 1 time a year, so

$n = 1$. Also, $P = \$1800$, $r = 2\% = 0.02$, and $t = 3$ years.

$$A = P\left(1 + \frac{r}{n}\right)^{n \cdot t}$$

$$= 1800\left(1 + \frac{0.02}{1}\right)^{1 \cdot 3}$$

$$= 1800(1.02)^3$$

$$\approx 1910.17 \qquad \text{Round to 2 decimal places.}$$

> **Helpful Hint**
> Remember order of operations. **First** evaluate $(1.02)^3$, then multiply by 1800.

The total amount at the end of 3 years is $1910.17.

■ Work Practice 4

Practice 5

$5500 is invested at $6\frac{1}{4}\%$ compounded *daily* for 5 years. Find the total amount at the end of 5 years. (Use 1 year = 365 days.)

Example 5 Finding Total Amount Received from an Investment

$4000 is invested at 5.3% compounded quarterly for 10 years. Find the total amount at the end of 10 years.

Solution: "Compounded quarterly" means 4 times a year, so

$n = 4$. Also, $P = \$4000$, $r = 5.3\% = 0.053$, and $t = 10$ years.

$$A = P\left(1 + \frac{r}{n}\right)^{n \cdot t}$$

$$= 4000\left(1 + \frac{0.053}{4}\right)^{4 \cdot 10}$$

$$= 4000(1.01325)^{40}$$

$$\approx 6772.12$$

The total amount after 10 years is $6772.12.

■ Work Practice 5

Note: Part of the compound interest formula, $\left(1 + \frac{r}{n}\right)^{n \cdot t}$, is called the **compound interest factor.** Appendix A.7 contains a table of various calculated compound interest factors. Another way to calculate the total amount, A, in the compound interest

Answers

4. $3795.96 **5.** $7517.41

formula is to multiply the principal, P, by the appropriate compound interest factor found in Appendix A.7.

The Calculator Explorations box below shows how compound interest factors are calculated.

Objective C Calculating a Monthly Payment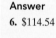

We conclude this section with a method to find the monthly payment on a loan.

Finding the Monthly Payment of a Loan

$$\text{monthly payment} = \frac{\text{principal} + \text{interest}}{\text{total number of payments}}$$

Example 6 Finding a Monthly Payment

Find the monthly payment on a $2000 loan for 2 years. The interest on the 2-year loan is $435.88.

Solution: First we determine the total number of monthly payments. The loan is for 2 years. Since there are 12 months per year, the number of payments is $2 \cdot 12$, or 24. Now we calculate the monthly payment.

$$\text{monthly payment} = \frac{\text{principal} + \text{interest}}{\text{total number of payments}}$$

$$\text{monthly payment} = \frac{\$2000 + \$435.88}{24}$$

$$\approx \$101.50.$$

The monthly payment is about $101.50.

■ Work Practice 6

Practice 6

Find the monthly payment on a $3000 3-year loan if the interest on the loan is $1123.58.

Answer

6. $114.54

Calculator Explorations **Compound Interest Factor**

A compound interest factor may be found by using your calculator and evaluating the formula

$$\textbf{compound interest factor} = \left(1 + \frac{r}{n}\right)^{n \cdot t}$$

where r is the interest rate, t is the time in years, and n is the number of times compounded per year. For example, the compound interest factor for 10 years at 8% compounded semiannually is about 2.19112. Let's find this factor by evaluating the compound interest factor formula when $r = 8\%$ or 0.08, $t = 10$, and $n = 2$ (compounded semiannually means 2 times per year). Thus,

$$\text{compound interest factor} = \left(1 + \frac{0.08}{2}\right)^{2 \cdot 10}$$

$$\text{or} \ \left(1 + \frac{0.08}{2}\right)^{20}$$

To evaluate, press the keys

$\boxed{(}\ \boxed{1}\ \boxed{+}\ \boxed{0.08}\ \boxed{\div}\ \boxed{2}\ \boxed{)}\ \boxed{y^x}\ \text{or}\ \boxed{\wedge}\ \boxed{20}\ \text{then}\ \boxed{=}\ \text{or}\ \boxed{\text{ENTER}}$.

The display will read $\boxed{2.1911231}$. Rounded to 5 decimal places, this is 2.19112.

Find the compound interest factors. Use the table in Appendix A.7 to check your answers. For Exercises 1–4, round to 5 decimal places. For Exercises 5 and 6, round to 2 decimal places.

1. 5 years, 9%, compounded quarterly
2. 15 years, 14%, compounded daily
3. 20 years, 11%, compounded annually
4. 1 year, 7%, compounded semiannually
5. Find the total amount after 4 years when $500 is invested at 6% compounded quarterly. (Multiply the appropriate compound interest factor by $500.)
6. Find the total amount for 19 years when $2500 is invested at 5% compounded daily.

Vocabulary, Readiness & Video Check

Use the choices below to fill in each blank. Choices may be used more than once.

total amount   simple   principal amount   compound

1. To calculate _____ interest, use $I = P \cdot R \cdot T$.

2. To calculate _____ interest, use $A = P\left(1 + \dfrac{r}{n}\right)^{n \cdot t}$.

3. _____ interest is computed not only on the original principal, but also on interest already earned in previous compounding periods.

4. When interest is computed on the original principal only, it is called _____ interest.

5. _____ (paid or received) = principal + interest.

6. The _____ is the money borrowed, loaned, or invested.

Martin-Gay Interactive Videos   *Watch the section lecture video and answer the following questions.*

See Video 6.7

Objective A 7. Complete this statement based on the lecture before ⊞ Example 1: Simple interest is charged on the _____ only. ▶

Objective B 8. In ⊞ Example 2, how often is the interest compounded and what number does this translate to in the formula? ▶

Objective C 9. In ⊞ Example 3, how was the denominator of 48 determined? ▶

6.7   Exercise Set MyMathLab® ▶

Objective A *Find the simple interest. See Examples 1 and 2.*

	Principal	Rate	Time
1.	$200	8%	2 years
3.	$160	11.5%	4 years
5.	$5000	10%	$1\frac{1}{2}$ years
7.	$375	18%	6 months
9.	$2500	16%	21 months

	Principal	Rate	Time
2.	$800	9%	3 years
4.	$950	12.5%	5 years
6.	$1500	14%	$2\frac{1}{4}$ years
8.	$775	15%	8 months
10.	$1000	10%	18 months

Solve. See Examples 1 through 3.

 11. A company borrows $162,500 for 5 years at a simple interest rate of 12.5%. Find the interest paid on the loan and the total amount paid back.

12. $265,000 is borrowed to buy a house. If the simple interest rate on the 30-year loan is 8.25%, find the interest paid on the loan and the total amount paid back.

13. A money market fund advertises a simple interest rate of 9%. Find the total amount received on an investment of $5000 for 15 months.

14. The Real Service Company takes out a 270-day (9-month) short-term, simple interest loan of $4500 to finance the purchase of some new equipment. If the interest rate is 14%, find the total amount that the company pays back.

15. Marsha borrows $8500 and agrees to pay it back in 4 years. If the simple interest rate is 17%, find the total amount she pays back.

16. An 18-year-old is given a high school graduation gift of $2000. If this money is invested at 8% simple interest for 5 years, find the total amount.

Objective B *Find the total amount in each compound interest account. See Examples 4 and 5.*

▶ **17.** $6150 is compounded semiannually at a rate of 14% for 15 years.

18. $2060 is compounded annually at a rate of 15% for 10 years.

19. $1560 is compounded daily at a rate of 8% for 5 years.

20. $1450 is compounded quarterly at a rate of 10% for 15 years.

21. $10,000 is compounded semiannually at a rate of 9% for 20 years.

22. $3500 is compounded daily at a rate of 8% for 10 years.

23. $2675 is compounded annually at a rate of 9% for 1 year.

24. $6375 is compounded semiannually at a rate of 10% for 1 year.

25. $2000 is compounded annually at a rate of 8% for 5 years.

26. $2000 is compounded semiannually at a rate of 8% for 5 years.

27. $2000 is compounded quarterly at a rate of 8% for 5 years.

28. $2000 is compounded daily at a rate of 8% for 5 years.

Objective C *Solve. See Example 6.*

29. A college student borrows $1500 for 6 months to pay for a semester of school. If the interest is $61.88, find the monthly payment.

30. Jim Tillman borrows $1800 for 9 months. If the interest is $148.90, find his monthly payment.

▶ **31.** $20,000 is borrowed for 4 years. If the interest on the loan is $10,588.70, find the monthly payment.

32. $105,000 is borrowed for 15 years. If the interest on the loan is $181,125, find the monthly payment.

Review

Find the perimeter of each figure. See Section 1.3.

△ **33.**

Rectangle | 6 yards
10 yards

△ **34.**

18 centimeters
Triangle
16 centimeters   12 centimeters

△ **35.**

Regular pentagon— All sides are same length | 7 meters

△ **36.**

Square | 21 miles

Concept Extensions

37. Explain how to look up a compound interest factor in the compound interest table.

38. Explain how to find the amount of interest in a compounded account.

39. Compare the following accounts: Account 1: $1000 is invested for 10 years at a simple interest rate of 6%. Account 2: $1000 is compounded semiannually at a rate of 6% for 10 years. Discuss how the interest is computed for each account. Determine which account earns more interest. Why?

Chapter 6 Group Activity

Fastest-Growing Occupations

According to U.S. Bureau of Labor Statistics projections, the careers listed below are the top ten fastest-growing jobs ranked by expected percent increase through the year 2020. (*Source:* Bureau of Labor Statistics)

	Occupation	Employment in 2010	Percent Change	Expected Employment in 2020
1	Personal care aides	861,000	70.5%	
2	Home health aides	1,017,700	69.4%	
3	Biomedical engineers	15,700	61.7%	
4	Helpers—masons	29,400	60.1%	
5	Helpers—carpenters	46,500	55.7%	
6	Veterinary technicians	80,200	52.0%	
7	Reinforcing iron/rebar workers	19,100	48.6%	
8	Physical therapist assistants	67,400	45.7%	
9	Helpers—plumbers	57,900	45.4%	
10	Meeting/event planners	71,600	43.7%	

What do most of these fast-growing occupations have in common? They require knowledge of math! For some careers, such as home health aides, event planners, and biomedical engineers, the ways math is used on the job may be obvious. For other occupations, the use of math may not be quite as apparent. However, tasks common to many jobs—filling in a time sheet, writing up an expense or mileage report, planning a budget, figuring a bill, ordering supplies, and even making a work schedule—all require math.

This activity may be completed by working in groups or individually.

1. List the top five occupations by order of employment figures for 2010.

2. Using the 2010 employment figures and the percent increase from 2010 to 2020, find the expected 2020 employment figure for each occupation listed in the table. Round to the nearest thousand.

3. List the top five occupations by order of employment figures for 2020. Did the order change at all from 2010? Explain.

Chapter 6 Vocabulary Check

Fill in each blank with one of the words or phrases listed below. Some choices may be used more than once.

percent   sales tax   is   0.01 $\frac{1}{100}$ amount of discount   percent of decrease   total price

base   of   amount   100%   compound interest   percent of increase   sale price   commission

1. In a mathematical statement, _____ usually means "multiplication."
2. In a mathematical statement, _____ means "equals."
3. _____ means "per hundred."
4. _____ is computed not only on the principal, but also on interest already earned in previous compounding periods.

5. In the percent proportion, $\dfrac{\rule{2cm}{0.4pt}}{\rule{2cm}{0.4pt}} = \dfrac{\text{percent}}{100}$.

6. To write a decimal or fraction as a percent, multiply by \rule{3cm}{0.4pt}.

7. The decimal equivalent of the % symbol is \rule{3cm}{0.4pt}.

8. The fraction equivalent of the % symbol is \rule{3cm}{0.4pt}.

9. The percent equation is \rule{3cm}{0.4pt} $\cdot$ percent $=$ \rule{3cm}{0.4pt}.

10. \rule{3cm}{0.4pt} $= \dfrac{\text{amount of decrease}}{\text{original amount}}$.

11. \rule{3cm}{0.4pt} $= \dfrac{\text{amount of increase}}{\text{original amount}}$.

12. \rule{3cm}{0.4pt} $=$ tax rate $\cdot$ purchase price.

13. \rule{3cm}{0.4pt} $=$ purchase price $+$ sales tax.

14. \rule{3cm}{0.4pt} $=$ commission rate $\cdot$ sales.

15. \rule{3cm}{0.4pt} $=$ discount rate $\cdot$ original price.

16. \rule{3cm}{0.4pt} $=$ original price $-$ amount of discount.

Helpful Hint

▶ Are you preparing for your test? Don't forget to take the Chapter 6 Test on page 430. Then check your answers at the back of the text and use the Chapter Test Prep Videos to see the fully worked-out solutions to any of the exercises you want to review.

6 Chapter Highlights

Definitions and Concepts	Examples
Section 6.1 Introduction to Percent	
Percent means "per hundred." The % symbol denotes percent.	$51\% = \dfrac{51}{100}$ 51 per 100
	$7\% = \dfrac{7}{100}$ 7 per 100
To write a percent as a decimal, replace the % symbol with its decimal equivalent, 0.01, and multiply.	$32\% = 32(0.01) = 0.32$
To write a decimal as a percent, multiply by 100%.	$0.08 = 0.08(100\%) = 08.\% = 8\%$
Section 6.2 Percents and Fractions	
To write a percent as a fraction, replace the % symbol with its fraction equivalent, $\dfrac{1}{100}$, and multiply.	$25\% = \dfrac{25}{100} = \dfrac{\overset{1}{\cancel{25}}}{4 \cdot \underset{1}{\cancel{25}}} = \dfrac{1}{4}$
To write a fraction as a percent, multiply by 100%.	$\dfrac{1}{6} = \dfrac{1}{6} \cdot 100\% = \dfrac{1}{6} \cdot \dfrac{100}{1}\% = \dfrac{100}{6}\% = 16\dfrac{2}{3}\%$

Definitions and Concepts	Examples

Section 6.3　Solving Percent Problems Using Equations

Three key words in the statement of a percent problem are

of, which means **multiplication** (·)

is, which means **equals** (=)

what (or some equivalent word or phrase), which stands for **the unknown number**

Solve:

$$6 \quad \text{is} \quad 12\% \quad \text{of} \quad \text{what number?}$$

$$6 \quad = \quad 12\% \quad \cdot \quad n$$

$6 = 0.12 \cdot n$　　Write 12% as a decimal.

$\dfrac{6}{0.12} = n$　　Divide 6 by 0.12, the number multiplied by n.

$50 = n$

Thus, 6 is 12% of 50.

Section 6.4　Solving Percent Problems Using Proportions

Percent Proportion

$$\dfrac{\text{amount}}{\text{base}} = \dfrac{\text{percent}}{100} \leftarrow \text{always } 100$$

or

$$\text{amount} \rightarrow \dfrac{a}{b} = \dfrac{p}{100} \leftarrow \text{percent}$$
$$\text{base} \rightarrow$$

Solve:

20.4 is what percent of 85?

amount　percent　base

$\text{amount} \rightarrow \dfrac{20.4}{85} = \dfrac{p}{100} \leftarrow \text{percent}$
$\text{base} \rightarrow$

$20.4 \cdot 100 = 85 \cdot p$　　Set cross products equal.

$2040 = 85 \cdot p$　　Multiply.

$\dfrac{2040}{85} = p$　　Divide 2040 by 85, the number multiplied by p.

$24 = p$　　Simplify.

Thus, 20.4 is 24% of 85.

Section 6.5　Applications of Percent

Percent of Increase

$$\text{percent of increase} = \dfrac{\text{amount of increase}}{\text{original amount}}$$

Percent of Decrease

$$\text{percent of decrease} = \dfrac{\text{amount of decrease}}{\text{original amount}}$$

A town's population of 16,480 decreased to 13,870 over a 12-year period. Find the percent decrease. Round to the nearest whole percent.

$\text{amount of decrease} = 16,480 - 13,870$

$= 2610$

$\text{percent of decrease} = \dfrac{\text{amount of decrease}}{\text{original amount}}$

$= \dfrac{2610}{16,480} \approx 0.16$

$= 16\%$

The town's population decreased by 16%.

Section 6.6　Percent and Problem Solving: Sales Tax, Commission, and Discount

Sales Tax and Total Price

$$\text{sales tax} = \text{sales tax rate} \cdot \text{purchase price}$$
$$\text{total price} = \text{purchase price} + \text{sales tax}$$

Find the sales tax and the total price on a purchase of $42 if the sales tax rate is 9%.

sales tax　=　sales tax rate　·　purchase price

$\text{sales tax} = 9\% \cdot \$42$

$= 0.09 \cdot \$42$

$= \$3.78$

(continued)

Definitions and Concepts	Examples

Section 6.6   Percent and Problem Solving: Sales Tax, Commission, and Discount (*continued*)

	The total price is

$$\text{total price} = \text{purchase price} + \text{sales tax}$$

$$\text{total price} = \$42 + \$3.78$$
$$= \$45.78$$

Commission

$$\text{commission} = \text{commission rate} \cdot \text{total sales}$$

A salesperson earns a commission of 3%. Find the commission from sales of $12,500 worth of appliances.

$$\text{commission} = \text{commission rate} \cdot \text{sales}$$

$$\text{commission} = 3\% \cdot \$12,500$$
$$= 0.03 \cdot \$12,500$$
$$= \$375$$

Discount and Sale Price

$$\text{amount of discount} = \text{discount rate} \cdot \text{original price}$$
$$\text{sale price} = \text{original price} - \text{amount of discount}$$

A suit is priced at $320 and is on sale today for 25% off. What is the sale price?

$$\text{amount of discount} = \text{discount rate} \cdot \text{original price}$$

$$\text{amount of discount} = 25\% \cdot \$320$$
$$= 0.25 \cdot \$320$$
$$= \$80$$

$$\text{sale price} = \text{original price} - \text{amount of discount}$$

$$\text{sale price} = \$320 - \$80$$
$$= \$240$$

The sale price is $240.

Section 6.7   Percent and Problem Solving: Interest

Simple Interest

$$\text{interest} = \text{principal} \cdot \text{rate} \cdot \text{time}$$

where the rate is understood to be per year.

Find the simple interest after 3 years on $800 at an interest rate of 5%.

$$\text{interest} = \text{principal} \cdot \text{rate} \cdot \text{time}$$

$$\text{interest} = \$800 \cdot 5\% \cdot 3$$
$$= \$800 \cdot 0.05 \cdot 3 \quad \text{Write 5\% as 0.05.}$$
$$= \$120 \quad \text{Multiply.}$$

The interest is $120.

Compound interest is computed not only on the principal, but also on interest already earned in previous compounding periods. (See Appendix A7 for various compound interest factors.)

$$A = P\left(1 + \frac{r}{n}\right)^{n \cdot t}$$

where n is the number of times compounded per year.

$800 is invested at 5% compounded quarterly for 10 years. Find the total amount at the end of 10 years.

$$A = \$800\left(1 + \frac{0.05}{4}\right)^{4 \cdot 10}$$
$$= \$800(1.0125)^{40}$$
$$\approx \$1314.90$$

(6.1) *Solve.*

1. In a survey of 100 adults, 37 preferred pepperoni on their pizzas. What percent preferred pepperoni?

2. A basketball player made 77 out of 100 attempted free throws. What percent of free throws was made?

Write each percent as a decimal.

3. 83%

4. 75%

5. 73.5%

6. 1.5%

7. 125%

8. 145%

9. 0.5%

10. 0.7%

11. 200%

12. 400%

13. 26.25%

14. 85.34%

Write each decimal as a percent.

15. 2.6

16. 1.02

17. 0.35

18. 0.055

19. 0.725

20. 0.252

21. 0.076

22. 0.085

23. 0.71

24. 0.65

25. 4

26. 9

(6.2) *Write each percent as a fraction or mixed number in simplest form.*

27. 1%

28. 10%

29. 25%

30. 8.5%

31. 10.2%

32. $16\frac{2}{3}\%$

33. $33\frac{1}{3}\%$

34. 110%

Write each fraction or mixed number as a percent.

35. $\frac{1}{5}$

36. $\frac{7}{10}$

37. $\frac{5}{6}$

38. $\frac{3}{5}$

39. $1\frac{1}{4}$

40. $1\frac{2}{3}$

41. $\frac{1}{16}$

42. $\frac{5}{8}$

(6.3) *Translate each to an equation and solve.*

43. 1250 is 1.25% of what number?

44. What number is $33\frac{1}{3}$% of 24,000?

45. 124.2 is what percent of 540?

46. 22.9 is 20% of what number?

47. What number is 40% of 7500?

48. 693 is what percent of 462?

(6.4) *Translate each to a proportion and solve.*

49. 104.5 is 25% of what number?

50. 16.5 is 5.5% of what number?

51. What number is 36% of 180?

52. 63 is what percent of 35?

53. 93.5 is what percent of 85?

54. What number is 33% of 500?

(6.5) *Solve.*

55. In a survey of 2000 people, it was found that 1320 have a microwave oven. Find the percent of people who own microwaves.

56. Of the 12,360 freshmen entering County College, 2000 are enrolled in basic college mathematics. Find the percent of entering freshmen who are enrolled in basic college mathematics. Round to the nearest whole percent.

57. The number of violent crimes in a city decreased from 675 to 534. Find the percent of decrease. Round to the nearest tenth of a percent.

58. The current charge for dumping waste in a local landfill is $16 per cubic foot. To cover new environmental costs, the charge will increase to $33 per cubic foot. Find the percent of increase.

59. This year the fund drive for a charity collected $215,000. Next year, a 4% decrease is expected. Find how much is expected to be collected in next year's drive.

60. A local union negotiated a new contract that increases the hourly pay 15% over last year's pay. The old hourly rate was $11.50. Find the new hourly rate rounded to the nearest cent.

(6.6) *Solve.*

61. If the sales tax rate is 5.5%, what is the total amount charged for a $250 coat?

62. Find the sales tax paid on a $25.50 purchase if the sales tax rate is 4.5%.

63. Russ James is a sales representative for a chemical company and is paid a commission rate of 5% on all sales. Find his commission if he sold $100,000 worth of chemicals last month.

64. Carol Sell is a sales clerk in a clothing store. She receives a commission of 7.5% on all sales. Find her commission for the week if her sales for the week were $4005. Round to the nearest cent.

65. A $3000 mink coat is on sale for 30% off. Find the discount and the sale price.

66. A $90 calculator is on sale for 10% off. Find the discount and the sale price.

(6.7) *Solve.*

67. Find the simple interest due on $4000 loaned for 4 months at 12% interest.

68. Find the simple interest due on $6500 loaned for 3 months at 20%.

69. Find the total amount in an account if $5500 is compounded annually at 12% for 15 years.

70. Find the total amount in an account if $6000 is compounded semiannually at 11% for 10 years.

71. Find the compound interest earned if $100 is compounded quarterly at 12% for 5 years.

72. Find the compound interest earned if $1000 is compounded quarterly at 18% for 20 years.

Mixed Review

Write each percent as a decimal.

73. 3.8%

74. 24.5%

75. 0.9%

Write each decimal as a percent.

76. 0.54

77. 95.2

78. 0.3

Write each percent as a fraction or mixed number in simplest form.

79. 47%

80. $6\frac{2}{5}\%$

81. 5.6%

Write each fraction or mixed number as a percent.

82. $\frac{3}{8}$

83. $\frac{2}{13}$

84. $\frac{6}{5}$

Translate each into an equation and solve.

85. 43 is 16% of what number?

86. 27.5 is what percent of 25?

87. What number is 36% of 1968?

88. 67 is what percent of 50?

Translate each into a proportion and solve.

89. 75 is what percent of 25?

90. What number is 16% of 240?

91. 28 is 5% of what number?

92. 52 is what percent of 16?

Solve.

93. The total number of cans in a soft drink machine is 300. If 78 soft drinks have been sold, find the percent of soft drink cans that have been sold.

94. A home valued at $96,950 last year has lost 7% of its value this year. Find the loss in value.

95. A dinette set sells for $568.00. If the sales tax rate is 8.75%, find the purchase price of the dinette set.

96. The original price of a video game is $23.00. It is on sale for 15% off. What is the amount of the discount?

97. A candy salesman makes a commission of $1.60 from each case of candy he sells. If a case of candy costs $12.80, what is his rate of commission?

98. Find the total amount due on a 6-month loan of $1400 at a simple interest rate of 13%.

99. $8800 is invested at 8% interest compounded quarterly. Find the total amount after 9 years.

100. Find the total amount due on a loan of $5500 for 9 years at 12.5% simple interest.

Chapter 6 | Test

Step-by-step test solutions are found on the Chapter Test Prep Videos. Where available: **MyMathLab®** or **You Tube™**

Answers

Write each percent as a decimal.

1. 85% **2.** 500% **3.** 0.8%

Write each decimal as a percent.

4. 0.056 **5.** 6.1 **6.** 0.39

Write each percent as a fraction or mixed number in simplest form.

7. 120% **8.** 38.5% **9.** 0.2%

Write each fraction or mixed number as a percent.

10. $\dfrac{11}{20}$ **11.** $\dfrac{3}{8}$ **12.** $1\dfrac{5}{9}$

Solve.

13. What number is 42% of 80? **14.** 0.6% of what number is 7.5?

15. 567 is what percent of 756?

1. _____

2. _____

3. _____

4. _____

5. _____

6. _____

7. _____

8. _____

9. _____

10. _____

11. _____

12. _____

13. _____

14. _____

15. _____

Solve. Round all dollar amounts to the nearest cent.

16. An alloy is 12% copper. How much copper is contained in 320 pounds of this alloy?

17. A farmer in Nebraska estimates that 20% of his potential crop, or $11,350, has been lost to a hard freeze. Find the total value of his potential crop.

18. If the local sales tax rate is 1.25%, find the total amount charged for a stereo system priced at $354.

19. A town's population increased from 25,200 to 26,460. Find the percent of increase.

20. A $120 framed picture is on sale for 15% off. Find the discount and the sale price.

21. Randy Nguyen is paid a commission rate of 4% on all sales. Find Randy's commission if his sales were $9875.

22. A sales tax of $1.53 is added to an item's price of $152.99. Find the sales tax rate. Round to the nearest whole percent.

23. Find the simple interest earned on $2000 saved for $3\frac{1}{2}$ years at an interest rate of 9.25%.

24. $1365 is compounded annually at 8%. Find the total amount in the account after 5 years.

25. A couple borrowed $400 from a bank at 13.5% simple interest for 6 months for car repairs. Find the total amount due the bank at the end of the 6-month period.

16. _____

17. _____

18. _____

19. _____

20. _____

21. _____

22. _____

23. _____

24. _____

25. _____

Answers

1. _____

2. _____

3. a. _____

   b. _____

   c. _____

4. a. _____

   b. _____

   c. _____

5. _____

6. _____

7. _____

8. _____

9. _____

10. _____

11. _____

12. _____

13. _____

14. _____

15. _____

16. _____

17. _____

18. _____

19. _____

20. _____

21. _____

22. _____

1. How many cases can be filled with 9900 cans of jalapeños if each case holds 48 cans? How many cans will be left over? Will there be enough cases to fill an order for 200 cases?

2. Multiply: 409×76

3. Write each fraction as a mixed number or a whole number.

    a. $\dfrac{30}{7}$ **b.** $\dfrac{16}{15}$ **c.** $\dfrac{84}{6}$

4. Write each mixed number as an improper fraction.

    a. $2\dfrac{5}{7}$ **b.** $10\dfrac{1}{10}$ **c.** $5\dfrac{3}{8}$

5. Use a factor tree to find the prime factorization of 80.

6. Find the area of the rectangle.

7 miles

17 miles

7. Write $\dfrac{10}{27}$ in simplest form.

8. Find the average of 28, 34, and 70.

9. Multiply and simplify: $\dfrac{23}{32} \cdot \dfrac{4}{7}$

10. Round 76,498 to the nearest ten.

11. Find the reciprocal of $\dfrac{11}{8}$.

12. Write the shaded part of the figure as an improper fraction and as a mixed number.

△ **13.** Find the perimeter of the rectangle.

$\frac{2}{15}$ inch

$\frac{4}{15}$ inch

14. Find $2 \cdot 5^2$

15. Find the LCM of 12 and 20.

16. Subtract $\dfrac{7}{9}$ from $\dfrac{10}{9}$.

17. Add: $\dfrac{2}{5} + \dfrac{4}{15}$

18. Find $\dfrac{2}{3}$ of 510.

19. Subtract: $7\dfrac{3}{14} - 3\dfrac{6}{7}$

20. Simplify: $9 \cdot \sqrt{25} - 6 \cdot \sqrt{4}$

Perform each indicated operation.

21. $\dfrac{1}{2} \div \dfrac{8}{7}$

22. $20\dfrac{4}{5} + 12\dfrac{7}{8}$

23. $\dfrac{2}{9} \cdot \dfrac{3}{11}$

24. $1\dfrac{7}{8} \cdot 3\dfrac{2}{5}$

Write each fraction as a decimal.

25. $\dfrac{8}{10}$

26. $\dfrac{9}{100}$

27. $\dfrac{87}{10}$

28. $\dfrac{48}{10,000}$

29. The price of a gallon of premium gasoline in Cross City is currently $3.1779. Round this to the nearest cent.

30. Subtract: $38 - 10.06$

31. Add: $763.7651 + 22.001 + 43.89$

32. Multiply: 12.483×100

33. Multiply: 23.6×0.78

34. Multiply: 76.3×1000

Divide.

35. $\dfrac{786.1}{1000}$

36. $0.5\overline{)0.638}$

37. $\dfrac{0.12}{10}$

38. $0.23\overline{)11.6495}$

39. Simplify: $723.6 \div 1000 \times 10$

40. Simplify: $\dfrac{3.19 - 0.707}{13}$

41. Write $\dfrac{1}{4}$ as a decimal.

42. Write $\dfrac{5}{9}$ as a decimal. Give an exact answer and a three-decimal-place approximation.

Write the following rates in Exercises 43 and 44 as fractions in simplest form.

43. 10 nails every 6 feet

44. 115 miles every 5 gallons

45. Is $\dfrac{4.1}{7} = \dfrac{2.9}{5}$ a true proportion?

46. Find each unit rate and decide on the better buy. Round the unit rates to three decimal places.
$0.93 for 18 flour tortillas
$1.40 for 24 flour tortillas

47. On a chamber of commerce map of Abita Springs, 5 miles corresponds to 2 inches. How many miles correspond to 7 inches?

48. Write each percent as a decimal.
a. 7% **b.** 200% **c.** 0.5%

49. Translate to an equation: What number is 25% of 0.008?

50. Write $\dfrac{3}{8}$ as a percent.

23. _____
24. _____
25. _____
26. _____
27. _____
28. _____
29. _____
30. _____
31. _____
32. _____
33. _____
34. _____
35. _____
36. _____
37. _____
38. _____
39. _____
40. _____
41. _____
42. _____
43. _____
44. _____
45. _____
46. _____
47. _____
48. a. _____
b. _____
c. _____
49. _____
50. _____

Measurement

The use of measurements is common in everyday life. A sales representative records the number of miles she has driven when she submits her travel expense report. A respiratory therapist measures the volume of air exhaled by a patient. A measurement is necessary in each case.

How Is the Crazy Horse Sculpture Progressing?

Only 8 miles from Mt. Rushmore, an American icon honoring four U.S. presidents, is another massive sculpture being carved into the Black Hills of South Dakota. This sculpture, of Lakota leader Crazy Horse, was officially started on June 3, 1948, and is the largest mountain carving in progress. It is now more than six decades in the making. As a comparison, many historians believe that it took about two decades to complete Egypt's Great Pyramid.

To compare the dimensions of these two sculptures, each U.S. president's head is 60 feet in height, while the head of Crazy Horse has a height of 87.5 feet. In fact, all heads on Mt. Rushmore will fit in Crazy Horse's head.

In Section 7.1, Exercises 83 and 84, we will explore measurements of the Crazy Horse memorial.

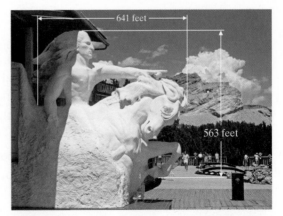

Dimensions of Sculpture, once Completed

Length: U.S. and Metric Systems of Measurement ▶

Objective A Defining and Converting U.S. System Units of Length ▶

In the United States, two systems of measurement are commonly used. They are the **United States (U.S.), or English, measurement system** and the **metric system.** The U.S. measurement system is familiar to most Americans. Units such as feet, miles, ounces, and gallons are used. However, the metric system is also commonly used in fields such as medicine, sports, international marketing, and certain physical sciences. We are accustomed to buying 2-liter bottles of soft drinks, watching televised coverage of the 100-meter dash at the Olympic Games, or taking a 200-milligram dose of pain reliever.

The U.S. system of measurement uses the **inch, foot, yard,** and **mile** to measure **length.** The following is a summary of equivalencies between units of length:

> **U.S. Units of Length**
>
> 12 inches (in.) = 1 foot (ft)
> 3 feet = 1 yard (yd)
> 36 inches = 1 yard
> 5280 feet = 1 mile (mi)

To convert from one unit of length to another, we will use **unit fractions.** We define a unit fraction to be a fraction that is equivalent to 1. Examples of unit fractions are as follows:

> **Unit Fractions**
>
> $\dfrac{12 \text{ in.}}{1 \text{ ft}} = 1$ or $\dfrac{1 \text{ ft}}{12 \text{ in.}} = 1$ (since 12 in. = 1 ft)
>
> $\dfrac{3 \text{ ft}}{1 \text{ yd}} = 1$ or $\dfrac{1 \text{ yd}}{3 \text{ ft}} = 1$ (since 3 ft = 1 yd)
>
> $\dfrac{5280 \text{ ft}}{1 \text{ mi}} = 1$ or $\dfrac{1 \text{ mi}}{5280 \text{ ft}} = 1$ (since 5280 ft = 1 mi)

Remember that multiplying a number by 1 does not change the value of the number.

Example 1 Convert 8 feet to inches.

Solution: We multiply 8 feet by a unit fraction that uses the equality 12 inches = 1 foot. The unit fraction should be in the form $\dfrac{\text{units to convert to}}{\text{original units}}$ or, in this case, $\dfrac{12 \text{ inches}}{1 \text{ foot}}$. We do this so that like units will divide out to 1, as shown.

$$8 \text{ ft} = \frac{8 \text{ ft}}{1} \cdot 1$$

$$= \frac{8 \text{ ft}}{1} \cdot \frac{12 \text{ in.}}{1 \text{ ft}} \qquad \text{Multiply by 1 in the form of } \frac{12 \text{ in.}}{1 \text{ ft}}.$$

$$= 8 \cdot 12 \text{ in.}$$

$$= 96 \text{ in.} \qquad \text{Multiply.}$$

(*Continued on next page*)

Objectives

A Define U.S. Units of Length and Convert from One Unit to Another. ▶

B Use Mixed U.S. Units of Length. ▶

C Perform Arithmetic Operations on U.S. Units of Length. ▶

D Define Metric Units of Length and Convert from One Unit to Another. ▶

E Perform Arithmetic Operations on Metric Units of Length. ▶

Practice 1

Convert 6 feet to inches.

Answer
1. 72 in.

Thus, 8 ft = 96 in., as shown in the diagram:

8 feet = 96 inches

■ Work Practice 1

Practice 2

Convert 8 yards to feet.

Helpful Hint

When converting from one unit to another, select a unit fraction with the properties below:

$$\frac{\text{units you are converting to}}{\text{original units}}$$

By using this unit fraction, the original units will divide out, as wanted.

Example 2   Convert 7 feet to yards.

Solution: We multiply by a unit fraction that compares 1 yard to 3 feet.

$$7 \text{ ft} = \frac{7 \text{ ft}}{1} \cdot 1$$

$$= \frac{7 \text{ ft}}{1} \cdot \frac{1 \text{ yd}}{3 \text{ ft}} \quad \leftarrow \text{Units to convert to}$$
$$\qquad\qquad\qquad \leftarrow \text{Original units}$$

$$= \frac{7}{3} \text{ yd}$$

$$= 2\frac{1}{3} \text{ yd} \qquad \text{Divide.}$$

Thus, $7 \text{ ft} = 2\frac{1}{3} \text{ yd}$, as shown in the diagram.

$7 \text{ feet} = 2\frac{1}{3} \text{ yards}$

■ Work Practice 2

Practice 3

Suppose the pelican's bill (in the photo) measures 18 inches. Convert 18 inches to feet, using decimals.

Example 3   Finding the Length of a Pelican's Bill

The Australian pelican has the longest bill, measuring from 13 to 18.5 inches long. The pelican in the photo has a 15-inch bill. Convert 15 inches to feet, using decimals in your final answer.

Solution:

$$15 \text{ in.} = \frac{15 \text{ in.}}{1} \cdot \frac{1 \text{ ft}}{12 \text{ in.}} \quad \leftarrow \text{Units to convert to}$$
$$\qquad\qquad\qquad\qquad \leftarrow \text{Original units}$$

$$= \frac{15}{12} \text{ ft}$$

$$= \frac{5}{4} \text{ ft} \qquad \text{Simplify } \frac{15}{12}.$$

$$= 1.25 \text{ ft} \qquad \text{Divide.}$$

Thus, 15 in. = 1.25 ft, as shown in the diagram.

15 inches = 1.25 ft

1 ft

$\frac{1}{4}$ or 0.25 ft

■ Work Practice 3

Answers

2. 24 ft   **3.** 1.5 ft

Objective B Using Mixed U.S. System Units of Length

Sometimes it is more meaningful to express a measurement of length with mixed units such as 1 ft and 5 in. We usually condense this and write 1 ft 5 in.

In Example 2, we found that 7 feet is the same as $2\frac{1}{3}$ yards. The measurement can also be written as a mixture of yards and feet. That is,

$$7 \text{ ft} = \underline{\hspace{1cm}} \text{ yd} \underline{\hspace{1cm}} \text{ ft}$$

Because 3 ft = 1 yd, we divide 3 into 7 to see how many whole yards are in 7 feet. The quotient is the number of yards, and the remainder is the number of feet.

$$
\begin{array}{r}
2 \text{ yd } 1 \text{ ft} \\
3\overline{)7} \\
\underline{-6} \\
1
\end{array}
$$

Thus, 7 ft = 2 yd 1 ft, as seen in the diagram:

```
|◄————————————— 7 ft —————————————►|
 ▓▓▓▓▓▓▓▓▓▓▓▓▓▓▓▓▓▓▓▓▓▓▓▓▓▓▓▓▓▓▓▓▓▓
|◄——— 1 yd ———►|◄——— 1 yd ———►|◄ 1 ft ►|
```

Example 4 Convert: 134 in. = _____ ft _____ in.

Solution: Because 12 in. = 1 ft, we divide 12 into 134. The quotient is the number of feet. The remainder is the number of inches. To see why we divide 12 into 134, notice that

$$134 \text{ in.} = \frac{134 \text{ in.}}{1} \cdot \frac{1 \text{ ft}}{12 \text{ in.}} = \frac{134}{12} \text{ ft}$$

$$
\begin{array}{r}
11 \text{ ft } 2 \text{ in} \\
12\overline{)134} \\
\underline{-12} \\
14 \\
\underline{-12} \\
2
\end{array}
$$

Thus, 134 in. = 11 ft 2 in.

■ Work Practice 4

Practice 4
Convert: 68 in. = _____ ft _____ in.

Example 5 Convert 3 feet 7 inches to inches.

Solution: First, we convert 3 feet to inches. Then we add 7 inches.

$$3 \text{ ft} = \frac{3 \text{ ft}}{1} \cdot \frac{12 \text{ in.}}{1 \text{ ft}} = 36 \text{ in.}$$

Then

$$3 \text{ ft } 7 \text{ in.} = 36 \text{ in.} + 7 \text{ in.} = 43 \text{ in.}$$

■ Work Practice 5

Practice 5
Convert 5 yards 2 feet to feet.

Answers
4. 5 ft 8 in. **5.** 17 ft

Objective C Performing Operations on U.S. System Units of Length

Finding sums or differences of measurements often involves converting units, as shown in the next example. Just remember that, as usual, only like units can be added or subtracted.

Practice 6

Add 4 ft 8 in. to 8 ft 11 in.

> **Example 6** Add 3 ft 2 in. and 5 ft 11 in.
>
> **Solution:** To add, we line up the similar units.
>
> $$\begin{array}{r} 3 \text{ ft} \;\; 2 \text{ in.} \\ + \; 5 \text{ ft} \; 11 \text{ in.} \\ \hline 8 \text{ ft} \; 13 \text{ in.} \end{array}$$
>
> Since 13 inches is the same as 1 ft 1 in., we have
>
> 8 ft 13 in. = 8 ft + 1 ft 1 in.
> = 9 ft 1 in.
>
> ■ Work Practice 6

✓**Concept Check** How could you estimate the following sum?

$$\begin{array}{r} 7 \text{ yd} \;\; 4 \text{ in.} \\ + \; 3 \text{ yd} \; 27 \text{ in.} \\ \hline \end{array}$$

Practice 7

Multiply 4 ft 7 in. by 4.

> **Example 7** Multiply 8 ft 9 in. by 3.
>
> **Solution:** By the distributive property, we multiply 8 ft by 3 and 9 in. by 3.
>
> $$\begin{array}{r} 8 \text{ ft} \;\; 9 \text{ in.} \\ \times \qquad\;\; 3 \\ \hline 24 \text{ ft} \; 27 \text{ in.} \end{array}$$
>
> Since 27 in. is the same as 2 ft 3 in., we simplify the product as
>
> 24 ft 27 in. = 24 ft + 2 ft 3 in.
> = 26 ft 3 in.
>
> ■ Work Practice 7

We divide in a similar manner as above.

Practice 8

A carpenter cuts 1 ft 9 in. from a board of length 5 ft 8 in. Find the remaining length of the board.

> **Example 8** Finding the Length of a Piece of Rope
>
> A rope of length 6 yd 1 ft has 2 yd 2 ft cut from one end. Find the length of the remaining rope.
>
> **Solution:** Subtract 2 yd 2 ft from 6 yd 1 ft.
>
> $$\begin{array}{rl} \text{beginning length} \rightarrow & 6 \text{ yd} \; 1 \text{ ft} \\ - \quad\;\; \text{amount cut} \rightarrow & -2 \text{ yd} \; 2 \text{ ft} \\ \hline \text{remaining length} \end{array}$$
>
> We cannot subtract 2 ft from 1 ft, so we borrow 1 yd from the 6 yd. One yard is converted to 3 ft and combined with the 1 ft already there.

Answers

6. 13 ft 7 in. **7.** 18 ft 4 in. **8.** 3 ft 11 in.

✓**Concept Check Answer**

round each to the nearest yard:
7 yd + 4 yd = 11 yd

Borrow 1 yd = 3 ft

5 yd + (1 yd)(3 ft)

$$
\begin{array}{rcl}
6\text{ yd } 1\text{ ft} & = & 5\text{ yd } 4\text{ ft} \\
-2\text{ yd } 2\text{ ft} & = & -2\text{ yd } 2\text{ ft} \\
\hline
& & 3\text{ yd } 2\text{ ft}
\end{array}
$$

The remaining rope is 3 yd 2 ft long.

▶ Work Practice 8

Objective D Defining and Converting Metric System Units of Length ▶

The basic unit of length in the metric system is the **meter.** A meter is slightly longer than a yard. It is approximately 39.37 inches long. Recall that a yard is 36 inches long.

1 yard = 36 inches

1 meter ≈ 39.37 inches

All units of length in the metric system are based on the meter. The following is a summary of the prefixes used in the metric system. Also shown are equivalencies between units of length. Like the decimal system, the metric system uses powers of 10 to define units.

Metric Units of Length
1 **kilo**meter (km) = 1000 meters (m)
1 **hecto**meter (hm) = 100 m
1 **deka**meter (dam) = 10 m
1 **meter** (m) = 1 m
1 **deci**meter (dm) = 1/10 m or 0.1 m
1 **centi**meter (cm) = 1/100 m or 0.01 m
1 **milli**meter (mm) = 1/1000 m or 0.001 m

The figure below will help you with decimeters, centimeters, and millimeters.

1 decimeter = $\frac{1}{10}$ meter   1 centimeter = $\frac{1}{100}$ meter   1 millimeter = $\frac{1}{1000}$ meter

Helpful Hint

Study the figure above for other equivalencies between metric units of length.

10 decimeters = 1 meter   10 millimeters = 1 centimeter
100 centimeters = 1 meter   10 centimeters = 1 decimeter
1000 millimeters = 1 meter

These same prefixes are used in the metric system for mass and capacity. The most commonly used measurements of length in the metric system are the **meter, millimeter, centimeter,** and **kilometer.**

✓**Concept Check** Is this statement reasonable? "The screen of a home television set has a 30-meter diagonal." Why or why not?

Being comfortable with the metric units of length means gaining a "feeling" for metric lengths, just as you have a "feeling" for the lengths of an inch, a foot, and a mile. To help you accomplish this, study the following examples:

A millimeter is about the thickness of a large paper clip.

A centimeter is about the width of a large paper clip.

A meter is slightly longer than a yard.

A kilometer is about two-thirds of a mile.

The width of this book is approximately 21.5 centimeters.

The distance between New York City and Philadelphia is about 160 kilometers.

$2\frac{1}{2}$ centimeters is about 1 inch.

1.7 meters

7 millimeters

19 centimeters

As with the U.S. system of measurement, unit fractions may be used to convert from one unit of length to another. For example, let's convert 1200 meters to kilometers. To do so, we will multiply by 1 in the form of the unit fraction

$$\frac{1 \text{ km}}{1000 \text{ m}} \quad \begin{array}{l} \leftarrow \text{Units to convert to} \\ \leftarrow \text{Original units} \end{array}$$

Unit fraction

$$1200 \text{ m} = \frac{1200 \text{ m}}{1} \cdot 1 = \frac{1200 \text{ m}}{1} \cdot \frac{1 \text{ km}}{1000 \text{ m}} = \frac{1200 \text{ km}}{1000} = 1.2 \text{ km}$$

The metric system does, however, have a distinct advantage over the U.S. system of measurement: the ease of converting from one unit of length to another. Since all units of length are powers of 10 of the meter, converting from one unit of length to another is as simple as moving the decimal point. Listing units of length in order from largest to smallest helps to keep track of how many places to move the decimal point when converting.

Let's again convert 1200 meters to kilometers. This time, to convert from meters to kilometers, we move along the chart shown, 3 units to the left, from meters to kilometers. This means that we move the decimal point 3 places to the left.

km	hm	dam	**m**	dm	cm	mm

3 units to the left

1200 m = 1.200 km	1000 m	200 m

3 places to the left |← 1 km →|← 0.2 km →|

✓**Concept Check Answer**

no; answers may vary

Thus, 1200 m = 1.2 km, as shown in the diagram.

Example 9 Convert 2.3 m to centimeters.

Solution: First we will convert by using a unit fraction.

$$2.3 \text{ m} = \frac{2.3 \text{ m}}{1} \cdot \overbrace{\frac{100 \text{ cm}}{1 \text{ m}}}^{\text{Unit fraction}} = 230 \text{ cm}$$

Now we will convert by listing the units of length in order from left to right and moving from meters to centimeters.

km   hm   dam   m   dm   cm   mm

2 units to the right

2.30 m = 230. cm

2 places to the right

With either method, we get 230 cm.

■ Work Practice 9

Practice 9
Convert 2.5 m to millimeters.

Example 10 Convert 450,000 mm to meters.

Solution: We list the units of length in order from left to right and move from millimeters to meters.

km   hm   dam   m   dm   cm   mm

3 units to the left

Thus, move the decimal point 3 places to the left.

450,000 mm = 450.000 m or 450 m

■ Work Practice 10

Practice 10
Convert 3500 m to kilometers.

✓**Concept Check** What is wrong with the following conversion of 150 cm to meters?

150.00 cm = 15,000 m

Objective E Performing Operations on Metric System Units of Length ▶

To add, subtract, multiply, or divide with metric measurements of length, we write all numbers using the same unit of length and then add, subtract, multiply, or divide as with decimals.

Example 11 Subtract 430 m from 1.3 km.

Solution: First we convert both measurements to kilometers or both to meters.

430 m = 0.43 km   or   1.3 km = 1300 m

       1.30 km   1300 m
      −0.43 km   −430 m
      ─────────   ─────────
       0.87 km   870 m

The difference is 0.87 km or 870 m.

■ Work Practice 11

Practice 11
Subtract 640 m from 2.1 km.

Answers
 9. 2500 mm   **10.** 3.5 km
11. 1.46 km or 1460 m

✓**Concept Check Answer**
decimal point should be moved two places to the left: 1.5 m

Practice 12

Multiply 18.3 hm by 5.

 Example 12   Multiply 5.7 mm by 4.

Solution: Here we simply multiply the two numbers. Note that the unit of measurement remains the same.

$$\begin{array}{r} 5.7 \text{ mm} \\ \times\ \ \ 4 \\ \hline 22.8 \text{ mm} \end{array}$$

■ Work Practice 12

Practice 13

A child was 55 centimeters at birth. Her adult height was 1.72 meters. Find how much she grew from birth to adult height.

Example 13   Finding a Person's Height

Fritz Martinson was 1.2 meters tall on his last birthday. Since then, he has grown 14 centimeters. Find his current height in meters.

Solution:

original height	→	1.20 m	
+ height grown	→	+ 0.14 m	(Since 14 cm = 0.14 m)
current height		1.34 m	

Fritz is now 1.34 meters tall.

■ Work Practice 13

Practice 14

Doris Blackwell is knitting a scarf that is currently 0.8 meter long. If she knits an additional 45 centimeters, how long will the scarf be?

Example 14   Finding a Crocodile's Length

A newly hatched Nile crocodile averages 26 centimeters in length. This type of crocodile normally grows 4.74 meters to reach its adult length. What is the adult length of this type of crocodile?

Solution:

original length	→	0.26 m	(since 26 cm = 0.26 m)
+ length grown	→	+ 4.74 m	
adult length		5.00 m	

The adult length is 5 meters.

■ Work Practice 14

Answers

12. 91.5 hm   **13.** 117 cm or 1.17 m   **14.** 125 cm or 1.25 m

Vocabulary, Readiness & Video Check

Use the choices below to fill in each blank. Some choices may be used more than once.

inches   yard   unit fraction
feet   meter

1. The basic unit of length in the metric system is the _____.

2. The expression $\dfrac{1 \text{ foot}}{12 \text{ inches}}$ is an example of a(n) _____.

3. A meter is slightly longer than a(n) _____.

4. One foot equals 12 _____.

5. One yard equals 3 _____.

6. One yard equals 36 _____.

7. One mile equals 5280 _____.

8. One foot equals $\dfrac{1}{3}$ _____.

Martin-Gay Interactive Videos   *Watch the section lecture video and answer the following questions.*

See Video 7.1

Objective A **9.** In ⊞ Example 3, what units are used in the denominator of the unit fraction and why was this decided? ▷

Objective B **10.** In ⊞ Example 4, how is a mixed unit similar to a mixed number? Use examples in your answer. ▷

Objective C **11.** In ⊞ Example 5, why is the sum of the addition problem not the final answer? Reference the sum in your answer. ▷

Objective D **12.** Based on the lecture before ⊞ Example 6, explain how to convert metric units to other metric units. ▷

Objective E **13.** What two answers did we get for ⊞ Example 8? Explain why both answers are correct. ▷

7.1 Exercise Set MyMathLab® ▷

Objective A *Convert each measurement as indicated. See Examples 1 through 3.*

▷ **1.** 60 in. to feet

2. 84 in. to feet

3. 12 yd to feet

4. 18 yd to feet

5. 42,240 ft to miles

6. 36,960 ft to miles

▷ **7.** $8\dfrac{1}{2}$ ft to inches

8. $12\dfrac{1}{2}$ ft to inches

▷ **9.** 10 ft to yards

10. 25 ft to yards

11. 6.4 mi to feet

12. 3.8 mi to feet

13. 162 in. to yd (Write answer as a decimal.)

14. 7216 yd to mi (Write answer as a decimal.)

15. 3 in. to ft (Write answer as a decimal.)

16. 129 in. to ft (Write answer as a decimal.)

Objective **B** *Convert each measurement as indicated. See Examples 4 and 5.*

17. 40 ft = _____ yd _____ ft

18. 100 ft = _____ yd _____ ft

19. 85 in. = _____ ft _____ in.

20. 47 in. = _____ ft _____ in.

21. 10,000 ft = _____ mi _____ ft

22. 25,000 ft = _____ mi _____ ft

▶ **23.** 5 ft 2 in. = _____ in.

24. 4 ft 11 in. = _____ in.

25. 8 yd 2 ft = _____ ft

26. 4 yd 1 ft = _____ ft

27. 2 yd 1 ft = _____ in.

28. 1 yd 2 ft = _____ in.

Objective **C** *Perform each indicated operation. Simplify the result if possible. See Examples 6 through 8.*

29. 3 ft 10 in. + 7 ft 4 in.

30. 12 ft 7 in. + 9 ft 11 in.

▶ **31.** 12 yd 2 ft + 9 yd 2 ft

32. 16 yd 2 ft + 8 yd 2 ft

33. 22 ft 8 in. − 16 ft 3 in.

34. 15 ft 5 in. − 8 ft 2 in.

35. 18 ft 3 in. − 10 ft 9 in.

36. 14 ft 8 in. − 3 ft 11 in.

37. 28 ft 8 in. ÷ 2

38. 34 ft 6 in. ÷ 2

39. 16 yd 2 ft × 5

40. 15 yd 1 ft × 8

Objective **D** *Convert as indicated. See Examples 9 and 10.*

41. 60 m to centimeters

42. 46 m to centimeters

43. 40 mm to centimeters

44. 14 mm to centimeters

45. 500 m to kilometers

46. 400 m to kilometers

47. 1700 mm to meters

48. 6400 mm to meters

▶ **49.** 1500 cm to meters

50. 6400 cm to meters

51. 0.42 km to centimeters

52. 0.95 km to centimeters

53. 7 km to meters

54. 5 km to meters

55. 8.3 cm to millimeters

56. 4.6 cm to millimeters

57. 20.1 mm to decimeters

58. 140.2 mm to decimeters

▶ **59.** 0.04 m to millimeters

60. 0.2 m to millimeters

Objective E *Perform each indicated operation. Remember to insert units when writing your answers. See Examples 11 through 14.*

61. 8.6 m + 0.34 m

62. 14.1 cm + 3.96 cm

63. 2.9 m + 40 mm

64. 30 cm + 8.9 m

▶ **65.** 24.8 mm − 1.19 cm

66. 45.3 m − 2.16 dam

67. 15 km − 2360 m

68. 14 cm − 15 mm

69. 18.3 m × 3

70. 14.1 m × 4

71. 6.2 km ÷ 4

72. 9.6 m ÷ 5

Objectives A C D E **Mixed Practice** *Solve. Remember to insert units when writing your answers. For Exercises 73 through 82, complete the charts. See Examples 1 through 14.*

		Yards	Feet	Inches
73.	Chrysler Building in New York City		1046	
74.	4-story building			792
75.	Python length		35	
76.	Ostrich height			108

Complete the chart.

		Meters	Millimeters	Kilometers	Centimeters
77.	Length of elephant	5			
78.	Height of grizzly bear	3			
79.	Tennis ball diameter				6.5
80.	Golf ball diameter				4.6
81.	Distance from London to Paris			342	
82.	Distance from Houston to Dallas			396	

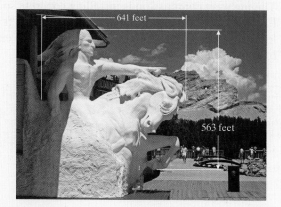

641 feet

563 feet

83. The total width of the Crazy Horse carving is 641 feet. Convert this width to
a. yards
b. inches

84. The total height of the Crazy Horse carving is 563 feet. Convert this height to
a. yards
b. inches

85. The National Zoo maintains a small patch of bamboo, which it grows as a food supply for its pandas. Two weeks ago, the bamboo was 6 ft 10 in. tall. Since then, the bamboo has grown 3 ft 8 in. How tall is the bamboo now?

86. While exploring in the Marianas Trench, a submarine probe was lowered to a point 1 mile 1400 feet below the ocean's surface. Later it was lowered an additional 1 mile 4000 feet below this point. How far was the probe below the surface of the Pacific?

87. At its deepest point, the Grand Canyon of the Colorado River in Arizona is about 6000 ft. The Grand Canyon of the Yellowstone River, which is in Yellowstone National Park in Wyoming, is at most 900 feet deep. How much deeper is the Grand Canyon of the Colorado River than the Grand Canyon of the Yellowstone River? (*Source:* National Park Service)

88. The Black Canyon of the Gunnison River is only 1150 ft wide at its narrowest point. At its narrowest, the Grand Canyon of the Yellowstone is $\frac{1}{2}$ mile wide.

Find the difference in width between the Grand Canyon of the Yellowstone and the Black Canyon of the Gunnison. (*Note:* Notice that the dimensions are different.) (*Source:* National Park Service)

89. The tallest man in the world is recorded as Robert Pershing Wadlow of Alton, Illinois. Born in 1918, he measured 8 ft 11 in. at his tallest. The shortest man in the world is Chandra Bahadur Dangi of Nepal, who measures 21.5 in. How many times taller than Chandra was Robert? Round to one decimal place. (*Source: Guinness World Records*)

90. A 3.4-m rope is attached to a 5.8-m rope. However, when the ropes are tied, 8 cm of length is lost to form the knot. What is the length of the tied ropes?

91. The length of one of the Statue of Liberty's hands is 16 ft 5 in. One of the statue's eyes is 2 ft 6 in. across. How much longer is a hand than the width of an eye? (*Source:* National Park Service)

92. The width of the Statue of Liberty's head from ear to ear is 10 ft. The height of the statue's head from chin to cranium is 17 ft 3 in. How much taller is the statue's head than its width? (*Source:* National Park Service)

93. The ice on a pond is 5.33 cm thick. For safe skating, the owner of the pond insists that it must be 80 mm thick. How much thicker must the ice be before skating is allowed?

94. The sediment on the bottom of the Towamencin Creek is normally 14 cm thick, but the recent flood washed away 22 mm of sediment. How thick is it now?

95. The Amana Corporation stacks up its microwave ovens in a distribution warehouse. Each stack is 1 ft 9 in. wide. How far from the wall would 9 of these stacks extend?

96. The highway commission is installing concrete sound barriers along a highway. Each barrier is 1 yd 2 ft long. Find the total length of 25 barriers placed end to end.

1 ft 9 in.

1 yd 2ft

97. A carpenter needs to cut a board into thirds. If the board is 9 ft 3 in. long originally, how long will each cut piece be?

98. A wall is erected exactly halfway between two buildings that are 192 ft 8 in. apart. If the wall is 8 in. wide, how far is it from the wall to either of the buildings?

99. An art class is learning how to make kites. The two sticks used for each kite have lengths of 1 m and 65 cm. What total length of wood must be ordered for the sticks if 25 kites are to be built?

100. The total pages of a hardbound economics text are 3.1 cm thick. The front and back covers are each 2 mm thick. How high would a stack of 10 of these texts be?

101. A logging firm needs to cut a 67-m-long redwood log into 20 equal pieces before loading it onto a truck for shipment. How long will each piece be?

102. An 18.3-m-tall flagpole is mounted on a 65-cm-high pedestal. How far is the top of the flagpole from the ground?

103. A 2.15-m-long sash cord has become frayed at both ends so that 1 cm is trimmed from each end. How long is the remaining cord?

104. A 112.5-foot-tall dead pine tree is removed by starting at the top and cutting off 9-foot-long sections. How many 9-foot sections are removed?

105. The longest truck in the world is operated by Gould Transport in Australia, and is the 182-ft Road Train. How many *yards* long are 2 of these trucks? (*Source: Guinness World Records*)

106. Three hundred fifty thousand people daily see the large Coca-Cola sign in the Tokyo Ginza shopping district. It is in the shape of a rectangle whose length is 31 yards and whose width is 49 feet. Find the area of the sign in square feet. (*Source:* Coca-Cola Company) (*Hint:* Recall that the area of a rectangle is the product length times width.)

107. A floor tile is 22.86 cm wide. How many tiles in a row are needed to cross a room 3.429 m wide?

108. A standard postcard is 1.6 times longer than it is wide. If it is 9.9 cm wide, what is its length?

Review

Write each decimal as a fraction and each fraction as a decimal. See Section 4.6.

109. 0.21   **110.** 0.86   **111.** $\dfrac{13}{100}$ **112.** $\dfrac{47}{100}$ **113.** $\dfrac{1}{4}$ **114.** $\dfrac{3}{20}$

Concept Extensions

Determine whether the measurement in each statement is reasonable.

115. The width of a twin-size bed is 20 meters.

116. A window measures 1 meter by 0.5 meter.

117. A drinking glass is made of glass 2 millimeters thick.

118. A paper clip is 4 kilometers long.

119. The distance across the Colorado River is 50 kilometers.

120. A model's hair is 30 centimeters long.

Estimate each sum or difference. See the first Concept Check in this section.

121.   5 yd   2 in.
     + 7 yd 30 in.
     ⎯⎯⎯⎯⎯

122.   45 ft   1 in.
     − 10 ft 11 in.
     ⎯⎯⎯⎯⎯

123. Using a unit other than the foot, write a length that is equivalent to 4 feet. (*Hint:* There are many possibilities.)

124. Using a unit other than the meter, write a length that is equivalent to 7 meters. (*Hint:* There are many possibilities.)

125. To convert from meters to centimeters, the decimal point is moved two places to the right. Explain how this relates to the fact that the prefix *centi* means $\dfrac{1}{100}$.

126. Explain why conversions in the metric system are easier to make than conversions in the U.S. system of measurement.

127. An advertisement sign outside Fenway Park in Boston measures 18.3 m by 18.3 m. What is the area of this sign?

7.2   # Weight and Mass: U.S. and Metric Systems of Measurement

Objectives

A Define U.S. Units of Weight and Convert from One Unit to Another.

B Perform Arithmetic Operations on U.S. Units of Weight.

C Define Metric Units of Mass and Convert from One Unit to Another.

D Perform Arithmetic Operations on Metric Units of Mass.

Objective A   Defining and Converting U.S. System Units of Weight

Whenever we talk about how heavy an object is, we are concerned with the object's **weight.** We discuss weight when we refer to a 12-ounce box of Rice Krispies, a 15-pound tabby cat, or a barge hauling 24 tons of garbage.

12 ounces

15 pounds

24 tons of garbage

The most common units of weight in the U.S. measurement system are the **ounce,** the **pound,** and the **ton.** The following is a summary of equivalencies between units of weight:

U.S. Units of Weight	Unit Fractions
16 ounces (oz) = 1 pound (lb)	$\dfrac{16 \text{ oz}}{1 \text{ lb}} = \dfrac{1 \text{ lb}}{16 \text{ oz}} = 1$
2000 pounds = 1 ton	$\dfrac{2000 \text{ lb}}{1 \text{ ton}} = \dfrac{1 \text{ ton}}{2000 \text{ lb}} = 1$

✓**Concept Check** If you were describing the weight of a fully loaded semitrailer, which type of unit would you use: ounce, pound, or ton? Why?

Unit fractions that equal 1 are used to convert between units of weight in the U.S. system. When converting using unit fractions, recall that the numerator of a unit fraction should contain the units we are converting to and the denominator should contain the original units.

Example 1 Convert 9000 pounds to tons.

Solution: We multiply 9000 lb by a unit fraction that uses the equality

2000 pounds = 1 ton.

Remember, the unit fraction should be $\dfrac{\text{units to convert to}}{\text{original units}}$ or $\dfrac{1 \text{ ton}}{2000 \text{ lb}}$.

$$9000 \text{ lb} = \frac{9000 \text{ lb}}{1} \cdot 1 = \frac{9000 \text{ lb}}{1} \cdot \frac{1 \text{ ton}}{2000 \text{ lb}} = \frac{9000 \text{ tons}}{2000} = \frac{9}{2} \text{ tons or } 4\frac{1}{2} \text{ tons}$$

2000 lb   2000 lb   2000 lb   2000 lb   1000 lb

1 ton   1 ton   1 ton   1 ton $\frac{1}{2}$ ton

$9000 \text{ lb} = 4\frac{1}{2} \text{ tons}$

 Work Practice 1

Example 2 Convert 3 pounds to ounces.

Solution: We multiply by the unit fraction $\dfrac{16 \text{ oz}}{1 \text{ lb}}$ to convert from pounds to ounces.

$$3 \text{ lb} = \frac{3 \text{ lb}}{1} \cdot 1 = \frac{3 \text{ lb}}{1} \cdot \frac{16 \text{ oz}}{1 \text{ lb}} = 3 \cdot 16 \text{ oz} = 48 \text{ oz}$$

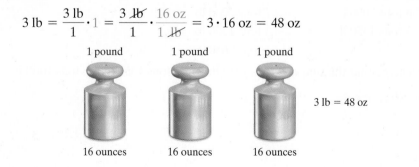

1 pound   1 pound   1 pound

$3 \text{ lb} = 48 \text{ oz}$

16 ounces   16 ounces   16 ounces

 Work Practice 2

Practice 1

Convert 6500 pounds to tons.

Practice 2

Convert 72 ounces to pounds.

Answers

1. $3\frac{1}{4}$ tons   **2.** $4\frac{1}{2}$ lb

✓**Concept Check Answer**

ton

As with length, it is sometimes useful to simplify a measurement of weight by writing it in terms of mixed units.

Practice 3

Convert:

47 ounces = _____ lb_____ oz

Example 3   Convert: 33 ounces = _____ lb_____ oz

Solution: Because 16 oz = 1 lb, divide 16 into 33 to see how many pounds are in 33 ounces. The quotient is the number of pounds, and the remainder is the number of ounces. To see why we divide 16 into 33, notice that

$$33 \text{ oz} = 33 \, \cancel{oz} \cdot \frac{1 \text{ lb}}{16 \, \cancel{oz}} = \frac{33}{16} \text{ lb}$$

$$\begin{array}{r} 2 \text{ lb } 1 \text{ oz} \\ 16\overline{)33} \\ -32 \\ \hline 1 \end{array}$$

Thus, 33 ounces is the same as 2 lb 1 oz.

16 ounces   16 ounces   1 ounce

33 oz = 2 lb 1 oz

1 pound   1 pound   1 ounce

■ Work Practice 3

Objective B Performing Operations on U.S. System Units of Weight ▶

Performing arithmetic operations on units of weight works the same way as performing arithmetic operations on units of length.

Practice 4

Subtract 5 tons 1200 lb from 8 tons 100 lb.

Example 4   Subtract 3 tons 1350 lb from 8 tons 1000 lb.

Solution: To subtract, we line up similar units.

$$\begin{array}{r} 8 \text{ tons } 1000 \text{ lb} \\ -3 \text{ tons } 1350 \text{ lb} \end{array}$$

Since we cannot subtract 1350 lb from 1000 lb, we borrow 1 ton from the 8 tons. To do so, we write 1 ton as 2000 lb and combine it with the 1000 lb.

7 tons + (1 ton) 2000 lb

$$\begin{array}{rcl} 8 \text{ tons } 1000 \text{ lb} & = & 7 \text{ tons } 3000 \text{ lb} \\ -3 \text{ tons } 1350 \text{ lb} & = & -3 \text{ tons } 1350 \text{ lb} \\ \hline & & 4 \text{ tons } 1650 \text{ lb} \end{array}$$

To check, see that the sum of 4 tons 1650 lb and 3 tons 1350 lb is 8 tons 1000 lb.

■ Work Practice 4

Answers
3. 2 lb 15 oz   **4.** 2 tons 900 lb

Example 5 Divide 9 lb 6 oz by 2.

Solution: We divide each of the units by 2.

$$
\begin{array}{r}
4\ \text{lb}\quad 11\ \text{oz}\\
2\overline{)\ 9\ \text{lb}\quad\ 6\ \text{oz}}\\
\underline{-8}\\
1\ \text{lb} = 16\ \text{oz}\\
\overline{22\ \text{oz}}\quad
\end{array}
$$

Divide 2 into 22 oz to get 11 oz.

To check, multiply 4 pounds 11 ounces by 2. The result is 9 pounds 6 ounces.

■ Work Practice 5

Example 6 Finding the Weight of a Child

Bryan weighed 8 lb 8 oz at birth. By the time he was 1 year old, he had gained 11 lb 14 oz. Find his weight at age 1 year.

Solution:

$$
\begin{array}{lll}
\text{birth weight} & \rightarrow & 8\ \text{lb}\ \ 8\ \text{oz}\\
\underline{+\ \text{weight gained}} & \rightarrow & \underline{+\ 11\ \text{lb}\ 14\ \text{oz}}\\
\text{total weight} & \rightarrow & 19\ \text{lb}\ 22\ \text{oz}
\end{array}
$$

Since 22 oz equals 1 lb 6 oz,

19 lb 22 oz = 19 lb + 1 lb 6 oz

= 20 lb 6 oz

Bryan weighed 20 lb 6 oz on his first birthday.

■ Work Practice 6

Objective C Defining and Converting Metric System Units of Mass

In scientific and technical areas, a careful distinction is made between **weight** and **mass. Weight** is really a measure of the pull of gravity. The farther from Earth an object gets, the less it weighs. However, **mass** is a measure of the amount of substance in the object and does not change. Astronauts orbiting Earth weigh much less than they weigh on Earth, but they have the same mass in orbit as they do on Earth. Here on Earth, weight and mass are the same, so either term may be used.

The basic unit of mass in the metric system is the **gram.** It is defined as the mass of water contained in a cube 1 centimeter (cm) on each side.

The following examples may help you get a feeling for metric masses:

A tablet contains 200 milligrams of ibuprofen.

A large paper clip weighs approximately 1 gram.

Practice 5
Divide 5 lb 8 oz by 4.

Practice 6
A 5-lb 14-oz batch of cookies is packed into a 6-oz container before it is mailed. Find the total weight.

Answers
5. 1 lb 6 oz **6.** 6 lb 4 oz

A box of crackers weighs 453 grams.

A kilogram is slightly over 2 pounds. An adult woman may weigh 60 kilograms.

The prefixes for units of mass in the metric system are the same as for units of length, as shown in the following table:

Metric Units of Mass
1 **kilo**gram (kg) = 1000 grams (g)
1 **hecto**gram (hg) = 100 g
1 **deka**gram (dag) = 10 g
1 gram (g) = 1 g
1 **deci**gram (dg) = 1/10 g or 0.1 g
1 **centi**gram (cg) = 1/100 g or 0.01 g
1 **milli**gram (mg) = 1/1000 g or 0.001 g

✓**Concept Check** True or false? A decigram is larger than a dekagram. Explain.

The **milligram,** the **gram,** and the **kilogram** are the three most commonly used units of mass in the metric system.

As with lengths, all units of mass are powers of 10 of the gram, so converting from one unit of mass to another only involves moving the decimal point. To convert from one unit of mass to another in the metric system, list the units of mass in order from largest to smallest.

Let's convert 4300 milligrams to grams. To convert from milligrams to grams, we move along the list 3 units to the left.

kg   hg   dag   **g**   dg   cg   **mg**

3 units to the left

This means that we move the decimal point 3 places to the left to convert from milligrams to grams.

4300 mg = 4.3 g

Don't forget, the same conversion can be done with unit fractions.

$$4300 \text{ mg} = \frac{4300 \text{ mg}}{1} \cdot 1 = \frac{4300 \text{ mg}}{1} \cdot \frac{0.001 \text{ g}}{1 \text{ mg}}$$

$$= 4300 \cdot 0.001 \text{ g}$$

$$= 4.3 \text{ g} \quad \text{To multiply by 0.001, move the decimal point 3 places to the left.}$$

To see that this is reasonable, study the diagram:

1000 mg	1000 mg	1000 mg	1000 mg	300 mg
1 g	1 g	1 g	1 g	0.3 g

4300 mg = 4.3 g

✓**Concept Check Answer**

false

Thus, 4300 mg = 4.3 g

Example 7 Convert 3.2 kg to grams.

Solution: First we convert by using a unit fraction.

$$3.2 \text{ kg} = 3.2 \text{ kg} \cdot 1 = 3.2 \, \overline{kg} \cdot \frac{\overbrace{1000 \text{ g}}^{\text{Unit fraction}}}{1 \, \overline{kg}} = 3200 \text{ g}$$

Now let's list the units of mass in order from left to right and move from kilograms to grams.

kg   hg   dag   g   dg   cg   mg

3 units to the right

3.200 kg = 3200. g

3 places to the right

1 kg   1 kg   1 kg   0.2 kg

3.2 kg = 3200 g

1000 g   1000 g   1000 g   200 g

■ **Work Practice 7**

Example 8 Convert 2.35 cg to grams.

Solution: We list the units of mass in a chart and move from centigrams to grams.

kg   hg   dag   g   dg   cg   mg

2 units to the left

02.35 cg = 0.0235 g

2 places to the left

■ **Work Practice 8**

Objective D Performing Operations on Metric System Units of Mass

Arithmetic operations can be performed with metric units of mass just as we performed operations with metric units of length. We convert each number to the same unit of mass and add, subtract, multiply, or divide as with decimals.

Example 9 Subtract 5.4 dg from 1.6 g.

Solution: We convert both numbers to decigrams or to grams before subtracting.

5.4 dg = 0.54 g   or   1.6 g = 16 dg

$$\begin{array}{r} 1.60 \text{ g} \\ - \, 0.54 \text{ g} \\ \hline 1.06 \text{ g} \end{array} \qquad \begin{array}{r} 16.0 \text{ dg} \\ - \, 5.4 \text{ dg} \\ \hline 10.6 \text{ dg} \end{array}$$

The difference is 1.06 g or 10.6 dg.

■ **Work Practice 9**

Practice 7

Convert 3.41 g to milligrams.

Practice 8

Convert 56.2 cg to grams.

Practice 9

Subtract 3.1 dg from 2.5 g.

Answers

7. 3410 mg   **8.** 0.562 g

9. 2.19 g or 21.9 dg

Practice 10

Twenty-four bags of cement weigh a total of 550 kg. Find the average weight of 1 bag, rounded to the nearest kilogram.

Example 10 Calculating Allowable Weight in an Elevator

An elevator has a weight limit of 1400 kg. A sign posted in the elevator indicates that the maximum capacity of the elevator is 17 persons. What is the average allowable weight for each passenger, rounded to the nearest kilogram?

Solution: To solve, notice that the total weight of 1400 kilograms ÷ 17 = average weight.

$$
\begin{array}{r}
82.3 \text{ kg} \approx 82 \text{ kg} \\
17\overline{)\ 1400.0 \text{ kg}} \\
-136 \\
\hline
40 \\
-34 \\
\hline
6\,0 \\
-5\,1 \\
\hline
9
\end{array}
$$

Each passenger can weigh an average of 82 kg. (Recall that a kilogram is slightly over 2 pounds, so 82 kilograms is over 164 pounds.)

■ Work Practice 10

Answer

10. 23 kg

Vocabulary, Readiness & Video Check

Use the choices below to fill in each blank.

mass   weight   gram

1. _____ is a measure of the amount of substance in an object. This measure does not change.

2. _____ is the measure of the pull of gravity.

3. The basic unit of mass in the metric system is the _____ .

Fill in these blanks with the correct number. Choices for these blanks are not shown in the list of terms above.

4. One pound equals _____ ounces.

5. One ton equals _____ pounds.

Convert without pencil or paper.

6. 3 tons to pounds   7. 32 ounces to pounds   8. 3 pounds to ounces   9. 4000 pounds to tons   10. 1 ton to pounds

Martin-Gay Interactive Videos

Watch the section lecture video and answer the following questions.

See Video 7.2

Objective A 11. In ▦ Example 2, what units are used in the numerator of the unit fraction and why was this decided? ▶

Objective B 12. In ▦ Example 4, explain the first step taken to solve the problem. ▶

Objective C 13. In ▦ Example 5, how many places is the decimal point moved and in what direction? What is the final conversion? ▶

Objective D 14. What is the answer to ▦ Example 7 in decigrams? ▶

7.2 Exercise Set MyMathLab ▶

Objective A *Convert as indicated. See Examples 1 through 3.*

1. 2 pounds to ounces

2. 5 pounds to ounces

3. 5 tons to pounds

4. 7 tons to pounds

5. 18,000 pounds to tons

6. 28,000 pounds to tons

▶ **7.** 60 ounces to pounds

8. 90 ounces to pounds

9. 3500 pounds to tons

10. 11,000 pounds to tons

11. 12.75 pounds to ounces

12. 9.5 pounds to ounces

▶ **13.** 4.9 tons to pounds

14. 8.3 tons to pounds

15. $4\frac{3}{4}$ pounds to ounces

16. $9\frac{1}{8}$ pounds to ounces

17. 2950 pounds to the nearest tenth of a ton

18. 51 ounces to the nearest tenth of a pound

19. $\frac{4}{5}$ oz to pounds

20. $\frac{1}{4}$ oz to pounds

21. $5\frac{3}{4}$ lb to ounces

22. $2\frac{1}{4}$ lb to ounces

23. 10 lb 1 oz to ounces

24. 7 lb 6 oz to ounces

▶ **25.** 89 oz = _____ lb _____ oz

26. 100 oz = _____ lb _____ oz

Objective B *Perform each indicated operation. See Examples 4 through 6.*

27. 34 lb 12 oz + 18 lb 14 oz

28. 6 lb 10 oz + 10 lb 8 oz

29. 3 tons 1820 lb + 4 tons 930 lb

30. 1 ton 1140 lb + 5 tons 1200 lb

31. 5 tons 1050 lb − 2 tons 875 lb

32. 4 tons 850 lb − 1 ton 260 lb

▶ **33.** 12 lb 4 oz − 3 lb 9 oz

34. 45 lb 6 oz − 26 lb 10 oz

35. 5 lb 3 oz × 6

36. 2 lb 5 oz × 5

37. 6 tons 1500 lb ÷ 5

38. 5 tons 400 lb ÷ 4

Objective C Convert as indicated. See Examples 7 and 8.

39. 500 g to kilograms   **40.** 820 g to kilograms   ▶ **41.** 4 g to milligrams   **42.** 9 g to milligrams

43. 25 kg to grams   **44.** 18 kg to grams   **45.** 48 mg to grams   **46.** 112 mg to grams

▶ **47.** 6.3 g to kilograms   **48.** 4.9 g to kilograms   **49.** 15.14 g to milligrams   **50.** 16.23 g to milligrams

51. 6.25 kg to grams   **52.** 3.16 kg to grams   **53.** 35 hg to centigrams   **54.** 4.26 cg to dekagrams

Objective D Perform each indicated operation. Remember to insert units when writing your answers. See Examples 9 and 10.

55. 3.8 mg + 9.7 mg   **56.** 41.6 g + 9.8 g   **57.** 205 mg + 5.61 g   **58.** 2.1 g + 153 mg

▶ **59.** 9 g − 7150 mg   **60.** 6.13 g − 418 mg   **61.** 1.61 kg − 250 g   **62.** 4 kg − 2410 g

63. 5.2 kg × 2.6   **64.** 4.8 kg × 9.3   **65.** 17 kg ÷ 8   **66.** 8.25 g ÷ 6

Objectives A B C D Mixed Practice Solve. Remember to insert units when writing your answers. For Exercises 67 through 74, complete the chart. See Examples 1 through 10.

	Object	Tons	Pounds	Ounces
67.	Statue of Liberty–weight of copper sheeting	100		
68.	Statue of Liberty–weight of steel	125		
69.	A 12-inch cube of osmium (heaviest metal)		1345	
70.	A 12-inch cube of lithium (lightest metal)		32	

	Object	Grams	Kilograms	Milligrams	Centigrams
71.	Capsule of amoxicillin (antibiotic)			500	
72.	Tablet of Topamax (epilepsy and migraine uses)			25	
73.	A six-year-old boy		21		
74.	A golf ball	45			

75. A can of 7-Up weighs 336 grams. Find the weight in kilograms of 24 cans.

76. Guy Green normally weighs 73 kg, but he lost 2800 grams after being sick with the flu. Find Guy's new weight.

77. Sudafed is a decongestant that comes in two strengths. Regular strength contains 60 mg of medication. Extra strength contains 0.09 g of medication. How much extra medication is in the extra-strength tablet?

78. A small can of Planters sunflower seeds weighs 177 g. If each can contains 6 servings, find the weight of one serving.

79. Doris Johnson has two open containers of Uncle Ben's rice. If she combines 1 lb 10 oz from one container with 3 lb 14 oz from the other container, how much total rice does she have?

80. Dru Mizel maintains the records of the amount of coal delivered to his department in the steel mill. In January, 3 tons 1500 lb were delivered. In February, 2 tons 1200 lb were delivered. Find the total amount delivered in these two months.

81. Elizabeth Hamtini was amazed when she grew a 28-lb 10-oz zucchini in her garden, but later she learned that the heaviest zucchini ever grown weighed 64 lb 8 oz in Llanharry, Wales, by B. Lavery in 1990. How far below the record weight was Carla's zucchini? (*Source: Guinness World Records*)

82. The heaviest baby born in good health weighed an incredible 22 lb 8 oz. He was born in Italy in September 1955. How much heavier is this than a 7-lb 12-oz baby? (*Source: Guinness World Records*)

83. The smallest baby born in good health weighed only 8.6 ounces, less than a can of soda. She was born in Chicago in December 2004. How much lighter was she than an average baby, who weighs about 7 lb 8 ounces?

84. A large bottle of Hire's Root Beer weighs 1900 grams. If a carton contains 6 large bottles of root beer, find the weight in kilograms of 5 cartons.

85. Three milligrams of preservatives are added to a 0.5-kg box of dried fruit. How many milligrams of preservatives are in 3 cartons of dried fruit if each carton contains 16 boxes?

86. One box of Swiss Miss Cocoa Mix weighs 0.385 kg, but 39 grams of this weight is the packaging. Find the actual weight of the cocoa in 8 boxes.

87. A carton of 12 boxes of Quaker Oats Oatmeal weighs 6.432 kg. Each box includes 26 grams of packaging material. What is the actual weight of the oatmeal in the carton?

88. The supermarket prepares hamburger in 85-gram market packages. When Leo Gonzalas gets home, he divides the package in half before refrigerating the meat. How much will each package weigh?

89. The Shop 'n Bag supermarket chain ships hamburger meat by placing 10 packages of hamburger in a box, with each package weighing 3 lb 4 oz. How much will 4 boxes of hamburger weigh?

90. The Quaker Oats Company ships its 1-lb 2-oz boxes of oatmeal in cartons containing 12 boxes of oatmeal. How much will 3 such cartons weigh?

91. A carton of Del Monte Pineapple weighs 55 lb 4 oz, but 2 lb 8 oz of this weight is due to packaging. Find the actual weight of the pineapple in 4 cartons.

92. The Hormel Corporation ships cartons of canned ham weighing 43 lb 2 oz each. Of this weight, 3 lb 4 oz is due to packaging. Find the actual weight of the ham found in 3 cartons.

Review

Write each fraction as a decimal. See Section 4.6.

93. $\dfrac{4}{25}$ **94.** $\dfrac{3}{5}$ **95.** $\dfrac{7}{8}$ **96.** $\dfrac{3}{16}$

Concept Extensions

Determine whether the measurement in each statement is reasonable.

97. The doctor prescribed a pill containing 2 kg of medication.

98. A full-grown cat weighs approximately 15 g.

99. A bag of flour weighs 4.5 kg.

100. A staple weighs 15 mg.

101. A professor weighs less than 150 g.

102. A car weighs 2000 mg.

103. Use a unit other than centigram and write a mass that is equivalent to 25 centigrams. (*Hint:* There are many possibilities.)

104. Use a unit other than pound and write a weight that is equivalent to 4000 pounds. (*Hint:* There are many possibilities.)

True or false? See the second Concept Check in this section.

105. A kilogram is larger than a gram.

106. A decigram is larger than a milligram.

107. Why is the decimal point moved to the right when grams are converted to milligrams?

108. To change 8 pounds to ounces, multiply by 16. Why is this the correct procedure?

7.3 **Capacity: U.S. and Metric Systems of Measurement** ▶

Objectives

A Define U.S. Units of Capacity and Convert from One Unit to Another. ▶

B Perform Arithmetic Operations on U.S. Units of Capacity. ▶

C Define Metric Units of Capacity and Convert from One Unit to Another. ▶

D Perform Arithmetic Operations on Metric Units of Capacity. ▶

Objective A Defining and Converting U.S. System Units of Capacity ▶

Units of **capacity** are generally used to measure liquids. The number of gallons of gasoline needed to fill a gas tank in a car, the number of cups of water needed in a bread recipe, and the number of quarts of milk sold each day at a supermarket are all examples of using units of capacity. The following summary shows equivalencies between units of capacity:

U.S. Units of Capacity

8 fluid ounces (fl oz) = 1 cup (c)

2 cups = 1 pint (pt)

2 pints = 1 quart (qt)

4 quarts = 1 gallon (gal)

Just as with units of length and weight, we can form unit fractions to convert between different units of capacity. For instance,

$$\frac{2 \text{ c}}{1 \text{ pt}} = \frac{1 \text{ pt}}{2 \text{ c}} = 1 \quad \text{and} \quad \frac{2 \text{ pt}}{1 \text{ qt}} = \frac{1 \text{ qt}}{2 \text{ pt}} = 1$$

Example 1 Convert 9 quarts to gallons.

Solution: We multiply by the unit fraction $\frac{1 \text{ gal}}{4 \text{ qt}}$.

$$9 \text{ qt} = \frac{9 \text{ qt}}{1} \cdot 1$$

$$= \frac{9 \text{ qt}}{1} \cdot \frac{1 \text{ gal}}{4 \text{ qt}}$$

$$= \frac{9 \text{ gal}}{4}$$

$$= 2\frac{1}{4} \text{ gal}$$

Thus, 9 quarts is the same as $2\frac{1}{4}$ gallons, as shown in the diagram:

9 quarts
$= 2\frac{1}{4}$ gal

1 gallon   1 gallon $\frac{1}{4}$ gallon

■ Work Practice 1

Practice 1

Convert 43 pints to quarts.

Example 2 Convert 14 cups to quarts.

Solution: Our equivalency table contains no direct conversion from cups to quarts. However, from this table we know that

$$1 \text{ qt} = 2 \text{ pt} = \frac{2 \text{ pt}}{1} \cdot 1 = \frac{2 \text{ pt}}{1} \cdot \frac{2 \text{ c}}{1 \text{ pt}} = 4 \text{ c}$$

so 1 qt = 4 c. Now we have the unit fraction $\frac{1 \text{ qt}}{4 \text{ c}}$. Thus,

$$14 \text{ c} = \frac{14 \text{ c}}{1} \cdot 1 = \frac{14 \text{ c}}{1} \cdot \frac{1 \text{ qt}}{4 \text{ c}} = \frac{14 \text{ qt}}{4} = \frac{7}{2} \text{ qt} \quad \text{or} \quad 3\frac{1}{2} \text{ qt}$$

14 cups
$= 3\frac{1}{2}$ qt

1 quart   1 quart   1 quart $\frac{1}{2}$ quart

■ Work Practice 2

Practice 2

Convert 26 quarts to cups.

✓**Concept Check** If 50 cups is converted to quarts, will the equivalent number of quarts be less than or greater than 50? Explain.

✓**Concept Check Answer**

less than 50

Objective B Performing Operations on U.S. System Units of Capacity ▶

As is true of units of length and weight, units of capacity can be added, subtracted, multiplied, and divided.

Practice 3

Subtract 2 qt from 1 gal 1 qt.

Example 3 Subtract 3 qt from 4 gal 2 qt.

Solution: To subtract, we line up similar units.

$$\begin{array}{r} 4 \text{ gal } 2 \text{ qt} \\ - \qquad 3 \text{ qt} \\ \hline \end{array}$$

We cannot subtract 3 qt from 2 qt. We need to borrow 1 gallon from the 4 gallons, convert it to 4 quarts, and then combine it with the 2 quarts.

$$\begin{array}{rclcr} 4 \text{ gal } 2 \text{ qt} & = & & 3 \text{ gal } 6 \text{ qt} \\ - \qquad 3 \text{ qt} & = & - & \qquad 3 \text{ qt} \\ \hline & & & 3 \text{ gal } 3 \text{ qt} \end{array}$$

To check, see that the sum of 3 gal 3 qt and 3 qt is 4 gal 2 qt.

■ Work Practice 3

Practice 4

Divide 6 gal 3 qt by 2.

Example 4 Divide 3 gal 2 qt by 2.

Solution: We divide each unit of capacity by 2.

$$\begin{array}{r} 1 \text{ gal} \qquad 3 \text{ qt} \\ 2\overline{)3 \text{ gal} \qquad 2 \text{ qt}} \\ \underline{-2} \qquad\qquad\qquad \\ 1 \text{ gal} = \underline{4 \text{ qt}} \\ 6 \text{ qt} \end{array}$$

Convert 1 gallon to 4 qt and add to 2 qt before continuing.

6 qt ÷ 2 = 3 qt

■ Work Practice 4

Practice 5

A large oil drum contains 15 gal 3 qt of oil. How much will be in the drum if an additional 4 gal 3 qt of oil is poured into it?

Example 5 Finding the Amount of Water in an Aquarium

An aquarium contains 6 gal 3 qt of water. If 2 gal 2 qt of water is added, what is the total amount of water in the aquarium?

Solution:

$$\begin{array}{rclr} \text{beginning water} & \rightarrow & & 6 \text{ gal } 3 \text{ qt} \\ + \qquad \text{water added} & \rightarrow & + & 2 \text{ gal } 2 \text{ qt} \\ \hline \text{total water} & \rightarrow & & 8 \text{ gal } 5 \text{ qt} \end{array}$$

Since 5 qt = 1 gal 1 qt, we have

$$\begin{array}{l} \overbrace{8 \text{ gal}} \quad \overbrace{5 \text{ qt}} \\ = 8 \text{ gal} + 1 \text{ gal } 1 \text{ qt} \\ = 9 \text{ gal } 1 \text{ qt} \end{array}$$

The total amount of water is 9 gal 1 qt.

■ Work Practice 5

Answers

3. 3 qt   **4.** 3 gal 1 qt 1 pt   **5.** 20 gal 2 qt

Objective C Defining and Converting Metric System Units of Capacity

Thus far, we know that the basic unit of length in the metric system is the meter and that the basic unit of mass in the metric system is the gram. What is the basic unit of capacity? The **liter.** By definition, a **liter** is the capacity or volume of a cube measuring 10 centimeters on each side.

The following examples may help you get a feeling for metric capacities:

One liter of liquid is slightly more than one quart.
Many soft drinks are packaged in 2-liter bottles.

The metric system was designed to be a consistent system. Once again, the prefixes for metric units of capacity are the same as for metric units of length and mass, as summarized in the following table:

Metric Units of Capacity
1 **kilo**liter (kl) = 1000 liters (L)
1 **hecto**liter (hl) = 100 L
1 **deka**liter (dal) = 10 L
1 liter (L) = 1 L
1 **deci**liter (dl) = 1/10 L or 0.1 L
1 **centi**liter (cl) = 1/100 L or 0.01 L
1 **milli**liter (ml) = 1/1000 L or 0.001 L

1 liter   1 quart

2 liters

The **milliliter** and the **liter** are the two most commonly used metric units of capacity.

Converting from one unit of capacity to another involves multiplying by powers of 10 or moving the decimal point to the left or to the right. Listing units of capacity in order from largest to smallest helps to keep track of how many places to move the decimal point when converting.

Let's convert 2.6 liters to milliliters. To convert from liters to milliliters, we move along the chart 3 units to the right.

kl   hl   dal   **L**   dl   cl   **ml**

3 units to the right

This means that we move the decimal point 3 places to the right to convert from liters to milliliters.

$$2.600 \, L = 2600. \, ml$$

This same conversion can be done with unit fractions.

$$2.6 \, L = \frac{2.6 \, L}{1} \cdot 1$$

$$= \frac{2.6 \, \cancel{L}}{1} \cdot \frac{1000 \, ml}{1 \, \cancel{L}}$$

$$= 2.6 \cdot 1000 \, ml$$

$$= 2600 \, ml \quad \text{To multiply by 1000, move the decimal point 3 places to the right.}$$

To visualize the result, study the diagram below:

2.6 L

1000 ml   1000 ml   600 ml = 2600 ml

Thus, 2.6 L = 2600 ml.

Practice 6

Convert 2100 ml to liters.

Example 6   Convert 3210 ml to liters.

Solution: Let's use the unit fraction method first.

Unit fraction

$$3210 \text{ ml} = \frac{3210 \text{ ml}}{1} \cdot 1 = 3210 \text{ ml} \cdot \frac{1 \text{ L}}{1000 \text{ ml}} = 3.21 \text{ L}$$

Now let's list the unit measures in order from left to right and move from milliliters to liters.

kl   hl   dal   L   dl   cl   ml

3 units to the left

3210 ml = 3.210 L, the same results as before and shown below in the diagram.

3 places to the left

1000 ml   1000 ml   1000 ml

210 ml

3210 ml

1 L   1 L   1 L   0.210 L = 3.210 L

■ Work Practice 6

Practice 7

Convert 2.13 dal to liters.

Example 7   Convert 0.185 dl to milliliters.

Solution: We list the unit measures in order from left to right and move from deciliters to milliliters.

kl   hl   dal   L   dl   cl   ml

2 units to the right

0.185 dl = 18.5 ml

2 places to the right

■ Work Practice 7

Objective D Performing Operations on Metric System Units of Capacity

As was true for length and weight, arithmetic operations involving metric units of capacity can also be performed. Make sure that the metric units of capacity are the same before adding or subtracting.

Answers
6. 2.1 L   **7.** 21.3 L

Example 8   Add 2400 ml to 8.9 L.

Solution: We must convert both to liters or both to milliliters before adding the capacities together.

$$2400 \text{ ml} = 2.4 \text{ L} \rightharpoondown \quad \text{or} \quad 8.9 \text{ L} = 8900 \text{ ml} \rightharpoondown$$

$$\begin{array}{r} 2.4 \text{ L} \leftharpoondown \\ +8.9 \text{ L} \\ \hline 11.3 \text{ L} \end{array} \qquad \begin{array}{r} 2400 \text{ ml} \\ +8900 \text{ ml} \leftharpoondown \\ \hline 11{,}300 \text{ ml} \end{array}$$

The total is 11.3 L or 11,300 ml. They both represent the same capacity.

■ Work Practice 8

✓**Concept Check** How could you estimate the following operation? Subtract 950 ml from 7.5 L.

Example 9   Finding the Amount of Medication a Person Has Received

A patient hooked up to an IV unit in the hospital is to receive 12.5 ml of medication every hour. How much medication does the patient receive in 3.5 hours?

Solution: We multiply 12.5 ml by 3.5.

$$\begin{array}{rcl} \text{medication per hour} & \rightarrow & 12.5 \text{ ml} \\ \times \qquad\qquad \text{hours} & \rightarrow & \times\ 3.5 \\ \hline \text{total medication} & & 625 \\ & & \underline{3750} \\ & & 43.75 \text{ ml} \end{array}$$

The patient receives 43.75 ml of medication.

■ Work Practice 9

Practice 8
Add 1250 ml to 2.9 L.

Practice 9
If 28.6 L of water can be pumped every minute, how much water can be pumped in 85 minutes?

Answers
8. 4150 ml or 4.15 L   **9.** 2431 L

✓**Concept Check Answer**
950 ml = 0.95 L; round 0.95 to 1;
7.5 − 1 = 6.5 L

Vocabulary, Readiness & Video Check

Use the choices below to fill in each blank. Some choices may be used more than once.

cups	pints	liter
quarts	fluid ounces	capacity

1. Units of _____ are generally used to measure liquids.
2. The basic unit of capacity in the metric system is the _____.
3. One cup equals 8 _____.
4. One quart equals 2 _____.
5. One pint equals 2 _____.
6. One quart equals 4 _____.
7. One gallon equals 4 _____.

Convert as indicated without pencil and paper or calculator.

8. 2 c to pints   **9.** 4 c to pints   **10.** 4 qt to gallons   **11.** 8 qt to gallons   **12.** 2 pt to quarts

13. 6 pt to quarts   **14.** 8 fl oz to cups   **15.** 24 fl oz to cups   **16.** 3 pt to cups

Martin-Gay Interactive Videos   *Watch the section lecture video and answer the following questions.*

See Video 7.3

Objective A 17. Complete this statement based on ⊞ Example 1: When using a unit fraction, we are not changing the _____, we are changing the _____. ▶

Objective B 18. In ⊞ Example 4, explain the first step taken to solve the problem. ▶

Objective C 19. In ⊞ Example 5, how many places is the decimal point moved and in what direction? What is the final conversion? ▶

Objective D 20. What is the answer to ⊞ Example 7 in dekaliters? ▶

7.3 **Exercise Set** MyMathLab® ▶

Objective A *Convert each measurement as indicated. See Examples 1 and 2.*

1. 32 fluid ounces to cups   **2.** 16 quarts to gallons   **3.** 8 quarts to pints   **4.** 9 pints to quarts

▶ **5.** 14 quarts to gallons   **6.** 11 cups to pints   **7.** 80 fluid ounces to pints   **8.** 18 pints to gallons

9. 2 quarts to cups   **10.** 3 pints to fluid ounces   **11.** 120 fluid ounces to quarts   **12.** 20 cups to gallons

▶ **13.** 42 cups to quarts   **14.** 7 quarts to cups   **15.** $4\frac{1}{2}$ pints to cups   **16.** $6\frac{1}{2}$ gallons to quarts

17. 5 gal 3 qt to quarts   **18.** 4 gal 1 qt to quarts   **19.** $\frac{1}{2}$ cup to pints   **20.** $\frac{1}{2}$ pint to quarts

▶ **21.** 58 qt = _____ gal _____ qt   **22.** 70 qt = _____ gal _____ qt

23. 39 pt = _____ gal _____ qt _____ pt   **24.** 29 pt = _____ gal _____ qt _____ pt

25. $2\frac{3}{4}$ gallons to pints   **26.** $3\frac{1}{4}$ quarts to cups

Objective B *Perform each indicated operation. See Examples 3 through 5.*

27. 5 gal 3 qt + 7 gal 3 qt

28. 2 gal 2 qt + 9 gal 3 qt

29. 1 c 5 fl oz + 2 c 7 fl oz

30. 2 c 3 fl oz + 2 c 6 fl oz

▶ **31.** 3 gal − 1 gal 3 qt

32. 2 pt − 1 pt 1 c

33. 3 gal 1 qt − 1 qt 1 pt

34. 3 qt 1 c − 1 c 4 fl oz

35. 8 gal 2 qt × 2

36. 6 gal 1 pt × 2

37. 9 gal 2 qt ÷ 2

38. 5 gal 6 fl oz ÷ 2

Objective C *Convert as indicated. See Examples 6 and 7.*

39. 5 L to milliliters

40. 8 L to milliliters

41. 0.16 L to kiloliters

42. 0.127 L to kiloliters

▶ **43.** 5600 ml to liters

44. 1500 ml to liters

45. 3.2 L to centiliters

46. 1.7 L to centiliters

47. 410 L to kiloliters

48. 250 L to kiloliters

49. 64 ml to liters

50. 39 ml to liters

▶ **51.** 0.16 kl to liters

52. 0.48 kl to liters

53. 3.6 L to milliliters

54. 1.9 L to milliliters

Objective D *Perform each indicated operation. Remember to insert units when writing your answers. See Examples 8 and 9.*

55. 3.4 L + 15.9 L

56. 18.5 L + 4.6 L

▶ **57.** 2700 ml + 1.8 L

58. 4.6 L + 1600 ml

59. 8.6 L − 190 ml

60. 4.8 L − 283 ml

61. 17,500 ml − 0.9 L

62. 6850 ml − 0.3 L

63. 480 ml × 8

64. 290 ml × 6

65. 81.2 L ÷ 0.5

66. 5.4 L ÷ 3.6

Objectives A B C D Mixed Practice *Solve. Remember to insert units when writing your answers. For Exercises 67 through 70, complete the chart. See Examples 1 through 9.*

	Capacity	Cups	Gallons	Quarts	Pints
67.	An average-size bath of water		21		
68.	A dairy cow's daily milk yield				38
69.	Your kidneys filter about this amount of blood every minute	4			
70.	The amount of water needed in a punch recipe	2			

71. Mike Schaferkotter drank 410 ml of Mountain Dew from a 2-liter bottle. How much Mountain Dew remains in the bottle?

72. The Werners' Volvo has a 54.5-L gas tank. Only 38 deciliters of gasoline still remain in the tank. How many liters will be needed to fill it?

73. Margie Phitts added 354 ml of Prestone gas treatment to the 18.6 L of gasoline in her car's tank. Find the total amount of gasoline in the tank.

74. Chris Peckaitis wishes to share a 2-L bottle of Coca-Cola equally with 7 of his friends. How much will each person get?

75. A garden tool engine requires a 30-to-1 gas-to-oil mixture. This means that $\frac{1}{30}$ of a gallon of oil should be mixed with 1 gallon of gas. Convert $\frac{1}{30}$ gallon to fluid ounces. Round to the nearest tenth.

76. Henning's Supermarket sells homemade soup in 1 qt 1 pt containers. How much soup is contained in three such containers?

77. Can 5 pt 1 c of fruit punch and 2 pt 1 c of ginger ale be poured into a 1-gal container without it overflowing?

78. Three cups of prepared Jell-O are poured into 6 dessert dishes. How many fluid ounces of Jell-O are in each dish?

79. Stanley Fisher paid $14 to fill his car with 44.3 liters of gasoline. Find the price per liter of gasoline to the nearest thousandth of a dollar.

80. A student carelessly misread the scale on a cylinder in the chemistry lab and added 40 cl of water to a mixture instead of 40 ml. Find the excess amount of water.

Review

Write each fraction in simplest form. See Section 2.3.

81. $\frac{20}{25}$ **82.** $\frac{75}{100}$ **83.** $\frac{27}{45}$ **84.** $\frac{56}{60}$ **85.** $\frac{72}{80}$ **86.** $\frac{18}{20}$

Concept Extensions

Determine whether the measurement in each statement is reasonable.

87. Clair took a dose of 2 L of cough medicine to cure her cough.

88. John drank 250 ml of milk for lunch.

89. Jeannie likes to relax in a tub filled with 3000 ml of hot water.

90. Sarah pumped 20 L of gasoline into her car yesterday.

Solve. See the Concept Checks in this section.

91. If 70 pints are converted to gallons, will the equivalent number of gallons be less than or greater than 70? Explain why.

92. If 30 gallons are converted to quarts, will the equivalent number of quarts be less than or greater than 30? Explain why.

93. Explain how to estimate the following operation: Add 986 ml to 6.9 L.

94. Explain how to borrow in order to subtract 1 gal 2 qt from 3 gal 1 qt.

95. Find the number of fluid ounces in 1 gallon.

96. Find the number of fluid ounces in 1.5 gallons.

A cubic centimeter (cc) is the amount of space that a volume of 1 ml occupies. Because of this, we will say that 1 cc = 1 ml.

A common syringe is one with a capacity of 3 cc. Use the diagram and give the measurement indicated by each arrow.

97. B   **98.** A   **99.** D   **100.** C

In order to measure small dosages, such as for insulin, u-100 syringes are used. For these syringes, 1 cc has been divided into 100 equal units (u). Use the diagram and give the measurement indicated by each arrow in units (u) and then in cubic centimeters. Use 100 u = 1 cc.

101. B   **102.** A

103. D   **104.** C

Length, Weight, and Capacity

Convert each measurement as indicated.

Length

1. 36 in. = _____ ft

2. 10,560 ft = _____ mi

3. 20 ft = _____ yd

4. $6\frac{1}{3}$ yd = _____ ft

5. 2.1 mi = _____ ft

6. 3.2 ft = _____ in.

7. 30 m = _____ cm

8. 24 mm = _____ cm

9. 2000 mm = _____ m

10. 1800 cm = _____ m

11. 7.2 cm = _____ mm

12. 600 m = _____ km

Weight or Mass

13. $7\frac{1}{2}$ tons = _____ lb

14. 11,000 lb = _____ tons

15. 8.5 lb = _____ oz

16. 72 oz = _____ lb

1. _____
2. _____
3. _____
4. _____
5. _____
6. _____
7. _____
8. _____
9. _____
10. _____
11. _____
12. _____
13. _____
14. _____
15. _____
16. _____

17. _____

18. _____

19. _____

20. _____

21. _____

22. _____

23. _____

24. _____

25. _____

26. _____

27. _____

28. _____

29. _____

30. _____

31. _____

32. _____

33. _____

34. _____

35. _____

36. _____

37. _____

38. _____

17. 104 oz = _____ lb

18. 5 lb = _____ oz

19. 28 kg = _____ g

20. 1400 mg = _____ g

21. 5.6 g = _____ kg

22. 6 kg = _____ g

23. 670 mg = _____ g

24. 3.6 g = _____ kg

Capacity

25. 6 qt = _____ pt

26. 5 pt = _____ qt

27. 14 qt = _____ gal

28. 17 c = _____ pt

29. $3\frac{1}{2}$ pt = _____ c

30. 26 qt = _____ gal

31. 7 L = _____ ml

32. 350 L = _____ kl

33. 47 ml = _____ L

34. 0.97 kl = _____ L

35. 0.126 kl = _____ L

36. 75 ml = _____ L

37. $\frac{1}{2}$ c = _____ fl oz

38. $\frac{3}{4}$ gal = _____ c

7.4 Conversions Between the U.S. and Metric Systems

Objective A Converting Between the U.S. and Metric Systems

The metric system probably had its beginnings in France in the 1600s, but it was the Metric Act of 1866 that made the use of this system legal (but not mandatory) in the United States. Other laws have followed that allow for a slow, but deliberate, transfer to the modernized metric system. In April 2001, for example, the U.S. Stock Exchanges completed their change to decimal trading instead of fractions. By the end of 2009, all products sold in Europe (with some exceptions) were required to have only metric units on their labels. (*Source:* U.S. Metric Association and National Institute of Standards and Technology)

You may be surprised at the number of everyday items we use that are already manufactured in metric units. We easily recognize 1L and 2L soda bottles, but what about the following?

Pencil leads (0.5 mm or 0.7 mm)

Camera film (35 mm)

Sporting events (5-km or 10-km races)

Medicines (500-mg capsules)

Labels on retail goods (dual-labeled since 1994)

Since the United States has not completely converted to the metric system, we need to practice converting from one system to the other. Below is a table of mostly approximate conversions.

Length:		Capacity:		Weight (mass):	
Metric	U.S. System	Metric	U.S. System	Metric	U.S. System
1 m ≈ 1.09 yd		1 L ≈ 1.06 qt		1 kg ≈ 2.20 lb	
1 m ≈ 3.28 ft		1 L ≈ 0.26 gal		1 g ≈ 0.04 oz	
1 km ≈ 0.62 mi		3.79 L ≈ 1 gal		0.45 kg ≈ 1 lb	
2.54 cm = 1 in.		0.95 L ≈ 1 qt		28.35 g ≈ 1 oz	
0.30 m ≈ 1 ft		29.57 ml ≈ 1 fl oz			
1.61 km ≈ 1 mi					

There are many ways to perform these metric-to-U.S. conversions. We will do so by using unit fractions.

Length
1 yard
1 meter

Capacity
1 quart   1 liter

Weight (Mass)
1 pound   1 kilogram

Example 1 Compact Discs

Standard-sized compact discs are 12 centimeters in diameter. Convert this length to inches. Round the result to two decimal places. (*Source:* usByte.com)

Solution: From our length conversion table, we know that 2.54 cm = 1 in. This fact gives us two unit fractions: $\frac{2.54 \text{ cm}}{1 \text{ in.}}$ and $\frac{1 \text{ in.}}{2.54 \text{ cm}}$. We use the unit fraction with cm in the denominator so that these units divide out.

$$12 \text{ cm} = \frac{12 \text{ cm}}{1} \cdot 1 = \frac{12 \cancel{\text{ cm}}}{1} \cdot \frac{1 \text{ in.} \quad \leftarrow \text{Units to convert to}}{2.54 \cancel{\text{ cm}} \quad \leftarrow \text{Original units}}$$

$$= \frac{12 \text{ in.}}{2.54}$$

$$\approx 4.72 \text{ in.} \quad \text{Divide.}$$

(*Continued on next page*)

Practice 1

The center hole of a standard-sized compact disc is 1.5 centimeters in diameter. Convert this length to inches. Round the result to 2 decimal places.

Answer
1. 0.59 in.

469

Thus, the diameter of a standard compact disc is exactly 12 cm or approximately 4.72 inches. For a dimension this size, you can use a ruler to check. Another method is to approximate. Our result, 4.72 in., is close to 5 inches. Since 1 in. is about 2.5 cm, then 5 in. is about 5(2.5 cm) = 12.5 cm, which is close to 12 cm.

▪ Work Practice 1

Practice 2

A full-grown human heart weighs about 8 ounces. Convert this weight to grams. If necessary, round your result to the nearest tenth of a gram.

Example 2   Liver

The liver is your largest internal organ. It weighs about 3.5 pounds in a grown man. Convert this weight to kilograms. Round to the nearest tenth. (*Source: Some Body!* by Dr. Pete Rowan)

Unit fraction

Solution: $3.5 \text{ lb} \approx \dfrac{3.5 \text{ lb}}{1} \cdot \dfrac{0.45 \text{ kg}}{1 \text{ lb}} = 3.5(0.45 \text{ kg}) \approx 1.6 \text{ kg}$

Thus 3.5 pounds are approximately 1.6 kilograms. From the table of conversions, we know that 1 kg ≈ 2.2 lb. So that means 0.5 kg ≈ 1.1 lb and after adding, we have 1.5 kg ≈ 3.3 lb. Our result is reasonable.

▪ Work Practice 2

Practice 3

Convert 237 ml to fluid ounces. Round to the nearest whole fluid ounce.

Example 3   Postage Stamp

Australia converted to the metric system in 1973. In that year, four postage stamps were issued to publicize this conversion. One such stamp is shown. Let's check the mathematics on the stamp by converting 7 fluid ounces to milliliters. Round to the nearest hundred.

Unit fraction

Solution: $7 \text{ fl oz} \approx \dfrac{7 \text{ fl oz}}{1} \cdot \dfrac{29.57 \text{ ml}}{1 \text{ fl oz}} = 7(29.57 \text{ ml}) = 206.99 \text{ ml}$

Rounded to the nearest hundred, 7 fl oz ≈ 200 ml.

▪ Work Practice 3

Answers

2. 226.8 g   **3.** 8 fl oz

Vocabulary, Readiness & Video Check

Martin-Gay Interactive Videos

Watch the section lecture video and answer the following questions.

Objective A   **1.** Write two conversions that may be used to solve ▯ Example 2. ▶

2. Why isn't 0.1125 kg the final answer to ▯ Example 3? ▶

See Video 7.4 🍒

7.4 Exercise Set MyMathLab®

Note: Because approximations are used, your answers may vary slightly from the answers given in the back of the book.

Objective A *Convert as indicated. If necessary, round answers to two decimal places. See Examples 1 through 3.*

1. 756 milliliters to fluid ounces

2. 18 liters to quarts

▶ **3.** 86 inches to centimeters

4. 86 miles to kilometers

5. 1000 grams to ounces

6. 100 kilograms to pounds

7. 93 kilometers to miles

8. 9.8 meters to feet

▶ **9.** 14.5 liters to gallons

10. 150 milliliters to fluid ounces

11. 30 pounds to kilograms

12. 15 ounces to grams

Fill in the chart. Give exact answers or round to 1 decimal place. See Examples 1 through 3.

		Meters	Yards	Centimeters	Feet	Inches
13.	The height of a woman				5	
14.	Statue of Liberty length of nose	1.37				
15.	Leaning Tower of Pisa		60			
16.	Blue whale		36			

Solve. If necessary, round answers to two decimal places. See Examples 1 through 3.

17. The balance beam for female gymnasts is 10 centimeters wide. Convert this width to inches.

18. In men's gymnastics, the rings are 250 centimeters from the floor. Convert this height to inches, then to feet.

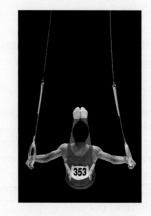

19. In many states, the maximum speed limit for recreational vehicles is 50 miles per hour. Convert this to kilometers per hour.

20. In some states, the speed limit is 70 miles per hour. Convert this to kilometers per hour.

21. Ibuprofen comes in 200-milligram tablets. Convert this to ounces. (Round your answer to this exercise to 3 decimal places.)

22. Vitamin C tablets come in 500-milligram caplets. Convert this to ounces.

23. A stone is a unit in the British customary system. Use the conversion 14 pounds = 1 stone to check the equivalencies in this 1973 Australian stamp. Is 100 kilograms approximately 15 stone 10 pounds?

24. Convert 5 feet 11 inches to centimeters and check the conversion on this 1973 Australian stamp. Is it correct?

25. The Monarch butterfly migrates annually between the northern United States and central Mexico. The trip is about 4500 km long. Convert this to miles.

26. There is a species of African termite that builds nests up to 18 ft high. Convert this to meters.

27. A $3\frac{1}{2}$-inch diskette is not really $3\frac{1}{2}$ inches. To find its actual width, convert this measurement to centimeters, then to millimeters. Round the result to the nearest ten.

28. The average two-year-old is 84 centimeters tall. Convert this to feet and inches.

29. For an average adult, the weight of the right lung is greater than the weight of the left lung. If the right lung weighs 1.5 pounds and the left lung weighs 1.25 pounds, find the difference in grams. (*Source: Some Body!*)

30. The skin of an average adult weighs 9 pounds and is the heaviest organ. Find the weight in grams. (*Source: Some Body!*)

31. A fast sneeze has been clocked at about 167 kilometers per hour. Convert this to miles per hour. Round to the nearest whole.

32. A Boeing 747 has a cruising speed of about 980 kilometers per hour. Convert this to miles per hour. Round to the nearest whole.

33. The General Sherman giant sequoia tree has a diameter of about 8 meters at its base. Convert this to feet. (*Source: Fantastic Book of Comparisons*)

34. The largest crater on the near side of the moon is Billy Crater. It has a diameter of 303 kilometers. Convert this to miles. (*Source: Fantastic Book of Comparisons*)

35. The total length of the track on a CD is about 4.5 kilometers. Convert this to miles. Round to the nearest whole mile.

36. The distance between Mackinaw City, Michigan, and Cheyenne, Wyoming, is 2079 kilometers. Convert this to miles. Round to the nearest whole mile.

37. A doctor orders a dosage of 5 ml of medicine every 4 hours for 1 week. How many fluid ounces of medicine should be purchased? Round up to the next whole fluid ounce.

38. A doctor orders a dosage of 12 ml of medicine every 6 hours for 10 days. How many fluid ounces of medicine should be purchased? Round up to the next whole fluid ounce.

Without actually converting, choose the most reasonable answer.

39. This math book has a height of about

_____ .

    a. 28 mm   **b.** 28 cm

    c. 28 m   **d.** 28 km

40. A mile is _____ a kilometer.

    a. shorter than   **b.** longer than

    c. the same length as

41. A liter has _____ capacity than a quart.
   a. less   **b.** greater
   c. the same

42. A foot is _____ a meter.
   a. shorter than   **b.** longer than
   c. the same length as

43. A kilogram weighs _____ a pound.
   a. the same as   **b.** less than
   c. more than

44. A football field is 100 yards, which is about
_____.
   a. 9 m   **b.** 90 m
   c. 900 m   **d.** 9000 m

45. An $8\frac{1}{2}$-ounce glass of water has a capacity of about
_____.
   a. 250 L   **b.** 25 L
   c. 2.5 L   **d.** 250 ml

46. A 5-gallon gasoline can has a capacity of about
_____.
   a. 19 L   **b.** 1.9 L
   c. 19 ml   **d.** 1.9 ml

47. The weight of an average man is about
_____.
   a. 700 kg   **b.** 7 kg
   c. 0.7 kg   **d.** 70 kg

48. The weight of a pill is about _____.
   a. 200 kg   **b.** 20 kg
   c. 2 kg   **d.** 200 mg

Review

Perform the indicated operations. See Section 1.9.

49. $6 \cdot 4 + 5 \div 1$

50. $10 \div 2 + 9(8)$

51. $\dfrac{10 + 8}{10 - 8}$

52. $\dfrac{14 + 1}{5(3)}$

53. $3 + 5(19 - 17) - 8$

54. $1 + 4(19 - 9) + 5$

55. $3[(1 + 5) \cdot (8 - 6)]$

56. $5[(18 - 8) - 9]$

Concept Extensions

Body surface area (BSA) is often used to calculate dosages for some drugs. BSA is calculated in square meters using a person's weight and height.

$$BSA = \sqrt{\frac{(\text{weight in kg}) \times (\text{height in cm})}{3600}}$$

For Exercises 57 through 62, calculate the BSA for each person. Round to the nearest hundredth. You will need to use the square root key on your calculator.

57. An adult whose height is 182 cm and weight is 90 kg.

58. An adult whose height is 157 cm and weight is 63 kg.

59. A child whose height is 50 in. and weight is 40 kg. (*Hint:* Don't forget to first convert inches to centimeters)

60. A child whose height is 26 in. and weight is 13 kg.

61. An adult whose height is 60 in. and weight is 150 lb.

62. An adult whose height is 69 in. and weight is 172 lb.

Solve.

63. Suppose the adult from Exercise 57 is to receive a drug that has a recommended dosage range of 10–12 mg per sq meter. Find the dosage range for the adult.

64. Suppose the child from Exercise 60 is to receive a drug that has a recommended dosage of 30 mg per sq meter. Find the dosage for the child.

65. A handball court is a rectangle that measures 20 meters by 40 meters. Find its area in square meters and square feet.

66. A backpack measures 16 inches by 13 inches by 5 inches. Find the volume of a box with these dimensions. Find the volume in cubic inches and cubic centimeters. Round the cubic centimeters to the nearest whole cubic centimeter.

7.5 Temperature: U.S. and Metric Systems of Measurement

Objectives

A Convert Temperatures from Degrees Celsius to Degrees Fahrenheit.

B Convert Temperatures from Degrees Fahrenheit to Degrees Celsius.

When Gabriel Fahrenheit and Anders Celsius independently established units for temperature scales, each based his unit on the heat of water the moment it boils compared to the moment it freezes. One degree Celsius is 1/100 of the difference in heat. One degree Fahrenheit is 1/180 of the difference in heat. Celsius arbitrarily labeled the temperature at the freezing point at 0°C, making the boiling point 100°C; Fahrenheit labeled the freezing point 32°F, making the boiling point 212°F. Water boils at 212°F and 100°C.

By comparing the two scales in the figure, we see that a 20°C day is as warm as a 68°F day. Similarly, a sweltering 104°F day in the Mojave Desert corresponds to a 40°C day.

✓**Concept Check** Which of the following statements is correct? Explain.

a. 6°C is below the freezing point of water.

b. 6°F is below the freezing point of water.

Objective A Converting Degrees Celsius to Degrees Fahrenheit ▶

To convert from Celsius temperatures to Fahrenheit temperatures, we can use either of the equations in the box below.

Converting Celsius to Fahrenheit

$$F = \frac{9}{5} \cdot C + 32 \qquad \text{or} \qquad F = 1.8 \cdot C + 32$$

(To convert to Fahrenheit temperature, multiply the Celsius temperature by $\frac{9}{5}$ or 1.8, and then add 32.)

In these equations, we use the symbol F to represent degrees Fahrenheit and the symbol C to represent degrees Celsius.

✓**Concept Check Answer**

b

Example 1   Convert 15°C to degrees Fahrenheit.

Solution: $F = \dfrac{9}{5} \cdot C + 32$

$\qquad = \dfrac{9}{5} \cdot 15 + 32$ Replace C with 15.

$\qquad = 27 + 32$ Simplify.

$\qquad = 59$ Add.

Thus, 15°C is equivalent to 59°F.

■ Work Practice 1

Practice 1
Convert 50°C to degrees Fahrenheit.

Example 2   Convert 29°C to degrees Fahrenheit.

Solution: $F = 1.8 \cdot C + 32$

$\qquad = 1.8 \cdot 29 + 32$ Replace C with 29.

$\qquad = 52.2 + 32$ Multiply 1.8 by 29.

$\qquad = 84.2$ Add.

Therefore, 29°C is the same as 84.2°F.

■ Work Practice 2

Practice 2
Convert 18°C to degrees Fahrenheit.

Objective B Converting Degrees Fahrenheit to Degrees Celsius ▶

To convert from Fahrenheit temperatures to Celsius temperatures, see the box below. The symbol C represents degrees Celsius and the symbol F represents degrees Fahrenheit.

Converting Fahrenheit to Celsius

$$C = \dfrac{5}{9}(F - 32)$$

(To convert to Celsius temperature, subtract 32 from the Fahrenheit temperature, and then multiply by $\dfrac{5}{9}$.)

Example 3   Convert 59°F to degrees Celsius.

Solution: We evaluate the formula $C = \dfrac{5}{9}(F - 32)$ when F is 59.

$C = \dfrac{5}{9}(F - 32)$

$\quad = \dfrac{5}{9} \cdot (59 - 32)$ Replace F with 59.

$\quad = \dfrac{5}{9} \cdot (27)$ Subtract inside parentheses.

$\quad = 15$ Multiply.

Therefore, 59°F is the same temperature as 15°C.

■ Work Practice 3

(Compare the results of Examples 1 and 3.)

Practice 3
Convert 68°F to degrees Celsius.

Answers
1. 122°F   **2.** 64.4°F   **3.** 20°C

Practice 4

Convert 113°F to degrees Celsius. If necessary, round to the nearest tenth of a degree.

Example 4 Convert 114°F to degrees Celsius. If necessary, round to the nearest tenth of a degree.

Solution:
$$C = \frac{5}{9}(F - 32)$$

$$= \frac{5}{9}(114 - 32) \quad \text{Replace F with 114.}$$

$$= \frac{5}{9} \cdot (82) \quad \text{Subtract inside parentheses.}$$

$$\approx 45.6 \quad \text{Multiply.}$$

Therefore, 114°F is approximately 45.6°C.

■ Work Practice 4

Practice 5

During a bout with the flu, Albert's temperature reaches 102.8°F. What is his temperature measured in degrees Celsius? Round to the nearest tenth of a degree.

Example 5 Body Temperature

Normal body temperature is 98.6°F. What is this temperature in degrees Celsius?

Solution: We evaluate the formula $C = \frac{5}{9}(F - 32)$ when F is 98.6.

$$C = \frac{5}{9}(F - 32)$$

$$= \frac{5}{9}(98.6 - 32) \quad \text{Replace F with 98.6.}$$

$$= \frac{5}{9} \cdot (66.6) \quad \text{Subtract inside parentheses.}$$

$$= 37 \quad \text{Multiply.}$$

Therefore, normal body temperature is 37°C.

■ Work Practice 5

Answers

4. 45°C   **5.** 39.3°C

✓ **Concept Check Answer**

She used the conversion for Celsius to Fahrenheit instead of Fahrenheit to Celsius.

✓**Concept Check** Clarissa must convert 40°F to degrees Celsius. What is wrong with her work shown below?

$$F = 1.8 \cdot C + 32$$
$$F = 1.8 \cdot 40 + 32$$
$$F = 72 + 32$$
$$F = 104$$

Vocabulary, Readiness & Video Check

Martin-Gay Interactive Videos   *Watch the section lecture video and answer the following questions.*

Objective A 1. Which version of the formula is used to solve ▥ Example 1? What is the replacement value for C? ▶

Objective B 2. In ▥ Example 2, what is the replacement value for F? What is the final conversion? ▶

See Video 7.5 🍎

7.5 Exercise Set MyMathLab® ▶

Objectives A B Mixed Practice *Convert as indicated. When necessary, round to the nearest tenth of a degree. See Examples 1 through 5.*

1. 41°F to degrees Celsius

2. 68°F to degrees Celsius

3. 104°F to degrees Celsius

▶ **4.** 77°F to degrees Celsius

5. 60°C to degrees Fahrenheit

6. 80°C to degrees Fahrenheit

7. 115°C to degrees Fahrenheit

8. 35°C to degrees Fahrenheit

9. 62°F to degrees Celsius

10. 182°F to degrees Celsius

11. 142.1°F to degrees Celsius

12. 43.4°F to degrees Celsius

13. 92°C to degrees Fahrenheit

14. 75°C to degrees Fahrenheit

15. 16.3°C to degrees Fahrenheit

16. 48.6°C to degrees Fahrenheit

17. The hottest temperature ever recorded in New Mexico was 122°F. Convert this temperature to degrees Celsius. (*Source:* National Climatic Data Center)

18. The hottest temperature ever recorded in Rhode Island was 104°F. Convert this temperature to degrees Celsius. (*Source:* National Climatic Data Center)

▶ **19.** A weather forecaster in Caracas predicts a high temperature of 27°C. Find this measurement in degrees Fahrenheit.

20. While driving to work, Alan Olda notices a temperature of 22°C flash on the local bank's temperature display. Find the corresponding temperature in degrees Fahrenheit.

21. Water boils at 212°F. Find this temperature in degrees Celsius.

22. Water freezes at 0°C. Find this temperature in degrees Fahrenheit.

23. Najib Tan is running a fever of 100.2°F. Find his temperature as it would be shown on a Celsius thermometer.

24. William Saylor generally has a subnormal temperature of 98.2°F. Find what this temperature would be on a Celsius thermometer.

25. In a European cookbook, a recipe requires the ingredients for caramels to be heated to 118°C, but the cook has access only to a Fahrenheit thermometer. Find the temperature in degrees Fahrenheit that should be used to make the caramels.

26. The ingredients for divinity should be heated to 127°C, but the candy thermometer that Myung Kim has is calibrated to degrees Fahrenheit. Find how hot he should heat the ingredients.

27. The surface temperature of Venus can reach 864°F. Find this temperature in degrees Celsius.

28. The temperature of Earth's core is estimated to be 4000°C. Find the corresponding temperature in degrees Fahrenheit.

29. At Mack Trucks' headquarters, the room temperature is to be set at 70°F, but the thermostat is calibrated in degrees Celsius. Find the temperature to be set.

30. The computer room at Merck, Sharp, and Dohm is normally cooled to 66°F. Find the corresponding temperature in degrees Celsius.

Review

Find the perimeter of each figure. Recall that the perimeter of a figure is the distance around the figure. See Section 1.3.

△ **31.**   3 in.

    3 in. | Square

△ **32.**   25 m

    6 m | Rectangle

△ **33.**   4 cm   3 cm
    Triangle
       5 cm

△ **34.**   3 ft   3 ft
    Pentagon
    3 ft   3 ft
        3 ft

△ **35.**   2 ft 8 in.

    1 ft 6 in. | Rectangle

△ **36.**   2.6 m

    2.6 m | Square

Concept Extensions

Solve. See the first Concept Check in this section. True or False.

37. 10°F is above the freezing point of water.

38. 10°C is above the freezing point of water.

39. 102°C is above the boiling point of water.

40. 102°F is above the boiling point of water.

41. On July 19, 1996, at the Naka Fusion Research Establishment in Nakamachi, Ibaraki, Japan, the highest temperature produced in a laboratory was achieved. This temperature was approximately 936,000,000°F. Convert this temperature to degrees Celsius. (*Note:* This is almost 30 times the temperature at the center of the sun.) (*Source: Guinness Book of Records*)

42. The hottest-burning substance known is carbon sub-nitride. Its flame at one atmospheric pressure reaches 9010°F. Convert this temperature to degrees Celsius. (*Source: Guinness Book of Records*)

Determine whether the measurement in each statement is reasonable by answering yes or no.

43. A 72°F room feels comfortable.

44. Water heated to 110°F will boil.

45. Josiah has a fever if a thermometer shows his temperature to be 40°F.

46. An air temperature of 20°F on a Vermont ski slope can be expected in the winter.

47. When the temperature is 30°C outside, an overcoat is needed.

48. An air-conditioned room at 60°C feels quite chilly.

49. Barbara has a fever when a thermometer records her temperature at 40°C.

50. Water cooled to 32°C will freeze.

51. In your own words, describe how to convert from degrees Celsius to degrees Fahrenheit.

52. In your own words, describe how to convert from degrees Fahrenheit to degrees Celsius.

Energy: U.S. and Metric Systems of Measurement ▶

Many people think of energy as a concept that involves movement or activity. However, **energy** is defined as "the capacity to do work." Often energy is stored, awaiting use at some later point in time.

Objective A Defining and Using U.S. System Units of Energy ▶

In the U.S. system of measurement, energy is commonly measured in foot-pounds. One **foot-pound (ft-lb)** is the amount of energy needed to lift a 1-pound object a distance of 1 foot. To determine the amount of energy necessary to move a 50-pound weight a distance of 100 feet, we simply multiply these numbers. That is,

50 pounds · 100 feet = 5000 ft-lb of energy

Example 1 Finding the Amount of Energy Needed to Move a Carton

An employee for the Jif Peanut Butter company must lift a carton of peanut butter jars 16 feet to the top of the warehouse. In the carton are 24 jars, each of which weighs 1.125 pounds. How much energy is required to lift the carton?

Solution: First we determine the weight of the carton.

weight of carton = weight of a jar · number of jars
= 1.125 pounds · 24
= 27 pounds

Thus, the carton weighs 27 pounds.
To find the energy needed to lift the 27-pound carton, we multiply the weight times the distance.

energy = 27 pounds · 16 feet = 432 ft-lb

Thus, 432 ft-lb of energy are required to lift the carton.

▬ Work Practice 1

✓ **Concept Check** Suppose you would like to find how many foot-pounds of energy are needed to lift an object weighing 12 ounces a total of 14 yards. What adjustments should you make before computing the answer? Next, compute the answer.

Another form of energy is heat. In the U.S. system of measurement, heat is measured in **British Thermal Units (BTU).** A BTU is the amount of heat required to raise the temperature of 1 pound of water 1 degree Fahrenheit. To relate British Thermal Units to foot-pounds, we need to know that

1 BTU = 778 ft-lb

Objectives

A Define and Use U.S. Units of Energy and Convert from One Unit to Another. ▶

B Define and Use Metric Units of Energy. ▶

Practice 1

Three bales of cardboard must be lifted 340 feet. If each bale weighs 63 pounds, find the amount of work required to lift the cardboard.

Answer
1. 64,260 ft-lb

✓ **Concept Check Answer**
convert 12 ounces to 0.75 pound and 14 yards to 42 feet; answer: 31.5 ft-lb

479

Practice 2

The FC-5 model furnace produced by Raywall uses 17,065 BTU every hour. Convert this energy requirement to foot-pounds.

70 calories

Practice 3

It takes 30 calories each hour for Alan to fly a kite. How many calories will he use if he flies his kite for 2 hours?

Practice 4

It takes 200 calories for Melanie to play Frisbee for an hour. How many calories will she use playing Frisbee for an hour each day for 5 days?

Answers
2. 13,276,570 ft-lb   **3.** 60 cal
4. 1000 cal

Example 2 Converting BTU to Foot-Pounds

The Raywall Company produces several different furnace models. Their FC-4 model requires 13,652 BTU every hour to operate. Convert the required energy to foot-pounds.

Solution:

To convert BTU to foot-pounds, we multiply by the unit fraction $\dfrac{778 \text{ ft-lb}}{1 \text{ BTU}}$.

$$13{,}652 \text{ BTU} = 13{,}652 \; \cancel{\text{BTU}} \cdot \overbrace{\frac{778 \text{ ft-lb}}{1 \; \cancel{\text{BTU}}}}^{\text{Unit fraction}}$$

$$= 10{,}621{,}256 \text{ ft-lb}$$

Thus, 13,652 BTU is equivalent to 10,621,256 ft-lb.

■ Work Practice 2

Objective B Defining and Using Metric System Units of Energy ▶

In the metric system, heat is measured in calories. A **calorie (cal)** is the amount of heat required to raise the temperature of 1 kilogram of water 1 degree Celsius.

The fact that an apple contains 70 calories means that 70 calories of heat energy are stored in our bodies whenever we eat an apple. This energy is stored in fat tissue and is burned (or "oxidized") by our bodies when we require energy to do work. We need 20 calories each hour just to stand still. This means that 20 calories of heat energy will be burned by our bodies each hour that we spend standing.

Example 3 Finding the Number of Calories Needed

It takes 20 calories for Jim to stand for 1 hour. How many calories does he use when standing for 3 hours at a crowded party?

Solution: We multiply the number of calories used in 1 hour by the number of hours spent standing.

total calories $= 20 \cdot 3 = 60$ calories

Therefore, Jim uses 60 calories to stand for 3 hours at the party.

■ Work Practice 3

Example 4 Finding the Number of Calories Needed

It takes 115 calories for Kathy to walk slowly for 1 hour. How many calories does she use walking slowly for 1 hour a day for 6 days?

Solution: We multiply the total number of calories used in 1 hour each day by the number of days.

total calories $= 115 \cdot 6 = 690$ calories

Therefore, Kathy uses 690 calories walking slowly for 1 hour for 6 days.

■ Work Practice 4

Example 5 Finding the Number of Calories Needed

It requires 100 calories to play a game of cards for an hour. If Jason plays poker for 1.5 hours each day for 5 days, how many calories are required?

Solution: We first determine the number of calories Jason uses each day to play poker.

calories used each day $= 100(1.5) = 150$ calories

Then we multiply the number of calories used each day by the number of days.

calories used for 5 days $= 150 \cdot 5 = 750$ calories

Thus, Jason uses 750 calories to play poker for 1.5 hours each day for 5 days.

■ Work Practice 5

Calories burned during exercise are affected by your body weight, the intensity of your workout, personal metabolism, and other factors. For example, the table below is for a 155-pound person.

1 Hour of Activity	Calories Burned
Raking lawn	281
Cross-country hiking	422
Golf (pulling clubs)	352
Ballroom dancing	387
Fishing	281
Playing catch football	141
Walking briskly	281
Running in general	563
Cleaning house	246

Practice 5

To play volleyball for an hour requires 300 calories. If Martha plays volleyball 1.25 hours each day for 4 days, how many calories does Martha use?

Answer

5. 1500 calories

Vocabulary, Readiness & Video Check

Use the choices below to fill in each blank.

calorie   energy   British Thermal Unit   foot-pound

1. _____ is defined as "the capacity to do work."

2. The amount of energy needed to lift a 1-pound object a distance of 1 foot is called a(n) _____.

3. The amount of heat required to raise the temperature of 1 pound of water 1 degree Fahrenheit is called a(n) _____.

4. The amount of heat required to raise the temperature of 1 kilogram of water 1 degree Celsius is called a(n) _____.

Solve mentally, without use of a pencil and paper or calculator.

5. How many foot-pounds of energy are needed to lift a 6-pound object 5 feet?

6. How many foot-pounds of energy are needed to lift a 10-pound object 4 feet?

7. How many foot-pounds of energy are needed to lift a 3-pound object 20 feet?

8. How many foot-pounds of energy are needed to lift a 5-pound object 9 feet?

9. If 30 calories are burned by the body in 1 hour, how many calories are burned (at the same rate) in 3 hours?

10. If 15 calories are burned by the body in 1 hour, how many calories are burned (at the same rate) in 2 hours?

11. If 20 calories are burned by the body in 1 hour, how many calories are burned (at the same rate) in $\frac{1}{4}$ of an hour?

12. If 50 calories are burned by the body in 1 hour, how many calories are burned (at the same rate) in $\frac{1}{2}$ of an hour?

Martin-Gay Interactive Videos

Watch the section lecture video and answer the following questions.

Objective A 13. Complete this statement based on the lecture before ▣ Example 2: In the _____ system of measurement, heat is measured in _____ Thermal Units (BTU). A BTU is the amount of heat required to raise the temperature of 1 _____ of water 1 degree _____. ▶

Objective B 14. Complete this statement based on the lecture before ▣ Example 3: In the _____ system of measurement, heat is measured in _____ (cal). A cal is the amount of heat required to raise the temperature of 1 _____ of water 1 degree _____. ▶

See Video 7.6

7.6 Exercise Set MyMathLab®

Objective A *Solve. See Examples 1 and 2.*

▶ 1. How much energy is required to lift a 3-pound math textbook 380 feet up a hill?

2. How much energy is required to lift a 20-pound sack of potatoes 55 feet?

3. How much energy is required to lift a 168-pound person 22 feet?

4. How much energy is needed to lift a 2250-pound car a distance of 45 feet?

5. How many foot-pounds of energy are needed to take 2.5 tons of topsoil 85 feet from the pile delivered by the nursery to the garden?

6. How many foot-pounds of energy are needed to lift 4.25 tons of coal 16 feet into a new coal bin?

▶ 7. Convert 30 BTU to foot-pounds.

8. Convert 50 BTU to foot-pounds.

9. Convert 1000 BTU to foot-pounds.

10. Convert 10,000 BTU to foot-pounds.

11. A 20,000-BTU air conditioner requires how many foot-pounds of energy to operate?

12. A 24,000-BTU air conditioner requires how many foot-pounds of energy to operate?

13. The Raywall model FC-10 heater uses 34,130 BTU each hour to operate. How many foot-pounds of energy does it use each hour?

14. The Raywall model FC-12 heater uses 40,956 BTU each hour to operate. How many foot-pounds of energy does it use each hour?

15. 8,000,000 ft-lb is equivalent to how many BTU, rounded to the nearest whole number?

16. 450,000 ft-lb is equivalent to how many BTU, rounded to the nearest whole number?

Objective B *Solve. See Examples 3 through 5.*

17. While walking slowly, Janie Gaines burns 115 calories each hour. How many calories does she burn if she walks slowly for an hour every day of the week?

18. Dancing burns 270 calories per hour. How many calories are needed to go dancing an hour a night for 3 nights?

19. Approximately 300 calories are burned each hour skipping rope. How many calories are required to skip rope $\frac{1}{2}$ of an hour each day for 5 days?

20. Ebony Jordan burns 360 calories per hour while riding her stationary bike. How many calories does she burn when she rides her bicycle $\frac{1}{4}$ of an hour each day for 6 days?

21. Julius Davenport goes through a rigorous exercise routine each day. He burns calories at a rate of 720 calories per hour. How many calories does he need to exercise 20 minutes per day, 6 days a week?

22. A roller skater can easily use 325 calories per hour while skating. How many calories are needed to roller skate 75 minutes per day for 3 days?

23. A casual stroll burns 165 calories per hour. How long will it take to stroll off the 425 calories contained in a hamburger, to the nearest tenth of an hour?

24. Even when asleep, the body burns 15 calories per hour. How long must a person sleep to burn off the calories in an 80-calorie orange, to the nearest tenth of an hour?

25. One pound of body weight is lost whenever 3500 calories are burned off. If walking briskly burns 200 calories for each mile walked, how far must Sheila Osby walk to lose 1 pound?

26. Bicycling can burn as much as 500 calories per hour. How long must a person ride a bicycle at this rate to use up the 3500 calories needed to lose 1 pound?

Review

Subtract. See Section 1.4.

27. 90 − 17   **28.** 90 − 49   **29.** 180 − 17   **30.** 180 − 49   **31.** 180 − 161   **32.** 180 − 101

Concept Extensions

33. A 123.9-pound pile of prepacked canned goods must be lifted 9 inches to permit a door to close. How much energy is needed to do the job?

34. A 14.3-pound framed picture must be lifted 6 feet 3 inches. How much energy is required to move the picture?

35. 6400 ft-lb of energy were needed to lift an anvil 25 feet. Find the weight of the anvil.

36. 825 ft-lb of energy were needed to lift 40 pounds of apples to a new container. How far were the apples moved?

37. In your own words, define calorie.

Chapter 7 Group Activity

Speeds

Section 7.1

A speed measures how far something travels in a given unit of time. You already learned in Section 5.2 that the speed 55 miles per hour is a rate that can be written as $\frac{55 \text{ miles}}{1 \text{ hour}}$. Just as there are different units of measurement for length or distance, there are different units of measurement for speed as well. It is also possible to perform unit conversions on speeds. Before we learn about converting speeds, we will review units of time. The following is a summary of equivalencies between various units of time.

Units of Time	Unit Fractions
60 seconds (s) = 1 minute (min)	$\frac{60 \text{ s}}{1 \text{ min}} = \frac{1 \text{ min}}{60 \text{ s}} = 1$
60 minutes = 1 hour (h)	$\frac{60 \text{ min}}{1 \text{ h}} = \frac{1 \text{ h}}{60 \text{ min}} = 1$
3600 seconds = 1 hour	$\frac{3600 \text{ s}}{1 \text{ h}} = \frac{1 \text{ h}}{3600 \text{ s}}$

Here are some common speeds.

Speeds

Miles per hour (mph)
Miles per minute (mi/min)
Miles per second (mi/s)
Feet per second (ft/s)
Feet per minute (ft/min)
Kilometers per hour (kmph or km/h)
Kilometers per second (kmps or km/s)
Meters per second (m/s)
Knots

To convert from one speed to another, unit fractions may be used. To convert from mph to ft/s first write the original speed as a unit rate. Then multiply by a unit fraction that relates miles to feet and by a unit fraction that relates hours to seconds. The unit fractions should be written so that like units will divide out. For example, to convert 55 mph to ft/s:

$$55 \text{ mph} = \frac{55 \text{ miles}}{1 \text{ hour}} = \frac{55 \text{ miles}}{1 \text{ hour}} \cdot \frac{5280 \text{ ft}}{1 \text{ mile}} \cdot \frac{1 \text{ hour}}{3600 \text{ s}}$$

$$= \frac{55 \cdot 5280 \text{ ft}}{3600 \text{ s}}$$

$$= \frac{290{,}400 \text{ ft}}{3600 \text{ s}}$$

$$= 80\frac{2}{3} \text{ ft/s}$$

Group Activity

1. Research the current world land speed record. Convert the speed from mph to feet per second.

2. Research the current world water speed record. Convert from mph to knots.

3. Research and then describe the Beaufort Wind Scale, its origins, and how it is used. Give the scale keyed to both miles per hour and knots. Why would both measures be useful?

> **Helpful Hint**
>
> A **knot** is 1 nautical mile per hour and is a measure of speed used for ships.
>
> 1 nautical mile (nmi) ≈ 1.15 miles (mi)
> 1 nautical mile (nmi) ≈ 6076.12 feet (ft)

Chapter 7 Vocabulary Check

Fill in each blank with one of the words or phrases listed below.

mass	unit fractions	gram	energy	weight
meter	liter		calorie	British Thermal Unit

1. _____ is a measure of the pull of gravity.

2. _____ is a measure of the amount of substance in an object. This measure does not change.

3. The basic unit of length in the metric system is the _____.

4. To convert from one unit of length to another, _____ may be used.

5. A(n) _____ is the basic unit of mass in the metric system.

6. _____ is the capacity to do work.

7. In the U.S. system of measurement, a(n) _____ is the amount of heat required to raise the temperature of 1 pound of water 1 degree Fahrenheit.

8. The _____ is the basic unit of capacity in the metric system.

9. In the metric system, a(n) _____ is the amount of heat required to raise the temperature of 1 kilogram of water 1 degree Celsius.

> **Helpful Hint**
>
> ▶ Are you preparing for your test? Don't forget to take the Chapter 7 Test on page 493. Then check your answers at the back of the text and use the Chapter Test Prep Videos to see the fully worked-out solutions to any of the exercises you want to review.

7 Chapter Highlights

Definitions and Concepts	Examples
Section 7.1 Length: U.S. and Metric Systems of Measurement	

Definitions and Concepts	Examples
To convert from one unit of length to another, multiply by a **unit fraction** in the form $$\frac{\text{units to convert to}}{\text{original units}}.$$ **Length: U.S. System of Measurement** $$12 \text{ inches (in.)} = 1 \text{ foot (ft)}$$ $$3 \text{ feet} = 1 \text{ yard (yd)}$$ $$5280 \text{ feet} = 1 \text{ mile (mi)}$$ The basic unit of length in the metric system is the **meter.** A meter is slightly longer than a yard.	$$\frac{12 \text{ inches}}{1 \text{ foot}}, \frac{1 \text{ foot}}{12 \text{ inches}}, \frac{3 \text{ feet}}{1 \text{ yard}}$$ Convert 6 feet to inches. $$6 \text{ ft} = \frac{6 \text{ ft}}{1} \cdot 1$$ $$= \frac{6 \text{ ft}}{1} \cdot \frac{12 \text{ in.}}{1 \text{ ft}} \begin{array}{l} \leftarrow \text{units to convert to} \\ \leftarrow \text{original units} \end{array}$$ $$= 6 \cdot 12 \text{ in.}$$ $$= 72 \text{ in.}$$ Convert 3650 centimeters to meters. $$3650 \text{ cm} = 3650 \text{ cm} \cdot 1$$ $$= \frac{3650 \text{ cm}}{1} \cdot \frac{0.01 \text{ m}}{1 \text{ cm}} = 36.5 \text{ m}$$

Length: Metric System of Measurement

Metric Unit of Length
1 **kilo**meter (km) = 1000 meters (m)
1 **hecto**meter (hm) = 100 m
1 **deka**meter (dam) = 10 m
1 meter (m) = 1 m
1 **deci**meter (dm) = 1/10 m or 0.1 m
1 **centi**meter (cm) = 1/100 m or 0.01 m
1 **milli**meter (mm) = 1/1000 m or 0.001 m

or

km   hm   dam   m   dm   cm   mm

2 units to the left

$$3650 \text{ cm} = 36.5 \text{ m}$$

2 places to the left

Definitions and Concepts	Examples

Section 7.2   Weight and Mass: U.S. and Metric Systems of Measurement

Weight is really a measure of the pull of gravity. **Mass** is a measure of the amount of substance in an object and does not change.

Weight: U.S. System of Measurement

   16 ounces (oz) = 1 pound (1b)

    2000 pounds = 1 ton

A **gram** is the basic unit of mass in the metric system. It is the mass of water contained in a cube 1 centimeter on each side. A paper clip weighs about 1 gram.

Mass: Metric System of Measurement

Metric Unit of Mass
1 kilogram (kg) = 1000 grams (g)
1 hectogram (hg) = 100 g
1 dekagram (dag) = 10 g
1 gram (g) = 1 g
1 decigram (dg) = 1/10 g or 0.1 g
1 centigram (cg) = 1/100 g or 0.01 g
1 milligram (mg) = 1/1000 g or 0.001 g

Convert 5 pounds to ounces.

$$5 \text{ lb} = 5 \text{ lb} \cdot 1 = \frac{5 \text{ lb}}{1} \cdot \frac{16 \text{ oz}}{1 \text{ lb}} = 80 \text{ oz}$$

Convert 260 grams to kilograms.

$$260 \text{ g} = \frac{260 \text{ g}}{1} \cdot 1 = \frac{260 \text{ g}}{1} \cdot \frac{1 \text{ kg}}{1000 \text{ g}} = 0.26 \text{ kg}$$

or

kg   hg   dag   g   dg   cg   mg

3 units to the left

260 g = 0.260 kg

3 places to the left

Section 7.3   Capacity: U.S. and Metric Systems of Measurement

Capacity: U.S. System of Measurement

   8 fluid ounces (fl oz) = 1 cup (c)

     2 cups = 1 pint (pt)

     2 pints = 1 quart (qt)

     4 quarts = 1 gallon (gal)

The **liter** is the basic unit of capacity in the metric system. It is the capacity or volume of a cube measuring 10 centimeters on each side. A liter of liquid is slightly more than 1 quart.

Capacity: Metric System of Measurement

Metric Unit of Capacity
1 kiloliter (kl) = 1000 liters (L)
1 hectoliter (hl) = 100 L
1 dekaliter (dal) = 10 L
1 liter (L) = 1 L
1 deciliter (dl) = 1/10 L or 0.1 L
1 centiliter (cl) = 1/100 L or 0.01 L
1 milliliter (ml) = 1/1000 L or 0.001 L

Convert 5 pints to gallons.

   1 gal = 4 qt = 8 pt

$$5 \text{ pt} = 5 \text{ pt} \cdot 1 = \frac{5 \text{ pt}}{1} \cdot \frac{1 \text{ gal}}{8 \text{ pt}} = \frac{5}{8} \text{ gal}$$

Convert 1.5 liters to milliliters.

$$1.5 \text{ L} = \frac{1.5 \text{ L}}{1} \cdot 1 = \frac{1.5 \text{ L}}{1} \cdot \frac{1000 \text{ ml}}{1 \text{ L}} = 1500 \text{ ml}$$

or

kl   hl   dal   L   dl   cl   ml

3 units to the right

1.500 L = 1500 ml

3 places to the right

Definitions and Concepts	Examples

Section 7.4   Conversions Between the U.S. and Metric Systems

To convert between systems, use approximate unit fractions from Section 7.4.	Convert 7 feet to meters. $$7 \text{ ft} \approx \frac{7 \cancel{\text{ft}}}{1} \cdot \frac{0.30 \text{ m}}{1 \cancel{\text{ft}}} = 2.1 \text{ m}$$ Convert 8 liters to quarts. $$8 \text{ L} \approx \frac{8 \cancel{\text{L}}}{1} \cdot \frac{1.06 \text{ qt}}{1 \cancel{\text{L}}} = 8.48 \text{ qt}$$ Convert 363 grams to ounces. $$363 \text{ g} \approx \frac{363 \cancel{\text{g}}}{1} \cdot \frac{0.04 \text{ oz}}{1 \cancel{\text{g}}} = 14.52 \text{ oz}$$

Section 7.5   Temperature: U.S. and Metric Systems of Measurement

To Convert from Celsius Temperature to Fahrenheit Temperature $$F = \frac{9}{5} \cdot C + 32 \quad \text{or} \quad F = 1.8 \cdot C + 32$$ **To Convert from Fahrenheit Temperature to Celsius Temperature** $$C = \frac{5}{9} \cdot (F - 32)$$	Convert $35°C$ to degrees Fahrenheit. $$F = \frac{9}{5} \cdot 35 + 32 = 63 + 32 = 95$$ $35°C = 95°F$ Convert $50°F$ to degrees Celsius. $$C = \frac{5}{9} \cdot (50 - 32) = \frac{5}{9} \cdot 18 = 10$$ $50°F = 10°C$

Section 7.6   Energy: U.S. and Metric Systems of Measurement

Energy is the capacity to do work. In the U.S. system of measurement, a **foot-pound** (ft-lb) is the amount of energy needed to lift a 1-pound object a distance of 1 foot. In the U.S. system of measurement, a **British Thermal Unit** (BTU) is the amount of heat required to raise the temperature of 1 pound of water 1 degree Fahrenheit. (1 BTU = 778 ft-lb) In the metric system, a **calorie** is the amount of heat required to raise the temperature of 1 kilogram of water 1 degree Celsius.	How much energy is needed to lift a 20-pound object 35 feet? $$20 \text{ pounds} \cdot 35 \text{ feet} = 700 \text{ foot-pounds of energy}$$ Convert 50 BTU to foot-pounds. $$50 \text{ BTU} = 50 \cancel{\text{BTU}} \cdot \frac{778 \text{ ft-lb}}{1 \cancel{\text{BTU}}} = 38,900 \text{ ft-lb}$$ A stationary bicyclist uses 350 calories in 1 hour. How many calories will the bicyclist use in $\frac{1}{2}$ of an hour? $$\text{calories used} = 350 \cdot \frac{1}{2} = 175 \text{ calories}$$

(7.1) *Convert.*

1. 108 in. to feet

2. 72 ft to yards

3. 1.5 mi to feet

4. $\frac{1}{2}$ yd to inches

5. 52 ft = ____ yd ____ ft

6. 46 in. = ____ ft ____ in.

7. 42 m to centimeters

8. 82 cm to millimeters

9. 12.18 mm to meters

10. 2.31 m to kilometers

Perform each indicated operation.

11. 4 yd 2 ft + 16 yd 2 ft

12. 12 ft 1 in. − 4 ft 8 in.

13. 8 ft 3 in. × 5

14. 7 ft 4 in. ÷ 2

15. 8 cm + 15 mm

16. 4 m − 126 cm

17. 8.62 m × 4

18. 19.6 km ÷ 8

Solve.

19. A bolt of cloth contains 333 yd 1 ft of cotton ticking. Find the amount of material that remains after 163 yd 2 ft is removed from the bolt.

20. The local ambulance corps plans to award 20 framed certificates of valor to some of its outstanding members. If each frame requires 6 ft 4 in. of framing material, how much material is needed for all the frames?

21. The trip from Philadelphia to Washington, D.C., is 217 km each way. Four friends agree to share the driving equally. How far must each drive on this round-trip vacation?

△ **22.** The college has ordered that NO SMOKING signs be placed above the doorway of each classroom. Each sign is 0.8 m long and 30 cm wide. Find the area of each sign. (*Hint:* Recall that the area of a rectangle = width · length.)

0.8 meter

30 centimeters

(7.2) *Convert.*

23. 66 oz to pounds

24. 2.3 tons to pounds

25. 52 oz = ____ lb ____ oz

26. 10,300 lb = ____ tons ____ lb

27. 27 mg to grams

28. 40 kg to grams

29. 2.1 hg to dekagrams

30. 0.03 mg to decigrams

Perform each indicated operation.

31. 6 lb 5 oz − 2 lb 12 oz

32. 5 tons 1600 lb + 4 tons 1200 lb

33. 6 tons 2250 lb ÷ 3

34. 8 lb 6 oz × 4

35. 4.3 mg × 5

36. 4.8 kg − 4200 g

Solve.

37. Donshay Berry ordered 1 lb 12 oz of soft-center candies and 2 lb 8 oz of chewy-center candies for his party. Find the total weight of the candy ordered.

38. Four local townships jointly purchase 38 tons 300 lb of cinders to spread on their roads during an ice storm. Determine the weight of the cinders each township receives if they share the purchase equally.

39. Linda Holden ordered 8.3 kg of whole wheat flour from the health-food store, but she received 450 g less. How much flour did she actually receive?

40. Eight friends spent a weekend in the Poconos tapping maple trees and preparing 9.3 kg of maple syrup. Find the weight each friend receives if they share the syrup equally.

(7.3) *Convert.*

41. 16 pints to quarts

42. 40 fluid ounces to cups

43. 3 qt 1 pt to pints

44. 18 quarts to cups

45. 9 pt = ____qt____pt

46. 15 qt = ____gal____qt

47. 3.8 L to milliliters

48. 4.2 ml to deciliters

49. 14 hl to kiloliters

50. 30.6 L to centiliters

Perform each indicated operation.

51. 1 qt 1 pt + 3 qt 1 pt

52. 3 gal 2 qt × 2

53. 0.946 L − 210 ml

54. 6.1 L + 9400 ml

Solve.

55. Carlos Perez prepares 4 gal 2 qt of iced tea for a block party. During the first 30 minutes of the party, 1 gal 3 qt of the tea is consumed. How much iced tea remains?

56. A recipe for soup stock calls for 1 c 4 fl oz of beef broth. How much should be used if the recipe is cut in half?

57. Each bottle of Kiwi liquid shoe polish holds 85 ml of the polish. Find the number of liters of shoe polish contained in 8 boxes if each box contains 16 bottles.

58. Ivan Miller wants to pour three separate containers of saline solution into a single vat with a capacity of 10 liters. Will 6 liters of solution in the first container combined with 1300 milliliters in the second container and 2.6 liters in the third container fit into the larger vat?

(7.4) *Note: Because approximations are used in this section, your answers may vary slightly from the answers given in the back of the book.*

Convert as indicated. If necessary, round to two decimal places.

59. 7 meters to feet

60. 11.5 yards to meters

61. 17.5 liters to gallons

62. 7.8 liters to quarts

63. 15 ounces to grams

64. 23 pounds to kilograms

65. A 100-meter dash is being held today. How many yards is this?

66. If a person weighs 82 kilograms, how many pounds is this?

67. How many quarts are contained in a 3-liter bottle of cola?

68. A compact disc is 1.2 mm thick. Find the height (in inches) of 50 discs.

(7.5) *Convert. Round to the nearest tenth of a degree, if necessary.*

69. 245°C to degrees Fahrenheit

70. 160°C to degrees Fahrenheit

71. 42°C to degrees Fahrenheit

72. 93.2°F to degrees Celsius

73. 41.3°F to degrees Celsius

74. 80°F to degrees Celsius

Solve. Round to the nearest tenth of a degree, if necessary.

75. A sharp dip in the jet stream caused the temperature in New Orleans to drop to 35°F. Find the corresponding temperature in degrees Celsius.

76. The recipe for meat loaf calls for a 165°C oven. Find the setting used if the oven has a Fahrenheit thermometer.

(7.6) *Solve.*

77. How many foot-pounds of energy are needed to lift a 5.6-pound radio a distance of 12 feet?

78. How much energy is required to lift a 21-pound carton of Rice-A-Roni a distance of 6.5 feet?

79. How much energy is used when a 1.2-ton pile of sand is lifted 15 yards?

80. The energy required to operate a 12,000-BTU air conditioner is equivalent to how many foot-pounds?

81. Convert 2,000,000 foot-pounds to BTU, rounded to the nearest hundred.

82. Kip Yates burns off 450 calories each hour he plays handball. How many calories does he use to play handball for $2\frac{1}{2}$ hours?

83. Qwanetta Sesson uses 210 calories each hour she spends mowing the grass. Find the number of calories needed to mow the grass if she spends 3 hours mowing each week for 24 weeks.

84. Four ounces of sirloin steak contain 420 calories. If Edith Lutrell uses 180 calories each hour she walks, how long must she walk to burn off the calories from this steak?

Mixed Review

Convert the following.

85. 2.5 mi to feet

86. 6.25 ft to inches

87. 23,760 ft to miles

88. 129 in. to feet

89. 8200 lb = ____ tons ____ lb

90. 4300 lb = ____ tons ____ lb

91. 5 m to centimeters

92. 286 mm to kilometers

93. 1400 mg to grams

94. 240 mg to grams

95. 6.75 gallons to quarts

96. 5.25 gallons to quarts

97. 8.5 pints to cups

98. 6.25 pints to cups

99. 86°C to degrees Fahrenheit

100. 15°C to degrees Fahrenheit

101. 51.8°F to degrees Celsius

102. 82.4°F to degrees Celsius

Perform the indicated operations and simplify.

103. 9.3 km − 183 m

104. 8.6 km − 247 m

105. 7.4 L + 6500 ml

106. 35 L + 700 ml

107. 9.3 g − 1200 mg

108. 3.4 g − 1800 mg

109. 6.3 kg × 8

110. 3.2 kg × 4

111. 3 gal 1 qt + 4 gal 2 qt

112. 6 gal 1 qt + 2 gal 1 qt

113. 4100 mm − 3 dm

114. 6300 mm − 5 dm

115. 4.5 tons ÷ 2

116. 6.75 tons ÷ 3

Convert.

1. 280 in. = ____ ft ____ in.

2. $2\frac{1}{2}$ gal to quarts

3. 30 oz to pounds

4. 2.8 tons to pounds

5. 38 pt to gallons

6. 40 mg to grams

7. 2.4 kg to grams

8. 3.6 cm to millimeters

9. 4.3 dg to grams

10. 0.83 L to milliliters

Perform each indicated operation.

11. 3 qt 1 pt + 2 qt 1 pt

12. 8 lb 6 oz − 4 lb 9 oz

13. 2 ft 9 in. × 3

14. 5 gal 2 qt ÷ 2

15. 8 cm − 14 mm

16. 1.8 km + 456 m

Convert. Round to the nearest tenth of a degree, if necessary.

17. 84°F to degrees Celsius

18. 12.6°C to degrees Fahrenheit

19. The sugar maples in front of Bette MacMillan's house are 8.4 meters tall. Because they interfere with the phone lines, the telephone company plans to remove the top third of the trees. How tall will the maples be after they are shortened?

20. A total of 15 gal 1 qt of oil has been removed from a 20-gallon drum. How much oil still remains in the container?

21. The engineer in charge of bridge construction said that the span of a certain bridge would be 88 m. But the actual construction required it to be 340 cm longer. Find the span of the bridge, in meters.

22. If 2 ft 9 in. of material is used to manufacture one scarf, how much material is needed for 6 scarves?

1. _____

2. _____

3. _____

4. _____

5. _____

6. _____

7. _____

8. _____

9. _____

10. _____

11. _____

12. _____

13. _____

14. _____

15. _____

16. _____

17. _____

18. _____

19. _____

20. _____

21. _____

22. _____

23. _____

24. _____

25. _____

26. _____

27. _____

28. _____

29. _____

30. _____

31. _____

32. _____

33. _____

23. The Vietnam Veterans Memorial, inscribed with the names of 58,226 deceased and missing U.S. soldiers from the Vietnam War, is located on the National Mall in Washington, D.C. This memorial is formed from two straight sections of wall that meet at an angle at the center of the monument. Each wall is 246 ft 9 in. long. What is the total length of the Vietnam Veterans Memorial's wall? (_Source:_ National Park Service)

24. Each panel making up the wall of the Vietnam Veterans Memorial is 101.6 cm wide. There are a total of 148 panels making up the wall. What is the total length of the wall in meters? (_Source:_ National Park Service)

25. The hottest temperature ever recorded was 134°F in Death Valley at Furnace Creek Ranch, CA, on July 10, 1913. Convert this temperature to degrees Celsius. Round your answer to the nearest tenth of a degree. (_Source:_ National Climatic Data Center)

26. The doctors are quite concerned about Lucia Gillespie, who is running a 41°C fever. Find Lucia's temperature in degrees Fahrenheit.

27. The largest ice cream sundae ever made in the United States was assembled in Anaheim, California, in 1985. This giant sundae used 4667 gallons of ice cream. How many pints of ice cream were used?

28. A piece of candy weighs 5 grams. How many ounces is this?

29. A 5-kilometer race is being held today. How many miles is this?

30. A 5-gallon container holds how many liters?

31. Phillipe Jordaine must lift a 48.5-pound carton of canned tomatoes a distance of 14 feet. Find the energy required to move the carton.

32. Energy used by a 26,000-BTU heater is equivalent to how many foot-pounds?

33. Robin Nestle burns 180 calories each hour when she swims. How many calories does she use if she swims 1 hour per day for 5 days?

1. Add: $1647 + 246 + 32 + 85$

2. Subtract: $2000 - 469$

3. Find the prime factorization of 945.

4. Find the area of the rectangle.

17 in.

9 in.

5. Find the LCM of 11 and 33.

6. Subtract: $\dfrac{8}{21} - \dfrac{2}{9}$

7. Add: $3\dfrac{4}{5} + 1\dfrac{4}{15}$

8. Multiply: $2\dfrac{1}{2} \cdot 4\dfrac{2}{15}$

Write each decimal as a fraction or mixed number in simplest form.

9. 0.125

10. 1.2

11. 105.083

12. Evaluate: $\left(\dfrac{2}{3}\right)^3$

13. Insert $<$, $>$, or $=$ to form a true statement.
0.052   0.236

14. Evaluate: $30 \div 6 \cdot 5$

15. Subtract: $85 - 17.31$. Check your answer.

16. Add: $27.9 + 8.07 + 103.261$

Multiply.

17. 42.1×0.1

18. 186.04×1000

19. 9.2×0.001

20. Find the average of 6.8, 9.7, and 0.9.

21. Divide: $8.32 \div 32$. Check your answer.

22. Add: $\dfrac{3}{10} + \dfrac{3}{4}$

23. Write $2\dfrac{3}{16}$ as a decimal.

24. Round 7.2846 to the nearest tenth.

25. Write $\dfrac{2}{3}$ as a decimal.

26. Simplify: $\dfrac{0.12 + 0.96}{0.5}$

Answers

1. _____

2. _____

3. _____

4. _____

5. _____

6. _____

7. _____

8. _____

9. _____

10. _____

11. _____

12. _____

13. _____

14. _____

15. _____

16. _____

17. _____

18. _____

19. _____

20. _____

21. _____

22. _____

23. _____

24. _____

25. _____

26. _____

27. _____

28. _____

29. _____

30. _____

31. _____

32. _____

33. _____

34. _____

35. _____

36. _____

37. _____

38. _____

39. _____

40. _____

41. _____

42. _____

43. _____

44. _____

45. _____

46. _____

47. _____

48. _____

49. _____

50. _____

27. Write the ratio of 12 to 17 using fractional notation.

28. Write the ratio of $2\frac{2}{3}$ to $5\frac{1}{9}$ as a ratio of whole numbers using fractional notation.

29. Write "337.5 miles every 15 gallons of gas" as a unit rate.

30. A square is 9 inches by 9 inches. Find the ratio of a side to its perimeter.

31. Find the unknown number n. $\dfrac{7}{n} = \dfrac{6}{5}$

32. A recipe that makes 2 piecrusts calls for 3 cups of flour. How much flour is needed to make 5 piecrusts?

△ **33.** A 50-pound bag of fertilizer covers 2400 square feet of lawn. How many bags of fertilizer are needed to cover a town square containing 15,360 square feet of lawn? Round the answer up to the nearest whole bag.

34. Write 23% as a fraction.

35. Write 23% as a decimal.

36. Write $\dfrac{7}{8}$ as a percent.

37. Write $\dfrac{1}{12}$ as a percent. Round to the nearest hundredth percent.

38. 108 is what percent of 450?

39. What number is 35% of 40?

40. Convert 4 gallons to pints.

41. Translate to a proportion. What percent of 30 is 75?

42. Convert 8.6 meters to centimeters.

43. In response to a decrease in sales, a company with 1500 employees reduces the number of employees to 1230. What is the percent of decrease?

44. Convert 13,000 pounds to tons.

45. An electric rice cooker that normally sells for $65 is on sale for 25% off. What is the amount of discount and what is the sale price?

46. Convert $3\frac{1}{4}$ pounds to ounces.

47. Find the simple interest after 2 years on $500 at an interest rate of 12%.

48. The number of faculty at a local community college was recently increased from 240 to 276. What is the percent of increase?

49. Convert 9000 pounds to tons.

50. Convert 25° Celsius to degrees Fahrenheit.

Geometry

A Swiss company created the Zaugg Pipe Monster (shown above) specifically for building superpipes.

What Is a Superpipe?

For winter sports, the term *superpipe* is used to describe a halfpipe built of snow that has walls 22 feet high from the flat bottom on both sides. The length of a superpipe ranges from 400 feet to 600 feet.

Halfpipes in snow were originally formed by hand tools or with heavy machinery. The current method of halfpipe cutting and grooming is by use of a Zaugg Pipe Monster, shown above. Because of the high expense of constructing and maintaining them, there are very few true superpipes. During the 2013–2014 northern hemisphere winter, only 14 superpipes existed globally.

Throughout this chapter, we work with lines, angles, circumferences, and other geometric concepts that give us an appreciation of the work that goes into constructing a halfpipe.

1998 Olympics Nagano, Japan — 394 feet, 11.5 feet, 49 feet

2014 Olympics Sochi, Russia — 591 feet, 22 feet, 66 feet

The word *geometry* is formed from the Greek words *geo*, meaning Earth, and *metron*, meaning measure. Geometry literally means to measure the Earth. In this chapter we learn about various geometric figures and their properties such as perimeter, area, and volume. Knowledge of geometry can help us solve practical problems in real-life situations. For instance, knowing certain measures of a circular swimming pool allows us to calculate how much water it can hold.

497

Objectives

A Identify Lines, Line Segments, Rays, and Angles. ▷

B Classify Angles as Acute, Right, Obtuse, or Straight. ▷

C Identify Complementary and Supplementary Angles. ▷

D Find Measures of Angles. ▷

Objective A Identifying Lines, Line Segments, Rays, and Angles ▷

Let's begin with a review of two important concepts—space and plane.

Space extends in all directions indefinitely. Examples of objects in space are houses, grains of salt, bushes, your *Basic College Mathematics* textbook, and you.

A **plane** is a flat surface that extends indefinitely. Surfaces like a plane are a classroom floor or a blackboard or whiteboard.

Plane

The most basic concept of geometry is the idea of a point in space. A **point** has no length, no width, and no height, but it does have location. We represent a point by a dot, and we usually label points with capital letters.

P

Point *P*

A **line** is a set of points extending indefinitely in two directions. A line has no width or height, but it does have length. We can name a line by any two of its points or by a single lowercase letter. A **line segment** is a piece of a line with two endpoints.

Line *AB*, $\overleftrightarrow{AB}$, or line *l**   Line segment *AB* or $\overline{AB}$

A **ray** is a part of a line with one endpoint. A ray extends indefinitely in one direction. An **angle** is made up of two rays that share the same endpoint. The common endpoint is called the **vertex.**

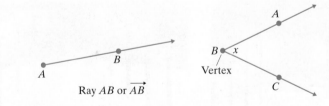

Ray *AB* or $\overrightarrow{AB}$

The angle in the figure above can be named

$\angle ABC$ $\angle CBA$ $\angle B$ or $\angle x$

   ↑   ↑

The vertex is the middle point.

Rays *BA* and *BC* are **sides** of the angle.

*Although line *l* is also line *BA* or $\overleftrightarrow{BA}$, we will use only one order of points to name a line or line segment.

Helpful Hint

Naming an Angle
When there is no confusion as to what angle is being named, you may use the vertex alone.

Name of ∠B is all right.
There is no confusion. ∠B means ∠1.

Name of ∠B is *not* all right.
There is confusion. Does ∠B mean
∠1, ∠2, ∠3, or ∠4?

 Example 1   Identify each figure as a line, a ray, a line segment, or an angle. Then name the figure using the given points.

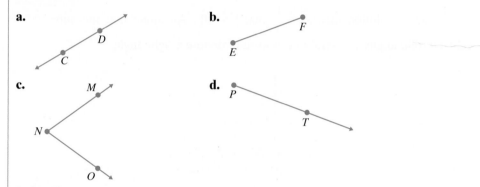

a.

b.

c.

d.

Solution:

Figure (a) extends indefinitely in two directions. It is line *CD* or $\overleftrightarrow{CD}$.
Figure (b) has two endpoints. It is line segment *EF* or $\overline{EF}$.
Figure (c) has two rays with a common endpoint. It is ∠*MNO*, ∠*ONM*, or ∠*N*.
Figure (d) is part of a line with one endpoint. It is ray *PT* or $\overrightarrow{PT}$.

 Work Practice 1

Practice 1

Identify each figure as a line, a ray, a line segment, or an angle. Then name the figure using the given points.

a.   **b.**

c.

d.

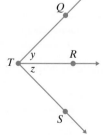 **Example 2**   List other ways to name ∠*y*.

Solution: Two other ways to name ∠*y* are ∠*QTR* and ∠*RTQ*. We may *not* use the vertex alone to name this angle because three different angles have *T* as their vertex.

Work Practice 2

Practice 2

Use the figure in Example 2 to list other ways to name ∠*z*.

Answers
1. a. ray; ray *AB* or $\overrightarrow{AB}$ **b.** line segment; line segment *RS* or $\overline{RS}$
c. line; line *EF* or $\overleftrightarrow{EF}$ **d.** angle; ∠*TVH* or ∠*HVT* or ∠*V*
2. ∠*RTS*, ∠*STR*

Objective B Classifying Angles as Acute, Right, Obtuse, or Straight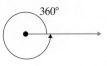

An angle can be measured in **degrees.** The symbol for degrees is a small, raised circle, °. There are 360° in a full revolution, or a full circle.

$\frac{1}{2}$ of a revolution measures $\frac{1}{2}(360°) = 180°$. An angle that measures 180° is called a **straight angle.**

$\angle RST$ is a straight angle.

$\frac{1}{4}$ of a revolution measures $\frac{1}{4}(360°) = 90°$. An angle that measures 90° is called a **right angle.** The symbol ∟ is used to denote a right angle.

$\angle ABC$ is a right angle.

An angle whose measure is between 0° and 90° is called an **acute angle.**

Acute angles

An angle whose measure is between 90° and 180° is called an **obtuse angle.**

Obtuse angles

Practice 3

Classify each angle as acute, right, obtuse, or straight.

a.   **b.**

c.   **d.**

Example 3 Classify each angle as acute, right, obtuse, or straight.

a.   **b.**

c.   **d.**

Solution:

a. $\angle R$ is a right angle, denoted by ∟. It measures 90°.
b. $\angle S$ is a straight angle. It measures 180°.
c. $\angle T$ is an acute angle. It measures between 0° and 90°.
d. $\angle Q$ is an obtuse angle. It measures between 90° and 180°.

▮ Work Practice 3

Let's look at $\angle B$ below, whose measure is 62°.

There is a shorthand notation for writing the measure of this angle. To write "The measure of $\angle B$ is 62°," we can write

$$m\angle B = 62°.$$

By the way, note that $\angle B$ is an acute angle because $m\angle B$ is between 0° and 90°.

Objective C Identifying Complementary and Supplementary Angles ▶

Two angles that have a sum of 90° are called **complementary angles.** We say that each angle is the **complement** of the other.

$\angle R$ and $\angle S$ are complementary angles because

$$m\angle R + m\angle S = 60° + 30° = 90°$$

Complementary angles
60° + 30° = 90°

Two angles that have a sum of 180° are called **supplementary angles.** We say that each angle is the **supplement** of the other.

$\angle M$ and $\angle N$ are supplementary angles because

$$m\angle M + m\angle N = 125° + 55° = 180°$$

Supplementary angles
125° + 55° = 180°

Example 4   Find the complement of a 48° angle.

Solution: Two angles that have a sum of 90° are complementary. This means that the complement of an angle that measures 48° is an angle that measures $90° - 48° = 42°$.

▮ Work Practice 4

Practice 4

Find the complement of a 29° angle.

Answer
4. 61°

Practice 5

Find the supplement of a 67° angle.

Example 5 Find the supplement of a 107° angle.

Solution: Two angles that have a sum of 180° are supplementary. This means that the supplement of an angle that measures 107° is an angle that measures 180° − 107° = 73°.

■ Work Practice 5

✓**Concept Check** True or false? The supplement of a 48° angle is 42°. Explain.

Objective D Finding Measures of Angles ▶

Measures of angles can be added or subtracted to find measures of related angles.

Practice 6

a. Find the measure of ∠y.

b. Find the measure of ∠x.

c. Classify ∠x and ∠y as acute, obtuse, or right angles.

Example 6 Find the measure of ∠x. Then classify ∠x as an acute, obtuse, or right angle.

Solution: $m\angle x = m\angle QTS - m\angle RTS$
$= 87° - 52°$
$= 35°$

Thus, the measure of ∠x (m∠x) is 35°.
Since ∠x measures between 0° and 90°, it is an acute angle.

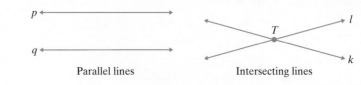

■ Work Practice 6

Two lines in a plane can be either parallel or intersecting. **Parallel lines** never meet. **Intersecting lines** meet at a point. The symbol ∥ is used to indicate "is parallel to." For example, in the figure, p ∥ q.

Parallel lines   Intersecting lines

Some intersecting lines are perpendicular. Two lines are **perpendicular** if they form right angles when they intersect. The symbol ⊥ is used to denote "is perpendicular to." For example, in the figure below, m ⊥ n.

Perpendicular lines

When two lines intersect, four angles are formed. Two angles that are opposite each other are called **vertical angles.** Vertical angles have the same measure.

Two angles that share a common side are called **adjacent angles.** Adjacent angles formed by intersecting lines are supplementary. That is, the sum of their measures is 180°.

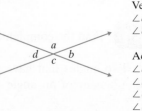

Vertical angles:
∠a and ∠c
∠d and ∠b

Adjacent angles:
∠a and ∠b
∠b and ∠c
∠c and ∠d
∠d and ∠a

Answers

5. 113° **6. a.** 44° **b.** 28° **c.** both acute

✓**Concept Check Answer**

false; the *complement* of a 48° angle is 42°; the *supplement* of a 48° angle is 132°

Here are a few real-life examples of the lines we just discussed.

Parallel lines

Vertical angles

Perpendicular lines

Example 7 Find the measures of $\angle x$, $\angle y$, and $\angle z$ if the measure of $\angle t$ is 42°.

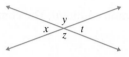

Solution: Since $\angle t$ and $\angle x$ are vertical angles, they have the same measure, so $\angle x$ measures 42°.

Since $\angle t$ and $\angle y$ are adjacent angles, their measures have a sum of 180°. So $\angle y$ measures $180° - 42° = 138°$.

Since $\angle y$ and $\angle z$ are vertical angles, they have the same measure. So $\angle z$ measures 138°.

▪ Work Practice 7

A line that intersects two or more lines at different points is called a **transversal.** Line l is a transversal that intersects lines m and n. The eight angles formed have special names. Some of these names are:

Corresponding angles: $\angle a$ and $\angle e$, $\angle c$ and $\angle g$, $\angle b$ and $\angle f$, $\angle d$ and $\angle h$

Alternate interior angles: $\angle c$ and $\angle f$, $\angle d$ and $\angle e$

When two lines cut by a transversal are *parallel,* the following statement is true:

Parallel Lines Cut by a Transversal

If two parallel lines are cut by a transversal, then the measures of **corresponding angles are equal** and the measures of the **alternate interior angles are equal.**

Example 8 Given that $m \parallel n$ and that the measure of $\angle w$ is 100°, find the measures of $\angle x$, $\angle y$, and $\angle z$.

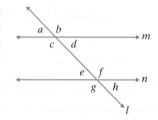

Solution:

$m\angle x = 100°$	$\angle x$ and $\angle w$ are vertical angles.
$m\angle z = 100°$	$\angle x$ and $\angle z$ are corresponding angles.
$m\angle y = 180° - 100° = 80°$	$\angle z$ and $\angle y$ are supplementary angles.

▪ Work Practice 8

Practice 7

Find the measures of $\angle a$, $\angle b$, and $\angle c$.

Practice 8

Given that $m \parallel n$ and that the measure of $\angle w = 45°$, find the measures of all the angles shown.

Answers

7. $m\angle a = 109°$; $m\angle b = 71°$; $m\angle c = 71°$

8. $m\angle x = 45°$; $m\angle y = 45°$; $m\angle z = 135°$; $m\angle a = 135°$; $m\angle b = 135°$; $m\angle c = 135°$; $m\angle d = 45°$

Vocabulary, Readiness & Video Check

Use the choices below to fill in each blank.

acute	straight	degrees	adjacent	parallel	intersecting
obtuse	space	plane	point	vertical	vertex
right	angle	ray	line	perpendicular	transversal

1. A(n) _____ is a flat surface that extends indefinitely.
2. A(n) _____ has no length, no width, and no height.
3. _____ extends in all directions indefinitely.
4. A(n) _____ is a set of points extending indefinitely in two directions.
5. A(n) _____ is part of a line with one endpoint.
6. A(n) _____ is made up of two rays that share a common endpoint. The common endpoint is called the _____.
7. A(n) _____ angle measures 180°.
8. A(n) _____ angle measures 90°.
9. A(n) _____ angle measures between 0° and 90°.
10. A(n) _____ angle measures between 90° and 180°.
11. _____ lines never meet and _____ lines meet at a point.
12. Two intersecting lines are _____ if they form right angles when they intersect.
13. An angle can be measured in _____.
14. A line that intersects two or more lines at different points is called a(n) _____.
15. When two lines intersect, four angles are formed. The angles that are opposite each other are called _____ angles.
16. Two angles that share a common side are called _____ angles.

Martin-Gay Interactive Videos   *Watch the section lecture video and answer the following questions.*

Objective A 17. In the lecture after ⊞ Example 2, what are the four ways we can name the angle shown? ▶

Objective B 18. In the lecture before ⊞ Example 3, what type of angle forms a line? What is its measure? ▶

Objective C 19. What calculation is used to find the answer to ⊞ Example 6? ▶

Objective D 20. In the lecture before ⊞ Example 7, two lines in a plane that aren't parallel must what? ▶

See Video 8.1

8.1 Exercise Set MyMathLab® ▶

Objective A *Identify each figure as a line, a ray, a line segment, or an angle. Then name the figure using the given points. See Examples 1 and 2.*

1.
2.
3.
4.

5.   **6.**   ▶ **7.**   **8.**

List two other ways to name each angle. See Example 2.

9. ∠x

10. ∠w

11. ∠z

12. ∠y

Objective **B** *Classify each angle as acute, right, obtuse, or straight. See Example 3.*

13.   **14.**   ▶ **15.**   **16.**

▶ **17.**   **18.**   **19.**   **20.**

Objective **C** *Find each complementary or supplementary angle as indicated. See Examples 4 and 5.*

▶ **21.** Find the complement of a 23° angle.   **22.** Find the complement of a 77° angle.

▶ **23.** Find the supplement of a 17° angle.   **24.** Find the supplement of a 77° angle.

25. Find the complement of a 58° angle.   **26.** Find the complement of a 22° angle.

27. Find the supplement of a 150° angle.   **28.** Find the supplement of a 130° angle.

29. Identify the pairs of complementary angles.   **30.** Identify the pairs of complementary angles.

31. Identify the pairs of supplementary angles.

32. Identify the pairs of supplementary angles.

Objective **D** *Find the measure of ∠x in each figure. See Example 6.*

33.   **34.**   **35.**   **36.**

Find the measures of angles x, y, and z in each figure. See Examples 7 and 8.

37.   **38.**   ▶ **39.**   **40.**

41. *m* ∥ *n*   **42.** *m* ∥ *n*   ▶ **43.** *m* ∥ *n*   **44.** *m* ∥ *n*

Objectives **A** **D** **Mixed Practice** *Find two other ways of naming each angle. See Example 2.*

45. ∠*x*

46. ∠*y*

47. ∠*z*

48. ∠*ABE* (just name one other way)

Find the measure of each angle in the figure above. See Example 6.

▶ **49.** ∠*ABC*   **50.** ∠*EBD*   **51.** ∠*CBD*   **52.** ∠*CBA*

▶ **53.** ∠*DBA*   **54.** ∠*EBC*   **55.** ∠*CBE*   **56.** ∠*ABE*

Review

Perform each indicated operation. See Sections 2.4, 2.5, 3.3, and 3.4.

57. $\frac{7}{8} + \frac{1}{4}$

58. $\frac{7}{8} - \frac{1}{4}$

59. $\frac{7}{8} \cdot \frac{1}{4}$

60. $\frac{7}{8} \div \frac{1}{4}$

61. $3\frac{1}{3} - 2\frac{1}{2}$

62. $3\frac{1}{3} + 2\frac{1}{2}$

63. $3\frac{1}{3} \div 2\frac{1}{2}$

64. $3\frac{1}{3} \cdot 2\frac{1}{2}$

Concept Extensions

65. The angle between the two walls of the Vietnam Veterans Memorial in Washington, D.C., is 125.2°. Find the supplement of this angle. (*Source:* National Park Service)

66. The faces of Khafre's Pyramid at Giza, Egypt, are inclined at an angle of 53.13°. Find the complement of this angle. (*Source:* PBS *NOVA* Online)

Answer true or false for Exercises 67 through 70. See the Concept Check in this section. If false, explain why.

67. The complement of a 100° angle is an 80° angle.

68. It is possible to find the complement of a 120° angle.

69. It is possible to find the supplement of a 120° angle.

70. The supplement of a 5° angle is a 175° angle.

71. If lines *m* and *n* are parallel, find the measures of angles *a* through *e*.

△ **72.** Below is a rectangle. List which segments, if extended, would be parallel lines.

73. Can two supplementary angles both be acute? Explain why or why not.

74. In your own words, describe how to find the complement and the supplement of a given angle.

75. Find two complementary angles with the same measure.

76. Is the figure below possible? Why or why not?

△ **8.2**   **Plane Figures and Solids** ▶

Objectives

A Identify Plane Figures. ▶

B Identify Solids. ▶

In order to prepare for the sections ahead in this chapter, we first review plane figures and solids.

Objective A Identifying Plane Figures ▶

Recall from Section 8.1 that a **plane** is a flat surface that extends indefinitely.

Plane

A **plane figure** is a figure that lies on a plane. Plane figures, like planes, have length and width but no thickness or depth.

A **polygon** is a closed plane figure that basically consists of three or more line segments that meet at their endpoints.

A **regular polygon** is one whose sides are all the same length and whose angles are the same measure.

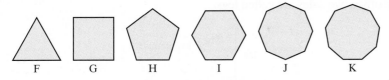

F   G   H   I   J   K

A polygon is named according to the number of its sides.

Polygons		
Number of Sides	**Name**	**Figure Examples**
3	Triangle	A, F
4	Quadrilateral	B, E, G
5	Pentagon	H
6	Hexagon	I
7	Heptagon	C
8	Octagon	J
9	Nonagon	K
10	Decagon	D

Some triangles and quadrilaterals are given special names, so let's study these polygons further. We begin with triangles.

The sum of the measures of the angles of a triangle is 180°.

$$m\angle x + m\angle y + m\angle z = 180°$$

Example 1 Find the measure of $\angle a$.

Solution: Since the sum of the measures of the three angles is 180°, we have

measure of $\angle a$, or $m\angle a = 180° - 95° - 35° = 50°$

To check, see that $95° + 35° + 50° = 180°$.

■ Work Practice 1

We can classify triangles according to the lengths of their sides. (We will use tick marks to denote the sides and angles of a figure that are equal.)

Equilateral triangle

All three sides are the same length. Also, all three angles have the same measure.

Isosceles triangle

Two sides are the same length. Also, the angles opposite the equal sides have equal measure.

Scalene triangle

No sides are the same length. No angles have the same measure.

Practice 1

Find the measure of $\angle x$.

Answer

1. 45°

One other important type of triangle is a right triangle. A **right triangle** is a triangle with a right angle. The side opposite the right angle is called the **hypotenuse,** and the other two sides are called **legs.**

Copyright 2015 Pearson Education, Inc.

Practice 2

Find the measure of ∠y.

Example 2 Find the measure of ∠b.

Solution: We know that the measure of the right angle, ∟, is 90°. Since the sum of the measures of the angles is 180°, we have

measure of ∠b, or $m\angle b = 180° - 90° - 30° = 60°$

▪ Work Practice 2

Helpful Hint

From the previous example, can you see that in a right triangle, the sum of the other two acute angles is 90°? This is because

$$90° \ + \ 90° \ = \ 180°$$
↑　　　↑　　　↑
right　　sum of　　sum of
angle's　other two　angles'
measure　angles'　measures
　　　measures

Now we review some special quadrilaterals. A **parallelogram** is a special quadrilateral with opposite sides parallel and equal in length.

A **rectangle** is a special **parallelogram** that has four right angles.

A **square** is a special **rectangle** that has all four sides equal in length.

Answer

2. 65°

A **rhombus** is a special **parallelogram** that has all four sides equal in length.

A **trapezoid** is a quadrilateral with exactly one pair of opposite sides parallel.

parallel
sides

✓**Concept Check** True or false? All quadrilaterals are parallelograms. Explain.

In addition to triangles, quadrilaterals, and other polygons, circles are also plane figures. A **circle** is a plane figure that consists of all points that are the same fixed distance from a point c. The point c is called the **center** of the circle. The **radius** of a circle is the distance from the center of the circle to any point on the circle. The **diameter** of a circle is the distance across the circle passing through the center. Notice that the diameter is twice the radius, and the radius is half the diameter.

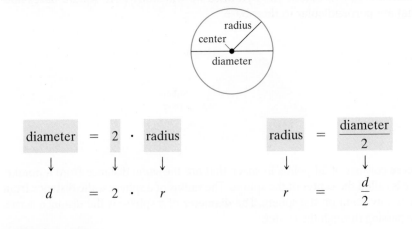

radius
center
diameter

diameter $= 2 \cdot$ radius   radius $= \dfrac{\text{diameter}}{2}$

$\quad\downarrow\qquad\quad\downarrow\quad\ \downarrow \qquad\qquad\qquad \downarrow\qquad\quad \downarrow$

$\quad d \quad = 2 \cdot \quad r \qquad\qquad\qquad r \quad = \dfrac{d}{2}$

| **Example 3** | Find the diameter of the circle. |

5 cm

Solution: The diameter is twice the radius.

$d = 2 \cdot r$

$d = 2 \cdot 5 \text{ cm} = 10 \text{ cm}$

The diameter is 10 centimeters.

■ Work Practice 3

Objective B Identifying Solid Figures ▶

Recall from Section 8.1 that space extends in all directions indefinitely.
   A **solid** is a figure that lies in space. Solids have length, width, and height or depth.

Practice 3
Find the radius of the circle.

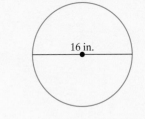

16 in.

A **rectangular solid** is a solid that consists of six sides, or faces, all of which are rectangles.

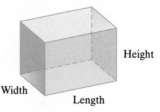

Height

Width

Length

A **cube** is a rectangular solid whose six sides are squares.

A **pyramid** is shown below. The pyramids we will study have square bases and heights that are perpendicular to their base.

Height

Square base

A **sphere** consists of all points in space that are the same distance from a point c. The point c is called the **center** of the sphere. The **radius** of a sphere is the distance from the center to any point on the sphere. The **diameter** of a sphere is the distance across the sphere passing through the center.

←— Diameter —→

Radius

Center

The radius and diameter of a sphere are related in the same way as the radius and diameter of a circle.

$$d = 2 \cdot r \quad \text{or} \quad r = \frac{d}{2}$$

Practice 4

Find the diameter of the sphere.

7 mi

Answer

4. 14 mi

Example 4 Find the radius of the sphere.

Solution: The radius is half the diameter.

$$r = \frac{d}{2}$$

$$r = \frac{36 \text{ feet}}{2} = 18 \text{ feet}$$

The radius is 18 feet.

36 ft

■ Work Practice 4

The **cylinders** we will study have bases that are in the shape of circles and heights that are perpendicular to their base.

Height

Circular base

The **cones** we will study have bases that are circles and heights that are perpendicular to their base.

Height

Circular base

Vocabulary, Readiness & Video Check

Martin-Gay Interactive Videos   *Watch the section lecture video and answer the following questions.*

Objective A **1.** From the lecture after ⊞ Example 2, since all angles of an equilateral triangle have the same measure, what is the measure of each angle? ▶

Objective B **2.** What solid is identified in ⊞ Example 6? What two real-life examples of the solid are given? ▶

See Video 8.2

8.2 Exercise Set MyMathLab® ▶

Objective A *Identify each polygon. See the table at the beginning of this section.*

1.   2.   3.   4.

5.   6.   7.   8.

Classify each triangle as equilateral, isosceles, or scalene. Also identify any triangles that are also right triangles. See the triangle classification after Example 1.

9.   10.   11.

12.   **13.**   **14.**

Find the measure of ∠x in each figure. See Examples 1 and 2.

◉15.

70°

85°   x

16.

x

112°

28°

17.

95°

72°   x

18.

x

80°
65°

◉19.

x

50°

20.

x

20°

Fill in each blank.

21. Twice the radius of a circle is its _____.

22. A rectangle with all four sides equal is a(n) _____.

23. A parallelogram with four right angles is a(n) _____.

24. Half the diameter of a circle is its _____.

◉25. A quadrilateral with opposite sides parallel is a(n) _____.

26. A quadrilateral with exactly one pair of opposite sides parallel is a(n) _____.

27. The side opposite the right angle of a right triangle is called the _____.

28. A triangle with no equal sides is a(n) _____.

Find the unknown diameter or radius in each figure. See Example 3.

◉29.

d = ?
r = 7 m

30.

r = 3 mi

d = ?

31.

r = ?
d = 29 cm

32.

r = ?
d = 13 ft

33.

d = ?   r = 20.3 cm

34.

d = ?

r = 7.8 in.

35.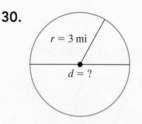

d = 168 in.

r = ?

Largest pumpkin pie
(*Source:* Circleville, Ohio, Pumpkin Festival)

36.

d = 2.6 m

r = ?

Largest cereal bowl
(*Source: Guinness World Records*)

Objective B *Identify each solid.*

37.

38.

39.

40.

41.

42.

Identify the basic shape of each item.

43.

44.

45.

46.

47.

48.

49.

50.

Find each unknown radius or diameter. See Example 4.

51. The radius of a sphere is 7.4 inches. Find its diameter.

52. The radius of a sphere is 5.8 meters. Find its diameter.

53. Find the radius of the sphere.

54. Find the radius of the sphere.

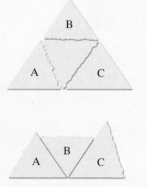

55. Saturn has a radius of approximately 36,184 miles. What is its diameter?

56. A sphere-shaped wasp nest found in Japan had a radius of approximately 15 inches. What was its diameter? (*Source: Guinness World Records*)

Review

Perform each indicated operation. See Sections 1.3, 1.6, 4.3, and 4.4.

57. $2(18) + 2(36)$

58. $4(87)$

59. $4(3.14)$

60. $2(7.8) + 2(9.6)$

Concept Extensions

Determine whether each statement is true or false. See the Concept Check in this section.

61. A square is also a rhombus.

62. A square is also a regular polygon.

63. A rectangle is also a parallelogram.

64. A trapezoid is also a parallelogram.

65. A pentagon is also a quadrilateral.

66. A rhombus is also a parallelogram.

67. Is an isosceles right triangle possible? If so, draw one.

68. In your own words, explain whether a square is also a rhombus.

69. The following demonstration is credited to the mathematician Pascal, who is said to have developed it as a young boy.

Cut a triangle from a piece of paper. The length of the sides and the size of the angles are unimportant. Tear the points off the triangle as shown in the top right figure.

Place the points of the triangle together, as shown in the bottom right figure. Notice that a straight line is formed. What was Pascal trying to show?

△ 8.3 Perimeter ▷

Objective A Using Formulas to Find Perimeters ▷

Recall from Section 1.3 that the perimeter of a polygon is the distance around the polygon. This means that the perimeter of a polygon is the sum of the lengths of its sides.

Objectives

A Use Formulas to Find Perimeters. ▷

B Use Formulas to Find Circumferences. ▷

Example 1 Find the perimeter of the rectangle below.

5 inches

9 inches

Solution:

$$perimeter = 9 \text{ inches} + 9 \text{ inches} + 5 \text{ inches} + 5 \text{ inches}$$
$$= 28 \text{ inches}$$

■ Work Practice 1

Notice that the perimeter of the rectangle in Example 1 can be written as $2 \cdot (9 \text{ inches}) + 2 \cdot (5 \text{ inches})$.

↑ length   ↑ width

In general, we can say that the perimeter of a rectangle is always

$$2 \cdot length + 2 \cdot width$$

As we have just seen, the perimeters of some special figures such as rectangles form patterns. These patterns are given as **formulas.** The formula for the perimeter of a rectangle is shown next:

Perimeter of a Rectangle

$$perimeter = 2 \cdot length + 2 \cdot width$$

In symbols, this can be written as

$$P = 2 \cdot l + 2 \cdot w$$

length

width   width

length

Example 2 Find the perimeter of a rectangle with a length of 11 inches and a width of 3 inches.

11 in.

3 in.

Solution: We use the formula for perimeter and replace the letters by their known lengths.

$$P = 2 \cdot l + 2 \cdot w$$
$$= 2 \cdot 11 \text{ in.} + 2 \cdot 3 \text{ in.} \quad \text{Replace } l \text{ with 11 in. and } w \text{ with 3 in.}$$
$$= 22 \text{ in.} + 6 \text{ in.}$$
$$= 28 \text{ in.}$$

The perimeter is 28 inches.

■ Work Practice 2

Practice 1

a. Find the perimeter of the rectangle.

15 meters

12 meters

b. Find the perimeter of the rectangular lot shown below:

60 feet

80 feet

Practice 2

Find the perimeter of a rectangle with a length of 22 centimeters and a width of 10 centimeters.

Answers

1. a. 54 m   **b.** 280 ft

2. 64 cm

Recall that a square is a special rectangle with all four sides the same length. The formula for the perimeter of a square is shown next:

Perimeter of a Square

Perimeter = side + side + side + side

= 4 · side

In symbols,

$P = 4 \cdot s$

Practice 3

Find the perimeter of a square tabletop if each side is 5 feet long.

Example 3 Finding the Perimeter of a Field

How much fencing is needed to enclose a square field 50 yards on a side?

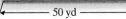

Solution: To find the amount of fencing needed, we find the distance around, or perimeter. The formula for the perimeter of a square is $P = 4 \cdot s$. We use this formula and replace s by 50 yards.

$P = 4 \cdot s$

$= 4 \cdot 50 \text{ yd}$

$= 200 \text{ yd}$

The amount of fencing needed is 200 yards.

■ Work Practice 3

The formula for the perimeter of a triangle with sides of lengths a, b, and c is given next:

Perimeter of a Triangle

Perimeter = side a + side b + side c

In symbols,

$P = a + b + c$

Practice 4

Find the perimeter of a triangle if the sides are 5 centimeters, 10 centimeters, and 6 centimeters in length.

Example 4 Find the perimeter of a triangle if the sides are 3 inches, 7 inches, and 6 inches.

Answers

3. 20 ft **4.** 21 cm

Solution: The formula for the perimeter is $P = a + b + c$, where a, b, and c are the lengths of the sides. Thus,

$$P = a + b + c$$
$$= 3 \text{ in.} + 7 \text{ in.} + 6 \text{ in.}$$
$$= 16 \text{ in.}$$

The perimeter of the triangle is 16 inches.

▧ **Work Practice 4**

Recall that to find the perimeter of other polygons, we find the sum of the lengths of their sides.

Example 5 Find the perimeter of the trapezoid shown below:

Solution: To find the perimeter, we find the sum of the lengths of its sides.

$$\text{perimeter} = 3 \text{ cm} + 2 \text{ cm} + 6 \text{ cm} + 3 \text{ cm} = 14 \text{ cm}$$

The perimeter is 14 centimeters.

▧ **Work Practice 5**

Example 6 Finding the Perimeter of a Room

Find the perimeter of the room shown below:

Solution: To find the perimeter of the room, we first need to find the lengths of all sides of the room.

This side
must measure
15 feet − 9 feet = 6 feet

This side
must measure
10 feet − 7 feet = 3 feet

(Continued on next page)

Practice 5

Find the perimeter of the trapezoid shown.

Practice 6

Find the perimeter of the room shown.

Answers
5. 23 km   **6.** 92 m

Now that we know the measures of all sides of the room, we can add the measures to find the perimeter.

9 ft
3 ft
10 ft
6 ft
7 ft
15 ft

$$\text{perimeter} = 10 \text{ ft} + 9 \text{ ft} + 3 \text{ ft} + 6 \text{ ft} + 7 \text{ ft} + 15 \text{ ft}$$
$$= 50 \text{ ft}$$

The perimeter of the room is 50 feet.

■ Work Practice 6

Practice 7

A rectangular lot measures 60 feet by 120 feet. Find the cost to install fencing around the lot if the cost of fencing is $1.90 per foot.

Example 7 Calculating the Cost of Wallpaper Border

A rectangular room measures 10 feet by 12 feet. Find the cost to hang a wallpaper border on the walls close to the ceiling if the cost of the wallpaper border is $1.09 per foot.

Solution: First we find the perimeter of the room.

$$P = 2 \cdot l + 2 \cdot w$$
$$= 2 \cdot 12 \text{ ft} + 2 \cdot 10 \text{ ft} \quad \text{Replace } l \text{ with 12 feet and } w \text{ with 10 feet.}$$
$$= 24 \text{ ft} + 20 \text{ ft}$$
$$= 44 \text{ ft}$$

The cost of the wallpaper is

$$\text{cost} = \$1.09 \cdot 44 \text{ ft} = 47.96$$

The cost of the wallpaper is $47.96.

■ Work Practice 7

Objective B Using Formulas to Find Circumferences

Recall from Section 4.4 that the distance around a circle is called the **circumference**. This distance depends on the radius or the diameter of the circle.

The formulas for circumference are shown next:

Circumference of a Circle

radius
center
diameter

Circumference = $2 \cdot \pi \cdot$ radius   or   Circumference = $\pi \cdot$ diameter

In symbols,

$$C = 2 \cdot \pi \cdot r \quad \text{or} \quad C = \pi \cdot d,$$

where $\pi \approx 3.14$ or $\pi \approx \dfrac{22}{7}$.

Answer

7. $684

To better understand circumference and π (pi), try the following experiment. Take any can and measure its circumference and its diameter.

The can in the figure above has a circumference of 23.5 centimeters and a diameter of 7.5 centimeters. Now divide the circumference by the diameter.

$$\frac{\text{circumference}}{\text{diameter}} = \frac{23.5 \text{ cm}}{7.5 \text{ cm}} \approx 3.13$$

Try this with other sizes of cylinders and circles–you should always get a number close to 3.1. The exact ratio of circumference to diameter is π. (Recall that $\pi \approx 3.14$ or $\pi \approx \frac{22}{7}$.)

Example 8   Finding Circumference of Spa

Mary Catherine Dooley plans to install a border of new tiling around the circumference of her circular spa. If her spa has a diameter of 14 feet, find its exact circumference. Then use the approximation 3.14 for π to approximate the circumference.

14 feet

Solution: Because we are given the diameter, we use the formula $C = \pi \cdot d$.

$C = \pi \cdot d$

   $= \pi \cdot 14 \text{ ft}$ Replace d with 14 feet.

   $= 14\pi \text{ ft}$

The circumference of the spa is *exactly* 14π feet. By replacing π with the *approximation* 3.14, we find that the circumference is *approximately* 14 feet $\cdot$ 3.14 = 43.96 feet.

■ Work Practice 8

✓**Concept Check**   The distance around which figure is greater: a square with side length 5 inches or a circle with radius 3 inches?

Practice 8

a. An irrigation device waters a circular region with a diameter of 20 yards. Find the exact circumference of the watered region, then use $\pi \approx 3.14$ to give an approximation.

20 yd

b. A manufacturer of clocks is designing a new model. To help the designer calculate the cost of materials to make the new clock, calculate the circumference of a clock with a face diameter of 12 inches. Give an exact circumference; then use $\pi \approx 3.14$ to approximate.

Answers

8. a. exactly 20π yd ≈ 62.8 yd

b. exactly 12π in. ≈ 37.68 in.

✓**Concept Check Answer**

a square with side length 5 in.

Vocabulary, Readiness & Video Check

Use the choices below to fill in each blank.

circumference   radius π $\dfrac{22}{7}$

diameter   perimeter   3.14

1. The ——————— of a polygon is the sum of the lengths of its sides.

2. The distance around a circle is called the ———————.

3. The exact ratio of circumference to diameter is ———————.

4. The diameter of a circle is double its ———————.

5. Both ——————— and ——————— are approximations for π.

6. The radius of a circle is half its ———————.

Martin-Gay Interactive Videos   Watch the section lecture video and answer the following questions.

Objective A   **7.** In ▯ Example 1, how can the perimeter be found if we forget the formula? ▶

Objective B   **8.** From the lecture before ▯ Example 6, circumference is a special name for what? ▶

See Video 8.3

8.3 Exercise Set MyMathLab® ▶

Objective A   *Find the perimeter of each figure. See Examples 1 through 6.*

1.
15 ft   Rectangle
17 ft

2.
Rectangle   14 m
5 m

3.
Parallelogram   25 cm
35 cm

4.
Parallelogram
3 yd
2 yd

5.
5 in.   7 in.
9 in.

6.
5 units   11 units
10 units

7.
10 ft   8 ft
7 ft   8 ft
15 ft

8.
10 m   4 m
10 m
13 m   9 m
20 m

Find the perimeter of each regular polygon. (The sides of a regular polygon have the same length.)

9. 14 inches

10. 50 m

11. 31 cm

12. 15 yd

Solve. See Examples 1 through 7.

13. A polygon has sides of length 5 feet, 3 feet, 2 feet, 7 feet, and 4 feet. Find its perimeter.

14. A triangle has sides of length 8 inches, 12 inches, and 10 inches. Find its perimeter.

15. A line-marking machine lays down lime powder to mark both foul lines on a baseball field. If each foul line for this field measures 312 feet, how many feet of lime powder will be deposited?

16. A baseball diamond has 4 sides, with each side length 90 feet. If a baseball player hits a home run, how far does the player run (home plate, around the bases, then back to home plate)?

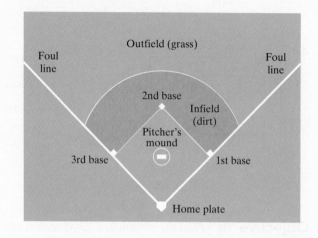

17. If a football field is 53 yards wide and 120 yards long, what is the perimeter?

18. A stop sign has eight equal sides of length 12 inches. Find its perimeter.

53 yd

120 yd

19. A metal strip is being installed around a workbench that is 8 feet long and 3 feet wide. Find how much stripping is needed for this project.

20. Find how much fencing is needed to enclose a rectangular garden 70 feet by 21 feet.

21. If the stripping in Exercise 19 costs $2.50 per foot, find the total cost of the stripping.

22. If the fencing in Exercise 20 costs $2 per foot, find the total cost of the fencing.

23. A regular octagon has a side length of 9 inches. Find its perimeter.

24. A regular pentagon has a side length of 14 meters. Find its perimeter.

▶ **25.** Find the perimeter of the top of a square compact disc case if the length of one side is 7 inches.

26. Find the perimeter of a square ceramic tile with a side of length 3 inches.

27. A rectangular room measures 10 feet by 11 feet. Find the cost of installing a strip of wallpaper around the room if the wallpaper costs $0.86 per foot.

28. A rectangular house measures 85 feet by 70 feet. Find the cost of installing gutters around the house if the cost is $2.36 per foot.

Find the perimeter of each figure. See Example 6.

29.
17 m
28 m
20 m
20 m

30.
13 in.
6 in.
13 in.
30 in.

▶ **31.**
3 ft   4 ft
5 ft
6 ft
15 ft

32.
16 cm
2 cm
11 cm
4 cm
3 cm
9 cm

33.
18 cm
5 cm
9 cm
14 cm

34.
22 km
12 km
5 km
6 km

Objective B *Find the circumference of each circle. Give the exact circumference and then an approximation. Use π ≈ 3.14. See Example 8.*

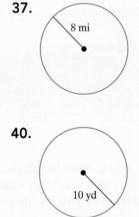

35.
17 cm

36.
2.5 in.

37.
8 mi

38.
50 ft

▶ **39.**
26 m

40.
10 yd

41. Wyley Robinson just bought a trampoline for his children to use. The trampoline has a diameter of 15 feet. If Wyley wishes to buy netting to go around the outside of the trampoline, how many feet of netting does he need?

42. The largest round barn in the world is located at the Marshfield Fairgrounds in Wisconsin. The barn has a diameter of 150 ft. What is the circumference of the barn? (*Source: The Milwaukee Journal Sentinel*)

43. Meteor Crater, near Winslow, Arizona, is 4000 feet in diameter. Approximate the distance around the crater. Use 3.14 for π. (*Source: The Handy Science Answer Book*)

44. The largest pearl, the *Pearl of Lao-tze*, has a diameter of $5\frac{1}{2}$ inches. Approximate the distance around the pearl. Use $\frac{22}{7}$ for π. (*Source: The Guinness Book of World Records*)

Objectives A B **Mixed Practice** *Find the distance around each figure. For circles, give the exact circumference and then an approximation. Use $\pi \approx 3.14$. See Examples 1 through 8.*

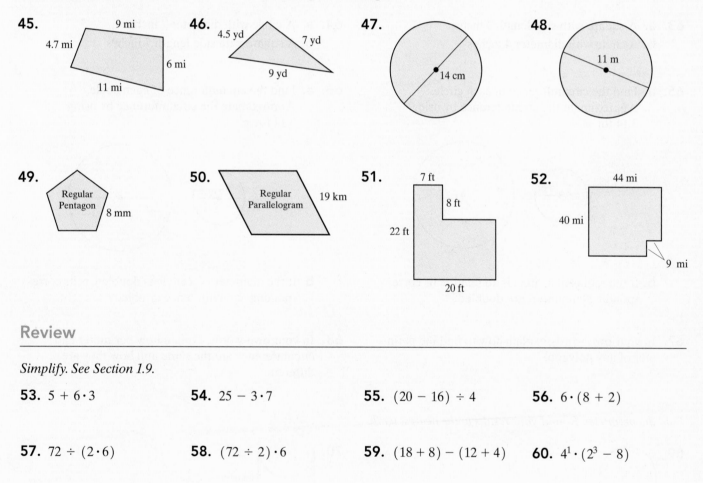

45. 9 mi, 4.7 mi, 6 mi, 11 mi

46. 4.5 yd, 7 yd, 9 yd

47. 14 cm

48. 11 m

49. Regular Pentagon, 8 mm

50. Regular Parallelogram, 19 km

51. 7 ft, 8 ft, 22 ft, 20 ft

52. 44 mi, 40 mi, 9 mi

Review

Simplify. See Section 1.9.

53. $5 + 6 \cdot 3$

54. $25 - 3 \cdot 7$

55. $(20 - 16) \div 4$

56. $6 \cdot (8 + 2)$

57. $72 \div (2 \cdot 6)$

58. $(72 \div 2) \cdot 6$

59. $(18 + 8) - (12 + 4)$

60. $4^1 \cdot (2^3 - 8)$

Concept Extensions

There are a number of factors that determine the dimensions of a rectangular soccer field. Use the table below to answer Exercises 61 and 62.

Soccer Field Width and Length		
Age	**Width Min–Max**	**Length Min–Max**
Under 6/7:	15–20 yards	25–30 yards
Under 8:	20–25 yards	30–40 yards
Under 9:	30–35 yards	40–50 yards
Under 10:	40–50 yards	60–70 yards
Under 11:	40–50 yards	70–80 yards
Under 12:	40–55 yards	100–105 yards
Under 13:	50–60 yards	100–110 yards
International:	70–80 yards	110–120 yards

61. a. Find the minimum length and width of a soccer field for 8-year-old children. (Carefully consider the age.)

  b. Find the perimeter of this field.

62. a. Find the maximum length and width of a soccer field for 12-year-old children.

  b. Find the perimeter of this field.

Solve. See the Concept Check in this section. Choose the figure that has the greater distance around.

63. a. A square with side length 3 inches

  b. A circle with diameter 4 inches

64. a. A circle with diameter 7 inches

  b. A square with side length 7 inches

65. a. Find the circumference of each circle. Approximate the circumference by using 3.14 for π.

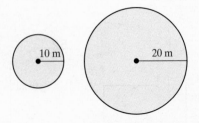

  b. If the radius of a circle is doubled, is its corresponding circumference doubled?

66. a. Find the circumference of each circle. Approximate the circumference by using 3.14 for π.

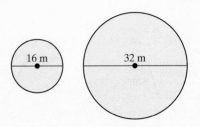

  b. If the diameter of a circle is doubled, is its corresponding circumference doubled?

67. In your own words, explain how to find the perimeter of any polygon.

68. In your own words, explain how perimeter and circumference are the same and how they are different.

Find the perimeter. Round your results to the nearest tenth.

69.

6 meters

6 meters

70.

6 meters

6 meters

71.

5 m

22 m

72.

← 5 feet →

7 feet

73. The perimeter of this rectangle is 31 feet. Find its width.

9 ft

?

74. The perimeter of this square is 18 inches. Find the length of a side.

△ 8.4   Area ▶

Objective A   Finding Areas of Geometric Figures ▶

Recall that area measures the amount of surface of a region. Thus far, we know how to find the area of a rectangle and a square. These formulas, as well as formulas for finding the areas of other common geometric figures, are given next:

Objective

A Find the Areas of Geometric Figures. ▶

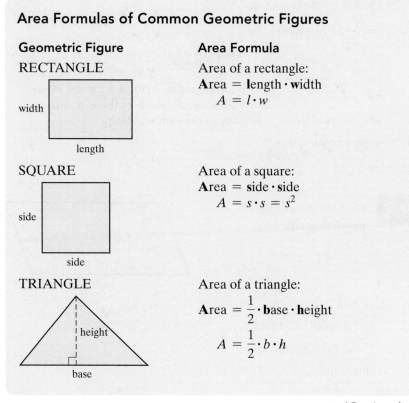

Area Formulas of Common Geometric Figures

Geometric Figure

Area Formula

RECTANGLE

width

length

Area of a rectangle:
Area = length · width
$$A = l \cdot w$$

SQUARE

side

side

Area of a square:
Area = side · side
$$A = s \cdot s = s^2$$

TRIANGLE

height

base

Area of a triangle:
$$\textbf{Area} = \frac{1}{2} \cdot \textbf{base} \cdot \textbf{height}$$
$$A = \frac{1}{2} \cdot b \cdot h$$

(Continued on next page)

Area Formulas of Common Geometric Figures (*continued*)

Geometric Figure   **Area Formula**

PARALLELOGRAM   Area of a parallelogram:
                            Area = **b**ase · **h**eight
                                $A = b \cdot h$

TRAPEZOID   Area of a trapezoid:
                                $\textbf{A}\text{rea} = \frac{1}{2} \cdot (\text{one } \textbf{b}\text{ase} + \text{other } \textbf{B}\text{ase}) \cdot \textbf{h}\text{eight}$

                                $A = \frac{1}{2} \cdot (b + B) \cdot h$

Use these formulas for the following examples.

Helpful Hint

Area is always measured in square units.

Practice 1

Find the area of the triangle.

$6\frac{1}{4}$ in.

8 in.

Example 1 Find the area of the triangle.

8 cm

14 cm

Solution: $A = \frac{1}{2} \cdot b \cdot h$

$= \frac{1}{2} \cdot 14 \text{ cm} \cdot 8 \text{ cm}$

$= \frac{\overset{1}{\cancel{2}} \cdot 7 \cdot 8}{\underset{1}{\cancel{2}}} \text{ sq cm}$

$= 56 \text{ square cm}$

The area is 56 square centimeters.

Helpful Hint You may see 56 sq cm, for example, written with the notation 56 cm². Both of these notations mean the same quantity.

■ Work Practice 1

Practice 2

Find the area of the square.

4.2 yd

4.2 yd

Example 2 Find the area of the parallelogram.

1.5 mi

3.4 mi

Solution: $A = b \cdot h$

$= 3.4 \text{ miles} \cdot 1.5 \text{ miles}$

$= 5.1 \text{ square miles}$

The area is 5.1 square miles.

■ Work Practice 2

Answers
1. 25 sq in.   **2.** 17.64 sq yd

Helpful Hint

When finding the area of figures, be sure all measurements are changed to the same unit before calculations are made.

Example 3 Find the area of the figure.

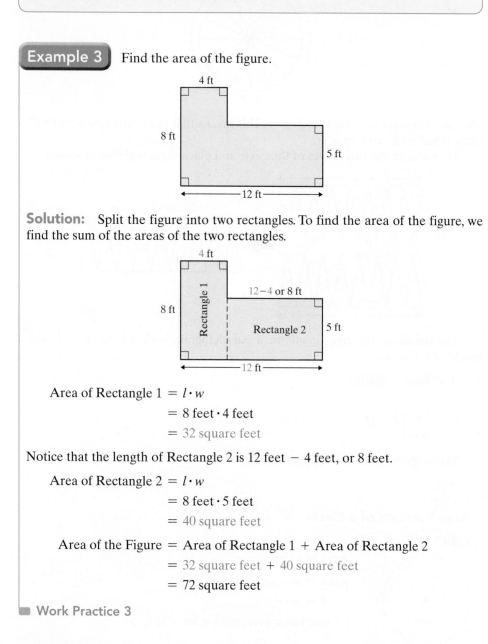

Solution: Split the figure into two rectangles. To find the area of the figure, we find the sum of the areas of the two rectangles.

$$\text{Area of Rectangle 1} = l \cdot w$$
$$= 8 \text{ feet} \cdot 4 \text{ feet}$$
$$= 32 \text{ square feet}$$

Notice that the length of Rectangle 2 is 12 feet − 4 feet, or 8 feet.

$$\text{Area of Rectangle 2} = l \cdot w$$
$$= 8 \text{ feet} \cdot 5 \text{ feet}$$
$$= 40 \text{ square feet}$$

$$\text{Area of the Figure} = \text{Area of Rectangle 1} + \text{Area of Rectangle 2}$$
$$= 32 \text{ square feet} + 40 \text{ square feet}$$
$$= 72 \text{ square feet}$$

■ Work Practice 3

Helpful Hint

The figure in Example 3 can also be split into two rectangles as shown:

Practice 3

Find the area of the figure.

Answer
3. 396 sq m

To better understand the formula for area of a circle, try the following. Cut a circle into many pieces as shown:

The circumference of a circle is $2 \cdot \pi \cdot r$. This means that the circumference of half a circle is half of $2 \cdot \pi \cdot r$, or $\pi \cdot r$.

Then unfold the two halves of the circle and place them together as shown:

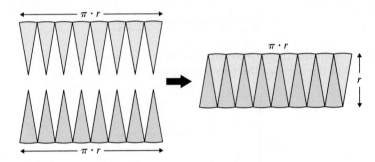

The figure on the right is almost a parallelogram with a base of $\pi \cdot r$ and a height of r. The area is

$$A = \boxed{\text{base}} \cdot \boxed{\text{height}}$$
$$= (\pi \cdot r) \cdot r$$
$$= \pi \cdot r^2$$

This is the formula for area of a circle.

Area Formula of a Circle

CIRCLE

Area of a circle:

Area $= \pi \cdot (\text{radius})^2$

$A = \pi \cdot r^2$

(A fraction approximation for π is $\dfrac{22}{7}$.)

(A decimal approximation for π is 3.14.)

Practice 4

Find the area of the given circle. Find the exact area and an approximation. Use 3.14 as an approximation for π.

7 cm

Example 4 Find the area of a circle with a radius of 3 feet. Find the exact area and an approximation. Use 3.14 as an approximation for π.

3 ft

Solution: We let $r = 3$ ft and use the formula.

$$A = \pi \cdot r^2$$
$$= \pi \cdot (3 \text{ ft})^2$$
$$= \pi \cdot 9 \text{ square ft, or } 9 \cdot \pi \text{ square ft}$$

Answer

4. 49π sq cm ≈ 153.86 sq cm

To approximate this area, we substitute 3.14 for π.

$$9 \cdot \pi \text{ square feet} \approx 9 \cdot 3.14 \text{ square feet}$$
$$= 28.26 \text{ square feet}$$

The *exact* area of the circle is 9π square feet, which is *approximately* 28.26 square feet.

■ **Work Practice 4**

✓**Concept Check** Use diagrams to decide which figure would have a larger area: a circle of diameter 10 inches or a square 10 inches long on each side.

✓**Concept Check Answer**
a square 10 in. long on each side

Vocabulary, Readiness & Video Check

Martin-Gay Interactive Videos   Watch the section lecture video and answer the following question.

Objective A 1. What formula was used twice and why did we use it twice to solve ▣ Example 3? ▶

See Video 8.4

8.4 Exercise Set MyMathLab® ▶

Objective A *Find the area of the geometric figure. If the figure is a circle, give an exact area and then use the given* **approximation** *for π to approximate the area. See Examples 1 through 4.*

▶**1.**

2 m   Rectangle
3.5 m

2.

2.75 ft   Rectangle
7 ft

▶**3.**

3 yd
$6\frac{1}{2}$ yd

4.

5 ft
$4\frac{1}{2}$ ft

5.

6 yd
5 yd

6.

5 ft   7 ft

▶**7.** Use 3.14 for π.

$d = 3$ in.

8. Use $\frac{22}{7}$ for π.

$r = 2$ cm

9. Square   4.2 ft

10. Square   2.6 m

11. 5 m
Trapezoid
4 m
9 m

12. Trapezoid
6 in.
5 in. $8\frac{1}{2}$ in.

13. 4 yd
4 yd   Trapezoid
7 yd

14. 10 ft
3 ft   Trapezoid
5 ft

15. 7 ft
Parallelogram
$5\frac{1}{4}$ ft

16. Parallelogram $4\frac{1}{4}$ cm
3 cm

17. $4\frac{1}{2}$ in.   Parallelogram
5 in.

18. 4 m
6 m
Parallelogram

19. 2 cm
$1\frac{1}{2}$ cm $1\frac{1}{2}$ cm
3 cm
7 cm

20. 6 km
4 km
5 km
10 km

21. 5 mi
10 mi
3 mi
17 mi

22. 25 cm
15 cm   12 cm
5 cm

23. 5 cm
3 cm

24. 4 in.
5 in.

25. Use $\dfrac{22}{7}$ for π.
$r = 6$ in.

26. Use 3.14 for π.
$d = 5$ m

Solve. See Examples 1 through 4.

27. A $10\frac{1}{2}$-foot by 16-foot concrete wall is to be built using concrete blocks. Find the area of the wall.

28. The floor of Terry's attic is 24 feet by 35 feet. Find how many square feet of insulation are needed to cover the attic floor.

29. The world's largest U.S. flag is the "Superflag," which measures 505 feet by 255 feet. Find its area. (*Source:* Superflag.com)

255 feet

505 feet

30. The world's largest illuminated indoor advertising sign is located in the Dubai International Airport in Dubai, UAE. It measures 28.0 meters in length by 6.2 meters in height. Find its area. (*Source:* World Record Academy)

28 meters

6.2 meters

31. The face of a watch has a diameter of 2 centimeters. What is its area? Give an exact answer then an approximation using 3.14 for π.

2 cm

32. The world's largest commercially available pizza is sold by Big Mama's & Papa's Pizzeria in Los Angeles, CA. This huge square pizza, called "The Giant Sicilian," measures 54 inches on each side and sells for $199.99 plus tax. Find the area of the top of the pizza. (*Source:* Guinness World Records, Big Mama's & Big Papa's Pizzeria Inc.)

54 in.

33. One side of a concrete block measures 8 inches by 16 inches. Find the area of the side in square inches. Find the area in square feet (144 sq in. = 1 sq ft).

34. A standard *double* roll of wallpaper is $6\frac{5}{6}$ feet wide and 33 feet long. Find the area of the *double* roll.

35. A picture frame measures 20 inches by $25\frac{1}{2}$ inches. Find how many square inches of glass the frame requires.

36. A mat to go under a tablecloth is made to fit a round dining table with a 4-foot diameter. Approximate how many square feet of mat there are. Use 3.14 as an approximation for π.

37. A drapery panel measures 6 feet by 7 feet. Find how many square feet of material are needed for *four* panels.

38. A page in a book measures 27.5 centimeters by 20.5 centimeters. Find its area.

39. Find how many square feet of land are in the plot shown:

90 feet
80 feet
140 feet

40. For Gerald Gomez to determine how much grass seed he needs to buy, he must know the size of his yard. Use the drawing to determine how many square feet are in his yard.

96 feet
48 feet
48 feet   24 feet
132 feet

41. The outlined part of the roof shown is in the shape of a trapezoid and needs to be shingled. The number of shingles to buy depends on the area.

a. Use the dimensions given to find the area of the outlined part of the roof to the nearest whole square foot.

36 ft
$12\frac{1}{2}$ ft   25 ft

b. Shingles are packaged in a unit called a "square." If a "square" covers 100 square feet, how many whole squares need to be purchased to shingle this part of the roof?

42. The entire side of the building shaded in the drawing is to be bricked. The number of bricks to buy depends on the area.

a. Find the area.

4 feet
12 feet   12 feet
8 feet

b. If the side area of each brick (including mortar room) is $\frac{1}{6}$ square foot, find the number of bricks needed to brick the end of the building.

Review

Find the perimeter or circumference of each geometric figure. See Section 8.3.

43. Give the exact circumference and an approximation. Use 3.14 for π.

14 in.

44.

4 cm
5 cm
Rectangle

45.

3 ft $3\frac{1}{2}$ ft
4 ft
6 ft
$8\frac{1}{2}$ ft

46.

6 mi
$9\frac{1}{4}$ mi
$7\frac{1}{2}$ mi
12 mi

47. $2\frac{1}{8}$ ft

Regular hexagon

48.

Equilateral triangle

3 in.

Concept Extensions

Given the following situations, tell whether you are more likely to be concerned with area or perimeter.

49. ordering fencing to fence a yard

50. ordering grass seed to plant in a yard

51. buying carpet to install in a room

52. buying gutters to install on a house

53. ordering paint to paint a wall

54. ordering baseboards to install in a room

55. buying a wallpaper border to go on the walls around a room

56. buying fertilizer for your yard

57. A pizza restaurant recently advertised two specials. The first special was a 12-inch diameter pizza for $10. The second special was two 8-inch diameter pizzas for $9. Determine the better buy. (*Hint:* First find and compare the areas of the pizzas in the two specials. Then find a price per square inch for the pizzas in both specials.)

58. Find the approximate area of the state of Utah.

105 mi

70 mi

350 mi

Utah

270 mi

59. Find the area of a rectangle that measures 2 *feet* by 8 *inches*. Give the area in square feet and in square inches.

60. In your own words, explain why perimeter is measured in units and area is measured in square units. (*Hint:* See Section 1.6 for an introduction on the meaning of area.)

61. Find the area of the shaded region. Use the approximation 3.14 for π.

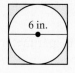

6 in.

62. Estimate the cost of a piece of carpet for a rectangular room 10 feet by 15 feet. The cost of the carpet is $6.50 per yard.

63. The largest pumpkin pie was made for the 100th anniversary of the Circleville, Ohio, Pumpkin Festival in October 2008. The pie had a diameter of 168 inches. Find the exact area of the top of the pie, and an approximation. Use $\pi \approx 3.14$. (*Source:* Circleville, Ohio, Pumpkin Festival)

64. The largest cereal bowl in the world was made by Kellogg's South Africa in July 2007. The bowl had a 2.6-meter diameter. Calculate the exact area of the circular base of the bowl and an approximation. Use $\pi \approx 3.14$. (*Source: Guinness Book of World Records*)

Find the area of each figure. If needed, use $\pi \approx 3.14$ and round results to the nearest tenth.

65. Find the skating area.

66.

There are a number of factors that determine the dimensions of a rectangular soccer field. Use the table below to answer Exercises 67 and 68.

Soccer Field Width and Length		
Age	**Width Min–Max**	**Length Min–Max**
Under 6/7:	15–20 yards	25–30 yards
Under 8:	20–25 yards	30–40 yards
Under 9:	30–35 yards	40–50 yards
Under 10:	40–50 yards	60–70 yards
Under 11:	40–50 yards	70–80 yards
Under 12:	40–55 yards	100–105 yards
Under 13:	50–60 yards	100–110 yards
International:	70–80 yards	110–120 yards

67. a. Find the minimum length and width of a soccer field for 9-year-old children. (Carefully consider the age.)

    b. Find the area of this field.

68. a. Find the maximum length and width of a soccer field for 11-year-old children.

    b. Find the area of this field.

69. Do two rectangles with the same perimeter have the same area? To see, find the perimeter and the area of each rectangle.

6 in.   8 in.   3 in.   11 in.

8.5 Volume

Objective A Finding Volumes of Solids

Objective

A Find the Volumes of Solids.

Volume is a measure of the space of a region. The volume of a box or can, for example, is the amount of space inside. Volume can be used to describe the amount of juice in a pitcher or the amount of concrete needed to pour a foundation for a house.

The volume of a solid is the number of **cubic units** in the solid. A cubic centimeter and a cubic inch are illustrated.

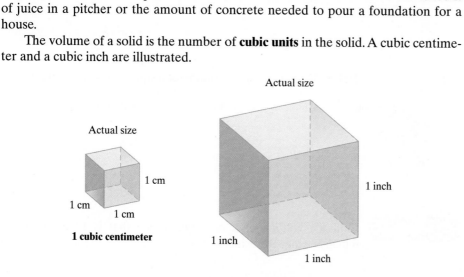

Actual size

Actual size

1 cm

1 cm

1 cm

1 cubic centimeter

1 inch

1 inch

1 inch

1 cubic inch

Formulas for finding the volumes of some common solids are given next:

Volume Formulas of Common Solids

Solid	**Volume Formulas**
RECTANGULAR SOLID height width length	Volume of a rectangular solid: **Volume = length · width · height** $V = l \cdot w \cdot h$
CUBE side side side	Volume of cube: **Volume = side · side · side** $V = s^3$
SPHERE radius	Volume of sphere: **Volume** $= \dfrac{4}{3} \cdot \pi \cdot (\textbf{radius})^3$ $V = \dfrac{4}{3} \cdot \pi \cdot r^3$
CIRCULAR CYLINDER height radius	Volume of a circular cylinder: **Volume** $= \pi \cdot (\textbf{radius})^2 \cdot \textbf{height}$ $V = \pi \cdot r^2 \cdot h$

(Continued on next page)

Volume Formulas of Common Solids (*continued*)

Solid	Volume Formulas
CONE	Volume of a cone: $$\mathbf{Volume} = \frac{1}{3} \cdot \pi \cdot (\mathbf{radius})^2 \cdot \mathbf{height}$$ $$V = \frac{1}{3} \cdot \pi \cdot r^2 \cdot h$$
SQUARE-BASED PYRAMID	Volume of a square-based pyramid: $$\mathbf{Volume} = \frac{1}{3} \cdot (\mathbf{side})^2 \cdot \mathbf{height}$$ $$V = \frac{1}{3} \cdot s^2 \cdot h$$

Practice 1

Find the volume of a rectangular box that is 5 feet long, 2 feet wide, and 4 feet deep.

Example 1   Find the volume of a rectangular box that is 12 inches long, 6 inches wide, and 3 inches high.

3 in.
6 in.   12 in.

Solution:

$$V = l \cdot w \cdot h$$
$$V = 12 \text{ in.} \cdot 6 \text{ in.} \cdot 3 \text{ in.} = 216 \text{ cubic in.}$$

The volume of the rectangular box is 216 cubic inches.

■ Work Practice 1

✓**Concept Check**   Juan is calculating the volume of the following rectangular solid. Find the error in his calculation.

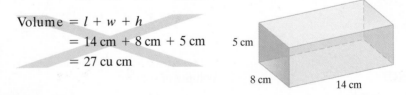

Volume $= l + w + h$
$= 14 \text{ cm} + 8 \text{ cm} + 5 \text{ cm}$
$= 27 \text{ cu cm}$

5 cm
8 cm   14 cm

Practice 2

Approximate the volume of a ball of radius $\frac{1}{2}$ centimeter. Use $\frac{22}{7}$ for π. Give an exact answer and an approximate answer.

Example 2   Approximate the volume of a ball of radius 3 inches. Use the approximation $\frac{22}{7}$ for π. Give an exact answer and an approximate answer.

3 in.

Answers

1. 40 cu ft   2. $\frac{1}{6}\pi$ cu cm $\approx \frac{11}{21}$ cu cm

✓**Concept Check Answer**

Volume $= l \cdot w \cdot h$
$= 14 \text{ cm} \cdot 8 \text{ cm} \cdot 5 \text{ cm}$
$= 560 \text{ cu cm}$

Solution:

$$V = \frac{4}{3} \cdot \pi \cdot r^3$$

$$= \frac{4}{3} \cdot \pi \, (3 \text{ in.})^3$$

$$= \frac{4}{3} \cdot \pi \cdot 27 \text{ cu in.}$$

$$= \frac{4 \cdot \pi \cdot \overset{1}{\cancel{3}} \cdot 9}{\underset{1}{\cancel{3}}} \text{ cu in.}$$

$$= 36\pi \text{ cu in.}$$

This is the exact volume. To approximate the volume, use the approximation $\frac{22}{7}$ for π.

$$V = 36\pi \text{ cu in.}$$

$$\approx 36 \cdot \frac{22}{7} \text{ cu in.} \quad \text{\small Replace } \pi \text{ with } \frac{22}{7}.$$

$$= \frac{36 \cdot 22}{7} \text{ cu in.}$$

$$= \frac{792}{7} \quad \text{or} \quad 113\frac{1}{7} \text{ cubic inches.}$$

The volume is *approximately* $113\frac{1}{7}$ cubic inches.

🔲 **Work Practice 2**

Example 3   Approximate the volume of a can that has a $3\frac{1}{2}$-inch radius and a height of 6 inches. Use $\frac{22}{7}$ for π. Give an exact volume and an approximate volume.

$3\frac{1}{2}$ in.

6 in.

Practice 3

Approximate the volume of a cylinder of radius 5 inches and height 7 inches. Use 3.14 for π. Give an exact answer and an approximate answer.

Solution: Using the formula for a circular cylinder, we have

$$V = \pi \cdot r^2 \cdot h$$

$$3\frac{1}{2} = \frac{7}{2}$$

$$= \pi \cdot \left(\frac{7}{2} \text{ in.}\right)^2 \cdot 6 \text{ in.}$$

$$= \pi \cdot \frac{49}{4} \text{ sq in.} \cdot 6 \text{ in.}$$

$$= \frac{\pi \cdot 49 \cdot \overset{1}{\cancel{2}} \cdot 3}{\underset{1}{\cancel{2} \cdot 2}} \text{ cu in.}$$

$$= 73\frac{1}{2}\pi \text{ cu in. or } 73.5\pi \text{ cu in.}$$

(Continued on next page)

(Continued on next page)

Answer

3. 175π cu in. ≈ 549.5 cu in.

This is the exact volume. To approximate the volume, use the approximation $\frac{22}{7}$ for π.

$$V = 73\frac{1}{2}\pi \approx \frac{147}{2} \cdot \frac{22}{7} \text{ cu in.} \quad \text{Replace } \pi \text{ with } \frac{22}{7}.$$

$$= \frac{21 \cdot \overset{1}{\cancel{7}} \cdot \overset{1}{\cancel{2}} \cdot 11}{\underset{1}{\cancel{2}} \cdot \underset{1}{\cancel{7}}} \text{ cu in.}$$

$$= 231 \text{ cubic in.}$$

The volume is approximately 231 cubic inches.

▣ Work Practice 3

Practice 4

Find the volume of a square-based pyramid that has a 3-meter side and a height of 5.1 meters.

5.1 m

3 m

Example 4 Approximate the volume of a cone that has a height of 14 centimeters and a radius of 3 centimeters. Use 3.14 for π. Give an exact answer and an approximate answer.

14 cm

3 cm

Solution: Using the formula for volume of a cone, we have

$$V = \frac{1}{3} \cdot \pi \cdot r^2 \cdot h$$

$$= \frac{1}{3} \cdot \pi \cdot (3 \text{ cm})^2 \cdot 14 \text{ cm} \quad \text{Replace } r \text{ with 3 cm and } h \text{ with 14 cm.}$$

$$= 42\pi \text{ cu cm}$$

Thus, 42π cubic centimeters is the exact volume. To approximate the volume, use the approximation 3.14 for π.

$$V \approx 42 \cdot 3.14 \text{ cu cm} \quad \text{Replace } \pi \text{ with 3.14.}$$

$$= 131.88 \text{ cu cm}$$

The volume is approximately 131.88 cubic centimeters.

▣ Work Practice 4

Answer

4. 15.3 cu m

Vocabulary, Readiness & Video Check

Use the choices below to fill in each blank. Some exercises are from Sections 8.3 and 8.4.

cubic	perimeter	volume
units	area	square

1. The measure of the amount of space inside a solid is its _____.

2. _____ measures the amount of surface of a region.

3. Volume is measured in _____ units.

4. Area is measured in _____ units.

5. The _____ of a polygon is the sum of the lengths of its sides.

6. Perimeter is measured in _____.

Martin-Gay Interactive Videos   *Watch the section lecture video and answer the following question.*

Objective A 7. In ▦ Examples 2 and 3, explain the difference in the two answers found for each. ▶

See Video 8.5

8.5 Exercise Set MyMathLab® ▶

Objective A *Find the volume of each solid. See Examples 1 through 4. Use $\frac{22}{7}$ for π.*

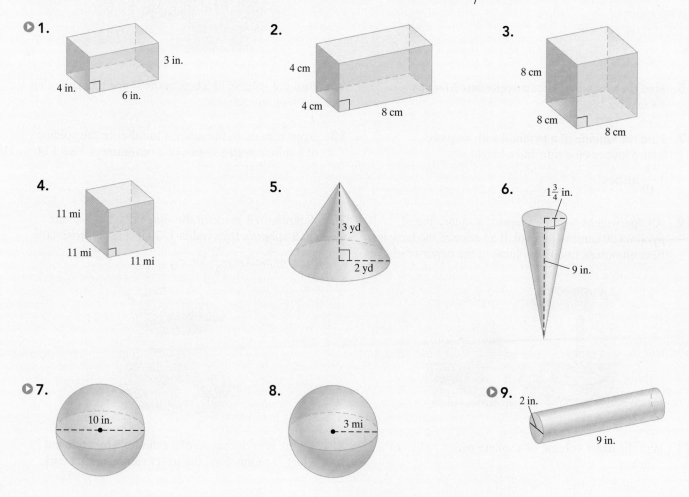

1. 3 in., 4 in., 6 in.

2. 4 cm, 4 cm, 8 cm

3. 8 cm, 8 cm, 8 cm

4. 11 mi, 11 mi, 11 mi

5. 3 yd, 2 yd

6. $1\frac{3}{4}$ in., 9 in.

7. 10 in.

8. 3 mi

9. 2 in., 9 in.

10. 10 ft   6 ft

11. 9 cm   5 cm

12. 15 m   7 m

Solve.

▶ **13.** Find the volume of a cube with edges of $1\frac{1}{3}$ inches.

$1\frac{1}{3}$ inches

14. A water storage tank is in the shape of a cone with the pointed end down. If the radius is 14 feet and the depth of the tank is 15 feet, approximate the volume of the tank in cubic feet. Use $\frac{22}{7}$ for π.

14 ft

15 ft

15. Find the volume of a rectangular box 2 feet by 1.4 feet by 3 feet.

16. Find the volume of a box in the shape of a cube that is 5 feet on each side.

17. Find the volume of a pyramid with a square base 5 inches on a side and a height of $1\frac{3}{10}$ inches.

18. Approximate to the nearest hundredth the volume of a sphere with a radius of 2 centimeters. Use 3.14 for π.

19. A paperweight is in the shape of a square-based pyramid 20 centimeters tall. If an edge of the base is 12 centimeters, find the volume of the paperweight.

20. A birdbath is made in the shape of a hemisphere (half-sphere). If its radius is 10 inches, approximate the volume. Use $\frac{22}{7}$ for π.

10 in.

21. Find the exact volume of a sphere with a radius of 7 inches.

22. A tank is in the shape of a cylinder 8 feet tall and 3 feet in radius. Find the exact volume of the tank.

23. Find the volume of a rectangular block of ice 2 feet by $2\frac{1}{2}$ feet by $1\frac{1}{2}$ feet.

24. Find the capacity (volume in cubic feet) of a rectangular ice chest with inside measurements of 3 feet by $1\frac{1}{2}$ feet by $1\frac{3}{4}$ feet.

25. In 2013, the largest free-floating soap bubble made with a wand had a diameter between 11 and 12 feet. Calculate the exact volume of a sphere with a diameter of 12 feet. (*Source:* Guinness World Records)

26. The largest inflatable beach ball was created in Poland in 2012. It has a diameter of just under 54 feet. Calculate the exact volume of a sphere with a diameter of 54 feet. (*Source:* Guinness World Records)

27. Find the exact volume of a waffle ice cream cone with a 3-in. diameter and a height of 7 inches.

28. A snow globe has a diameter of 6 inches. Find its exact volume. Then approximate its volume using 3.14 for π.

29. The largest cereal bowl in the world was made by Kellogg's South Africa in July 2007. The bowl had a 2.6-m diameter and a height of 1.5 m. Calculate the volume of cereal you could put into this bowl. Use $\pi \approx 3.14$ and round to the nearest hundredth of a cubic meter. (*Source: Guinness Book of World Records*)

30. Mount Fuji, in Japan, is considered the most beautiful composite volcano in the world. The mountain is in the shape of a cone whose height is about 3.5 kilometers and whose base radius is about 3 kilometers. Approximate the volume of Mt. Fuji in cubic kilometers. Use $\frac{22}{7}$ for π.

31. An ice cream cone with a 4-centimeter diameter and 3-centimeter depth is filled exactly level with the top of the cone. Approximate how much ice cream (in cubic centimeters) is in the cone. Use $\frac{22}{7}$ for π.

32. A child's toy is in the shape of a square-based pyramid 10 inches tall. If an edge of the base is 7 inches, find the volume of the toy.

The Space Cube is supposed to be the world's smallest computer, with dimensions of 2 inches by 2 inches by 2.2 inches.

33. Find the volume of the Space Cube.

34. Find the volume of an actual cube that measures 2 inches by 2 inches by 2 inches.

35. Find the volume of an actual cube that measures 2.2 inches by 2.2 inches by 2.2 inches.

36. Comment on the results of Exercises 33–35. Were you surprised when you compared volumes? Why or why not?

Review

Evaluate. See Section 1.9.

37. 5^2

38. 7^2

39. 3^2

40. 20^2

41. $1^2 + 2^2$

42. $5^2 + 3^2$

43. $4^2 + 2^2$

44. $1^2 + 6^2$

Concept Extensions

45. The Hayden Planetarium, at the Museum of Natural History in New York City, boasts a dome that has a diameter of 20 m. The dome is a hemisphere, or half a sphere. What is the volume enclosed by the dome at the Hayden Planetarium? Use 3.14 for π and round to the nearest hundredth. (*Source:* Hayden Planetarium)

hemisphere

46. The Adler Museum in Chicago recently added a new planetarium, its StarRider Theater, which has a diameter of 55 feet. Find the volume of its hemispheric (half a sphere) dome. Use 3.14 for π and round to the nearest hundredth. (*Source:* The Adler Museum)

47. Do two rectangular solids with the same volume have the same shape? To help, find the volume of each rectangular solid.

48. Do two rectangular solids with the same volume have the same surface area? To see, find the volume and surface area of each rectangular solid. Surface area is the area of the surface of the solid. To find the surface area of each rectangular solid, find the sum of the areas of the 6 rectangles that form each solid.

49. Two kennels are offered at a hotel. The kennels measure

a. 2′1″ by 1′8″ by 1′7″ and

b. 1′1″ by 2′ by 2′8″

What is the volume of each kennel rounded to the nearest tenth of a cubic foot? Which is larger?

50. The centerpiece of the New England Aquarium in Boston is its Giant Ocean Tank. This exhibit is a four-story cylindrical saltwater tank containing sharks, sea turtles, stingrays, and tropical fish. The radius of the tank is 16.3 feet and its height is 32 feet (assuming that a story is 8 feet). What is the volume of the Giant Ocean Tank? Use $\pi \approx 3.14$ and round to the nearest tenth of a cubic foot. (*Source:* New England Aquarium)

51. Find the volume of the figure below. Give an exact measure and then a whole number approximation.

2 in.

6 in.

52. Can you compute the volume of a rectangle? Why or why not?

Geometry Concepts

1. _____

△ **1.** Find the supplement and the complement of a 27° angle.

2. _____

Find the measures of angles x, y, and z in each figure.

3. _____

2.

3. $m \| n$

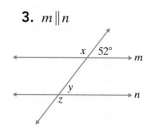

4. _____

4. Find the measure of ∠x. **5.** Find the diameter. **6.** Find the radius.

5. _____

6. _____

7. _____

For Exercises 7 through 11, find the perimeter (or circumference) and area of each figure. For the circle give an exact circumference and area. Then use $\pi \approx 3.14$ to approximate each. Don't forget to attach correct units.

8. _____

7. Square 5 m **8.** 4 ft, 3 ft, 5 ft **9.** 5 cm **10.** 11 mi Parallelogram 5 mi 4 mi

9. _____

10. _____

11. 8 cm, 3 cm, 7 cm, 17 cm

11. _____

12. The smallest cathedral is in High-landville, Missouri. The rectangular floor of the cathedral measures 14 feet by 17 feet. Find its perimeter and its area. (*Source: The Guinness Book of Records*)

12. _____

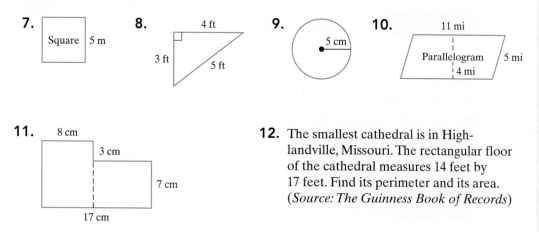

13. _____

Find the volume of each solid. Don't forget to attach correct units.

13. A cube with edges of 4 inches each.

14. A rectangular box 2 feet by 3 feet by 5.1 feet.

14. _____

15. A pyramid with a square base 10 centimeters on a side and a height of 12 centimeters.

16. A sphere with a diameter of 3 miles. Give the exact volume and then use $\pi \approx \dfrac{22}{7}$ to approximate.

15. _____

16. _____

Objectives

A Find the Square Root of a Number.

B Approximate Square Roots.

C Use the Pythagorean Theorem.

Objective A Finding Square Roots

The square of a number is the number times itself. For example:

The square of 5 is 25 because 5^2 or $5 \cdot 5 = 25$.
The square of 4 is 16 because 4^2 or $4 \cdot 4 = 16$.
The square of 10 is 100 because 10^2 or $10 \cdot 10 = 100$.

Recall from Chapter 1 that the reverse process of squaring is finding a **square root.** For example:

A square root of 16 is 4 because $4^2 = 16$.
A square root of 25 is 5 because $5^2 = 25$.
A square root of 100 is 10 because $10^2 = 100$.

We use the symbol $\sqrt{}$, called a **radical sign,** to name square roots. For example:

$\sqrt{16} = 4$ because $4^2 = 16$
$\sqrt{25} = 5$ because $5^2 = 25$

Square Root of a Number

A square root of a number a is a number b whose square is a. We use the radical sign $\sqrt{}$ to name square roots. In symbols,

$$\sqrt{a} = b \quad \text{if} \quad b^2 = a$$

Also,

$$\sqrt{0} = 0$$

Practice 1

Find each square root.
a. $\sqrt{100}$ **b.** $\sqrt{64}$
c. $\sqrt{121}$ **d.** $\sqrt{0}$

Example 1 Find each square root.

a. $\sqrt{49}$ **b.** $\sqrt{1}$ **c.** $\sqrt{81}$

Solution:
a. $\sqrt{49} = 7$ because $7^2 = 49$
b. $\sqrt{1} = 1$ because $1^2 = 1$
c. $\sqrt{81} = 9$ because $9^2 = 81$

■ Work Practice 1

Practice 2

Find: $\sqrt{\dfrac{1}{4}}$

Example 2 Find: $\sqrt{\dfrac{1}{36}}$

Solution: $\sqrt{\dfrac{1}{36}} = \dfrac{1}{6}$ because $\dfrac{1}{6} \cdot \dfrac{1}{6} = \dfrac{1}{36}$

■ Work Practice 2

Practice 3

Find: $\sqrt{\dfrac{9}{16}}$

Example 3 Find: $\sqrt{\dfrac{4}{25}}$

Solution: $\sqrt{\dfrac{4}{25}} = \dfrac{2}{5}$ because $\dfrac{2}{5} \cdot \dfrac{2}{5} = \dfrac{4}{25}$

■ Work Practice 3

Answers
1. a. 10 **b.** 8 **c.** 11 **d.** 0
2. $\dfrac{1}{2}$ **3.** $\dfrac{3}{4}$

Objective B Approximating Square Roots

Thus far, we have found square roots of perfect squares. Numbers like $\frac{1}{4}$, 36, $\frac{4}{25}$, and 1 are called **perfect squares** because their square root is a whole number or a fraction. A square root such as $\sqrt{5}$ cannot be written as a whole number or a fraction since 5 is not a perfect square.

Although $\sqrt{5}$ cannot be written as a whole number or a fraction, it can be approximated by estimating, by using a table (as in the appendix), or by using a calculator.

> **Example 4** Use Appendix A.6 or a calculator to approximate each square root to the nearest thousandth.
>
> **a.** $\sqrt{43} \approx 6.557$ is approximately
> **b.** $\sqrt{80} \approx 8.944$

■ Work Practice 4

Practice 4

Use Appendix A.6 or a calculator to approximate each square root to the nearest thousandth.
a. $\sqrt{21}$ **b.** $\sqrt{52}$

> **Helpful Hint**
>
> $\sqrt{80}$, above, is *approximately* 8.944. This means that if we multiply 8.944 by 8.944, the product is *close* to 80.
>
> $8.944 \times 8.944 \approx 79.995$

It is possible to approximate a square root to the nearest whole number without the use of a calculator or table. To do so, study the number line below and look for patterns.

$$\sqrt{25} \quad \sqrt{31} \; \sqrt{36} \; \sqrt{40} \qquad \sqrt{49} \qquad \sqrt{59} \; \sqrt{64} \quad \sqrt{72} \quad \sqrt{81}$$

$$5 \qquad\qquad 6 \qquad\qquad 7 \qquad\qquad 8 \qquad\qquad 9$$

Above the number line, notice that as the numbers under the radical signs increase, their values, and thus their placement on the number line, increase also.

> **Example 5** Without a calculator or table:
>
> **a.** Determine which two whole numbers $\sqrt{78}$ is between.
> **b.** Use part **a** to approximate $\sqrt{78}$ to the nearest whole.
>
> **Solution:**
>
> **a.** Review perfect squares and recall that $\sqrt{64} = 8$ and $\sqrt{81} = 9$. Since 78 is between 64 and 81, $\sqrt{78}$ is between $\sqrt{64}$ (or 8) and $\sqrt{81}$ (or 9).

> Thus, $\sqrt{78}$ is between 8 and 9.
> **b.** Since 78 is closer to 81, then (as our number line shows) $\sqrt{78}$ is closer to $\sqrt{81}$, or 9.

■ Work Practice 5

Practice 5

Without a calculator or table, approximate $\sqrt{62}$ to the nearest whole.

Objective C Using the Pythagorean Theorem

One important application of square roots has to do with right triangles. Recall that a **right triangle** is a triangle in which one of the angles is a right angle, or measures 90° (degrees). The **hypotenuse** of a right triangle is the side opposite the right angle.

Answers
4. a. 4.583 **b.** 7.211
5. 8

The **legs** of a right triangle are the other two sides. These are shown in the following figure. The right angle in the triangle is indicated by the small square drawn in that angle.

The following theorem is true for all right triangles:

> ### Pythagorean Theorem
>
> In any **right triangle,**
>
> $$(\text{leg})^2 + (\text{other leg})^2 = (\text{hypotenuse})^2$$
>
>

Using the Pythagorean theorem, we can use one of the following formulas to find an unknown length of a right triangle:

> ### Finding an Unknown Length of a Right Triangle
>
> $$\text{hypotenuse} = \sqrt{(\text{leg})^2 + (\text{other leg})^2}$$
>
> or
>
> $$\text{leg} = \sqrt{(\text{hypotenuse})^2 - (\text{other leg})^2}$$

Practice 6

Find the length of the hypotenuse of the given right triangle.

12 ft
16 ft

Example 6 Find the length of the hypotenuse of the given right triangle.

Solution: Since we are finding the hypotenuse, we use the formula

$$\text{hypotenuse} = \sqrt{(\text{leg})^2 + (\text{other leg})^2}$$

6 ft
8 ft

Putting the known values into the formula, we have

$$\begin{aligned}
\text{hypotenuse} &= \sqrt{(6)^2 + (8)^2} \qquad \text{The legs are 6 feet and 8 feet.}\\
&= \sqrt{36 + 64}\\
&= \sqrt{100}\\
&= 10
\end{aligned}$$

The hypotenuse is 10 feet long.

■ Work Practice 6

Practice 7

Approximate the length of the hypotenuse of the given right triangle. Round to the nearest whole unit.

7 km
9 km

Example 7 Approximate the length of the hypotenuse of the given right triangle. Round the length to the nearest whole unit.

Solution:

10 m

17 m

$$\begin{aligned}
\text{hypotenuse} &= \sqrt{(\text{leg})^2 + (\text{other leg})^2}\\
&= \sqrt{(17)^2 + (10)^2} \qquad \text{The legs are 10 meters and 17 meters.}\\
&= \sqrt{289 + 100}\\
&= \sqrt{389}\\
&\approx 20 \qquad\qquad \text{From Appendix A.6 or a calculator}
\end{aligned}$$

The hypotenuse is exactly $\sqrt{389}$ meters, which is approximately 20 meters.

■ Work Practice 7

Answers

6. 20 ft   **7.** 11 km

Example 8 Find the length of the leg in the given right triangle. Give the exact length and a two-decimal-place approximation.

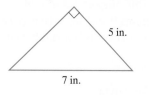

Solution: Notice that the hypotenuse measures 7 inches and the length of one leg measures 5 inches. Since we are looking for the length of the other leg, we use the formula

$$\text{leg} = \sqrt{(\text{hypotenuse})^2 - (\text{other leg})^2}$$

Putting the known values into the formula, we have

$$\text{leg} = \sqrt{(7)^2 - (5)^2} \quad \text{The hypotenuse is 7 inches, and the other leg is 5 inches.}$$
$$= \sqrt{49 - 25}$$
$$= \sqrt{24} \qquad \text{Exact answer}$$
$$\approx 4.90 \qquad \text{From Appendix A.6 or a calculator}$$

The length of the leg is exactly $\sqrt{24}$ inches, which is approximately 4.90 inches.

Work Practice 8

✓**Concept Check** The following lists are the lengths of the sides of two triangles. Which set forms a right triangle? Explain.
   a. 8, 15, 17   **b.** 24, 30, 40

△ **Example 9** Finding the Dimensions of a Park

An inner-city park is in the shape of a square that measures 300 feet on a side. A sidewalk is to be constructed along the diagonal of the park. Find the length of the sidewalk rounded to the nearest whole foot.

Solution: The diagonal is the hypotenuse of a right triangle, so we use the formula

$$\text{hypotenuse} = \sqrt{(\text{leg})^2 + (\text{other leg})^2}$$

Putting the known values into the formula we have

$$\text{hypotenuse} = \sqrt{(300)^2 + (300)^2} \quad \text{The legs are both 300 feet.}$$
$$= \sqrt{90,000 + 90,000}$$
$$= \sqrt{180,000}$$
$$\approx 424 \qquad \text{From Appendix A.6 or a calculator}$$

The length of the sidewalk is approximately 424 feet.

Work Practice 9

Practice 8

Find the length of the leg in the given right triangle. Give the exact length and a two-decimal-place approximation.

Practice 9

A football field is a rectangle measuring 100 yards by 53 yards. Draw a diagram and find the length of the diagonal of a football field to the nearest yard.

Answers
8. $\sqrt{72}$ ft ≈ 8.49 ft
9. 113 yd

✓**Concept Check Answer**
set (a) forms a right triangle

▦ Calculator Explorations   Finding Square Roots

To simplify or approximate square roots using a calculator, locate the key marked $\boxed{\sqrt{\ \ }}$.

To simplify $\sqrt{64}$, for example, press the keys

$\boxed{64}\ \boxed{\sqrt{\ \ }}$ or $\boxed{\sqrt{\ \ }}\ \boxed{64}$

The display should read $\boxed{8}$. Then

$\sqrt{64} = 8$

To *approximate* $\sqrt{10}$, press the keys

$\boxed{10}\ \boxed{\sqrt{\ \ }}$ or $\boxed{\sqrt{\ \ }}\ \boxed{10}$

The display should read $\boxed{3.16227766}$. This is an *approximation* for $\sqrt{10}$. A three-decimal-place approximation is

$\sqrt{10} \approx 3.162$

Is this answer reasonable? Since 10 is between perfect squares 9 and 16, $\sqrt{10}$ is between $\sqrt{9} = 3$ and $\sqrt{16} = 4$. Our answer is reasonable since 3.162 is between 3 and 4.

Simplify.

1. $\sqrt{1024}$
2. $\sqrt{676}$

Approximate each square root. Round each answer to the nearest thousandth.

3. $\sqrt{31}$
4. $\sqrt{19}$
5. $\sqrt{97}$
6. $\sqrt{56}$

Vocabulary, Readiness & Video Check

Use the choices below to fill in each blank. Some choices may be used more than once.

| squaring | Pythagorean theorem | radical | leg |
| hypotenuse | perfect squares | 10 | |

1. $\sqrt{100} =$ _____ because $10 \cdot 10 = 100$.

2. The _____ sign is used to denote the square root of a number.

3. The reverse process of _____ a number is finding a square root of a number.

4. The numbers 9, 1, and $\dfrac{1}{25}$ are called _____.

5. Label the parts of the right triangle. _____

6. The _____ can be used for right triangles.

Martin-Gay Interactive Videos   Watch the section lecture video and answer the following questions.

Objective A   7. From the lecture before ▤ Example 1, explain why $\sqrt{49} = 7$. ▶

Objective B   8. In ▤ Example 5, how do we know $\sqrt{15}$ is closer to 4 than to 3? ▶

Objective C   9. At the beginning of ▤ Example 6, what are we reminded about regarding the Pythagorean theorem? ▶

See Video 8.6

8.6 Exercise Set MyMathLab®

Objective A *Find each square root. See Examples 1 through 3.*

1. $\sqrt{4}$ **2.** $\sqrt{9}$ **3.** $\sqrt{121}$ **4.** $\sqrt{144}$

5. $\sqrt{\dfrac{1}{81}}$ **6.** $\sqrt{\dfrac{1}{64}}$ **7.** $\sqrt{\dfrac{16}{64}}$ **8.** $\sqrt{\dfrac{36}{81}}$

Objective B *Use Appendix A.6 or a calculator to approximate each square root. Round the square root to the nearest thousandth. See Examples 4 and 5.*

9. $\sqrt{3}$ **10.** $\sqrt{5}$ **11.** $\sqrt{15}$ **12.** $\sqrt{17}$

13. $\sqrt{47}$ **14.** $\sqrt{85}$ **15.** $\sqrt{26}$ **16.** $\sqrt{35}$

Determine what two whole numbers each square root is between without using a calculator or table. Then use a calculator or Appendix A.6 to check. See Example 5.

17. $\sqrt{38}$ **18.** $\sqrt{27}$ **19.** $\sqrt{101}$ **20.** $\sqrt{85}$

Objectives A B Mixed Practice *Find each square root. If necessary, round the square root to the nearest thousandth. See Examples 1 through 5.*

21. $\sqrt{256}$ **22.** $\sqrt{625}$ **23.** $\sqrt{92}$ **24.** $\sqrt{18}$

25. $\sqrt{\dfrac{49}{144}}$ **26.** $\sqrt{\dfrac{121}{169}}$ **27.** $\sqrt{71}$ **28.** $\sqrt{62}$

Objective C *Find the unknown length in each right triangle. If necessary, approximate the length to the nearest thousandth. See Examples 6 through 8.*

29. 5 in., 12 in., ?

30. ?, 15 ft, 36 ft

31. 10 cm, 12 cm, ?

32. 3 yd, 9 yd, ?

33. 22 m, 48 m, ?

34. 34 mi, 70 mi, ?

35. 108 mm, 45 mm, ?

36. ?, 27 km, 36 km

Sketch each right triangle and find the length of the side not given. If necessary, approximate the length to the nearest thousandth. (Each length is in units.) See Examples 6 through 8.

37. leg = 3, leg = 4

38. leg = 9, leg = 12

39. leg = 5, hypotenuse = 13

40. leg = 6, hypotenuse = 10

41. leg = 10, leg = 14

42. leg = 2, leg = 16

43. leg = 35, leg = 28

44. leg = 30, leg = 15

45. leg = 30, leg = 30

46. leg = 21, leg = 21

▶ **47.** hypotenuse = 2, leg = 1

48. hypotenuse = 9, leg = 8

49. leg = 7.5, leg = 4

50. leg = 12, leg = 22.5

Solve. See Example 9.

51. A standard city block is a square with each side measuring 100 yards. Find the length of the diagonal of a city block to the nearest hundredth yard.

52. A section of land is a square with each side measuring 1 mile. Find the length of the diagonal of the section of land to the nearest thousandth mile.

53. Find the height of the tree. Round the height to one decimal place.

54. Find the height of the antenna. Round the height to one decimal place.

55. The playing field for football is a rectangle that is 300 feet long by 160 feet wide. Find, to the nearest foot, the length of a straight-line run that started at one corner and went diagonally to end at the opposite corner.

56. A soccer field is in the shape of a rectangle and its dimensions depend on the age of the players. The dimensions of the soccer field below are the minimum dimensions for international play. Find the length of the diagonal of this rectangle. Round the answer to the nearest tenth of a yard.

Review

Find the value of n in each proportion. See Section 5.3.

57. $\dfrac{n}{6} = \dfrac{2}{3}$

58. $\dfrac{8}{n} = \dfrac{4}{8}$

59. $\dfrac{9}{11} = \dfrac{n}{55}$

60. $\dfrac{5}{6} = \dfrac{35}{n}$

61. $\dfrac{3}{n} = \dfrac{7}{14}$

62. $\dfrac{n}{9} = \dfrac{4}{6}$

Concept Extensions

Use the results of Exercises 17–20 and approximate each square root to the nearest whole without using a calculator or table. Then use a calculator or Appendix A.6 to check. See Example 5.

63. $\sqrt{38}$ **64.** $\sqrt{27}$ **65.** $\sqrt{101}$ **66.** $\sqrt{85}$

67. Without using a calculator, explain how you know that $\sqrt{105}$ is *not* approximately 9.875.

68. Without using a calculator, explain how you know that $\sqrt{27}$ is *not* approximately 3.296.

Does the set form the lengths of the sides of a right triangle? See the Concept Check in this section.

69. $25, 60, 65$ **70.** $20, 45, 50$

△**71.** Find the exact length of *x*. Then give a two-decimal-place approximation.

12 inches
10 inches
8 inches

8.7   Congruent and Similar Triangles

Objective A   Deciding Whether Two Triangles Are Congruent.

Objectives

A Deciding Whether Two Triangles Are Congruent.

B Find the Ratio of Corresponding Sides in Similar Triangles.

C Find Unknown Lengths of Sides in Similar Triangles.

Congruent angles are angles that have the same measure. Two triangles are **congruent** when they have the same shape and the same size. In congruent triangles, the measures of corresponding angles are equal and the lengths of corresponding sides are equal. The following triangles are congruent:

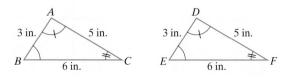

Since these triangles are congruent, the measures of corresponding angles are equal.

     Angles with equal measure: $\angle A$ and $\angle D$, $\angle B$ and $\angle E$, $\angle C$ and $\angle F$. Also, the lengths of corresponding sides are equal.

     Equal corresponding sides: $\overline{AB}$ and $\overline{DE}$, $\overline{BC}$ and $\overline{EF}$, $\overline{CA}$ and $\overline{FD}$

Any one of the following may be used to determine whether two triangles are congruent:

Congruent Triangles

Angle-Side-Angle (ASA)

If the measures of two angles of a triangle equal the measures of two angles of another triangle, and the lengths of the sides between each pair of angles are equal, the triangles are congruent.

For example, these two triangles are congruent by Angle-Side-Angle.

Side-Side-Side (SSS)

If the lengths of the three sides of a triangle equal the lengths of the corresponding sides of another triangle, the triangles are congruent.

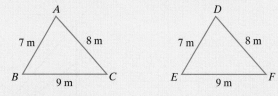

For example, these two triangles are congruent by Side-Side-Side.

Side-Angle-Side (SAS)

If the lengths of two sides of a triangle equal the lengths of corresponding sides of another triangle, and the measures of the angles between each pair of sides are equal, the triangles are congruent.

For example, these two triangles are congruent by Side-Angle-Side.

Practice 1

a. Determine whether triangle *MNO* is congruent to triangle *RQS*.

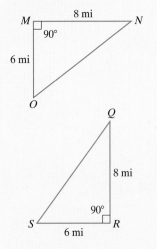

b. Determine whether triangle *GHI* is congruent to triangle *JKL*.

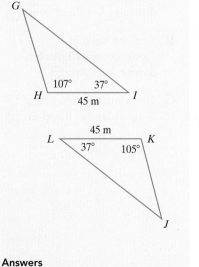

Example 1   Determine whether triangle *ABC* is congruent to triangle *DEF*.

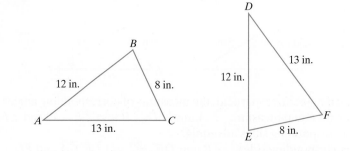

Solution: Since the lengths of all three sides of triangle *ABC* equal the lengths of all three sides of triangle *DEF*, the triangles are congruent.

■ Work Practice 1

Answers

1. a. congruent   **b.** not congruent

In Example 1, notice that as soon as we know that the two triangles are congruent, we know that all three corresponding angles are congruent.

Objective B Finding the Ratio of Corresponding Sides in Similar Triangles

Two triangles are **similar** when they have the same shape but not necessarily the same size. In similar triangles, the measures of corresponding angles are equal and corresponding sides are in proportion. The following triangles are similar:

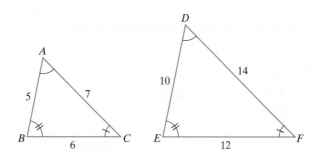

Since these triangles are similar, the measures of corresponding angles are equal.

Angles with equal measure: $\angle A$ and $\angle D$, $\angle B$ and $\angle E$, $\angle C$ and $\angle F$. Also, the lengths of corresponding sides are in proportion.

Sides in proportion: $\dfrac{AB}{DE} = \dfrac{BC}{EF} = \dfrac{CA}{FD}$ or, in this particular case,

$$\frac{AB}{DE} = \frac{5}{10} = \frac{1}{2}, \frac{BC}{EF} = \frac{6}{12} = \frac{1}{2}, \frac{CA}{FD} = \frac{7}{14} = \frac{1}{2}$$

The ratio of corresponding sides is $\dfrac{1}{2}$.

Example 2   Find the ratio of corresponding sides for the similar triangles *ABC* and *DEF*.

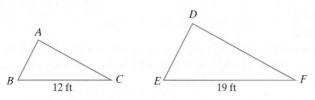

Solution: We are given the lengths of two corresponding sides. Their ratio is

$$\frac{12 \text{ feet}}{19 \text{ feet}} = \frac{12}{19}$$

 Work Practice 2

Objective C Finding Unknown Lengths of Sides in Similar Triangles

Because the ratios of lengths of corresponding sides are equal, we can use proportions to find unknown lengths in similar triangles.

Practice 2

Find the ratio of corresponding sides for the similar triangles *QRS* and *XYZ*.

Answer

2. $\dfrac{9}{13}$

Practice 3

Given that the triangles are similar, find the missing length n.

a.

b.

Example 3

Given that the triangles are similar, find the missing length n.

Solution: Since the triangles are similar, corresponding sides are in proportion. Thus, the ratio of 2 to 3 is the same as the ratio of 10 to n, or

$$\frac{2}{3} = \frac{10}{n}$$

To find the unknown length n, we set cross products equal.

$$\frac{2}{3} = \frac{10}{n}$$

$2 \cdot n = 3 \cdot 10$ Set cross products equal.

$2 \cdot n = 30$ Multiply.

$n = \dfrac{30}{2}$ Divide 30 by 2, the number multiplied by n.

$n = 15$

The missing length is 15 units.

■ Work Practice 3

✓ Concept Check

The following two triangles are similar. Which vertices of the first triangle appear to correspond to which vertices of the second triangle?

Many applications involve diagrams containing similar triangles. Surveyors, astronomers, and many other professionals continually use similar triangles in their work.

Practice 4

Tammy Shultz, a firefighter, needs to estimate the height of a burning building. She estimates the length of her shadow to be 8 feet long and the length of the building's shadow to be 60 feet long. Find the approximate height of the building if she is 5 feet tall.

Example 4   Finding the Height of a Tree

Mel Rose is a 6-foot-tall park ranger who needs to know the height of a particular tree. He measures the shadow of the tree to be 69 feet long when his own shadow is 9 feet long. Find the height of the tree.

Solution:

1. UNDERSTAND. Read and reread the problem. Notice that the triangle formed by the Sun's rays, Mel, and his shadow is similar to the triangle formed by the Sun's rays, the tree, and its shadow.

Answers

3. a. $n = 8$ **b.** $n = \dfrac{10}{3}$ or $3\dfrac{1}{3}$

4. approximately 37.5 ft

✓ Concept Check Answer

A corresponds to O; B corresponds to N; C corresponds to M

2. TRANSLATE. Write a proportion from the similar triangles formed.

$$\underbrace{\frac{\text{Mel's height}}{\text{height of tree}}}_{} \quad \overset{\rightarrow}{\rightarrow} \quad \frac{6}{n} = \frac{9}{69} \quad \overset{\leftarrow}{\leftarrow} \quad \underbrace{\frac{\text{length of Mel's shadow}}{\text{length of tree's shadow}}}_{}$$

$$\text{or } \frac{6}{n} = \frac{3}{23} \quad \text{Simplify } \frac{9}{69}. \text{ (ratio in lowest terms)}$$

3. SOLVE for n:

$$\frac{6}{n} = \frac{3}{23}$$

$$6 \cdot 23 = n \cdot 3 \quad \text{Set cross products equal.}$$

$$138 = n \cdot 3 \quad \text{Multiply.}$$

$$\frac{138}{3} = n \quad \text{Divide 138 by 3, the number multiplied by } n.$$

$$46 = n$$

4. INTERPRET. *Check* to see that replacing n with 46 in the proportion makes the proportion true. *State* your conclusion: The height of the tree is 46 feet.

▨ **Work Practice 4**

Vocabulary, Readiness & Video Check

Answer each question true or false.

1. Two triangles that have the same shape but not necessarily the same size are congruent.

2. Two triangles are congruent if they have the same shape and size.

3. Congruent triangles are also similar.

4. Similar triangles are also congruent.

5. For the two similar triangles, the ratio of corresponding sides is $\frac{5}{6}$.

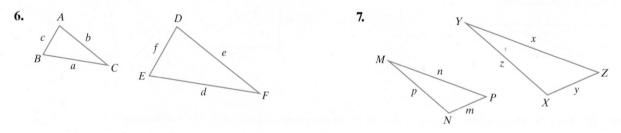

Each pair of triangles is similar. Name the congruent angles and the corresponding sides that are proportional.

6.

7.

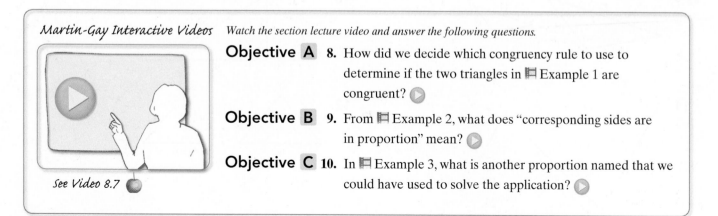

Martin-Gay Interactive Videos   Watch the section lecture video and answer the following questions.

See Video 8.7 🍎

Objective A   **8.** How did we decide which congruency rule to use to determine if the two triangles in ▥ Example 1 are congruent? ▶

Objective B   **9.** From ▥ Example 2, what does "corresponding sides are in proportion" mean? ▶

Objective C   **10.** In ▥ Example 3, what is another proportion named that we could have used to solve the application? ▶

8.7 Exercise Set MyMathLab® ▶

Objective A *Determine whether each pair of triangles is congruent. If congruent, state the reason why, such as SSS, SAS, or ASA. See Example 1.*

1.

5 in.   6 in.   7 in.

6 in.   7 in.   5 in.

2.

10 mi   14 mi   14 mi

14 mi   14 mi   10 mi

▶ **3.**

25 m   40 m   23 m

25 m   40 m   24 m

4.

21 cm   17 cm   7 cm

7 cm   21 cm   16 cm

5.

48°   30 m   42°

48°   42°   30 m

6.

15 yd   23°   36°

36°   23°   15 yd

7.

52 m   122°   98 m

52 m   122°   98 m

8.

36 km   93°   57 km

57 km   93°   36 km

Objective B *Find each ratio of the corresponding sides of the given similar triangles. See Example 2.*

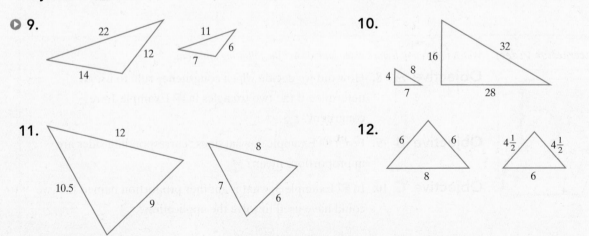

▶ **9.**

22   11   12   6   7   14

10.

16   32   4   8   7   28

11.

12   10.5   9   8   7   6

12.

6   6   8 $4\frac{1}{2}$ $4\frac{1}{2}$ 6

Objective C *Given that the pairs of triangles are similar, find the unknown length of the side labeled n.*
See Example 3.

13.

14.

⏵ **15.**

16.

17.

18.

19.

20.

21.

22.

23.

24.

25.

26.

27.

28.

Solve. For Exercises 29 and 30, the solutions have been started for you. See Example 4.

29. Given the following diagram, approximate the height of the observation deck in the Seattle Space Needle in Seattle, Washington. (*Source:* Seattle Space Needle)

Start the solution:

1. UNDERSTAND the problem. Reread it as many times as needed.
2. TRANSLATE into a proportion using the similar triangles formed. (Fill in the blanks.)

$$\begin{array}{ll} \text{height of} \\ \text{observation deck} \rightarrow \\ \text{height of pole} \rightarrow \end{array} \dfrac{n}{13} = \dfrac{\quad}{\quad} \begin{array}{l} \leftarrow \text{length of Space} \\ \quad \text{Needle shadow} \\ \leftarrow \text{length of pole} \\ \quad \text{shadow} \end{array}$$

3. SOLVE by setting cross products equal.
4. INTERPRET.

30. A fountain in Fountain Hills, Arizona, sits in a 28-acre lake and shoots up a column of water every hour. Based on the diagram below, what is the height of the fountain?

Start the solution:

1. UNDERSTAND the problem. Reread it as many times as needed.
2. TRANSLATE into a proportion using the similar triangles formed. (Fill in the blanks.)

$$\begin{array}{ll} \text{height of tree} \rightarrow \\ \text{height of fountain} \rightarrow \end{array} \dfrac{28}{n} = \dfrac{\quad}{\quad} \begin{array}{l} \leftarrow \text{length of tree} \\ \quad \text{shadow} \\ \leftarrow \text{length of fountain} \\ \quad \text{shadow} \end{array}$$

3. SOLVE by setting cross products equal.
4. INTERPRET.

31. Given the following diagram, approximate the height of the Chase Tower in Oklahoma City, Oklahoma. Here, we use x to represent the unknown number. (*Source:* Council on Tall Buildings and Urban Habitat)

32. The tallest tree standing today is a redwood located in the Humboldt Redwoods State Park near Ukiah, California. Given the following diagram, approximate its height. Here, we use x to represent the unknown number. (*Source: Guinness World Records*)

33. Samantha Black, a 5-foot-tall park ranger, needs to know the height of a tree. She notices that when the shadow of the tree is 48 feet long, her shadow is 4 feet long. Find the height of the tree.

34. Lloyd White, a firefighter, needs to estimate the height of a burning building. He estimates the length of his shadow to be 9 feet long and the length of the building's shadow to be 75 feet long. Find the approximate height of the building if he is 6 feet tall.

35. If a 30-foot tree casts an 18-foot shadow, find the length of the shadow cast by a 24-foot tree.

36. If a 24-foot flagpole casts a 32-foot shadow, find the length of the shadow cast by a 44-foot antenna. Round to the nearest tenth.

Review

Solve. See Section 5.3.

37. For the health of his fish, the owner of Pete's Sea World uses the standard that a 20-gallon tank should house only 19 neon tetras. Find the number of neon tetras that Pete would place into a 55-gallon tank.

38. A local package express deliveryman is traveling the city expressway at 45 mph when he is forced to slow down due to traffic ahead. His truck slows at the rate of 3 mph every 5 seconds. Find his speed 8 seconds after braking.

Solve. See Section 8.6.

39. Launch Umbilical Tower 1 is the name of the gantry used for the *Apollo* launch that took Neil Armstrong and Buzz Aldrin to the moon. Find the height of the gantry to the nearest whole foot.

430 feet

200 feet

40. Arena polo, popular in the United States and England, is played on a field that is 100 yards long and usually 50 yards wide. Find the length, to the nearest yard, of the diagonal of this field.

50 yards ? 100 yards

Perform the indicated operation. See Sections 4.3 through 4.5.

41. $3.6 + 0.41$

42. $3.6 - 0.41$

43. $(0.41)(3)$

44. $0.48 \div 3$

Concept Extensions

45. The print area on a particular page measures 7 inches by 9 inches. A printing shop is to copy the page and reduce the print area so that its length is 5 inches. What will its width be? Will the print now fit on a 3-by-5-inch index card?

46. The art sample for a banner measures $\frac{1}{3}$ foot in width by $1\frac{1}{2}$ feet in length. If the completed banner is to have a length of 9 feet, find its width.

Given that the pairs of triangles are similar, find the length of the side labeled n. Round your results to 1 decimal place.

47.

5.2   7.8   n   12.6

48.

11.6   n   20.8   58.7

49. In your own words, describe any differences in similar triangles and congruent triangles.

50. Describe a situation where similar triangles would be useful for a contractor building a house.

51. A triangular park is planned and waiting to be approved by the city zoning commission. A drawing of the park shows sides of length 5 inches, $7\frac{1}{2}$ inches, and $10\frac{5}{8}$ inches. If the scale on the drawing is $\frac{1}{4}$ in. = 10 ft, find the actual proposed dimensions of the park.

52. John and Robyn Costello draw a triangular deck on their house plans. Robyn measures sides of the deck drawing on the plans to be 3 inches, $4\frac{1}{2}$ inches, and 6 inches. If the scale on the drawing is $\frac{1}{4}$ in. = 1 foot, find the lengths of the sides of the deck they want built.

Chapter 8 Group Activity

The Cost of Road Signs

Sections 8.1, 8.2, 8.4

There are nearly 4 million miles of streets and roads in the United States. With streets, roads, and highways comes the need for traffic control, guidance, warning, and regulation. Road signs perform many of these tasks. Just in our routine travels, we see a wide variety of road signs every day. Think how many road signs must exist on the 4 million miles of roads in the United States. Have you ever wondered how much signs like these cost?

The cost of a road sign generally depends on the type of sign. Costs for several types of signs and signposts are listed in the table. Examples of various types of signs are shown below.

Road Sign Costs	
Type of Sign	**Cost**
Regulatory, warning, marker	$15–$18 per square foot
Large guide	$20–$25 per square foot
Type of Post	**Cost**
U-channel	$125–$200 each
Square tube	$10–$15 per foot
Steel breakaway posts	$15–$25 per foot

The cost of a sign is based on its area. For diamond, square, or rectangular signs, the area is found by multiplying the length (in feet) times the width (in feet). Then the area is multiplied by the cost per square foot. For signs with irregular shapes, costs are generally figured *as if* the sign were a rectangle, multiplying the height and width at the tallest and widest parts of the sign.

Group Activity

Locate four different kinds of road signs on or near your campus. Measure the dimensions of each sign, including the height of the post on which it is mounted. Using the cost data given in the table, find the minimum and maximum costs of each sign, including its post. Summarize your results in a table, and include a sketch of each sign.

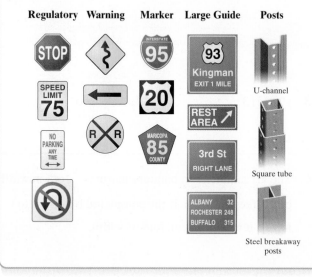

Chapter 8   Vocabulary Check

Fill in each blank with one of the words or phrases listed below.

transversal	line	congruent	hypotenuse	legs	acute
right	line segment	complementary	square root	vertical	supplementary
right triangle	volume	obtuse	vertex	ray	angle
similar	perimeter	area	straight	adjacent	

1. A(n) _____ is a triangle with a right angle. The side opposite the right angle is called the _____, and the other two sides are called _____.

2. A(n) _____ is a piece of a line with two endpoints.

3. Two angles that have a sum of 90° are called _____ angles.

4. A(n) _____ is a set of points extending indefinitely in two directions.

5. The _____ of a polygon is the distance around the polygon.

6. A(n) _____ is made up of two rays that share the same endpoint. The common endpoint is called the _____.

7. _____ triangles have the same shape and the same size.

8. _____ measures the amount of surface of a region.

9. A(n) _____ is a part of a line with one endpoint. A ray extends indefinitely in one direction.

10. A(n) _____ of a number a is a number b whose square is a.

11. A line that intersects two or more lines at different points is called a(n) _____.

12. A angle that measures 180° is called a(n) _____ angle.

13. The measure of the space of a solid is called its _____.

14. When two lines intersect, four angles are formed. The angles that are opposite each other are called _____ angles.

15. When two of the four angles from Exercise 14 share a common side, they are called _____ angles.

16. An angle whose measure is between 90° and 180° is called a(n) _____ angle.

17. An angle that measures 90° is called a(n) _____ angle.

18. An angle whose measure is between 0° and 90° is called a(n) _____ angle.

19. Two angles that have a sum of 180° are called _____ angles.

20. _____ triangles have exactly the same shape but not necessarily the same size.

Helpful Hint

▶ Are you preparing for your test? Don't forget to take the Chapter 8 Test on page 574. Then check your answers at the back of the text and use the Chapter Test Prep Videos to see the fully worked-out solutions to any of the exercises you want to review.

8 Chapter Highlights

Definitions and Concepts	Examples
Section 8.1 Lines and Angles	
A **line** is a set of points extending indefinitely in two directions. A line has no width or height, but it does have length. We name a line by any two of its points.	Line AB or $\overleftrightarrow{AB}$

(continued)

Definitions and Concepts	Examples

Section 8.1   Lines and Angles (continued)

A **line segment** is a piece of a line with two endpoints.

Line segment AB or $\overline{AB}$

A **ray** is a part of a line with one endpoint. A ray extends indefinitely in one direction.

Ray AB or $\overrightarrow{AB}$

An **angle** is made up of two rays that share the same endpoint. The common endpoint is called the **vertex.**

Angle ABC, $\angle ABC$, $\angle CBA$, or $\angle B$

An angle that measures 180° is called a **straight angle.**

$\angle RST$ is a straight angle.

An angle that measures 90° is called a **right angle.** The symbol ∟ is used to denote a right angle.

$\angle ABC$ is a right angle.

An angle whose measure is between 0° and 90° is called an **acute angle.**

Acute angles

An angle whose measure is between 90° and 180° is called an **obtuse angle.**

Obtuse angles

Two angles that have a sum of 90° are called **complementary angles.** We say that each angle is the **complement** of the other.

Complementary angles
60° + 30° = 90°

Two angles that have a sum of 180° are called **supplementary angles.** We say that each angle is the **supplement** of the other.

Supplementary angles
125° + 55° = 180°

Definitions and Concepts	Examples

Section 8.1   Lines and Angles (*continued*)

When two lines intersect, four angles are formed. Two of these angles that are opposite each other are called **vertical angles.** Vertical angles have the same measure.

Two of these angles that share a common side are called **adjacent angles.** Adjacent angles formed by intersecting lines are supplementary.

A line that intersects two or more lines at different points is called a **transversal.** Line l is a transversal that intersects lines m and n. The eight angles formed have special names. Some of these names are:

Corresponding angles: $\angle a$ and $\angle e$, $\angle c$ and $\angle g$, $\angle b$ and $\angle f$, $\angle d$ and $\angle h$

Alternate interior angles: $\angle c$ and $\angle f$, $\angle d$ and $\angle e$

Parallel Lines Cut by a Transversal

If two parallel lines are cut by a transversal, then the measures of **corresponding angles are equal** and the measures of **alternate interior angles are equal.**

Vertical angles:
  $\angle a$ and $\angle c$
  $\angle d$ and $\angle b$

Adjacent angles:
  $\angle a$ and $\angle b$
  $\angle b$ and $\angle c$
  $\angle c$ and $\angle d$
  $\angle d$ and $\angle a$

Section 8.2   Plane Figures and Solids

The **sum of the measures** of the angles of a triangle is 180°.

Find the measure of $\angle x$.

The measure of $\angle x = 180° - 85° - 45° = 50°$

A **right triangle** is a triangle with a right angle. The side opposite the right angle is called the **hypotenuse,** and the other two sides are called **legs.**

For a circle or a sphere:

diameter $= 2 \cdot$ radius

$d = 2 \cdot r$

radius $= \dfrac{\text{diameter}}{2}$

$r = \dfrac{d}{2}$

Find the diameter of the circle.

$d = 2 \cdot r$
$ = 2 \cdot 6 \text{ feet} = 12 \text{ feet}$

Definitions and Concepts	Examples

Section 8.3   Perimeter

Perimeter Formulas

Rectangle: $P = 2 \cdot l + 2 \cdot w$

Square: $P = 4 \cdot s$

Triangle: $P = a + b + c$

Circumference of a Circle: $C = 2 \cdot \pi \cdot r$ or $C = \pi \cdot d$,

where $\pi \approx 3.14$ or $\pi \approx \dfrac{22}{7}$

Find the perimeter of a rectangle with length 28 meters and width 15 meters.

$$P = 2 \cdot l + 2 \cdot w$$
$$= 2 \cdot 28 \text{ m} + 2 \cdot 15 \text{ m}$$
$$= 56 \text{ m} + 30 \text{ m}$$
$$= 86 \text{ m}$$

The perimeter is 86 meters.

Section 8.4   Area

Area Formulas

Rectangle: $A = l \cdot w$

Square: $A = s^2$

Triangle: $A = \dfrac{1}{2} \cdot b \cdot h$

Parallelogram: $A = b \cdot h$

Trapezoid: $A = \dfrac{1}{2} \cdot (b + B) \cdot h$

Circle: $A = \pi \cdot r^2$

Find the area of a square with side length 8 centimeters.

$$A = s^2$$
$$= (8 \text{ cm})^2$$
$$= 64 \text{ square centimeters}$$

The area of the square is 64 square centimeters.

Section 8.5   Volume

Volume Formulas

Rectangular Solid: $V = l \cdot w \cdot h$

Cube: $V = s^3$

Sphere: $V = \dfrac{4}{3} \cdot \pi \cdot r^3$

Right Circular Cylinder: $V = \pi \cdot r^2 \cdot h$

Cone: $V = \dfrac{1}{3} \cdot \pi \cdot r^2 \cdot h$

Square-Based Pyramid: $V = \dfrac{1}{3} \cdot s^2 \cdot h$

Find the volume of the sphere. Use $\dfrac{22}{7}$ for π.

4 in.

$$V = \dfrac{4}{3} \cdot \pi \cdot r^3$$

$$\approx \dfrac{4}{3} \cdot \dfrac{22}{7} \cdot (4 \text{ inches})^3$$

$$= \dfrac{4 \cdot 22 \cdot 64}{3 \cdot 7} \text{ cubic inches}$$

$$= \dfrac{5632}{21} \quad \text{or} \quad 268\dfrac{4}{21} \text{ cubic inches}$$

Definitions and Concepts	Examples

Section 8.6   Square Roots and the Pythagorean Theorem

Square Root of a Number

A **square root** of a number a is a number b whose square is a. We use the radical sign $\sqrt{}$ to name square roots.

$\sqrt{9} = 3, \sqrt{100} = 10, \sqrt{1} = 1$

Pythagorean Theorem

$$(\text{leg})^2 + (\text{other leg})^2 = (\text{hypotenuse})^2$$

Find the hypotenuse of the given triangle.

To Find an Unknown Length of a Right Triangle

$$\text{hypotenuse} = \sqrt{(\text{leg})^2 + (\text{other leg})^2}$$

$$\text{leg} = \sqrt{(\text{hypotenuse})^2 - (\text{other leg})^2}$$

$$\begin{aligned}
\text{hypotenuse} &= \sqrt{(\text{leg})^2 + (\text{other leg})^2} \\
&= \sqrt{(3)^2 + (8)^2} \quad \text{The legs are 3 and} \\
&\qquad\qquad\qquad\qquad \text{8 inches.} \\
&= \sqrt{9 + 64} \\
&= \sqrt{73} \text{ inches} \\
&\approx 8.5 \text{ inches}
\end{aligned}$$

Section 8.7   Congruent and Similar Triangles

Congruent triangles have the same shape and the same size. Corresponding angles are equal, and corresponding sides are equal.

Similar triangles have exactly the same shape but not necessarily the same size. Corresponding angles are equal, and the ratios of the lengths of corresponding sides are equal.

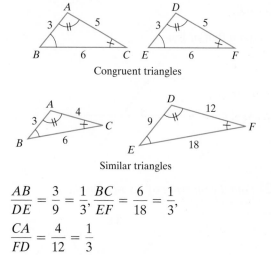

Congruent triangles

Similar triangles

$$\frac{AB}{DE} = \frac{3}{9} = \frac{1}{3}, \frac{BC}{EF} = \frac{6}{18} = \frac{1}{3},$$

$$\frac{CA}{FD} = \frac{4}{12} = \frac{1}{3}$$

Chapter 8   Review

(8.1) *Classify each angle as acute, right, obtuse, or straight.*

1.   **2.**   **3.** *C*   **4.**

5. Find the complement of a 25° angle.

6. Find the supplement of a 105° angle.

Find the measure of angle x in each figure.

7.

8.

9.

10.

11. Identify the pairs of supplementary angles.

12. Identify the pairs of complementary angles.

Find the measures of angles x, y, and z in each figure.

13.

14.
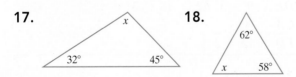

15. Given that $m \parallel n$.

16. Given that $m \parallel n$.

(8.2) *Find the measure of ∠x in each figure.*

17.

18.

19.

20.

Find the unknown diameter or radius as indicated.

21.

22.

23.

24.

Identify each solid.

25.

26.

27.

28.

Find the unknown radius or diameter as indicated.

29. The radius of a sphere is 9 inches. Find its diameter.

30. The diameter of a sphere is 4.7 meters. Find its radius.

Identify each regular polygon.

31.

32.

Identify each triangle as equilateral, isosceles, or scalene. Also identify any triangle that is a right triangle.

33.

34.

(8.3) *Find the perimeter of each figure.*

35.
27 m
Parallelogram $17\frac{1}{2}$ m

36.

11 cm   7.6 cm
12 cm

37.

7 m
8 m
5 m
10 m

38.
5 ft
4 ft
11 ft   3 ft
22 ft

Solve.

39. Find the perimeter of a rectangular sign that measures 6 feet by 10 feet.

40. Find the perimeter of a town square that measures 110 feet on a side.

Find the circumference of each circle. Use $\pi \approx 3.14$.

41.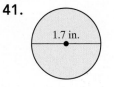

1.7 in.

42.

5 yd

(8.4) *Find the area of each figure. For the circles, find the exact area and then use π ≈ 3.14 to approximate the area.*

43.

12 ft

10 ft

36 ft

44.

14 m

20 m

45.

15 cm

40 cm

46.

9 yd

21 yd

47.

7 ft

48.

Square   9.1 m

49.

34 in.

7 in.

50.

64 cm

26 cm

32 cm

51.

4 m

3 m

12 m

13 m

52. The amount of sealant necessary to seal a driveway depends on the area. Find the area of a rectangular driveway 36 feet by 12 feet.

53. Find how much carpet is necessary to cover the floor of the room shown.

10 feet

13 feet

(8.5) *Find the volume of each solid. For Exercises 56 and 57, give an exact volume and an approximation.*

54.

$2\frac{1}{2}$ in.

$2\frac{1}{2}$ in.

$2\frac{1}{2}$ in.

55.

6 ft

2 ft   7 ft

56. Use π ≈ 3.14.

50 cm

20 cm

57. Use π ≈ $\frac{22}{7}$.

$\frac{1}{2}$ km

58. Find the volume of a pyramid with a square base 2 feet on a side and a height of 2 feet.

59. Approximate the volume of a tin can 8 inches high and 3.5 inches in radius. Use 3.14 for π.

60. A chest has 3 drawers. If each drawer has inside measurements of $2\frac{1}{2}$ feet by $1\frac{1}{2}$ feet by $\frac{2}{3}$ foot, find the total volume of the 3 drawers.

61. A cylindrical canister for a shop vacuum is 2 feet tall and 1 foot in *diameter*. Find its exact volume.

(8.6) *Simplify.*

62. $\sqrt{64}$

63. $\sqrt{144}$

64. $\sqrt{\dfrac{4}{25}}$

65. $\sqrt{\dfrac{1}{100}}$

Find the unknown length of each given right triangle. If necessary, round to the nearest tenth.

66. leg $= 12$, leg $= 5$

67. leg $= 20$, leg $= 21$

68. leg $= 9$, hypotenuse $= 14$

69. leg $= 124$, hypotenuse $= 155$

70. A baseball diamond is in the shape of a square and has sides of length 90 feet. Find the distance across the diamond from third base to first base, to the nearest tenth of a foot.

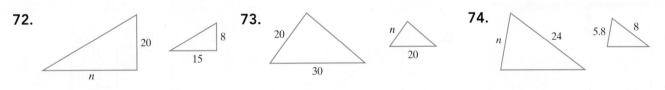

71. Find the height of the building rounded to the nearest tenth of a foot.

(8.7) *Given that the pairs of triangles are similar, find the unknown length n.*

72.

73.

74.

Solve.

75. A housepainter needs to estimate the height of a condominium. He estimates the length of his shadow to be 7 feet long and the length of the building's shadow to be 42 feet long. Find the approximate height of the building if the housepainter is $5\frac{1}{2}$ feet tall.

76. A toy company is making a triangular sail for a toy sailboat. The toy sail is to be the same shape as a real sailboat's sail. Use the following diagram to find the unknown lengths x and y.

Mixed Review

Find the following.

77. Find the supplement of a 72° angle.

78. Find the complement of a 1° angle.

Find the measure of angle x in each figure.

79.

80.

81.

82.

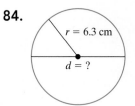

Find the unknown diameter or radius as indicated.

83.

$d = 26$ m
$r = ?$

84.

$r = 6.3$ cm
$d = ?$

Find the perimeter of each figure.

85.

Square $5\frac{1}{2}$ dm

86.

7 in.   11.2 in.
9.1 in.

87.

22 ft   11 ft
15 ft
42 ft
40 ft

Find the area of each figure. For the circle, find the exact area and then use $\pi = 3.14$ to approximate the area.

88.

43 m
42 m
13 m
14 m

89.

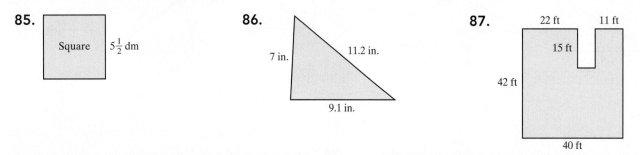

3 m

Find the volume of each solid.

90. Give an approximation using $\frac{22}{7}$ for π.

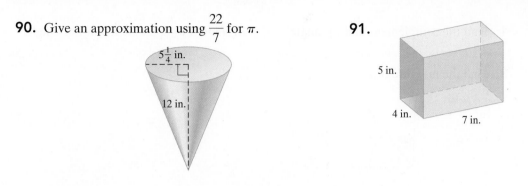

$5\frac{1}{4}$ in.

12 in.

91.

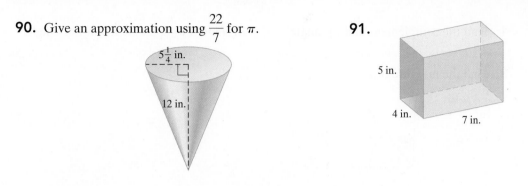

5 in.

4 in.   7 in.

Solve.

92. Find the volume of air in a rectangular room 15 feet by 12 feet with a 7-foot ceiling.

93. A mover has two boxes left for packing. Both are cubical, one 3 feet on a side and the other 1.2 feet on a side. Find their combined volume.

Simplify.

94. $\sqrt{1}$

95. $\sqrt{36}$

96. $\sqrt{\dfrac{16}{81}}$

Find the unknown length of each given right triangle. If necessary, round to the nearest tenth.

97. leg = 66, leg = 56

98. leg = 12, hypotenuse = 24

99. leg = 17, hypotenuse = 51

100. leg = 10, leg = 17

Given that the pairs of triangles are similar, find the unknown length n.

101.

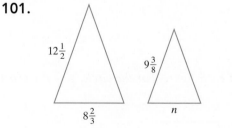

$12\frac{1}{2}$

$9\frac{3}{8}$

$8\frac{2}{3}$ n

102.

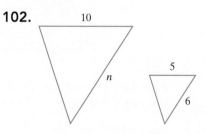

10

n

5

6

Answers

1. Find the complement of a 78° angle.

2. Find the supplement of a 124° angle.

3. Find the measure of $\angle x$.

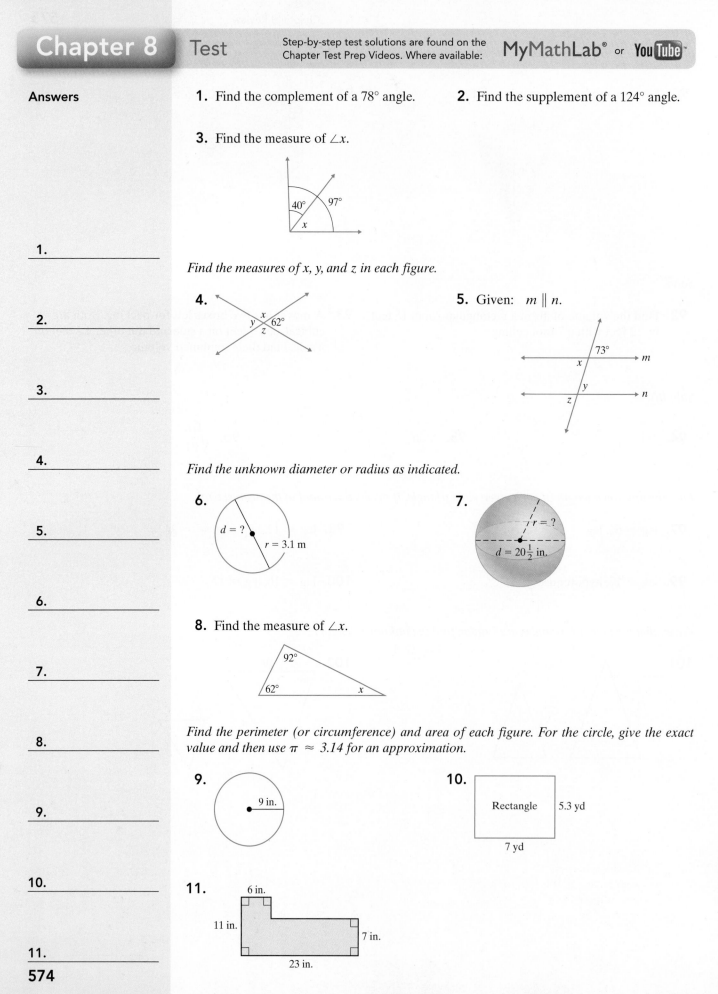

Find the measures of x, y, and z in each figure.

4.

5. Given: $m \parallel n$.

Find the unknown diameter or radius as indicated.

6.

7.

8. Find the measure of $\angle x$.

Find the perimeter (or circumference) and area of each figure. For the circle, give the exact value and then use $\pi \approx 3.14$ for an approximation.

9.

10.

Rectangle   5.3 yd

7 yd

11.

6 in.

11 in.

7 in.

23 in.

1. _____

2. _____

3. _____

4. _____

5. _____

6. _____

7. _____

8. _____

9. _____

10. _____

11. _____

Find the volume of each solid. For the cylinder, use $\pi \approx \dfrac{22}{7}$.

12.

5 in.

2 in.

13.

2 ft

3 ft   5 ft

Find each square root and simplify. Round the square root to the nearest thousandth if necessary.

14. $\sqrt{49}$ **15.** $\sqrt{79}$ **16.** $\sqrt{\dfrac{64}{100}}$

Solve.

17. Find the perimeter of a square photo with a side length of 4 inches.

18. How much soil is needed to fill a rectangular hole 3 feet by 3 feet by 2 feet?

19. Find how much baseboard is needed to go around a rectangular room that measures 18 feet by 13 feet. If baseboard costs $1.87 per foot, also calculate the total cost needed for materials.

20. Approximate to the nearest hundredth of a centimeter the length of the hypotenuse of a right triangle with legs of 4 centimeters each.

21. Vivian Thomas is going to put insecticide on her lawn to control grubworms. The lawn is a rectangle measuring 123.8 feet by 80 feet. The amount of insecticide required is 0.02 ounces per square foot. Find how much insecticide Vivian needs to purchase.

22. Given that the following triangles are similar, find the missing length n.

8

12

5

n

23. Tamara Watford, a surveyor, needs to estimate the height of a tower. She estimates the length of her shadow to be 4 feet long and the length of the tower's shadow to be 48 feet long. Find the approximate height of the tower if she is $5\dfrac{3}{4}$ feet tall.

$5\dfrac{3}{4}$ ft

4 ft   48 ft

12.

13.

14.

15.

16.

17.

18.

19.

20.

21.

22.

23.

Answers

1. _____

2. _____

3. _____

4. _____

5. _____

6. _____

7. _____

8. _____

9. _____

10. _____

11. _____

12. _____

13. _____

14. _____

15. _____

16. _____

17. _____

18. _____

19. _____

20. _____

21. _____

22. _____

1. Write the decimal 19.5023 in words.

2. Add: $\dfrac{7}{11} + \dfrac{1}{6}$

3. Round 736.2359 to the nearest tenth.

4. Round 736.2359 to the nearest hundred.

5. Add: $45 + 2.06$

6. Divide: $3\dfrac{1}{3} \div 1\dfrac{5}{6}$

Multiply.

7. 7.68×10

8. $\dfrac{7}{11} \cdot \dfrac{1}{6}$

9. 76.3×1000

10. $5\dfrac{1}{2} \cdot 2\dfrac{1}{11}$

11. Divide: $270.2 \div 7$. Check your answer.

12. Divide: $\dfrac{56.7}{100}$

13. Simplify: $0.5(8.6 - 1.2)$

14. Simplify: $\dfrac{5 + 2(8 - 3)}{30 \div 6 \cdot 5}$

15. Insert $<$, $>$, or $=$ to form a true statement. $\dfrac{1}{8}$ ___ 0.12

16. Insert $<$, $>$, or $=$ to form a true statement. $\dfrac{3}{4}$ ___ $\dfrac{13}{16}$

17. Write the ratio of 2.6 to 3.1 as a fraction in simplest form.

18. Find: $\dfrac{2}{9} + \dfrac{7}{15} - \dfrac{1}{3}$

19. Is $\dfrac{2}{3} = \dfrac{4}{6}$ a true proportion?

20. Solve for n: $\dfrac{7}{8} = \dfrac{n}{20}$

21. Since 2011, white has been the world's most popular color for cars. For 2013 model-year cars, 25 out of every 100 were painted white. What percent of model-year 2013 cars were white?

22. In a survey of 50 people, 34 people prefer taking pictures with digital cameras. What percent is this?

Write each percent as a fraction or mixed number in simplest form.

23. 1.9%   **24.** 26%   **25.** 125%   **26.** 560%

27. 85% of 300 is what number?

28. What percent of 16 is 2.4?

29. 20.8 is 40% of what number?

30. Find: $\left(7 - \sqrt{16}\right)^2$

31. Mr. Buccaran, the principal at Slidell High School, counted 31 freshmen absent during a particular day. If this is 4% of the total number of freshmen, how many freshmen are there at Slidell High School?

32. Flooring tiles cost $90 for a box with 40 tiles. Each tile is 1 square foot. Find the unit price in dollars per square foot.

33. Sherry Souter, a real estate broker for Wealth Investments, sold a house for $214,000 last week. If her commission rate is 1.5% of the selling price of the home, find the amount of her commission.

34. A student can complete 7 exercises in 6 minutes. At this rate, how many exercises can be completed in 30 minutes?

35. Convert 8 feet to inches.

36. 100 inches = ____ yd ____ ft ____ in.

37. Convert 3.2 kilograms to grams.

38. Convert 70 mm to meters.

39. Subtract 3 quarts from 4 gallons 2 quarts.

40. Write seventy thousand, fifty-two in standard form.

41. Convert 29°C to degrees Fahrenheit.

42. Write $\dfrac{1}{8}$ as a percent.

43. Find the measure of $\angle a$.

44. Find the perimeter of the triangle in Exercise 43.

45. Find the perimeter of the rectangle below.

46. Find the area of the rectangle in Exercise 45.

47. Find $\sqrt{\dfrac{4}{25}}$.

48. Find $\sqrt{\dfrac{9}{16}}$.

49. Find the ratio of corresponding sides for triangles *ABC* and *DEF*.

50. Use the figures in Exercise 49 and find the value of *x*.

23. _____	
24. _____	
25. _____	
26. _____	
27. _____	
28. _____	
29. _____	
30. _____	
31. _____	
32. _____	
33. _____	
34. _____	
35. _____	
36. _____	
37. _____	
38. _____	
39. _____	
40. _____	
41. _____	
42. _____	
43. _____	
44. _____	
45. _____	
46. _____	
47. _____	
48. _____	
49. _____	
50. _____	

Statistics and Probability

We often need to make decisions based on known statistics or the probability of an event occurring. For example, we decide whether or not to bring an umbrella to work based on the probability of rain. We choose an investment based on its mean, or average, return. We can predict which football team will win based on the trend in its previous wins and losses. This chapter reviews presenting data in a usable form on a graph and the basic ideas of statistics and probability.

Can We Experience Weightlessness?

Today, space travel, while still not commonplace, is well within everyone's comprehension. From landing on the Moon, to probes to Mars and beyond, the Hubble Space Telescope, and an International Space Station, there are many exciting explorations in space. There are even space tourists who pay big money to travel in space.

If space travel is not for you, but you'd like to experience weightlessness, there are commercial companies that offer rides on Boeing 727s with flight patterns similar to that shown below to simulate zero gravity.

In Section 9.1, Example 2, we explore lunar and planetary exploration since 1957.

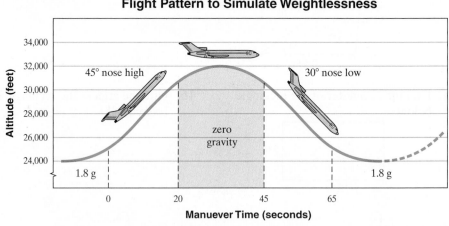

Flight Pattern to Simulate Weightlessness

Reading Pictographs, Bar Graphs, Histograms, and Line Graphs

Often data are presented visually in a graph. In this section, we practice reading several kinds of graphs, including pictographs, bar graphs, and line graphs.

Objective A Reading Pictographs

A **pictograph** such as the one below is a graph in which pictures or symbols are used. This type of graph contains a key that explains the meaning of the symbol used. An advantage of using a pictograph to display information is that comparisons can easily be made. A disadvantage of using a pictograph is that it is often hard to tell what fractional part of a symbol is shown. For example, in the pictograph below, Arabic shows a part of a symbol, but it's hard to read with any accuracy what fractional part of a symbol is shown.

Objectives

A Read Pictographs.

B Read and Construct Bar Graphs.

C Read and Construct Histograms.

D Read Line Graphs.

Example 1 Calculating Languages Spoken

The following pictograph shows the top eight most-spoken (primary) languages. Use this pictograph to answer the questions.

Top 8 Most-Spoken (Primary) Languages

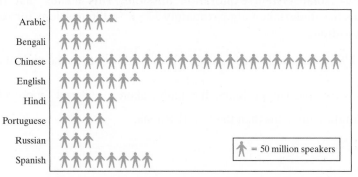

Source: www.ethnologue.com

a. Approximate the number of people who primarily speak Hindi.

b. Approximate how many more people primarily speak English than Hindi.

Solution:

a. Hindi corresponds to 5 symbols, and each symbol represents 50 million speakers. This means that the number of people who primarily speak Hindi is approximately $5 \cdot (50 \text{ million})$ or 250 million people. This can also be written as 250,000,000 people.

b. English shows $1\frac{1}{2}$ more symbols than Hindi. This means that $1\frac{1}{2} \cdot (50 \text{ million})$ or 75 million or 75,000,000 more people primarily speak English than Hindi.

■ Work Practice 1

Practice 1

Use the pictograph shown in Example 1 to answer the following questions:

a. Approximate the number of people who primarily speak Spanish.

b. Approximate how many more people primarily speak Spanish than Arabic.

Answers

1. a. 400 million or 400,000,000 people
b. 175 million or 175,000,000 people

Practice 2

Use the pictograph shown in Example 2 to answer the following questions:

a. Approximate the number of solar system exploration missions undertaken by the European Space Agency.

b. Approximate the total number of solar system exploration missions undertaken by the European Space Agency and Japan.

Example 2   Calculating Solar System Exploration

The following pictograph shows the approximate number of solar system exploration missions by various countries or space consortia from 1957 to the present day. Use this pictograph to answer the questions.

Solar System Exploration Missions

a. Approximate the number of solar system exploration missions undertaken by the United States.

b. Approximate how many more solar system exploration missions have been undertaken by the United States than by the USSR/Russia.

Solution:

a. The United States corresponds to 17 symbols, and each symbol represents 8 solar system exploration missions. This means that the United States has undertaken approximately $17 \cdot 8 = 136$ missions for solar system exploration.

b. The USSR/Russia shows $15\frac{1}{2}$ symbols, or $1\frac{1}{2}$ fewer than the United States. This means that the United States has undertaken $1\frac{1}{2} \cdot 8 = 12$ more solar system exploration missions than the USSR/Russia.

■ Work Practice 2

Objective B   Reading and Constructing Bar Graphs ▶

Another way to visually present data is with a **bar graph.** Bar graphs can appear with vertical bars or horizontal bars. Although we have studied bar graphs in previous sections, we now practice reading the height or length of the bars contained in a bar graph. An advantage to using bar graphs is that a scale is usually included for greater accuracy. Care must be taken when reading bar graphs, as well as other types of graphs—they may be misleading, as shown later in this section.

Answers

2. a. 12   **b.** 20

Example 3 Finding the Number of Endangered Species

The following bar graph shows the number of endangered species in the United States in 2013. Use this graph to answer the questions.

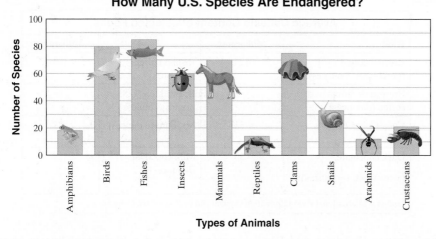

How Many U.S. Species Are Endangered?

Source: U.S. Fish and Wildlife Service

a. Approximate the number of endangered species that are clams.

b. Which category has the most endangered species?

Solution:

a. To approximate the number of endangered species that are clams, we go to the top of the bar that represents clams. From the top of this bar, we move horizontally to the left until the scale is reached. We read the height of the bar on the scale as approximately 75. There are approximately 75 clam species that are endangered, as shown.

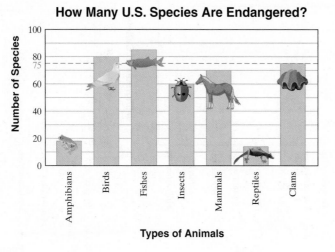

How Many U.S. Species Are Endangered?

Source: U.S. Fish and Wildlife Service

b. The most endangered species is represented by the tallest (longest) bar. The tallest bar corresponds to fishes.

■ Work Practice 3

Practice 3

Use the bar graph in Example 3 to answer the following questions:

a. Approximate the number of endangered species that are amphibians.

b. Which category shows the fewest endangered species?

Answers
3. a. 18 **b.** arachnids

Practice 4

Draw a vertical bar graph using the information in the table about selected states' electoral votes for President in the 2012, 2016, and 2020 presidential elections.

| Total Electoral Votes by Selected States ||
State	Electoral Votes
Texas	38
California	55
Florida	29
Nebraska	5
Indiana	11
Georgia	16

(*Source: U.S. Electoral College*)

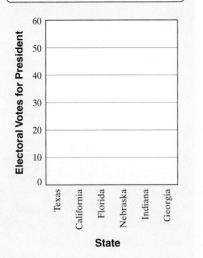

Next, we practice constructing a bar graph.

Example 4 Draw a vertical bar graph using the information in the table below that gives the caffeine content of selected foods.

| Average Caffeine Content of Selected Foods ||||
Food	Milligrams	Food	Milligrams
Brewed coffee (percolator, 8 ounces)	124	Instant coffee (8 ounces)	104
Brewed decaffeinated coffee (8 ounces)	3	Brewed tea (U.S. brands, 8 ounces)	64
Coca-Cola Classic (8 ounces)	31	Mr. Pibb (8 ounces)	27
Dark chocolate (semisweet, $1\frac{1}{2}$ ounces)	30	Milk chocolate (8 ounces)	9

(*Sources:* International Food Information Council and the Coca-Cola Company)

Solution: We draw and label a vertical line and a horizontal line as shown below on the left. These lines are also called axes. We place the different food categories along the horizontal axis. Along the vertical axis, we place a scale.

There are many choices of scales that would be appropriate. Notice that the milligrams range from a low of 3 to a high of 124. From this information, we use a scale that starts at 0 and then shows multiples of 20 so that the scale is not too cluttered. The scale stops at 140, the smallest multiple of 20 that will allow all milligrams to be graphed. It may also be helpful to draw horizontal lines along the scale markings to help draw the vertical bars at the correct height. The finished bar graph is shown below on the right.

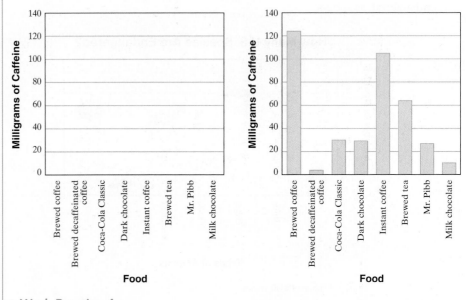

■ Work Practice 4

As mentioned previously, graphs can be misleading. Both graphs on the next page show the same information, but with different scales. Special care should be taken when forming conclusions from the appearance of a graph.

Answer

4.

Notice the ⧨ symbol on each vertical scale on the graphs below. This symbol alerts us that numbers are missing from that scale.

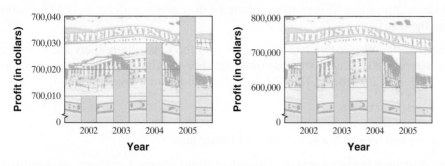

Are profits shown in the graphs above greatly increasing, or are they remaining about the same?

Objective C Reading and Constructing Histograms ▶

Suppose that the test scores of 36 students are summarized in the table below:

Student Scores	Frequency (Number of Students)
40–49	1
50–59	3
60–69	2
70–79	10
80–89	12
90–99	8

The results in the table can be displayed in a histogram. A **histogram** is a special bar graph. The width of each bar represents a range of numbers called a **class interval.** The height of each bar corresponds to how many times a number in the class interval occurs and is called the **class frequency.** The bars in a histogram lie side by side with no space between them.

Student Test Scores

Example 5 Reading a Histogram on Student Test Scores

Use the preceding histogram to determine how many students scored 50–59 on the test.

Solution: We find the bar representing 50–59. The height of this bar is 3, which means 3 students scored 50–59 on the test.

▶ Work Practice 5

Practice 5

Use the histogram on the left to determine how many students scored 80–89 on the test.

Answer

5. 12

Practice 6

Use the histogram above Example 5 to determine how many students scored less than 80 on the test.

Practice 7

Complete the frequency distribution table for the data below. Each number represents a credit card owner's unpaid balance for one month.

0	53	89	125
265	161	37	76
62	201	136	42

Class Intervals (Credit Card Balances)	Tally	Class Frequency (Number of Months)
$0–$49	____	____
$50–$99	____	____
$100–$149	____	____
$150–$199	____	____
$200–$249	____	____
$250–$299	____	____

Practice 8

Construct a histogram from the frequency distribution table above.

Answers

6. 16
7. table in class interval order:

Tally	Class Frequency (Number of Months)	Tally	Class Frequency (Number of Months)
				3			1
					4			1
			2			1

8.

Example 6 — Reading a Histogram on Student Test Scores

Use the preceding histogram to determine how many students scored 80 or above on the test.

Solution: We see that two different bars fit this description. There are 12 students who scored 80–89 and 8 students who scored 90–99. The sum of these two categories is 12 + 8 or 20 students. Thus, 20 students scored 80 or above on the test.

■ Work Practice 6

Now we will look at a way to construct histograms.

The daily high temperatures for 1 month in New Orleans, Louisiana, are recorded in the following list:

85°	90°	95°	89°	88°	94°
87°	90°	95°	92°	95°	94°
82°	92°	96°	91°	94°	92°
89°	89°	90°	93°	95°	91°
88°	90°	88°	86°	93°	89°

The data in this list have not been organized and can be hard to interpret. One way to organize the data is to place them in a **frequency distribution table.** We will do this in Example 7.

Example 7 — Completing a Frequency Distribution on Temperature

Complete the frequency distribution table for the preceding temperature data.

Solution: Go through the data and place a tally mark in the second column of the table next to the class interval. Then count the tally marks and write each total in the third column of the table.

Class Intervals (Temperatures)	Tally	Class Frequency (Number of Days)
82°–84°	|	1
85°–87°	|||	3
88°–90°	⪫⪫ ⪫⪫|	11
91°–93°	⪫⪫||	7
94°–96°	⪫⪫|||	8

■ Work Practice 7

Example 8 — Constructing a Histogram

Construct a histogram from the frequency distribution table in Example 7.

Solution:

■ Work Practice 8

✓**Concept Check** Which of the following sets of data is better suited to representation by a histogram? Explain.

Set 1		Set 2	
Grade on Final	# of Students	Section Number	Avg. Grade on Final
51–60	12	150	78
61–70	18	151	83
71–80	29	152	87
81–90	23	153	73
91–100	25		

Objective D Reading Line Graphs

Another common way to display information with a graph is by using a **line graph.** An advantage of a line graph is that it can be used to visualize relationships between two quantities. A line graph can also be very useful in showing a change over time.

Example 9 Reading Temperatures from a Line Graph

The following line graph shows the average daily temperature for each month in Omaha, Nebraska. Use this graph to answer the questions below.

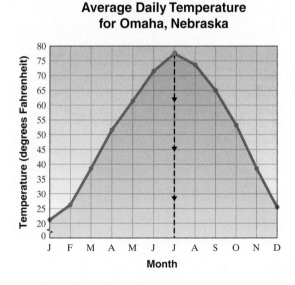

Source: National Climatic Data Center

a. During what month is the average daily temperature the highest?
b. During what month, from July through December, is the average daily temperature 65°F?
c. During what months is the average daily temperature less than 30°F?

Solution:

a. The month with the highest temperature corresponds to the highest point. This is the red point shown on the graph above. We follow this highest point downward to the horizontal month scale and see that this point corresponds to July.

(Continued on next page)

Practice 9

Use the temperature graph in Example 9 to answer the following questions:

a. During what month is the average daily temperature the lowest?

b. During what month is the average daily temperature 25°F?

c. During what months is the average daily temperature greater than 70°F?

Answers
9. a. January **b.** December
c. June, July, and August

✓ **Concept Check Answer**

Set 1; the grades are arranged in ranges of scores.

b. The months July through December correspond to the right side of the graph. We find the 65°F mark on the vertical temperature scale and move to the right until a point on the right side of the graph is reached. From that point, we move downward to the horizontal month scale and read the corresponding month. During the month of September, the average daily temperature is 65°F.

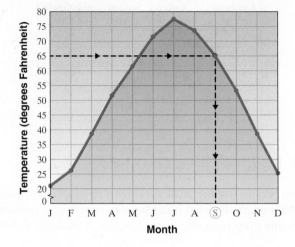

Source: National Climatic Data Center

c. To see what months the temperature is less than 30°F, we find what months correspond to points that fall below the 30°F mark on the vertical scale. These months are January, February, and December.

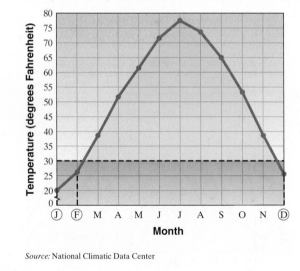

Source: National Climatic Data Center

■ Work Practice 9

Vocabulary, Readiness & Video Check

Fill in each blank with one of the choices below.

pictograph	bar	class frequency
histogram	line	class interval

1. A(n) _____ graph presents data using vertical or horizontal bars.

2. A(n) _____ is a graph in which pictures or symbols are used to visually present data.

3. A(n) _____ graph displays information with a line that connects data points.

4. A(n) _____ is a special bar graph in which the width of each bar represents a(n) _____ and the height of each bar represents the _____.

Martin-Gay Interactive Videos   Watch the section lecture video and answer the following questions.

Objective A **5.** From the pictograph in ⊞ Example 1, how would you approximate the number of wildfires for any given year? ▶

Objective B **6.** From ⊞ Example 5, what is one advantage of displaying data in a bar graph? ▶

Objective C **7.** Complete this statement based on the lecture before ⊞ Example 6: A histogram is a special kind of _____. ▶

Objective D **8.** From the line graph in ⊞ Examples 10–13, during which year(s) were total points scored greater than 40? ▶

See Video 9.1

9.1 Exercise Set MyMathLab® ▶

Objective A *The following pictograph shows the number of acres devoted to wheat production in selected states in 2013. Use this graph to answer Exercises 1 through 8. See Examples 1 and 2. (Source: National Agricultural Statistics Service)*

1. Which of the states shown planted the greatest quantity of acreage in wheat?

2. Which of the states shown planted the least amount of wheat acreage?

3. Approximate the number of acres of wheat planted in Oklahoma.

4. Approximate the number of acres of wheat planted in Kansas.

5. Which state plants about 2,500,000 acres of wheat?

6. Which states planted about 6,000,000 acres of wheat?

7. How many more acres of wheat were planted in Kansas than in Montana?

8. How many more acres of wheat were planted in Oklahoma than in Washington?

**Wheat Acreage
in Selected Top States, 2013**

States	
North Dakota	🌾 🌾 🌾 🌾 🌾 🌾
Kansas	🌾 🌾 🌾 🌾 🌾 🌾 🌾 🌾 🌾
Montana	🌾 🌾 🌾 🌾 🌾
South Dakota	🌾 🌾 🌾
Texas	🌾 🌾 🌾 🌾 🌾
Oklahoma	🌾 🌾 🌾 🌾 🌾
Colorado	🌾 🌾
Washington	🌾 🌾

🌾 = 1 million acres

The following pictograph shows the annual number of wildfires in the United States between 2006 and 2012. Use this graph to answer Exercises 9 through 16. See Examples 1 and 2. (Source: National Interagency Fire Center)

▶ **9.** Approximate the number of wildfires in 2012.

10. Approximately how many wildfires were there in 2006?

▶ **11.** Which year, of the years shown, had the most wildfires?

12. In what years were the number of wildfires greater than 72,000?

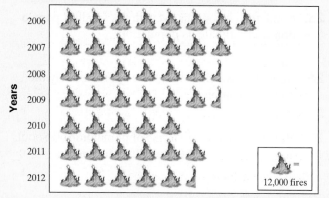

Wildfires in the United States

= 12,000 fires

13. What was the amount of increase in wildfires from 2010 to 2011?

14. What was the amount of decrease in wildfires from 2006 to 2012?

15. What was the average annual number of wildfires from 2010 to 2012? (*Hint:* How do you calculate the average?)

16. Give an explanation for the large number of wildfires in 2006.

Objective **B** *The National Weather Service has exacting definitions for hurricanes; they are tropical storms with winds in excess of 74 mph. The following bar graph shows the number of hurricanes, by month, that have made landfall on the mainland United States between 1851 and 2012. Use this graph to answer Exercises 17 through 22. See Example 3. (Source: NOAA: Hurricane Research Division)*

17. In which month did the most hurricanes make landfall in the United States?

18. In which month did the fewest hurricanes make landfall in the United States?

19. Approximate the number of hurricanes that made landfall in the United States during the month of August.

20. Approximate the number of hurricanes that made landfall in the United States in September.

21. In 2008, two hurricanes made landfall during the month of August. What fraction of all the 76 hurricanes that made landfall during August is this?

22. In 2012, only one hurricane, Hurricane Sandy, made landfall on the United States during the month of October. If there have been 51 hurricanes to make landfall in the month of October since 1851, approximately what percent of these occurred in 2012?

Hurricanes Making Landfall in the United States, by Month, 1851–2012

The following horizontal bar graph shows the recent population of the world's largest cities (including their suburbs). Use this graph to answer Exercises 23 through 28. See Example 3. (Source: CityPopulation)

23. Name the city with the largest population, and estimate its population.

24. Name the city whose population is between 26 million and 27 million, and estimate its population.

25. Name the city on this list with the smallest population, and estimate its population.

26. Name the two cities that have approximately the same population.

27. How much larger (in terms of population) is Seoul, South Korea, than Mexico City, Mexico?

28. How much larger (in terms of population) is Guangzhou, China, than Karachi, Pakistan?

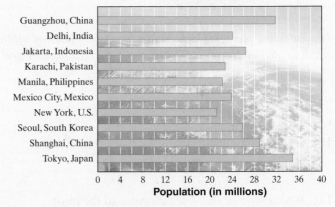

World's Largest Cities (including suburbs)

Use the information given to draw a vertical bar graph. Clearly label the bars. See Example 4.

29.

Fiber Content of Selected Foods

Food	Grams of Total Fiber
Kidney beans $\left(\frac{1}{2}\text{ c}\right)$	4.5
Oatmeal, cooked $\left(\frac{3}{4}\text{ c}\right)$	3.0
Peanut butter, chunky (2 tbsp)	1.5
Popcorn (1 c)	1.0
Potato, baked, with skin (1 med)	4.0
Whole wheat bread (1 slice)	2.5

(*Sources:* American Dietetic Association and National Center for Nutrition and Dietetics.)

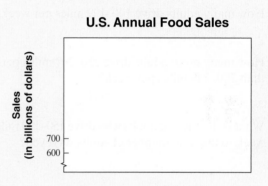

Fiber Content of Selected Foods

30.

U.S. Annual Food Sales

Year	Sales in Billions of Dollars
2007	1079
2008	1117
2009	1086
2010	1139
2011	1274
2012	1345

(*Source:* U.S. Department of Agriculture)

U.S. Annual Food Sales

31.

Best-Selling Albums of All Time (U.S. sales)

Album	Estimated Units Sold (in millions)
Led Zeppelin: *Led Zeppelin IV* (1971)	23
Eagles: *Their Greatest Hits* (1976)	29
Pink Floyd: *The Wall* (1979)	23
Michael Jackson: *Thriller* (1982)	29
Billy Joel: *Greatest Hits: Volumes I & II* (1985)	23

(*Source:* Recording Industry Association of America)

Best-Selling Albums of All Time (U.S. sales)

32.

Selected Worldwide Commercial Space Launches

Source	Total Commercial Space Launches 1990–2012
United States	156
Europe	153
Russia	141
China	23
Sea Launch*	39

*Sea Launch is an international venture involving 4 countries that uses its own launch facility outside national borders.

(*Source:* Bureau of Transportation Statistics)

Selected Worldwide Commercial Space Launches (1990–2012)

Objective C *The following histogram shows the number of miles that each adult, from a survey of 100 adults, drives per week. Use this histogram to answer Exercises 33 through 42. See Examples 5 and 6.*

33. How many adults drive 100–149 miles per week?

34. How many adults drive 200–249 miles per week?

35. How many adults drive fewer than 150 miles per week?

36. How many adults drive 200 miles or more per week?

37. How many adults drive 100–199 miles per week?

38. How many adults drive 150–249 miles per week?

39. How many more adults drive 250–299 miles per week than 200–249 miles per week?

40. How many more adults drive 0–49 miles per week than 50–99 miles per week?

41. What is the ratio of adults who drive 150–199 miles per week to the total number of adults surveyed?

42. What is the ratio of adults who drive 50–99 miles per week to the total number of adults surveyed?

The following histogram shows the projected population (in millions), by age groups, for the United States for the year 2020. Use this histogram to answer Exercises 43 through 50. For Exercises 45 through 48, estimate to the nearest whole million. See Examples 5 and 6.

43. What age range will be the largest population group in 2020?

44. What age range will be the smallest population group in 2020?

Projected U.S. 2020 Population by Age

45. How large is the population of 20- to 44-year-olds expected to be in 2020?

46. How large is the population of 45- to 64-year-olds expected to be in 2020?

47. How large is the population of those less than 4 years old expected to be in 2020?

48. How large is the population of 5- to 19-year-olds expected to be in 2020?

49. Which bar represents the age range you expect to be in during 2020?

50. How many more 20- to 44-year-olds are there expected to be than 45- to 64-year-olds in 2020?

The following list shows the golf scores for an amateur golfer. Use this list to complete the frequency distribution table to the right. See Example 7.

78	84	91	93	97
97	95	85	95	96
101	89	92	89	100

	Class Intervals (Scores)	Tally	Class Frequency (Number of Games)
▶ **51.**	70–79		
▶ **52.**	80–89		
▶ **53.**	90–99		
▶ **54.**	100–109		

Twenty-five people in a survey were asked to give their current checking account balances. Use the balances shown in the following list to complete the frequency distribution table to the right. See Example 7.

$53	$105	$162	$443	$109
$468	$47	$259	$316	$228
$207	$357	$15	$301	$75
$86	$77	$512	$219	$100
$192	$288	$352	$166	$292

	Class Intervals (Account Balances)	Tally	Class Frequency (Number of People)
55.	$0–$99		
56.	$100–$199		
57.	$200–$299		
58.	$300–$399		
59.	$400–$499		
60.	$500–$599		

▶ **61.** Use the frequency distribution table from Exercises 51 through 54 to construct a histogram. See Example 8.

Golf Scores

62. Use the frequency distribution table from Exercises 55 through 60 to construct a histogram. See Example 8.

Account Balances

Objective D *The following line graph shows the total points scored by both teams in the NFL Super Bowl from 2006 through 2013. Use this graph to answer Exercises 63 through 70. See Example 9. (Source: superbowlhistory.net)*

▶ **63.** Find the total points scored in the Super Bowl in 2009.

64. Find the total points scored in the Super Bowl in 2013.

▶ **65.** During which year(s) shown were the total points scored in the Super Bowl greater than 50?

66. During which year(s) shown was the total score in the Super Bowl the highest?

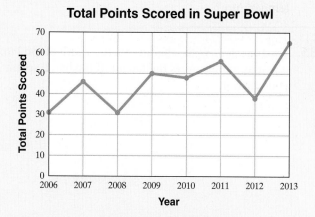

Total Points Scored in Super Bowl

67. During which year(s) shown was the total score in the Super Bowl the lowest?

68. Between 2011 and 2012, did the total score in the Super Bowl increase or decrease?

69. During which year(s) was the total score in the Super Bowl less than 40?

70. Between 2008 and 2009, did the total score in the Super Bowl increase or decrease?

Review

Find each percent. See Sections 6.3 and 6.4.

71. 30% of 12

72. 45% of 120

73. 10% of 62

74. 95% of 50

Write each fraction as a percent. See Section 6.2.

75. $\dfrac{1}{4}$

76. $\dfrac{2}{5}$

77. $\dfrac{17}{50}$

78. $\dfrac{9}{10}$

Concept Extensions

The following double line graph shows temperature highs and lows for a week. Use this graph to answer Exercises 79 through 84.

79. What was the high temperature reading on Thursday?

80. What was the low temperature reading on Thursday?

81. What day was the temperature the lowest? What was this low temperature?

82. What day of the week was the temperature the highest? What was this high temperature?

83. On what day of the week was the difference between the high temperature and the low temperature the greatest? What was this difference in temperature?

84. On what day of the week was the difference between the high temperature and the low temperature the least? What was this difference in temperature?

85. True or false? With a bar graph, the width of the bar is just as important as the height of the bar. Explain your answer.

86. Kansas plants about 24% of the wheat acreage in the United States. About how many acres of wheat are planted in the United States, according to the pictograph for Exercises 1 through 8? Round to the nearest million acre.

Objective A Reading Circle Graphs

In Exercise Set 6.1, the following **circle graph** was shown. This particular graph shows the favorite sport for 100 adults.

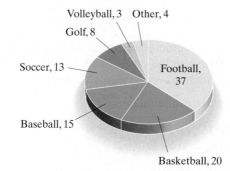

Volleyball, 3 Other, 4
Golf, 8
Soccer, 13
Football, 37
Baseball, 15
Basketball, 20

Each sector of the graph (shaped like a piece of pie) shows a category and the relative size of the category. In other words, the most popular sport is football, and it is represented by the largest sector.

Example 1 Find the ratio of adults preferring basketball to total adults. Write the ratio as a fraction in simplest form.

Solution: The ratio is

$$\frac{\text{people preferring basketball}}{\text{total adults}} = \frac{20}{100} = \frac{1}{5}$$

 Work Practice 1

Practice 1

Find the ratio of adults preferring golf to total adults. Write the ratio as a fraction in simplest form.

A circle graph is often used to show percents in different categories, with the whole circle representing 100%.

Example 2 Using a Circle Graph

The following graph shows the percent of visitors to the United States in 2012 from various regions. Using the circle graph shown, determine the percent of visitors who came to the United States from Mexico and Canada.

Solution: To find this percent, we add the percents corresponding to Mexico and Canada. The percent of visitors to the United States that came from Mexico and Canada is

21.3% + 34.1% = 55.4%

Work Practice 2

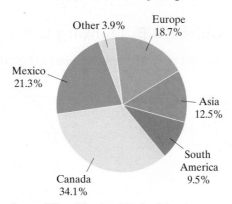

Visitors to U.S. by Region

Other 3.9%   Europe 18.7%
Mexico 21.3%
Asia 12.5%
South America 9.5%
Canada 34.1%

Source: Office of Travel and Tourism Industries

Practice 2

Using the circle graph shown in Example 2, determine the percent of visitors to the United States that came from Europe, Asia, and South America.

Answers

1. $\frac{2}{25}$ **2.** 40.7%

Helpful Hint

Since a circle graph represents a whole, the percents should add to 100% or 1. Notice this is true for Example 2.

Example 3 Finding Percent of Population

The U.S. Department of Commerce recorded 67 million international visitors to the United States in 2012. Use the circle graph from Example 2 and estimate the number of tourists that might have come from Europe.

Solution: We use the percent equation.

$$\text{amount} = \boxed{\text{percent}} \cdot \boxed{\text{base}}$$

$$\text{amount} = 0.187 \cdot 67{,}000{,}000$$

$$= 0.187(67{,}000{,}000)$$

$$= 12{,}529{,}000$$

Thus, 12,529,000 tourists might have come from Europe in 2012.

■ Work Practice 3

Practice 3

Use the information in Example 3 and the circle graph from Example 2 to estimate the number of tourists that might have come from Mexico in 2012.

✓**Concept Check** Can the following data be represented by a circle graph? Why or why not?

Responses to the Question, "In Which Activities Are You Involved?"	
Intramural sports	60%
On-campus job	42%
Fraternity/sorority	27%
Academic clubs	21%
Music programs	14%

Objective B Drawing Circle Graphs ▶

To draw a circle graph, we use the fact that a whole circle contains 360° (degrees).

360°

Answer

3. 14,271,000 tourists from Mexico

✓**Concept Check Answer**

no; the percents add up to more than 100%

Example 4 Drawing a Circle Graph for U.S. Armed Forces Personnel

The following table shows the percent of U.S. armed forces personnel that were in each branch of service in 2011. (*Source:* U.S. Department of Defense)

Branch of Service	Percent
Army	38
Navy	23
Marine Corps	14
Air Force	22
Coast Guard	3

Draw a circle graph showing this data.

Solution: First we find the number of degrees in each sector representing each branch of service. Remember that the whole circle contains 360°. (We will round degrees to the nearest whole.)

Sector	Degrees in Each Sector
Army	38% × 360° = 0.38 × 360° = 136.8° ≈ 137°
Navy	23% × 360° = 0.23 × 360° = 82.8° ≈ 83°
Marine Corps	14% × 360° = 0.14 × 360° = 50.4° ≈ 50°
Air Force	22% × 360° = 0.22 × 360° = 79.2° ≈ 79°
Coast Guard	3% × 360° = 0.03 × 360° = 10.8° ≈ 11°

Helpful Hint

Check your calculations by finding the sum of the degrees.

137° + 83° + 50° + 79° + 11° = 360°

The sum should be 360°. (It may vary only slightly because of rounding.)

Next we draw a circle and mark its center. Then we draw a line from the center of the circle to the circle itself.

To construct the sectors, we will use a **protractor.** A protractor measures the number of degrees in an angle. We place the hole in the protractor over the center of the circle. Then we adjust the protractor so that 0° on the protractor is aligned with the line that we drew.

It makes no difference which sector we draw first. To construct the "Army" sector, we find 137° on the protractor and mark our circle. Then we remove the protractor and use this mark to draw a second line from the center to the circle itself.

(*Continued on next page*)

Practice 4

Use the data shown to draw a circle graph.

Freshmen	30%
Sophomores	27%
Juniors	25%
Seniors	18%

Answer

4.

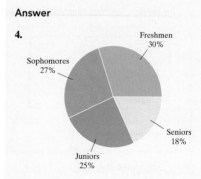

To construct the "Navy" sector, we follow the same procedure as above, except that we line up 0° with the second line we drew and mark the protractor at 83°.

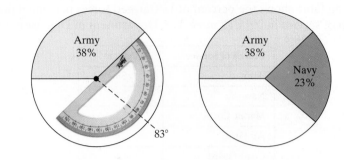

We continue in this manner until the circle graph is complete.

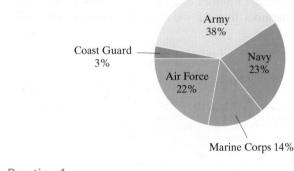

Coast Guard
3%

Marine Corps 14%

■ Work Practice 4

✓Concept Check True or false? The larger a sector in a circle graph, the larger the percent of the total it represents. Explain your answer.

Vocabulary, Readiness & Video Check

Use the choices below to fill in each blank.

sector   circle   100   360

1. In a(n) _____ graph, each section (shaped like a piece of pie) shows a category and the relative size of the category.

2. A circle graph contains pie-shaped sections, each called a(n) _____ .

3. The number of degrees in a whole circle is _____ .

4. If a circle graph has percent labels, the percents should add up to _____ .

Martin-Gay Interactive Videos

Watch the section lecture video and answer the following questions.

Objective A 5. From ⊟ Example 3, when a circle graph shows percents in different categories, what does the whole circle graph always represent? ▶

Objective B 6. From ⊟ Example 6, when looking at the sector degree measures of a circle graph, the whole circle graph corresponds to what degree measure? ▶

See Video 9.2 🔴

9.2 Exercise Set MyMathLab®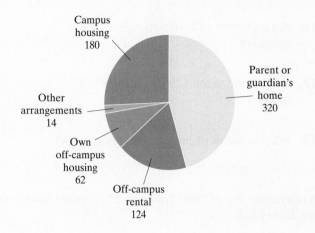

Objective **A** *The following circle graph is a result of surveying 700 college students. They were asked where they live while attending college. Use this graph to answer Exercises 1 through 6. Write all ratios as fractions in simplest form. See Example 1.*

1. Where do most of these college students live?

2. Besides the category "Other arrangements," where do the fewest of these college students live?

3. Find the ratio of students living in campus housing to total students.

4. Find the ratio of students living in off-campus rentals to total students.

5. Find the ratio of students living in campus housing to students living in a parent or guardian's home.

6. Find the ratio of students living in off-campus rentals to students living in a parent or guardian's home.

The following circle graph shows the percent of the land area of the continents of Earth. Use this graph for Exercises 7 through 14. See Example 2.

7. Which continent is the largest?

8. Which continent is the smallest?

9. What percent of the land on Earth is accounted for by Asia and Europe together?

10. What percent of the land on Earth is accounted for by North and South America?

Source: National Geographic Society

The total amount of land from the continents is approximately 57,000,000 square miles. Use the graph to find the area of the continents given in Exercises 11 through 14. See Example 3.

11. Asia   **12.** South America   **13.** Australia   **14.** Europe

The following circle graph shows the percent of the types of books available at Midway Memorial Library. Use this graph for Exercises 15 through 24. See Example 2.

▶ **15.** What percent of books are classified as some type of fiction?

16. What percent of books are nonfiction or reference?

▶ **17.** What is the second-largest category of books?

18. What is the third-largest category of books?

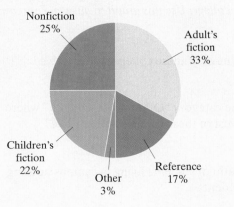

If this library has 125,600 books, find how many books are in each category given in Exercises 19 through 24. See Example 3.

▶ **19.** Nonfiction   **20.** Reference   **21.** Children's fiction

22. Adult's fiction   **23.** Reference or other **24.** Nonfiction or other

Objective **B** *Fill in the tables. Round to the nearest degree. Then draw a circle graph to represent the information given in each table. (Remember: The total of "Degrees in Sector" column should equal 360° or very close to 360° because of rounding.) See Example 4.*

▶ **25.**

Types of Apples Grown in Washington State		
Type of Apple	**Percent**	**Degrees in Sector**
Red Delicious	37%	
Golden Delicious	13%	
Fuji	14%	
Gala	15%	
Granny Smith	12%	
Other varieties	6%	
Braeburn	3%	
(*Source:* U.S. Apple Association)		

26.

Color Distribution of M&M's Milk Chocolate		
Color	**Percent**	**Degrees in Sector**
Blue	22.1%	
Orange	16.7%	
Green	16.7%	
Red	16.7%	
Brown	16.7%	
Yellow	11.1%	
(*Source:* M&M Mars)		

27.

Distribution of Large Dams by Continent		
Continent	Percent	Degrees in Sector
Europe	19%	
North America	32%	
South America	3%	
Asia	39%	
Africa	5%	
Australia	2%	

(*Source:* International Commission on Large Dams)

28.

Number of Times the "Are We There Yet?" Question Is Asked of Parents During Road Trips		
	Percent	Degrees in Sector
Never	20%	
Once	11%	
2–5 times	36%	
6–10 times	14%	
More than 10 times	19%	

(*Source:* KRC Research for Goodyear Tire & Rubber Co.)

Review

Write the prime factorization of each number. See Section 2.2.

29. 20

30. 25

31. 40

32. 16

33. 85

34. 105

Concept Extensions

The following circle graph shows the relative sizes of the great oceans. Use this graph for Exercises 35 through 40.

35. Without calculating, determine which ocean is the largest. How can you answer this question by looking at the circle graph?

36. Without calculating, determine which ocean is the smallest. How can you answer this question by looking at the circle graph?

Source: Philip's World Atlas

These oceans together make up 264,489,800 square kilometers of the Earth's surface. Find the square kilometers for each ocean.

37. Pacific Ocean

38. Atlantic Ocean

39. Indian Ocean

40. Arctic Ocean

The following circle graph summarizes the results of a survey of 2800 Internet users who make purchases online. Use this graph for Exercises 41 through 46. Round to the nearest whole.

41. How many of the survey respondents said that they spend $0–$15 online each month?

42. How many of the survey repondents said that they spend $15–$175 online each month?

43. How many of the survey respondents said that they spend $0 to $175 online each month?

44. How many of the survey respondents said that they spend $15 to over $175 online each month?

Online Spending per Month

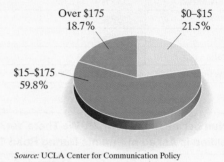

Over $175
18.7%

$0–$15
21.5%

$15–$175
59.8%

Source: UCLA Center for Communication Policy

45. Find the ratio of *number* of respondents who spend $0–$15 online to *number* of respondents who spend $15–$175 online. Write the ratio as a fraction with integers in the numerator and denominator.

46. Find the ratio of *percent* of respondents who spend $0–$15 online to *percent* of those who spend $15–$175. Write the ratio as a fraction with integers in the numerator and denominator.

See the Concept Checks in this section.

47. Can the data below be represented by a circle graph? Why or why not?

Responses to the Question, "What Classes Are You Taking?"	
Math	80%
English	72%
History	37%
Biology	21%
Chemistry	14%

48. True or false? The smaller a sector in a circle graph, the smaller the percent of the total it represents. Explain why.

Integrated Review

Reading Graphs

The following pictograph shows the six occupations with the largest estimated numerical increase in employment in the United States between 2010 and 2020. Use this graph to answer Exercises 1 through 4.

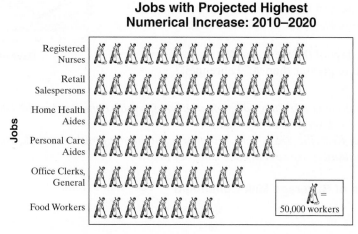

Jobs with Projected Highest Numerical Increase: 2010–2020

Source: Bureau of Labor Statistics

1. _____

1. Approximate the increase in the number of registered nurses from 2010 to 2020.

2. _____

2. Approximate the increase in the number of general office clerks from 2010 to 2020.

3. Which occupations are expected to show approximately the same increase in numbers of employees between the years shown?

3. _____

4. Which of the listed occupations is expected to show the least increase in numbers of employees between the years shown?

4. _____

The following bar graph shows the highest U.S. dams. Use this graph to answer Exercises 5 through 8.

5. _____

Highest U.S. Dams

Source: Committee on Register of Dams

6. _____

5. Name the U.S. dam with the greatest height and estimate its height.

6. Name the U.S. dam whose height is between 625 and 650 feet and estimate its height.

7. _____

7. Estimate how much higher the Hoover Dam is than the Glen Canyon Dam.

8. How many U.S. dams have heights over 700 feet?

8. _____

9. _____

10. _____

11. _____

12. _____

13. _____

14. _____

15. _____

16. _____

17. _____

18. _____

19. _____

20. _____

21. _____

22. _____

The following line graph shows the daily high temperatures for 1 week in Annapolis, Maryland. Use this graph to answer Exercises 9 through 12.

9. Name the day(s) of the week with the highest temperature and give that high temperature.

10. Name the day(s) of the week with the lowest temperature and give that low temperature.

11. On what days of the week was the temperature less than 90° Fahrenheit?

12. On what days of the week was the temperature greater than 90° Fahrenheit?

The following circle graph shows the type of beverage milk consumed in the United States. Use this graph for Exercises 13 through 16.

If a store in Kerrville, Texas, sells 200 quart containers of milk per week, estimate how many quart containers are sold in each category below.

Types of Beverage Milk Consumed

Source: U.S. Department of Agriculture

13. Whole milk

14. Skim milk

15. Buttermilk

16. Flavored reduced fat and skim milk

The following list shows weekly quiz scores for a student in basic college mathematics. Use this list to complete the frequency distribution table.

50	80	71	83	86
67	89	93	88	97
	53	90		
75	80	78	93	99

	Class Intervals (Scores)	Tally	Class Frequency (Number of Quizzes)
17.	50–59		
18.	60–69		
19.	70–79		
20.	80–89		
21.	90–99		

22. Use the table from Exercises 17 through 21 to construct a histogram.

Mean, Median, and Mode ⊙

Objective A Finding the Mean ⊙

Sometimes we want to summarize data by displaying them in a graph, but sometimes it is also desirable to be able to describe a set of data, or a set of numbers, by a single "middle" number. Three such **measures of central tendency** are the **mean,** the **median,** and the **mode.**

The most common measure of central tendency is the mean (sometimes called the "arithmetic mean" or the "average"). Recall that we first introduced finding the average of a list of numbers in Section 1.7.

> The **mean (average)** of a set of number items is the sum of the items divided by the number of items.
>
> $$\text{mean} = \frac{\text{sum of items}}{\text{number of items}}$$

Objectives

A Find the Mean of a List of Numbers. ⊙

B Find the Median of a List of Numbers. ⊙

C Find the Mode of a List of Numbers. ⊙

Example 1 Finding the Mean Time in an Experiment

Seven students in a psychology class conducted an experiment on mazes. Each student was given a pencil and asked to successfully complete the same maze. The timed results are below:

Student	Ann	Thanh	Carlos	Jesse	Melinda	Ramzi	Dayni
Time (Seconds)	13.2	11.8	10.7	16.2	15.9	13.8	18.5

a. Who completed the maze in the shortest time? Who completed the maze in the longest time?

b. Find the mean time.

c. How many students took longer than the mean time? How many students took shorter than the mean time?

Solution:

a. Carlos completed the maze in 10.7 seconds, the shortest time. Dayni completed the maze in 18.5 seconds, the longest time.

b. To find the mean (or average), we find the sum of the items and divide by 7, the number of items.

$$\text{mean} = \frac{13.2 + 11.8 + 10.7 + 16.2 + 15.9 + 13.8 + 18.5}{7}$$

$$= \frac{100.1}{7} = 14.3$$

c. Three students, Jesse, Melinda, and Dayni, had times longer than the mean time. Four students, Ann, Thanh, Carlos, and Ramzi, had times shorter than the mean time.

■ Work Practice 1

Practice 1

Find the mean of the following test scores: 87, 75, 96, 91, and 78.

✓**Concept Check** Estimate the mean of the following set of data:

5, 10, 10, 10, 10, 15

Often in college, the calculation of a **grade point average** (GPA) is a **weighted mean** and is calculated as shown in Example 2.

Answer

1. 85.4

✓**Concept Check Answer**

10

Practice 2

Find the grade point average if the following grades were earned in one semester.

Grade	Credit Hours
A	2
B	4
C	5
D	2
A	2

Example 2 Calculating Grade Point Average (GPA)

The following grades were earned by a student during one semester. Find the student's grade point average.

Course	Grade	Credit Hours
College mathematics	A	3
Biology	B	3
English	A	3
PE	C	1
Social studies	D	2

Solution: To calculate the grade point average, we need to know the point values for the different possible grades. The point values of grades commonly used in colleges and universities are given below:

A: 4, B: 3, C: 2, D: 1, F: 0

Now, to find the grade point average, we multiply the number of credit hours for each course by the point value of each grade. The grade point average is the sum of these products divided by the sum of the credit hours.

Course	Grade	Point Value of Grade	Credit Hours	Point Value × Credit Hours
College mathematics	A	4	3	12
Biology	B	3	3	9
English	A	4	3	12
PE	C	2	1	2
Social studies	D	1	2	2
		Totals:	12	37

$$\text{grade point average} = \frac{37}{12} \approx 3.08 \text{ rounded to two decimal places}$$

The student earned a grade point average of 3.08.

■ Work Practice 2

Objective B Finding the Median ▶

You may have noticed that a very low number or a very high number can affect the mean of a list of numbers. Because of this, you may sometimes want to use another measure of central tendency. A second measure of central tendency is called the **median.** The median of a list of numbers is not affected by a low or high number in the list.

> The **median** of a set of numbers in numerical order is the middle number. If the number of items is odd, the median is the middle number. If the number of items is even, the median is the mean of the two middle numbers.

Practice 3

Find the median of the list of numbers: 5, 11, 14, 23, 24, 35, 38, 41, 43

Example 3 Find the median of the following list of numbers:

25, 54, 56, 57, 60, 71, 98

Solution: Because this list is in numerical order, the median is the middle number, 57.

■ Work Practice 3

Answers

2. 2.67 **3.** 24

Example 4 Find the median of the following list of scores: 67, 91, 75, 86, 55, 91

Solution: First we list the scores in numerical order and then find the middle number.

55, 67, 75, 86, 91, 91

Since there is an even number of scores, there are two middle numbers, 75 and 86. The median is the mean of the two middle numbers.

$$\text{median} = \frac{75 + 86}{2} = 80.5$$

The median is 80.5.

■ Work Practice 4

> **Helpful Hint** Don't forget to write the numbers in order from smallest to largest before finding the median.

Objective C Finding the Mode ▷

The last common measure of central tendency is called the **mode.**

> The **mode** of a set of numbers is the number that occurs most often. (It is possible for a set of numbers to have more than one mode or to have no mode.)

Example 5 Find the mode of the list of numbers:

11, 14, 14, 16, 31, 56, 65, 77, 77, 78, 79

Solution: There are two numbers that occur the most often. They are 14 and 77. This list of numbers has two modes, 14 and 77.

■ Work Practice 5

Example 6 Find the median and the mode of the following set of numbers. These numbers were high temperatures for 14 consecutive days in a city in Montana.

76, 80, 85, 86, 89, 87, 82, 77, 76, 79, 82, 89, 89, 92

Solution: First we write the numbers in numerical order.

76, 76, 77, 79, 80, 82, 82, 85, 86, 87, 89, 89, 89, 92

Since there is an even number of items, the median is the mean of the two middle numbers, 82 and 85.

$$\text{median} = \frac{82 + 85}{2} = 83.5$$

The mode is 89, since 89 occurs most often.

■ Work Practice 6

✓**Concept Check** True or false? Every set of numbers *must* have a mean, median, and mode. Explain your answer.

Practice 4

Find the median of the list of scores:

36, 91, 78, 65, 95, 95, 88, 71

Practice 5

Find the mode of the list of numbers:

14, 10, 10, 13, 15, 15, 15, 17, 18, 18, 20

Practice 6

Find the median and the mode of the list of numbers:

26, 31, 15, 15, 26, 30, 16, 18, 15, 35

Answers
4. 83 **5.** 15 **6.** median: 22; mode: 15

✓**Concept Check Answer**
false; a set of numbers may have no mode

> **Helpful Hint**
>
> Don't forget that it is possible for a list of numbers to have no mode. For example, the list
>
> 2, 4, 5, 6, 8, 9
>
> has no mode. There is no number or numbers that occur more often than the others.

Vocabulary, Readiness & Video Check

Use the choices below to fill in each blank. Some choices may be used more than once.

mean	mode	grade point average
median	average	

1. Another word for "mean" is _____.

2. The number that occurs most often in a set of numbers is called the _____.

3. The _____ of a set of number items is $\dfrac{\text{sum of items}}{\text{number of items}}$.

4. The _____ of a set of numbers is the middle number. If the number of numbers is even, it is the _____ of the two middle numbers.

5. An example of weighted mean is a calculation of _____.

Martin-Gay Interactive Videos   *Watch the section lecture video and answer the following questions.*

Objective A   6. Why is the ≈ symbol used in ▣ Example 1? ⊙

Objective B   7. From ▣ Example 3, what is always the first step when finding the median of a set of data numbers? ⊙

Objective C   8. From ▣ Example 4, why do you think it is helpful to have data numbers in numerical order when finding the mode? ⊙

See Video 9.3

9.3 Exercise Set MyMathLab® ⊙

Objectives A B C Mixed Practice *For each set of numbers, find the mean, median, and mode. If necessary, round the mean to one decimal place. See Examples 1 and 3 through 6.*

1. 15, 23, 24, 18, 25

2. 45, 36, 28, 46, 52

3. 7.6, 8.2, 8.2, 9.6, 5.7, 9.1

4. 4.9, 7.1, 6.8, 6.8, 5.3, 4.9

5. 0.5, 0.2, 0.2, 0.6, 0.3, 1.3, 0.8, 0.1, 0.5

6. 0.6, 0.6, 0.8, 0.4, 0.5, 0.3, 0.7, 0.8, 0.1

7. 231, 543, 601, 293, 588, 109, 334, 268

8. 451, 356, 478, 776, 892, 500, 467, 780

The ten tallest buildings in the world, completed as of 2013, are listed in the following table. Use this table to answer Exercises 9 through 14. If necessary, round results to one decimal place. See Examples 1 and 3 through 6.

9. Find the mean height of the five tallest buildings.

Building	Height (in feet)
Burj Khalifa	2717
Makkah Royal Clock Tower Hotel	1972
Taipei 101	1667
Shanghai World Financial Center	1614
International Commerce Centre	1588
Petronas Tower 1	1483
Petronas Tower 2	1483
Zifeng Tower	1476
Willis Tower	1451
KK100	1449

(*Source:* Council on Tall Buildings and Urban Habitat)

10. Find the median height of the five tallest buildings.

11. Find the median height of the eight tallest buildings.

12. Find the mean height of the eight tallest buildings.

13. Given the building heights, explain how you know, without calculating, that the answer to Exercise 10 is greater than the answer to Exercise 11.

14. Given the building heights, explain how you know, without calculating, that the answer to Exercise 12 is less than the answer to Exercise 9.

For Exercises 15 through 18, the grades are given for a student for a particular semester. Find the grade point average. If necessary, round the grade point average to the nearest hundredth. See Example 2.

15.

Grade	Credit Hours
B	3
C	3
A	4
C	4

16.

Grade	Credit Hours
D	1
F	1
C	4
B	5

17.

Grade	Credit Hours
A	3
A	3
A	4
B	3
C	1

18.

Grade	Credit Hours
B	2
B	2
C	3
A	3
B	3

For Exercises 19 through 27, find the mean, median, and mode, as requested. See Examples 1 and 3 through 6. During an experiment, the following times (in seconds) were recorded:

7.8, 6.9, 7.5, 4.7, 6.9, 7.0.

19. Find the mean.

20. Find the median.

21. Find the mode.

In a mathematics class, the following test scores were recorded for a student: 93, 85, 89, 79, 88, 92.

22. Find the mean. Round to the nearest hundredth.

23. Find the median.

24. Find the mode.

The following pulse rates were recorded for a group of 15 students:

78, 80, 66, 68, 71, 64, 82, 71, 70, 65, 70, 75, 77, 86, 72.

25. Find the mean.

26. Find the median.

27. Find the mode.

28. How many pulse rates were higher than the mean?

29. How many pulse rates were lower than the mean?

Review

Write each fraction in simplest form. See Section 2.3.

30. $\dfrac{12}{20}$ **31.** $\dfrac{6}{18}$ **32.** $\dfrac{4}{36}$ **33.** $\dfrac{18}{30}$ **34.** $\dfrac{35}{100}$ **35.** $\dfrac{55}{75}$

Concept Extensions

Find the missing numbers in each set of numbers.

36. 16, 18, _____, _____, _____. The mode is 21. The median is 20.

37. _____, _____, _____, 40, _____. The mode is 35. The median is 37. The mean is 38.

38. Write a list of numbers for which you feel the median would be a better measure of central tendency than the mean.

39. Without making any computations, decide whether the median of the following list of numbers will be a whole number. Explain your reasoning.

36, 77, 29, 58, 43

9.4   Counting and Introduction to Probability

Objectives

A Use a Tree Diagram to Count Outcomes.

B Find the Probability of an Event.

Objective A   Using a Tree Diagram

In our daily conversations, we often talk about the likelihood or **probability** of a given result occurring. For example:

> The *chance* of thundershowers is 70 percent.
>
> What are the *odds* that the New Orleans Saints will go to the Super Bowl?
>
> What is the *probability* that you will finish cleaning your room today?

Each of these chance happenings—thundershowers, the New Orleans Saints playing in the Super Bowl, and cleaning your room today—is called an **experiment.** The possible results of an experiment are called **outcomes.** For example, flipping a coin is an experiment, and the possible outcomes are heads (H) or tails (T).

One way to picture the outcomes of an experiment is to draw a **tree diagram.** Each outcome is shown on a separate branch. For example, the outcomes of flipping a coin are

H

T

Heads   Tails

Example 1 Draw a tree diagram for tossing a coin twice. Then use the diagram to find the number of possible outcomes.

Solution:

First Coin Toss	Second Coin Toss	Outcomes

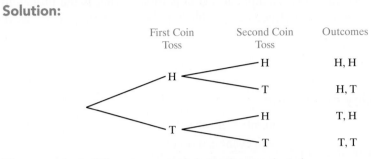

H
    H → H, H
    T → H, T

T
    H → T, H
    T → T, T

There are 4 possible outcomes when tossing a coin twice.

▪ **Work Practice 1**

Example 2 Draw a tree diagram for an experiment consisting of rolling a die and then tossing a coin. Then use the diagram to find the number of possible outcomes.

Die

Solution: Recall that a die has six sides and that each side represents a number, 1 through 6.

Roll a Die	Toss a Coin	Outcomes

1 — H → 1, H
   — T → 1, T

2 — H → 2, H
   — T → 2, T

3 — H → 3, H
   — T → 3, T

4 — H → 4, H
   — T → 4, T

5 — H → 5, H
   — T → 5, T

6 — H → 6, H
   — T → 6, T

There are 12 possible outcomes for rolling a die and then tossing a coin.

▪ **Work Practice 2**

    Any number of outcomes considered together is called an **event.** For example, when tossing a coin twice, H, H is an event. The event is tossing heads first and tossing heads second. Another event would be tossing tails first and then heads (T, H), and so on.

Practice 1

Draw a tree diagram for tossing a coin three times. Then use the diagram to find the number of possible outcomes.

Practice 2

Draw a tree diagram for an experiment consisting of tossing a coin and then rolling a die. Then use the diagram to find the number of possible outcomes.

Answers

1.

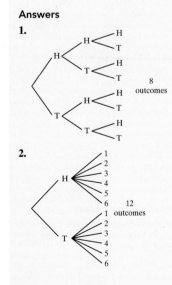

H
  H — H
    — T
  T — H
    — T

T
  H — H
    — T
  T — H
    — T

8 outcomes

2.

H
  1
  2
  3
  4
  5
  6

T
  1
  2
  3
  4
  5
  6

12 outcomes

Objective B Finding the Probability of an Event

As we mentioned earlier, the **probability of an event is a measure of the chance or likelihood of it occurring.** For example, if a coin is tossed, what is the probability that heads occurs? Since one of two equally likely possible outcomes is heads, the probability is $\frac{1}{2}$.

> **The Probability of an Event**
>
> $$\text{probability of an event} = \frac{\text{number of ways that the event can occur}}{\text{number of possible outcomes}}$$

Note from the definition of probability that the probability of an event is always between 0 and 1, inclusive (i.e., including 0 and 1). A probability of 0 means that an event won't occur, and a probability of 1 means that an event is certain to occur.

Practice 3

If a coin is tossed three times, find the probability of tossing tails, then heads, then tails (T, H, T).

Example 3 If a coin is tossed twice, find the probability of tossing heads on the first toss and then heads again on the second toss (H, H).

Solution: 1 way the event can occur

$$\overbrace{\text{H,T,   H,H,   T,H,   T,T}}^{}$$

4 possible outcomes

$$\text{probability} = \frac{1}{4} \quad \begin{array}{l}\text{Number of ways the event can occur}\\ \text{Number of possible outcomes}\end{array}$$

The probability of tossing heads and then heads is $\frac{1}{4}$.

■ Work Practice 3

Practice 4

If a die is rolled one time, find the probability of rolling a 2 or a 5.

Example 4 If a die is rolled one time, find the probability of rolling a 3 or a 4.

Solution: Recall that there are 6 possible outcomes when rolling a die.

2 ways that the event can occur

possible outcomes: 1, 2, 3, 4, 5, 6

6 possible outcomes

$$\text{probability of a 3 or a 4} = \frac{2}{6} \quad \begin{array}{l}\text{Number of ways the event can occur}\\ \text{Number of possible outcomes}\end{array}$$

$$= \frac{1}{3} \quad \text{Simplest form}$$

■ Work Practice 4

Answers

3. $\frac{1}{8}$ 4. $\frac{1}{3}$

✓ **Concept Check Answer**

The number of ways an event can occur can't be larger than the number of possible outcomes.

✓ **Concept Check** Suppose you have calculated a probability of $\frac{11}{9}$. How do you know that you have made an error in your calculation?

Example 5   Find the probability of choosing a red marble from a box containing 1 red, 1 yellow, and 2 blue marbles.

Solution: 1 way that event can occur

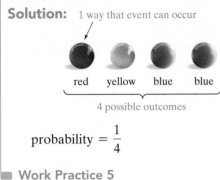

red   yellow   blue   blue

4 possible outcomes

$$\text{probability} = \frac{1}{4}$$

■ Work Practice 5

Practice 5

Use the diagram and information in Example 5 and find the probability of choosing a blue marble from the box.

Answer

5. $\frac{1}{2}$

Vocabulary, Readiness & Video Check

Use the choices below to fill in each blank. Choices may be used more than once.

0	probability	tree diagram
1	outcome	

1. A possible result of an experiment is called a(n) _____.

2. A(n) _____ shows each outcome of an experiment as a separate branch.

3. The _____ of an event is a measure of the likelihood of it occurring.

4. _____ is calculated by the number of ways that the event can occur divided by the number of possible outcomes.

5. A probability of _____ means that an event won't occur.

6. A probability of _____ means that an event is certain to occur.

Martin-Gay Interactive Videos

Watch the section lecture video and answer the following questions.

Objective A   **7.** In ▥ Example 1, how was the possible number of outcomes to the experiment determined from the tree diagram? ▶

Objective B   **8.** In ▥ Example 2, what is the probability of getting a 7? Explain this result. ▶

See Video 9.4 🍎

9.4 **Exercise Set** MyMathLab® ▶

Objective A *Draw a tree diagram for each experiment. Then use the diagram to find the number of possible outcomes. See Examples 1 and 2.*

▶ **1.** Choosing a letter in the word MATH, then a number (1, 2, or 3)

2. Choosing a number (1 or 2) and then a vowel (a, e, i, o, u)

Use the following spinners to draw a tree diagram and find the number of outcomes for Exercises 3–10.

Spinner A

Spinner B

3. Spinning Spinner A once

4. Spinning Spinner B once

5. Spinning Spinner B twice

6. Spinning Spinner A twice

7. Spinning Spinner A and then Spinner B

8. Spinning Spinner B and then Spinner A

9. Tossing a coin and then spinning Spinner B

10. Spinning Spinner A and then tossing a coin

Objective **B** *If a single die is tossed once, find the probability of each event. See Examples 3 through 5.*

11. A 5

12. A 9

13. A 1 or a 6

14. A 2 or a 3

15. An even number

16. An odd number

17. A number greater than 2

18. A number less than 6

Suppose the spinner shown is spun once. Find the probability of each event. See Examples 3 through 5.

19. The result of the spin is 2.

20. The result of the spin is 3.

21. The result of the spin is 1, 2, or 3.

22. The result of the spin is not 3.

23. The result of the spin is an odd number.

24. The result of the spin is an even number.

If a single choice is made from the bag of marbles shown, find the probability of each event. See Examples 3 through 5.

25. A red marble is chosen.

26. A blue marble is chosen.

27. A yellow marble is chosen.

28. A green marble is chosen.

29. A green or red marble is chosen.

30. A blue or yellow marble is chosen.

A new drug is being tested that is supposed to lower blood pressure. This drug was given to 200 people and the results are shown below. See Examples 3 through 5.

Lower Blood Pressure	Higher Blood Pressure	Blood Pressure Not Changed
152	38	10

31. If a person is testing this drug, what is the probability that his or her blood pressure will be higher?

32. If a person is testing this drug, what is the probability that his or her blood pressure will be lower?

33. If a person is testing this drug, what is the probability that his or her blood pressure will not change?

34. What is the sum of the answers to Exercises 31, 32, and 33? In your own words, explain why.

Review

Perform each indicated operation. See Sections 2.4, 2.5, and 3.3.

35. $\dfrac{1}{2} + \dfrac{1}{3}$ **36.** $\dfrac{7}{10} - \dfrac{2}{5}$ **37.** $\dfrac{1}{2} \cdot \dfrac{1}{3}$ **38.** $\dfrac{7}{10} \div \dfrac{2}{5}$ **39.** $5 \div \dfrac{3}{4}$ **40.** $\dfrac{3}{5} \cdot 10$

Concept Extensions

Recall that a deck of cards contains 52 cards. These cards consist of four suits (hearts, spades, clubs, and diamonds) of each of the following: 2, 3, 4, 5, 6, 7, 8, 9, 10, jack, queen, king, and ace. If a card is chosen from a deck of cards, find the probability of each event.

41. The king of hearts

42. The 10 of spades

43. A king

44. A 10

45. A heart

46. A club

47. A card in black ink

48. A queen or ace

Two dice are tossed. Find the probability of each sum of the dice. (Hint: Draw a tree diagram of the possibilities of two tosses of a die, and then find the sum of the numbers on each branch.)

49. A sum of 6

50. A sum of 10

51. A sum of 13

52. A sum of 2

Solve. See the Concept Check in this section.

53. In your own words, explain why the probability of an event cannot be greater than 1.

54. In your own words, explain when the probability of an event is 0.

Chapter 9 Group Activity

Sections 9.1, 9.3

This activity may be completed by working in groups or individually.

How often have you read an article in a newspaper or in a magazine that included results from a survey or poll? Surveys seem to have become very popular ways of getting feedback on anything from a political candidate, to a new product, to services offered by a health club. In this activity, you will conduct a survey and analyze the results.

1. Conduct a survey of 30 students in one of your classes. Ask each student to report his or her age.

2. Classify each age according to the following categories: under 20, 20 to 24, 25 to 29, 30 to 39, 40 to 49, and 50 or over. Tally the number of your survey respondents that fall into each category. Make a bar graph of your results. What does this graph tell you about the ages of your survey respondents?

3. Find the average age of your survey respondents.

4. Find the median age of your survey respondents.

5. Find the mode of the ages of your survey respondents.

6. Compare the mean, median, and mode of your age data. Are these measures similar? Which is largest? Which is smallest? If there is a noticeable difference between any of these measures, can you explain why?

Chapter 9 Vocabulary Check

Fill in each blank with one of the words or phrases listed below.

outcomes	bar	experiment	mean	tree diagram
pictograph	line	class interval	median	probability
histogram	circle	class frequency	mode	

1. A(n) _____ graph presents data using vertical or horizontal bars.

2. The _____ of a set of number items is $\dfrac{\text{sum of items}}{\text{number of items}}$.

3. The possible results of an experiment are the _____.

4. A(n) _____ is a graph in which pictures or symbols are used to visually present data.

5. The _____ of a set of numbers is the number that occurs most often.

6. A(n) _____ graph displays information with a line that connects data points.

7. The _____ of an ordered set of numbers is the middle number.

8. A(n) _____ is one way to picture and count outcomes.

9. A(n) _____ is an activity being considered, such as tossing a coin or rolling a die.

10. In a(n) _____ graph, each section (shaped like a piece of pie) shows a category and the relative size of the category.

11. The _____ of an event is $\dfrac{\text{number of ways that the event can occur}}{\text{number of possible outcomes}}$.

12. A(n) _____ is a special bar graph in which the width of each bar represents a(n) _____ and the height of each bar represents the _____.

▶ Are you preparing for your test? Don't forget to take the Chapter 9 Test on page 623. Then check your answers at the back of your text and use the Chapter Test Prep Videos to see the fully worked-out solutions to any of the exercises you want to review.

9 Chapter Highlights

Definitions and Concepts	Examples

Section 9.1 Reading Pictographs, Bar Graphs, Histograms, and Line Graphs

A **pictograph** is a graph in which pictures or symbols are used to visually present data.

A **line graph** displays information with a line that connects data points.

A **bar graph** presents data using vertical or horizontal bars.

The bar graph on the right shows the number of acres of wheat harvested in 2013 for leading states.

Wheat Production

Source: National Agricultural Statistics Service

1. Approximately how many acres of wheat were harvested in Kansas?

   9,500,000 acres

2. About how many more acres of wheat were harvested in North Dakota than South Dakota?

   $$\begin{array}{r} 6 \text{ million} \\ -2.5 \text{ million} \\ \hline 3.5 \text{ million} \end{array}\quad \text{or } 3,500,000 \text{ acres}$$

A **histogram** is a special bar graph in which the width of each bar represents a **class interval** and the height of each bar represents the **class frequency.** The histogram on the right shows student quiz scores.

1. How many students received a score of 6–10?

   4 students

2. How many students received a score of 11–20?

   $9 + 13 = 22$ students

Definitions and Concepts	Examples

Section 9.2   Reading Circle Graphs

In a **circle graph,** each section (shaped like a piece of pie) shows a category and the relative size of the category.

The circle graph on the right classifies tornadoes by wind speed.

Tornado Wind Speeds

110–205 mph
28%

Greater than
205 mph
2%

Less than
110 mph
70%

Source: National Oceanic and Atmospheric Administration

1. What percent of tornadoes have wind speeds of 110 mph or greater?

   $28\% + 2\% = 30\%$

2. If there were 1235 tornadoes in the United States in 1995, how many of these might we expect to have had wind speeds less than 110 mph? Find 70% of 1235.

   $70\%(1235) = 0.70(1235) = 864.5 \approx 865$

   Around 865 tornadoes would be expected to have had wind speeds of less than 110 mph.

Section 9.3   Mean, Median, and Mode

The **mean** (or **average**) of a set of number items is

$$\text{mean} = \frac{\text{sum of items}}{\text{number of items}}$$

Find the mean, median, and mode of the following set of numbers: 33, 35, 35, 43, 68, 68

$$\text{mean} = \frac{33 + 35 + 35 + 43 + 68 + 68}{6} = 47$$

The **median** of a set of numbers in numerical order is the middle number. If the number of items is even, the median is the mean of the two middle numbers.

The median is the mean of the two middle numbers, 35 and 43

$$\text{median} = \frac{35 + 43}{2} = 39$$

The **mode** of a set of numbers is the number that occurs most often. (A set of numbers may have no mode or more than one mode.)

There are two modes because there are two numbers that occur twice:

35 and 68

Section 9.4   Counting and Introduction to Probability

An **experiment** is an activity being considered, such as tossing a coin or rolling a die. The possible results of an experiment are the **outcomes.** A **tree diagram** is one way to picture and count outcomes.

Draw a tree diagram for tossing a coin and then choosing a number from 1 to 4.

Tossing a Coin	Choosing a Number	Outcomes
H	1	H, 1
	2	H, 2
	3	H, 3
	4	H, 4
T	1	T, 1
	2	T, 2
	3	T, 3
	4	T, 4

(continued)

Definitions and Concepts	Examples
Section 9.4 Counting and Introduction to Probability (*continued*)	
Any number of outcomes considered together is called an **event.** The **probability** of an event is a measure of the chance or likelihood of it occurring. $$\text{probability of an event} = \frac{\text{number of ways that the event can occur}}{\text{number of possible outcomes}}$$	Find the probability of tossing a coin twice and tails occurring each time. 1 way the event can occur HH, HT, TH, TT 4 possible outcomes $$\text{probability} = \frac{1}{4}$$

Chapter 9 Review

(9.1) *The following pictograph shows the number of new homes constructed from August 2012 to August 2013, by region. Use this graph to answer Exercises 1 through 6.*

New Home Construction

Northeast	
Midwest	
South	
West	

Each ⌂ represents 500,000 homes

Source: U.S. Census Bureau

1. How many new homes were constructed in the Midwest during the given year?

2. How many new homes were constructed in the West during the given year?

3. Which region had the most new homes constructed?

4. Which region had the fewest new homes constructed?

5. Which region(s) had 3,000,000 or more new homes constructed?

6. Which region(s) had fewer than 3,000,000 new homes constructed?

The following bar graph shows the percent of persons age 25 or over who completed four or more years of college. Use this graph to answer Exercises 7 through 10.

Four or More Years of College by Persons Age 25 or Over

Source: U.S. Census Bureau

7. Approximate the percent of persons who had completed four or more years of college in 1972.

8. What year shown had the greatest percent of persons completing four or more years of college?

9. What years shown had 20% or more of persons completing four or more years of college?

10. Describe any patterns you notice in this graph.

The following line graph shows the total number of Olympic medals awarded during the Summer Olympics between 1992 and 2012. Use this graph to answer Exercises 11 through 16.

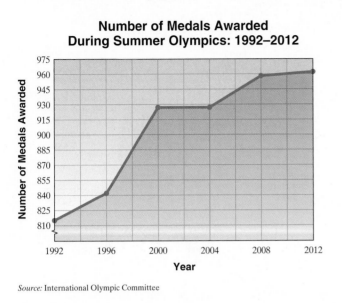

Number of Medals Awarded During Summer Olympics: 1992–2012

Source: International Olympic Committee

11. Approximate the number of medals awarded during the Summer Olympics of 2012.

12. Approximate the number of medals awarded during the Summer Olympics of 2000.

13. Between which two Summer Olympics did the number of medals awarded not change?

14. Between which two Summer Olympics did the number of medals awarded change the most?

15. How many more medals were awarded at the Summer Olympics of 1996 than at the Summer Olympics of 1992?

16. How many more medals were awarded at the Summer Olympics of 2012 than at the Summer Olympics of 1996?

The following histogram shows the hours worked per week by the employees of Southern Star Furniture. Use this histogram to answer Exercises 17 through 20.

Southern Star Furniture

17. How many employees work 41–45 hours per week?

18. How many employees work 21–25 hours per week?

19. How many employees work 30 hours or less per week?

20. How many employees work 36 hours or more per week?

Following is a list of monthly record high temperatures for New Orleans, Louisiana. Use this list to complete the frequency distribution table below.

83	96	101	92
85	100	92	102
89	101	87	84

	Class Intervals (Temperatures)	Tally	Class Frequency (Number of Months)
21.	80°–89°		
22.	90°–99°		
23.	100°–109°		

24. Use the table from Exercises 21–23 to draw a histogram.

(9.2) *The following circle graph shows a family's $4000 monthly budget. Use this graph to answer Exercises 25 through 30. Write all ratios as fractions in simplest form.*

25. What is the largest budget item?

26. What is the smallest budget item?

27. How much money is budgeted for the mortgage payment and utilities?

28. How much money is budgeted for savings and contributions?

29. Find the ratio of the mortage payment to the total monthly budget.

30. Find the ratio of food to the total monthly budget.

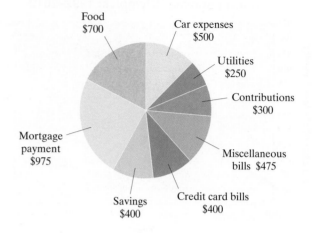

In 2013, there were 65 buildings over 1000 feet tall in the world. The following circle graph shows the percent of buildings over 1000 feet tall in the world by region in 2013. Use this graph to determine the number of tall buildings on each continent in Exercises 31 through 34. Round each answer to the nearest whole.

Source: Council on Tall Buildings and Urban Habitats

31. How many tall buildings were located in Asia?

32. How many tall buildings were located in North America?

33. How many tall buildings were located in the Middle East?

34. How many tall buildings were located in Europe?

(9.3) *Find the mean, median, and any mode(s) for each list of numbers. If necessary, round to the nearest tenth.*

35. 13, 23, 33, 14, 6

36. 45, 86, 21, 60, 86, 64, 45

37. 14,000, 20,000, 12,000, 20,000, 36,000, 45,000

38. 560, 620, 123, 400, 410, 300, 400, 780, 430, 450

For Exercises 39 and 40, the grades are given for a student for a particular semester. Find each grade point average. If necessary, round the grade point average to the nearest hundredth.

39.

Grade	Credit Hours
A	3
A	3
C	2
B	3
C	1

40.

Grade	Credit Hours
B	3
B	4
C	2
D	2
B	3

(9.4) *Draw a tree diagram for each experiment. Then use the diagram to determine the number of outcomes.*

Spinner 1   Spinner 2

41. Tossing a coin and then spinning Spinner 1

42. Spinning Spinner 2 and then tossing a coin

43. Spinning Spinner 1 twice

44. Spinning Spinner 2 twice

45. Spinning Spinner 1 and then Spinner 2

Find the probability of each event.

46. Rolling a 4 on a die

47. Rolling a 3 on a die

48. Spinning a 4 on the spinner

49. Spinning a 3 on the spinner

50. Spinning either a 1, 3, or 5 on the spinner

51. Spinning either a 2 or a 4 on the spinner

52. Rolling an even number on a die

53. Rolling a number greater than 3 on a die

Mixed Review

Find the mean, median, and any mode(s) for each list of numbers. If needed, round answers to two decimal places.

54. 73, 82, 95, 68, 54

55. 25, 27, 32, 98, 62

56. 750, 500, 427, 322, 500, 225

57. 952, 327, 566, 814, 327, 729

Given a bag containing 2 red marbles, 2 blue marbles, 3 yellow marbles, and 1 green marble, find the following:

58. The probability of choosing a blue marble from the bag

59. The probability of choosing a yellow marble from the bag

60. The probability of choosing a red marble from the bag

61. The probability of choosing a green marble from the bag

The following pictograph shows the money collected each week from a wrapping paper fund-raiser. Use this graph to answer Exercises 1 through 3.

Answers

Weekly Wrapping Paper Sales

1. How much money was collected during the second week?

2. During which week was the most money collected? How much money was collected during that week?

3. What was the total money collected for the fundraiser?

1. _____

The bar graph shows the normal monthly precipitation in centimeters for Chicago, Illinois. Use this graph to answer Exercises 4 through 6.

2. _____

Chicago Precipitation

Source: U.S. National Oceanic and Atmospheric Administration, *Climatography of the United States*, No. 81

3. _____

4. During which month(s) does Chicago normally have more than 9 centimeters of precipitation?

4. _____

5. During which month does Chicago normally have the least amount of precipitation? How much precipitation occurs during that month?

5. _____

6. _____

6. During which month(s) does 7 centimeters of precipitation normally occur?

7. _____

7. Use the information in the table to draw a bar graph. Clearly label each bar.

Most Common Blood Types	
Blood Type	**% of Population with This Blood Type**
O+	38%
A+	34%
B+	9%
O−	7%
A−	6%
AB+	3%
B−	2%
AB−	1%

Most Common Blood Types by Percent in the Population

Percent of Population

10

5

Blood Type

8. _____

The following line graph shows the average annual inflation rate in the United States for the years 2003–2013. Use this graph to answer Exercises 8 through 10.

9. _____

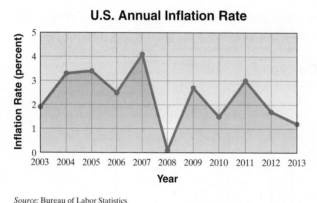

U.S. Annual Inflation Rate

Inflation Rate (percent)

Year

Source: Bureau of Labor Statistics

10. _____

8. Approximate the annual inflation rate in 2011.

9. During which of the years shown was the inflation rate greater than 3%?

10. During which sets of years was the inflation rate increasing?

11. _____

The result of a survey of 200 people is shown in the following circle graph. Each person was asked to tell his or her favorite type of music. Use this graph to answer Exercises 11 and 12.

Country
62

Rock
85

Jazz
44

Classical
9

11. Find the ratio of those who prefer rock music to the total number surveyed.

12. Find the ratio of those who prefer country music to those who prefer jazz.

12. _____

The following graph (from earlier in this chapter) shows the percent of visitors to the United States in 2012 by various regions. During 2012, there were approximately 66,700,000 foreign visitors to the United States. Use the graph to find how many people visited the United States in 2012 from the world regions given in Exercises 13 and 14. (Source: Office of Travel & Tourism Industries)

Visitors to U.S. by Region

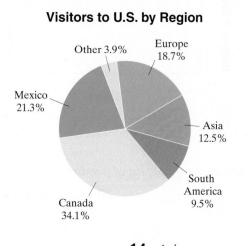

13. Canada

14. Asia

A professor measures the heights of the students in her class. The results are shown in the following histogram. Use this histogram to answer Exercises 15 and 16.

Student Heights

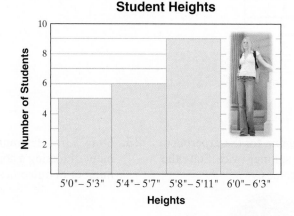

15. How many students are 5′8″–5′11″ tall?

16. How many students are 5′7″ or shorter?

17. The history test scores of 25 students are shown below. Use these scores to complete the frequency distribution table.

70	86	81	65	92
43	72	85	69	97
82	51	75	50	68
88	83	85	77	99
77	63	59	84	90

Class Intervals (Scores)	Tally	Class Frequency (Number of Students)
40–49		
50–59		
60–69		
70–79		
80–89		
90–99		

13. _____

14. _____

15. _____

16. _____

17. _____

18. _____

18. Use the results of Exercise 17 to draw a histogram.

19. _____

20. _____

Find the mean, median, and mode of each list of numbers.

19. 26, 32, 42, 43, 49

20. 8, 10, 16, 16, 14, 12, 12, 13

21. _____

Find the grade point average. If necessary, round to the nearest hundredth.

21.

Grade	Credit Hours
A	3
B	3
C	3
B	4
A	1

22. _____

22. Draw a tree diagram for the experiment of spinning the spinner twice. State the number of outcomes.

23. Draw a tree diagram for the experiment of tossing a coin twice. State the number of outcomes.

23. _____

24. _____

Suppose that the numbers 1 to 10 are each written on a scrap of paper and placed in a bag. You then select one number from the bag.

25. _____

24. What is the probability of choosing a 6 from the bag?

25. What is the probability of choosing a 3 or a 4 from the bag?

1. Simplify: $(8 - 6)^2 + 2^3 \cdot 3$

2. Simplify: $48 \div 8 \cdot 2$

3. Write $\dfrac{30}{108}$ in simplest form.

4. Subtract: $\dfrac{19}{40} - \dfrac{3}{10}$

5. Add: $1\dfrac{4}{5} + 4 + 2\dfrac{1}{2}$

6. Multiply: $5\dfrac{1}{3} \cdot 2\dfrac{1}{8}$

△ **7.** The formula for finding the area of a triangle is Area = $\dfrac{1}{2} \cdot$ base $\cdot$ height. Find the area of a triangle with height 3 feet and base length 5.6 feet.

8. Find the perimeter of a rectangle with length $3\dfrac{1}{2}$ meters and width $1\dfrac{1}{2}$ meters.

9. Write the ratio of $15 to $10 as a fraction in simplest form.

10. Write the ratio 14 inches to 3 feet as a fraction in simplest form.

Write each rate as a fraction in simplest form.

11. $2160 for 12 weeks

12. 340 miles every 5 hours

13. 360 miles on 16 gallons of gasoline

14. 78 files every 4 hours

Answers

1. _____

2. _____

3. _____

4. _____

5. _____

6. _____

7. _____

8. _____

9. _____

10. _____

11. _____

12. _____

13. _____

14. _____

15. _____

16. _____

17. _____

18. _____

19. _____

20. _____

21. _____

22. _____

23. _____

24. _____

25. _____

26. _____

27. _____

28. _____

29. _____

30. _____

31. _____

32. _____

15. Is $\dfrac{1\frac{1}{6}}{10\frac{1}{2}} = \dfrac{\frac{1}{2}}{4\frac{1}{2}}$ a true proportion?

16. Is $\dfrac{7.8}{3} = \dfrac{5.2}{2}$ a true proportion?

17. The standard dose of an antibiotic is 4 cc (cubic centimeters) for every 25 pounds (lb) of body weight. At this rate, find the standard dose for a 140-lb woman.

18. On a certain map, 2 inches represents 75 miles. How many miles are represented by 7 inches?

Write each percent as a decimal.

19. 4.6%

20. 0.29%

21. 190%

22. 452%

Write each percent as a fraction in simplest form.

23. 40%

24. 27%

25. $33\frac{1}{3}\%$

26. $61\frac{1}{7}\%$

27. Translate to an equation: Five is what percent of 20?

28. Translate to a proportion: Five is what percent of 20?

29. Find the sales tax and the total price on the purchase of an $85.50 atlas in a city where the sales tax rate is 7.5%.

30. A salesperson makes a 7% commission rate on her total sales. If her total sales are $23,000, what is her commission?

31. An accountant invested $2000 at a simple interest rate of 10% for 2 years. What total amount of money will she have from her investment in 2 years?

32. Find the mean (or average) of 28, 35, 40, and 32.

33. Convert 7 feet to yards.

34. Convert 2.5 tons to pounds.

35. Divide 9 lb 6 oz by 2.

36. Find: $1 + 3(10 - 8)$

37. Convert 2.35 cg to grams.

38. Convert 106 cm to millimeters.

39. Convert 3210 ml to liters.

40. Convert 5 m to centimeters.

41. Convert 15°C to degrees Fahrenheit.

42. Convert 41°F to degrees Celsius.

43. Find the complement of a 48° angle.

44. Find the supplement of a 48° angle.

45. Find: $\sqrt{\dfrac{1}{36}}$

46. Find: $\sqrt{\dfrac{1}{25}}$

47. Find the mode of the list of numbers:
11, 14, 14, 16, 31, 56, 65, 77, 77, 78, 79

48. Find the median of the numbers in Exercise 47.

49. If a coin is tossed twice, find the probability of tossing heads on the first toss and then heads again on the second toss.

50. A bag contains 3 red marbles and 2 blue marbles. Find the probability of choosing a red marble.

33. _____

34. _____

35. _____

36. _____

37. _____

38. _____

39. _____

40. _____

41. _____

42. _____

43. _____

44. _____

45. _____

46. _____

47. _____

48. _____

49. _____

50. _____

10

Signed Numbers

Check Your Progress

Vocabulary Check

Chapter Highlights

Chapter Review

Chapter Test

Cumulative Review

Thus far, we have studied whole numbers, fractions, and decimals. However, these numbers are not sufficient for representing many situations in real life. For example, to express 5 degrees below zero or $100 in debt, numbers less than 0 are needed. This chapter is devoted to signed numbers, which include numbers less than 0, and to operations on these numbers.

Is Pluto the Only Dwarf Planet?

You may know by now that Pluto is no longer classified as a planet, but according to the International Astronomical Union (IAU), it is a dwarf planet. Including Pluto, the IAU has named 5 dwarf planets and more are on the way. The dwarf planets thus far are named Ceres, Pluto, Makemake, Haumea, and Eris.

There are now 8 planets in the Solar System and 5 dwarf planets. Of course, the 8 planets are Mercury, Venus, Earth, Mars (the smaller, rocky ones), then Jupiter, Saturn, Uranus, and Neptune (the larger, low-density gas giants).

In Section 10.1, Example 23, we study the average surface temperature of planets.

Average Surface Temperature of Planets* and Dwarf Planets

Source: The World Almanac and Internet research

* For some planets, the temperature given is the temperature where the atmospheric pressure equals 1 Earth atmosphere.

Signed Numbers

Objective A Representing Real-Life Situations

Thus far in this text, all numbers have been 0 or greater than 0. Numbers greater than 0 are called **positive numbers.** However, sometimes situations exist that cannot be represented by a number greater than 0. For example,

0° — 5 degrees below 0°

Sea level

20 feet below sea level

To represent these situations, we need numbers less than 0.

Extending the number line to the left of 0 allows us to picture **negative numbers,** which are numbers that are less than 0.

Negative numbers | Zero | Positive numbers

−5 −4 −3 −2 −1 0 1 2 3 4 5

When a single + sign or no sign is in front of a number, the number is a positive number. When a single − sign is in front of a number, the number is a negative number. Together, we call positive numbers, negative numbers, and zero the **signed numbers.**

−5 indicates "negative five."

5 and +5 both indicate "positive five."

The number 0 is neither positive nor negative.

Now we have numbers to represent the situations previously mentioned.

5 degrees below 0   −5°
20 feet below sea level   −20 feet

Helpful Hint

A − sign, such as the one in −1, tells us that the number is to the left of 0 on a number line.

−1 is read "negative one."

A + sign or no sign tells us that a number lies to the right of 0 on a number line. For example, 3 and +3 both mean positive three.

Practice 1

a. The world's deepest bat colony spends each winter in a New York zinc mine at a depth of 3805 feet. Represent this position with a signed number. (*Source: Guinness Book of World Records,* 2009)

b. The tamarack tree, a kind of larch, commonly grows at the edge of the arctic tundra and survives winter temperatures of 85 degrees below zero Fahrenheit. Represent this temperature with a signed number in degrees Fahrenheit.

Practice 2

Graph the signed numbers $-5, 3, -3, -1\frac{3}{4}$, and -4.5 on a number line.

Example 1 Representing Depth with an Integer

The world's deepest cave is Krubera (or Voronja), in the country of Georgia, located by the Black Sea in Asia. It has been explored to a depth of 7208 feet below the surface of the Earth. Represent this position using a signed number. (*Source:* Speleogenesis.info)

Solution: If 0 represents the surface of the Earth, then 7208 feet below the surface can be represented by -7208.

■ Work Practice 1

Objective B Graphing Signed Numbers ▶

Example 2 Graph the signed numbers $-2\frac{1}{4}, -2, -\frac{1}{2}, 4$, and -3.8 on a number line.

Solution:

■ Work Practice 2

Some signed numbers are integers. The **integers** consist of 0, the natural numbers, and the opposites of the natural numbers. (A formal definition of opposites is given later in this section.) The integers are $\dots, -3, -2, -1, 0, 1, 2, 3, \dots$ and are graphed below.

In Example 2 above we graphed the signed numbers $-2\frac{1}{4}, -2, -\frac{1}{2}, 4, -3.8$. Of these signed numbers, -2 and 4 are integers; the rest are not integers.

Objective C Comparing Signed Numbers ▶

For any two numbers graphed on a number line, recall that the number to the **right** is the **greater number,** and the number to the **left** is the **smaller number.** Also, recall that the symbol $>$ means "is greater than" and the symbol $<$ means "is less than."

To illustrate, both -5 and -2 are graphed on the number line shown:

The graph of -5 is **to the left of** -2, so -5 **is less than** -2. We can write this as

$$-5 < -2$$

We can also write

$$-2 > -5$$

since -2 is **to the right** of -5, so -2 **is greater than** -5.

✓ **Concept Check** Is there a smallest negative number? Is there a largest positive number? Explain.

Answers

1. a. -3805 **b.** $-85°F$

2.

✓ **Concept Check Answer**

no; no

Examples Insert $<$ or $>$ between each pair of numbers to make a true statement.

3. -7 7 -7 is to the left of 7, so $-7 < 7$.
4. 0 -4 0 is to the right of -4, so $0 > -4$.
5. -9 -11 -9 is to the right of -11, so $-9 > -11$.
6. -2.9 -1 -2.9 is to the left of -1, so $-2.9 < -1$.
7. -6 0 -6 is to the left of 0, so $-6 < 0$.
8. 8.6 -8.6 8.6 is to the right of -8.6, so $8.6 > -8.6$.
9. $-1\dfrac{1}{4}$ $-2\dfrac{1}{2}$ $-1\dfrac{1}{4}$ is to the right of $-2\dfrac{1}{2}$, so $-1\dfrac{1}{4} > -2\dfrac{1}{2}$.

🔲 **Work Practice 3–9**

Practice 3–9

Insert $<$ or $>$ between each pair of numbers to make a true statement.
3. 8 -8
4. -11 0
5. -15 -14
6. 3 -4.6
7. 0 -2
8. -7.9 7.9
9. $-\dfrac{3}{8}$ $-1\dfrac{1}{7}$

Helpful Hint

If you think of $<$ and $>$ as arrowheads, notice that in a true statement the arrow always points to the smaller number.

$$5 > -4 \qquad -3 < -1$$
$$\uparrow \qquad\qquad \uparrow$$
smaller   smaller
number   number

Objective D Finding the Absolute Value of a Number ▶

The **absolute value** of a number is the number's distance from 0 on a number line. The symbol for absolute value is $|\ \ |$. For example, $|3|$ is read as "the absolute value of 3."

$|3| = 3$ because 3 is 3 units from 0.

$|-3| = 3$ because -3 is 3 units from 0.

Examples Find each absolute value.

10. $|-2| = 2$ because -2 is 2 units from 0.
11. $|5| = 5$ because 5 is 5 units from 0.
12. $|0| = 0$ because 0 is 0 units from 0.
13. $\left|-\dfrac{3}{4}\right| = \dfrac{3}{4}$ because $-\dfrac{3}{4}$ is $\dfrac{3}{4}$ unit from 0.
14. $|1.2| = 1.2$ because 1.2 is 1.2 units from 0.

 Work Practice 10–14

Practice 10–14

Find each absolute value.
10. $|6|$
11. $|-4|$
12. $|0|$
13. $\left|\dfrac{7}{8}\right|$
14. $|-3.4|$

Helpful Hint

Since the absolute value of a number is that number's *distance* from 0, the absolute value of a number is always 0 or positive. It is never negative.

$$|0| = 0 \qquad |-6| = 6$$
$$\uparrow \qquad\qquad \uparrow$$
zero   a positive number

Answers
3. $>$ **4.** $<$ **5.** $<$ **6.** $>$ **7.** $>$ **8.** $<$
9. $>$ **10.** 6   **11.** 4   **12.** 0   **13.** $\dfrac{7}{8}$
14. 3.4

Objective E Finding the Opposite of a Number

Two numbers that are the same distance from 0 on a number line but are on opposite sides of 0 are called **opposites.**

4 and −4 are opposites.

When two numbers are opposites, we say that each is the opposite of the other. Thus, **4 is the opposite of −4 and −4 is the opposite of 4.**

The phrase "the opposite of" is written in symbols as "−". For example,

the opposite of   9   is   −9

−   (9)   =   −9

the opposite of   −3   is   3

−   (−3)   =   3

Notice we just stated that

$-(-3) = 3$

In general, we have the following:

Opposites

If a is a number, then $-(-a) = a$.

Notice that because "the opposite of" is written as "−", to find the opposite of a number we place a "−" sign in front of the number.

Examples   Find the opposite of each number.

15. 12   The opposite of 12 is -12.
16. 1.7   The opposite of 1.7 is -1.7.
17. −1   The opposite of −1 is $-(-1)$ or 1.
18. 0   The opposite of 0 is -0 or 0.
19. $-\frac{10}{13}$ The opposite of $-\frac{10}{13}$ is $-\left(-\frac{10}{13}\right)$ or $\frac{10}{13}$.

> **Helpful Hint** Remember that 0 is neither positive nor negative.

Work Practice 15–19

✓**Concept Check** True or false? The number 0 is the only number that is its own opposite.

Examples Simplify.

20. $-(-4) = 4$ The opposite of negative 4 is 4.
21. $-|6| = -6$ The opposite of the absolute value of 6 is the opposite of 6, which is −6.
22. $-|-5| = -5$ The opposite of the absolute value of −5 is the opposite of 5, which is −5.

Work Practice 20–22

Practice 15–19

Find the opposite of each number.
15. −7   16. 0
17. $\frac{11}{15}$ 18. −9.6
19. 4

Practice 20–22

Simplify.
20. $-(-11)$
21. $-|7|$
22. $-|-2|$

Answers
15. 7 16. 0 17. $-\frac{11}{15}$ 18. 9.6
19. −4 20. 11 21. −7 22. −2

✓**Concept Check Answer**
true

Objective F Reading Bar Graphs Containing Signed Numbers

The bar graph below shows the average temperature (in degrees Fahrenheit) of the eight planets, excluding the newly classified "dwarf planet," Pluto. Notice that a negative temperature is illustrated by a bar below the horizontal line representing 0°F, and a positive temperature is illustrated by a bar above the horizontal line representing 0°F.

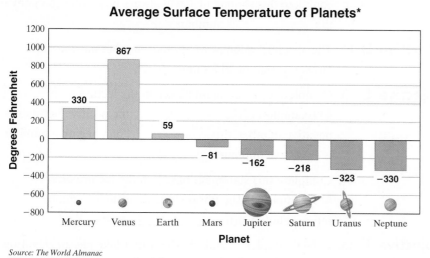

Average Surface Temperature of Planets*

Source: The World Almanac
* For some planets, the temperature given is the temperature where the atmospheric pressure equals 1 Earth atmosphere.

Example 23 Which planet has the lowest average temperature?

Solution: The planet with the lowest average temperature is the one that corresponds to the bar that extends the farthest in the negative direction (downward). Neptune has the lowest average temperature of −330°F.

▪ Work Practice 23

Practice 23

Which planet has the highest average temperature?

Answer
23. Venus

Vocabulary, Readiness & Video Check

Use the choices below to fill in each blank. Not all choices will be used.

opposites	absolute value	right	is less than
inequality symbols	negative	positive	left
signed	integers	is greater than	

1. The numbers . . . −3, −2, −1, 0, 1, 2, 3, . . . are called _____.

2. Positive numbers, negative numbers, and zero, together, are called _____ numbers.

3. The symbols "<" and ">" are called _____.

4. Numbers greater than 0 are called _____ numbers while numbers less than 0 are called _____ numbers.

5. The sign "<" means _____ and ">" means _____.

6. On a number line, the greater number is to the _____ of the lesser number.

7. A number's distance from 0 on a number line is the number's _____.

8. The numbers −5 and 5 are called _____.

Martin-Gay Interactive Videos   *Watch the section lecture video and answer the following questions.*

See Video 10.1

Objective A **9.** In ▐ Example 1, what application is used to represent a negative number? ▶

Objective B **10.** In ▐ Example 2, the tick marks are labeled with what numbers on the number line? ▶

Objective C **11.** From ▐ Example 4 and our knowledge of a number line, complete this statement: 0 will always be greater than any of the _____ numbers. ▶

Objective D **12.** The absolute value of what other number has the same answer as that of ▐ Example 6? ▶

Objective E **13.** Complete this statement based on the lecture after ▐ Example 11: The _____ of a positive number is a negative number; the _____ of a negative number is a positive number. ▶

14. Complete this statement based on the lecture after ▐ Example 11: The phrase "the opposite of" can be represented by a(n) _____ symbol. ▶

Objective F **15.** In ▐ Examples 14 and 15, what other lake has an elevation below sea level? ▶

10.1 Exercise Set MyMathLab® ▶

Objective A *Represent each quantity by a signed number. See Example 1.*

▶ **1.** A worker in a silver mine in Nevada works 1235 feet underground.

2. A scuba diver is swimming 25 feet below the surface of the water in the Gulf of Mexico.

3. The peak of Mount Elbert in Colorado is 14,433 feet above sea level. (*Source:* U.S. Geological Survey)

4. The lowest elevation in the United States is found at Death Valley, California, at an elevation of 282 feet below sea level. (*Source:* U.S. Geological Survey)

5. The record high temperature in Nevada is 125 degrees above zero Fahrenheit. (*Source:* National Climatic Data Center)

6. The record high temperature in Seattle, Washington, is 103 degrees above zero Fahrenheit. (*Source:* National Climatic Data Center)

7. The average depth of the Pacific Ocean is 13,215 feet below the surface of the ocean. (*Source:* Information Please Database)

8. The average depth of the Atlantic Ocean is 12,880 feet below the surface of the ocean. (*Source:* Information Please Database)

9. Barnes & Noble had an operating loss of $61,303,000 in fiscal year 2012. (*Source:* Barnes & Noble, Inc.)

10. U.S. wheat production decreased by 155,000,000 bushels between 2012 and 2013. (*Source:* U.S. Department of Agriculture)

11. Two divers are exploring the wreck of the *Andrea Doria,* south of Nantucket Island, Massachusetts. Guillermo is 160 feet below the surface of the ocean and Luigi is 147 feet below the surface. Represent each quantity by a signed number and determine who is deeper.

12. The temperature on one January day in Chicago was 10° below 0° Celsius. Represent this quantity by a signed number and tell whether this temperature is cooler or warmer than 5° below 0° Celsius.

13. Sales of video games fell by 11 percent from 2009 to 2011. Write a signed number to represent the percent of decrease in the sales of video games. (*Source:* Entertainment Software Association)

14. In 2012, the number of CDs shipped to music retailers reflected a 12 percent decrease from the previous year. Write a signed number to represent the percent of decrease in CDs shipped. (*Source:* Recording Industry Association of America)

Objective B *Graph the signed numbers in each list on a number line. See Example 2.*

▶15. $-3, 0, 4, -1\frac{1}{2}$

16. $-4, 0, 2, -3\frac{1}{4}$

17. $5, -2, -4.7$

18. $3, -1, -4, -2.1$

Objective C *Insert < or > between each pair of numbers to make a true statement. See Examples 3 through 9.*

19. 5   7

20. 16   10

21. 4   0

22. 8   0

▶23. -5 -7

24. -12 -10

▶25. 0 -3

26. 0 -7

27. -4.6 -2.7

28. -8.4 -1.6

29. $-1\frac{3}{4}$ 0

30. 0 $-\frac{7}{8}$

31. $\frac{1}{4}$ $-\frac{8}{11}$

32. $-\frac{1}{5}$ 6

▶33. $-2\frac{1}{2}$ $-\frac{9}{10}$

34. $-1\frac{1}{5}$ $-3\frac{1}{9}$

Objective D *Find each absolute value. See Examples 10 through 14.*

35. $|9|$

36. $|7|$

▶37. $|-8|$

38. $|-19|$

▶39. $|0|$

40. $|100|$

41. $|-5|$

42. $|-10|$

▶43. $|-8.1|$

44. $|-31.6|$

45. $\left|-\frac{1}{2}\right|$

46. $\left|\frac{9}{10}\right|$

47. $\left|-\frac{3}{8}\right|$

48. $\left|\frac{20}{23}\right|$

49. $|7.6|$

50. $|-0.6|$

Objective E *Find the opposite of each number. See Examples 15 through 19.*

▶ **51.** 5   **52.** 8   ▶ **53.** −4   **54.** −6   **55.** 23.6   **56.** 123.9

▶ **57.** $-\dfrac{9}{16}$ **58.** $-\dfrac{4}{9}$ **59.** −0.7   **60.** −4.4   **61.** $\dfrac{17}{18}$ **62.** $\dfrac{2}{3}$

Simplify. See Examples 20 through 22.

63. $|-7|$ **64.** $|-11|$ **65.** $-|20|$ **66.** $-|43|$ ▶ **67.** $-|-3|$ **68.** $-|-18|$

▶ **69.** $-(-8)$ **70.** $-(-7)$ **71.** $|-14|$ **72.** $-(-14)$ **73.** $-(-29)$ **74.** $-|-29|$

Objectives D E **Mixed Practice** *Fill in the charts. See Examples 10 through 19.*

	Number	Absolute Value of Number	Opposite of Number
75.	31		
77.			8.4

	Number	Absolute Value of Number	Opposite of Number
76.	−13		
78.			$-\dfrac{11}{18}$

Objective F *The bar graph shows the elevations of selected lakes. Use this graph for Exercises 79 through 82 (Source: U.S. Geological Survey). See Example 23.*

Elevations of Selected Lakes

Feet Above or Below Sea Level

Superior, North America 600; Ontario, North America 245; Caspian Sea, Asia-Europe −92; Maracaibo, South America 0; Great Bear, North America 512; Eyre, Australia −52; Vanern, Europe 144

▶ **79.** Which lake shown has the lowest elevation?

80. Which lake has an elevation at sea level?

▶ **81.** Which lake shown has the highest elevation?

82. Which lake shown has the second lowest elevation?

The following bar graph represents the boiling temperature, the temperature at which a substance changes from liquid to gas at standard atmospheric pressure. Use this graph to answer Exercises 83 through 86.

83. Which element has a positive boiling temperature closest to that of water?

84. Which element has the lowest boiling temperature?

85. Which element has a boiling temperature closest to −200°C?

86. Which element has an average boiling temperature closest to +300°C?

Review

Add. See Section 1.3.

87. $0 + 13$

88. $9 + 0$

89. $15 + 20$

90. $20 + 15$

91. $47 + 236 + 77$

92. $362 + 37 + 90$

Concept Extensions

Write the given numbers in order from least to greatest.

93. $2^2, -|3|, -(-5), -|-8|$

94. $|10|, 2^3, -|-5|, -(-4)$

95. $|-1|, -|-6|, -(-6), -|1|$

96. $1^4, -(-3), -|7|, |-20|$

97. $-(-2), 5^2, -10, -|-9|, |-12|$

98. $3^3, -|-11|, -(-10), -4, -|2|$

For Exercises 99 and 100, choose all numbers for x from each given list that make each statement true.

99. $|x| > 8$
   a. 0   **b.** −5   **c.** 8   **d.** −12

100. $|x| > 4$
   a. 0   **b.** 4   **c.** −1   **d.** −100

101. Evaluate: $-(-|-5|)$

102. Evaluate: $-(-|-(-7)|)$

Answer true or false for Exercises 103 through 107.

103. If $a > b$, then a must be a positive number.

104. The absolute value of a number is *always* a positive number.

105. A positive number is always greater than a negative number.

106. Zero is always less than a positive number.

107. The number $-a$ is always a negative number. (*Hint:* Read "$-$" as "the opposite of.")

108. Given the number line ⟵•—•—+—+—+—+⟶, is it true that $b < a$?
         $\quad a\ \ b\ -1\ \ 0\ \ 1$

109. Write in your own words how to find the absolute value of a signed number.

110. Explain how to determine which of two signed numbers is larger.

For Exercises 111 and 112, see the first Concept Check in this section.

111. Is there a largest negative number? If so, what is it?

112. Is there a smallest positive number? If so, what is it?

10.2 Adding Signed Numbers ▶

Objective A Adding Signed Numbers ▶

Objectives

A Add Signed Numbers. ▶

B Solve Problems by Adding Signed Numbers. ▶

Adding signed numbers can be visualized by using a number line. A positive number can be represented on the number line by an arrow of appropriate length pointing to the right, and a negative number by an arrow of appropriate length pointing to the left.

Both arrows represent 2 or +2. They both point to the right, and they are both 2 units long.

Both arrows represent −3. They both point to the left, and they are both 3 units long.

To add signed numbers such as $5 + (-2)$ on a number line, we start at 0 on the number line and draw an arrow representing 5. From the tip of this arrow, we draw another arrow representing −2. The tip of the second arrow ends at their sum, 3.

$$5 + (-2) = 3$$

To add $-1 + (-4)$ on the number line, we start at 0 and draw an arrow representing −1. From the tip of this arrow, we draw another arrow representing −4. The tip of the second arrow ends at their sum, −5.

$$-1 + (-4) = -5$$

Example 1 Add using a number line: $-3 + (-4)$

Solution:

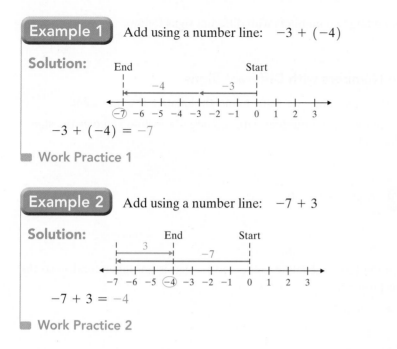

$$-3 + (-4) = -7$$

▪ Work Practice 1

Practice 1
Add using a number line:
$-5 + (-3)$

Example 2 Add using a number line: $-7 + 3$

Solution:

$$-7 + 3 = -4$$

▪ Work Practice 2

Practice 2
Add using a number line:
$5 + (-1)$

Using a number line each time we add two numbers can be time consuming. Instead, we can notice patterns in the previous examples and write rules for adding signed numbers.

Rules for adding signed numbers depend on whether we are adding numbers with the same sign or different signs. When adding two numbers with the same sign, notice that the sign of the sum is the same as the sign of the addends.

Adding Two Numbers with the Same Sign

Step 1: Add their absolute values.

Step 2: Use their common sign as the sign of the sum.

Example 3 Add: $-2 + (-21)$

Solution:

Step 1: $|-2| = 2, |-21| = 21,$ and $2 + 21 = 23.$

Step 2: Their common sign is negative, so the sum is negative:

$$-2 + (-21) = -23$$

▪ Work Practice 3

Practice 3
Add: $(-3) + (-9)$

Examples Add.

4. $-5 + (-11) = -16$

5. $2 + 6 = 8$

6. $-\dfrac{3}{8} + \left(-\dfrac{1}{8}\right) = -\dfrac{4}{8} = -\dfrac{\cancel{4}^{1}}{2 \cdot \cancel{4}_{1}} = -\dfrac{1}{2}$

7. $-1.2 + (-7.1) = -8.3$

▪ Work Practice 4–7

Practice 4–7
Add.
4. $-12 + (-3)$ **5.** $9 + 5$
6. $-\dfrac{4}{15} + \left(-\dfrac{1}{15}\right)$
7. $-8.3 + (-5.7)$

Answers

1. $-5 + (-3) = -8$

2. $5 + (-1) = 4$

3. -12 **4.** -15 **5.** 14 **6.** $-\dfrac{1}{3}$ **7.** -14

The rule for adding two numbers with different signs follows:

> ### Adding Two Numbers with Different Signs
> **Step 1:** Find the larger absolute value minus the smaller absolute value.
> **Step 2:** Use the sign of the number with the larger absolute value as the sign of the sum.

Practice 8

Add: $-3 + 9$

Example 8 Add: $-2 + 20$

Solution:

Step 1: $|-2| = 2$, $|20| = 20$, and $20 - 2 = 18$.

Step 2: 20 has the larger absolute value and its sign is an understood $+$ so the sum is positive.

$$-2 + 20 = +18 \text{ or } 18$$

◼ Work Practice 8

Practice 9

Add: $2 + (-8)$

Example 9 Add: $3 + (-7)$

Solution:

Step 1: $|3| = 3$, $|-7| = 7$, and $7 - 3 = 4$.

Step 2: -7 has the larger absolute value and its sign is $-$ so the sum is negative.

$$3 + (-7) = -4$$

◼ Work Practice 9

Practice 10–13

Add.

10. $-46 + 20$ **11.** $8.6 + (-6.2)$

12. $-\dfrac{3}{4} + \dfrac{1}{8}$ **13.** $-2 + 0$

Examples Add.

10. $-18 + 10 = -8$

11. $12.9 + (-8.6) = 4.3$

12. $-\dfrac{1}{2} + \dfrac{1}{6} = -\dfrac{3}{6} + \dfrac{1}{6} = -\dfrac{2}{6} = -\dfrac{1}{3}$

13. $0 + (-5) = -5$ The sum of 0 and any number is the number.

◼ Work Practice 10–13

Recall that numbers such as 7 and -7 are called opposites. In general, the sum of a number and its opposite is always 0.

$$7 + (-7) = 0 \qquad -26 + 26 = 0 \qquad 1008 + (-1008) = 0$$

opposites   opposites   opposites

> If a is a number, then
> $-a$ is its opposite. Also,
> $$\left.\begin{array}{l} a + (-a) = 0 \\ -a + a = 0 \end{array}\right\}$$ The sum of a number and its opposite is 0.

Examples Add.

14. $-21 + 21 = 0$
15. $36 + (-36) = 0$

Work Practice 14–15

Practice 14–15
Add.
14. $18 + (-18)$
15. $-64 + 64$

✓**Concept Check** What is wrong with the following calculation?

$5 + (-22) = 17$

In the following examples, we add three or more signed numbers. Remember that by the associative and commutative properties for addition, we may add numbers in any order that we wish. In Examples 16 and 17, let's add the numbers from left to right.

Example 16 Add: $(-3) + 4 + (-11)$

Solution: $(-3) + 4 + (-11) = 1 + (-11)$
$$= -10$$

Work Practice 16

Practice 16
Add: $6 + (-2) + (-15)$

Example 17 Add: $1 + (-10) + (-8) + 9$

Solution: $1 + (-10) + (-8) + 9 = -9 + (-8) + 9$
$$= -17 + 9$$
$$= -8$$

The sum will be the same if we add the numbers in any order. To see this, let's add the positive numbers together and then the negative numbers together first.

$$1 + 9 = 10 \quad \text{Add the positive numbers.}$$
$$(-10) + (-8) = -18 \quad \text{Add the negative numbers.}$$
$$10 + (-18) = -8 \quad \text{Add these results.}$$

The sum is -8.

Work Practice 17

Practice 17
Add:
$5 + (-3) + 12 + (-14)$

Helpful Hint Don't forget that addition is commutative and associative. In other words, numbers may be added in any order.

Objective B Solving Problems by Adding Signed Numbers ▶

Next, we practice solving problems that require adding signed numbers.

Answers
14. 0 **15.** 0 **16.** −11 **17.** 0

✓**Concept Check Answer**
$5 + (-22) = -17$

Practice 18

If the temperature was $-7°$ Fahrenheit at 6 a.m., and it rose 4 degrees by 7 a.m. and then rose another 7 degrees in the hour from 7 a.m. to 8 a.m., what was the temperature at 8 a.m.?

Answer

18. $4°F$

Example 18   Calculating Temperature

In Philadelphia, Pennsylvania, the record extreme high temperature is $104°F$. Decrease this temperature by 111 degrees, and the result is the record extreme low temperature. Find this temperature. (*Source:* National Climatic Data Center)

Solution:

In words: | extreme low temperature | = | extreme high temperature | + | decrease of 111° |

$$\text{extreme low temperature} = 104 + (-111)$$

$$= -7$$

The record extreme low temperature in Philadelphia, Pennsylvania, is $-7°F$.

■ Work Practice 18

🖩 Calculator Explorations   Entering Negative Numbers

To enter a negative number on a calculator, find the key marked $\boxed{+/-}$. (Some calculators have a key marked $\boxed{CHS}$ and some calculators have a special key $\boxed{(-)}$ for entering a negative sign.) To enter the number -2, for example, press the keys $\boxed{2}$ $\boxed{+/-}$. The display will read $\boxed{-2}$.

To find $-32 + (-131)$, press the keys

$\boxed{32}$ $\boxed{+/-}$ $\boxed{+}$ $\boxed{131}$ $\boxed{+/-}$ $\boxed{=}$ or

$\boxed{(-)}$ $\boxed{32}$ $\boxed{+}$ $\boxed{(-)}$ $\boxed{131}$ $\boxed{ENTER}$

The display will read $\boxed{-163}$.

Thus $-32 + (-131) = -163$.

Use a calculator to perform each indicated operation.

1. $-256 + 97$

2. $811 + (-1058)$

3. $6(15) + (-46)$

4. $-129 + 10(48)$

5. $-108,650 + (-786,205)$

6. $-196,662 + (-129,856)$

Vocabulary, Readiness & Video Check

Use the choices below to fill in each blank. Not all choices will be used.

$-a$ a 0   commutative   associative

1. If n is a number, then $-n + n = $ _____.

2. Since $x + n = n + x$, we say that addition is _____.

3. If a is a number, then $-(-a) = $ _____.

4. Since $n + (x + a) = (n + x) + a$, we say that addition is _____.

Martin-Gay Interactive Videos   *Watch the section lecture video and answer the following questions.*

Objective A **5.** What is the sign of the sum in ⊞ Example 5 and why? ▶

Objective B **6.** What does the answer to ⊞ Example 9, −231, mean in the context of the application? ▶

See Video 10.2 🍎

10.2 Exercise Set MyMathLab® ▶

Objective A *Add using a number line. See Examples 1 and 2.*

▶**1.** $-1 + (-6)$

2. $-6 + (-5)$

▶**3.** $-4 + 7$

4. $10 + (-3)$

5. $-13 + 7$

6. $9 + (-4)$

Add. See Examples 3 through 15.

7. $23 + 12$ **8.** $15 + 42$ ▶**9.** $-6 + (-2)$ **10.** $-5 + (-4)$ **11.** $-43 + 43$

12. $-62 + 62$ ▶**13.** $6 + (-2)$ **14.** $8 + (-3)$ **15.** $-6 + 8$ **16.** $-8 + 12$

17. $3 + (-5)$ ▶**18.** $5 + (-9)$ **19.** $-2 + (-9)$ **20.** $-6 + (-1)$ **21.** $-7 + 7$

22. $-10 + 10$ **23.** $12 + (-5)$ **24.** $24 + (-10)$ **25.** $-6 + 3$ **26.** $-10 + 2$

27. $-12 + 3$ **28.** $-15 + 5$ **29.** $-12 + (-12)$ **30.** $-23 + (-23)$ **31.** $-25 + (-32)$

32. $-45 + (-90)$ **33.** $56 + (-26)$ **34.** $89 + (-37)$ **35.** $-37 + 57$ **36.** $-25 + 65$

37. $-123 + (-100)$ **38.** $-500 + (-230)$ **39.** $-42 + 193$ **40.** $-64 + 164$ **41.** $-6.3 + (-2.2)$

42. $-9.1 + (-4.6)$ ▶ **43.** $-10.7 + 15.3$ **44.** $-19.6 + 23.1$ **45.** $-\dfrac{7}{11} + \left(-\dfrac{1}{11}\right)$

46. $-\dfrac{9}{13} + \left(-\dfrac{1}{13}\right)$ ▶ **47.** $-\dfrac{2}{3} + \left(-\dfrac{1}{6}\right)$ **48.** $-\dfrac{3}{8} + \left(-\dfrac{1}{4}\right)$ **49.** $-\dfrac{4}{5} + \dfrac{1}{10}$

50. $-\dfrac{9}{16} + \dfrac{3}{8}$ **51.** $-2\dfrac{1}{4} + 7\dfrac{7}{8}$ **52.** $-4\dfrac{1}{5} + 6\dfrac{9}{10}$

Add. See Examples 16 and 17.

▶ **53.** $-4 + 2 + (-5)$ **54.** $-1 + 5 + (-8)$ **55.** $-5.2 + (-7.7) + (-11.7)$

56. $-10.3 + (-3.2) + (-2.7)$ **57.** $12 + (-4) + (-4) + 12$ **58.** $18 + (-9) + 5 + (-2)$

59. $(-10) + 14 + 25 + (-16)$ **60.** $34 + (-12) + (-11) + 213$

Objective B Translating *Solve. See Example 18.*

61. Find the sum of -8 and 25.

62. Find the sum of -30 and 10.

63. Find the sum of -31, -9, and 30.

64. Find the sum of -49, -2, and 40.

▶ **65.** Suppose a deep-sea diver dives from the surface to 215 feet below the surface. He then dives down 16 more feet. Use positive and negative numbers (or 0) to represent this situation. Then find the diver's present depth.

66. Suppose a diver dives from the surface to 248 meters below the surface and then swims up 6 meters, down 17 meters, down another 24 meters, and then up 23 meters. Use positive and negative numbers (or 0) to represent this situation. Then find the diver's depth after these movements.

In some card games, it is possible to have positive and negative scores. The table shows the scores for two teams playing a series of four games. Use this table to answer Exercises 67 and 68.

	Game 1	Game 2	Game 3	Game 4
Team 1	-2	-13	20	2
Team 2	5	11	-7	-3

67. Find each team's total score after four games. If the winner is the team with the greater score, find the winning team.

68. Find each team's total score after three games. If the winner is the team with the greater score, which team was winning after three games?

In golf, it is possible to have positive and negative scores. The following table shows the results of the eighteen-hole Round 1 for Inbee Park and Angela Stanford at the 2013 U.S. Women's Open in Southampton, NY. Use the table to answer Exercises 69 through 72.

Player/Hole	1	2	3	4	5	6	7	8	9	10	11	12	13	14	15	16	17	18
Park	−1	−1	0	−1	0	+1	0	−1	0	−1	0	0	0	−1	0	0	0	0
Stanford	+2	+1	0	0	0	−1	0	+1	0	−1	0	−1	0	0	0	0	0	0

(*Source:* Ladies' Professional Golf Association)

69. Find the score for each athlete after playing 9 holes.

70. In golf, the lower score is the winner. Use the result of Exercise 69 to determine who was winning after playing 9 holes of golf.

71. Find the total score for each athlete shown in Round 1.

72. In golf, the lower score is the winner. Use the result of Exercise 71 to determine who won Round 1.

The following bar graph shows the yearly net income for Amazon.com, Inc. Net income is one indication of a company's health. It measures revenue (money taken in) minus cost (money spent). Use this graph to answer Exercises 73 through 76. (Source: Amazon.com, Inc.)

73. What was the net income (in dollars) for Amazon.com, Inc. in 2012?

74. What was the net income (in dollars) for Amazon.com, Inc. in 2010?

75. Find the total net income for the years 2009 and 2010.

76. Find the total net income for all the years shown.

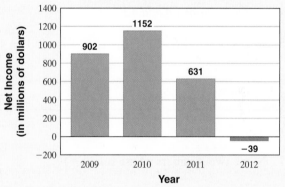

Amazon.com, Inc. Net Income

77. The temperature at 4 p.m. on February 2 was −10° Celsius. By 11 p.m. the temperature had risen 12 degrees. Find the temperature at 11 p.m.

78. In some card games, it is possible to have both positive and negative scores. After four rounds of play, Michelle had scores of 14, −5, −8, and 7. What was her total score for the game?

A small business company reports the following net incomes. Use this table to answer Exercises 79 and 80.

Year	Net Income (in dollars)
2006	−$10,412
2007	−$1786
2008	$15,395
2009	$31,418

79. Find the sum of the net incomes for 2007 and 2008.

80. Find the sum of the net incomes for all four years shown.

81. The all-time record low temperature for Wisconsin is −55°F. West Virginia's all-time record low temperature is 8°F higher than Wisconsin's record low. What is West Virginia's record low temperature? (*Source:* National Climatic Data Center)

82. The all-time record low temperature for Oklahoma is −27°F. In Georgia, the lowest temperature ever recorded is 10°F higher than Oklahoma's all-time low temperature. What is the all-time record low temperature for Georgia? (*Source:* National Climatic Data Center)

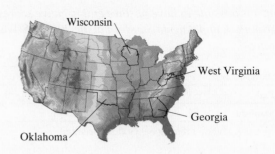

83. The deepest spot in the Pacific Ocean is the Mariana Trench, which has an elevation of 10,924 meters below sea level. The bottom of the Pacific's Aleutian Trench has an elevation 3245 meters higher than that of the Mariana Trench. Use a negative number to represent the depth of the Aleutian Trench. (*Source:* Defense Mapping Agency)

84. The deepest spot in the Atlantic Ocean is the Puerto Rico Trench, which has an elevation of 8605 meters below sea level. The bottom of the Atlantic's Cayman Trench has an elevation 1070 meters above the level of the Puerto Rico Trench. Use a negative number to represent the depth of the Cayman Trench. (*Source:* Defense Mapping Agency)

Review

Subtract. See Sections 1.4 and 4.3.

85. $44 - 0$ **86.** $91 - 0$ **87.** $76.1 - 4.09$ **88.** $93.7 - 10.08$ **89.** $200 - 59$ **90.** $400 - 18$

Concept Extensions

Solve.

91. Name two numbers whose sum is −17.

92. Name two numbers whose sum is −30.

Each calculation below is incorrect. Find the error and correct it. See the Concept Check in this section.

93. $7 + (-10) \overset{?}{=} 17$

94. $-4 + 14 \overset{?}{=} -18$

95. $-10 + (-12) \overset{?}{=} -120$

96. $-15 + (-17) \overset{?}{=} 32$

For Exercises 97 through 100, determine whether each statement is true or false.

97. The sum of two negative numbers is always a negative number.

98. The sum of two positive numbers is always a positive number.

99. The sum of a positive number and a negative number is always a negative number.

100. The sum of zero and a negative number is always a negative number.

101. In your own words, explain how to add two negative numbers.

102. In your own words, explain how to add a positive number and a negative number.

Subtracting Signed Numbers ⊙

In Section 10.1 we discussed the opposite of a number.

> The opposite of 3 is -3.
>
> The opposite of -6.7 is 6.7.
>
> Notice the pattern when we find the sum of these opposites.
>
> $3 + (-3) = 0$ and $-6.7 + 6.7 = 0$

This pattern is always true. That is, the sum of a number and its opposite is always 0. Because of this, the opposite of a number is also called the **additive inverse** of a number. In this section, we use opposites (or additive inverses) to subtract signed numbers.

Objectives

A Subtract Signed Numbers. ⊙

B Add and Subtract Signed Numbers. ⊙

C Solve Problems by Subtracting Signed Numbers. ⊙

Objective A Subtracting Signed Numbers ⊙

To subtract signed numbers, we write the subtraction problem as an addition problem. To see how we do this, study the examples below:

$$10 - 4 = 6$$
$$10 + (-4) = 6$$

Since both expressions simplify to 6, this means that

$$10 - 4 = 10 + (-4) = 6$$

Also,

$$3 - 2 = 3 + (-2) = 1$$
$$15 - 1 = 15 + (-1) = 14$$

Thus, to subtract two numbers, we add the first number to the opposite (also called the **additive inverse**) of the second number.

Subtracting Two Numbers

If a and b are numbers, then $a - b = a + (-b)$.

Examples Subtract.

subtraction	=	first number	+	opposite of the second number		
1. $8 - 5$	=	8	+	(-5)	=	3
2. $-4 - 10$	=	-4	+	(-10)	=	-14
3. $6 - (-5)$	=	6	+	5	=	11
4. $-11 - (-7)$	=	-11	+	7	=	-4

▨ Work Practice 1–4

Practice 1–4

Subtract.

1. $12 - 7$ **2.** $-6 - 4$

3. $11 - (-14)$ **4.** $-9 - (-1)$

Practice 5–9

Subtract.

5. $5 - 9$ **6.** $-12 - 4$

7. $-2 - (-7)$ **8.** $-10.5 - 14.3$

9. $\dfrac{5}{13} - \dfrac{12}{13}$

Examples Subtract.

5. $-10 - 5 = -10 + (-5) = -15$

6. $8 - 15 = 8 + (-15) = -7$

7. $-4 - (-5) = -4 + 5 = 1$

8. $-1.7 - 6.2 = -1.7 + (-6.2) = -7.9$

9. $-\dfrac{10}{11} - \left(-\dfrac{3}{11}\right) = -\dfrac{10}{11} + \dfrac{3}{11} = -\dfrac{7}{11}$

■ Work Practice 5–9

Helpful Hint

To visualize subtraction, try the following:

The difference between $5°F$ and $-2°F$ can be found by subtracting. That is,

$$5 - (-2) = 5 + 2 = 7$$

Can you visually see from the thermometer on the right that there are actually 7 degrees between $5°F$ and $-2°F$?

✓**Concept Check** What is wrong with the following calculation?

$$-7 - (-3) = -10$$

Practice 10

Subtract 5 from -10.

Example 10 Subtract 7 from -3.

Solution: To subtract 7 *from* -3, we find

$$-3 - 7 = -3 + (-7) = -10$$

■ Work Practice 10

Objective B Adding and Subtracting Signed Numbers

If a problem involves adding or subtracting more than two signed numbers, we rewrite differences as sums and add. Recall that by associative and commutative properties, we may add numbers in any order. In Examples 11 and 12, we will add from left to right.

Practice 11

Simplify: $-4 - 3 - 7 - (-5)$

Example 11 Simplify: $7 - 8 - (-5) - 1$

Solution:

$$
\begin{aligned}
7 - 8 - (-5) - 1 &= 7 + (-8) + 5 + (-1) \\
&= -1 + 5 + (-1) \\
&= 4 + (-1) \\
&= 3
\end{aligned}
$$

■ Work Practice 11

Answers

5. -4 **6.** -16 **7.** 5 **8.** -24.8

9. $-\dfrac{7}{13}$ **10.** -15 **11.** -9

✓**Concept Check Answer**

$-7 - (-3) = -4$

Example 12 Simplify: $7 + (-12) - 3 - (-8)$

Solution:

$$7 + (-12) - 3 - (-8) = \underbrace{7 + (-12)} + (-3) + 8$$
$$= \underbrace{-5 + (-3)} + 8$$
$$= \underbrace{-8 \qquad + 8}$$
$$= 0$$

◼ Work Practice 12

Objective C Solving Problems by Subtracting Signed Numbers ▶

Solving problems often requires subtraction of signed numbers.

Example 13 Finding a Change in Elevation

The highest point in the United States is the top of Mount McKinley, at a height of 20,237 feet above sea level. The lowest point is Death Valley, California, which is 282 feet below sea level. How much higher is Mount McKinley than Death Valley? (*Source:* U.S. Geological Survey)

Solution:

1. **UNDERSTAND.** Read and reread the problem. To find "how much higher," we subtract. Don't forget that since Death Valley is 282 feet *below* sea level, we represent its height by -282. Draw a diagram to help visualize the problem.

2. **TRANSLATE.**

In words:
how much higher is Mt. McKinley	=	height of Mt. McKinley	minus	height of Death Valley
↓	↓	↓	↓	↓

Translate:
how much higher is Mt. McKinley	=	20,237	−	(−282)

3. **SOLVE:**

$$20,237 - (-282) = 20,237 + 282 = 20,519$$

4. **INTERPRET.** Check and state your conclusion: Mount McKinley is 20,519 feet higher than Death Valley.

(Note: A new digital survey measured the elevation as 20,237 feet, somewhat lower than the 20,320 feet surveyed in 1952.)

◼ Work Practice 13

Practice 12
Simplify:
$3 + (-5) - 6 - (-4)$

Practice 13
The highest point in Asia is the top of Mount Everest, at a height of 29,028 feet above sea level. The lowest point is the Dead Sea, which is 1312 feet below sea level. How much higher is Mount Everest than the Dead Sea? (*Source:* National Geographic Society)

Answers
12. −4 **13.** 30,340 ft

Vocabulary, Readiness & Video Check

Multiple choice: Select the correct lettered response following each exercise.

1. It is true that $a - b =$ _____.
   a. $b - a$ **b.** $a + (-b)$ **c.** $a + b$

2. The opposite of *n* is _____.
   a. $-n$ **b.** $-(-n)$ **c.** n

3. The expression $-10 - (-14)$ equals _____.
   a. $10 - 14$ **b.** $-10 - 14$ **c.** $-14 - 10$ **d.** $-10 + 14$

4. The expression $-5 - 10$ equals _____.
   a. $5 - 10$ **b.** $5 + 10$ **c.** $-5 + (-10)$ **d.** $10 - 5$

Subtract.

5. $5 - 5$ **6.** $7 - 7$ **7.** $6.2 - 6.2$ **8.** $1.9 - 1.9$

Martin-Gay Interactive Videos   *Watch the section lecture video and answer the following questions.*

Objective A **9.** In ▥ Example 2, why did subtracting -8 become adding 8? ▶

Objective B **10.** In ▥ Example 7, if both subtractions were rewritten as equivalent additions in one step, what would the resulting expression be? ▶

Objective C **11.** What does the answer to ▥ Example 8, 263, mean in the context of the application? ▶

See Video 10.3 🍎

10.3 **Exercise Set** MyMathLab® ▶

Objective A *Subtract. See Examples 1 through 9.*

  1. $-5 - (-5)$ **2.** $-6 - (-6)$ **3.** $8 - 3$ **4.** $5 - 2$ ▶**5.** $3 - 8$

  6. $2 - 5$ **7.** $7 - (-7)$ **8.** $12 - (-12)$ ▶**9.** $-5 - (-8)$ **10.** $-25 - (-25)$

▶**11.** $-14 - 4$ **12.** $-2 - 42$ **13.** $2 - 16$ **14.** $8 - 9$ **15.** $2.2 - 5.5$

 16. $1.7 - 6.3$ **17.** $3.62 - (-0.4)$ **18.** $8.44 - (-0.2)$ **19.** $-\dfrac{3}{10} - \left(-\dfrac{7}{10}\right)$ **20.** $-\dfrac{2}{11} - \left(-\dfrac{5}{11}\right)$

▶**21.** $\dfrac{2}{5} - \dfrac{7}{10}$ **22.** $\dfrac{3}{16} - \dfrac{7}{8}$ **23.** $\dfrac{1}{2} - \left(-\dfrac{1}{3}\right)$ **24.** $\dfrac{1}{7} - \left(-\dfrac{1}{4}\right)$

Translating *Translate each phrase, then simplify. See Example 10.*

25. Subtract 18 from −20.

26. Subtract 10 from −22.

27. Find the difference of −20 and −3.

28. Find the difference of −8 and −13.

◐ 29. Subtract −11 from 2.

30. Subtract −50 from −50.

Mixed Practice (*Sections 10.2, 10.3*) *Add or subtract as indicated.*

31. −21 + (−17)

32. −35 + (−11)

33. 9 − 20

34. 7 − 30

35. −4.9 − 7.8

36. −10.5 − 6.8

37. $\frac{4}{7} + \left(-\frac{1}{7}\right)$

38. $\frac{9}{13} + \left(-\frac{5}{13}\right)$

Objective B *Simplify. See Examples 11 and 12.*

39. 7 − 3 − 2

40. 8 − 4 − 1

41. 12 − 5 − 7

42. 30 − 7 − 12

◐ 43. −5 − 8 − (−12)

44. −10 − 6 − (−9)

45. −10 + (−5) − 12

46. −15 + (−8) − 4

47. 12 − (−34) + (−6)

48. 23 − (−17) + (−9)

49. 19 − 14 + (−6) + (−50)

50. 28 − 16 + (−14) + (−1)

Objective C *Solve. See Example 13.*

The bar graph below shows the monthly average temperature in Fairbanks, Alaska. Notice that a negative temperature is illustrated by a bar below the horizontal line representing 0°F. Use this graph to answer Exercises 51 through 54.

51. Find the difference in temperature between the months of March and February.

52. Find the difference in temperature between the months of November and December.

53. Find the difference in temperature between the two months with the lowest temperatures.

54. Find the difference in temperature between the month with the warmest temperature and the month with the coldest temperature.

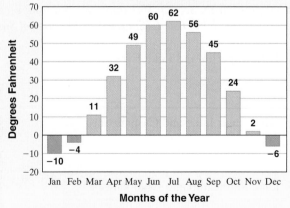

Monthly Average Temperatures in Fairbanks, AK

Source: National Climatic Data Center

55. The coldest temperature ever recorded on Earth was −129°F in Antarctica. The warmest temperature ever recorded was 134°F in Death Valley. How many degrees warmer is 134°F than −129°F? (*Source:* World Meteorological Organization) (*Note:* The former record of 136°F in the Sahara Desert has been disqualified.)

56. The coldest temperature ever recorded in the United States was −80°F in Alaska. The warmest temperature ever recorded was 134°F in California. How many degrees warmer is 134°F than −80°F? (*Source: The World Almanac*)

Solve.

57. Aaron Aiken has $125 in his checking account. He writes a check for $117, makes a deposit of $45, and then writes another check for $69. Find the balance in his account. (Write the amount as a signed number.)

58. A woman received a statement of her charge account at Old Navy. She spent $93 on purchases last month. She returned an $18 top because she didn't like the color. She also returned a $26 nightshirt because it was damaged. What does she actually owe on her account?

59. The temperature one morning is −4° Celsius at 6 a.m. If the temperature drops 3 degrees by 7 a.m., rises 4 degrees between 7 a.m. and 8 a.m., and then drops 7 degrees between 8 a.m. and 9 a.m., find the temperature at 9 a.m.

60. Mauna Kea in Hawaii has an elevation of 13,796 feet above sea level. The Mid-America Trench in the Pacific Ocean has an elevation of 21,857 feet below sea level. Find the difference in elevation between those two points. (*Source:* National Geographic Society and Defense Mapping Agency)

Some places on Earth lie below sea level, which is the average level of the surface of the oceans. Use this diagram to answer Exercises 61 through 64. (Source: Fantastic Book of Comparisons, Russell Ash)

61. Find the difference in elevation between Death Valley and Qattâra Depression.

62. Find the difference in elevation between the Danakil and Turfan Depressions.

63. Find the difference in elevation between the two lowest elevations shown.

64. Find the difference in elevation between the highest elevation shown and the lowest elevation shown.

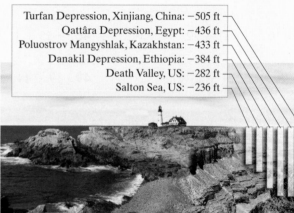

Turfan Depression, Xinjiang, China: −505 ft
Qattâra Depression, Egypt: −436 ft
Poluostrov Mangyshlak, Kazakhstan: −433 ft
Danakil Depression, Ethiopia: −384 ft
Death Valley, US: −282 ft
Salton Sea, US: −236 ft

The bar graph from Section 10.1 shows elevations of selected lakes. For Exercises 65 through 68, find the difference in elevation for the lakes listed. (Source: U.S. Geological Survey)

65. Lake Superior and Lake Eyre

66. Great Bear Lake and Caspian Sea

67. Lake Maracaibo and Lake Vanern

68. Lake Eyre and Caspian Sea

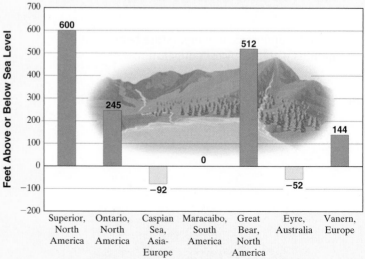

Elevations of Selected Lakes

Feet Above or Below Sea Level

700
600 — 600 (Superior, North America)
512 (Great Bear, North America)
500
400
300
245 (Ontario, North America)
200
144 (Vanern, Europe)
100
0 (Maracaibo, South America)
−92 (Caspian Sea, Asia-Europe)
−52 (Eyre, Australia)
−100
−200

Superior, North America | Ontario, North America | Caspian Sea, Asia-Europe | Maracaibo, South America | Great Bear, North America | Eyre, Australia | Vanern, Europe

Solve.

69. The average surface temperature of the hottest planet, Venus, is 867°F, while the average surface temperature of the coldest planet, Neptune, is −330°F. Find the difference in temperatures.

70. The average surface temperature of Mercury is 330°F, while the average surface temperature of Jupiter is −162°F. Find the difference in temperatures.

The difference between a country's exports and imports is called the country's trade balance.

71. In 2012, the United States had $2211 billion in exports and $2745 billion in imports. What was the U.S. trade balance in 2012? (*Source:* U.S. Census Bureau)

72. In 2012, the United States exported 25 million barrels of crude oil and imported 3121 million barrels of crude oil. What was the U.S. trade balance for crude oil in 2012? (*Source:* U.S. Energy Information Administration)

Review

Multiply. See Section 1.6.

73. $8 \cdot 0$

74. $0 \cdot 8$

75. $1 \cdot 8$

76. $8 \cdot 1$

77. $\begin{array}{r} 23 \\ \times\ 46 \\ \hline \end{array}$

78. $\begin{array}{r} 51 \\ \times\ 89 \\ \hline \end{array}$

Concept Extensions

Solve.

79. Name two numbers whose difference is −3.

80. Name two numbers whose difference is −10.

Each calculation below is incorrect. Find the error and correct it. See the Concept Check in this section.

81. $9 - (-7) \stackrel{?}{=} 2$

82. $-4 - 8 \stackrel{?}{=} 4$

83. $10 - 30 \stackrel{?}{=} 20$

84. $-3 - (-10) \stackrel{?}{=} -13$

Simplify. (Hint: Find the absolute values first.)

85. $|-3| - |-7|$

86. $|-12| - |-5|$

87. $|-5| - |5|$

88. $|-8| - |8|$

89. $|-15| - |-29|$

90. $|-23| - |-42|$

For Exercises 91 and 92, determine whether each statement is true or false.

91. $|-8 - 3| = 8 - 3$

92. $|-2 - (-6)| = |-2| - |-6|$

93. In your own words, explain how to subtract one signed number from another.

94. A student explains to you that the first step to simplify $8 + 12 \cdot 5 - 100$ is to add 8 and 12. Is the student correct? Explain why or why not.

Signed Numbers

Answers

1. _____

2. _____

3. _____

4. _____

5. _____

6. _____

7. _____

8. _____

9. _____

10. _____

11. _____

12. _____

13. _____

14. _____

15. _____

16. _____

17. _____

18. _____

19. _____

20. _____

21. _____

Represent each quantity by a signed number.

1. The peak of Mount Everest in Asia is 29,028 feet above sea level. (*Source: U.S. Geological Survey*)

2. The Mariana Trench in the Pacific Ocean is 35,840 feet below sea level. (*Source: The World Almanac,* 1998)

3. The deepest hole ever drilled in the Earth's crust is in Russia and its depth is over 7 miles below sea level. (*Source: Fantastic Book of Comparisons,* 1999)

Surface of the earth

Over 7 miles

4. The peak of K2 in Asia is 28,250 feet above sea level. (*Source:* National Geographic Society)

5. Graph the signed numbers on the given number line. $-4, 0, -1.5, 3\frac{1}{4}$

$$\xleftarrow{\hspace{0.3cm}}\overset{\begin{array}{ccccccccccc}-5&-4&-3&-2&-1&0&1&2&3&4&5\end{array}}{\rule{6cm}{0.4pt}}\xrightarrow{\hspace{0.3cm}}$$

Insert < or > between each pair of numbers to make a true statement.

6. 0 -3

7. 7   0

8. $-1\frac{1}{2}$ $1\frac{1}{4}$

9. -2 -7.6

10. -15 -5

11. -4 -40

12. 3.9 -3.9

13. $-\frac{1}{2}$ $-\frac{1}{3}$

Simplify.

14. $|-1|$

15. $|1|$

16. $|0|$

17. $-|-4|$

18. $|-8.6|$

19. $|100.3|$

20. $-\left(-\frac{3}{4}\right)$

21. $-\left|-\frac{3}{4}\right|$

Find the opposite of each number.

22. 6 **23.** −13 **24.** 89.1 **25.** $-\dfrac{2}{9}$

Add or subtract as indicated.

26. −7 + 12 **27.** −9.2 + (−11.6) **28.** $\dfrac{5}{9} + \left(-\dfrac{1}{3}\right)$

29. 1 − 3 **30.** $\dfrac{3}{8} - \left(-\dfrac{3}{8}\right)$ **31.** −2.6 − 1.4

32. −7 + (−2.6) **33.** −14 + 8 **34.** −8 − (−20)

35. 18 − (−102) **36.** 8.65 − 12.09

37. $-\dfrac{4}{5} - \dfrac{3}{10}$ **38.** −8 + (−6) + 20

39. −11 − 7 − (−19) **40.** −4 + (−8) − 16 − (−9)

Translate each phrase, then simplify.

41. Subtract 26 from 14. **42.** Subtract −8 from −12.

43. Find the sum of −17 and −27. **44.** Add −3.6 and 2.1

Solve.

45. The coldest temperature ever recorded in Europe was −73°F in Ust'Shchugor, Russia. The warmest temperature ever recorded in Europe was 118°F in Elefsina, Greece. How many degrees warmer is 118°F than −73°F? (*Source:* World Meteorological Organization)

46. The coldest temperature ever recorded in Asia was −90°F in Oimekon, Russia. The warmest temperature ever recorded in Asia was 129°F in Tirat Tsvi, Israel. How many degrees warmer is 129°F than −90°F? (*Source:* World Meteorological Organization)

22. _____

23. _____

24. _____

25. _____

26. _____

27. _____

28. _____

29. _____

30. _____

31. _____

32. _____

33. _____

34. _____

35. _____

36. _____

37. _____

38. _____

39. _____

40. _____

41. _____

42. _____

43. _____

44. _____

45. _____

46. _____

Multiplying and Dividing Signed Numbers ▶

Multiplying and dividing signed numbers is similar to multiplying and dividing whole numbers. One difference is that we need to determine whether the result is a positive number or a negative number.

Objectives

A Multiply Signed Numbers. ▶

B Divide Signed Numbers. ▶

C Solve Problems by Multiplying and Dividing Signed Numbers. ▶

Objective A Multiplying Signed Numbers ▶

Consider the following pattern of products:

First factor decreases by 1 each time.

$$3 \cdot 2 = 6$$
$$2 \cdot 2 = 4$$
$$1 \cdot 2 = 2$$
$$0 \cdot 2 = 0$$

Product decreases by 2 each time.

This pattern can be continued, as follows:

$$-1 \cdot 2 = -2$$
$$-2 \cdot 2 = -4$$
$$-3 \cdot 2 = -6$$

This suggests that the product of a negative number and a positive number is a negative number.

What is the sign of the product of two negative numbers? To find out, we form another pattern of products. Again, we decrease the first factor by 1 each time, but this time the second factor is negative.

$$2 \cdot (-3) = -6$$
$$1 \cdot (-3) = -3$$
$$0 \cdot (-3) = 0$$

Product increases by 3 each time.

This pattern continues as

$$-1 \cdot (-3) = 3$$
$$-2 \cdot (-3) = 6$$
$$-3 \cdot (-3) = 9$$

This suggests that the product of two negative numbers is a positive number. Thus, we can determine the sign of a product when we know the signs of the factors.

Multiplying Numbers

The product of two numbers having the same sign is a positive number.

The product of two numbers having different signs is a negative number.

Product of Like Signs

$$(+)(+) = +$$
$$(-)(-) = +$$

Product of Different Signs

$$(-)(+) = -$$
$$(+)(-) = -$$

Practice 1–5

Multiply.

1. $-2 \cdot 6$ **2.** $-4(-3)$

3. $0 \cdot (-10)$ **4.** $\left(\dfrac{3}{7}\right)\left(-\dfrac{1}{3}\right)$

5. $-4(-3.2)$

Answers

1. -12 **2.** 12 **3.** 0 **4.** $-\dfrac{1}{7}$ **5.** 12.8

Examples Multiply.

1. $-7 \cdot 3 = -21$ **2.** $-2(-5) = 10$

3. $0 \cdot (-4) = 0$ **4.** $\left(-\dfrac{1}{2}\right)\left(-\dfrac{2}{3}\right) = \dfrac{1 \cdot \overset{1}{\cancel{2}}}{\underset{1}{\cancel{2}} \cdot 3} = \dfrac{1}{3}$

5. $-3(1.2) = -3.6$

■ Work Practice 1–5

Recall that by the associative and commutative properties for multiplication, we may multiply numbers in any order that we wish. In Examples 6 and 7, we multiply from left to right.

Examples Multiply.

6. $7(-6)(-2) = -42(-2)$
$\qquad\qquad = 84$

7. $(-2)(-3)(-4) = 6(-4)$
$\qquad\qquad\quad = -24$

8. $(-1)(-2)(-3)(-4) = -1(-24)$ We have -24 from Example 7.
$\qquad\qquad\qquad\qquad = 24$

■ Work Practice 6–8

Practice 6–8

Multiply.
6. $7(-2)(-4)$
7. $(-5)(-6)(-1)$
8. $(-2)(-5)(-6)(-1)$

✓**Concept Check** What is the sign of the product of five negative numbers? Explain.

Helpful Hint

Have you noticed a pattern when multiplying signed numbers?

If we let $(-)$ represent a negative number and $(+)$ represent a positive number, then

$$(-)(-) = (+)$$
$$(-)(-)(-) = (-)$$
$$(-)(-)(-)(-) = (+)$$
$$(-)(-)(-)(-)(-) = (-)$$

The product of an even number of negative numbers is a positive result.

The product of an odd number of negative numbers is a negative result.

Recall from our study of exponents that $2^3 = 2 \cdot 2 \cdot 2 = 8$. We can now work with bases that are negative numbers. For example,

$$(-2)^3 = \underbrace{(-2)(-2)(-2)}_{3 \text{ factors of } -2} = -8$$

Example 9 Evaluate: $(-5)^2$

Solution: Remember that $(-5)^2$ means 2 factors of -5.

$$(-5)^2 = (-5)(-5) = 25$$

■ Work Practice 9

Practice 9

Evaluate: $(-3)^4$

Notice in Example 9 the parentheses around -5 in $(-5)^2$. With these parentheses, -5 is the base that is squared. Without parentheses, such as -7^2, only the 7 is squared, as shown next.

Example 10 Evaluate: -7^2

Solution: Remember that without parentheses, only the 7 is squared.

$$-7^2 = -(7 \cdot 7) = -49$$

■ Work Practice 10

Practice 10

Evaluate: -9^2

Answers
6. 56 **7.** -30 **8.** 60 **9.** 81 **10.** -81

✓**Concept Check Answer**
negative

Helpful Hint

Make sure you understand the difference between Examples 9 and 10.

→ parentheses, so −5 is squared

$$(-5)^2 = (-5)(-5) = 25$$

→ no parentheses, so only the 7 is squared

$$-7^2 = -(7 \cdot 7) = -49$$

Objective B   Dividing Signed Numbers ▶

Division of signed numbers is related to multiplication of signed numbers. The sign rules for division can be discovered by writing a related multiplication problem. For example,

$$\frac{6}{2} = 3 \qquad \text{because } 3 \cdot 2 = 6$$

$$\frac{-6}{2} = -3 \qquad \text{because } -3 \cdot 2 = -6$$

$$\frac{6}{-2} = -3 \qquad \text{because } -3 \cdot (-2) = 6$$

$$\frac{-6}{-2} = 3 \qquad \text{because } 3 \cdot (-2) = -6$$

Helpful Hint

Just as for whole numbers, division can be checked by multiplication.

Dividing Numbers

The quotient of two numbers having the same sign is a positive number.

The quotient of two numbers having different signs is a negative number.

Quotient of Like Signs

$$\frac{(+)}{(+)} = + \qquad \frac{(-)}{(-)} = +$$

Quotient of Different Signs

$$\frac{(+)}{(-)} = - \qquad \frac{(-)}{(+)} = -$$

Practice 11–14

Divide.

11. $\dfrac{28}{-7}$ **12.** $-18 \div (-2)$

13. $\dfrac{-4.6}{0.2}$ **14.** $\dfrac{3}{5} \div \left(-\dfrac{2}{7}\right)$

Examples   Divide.

11. $\dfrac{-12}{6} = -2$

12. $-20 \div (-4) = 5$

13. $\dfrac{1.2}{-0.6} = -2 \qquad 0.6\overline{)1.2}$

14. $\dfrac{7}{9} \div \left(-\dfrac{2}{5}\right) = \dfrac{7}{9} \cdot \left(-\dfrac{5}{2}\right) = -\dfrac{7 \cdot 5}{9 \cdot 2} = -\dfrac{35}{18} \text{ or } -1\dfrac{17}{18}$

■ Work Practice 11–14

Answers

11. −4   **12.** 9   **13.** −23

14. $-\dfrac{21}{10}$ or $-2\dfrac{1}{10}$

✓ **Concept Check Answer**

$\dfrac{1}{2} \div (-3) = \dfrac{1}{2} \cdot \left(-\dfrac{1}{3}\right) = -\dfrac{1}{6}$

✓ **Concept Check** Find the error in the following computation:

$$\frac{1}{2} \div (-3) = \frac{1}{2} \cdot \frac{1}{3} = \frac{1}{6}$$

Examples Divide, if possible.

15. $\dfrac{0}{-5} = 0$ because $0 \cdot -5 = 0$

16. $\dfrac{-7}{0}$ is undefined because there is no number that gives a product of -7 when multiplied by 0

■ Work Practice 15–16

Let's take a moment and notice a pattern from earlier work in this section.

$$\frac{-6}{2} = -3, \quad \frac{6}{-2} = -3, \quad \text{and} \quad -\frac{6}{2} = -3$$

Since $\dfrac{-6}{2}, \dfrac{6}{-2},$ and $-\dfrac{6}{2}$ all equal -3, then they must equal each other. Thus,

$$\frac{-6}{2} = \frac{6}{-2} = -\frac{6}{2}.$$

In general, this shows us three equivalent ways to write a negative fraction.

> The fraction $-\dfrac{a}{b} = \dfrac{-a}{b} = \dfrac{a}{-b}$

✓**Concept Check** Write $-\dfrac{4}{7}$ in two other equivalent ways.

Objective C Solving Problems by Multiplying and Dividing Signed Numbers ▶

Many real-life problems involve multiplication and division of signed numbers.

Example 17 Calculating a Total Golf Score

A professional golfer finished seven strokes under par (-7) for each of three days of a tournament. What was her total score for the tournament?

Solution:

1. UNDERSTAND. Read and reread the problem. Although the key word is "total," since this is repeated addition of the same number, we multiply.
2. TRANSLATE.

In words:
golfer's total score	=	number of days	·	score each day
↓	↓	↓	↓	↓

Translate:
| golfer's total | = | 3 | · | (-7) |

3. SOLVE: $3 \cdot (-7) = -21$
4. INTERPRET. Check and state your conclusion: The golfer's total score was -21, or 21 strokes under par.

■ Work Practice 17

Practice 15–16

Divide, if possible.

15. $\dfrac{-1}{0}$ **16.** $\dfrac{0}{-2}$

Practice 17

A card player had a score of -13 for each of four games. Find the total score.

Answers

15. undefined **16.** 0 **17.** -52

✓**Concept Check Answer**

$\dfrac{-4}{7}, \dfrac{4}{-7}$

Vocabulary, Readiness & Video Check

Use the choices below to fill in each blank. Choices may be used more than once.

negative   0
positive   undefined

1. The product of a negative number and a positive number is a(n) _____ number.
2. The product of two negative numbers is a(n) _____ number.
3. The quotient of two negative numbers is a(n) _____ number.
4. The quotient of a negative number and a positive number is a(n) _____ number.
5. The product of a negative number and zero is _____.
6. The quotient of 0 and a negative number is _____.
7. The quotient of a negative number and 0 is _____.

Martin-Gay Interactive Videos   *Watch the section lecture video and answer the following questions.*

Objective A 8. When multiplying two signed numbers in ▣ Examples 1–5, what was done first? ▶

Objective B 9. Complete this statement based on the lecture before ▣ Example 6: We can find out about sign rules for division because we know sign rules for _____. ▶

Objective C 10. In ▣ Example 11, what phrase told us that the example will involve a negative number? ▶

See Video 10.4 🍎

10.4 Exercise Set MyMathLab® ▶

Objective A *Multiply. See Examples 1 through 5.*

1. $-2(-3)$
2. $5(-3)$
▶ 3. $-4(9)$
4. $-7(-2)$
5. $(2.6)(-1.2)$

6. $-0.3(5.6)$
7. $0(-14)$
8. $-6(0)$
▶ 9. $-\dfrac{3}{5}\left(-\dfrac{2}{7}\right)$
10. $-\dfrac{3}{4}\left(\dfrac{9}{10}\right)$

Multiply. See Examples 6 through 8.

11. $6(-4)(2)$
12. $8(-3)(3)$
13. $-1(-2)(-4)$
14. $-2(-5)(-4)$

▶ 15. $-4(4)(-5)$
16. $-2(3)(-7)$
17. $10(-5)(-1)(-3)$
18. $2(-1)(3)(-2)$

Evaluate. See Examples 9 and 10.

19. $(-2)^2$

20. $(-2)^4$

21. -5^2

22. -3^2

23. $(-3)^3$

24. $(-1)^4$

25. -2^3

26. -1^3

27. $(-9)^2$

28. $(-4)^3$

29. $\left(-\dfrac{5}{11}\right)^2$

30. $\left(-\dfrac{3}{7}\right)^2$

Objective B *Divide. See Examples 11 through 16.*

31. $-24 \div 6$

32. $90 \div (-9)$

33. $\dfrac{-30}{6}$

34. $\dfrac{56}{-8}$

35. $\dfrac{-88}{-11}$

36. $\dfrac{-32}{4}$

37. $\dfrac{0}{14}$

38. $\dfrac{-13}{0}$

39. $\dfrac{39}{-3}$

40. $\dfrac{-24}{-12}$

41. $\dfrac{7.8}{-0.3}$

42. $\dfrac{1.21}{-1.1}$

43. $-\dfrac{5}{12} \div \left(-\dfrac{1}{5}\right)$

44. $-\dfrac{3}{8} \div \left(-\dfrac{2}{7}\right)$

45. $\dfrac{100}{-20}$

46. $\dfrac{240}{-40}$

47. $450 \div (-9)$

48. $480 \div (-8)$

49. $\dfrac{-12}{-4}$

50. $\dfrac{-36}{-3}$

51. $\dfrac{-120}{0.4}$

52. $\dfrac{-200}{2.5}$

53. $-\dfrac{8}{15} \div \dfrac{2}{3}$

54. $-\dfrac{1}{6} \div \dfrac{7}{18}$

Objectives A B Mixed Practice *Perform the indicated operation. See Examples 1 through 16.*

55. $-12(0)$

56. $0(-100)$

57. $(-5)^3$

58. $(-3)^2$

59. $\dfrac{430}{-8.6}$

60. $\dfrac{-423}{4.7}$

61. $-1(2)(7)(-3.1)$

62. $-2(3)(5)(-6.2)$

63. $\dfrac{4}{3}\left(-\dfrac{8}{9}\right)$

64. $-\dfrac{1}{9}\left(-\dfrac{1}{3}\right)$

65. $\dfrac{4}{3} \div \left(-\dfrac{8}{9}\right)$

66. $-\dfrac{1}{9} \div \left(-\dfrac{1}{3}\right)$

Objective C Translating *Translate each phrase, then simplify. See Example 17.*

67. Find the quotient of -54 and 9.

68. Find the quotient of -63 and -3.

69. Find the product of -42 and -6.

70. Find the product of -49 and 5.

Solve. See Example 17.

71. A football team lost four yards on each of three consecutive plays. Represent the total loss as a product of signed numbers and find the total loss.

72. Joe Norstrom lost $400 on each of seven consecutive days in the stock market. Represent his total loss as a product of signed numbers and find his total loss.

73. A deep-sea diver must move up or down in the water in short steps in order to keep from getting a physical condition called the "bends." Suppose a diver moves down from the surface in five steps of 20 feet each. Represent his total movement as a product of signed numbers and find the product.

74. A weather forecaster predicts that the temperature will drop five degrees each hour for the next six hours. Represent this drop as a product of signed numbers and find the total drop in temperature.

The graph shows melting points in degrees Celsius of selected elements. Use this graph to answer Exercises 75 through 78.

75. The melting point of nitrogen is 3 times the melting point of radon. Find the melting point of nitrogen.

76. The melting point of rubidium is −1 times the melting point of mercury. Find the melting point of rubidium.

77. The melting point of argon is −3 times the melting point of potassium. Find the melting point of argon.

78. The melting point of strontium is −11 times the melting point of radon. Find the melting point of strontium.

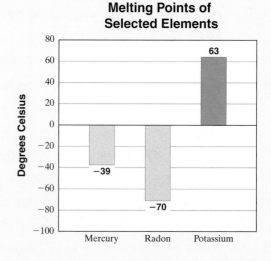

Melting Points of Selected Elements

79. During a quarter of 2013, Eastman Kodak posted a net income of −$224 million. If this continued, what would Eastman Kodak's net income have been after four quarters? (*Source:* EastmanKodak)

80. During a quarter of 2013, Amazon.com posted a net income of −$41 million. If this continued, what would Amazon.com's net income have been after four quarters? (*Source:* Amazon.com, Inc.)

81. In 1987, there were only 27 California condors in the entire world. Thanks to conservation efforts, in 2013 there were 420 California condors. (*Source:* California Department of Fish and Game)
  a. Find the change in the numbers of California condors from 1987 to 2013.
  b. Find the average change per year in the California condor population over the period in part a. Round to the nearest whole.

82. In November 2012, the average price for a gallon of regular gasoline in the United States was $3.492. In November 2013, this had dropped to $3.265 per gallon. (*Source:* U.S. Energy Information Administration)
  a. Find the change in the average price per gallon of regular gasoline from November 2012 to November 2013.
  b. Find the average change per month in the price of a gallon of regular gasoline over this 12-month period. Round to the nearest cent.

Review

Perform each indicated operation. See Section 1.9.

83. $(11 - 8) \cdot (200 - 100)$

84. $(2 \cdot 3)^2$

85. $(3 \cdot 5)^2$

86. $(12 - 3) \cdot (18 - 10)$

87. $90 + 12^2 - 5^3$

88. $3 \cdot (7 - 4) + 2 \cdot 5^2$

89. $12 \div 4 - 2 + 7$

90. $12 \div (4 - 2) + 7$

Concept Extensions

Mixed Practice (Sections 10.2, 10.3, 10.4) *Perform the indicated operations.*

91. $-87 \div 3$

92. $-9(-10)$

93. $-9 - 10$

94. $-4 + (-3) + 21$

95. $-4 - 15 - (-11)$

96. $-16 - (-2)$

In Exercises 97 through 100, determine whether each statement is true or false.

97. The product of two negative numbers is always a negative number.

98. The product of a positive number and a negative number is always a negative number.

99. The quotient of two negative numbers is always a positive number.

100. The quotient of zero and a negative number is always zero.

Solve. For Exercises 101 and 102, see the first Concept Check in this section.

101. What is the sign of the product of seven negative numbers?

102. What is the sign of the product of ten negative numbers?

For Exercises 103 and 104, write the fractions in two other equivalent ways by different placements of the negative sign. See the third Concept Check in this section.

103. $-\dfrac{20}{41}$

104. $-\dfrac{8}{17}$

Without actually finding the product, write the list of numbers in Exercises 105 and 106 in order from least to greatest. For help, see a Helpful Hint box in this section.

105. $(-2)^{12}, (-2)^{17}, (-5)^{12}, (-5)^{17}$

106. $(-1)^{50}, (-1)^{55}, 0^{15}, (-7)^{20}, (-7)^{23}$

107. In your own words, explain how to divide two signed numbers.

108. In your own words, explain how to multiply two signed numbers.

Objectives

A Simplify Expressions by Using the Order of Operations.

B Find the Average of a List of Numbers.

Objective A Simplifying Expressions

We first discussed the order of operations in Chapter 1. In this section, you are given an opportunity to practice using the order of operations when expressions contain signed numbers. The rules for the order of operations from Section 1.9 are repeated here.

Order of Operations

1. Perform all operations within parentheses ( ), brackets [ ], or other grouping symbols such as square roots or fraction bars.
2. Evaluate any expressions with exponents.
3. Multiply or divide in order from left to right.
4. Add or subtract in order from left to right.

Practice 1

Simplify: $20 + 50 + (-4)^3$

Example 1 Simplify: $60 + 30 + (-2)^3$

Solution: $60 + 30 + (-2)^3 = 60 + 30 + (-8)$ Write $(-2)^3$ as -8.

$$= 90 + (-8) \qquad \text{Add from left to right.}$$
$$= 82$$

■ Work Practice 1

Practice 2

Simplify: $\dfrac{25}{5(-1)}$

Example 2 Simplify: $\dfrac{-6(2)}{-3}$

Solution: First we multiply -6 and 2. Then we divide.

$$\frac{-6(2)}{-3} = \frac{-12}{-3}$$
$$= 4$$

■ Work Practice 2

Practice 3

Simplify: $\dfrac{-18 + 6}{-3 - 1}$

Example 3 Simplify: $\dfrac{12 - 16}{-1 + 3}$

Solution: We simplify above and below the fraction bar separately. Then we divide.

$$\frac{12 - 16}{-1 + 3} = \frac{-4}{2}$$
$$= -2$$

■ Work Practice 3

Practice 4

Simplify: $-2^3 + (-4)^2 + 1^5$

Example 4 Simplify: $-4^2 + (-3)^2 - 1^3$

Solution:

$$-4^2 + (-3)^2 - 1^3 = -16 + 9 - 1 \qquad \text{Simplify expressions with exponents.}$$
$$= -7 - 1 \qquad \text{Add or subtract from left to right.}$$
$$= -8$$

■ Work Practice 4

Answers

1. 6 **2.** −5 **3.** 3 **4.** 9

Example 5 Simplify: $3(4 - 7) + (-2) - \sqrt{25}$

Solution:

$$3(4 - 7) + (-2) - \sqrt{25} = 3(-3) + (-2) - 5 \quad \text{Simplify inside parentheses and replace } \sqrt{25} \text{ with 5.}$$
$$= -9 + (-2) - 5 \quad \text{Multiply.}$$
$$= -11 - 5 \quad \text{Add or subtract from left to right.}$$
$$= -16$$

■ Work Practice 5

Practice 5

Simplify:
$2(2 - 8) + (-12) - \sqrt{9}$

Example 6 Simplify: $(-3) \cdot |-5| - (-2) + 4^2$

Solution:

$$(-3) \cdot |-5| - (-2) + 4^2 = (-3) \cdot 5 - (-2) + 4^2 \quad \text{Write } |-5| \text{ as 5.}$$
$$= (-3) \cdot 5 - (-2) + 16 \quad \text{Write } 4^2 \text{ as 16.}$$
$$= -15 - (-2) + 16 \quad \text{Multiply.}$$
$$= -13 + 16 \quad \text{Add or subtract from left to right.}$$
$$= 3$$

■ Work Practice 6

Practice 6

Simplify:
$(-5) \cdot |-4| + (-3) + 2^3$

Example 7 Simplify: $-2[-3 + 2(-1 + 6)] - 5.3$

Solution: Here we begin with the innermost set of parentheses.

$$-2[-3 + 2(-1 + 6)] - 5.3 = -2[-3 + 2(5)] - 5.3 \quad \text{Write } -1 + 6 \text{ as 5.}$$
$$= -2[-3 + 10] - 5.3 \quad \text{Multiply.}$$
$$= -2(7) - 5.3 \quad \text{Add.}$$
$$= -14 - 5.3 \quad \text{Multiply.}$$
$$= -19.3 \quad \text{Subtract.}$$

■ Work Practice 7

Practice 7

Simplify:
$-4[-2 + 5(-3 + 5)] - 7.9$

✔**Concept Check** True or false? Explain your answer. The result of

$$-4(3 - 7) - 8(9 - 6)$$

is positive because there are four negative signs.

Example 8 Simplify: $\left(\dfrac{1}{6} - \dfrac{5}{6}\right) \cdot \dfrac{3}{7}$

Solution:

$$\left(\frac{1}{6} - \frac{5}{6}\right) \cdot \frac{3}{7} = -\frac{2}{3} \cdot \frac{3}{7} \qquad \frac{1}{6} - \frac{5}{6} = -\frac{4}{6} = -\frac{2}{3}$$
$$= -\frac{2 \cdot \overset{1}{\cancel{3}}}{\underset{1}{\cancel{3}} \cdot 7} \qquad \text{Multiply.}$$
$$= -\frac{2}{7} \qquad \text{Simplify.}$$

■ Work Practice 8

Practice 8

Simplify: $\dfrac{5}{8} \div \left(\dfrac{1}{4} - \dfrac{3}{4}\right)$

Answers

5. -27 **6.** -15 **7.** -39.9 **8.** $-\dfrac{5}{4}$

✔**Concept Check Answer**

false; $-4(3 - 7) - 8(9 - 6) = -8$

Objective B Review Finding Averages ▶

Recall from Chapter 1 that the average of a list of numbers is

$$\text{average} = \frac{\text{sum of numbers}}{\textit{number} \text{ of numbers}}$$

Practice 9

Find the average of the temperatures for the months October through April.

Example 9 The graph shows some monthly normal temperatures for Barrow, Alaska. Use this graph to find the average of the temperatures for the months January through May.

Monthly Normal Temperatures for Barrow, Alaska

Solution: By reading the graph, we have

$$\text{average} = \frac{-14 + (-16) + (-14) + (-1) + 20}{5}$$

There are 5 months from January through May.

$$= \frac{-25}{5}$$

$$= -5$$

The average of the temperatures is $-5°F$.

■ Work Practice 9

Answer

9. $-6°F$

📱 Calculator Explorations Simplifying an Expression Containing a Fraction Bar

Recall that even though most calculators follow the order of operations, parentheses must sometimes be inserted. For example, to simplify $\dfrac{-8 + 6}{-2}$ on a calculator, enter parentheses about the expression above the fraction bar so that it is simplified separately.

To simplify $\dfrac{-8 + 6}{-2}$, press the keys

[( ] [8] [+/−] [+] [6] [ )] [÷] [2] [+/−] [=] or
[( ] [(−)] [8] [+] [6] [ )] [÷] [(−)] [2] [ENTER].

The display will read ▢ 1 ▢.

Thus, $\dfrac{-8 + 6}{-2} = 1$.

Use a calculator to simplify.

1. $\dfrac{-120 - 360}{-10}$

2. $\dfrac{4750}{-2 + (-17)}$

3. $\dfrac{-316 + (-458)}{28 + (-25)}$

4. $\dfrac{-234 + 86}{-18 + 16}$

Vocabulary, Readiness & Video Check

Use the choices below to fill in each blank.

average   subtraction   division
addition   multiplication

1. To simplify $-2 \div 2 \cdot (3)$, which operation should be performed first? _____

2. To simplify $-9 - 3 \cdot 4$, which operation should be performed first? _____

3. The _____ of a list of numbers is $\dfrac{sum\ of\ numbers}{number\ of\ numbers}$.

4. To simplify $5[-9 + (-3)] \div 4$, which operation should be performed first? _____

5. To simplify $-2 + 3(10 - 12) \cdot (-8)$, which operation should be performed first? _____

Martin-Gay Interactive Videos   *Watch the section lecture video and answer the following questions.*

Objective A 6. In ⊟ Example 3, what two roles does the fraction bar play? ▶

Objective B 7. From the lecture before ⊟ Example 8, explain why finding the average is a good example of an application for this section. ▶

See Video 10.5

10.5 Exercise Set MyMathLab® ▶

Objective A *Simplify. See Examples 1 through 8.*

▶**1.** $-1(-2) + 1$

2. $3 + (-8) \div 2$

3. $\dfrac{-8(-5)}{10}$

4. $\dfrac{-9(-6)}{18}$

5. $9 - 12 - 4$

6. $10 - 23 - 12$

7. $4 + 3(-6)$

8. $8 + 4(-3)$

▶**9.** $\dfrac{4}{9}\left(\dfrac{2}{10} - \dfrac{7}{10}\right)$

10. $\dfrac{2}{5}\left(\dfrac{3}{8} - \dfrac{4}{8}\right)$

11. $-10 + 4 \div 2$

12. $-12 + 6 \div 3$

13. $25 \div (-5) + \sqrt{81}$

14. $28 \div (-7) + \sqrt{25}$

15. $\dfrac{-|-5| + 3}{7 - 9}$

16. $\dfrac{-|-11| + 7}{7 - 9}$

▶**17.** $\dfrac{24}{10 + (-4)}$

18. $\dfrac{88}{-8 - 3}$

19. $5(-3) - (-12)$

20. $7(-4) - (-6)$

21. $-19 - 12(3)$

22. $-24 - 14(2)$

23. $\dfrac{-23 + 7}{14 + (-10)}$

24. $\dfrac{-27 + 6}{13 + (-10)}$

25. $[8 + (-4)]^2$ **26.** $[9 + (-2)]^3$ ◗**27.** $3^3 - 12$ **28.** $5^2 - 100$

29. $(3 - 12) \div 3$ **30.** $(12 - 19) \div 7$ **31.** $5 + 2^3 - 4^2$ **32.** $12 + 5^2 - 2^4$

33. $(5 - 9)^2 \div (4 - 2)^2$ **34.** $(2 - 7)^2 \div (4 - 3)^4$ ◗**35.** $|8 - 24| \cdot (-2) \div (-2)$

36. $|3 - 15| \cdot (-4) \div (-16)$ **37.** $(-12 - 20) \div 16 - 25$ **38.** $(-20 - 5) \div 5 - 15$

39. $5(5 - 2) + (-5)^2 - \sqrt{36}$ **40.** $3(8 - 3) + (-4)^2 - \sqrt{100}$ **41.** $(0.2 - 0.7)(0.6 - 1.9)$

42. $(0.4 - 1.2)(0.8 - 1.7)$ ◗**43.** $2 - 7 \cdot 6 - 19$ **44.** $4 - 12 \cdot 8 - 17$

45. $(-36 \div 6) - (4 \div 4)$ **46.** $(-4 \div 4) - (8 \div 8)$ **47.** $\left(\dfrac{1}{2}\right)^2 - \left(\dfrac{1}{3}\right)^2$

48. $\left(\dfrac{1}{4}\right)^2 - \left(\dfrac{1}{2}\right)^2$ **49.** $(-5)^2 - 6^2$ **50.** $(-4)^4 - (5)^4$

51. $(10 - 4^2)^2$ **52.** $(11 - 3^2)^3$ **53.** $2(8 - 10)^2 - 5(1 - 6)^2$

54. $-3(4 - 8)^2 + 5(14 - 16)^3$ **55.** $3(-10) \div [5(-3) - 7(-2)]$ **56.** $12 - [7 - (3 - 6)] + (2 - 3)^3$

◗**57.** $\dfrac{(-7)(-3) - 4(3)}{3[7 \div (3 - 10)]}$ **58.** $\dfrac{10(-1) - (-2)(-3)}{2[-8 \div (-2 - 2)]}$ **59.** $(0.2)^2 - (1.5)^2$

60. $(1.3)^2 - (2.2)^2$ **61.** $-3^2 + (-2)^3 - 4^2$ **62.** $-4^2 + (-3)^2 - 2^2$

63. $\dfrac{|8 - 13| + \sqrt{81}}{-5(3) - (-5)}$ **64.** $\dfrac{|10 - 6| + \sqrt{36}}{-4(5) - (-12)}$

Objective B *Find the average of each list of numbers. See Example 9.*

▶ **65.** −10, 8, −4, 2, 7, −5, −12

66. −18, −8, −1, −1, 0, 4

67. −17, −26, −20, −13

68. −40, −20, −10, −15, −5

Let's use the graph from Example 9, which shows some monthly normal temperatures for Barrow, Alaska. Use this for Exercises 69 through 72.

Monthly Normal Temperatures for Barrow, Alaska

69. Find the average of the temperatures for the months December through March.

70. Find the average of the temperatures for the months October through January.

71. Can the average of all these temperatures be greater than the highest temperature, 20? Explain why or why not.

72. Can the average of all these temperatures be less than the lowest temperature, −16? Explain why or why not.

Review

Perform each indicated operation. See Sections 1.3, 1.4, 1.6, and 1.7.

73. 45 · 90

74. 90 ÷ 45

75. 90 − 45

76. 45 + 90

Find the perimeter of each figure. See Section 8.3.

△ **77.** Square — 8 in.

△ **78.** Parallelogram — 5 cm, 3 cm

△ **79.** Rectangle — 6 ft, 9 ft

△ **80.** Triangle — 17 m, 23 m, 32 m

Concept Extensions

Insert parentheses where needed so that each expression evaluates to the given number.

81. 2 · 7 − 5 · 3; evaluates to 12

82. 7 · 3 − 4 · 2; evaluates to 34

83. −6 · 10 − 4; evaluates to −36

84. 2 · 8 ÷ 4 − 20; evaluates to −36

85. Are parentheses necessary in the expression $3 + (4 \cdot 5)$? Explain your answer.

86. Are parentheses necessary in the expression $(3 + 4) \cdot 5$? Explain your answer.

87. Discuss the effect parentheses have in an exponential expression. For example, what is the difference between $(-6)^2$ and -6^2?

88. Discuss the effect parentheses have in an exponential expression. For example, what is the difference between $(2 \cdot 4)^2$ and $2 \cdot 4^2$?

Evaluate.

89. $(-12)^4$

90. $(-17)^6$

Chapter 10 Group Activity

Magic Squares

Sections 10.1–10.3

A magic square is a set of numbers arranged in a square table so that the sum of the numbers in each column, row, and diagonal is the same. For instance, in the magic square below, the sum of each column, row, and diagonal is 15. Notice that no number is used more than once in the magic square.

2	9	4
7	5	3
6	1	8

The properties of magic squares have been known for a very long time and once were thought to be good luck charms. The ancient Egyptians and Greeks understood their patterns. A magic square even made it into a famous work of art. The engraving titled *Melencolia I*, created by German artist Albrecht Dürer in 1514, features the following four-by-four magic square on the building behind the central figure.

16	3	2	13
5	10	11	8
9	6	7	12
4	15	14	1

Exercises

1. Verify that what is shown in the Dürer engraving is, in fact, a magic square. What is the common sum of the columns, rows, and diagonals?

2. Negative numbers can also be used in magic squares. Complete the following magic square:

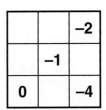

3. Use the numbers $-16, -12, -8, -4, 0, 4, 8, 12,$ and 16 to form a magic square:

Chapter 10 Vocabulary Check

Fill in each blank with one of the words or phrases listed below.

signed   positive   opposites   negative   absolute value   additive inverse

integers   inequality symbols   average   is less than   is greater than

1. Two numbers that are the same distance from 0 on a number line but are on opposite sides of 0 are called _____.

2. Together, positive numbers, negative numbers, and 0 are called _____ numbers.

3. The _____ of a number is that number's distance from 0 on a number line.

4. The _____ are . . . , −3, −2, −1, 0, 1, 2, 3, ...

5. The opposite of a number is also called its _____.

6. The _____ numbers are numbers less than zero.

7. The _____ numbers are numbers greater than zero.

8. The symbols "<" and ">" are called _____.

9. The _____ of a list of numbers is $\dfrac{sum\ of\ numbers}{number\ of\ numbers}$.

10. The sign "<" means _____ and ">" means _____.

> **Helpful Hint**
>
> ▶ Are you preparing for your test? Don't forget to take the Chapter 10 Test on page 678. Then check your answers at the back of the text and use the Chapter Test Prep Videos to see the fully worked-out solutions to any of the exercises you want to review.

10 Chapter Highlights

Definitions and Concepts	Examples
Section 10.1 Signed Numbers	
Together, positive numbers, negative numbers, and 0 are called **signed numbers.**	−432, −10, 0, 15
The **integers** are . . . , −3, −2, −1, 0, 1, 2, 3, ...	
The **absolute value** of a number is that number's distance from 0 on a number line. The symbol for absolute value is | |.	$\|-2\| = 2$ $\|2\| = 2$
Two numbers that are the same distance from 0 on a number line but are on opposite sides of 0 are called **opposites.**	5 and −5 are opposites.
If a is a number, then $-(-a) = a$.	$-(-11) = 11$. Do not confuse with $-\|-3\| = -3$

Definitions and Concepts	Examples

Section 10.2   Adding Signed Numbers

Adding Two Numbers with the Same Sign

Step 1:   Add their absolute values.

Step 2:   Use their common sign as the sign of the sum.

Adding Two Numbers with Different Signs

Step 1:   Find the larger absolute value minus the smaller absolute value.

Step 2:   Use the sign of the number with the larger absolute value as the sign of the sum.

Add:

$$-2 + (-3) = -5$$
$$-7 + (-15) = -22$$
$$-1.2 + (-5.7) = -6.9$$
$$-6 + 4 = -2$$
$$17 + (-12) = 5$$
$$-\frac{4}{11} + \frac{1}{11} = -\frac{3}{11}$$
$$-32 + (-2) + 14 = -34 + 14$$
$$= -20$$

Section 10.3   Subtracting Signed Numbers

Subtracting Two Numbers

If a and b are numbers, then $a - b = a + (-b)$.

Subtract:

$$-35 - 4 = -35 + (-4) = -39$$
$$3 - 8 = 3 + (-8) = -5$$
$$-7.8 - (-10.2) = -7.8 + 10.2 = 2.4$$
$$7 - 20 - 18 - (-3) = 7 + (-20) + (-18) + 3$$
$$= -13 + (-18) + 3$$
$$= -31 + 3$$
$$= -28$$

Section 10.4   Multiplying and Dividing Signed Numbers

Multiplying Signed Numbers

The product of two numbers having the same sign is a positive number.

The product of two numbers having different signs is a negative number.

Multiply:

$$(-7)(-6) = 42$$
$$9(-4) = -36$$
$$-3(0.7) = -2.1$$

Evaluate:

$$(-3)^2 = (-3)(-3) = 9$$
$$\left(-\frac{2}{3}\right)^2 = \left(-\frac{2}{3}\right)\left(-\frac{2}{3}\right) = \frac{4}{9}$$
$$-3^2 = -(3 \cdot 3) = -9$$

Dividing Signed Numbers

The quotient of two numbers having the same sign is a positive number.

The quotient of two numbers having different signs is a negative number.

Divide:

$$-100 \div (-10) = 10$$
$$\frac{14}{-2} = -7, \frac{-3.6}{-0.3} = 12, \frac{0}{-3} = 0, \frac{22}{0} \text{ is undefined.}$$

Definitions and Concepts	Examples
Section 10.5 Order of Operations	

Definitions and Concepts	Examples
Order of Operations 1. Perform all operations within parentheses ( ), brackets [ ], or other grouping symbols such as square roots or fraction bars. 2. Evaluate any expressions with exponents. 3. Multiply or divide in order from left to right. 4. Add or subtract in order from left to right.	Simplify: $$3 + 2 \cdot (-5) = 3 + (-10)$$ $$= -7$$ $$\frac{-2(5-7)}{-7 + |-3|} = \frac{-2(-2)}{-7 + 3}$$ $$= \frac{4}{-4}$$ $$= -1$$

Chapter 10 Review

(10.1) *Represent each quantity by a signed number.*

1. A gold miner is working 1435 feet below the surface of Earth.

2. A mountain peak is 7562 meters above sea level.

Graph each number on a number line.

3. $-2, 4, -3.5, 0$

4. $-7, -1.6, 3, -4\frac{1}{3}$

Insert $<$ or $>$ between each pair of numbers to form a true statement.

5. $-18 \quad -20$

6. $-5\frac{1}{3} \quad -4\frac{7}{8}$

7. $-12.3 \quad -19.8$

Find each absolute value.

8. $|-12|$

9. $|0|$

10. $\left|-\frac{7}{8}\right|$

Find the opposite of each number.

11. -12

12. $\frac{1}{2}$

Simplify.

13. $-(-3.9)$

14. $-|-7|$

Determine whether each statement is true or false.

15. A negative number is always less than a positive number.

16. The absolute value of a number is always 0 or a positive number.

(10.2) *Add.*

17. $5 + (-3)$

18. $18 + (-4)$

19. $-12 + 16$

20. $-23 + 40$

21. $-8 + (-15)$

22. $-5 + (-17)$

23. $-2.4 + 0.3$

24. $-8.9 + 1.9$

25. $\frac{2}{3} + \left(-\frac{2}{5}\right)$

26. $-\frac{8}{9} + \frac{1}{3}$

27. $-43 + (-108)$

28. $-100 + (-506)$

29. The temperature at 5 a.m. on a day in January was $-15°$ Celsius. By 6 a.m. the temperature had fallen 5 degrees. Use a signed number to represent the temperature at 6 a.m.

30. A diver starts out at 127 feet below the surface and then swims downward another 23 feet. Use a signed number to represent the diver's current depth.

31. During the 2013 Canadian Women's Open golf tournament, the winner, Lydia Ko, had scores of -5, -1, -3, and -6 over four rounds of golf. What was her total score for the tournament? (*Source*: Ladies Professional Golf Association)

32. During the 2013 Shriners Hospitals for Children Open golf tournament, the winner, Webb Simpson, had a score of -24. The fourth-place finisher, Charley Hoffman, had a score that was 7 points more than the winning score. What was Charley Hoffman's score in the 2013 Shriners Hospitals for Children Open? (*Source:* Professional Golf Association)

(10.3) *Subtract.*

33. $4 - 12$

34. $-12 - 14$

35. $-\frac{3}{4} - \frac{1}{2}$

36. $-\frac{2}{5} - \frac{7}{10}$

37. $7 - (-13)$

38. $-6 - (-14)$

39. $-(-5) - 12 - (-3)$

40. $-16 - 16 - (-4) - (-12)$

41. $-1.7 - (-2.9)$

42. $-0.5 - (-1.2)$

43. $\frac{4}{9} - \frac{7}{9} - \left(-\frac{1}{18}\right)$

44. $\frac{3}{7} - \frac{5}{7} - \left(-\frac{1}{14}\right)$

45. Josh Weidner has $142 in his checking account. He writes a check for $125, makes a deposit for $43, and then writes another check for $85. Represent the balance in his account by a signed number.

46. If the elevation of Lake Superior is 600 feet above sea level and the elevation of the Caspian Sea is 92 feet below sea level, find the difference in elevation. Represent the difference by a signed number.

(10.4) *Perform the indicated operations.*

47. $(-11)^2$

48. $(-3)^4$

49. -3^4

50. -11^2

51. $\left(-\frac{4}{5}\right)^3$

52. $\left(-\frac{2}{3}\right)^3$

53. $(-3)(-7)$

54. $-6 \cdot 3$

55. $\frac{-78}{3}$

56. $\frac{38}{-2}$

57. $(-4)(-1)(-9)$

58. $(-2)(-2)(-10)$

59. $-\frac{2}{3} \cdot \frac{7}{8}$

60. $-0.5(-12)$

61. $-2 \div 0$

62. $0 \div (-61)$

63. $\frac{-0.072}{-0.06}$

64. $-\frac{8}{25} \div \left(-\frac{2}{5}\right)$

(10.5) *Find the average of each list of numbers.*

65. $94, -85, 77, 0, 30,$ and -26

66. $-20, -17, 5, 9,$ and -12

Simplify.

67. $-10 + 3(-2)$

68. $3(-12) - \sqrt{64}$

69. $5.2 + \sqrt{36} \div (-0.3)$

70. $-6 + (-10) \div (-0.2)$

71. $|-16| + (-3) \cdot 12 \div 4$

72. $-(-4) \cdot |-3| - 5$

73. $4^3 - (8 - 3)^2$

74. $2^3 - (9 - 2)^2$

75. $\left(-\dfrac{1}{3}\right)^2 - \dfrac{8}{3}$

76. $\left(-\dfrac{1}{10}\right)^2 - \dfrac{9}{10}$

77. $\dfrac{(-4)(-3) - (-2)(-1)}{-10 + 5}$

78. $\dfrac{4(12 - 18)}{-10 \div (-2 - 3)}$

Mixed Review

Perform the indicated operations.

79. $(-4)^2$

80. -4^2

81. $-6 + (-9)$

82. $-16 - 3$

83. $-4(-12)$

84. $\dfrac{84}{-4}$

85. $-7.6 - (-9.7)$

86. $-\dfrac{9}{20} + \dfrac{4}{20}$

87. $-\dfrac{15}{14} \div \left(-\dfrac{25}{28}\right)$

88. $\dfrac{3}{11} \cdot \left(-\dfrac{22}{4}\right)$

89. $\dfrac{7}{8} - \dfrac{9}{10}$

90. $5(-0.27)$

91. Joe owed his mother $32. He gave her $23. Write his financial situation as a signed number.

92. The temperature at noon on a Monday in December was $-11°C$. By noon on Tuesday, it had warmed by $17°C$. What was the temperature at noon on Tuesday?

93. The top of the mountain has an altitude of 12,923 feet. The bottom of the valley is 195 feet below sea level. Find the difference between these two elevations.

94. Wednesday's lowest temperature was $-18°C$. The cold weather continued and by Friday it had dropped another $9°C$. What was the low temperature on Friday?

Simplify.

95. $(3 - 7)^2 \div (6 - 4)^3$

96. $(4 + 6)^2 \div (2 - 7)^2$

97. $3(4 + 2) + (-6) - 3^2$

98. $4(5 - 3) - (-2) + 3^3$

99. $2 - 4 \cdot 3 + \sqrt{25}$

100. $4 - 6 \cdot 5 + \sqrt{1}$

101. $\dfrac{-|-14| - 6}{7 + 2(-3)}$

102. $5(7 - 6)^3 - 4(2 - 3)^2 + 2^4$

Chapter 10 Test

Step-by-step test solutions are found on the Chapter Test Prep Videos. Where available: MyMathLab® or YouTube™

Answers

1. Graph each number on a number line.

$$-5, 0, 2, -6.5, -1\frac{2}{3}$$

1. _____

2. _____

2. Find the opposite of $\frac{2}{3}$.

3. _____

4. _____

Simplify.

3. $|-2.9|$ **4.** $-(-98)$ **5.** $-|-98|$

5. _____

6. _____

Insert $<$ or $>$ between each pair of numbers to form a true statement.

7. _____

6. $-17 \quad -19$ **7.** $0 \quad -2.5$

8. _____

9. _____

Simplify each expression.

10. _____

8. $-5 + 8$ **9.** $18 - 24$ **10.** $0.5(-20)$

11. _____

12. _____

11. $-16 \div (-4)$ **12.** $-\frac{3}{11} + \left(-\frac{5}{22}\right)$ **13.** $-7 - (-19)$

13. _____

14. _____

15. _____

14. $-5(-13)$ **15.** $\frac{-2.5}{-0.5}$ **16.** $|-25| + (-13)$

16. _____

17. _____

17. $-8 + 9 \div (-3)$ **18.** $-7 + (-32) - 12 + \sqrt{25}$

18. _____

19. _____

19. $(-5)^3 - 24 \div (-3)$ **20.** $(5 - 9)^2 \cdot (8 - 6)^3$

20. _____

21. $\dfrac{-3(-2) + 12}{-1(-4 - 5)}$

22. $\dfrac{|25 - 30|^2}{2(-6) + 7}$

23. $\left(\dfrac{5}{9} - \dfrac{7}{9}\right)^2 + \left(-\dfrac{2}{9}\right)$

24. Find the sum of -7, -9, and 35.

25. Subtract 43 from 40.

26. Find the product of $-\dfrac{1}{10}$ and $\dfrac{2}{9}$.

27. Find the quotient of $-\dfrac{1}{10}$ and $\dfrac{2}{9}$.

28. A diver starts at sea level and then makes 4 successive descents of 22 feet. After the descents, what is her elevation?

29. Aaron Hawn has $129 in his checking account. He writes a check for $79, withdraws $40 from an ATM, and then deposits $35. Represent the new balance in his account by an integer.

30. Mt. Washington in New Hampshire has an elevation of 6288 feet above sea level. The Romanche Gap in the Atlantic Ocean has an elevation of 25,354 feet below sea level. Represent the difference in elevation between these two points by an integer. (*Source:* National Geographic Society and Defense Mapping Agency)

31. Lake Baykal in Siberian Russia is the deepest lake in the world, with a maximum depth of 5315 feet. The elevation of the lake's surface is 1495 feet above sea level. What is the elevation (with respect to sea level) of the deepest point in the lake? (*Source:* U.S. Geological Survey)

1495 feet above sea level

Sea level — 1495 feet

5315 feet

? elevation

32. Find the average of -12, -13, 0, 9.

21. _____

22. _____

23. _____

24. _____

25. _____

26. _____

27. _____

28. _____

29. _____

30. _____

31. _____

32. _____

Answers

1. _____

2. _____

3. _____

4. _____

5. _____

6. _____

7. _____

8. _____

9. _____

10. _____

11. _____

12. _____

13. _____

14. _____

15. _____

16. _____

17. _____

18. _____

19. _____

20. _____

680

1. Multiply: 631×125

2. Multiply: $\dfrac{5}{8} \cdot \dfrac{10}{11}$

3. Divide: $\dfrac{2}{5} \div \dfrac{1}{2}$

4. Divide: $2124 \div 9$

5. Add: $\dfrac{2}{3} + \dfrac{1}{7}$

6. Subtract: $9\dfrac{2}{7} - 7\dfrac{1}{2}$

For Exercises 7 through 9, write each decimal in standard form.

7. Forty-eight and twenty-six hundredths

8. Eight hundredths

9. Six and ninety-five thousandths

10. Multiply: 563.21×100

11. Subtract: $3.5 - 0.068$

12. Divide: $0.27 \div 0.02$

13. Simplify: $\dfrac{5.68 + (0.9)^2 \div 100}{0.2}$

14. Simplify: $50 \div 5 \cdot 2$

15. Write "\$31,500 every 7 months" as a unit rate.

16. Write "300 miles every 5 hours" as a unit rate.

Find the value of the unknown number n.

17. $\dfrac{51}{34} = \dfrac{3}{n}$

18. $\dfrac{7}{8} = \dfrac{6}{n}$

19. 46 out of every 100 college students live at home. What percent of students live at home? (*Source:* Independent Insurance Agents of America)

20. A basketball player made 4 out of 5 free throws. What percent of free throws were made?

Write each fraction, mixed, or whole number as a percent.

21. $\dfrac{9}{20}$

22. $\dfrac{53}{50}$

23. $1\dfrac{1}{2}$

24. 5

Solve.

25. 13 is $6\dfrac{1}{2}$% of what number?

26. What is 110% of 220?

27. Translate to a proportion. 101 is what percent of 200?

28. Translate to an equation. 101 is what percent of 200?

29. Ivan Borski borrowed $2400 at 10% simple interest for 8 months to buy a used Toyota Corolla. Find the simple interest he paid.

30. C. J. Dufour wants to buy a digital camera. She has $762 in her savings account. If the camera costs $237, how much money will she have in her account after buying the camera?

31. Add 3 ft 2 in. and 5 ft 11 in.

32. Find the perimeter of the square.

17 meters

33. The normal body temperature is 98.6°F. What is this temperature in degrees Celsius?

34. Convert 7.2 meters to centimeters.

35. Find the supplement of a 107° angle.

36. Find the complement of a 34° angle.

37. Find the measure of $\angle b$.

b

30°

△ **38.** Find the measure of $\angle a$.

108°

a 23°

21. _____

22. _____

23. _____

24. _____

25. _____

26. _____

27. _____

28. _____

29. _____

30. _____

31. _____

32. _____

33. _____

34. _____

35. _____

36. _____

37. _____

38. _____

△ **39.** Find the area of the parallelogram.

1.5 miles

3.4 miles

△ **40.** Find the area of the circle. Give an exact area, then use 3.14 for π to approximate the area.

5 m

△ **41.** Approximate the volume of a ball of radius 3 inches. Use the approximation $\frac{22}{7}$ for π. Give an exact answer and an approximate answer.

3 inches

△ **42.** Find the volume of the box.

3 cm

2 cm   4 cm

43. Find the median of the list of numbers: $25, 54, 56, 57, 60, 71, 98$

44. Find the mean or average of $36, 25, 18,$ and 19.

45. If a die is rolled one time, find the probability of rolling a 3 or a 4.

46. Subtract: $-9 - (-4.1)$

47. Add: $-2 + (-21)$

48. Simplify:

a. $(-10)^3$

b. -10^3

49. Simplify: $60 + 30 + (-2)^3$

50. Simplify: $\dfrac{\sqrt{16} - (-8)}{9 - 12}$

39. _____

40. _____

41. _____

42. _____

43. _____

44. _____

45. _____

46. _____

47. _____

48. a. _____

b. _____

49. _____

50. _____

Introduction to Algebra

The Top 4 Social Networking Sites Have Changed!

A social network site or service consists of a profile or representation of each user, his or her social links, and a variety of other services, including a way for users to interact over the Internet. Although Google is currently the most popular Internet site, with an estimated 1,000,000,000 unique monthly visitors, social networking giant Facebook is the third most popular Internet site and tops among social networking sites. Some social networking sites concentrate on entertainment, some on career-minded professionals, and some, like Twitter, let members send out short, 140-character messages called "tweets."

In Section 11.5, Exercises 65 and 66, we explore numbers of personal computers in use in select countries.

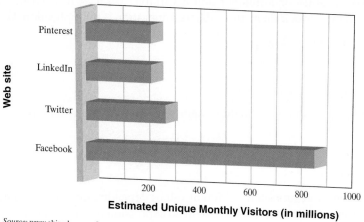

Top 4 Social Networking Sites

Source: www.ebizmba.com, September 2014

In this chapter we make the transition from arithmetic to algebra. In algebra, letters are used to stand for unknown quantities. Using variables is a very powerful tool for solving problems that cannot be solved with arithmetic alone. This chapter introduces variables, algebraic expressions, and solving variable equations.

Objectives

A Evaluate Algebraic Expressions for Given Replacement Values for the Variables.

B Use Properties of Numbers to Combine Like Terms.

C Use Properties of Numbers to Multiply Expressions.

D Simplify Expressions by Multiplying and Then Combining Like Terms.

E Find the Perimeter and Area of Figures.

Objective A Evaluating Algebraic Expressions

Perhaps the most important quality of mathematics is that it is a science of pattern. Communicating about patterns is made possible by using a letter to represent all the numbers fitting a pattern. We call such a letter a **variable.** For example, in Section 1.3 we presented the addition property of 0, which states that the sum of 0 and any whole number is that number. We might write

$$0 + 1 = 1$$
$$0 + 2 = 2$$
$$0 + 3 = 3$$
$$0 + 4 = 4$$
$$0 + 5 = 5$$
$$0 + 6 = 6$$
$$\vdots$$

continuing indefinitely. This is a pattern, and all whole numbers fit the pattern. We can communicate this pattern for all whole numbers by letting a letter, such as a, represent all whole numbers. We can then write

$$0 + a = a$$

> **Helpful Hint**
>
> Variables have been used in previous chapters, although we have not called them that. For example, in the ratio and proportion chapter, we wrote equations such as
>
> $$\frac{4}{n} = \frac{6}{12}$$
>
> Here, the letter n is a variable.

Learning to use variable notation is a primary goal of learning **algebra.** We now take some important beginning steps in learning to use variable notation.

A combination of numbers, letters (variables), and operation symbols is called an **algebraic expression** or simply an **expression.** For example,

$$3 + x, \qquad 5 \cdot y, \qquad \text{and} \qquad 2 \cdot z - 1 + x$$

are expressions.

If two variables or a number and a variable are next to each other, with no operation sign between them, the indicated operation is multiplication. For example,

$$2x \qquad \text{means} \qquad 2 \cdot x$$

and

$$xy \text{ or } x(y) \qquad \text{means} \qquad x \cdot y$$

Also, the meaning of an exponent remains the same when the base is a variable. For example,

$$x^2 = \underbrace{x \cdot x}_{2 \text{ factors of } x} \quad \text{and} \quad y^5 = \underbrace{y \cdot y \cdot y \cdot y \cdot y}_{5 \text{ factors of } y}$$

Algebraic expressions have different values depending on the replacement values for the variable(s). Replacing a variable in an expression by a number and then finding the value of the expression is called **evaluating the expression.** When finding the value of an expression, remember to follow the order of operations.

Example 1   Evaluate: $3x - 7$ when $x = 2$

Solution:

$$3x - 7 = 3 \cdot 2 - 7 \quad \text{Replace } x \text{ with 2.}$$
$$= 6 - 7 \quad \text{Multiply.}$$
$$= -1 \quad \text{Subtract.}$$

■ Work Practice 1

Example 2   Evaluate: $2x + y$ when $x = 8$ and $y = -7$

Solution: Replace x with 8 and y with -7 in $2x + y$.

$$2x + y = 2 \cdot 8 + (-7) \quad \text{Replace } x \text{ with 8 and } y \text{ with } -7.$$
$$= 16 + (-7) \quad \text{Multiply first because of the order of operations.}$$
$$= 9 \quad \text{Add.}$$

■ Work Practice 2

Example 3   Evaluate: $\dfrac{3m - 2n}{-2q}$ when $m = 8, n = 4,$ and $q = 1$

Solution:

$$\frac{3m - 2n}{-2q} = \frac{3 \cdot 8 - 2 \cdot 4}{-2 \cdot 1} \quad \text{Replace } m \text{ with 8, } n \text{ with 4, and } q \text{ with 1.}$$
$$= \frac{24 - 8}{-2} \quad \text{Multiply.}$$
$$= \frac{16}{-2} \quad \text{Subtract in the numerator.}$$
$$= -8 \quad \text{Divide.}$$

■ Work Practice 3

Example 4   Evaluate: $8 - (6a - 5)$ when $a = -3$

Solution:

$$8 - (6a - 5) = 8 - (6 \cdot (-3) - 5) \quad \text{Replace } a \text{ with } -3.$$
$$= 8 - (-18 - 5) \quad \text{Multiply.}$$
$$= 8 - (-23) \quad \text{Simplify inside the parentheses.}$$
$$= 8 + 23$$
$$= 31 \quad \text{Add.}$$

■ Work Practice 4

Example 5   Evaluate: $x^3 - 1.1y$ when $x = 4$ and $y = -1$

Solution:

$$x^3 - 1.1y = 4^3 - 1.1(-1) \quad \text{Replace } x \text{ with 4 and } y \text{ with } -1.$$
$$= 64 - 1.1(-1) \quad \text{Evaluate } 4^3.$$
$$= 64 - (-1.1) \quad \text{Multiply.}$$
$$= 64 + 1.1$$
$$= 65.1 \quad \text{Add.}$$

■ Work Practice 5

Practice 1

Evaluate: $5x - 12$ when $x = 2$

Practice 2

Evaluate: $5x - y$ when $x = 2$ and $y = -3$

Practice 3

Evaluate: $\dfrac{5r - 2s}{-3q}$ when $r = 3, s = 3,$ and $q = 1$

Practice 4

Evaluate: $13 - (3a + 8)$ when $a = -2$

Practice 5

Evaluate: $a^2 - 0.7b$ when $a = 6$ and $b = -2$

Answers

1. -2 **2.** 13   **3.** -3 **4.** 11   **5.** 37.4

Practice 6

The formula for finding the perimeter of a rectangle is $P = 2 \cdot l + 2 \cdot w$ (which we now know can be written as $P = 2l + 2w$). Find the perimeter of a rectangular garden that is 25 meters wide and 40 meters long.

40 meters   25 meters

Example 6   Finding the Area of a Rectangle

The formula for finding the area of a rectangle is $A = l \cdot w$ (which we now know can be written as $A = lw$), where l is the length of the rectangle and w is the width. Find the area of a rectangular floor that is 45 feet long and 30 feet wide.

30 feet

45 feet

Solution:

$$A = \quad l \quad \cdot \quad w$$
$$\quad\quad \downarrow \quad\quad\quad \downarrow$$
$$= (45 \text{ feet}) \cdot (30 \text{ feet}) \quad \text{Replace } l \text{ with 45 feet and } w \text{ with 30 feet.}$$
$$= 1350 \text{ square feet}$$

The area of the floor is 1350 square feet.

▪ Work Practice 6

Objective B Combining Like Terms ▶

The addends of an algebraic expression are called the **terms** of the expression.

$$x + 3$$
⎿___⎤ 2 terms

$$3y^2 + (-6y) + 4$$
⎿___⎤ 3 terms

A term that is only a number has a special name. It is called a **constant term,** or simply a **constant.** A term that contains a variable is called a **variable term.**

$$x \quad + \quad 3 \qquad\qquad 3y^2 + (-6y) + \quad 4$$
↑   ↑   ↑   ↑   ↑

| variable term | constant term | variable terms | constant term |

Helpful Hint Recall that $1 \cdot$ any number $=$ that number. This means that

$1 \cdot x = x$ or that $1x = x$.

Thus x can always be replaced by $1x$ or $1 \cdot x$.

The number factor of a variable term is called the **numerical coefficient.** A numerical coefficient of 1 is usually not written.

$$5x \qquad\qquad x \text{ or } 1x \qquad\qquad 3y^2 \qquad\qquad -6y$$
↑   ↑   ↑   ↑

| Numerical coefficient is 5. | Understood numerical coefficient is 1. | Numerical coefficient is 3. | Numerical coefficient is −6. |

Terms with the same variable factors, except that they may have different numerical coefficients, are called **like terms.**

Like Terms	Unlike Terms
$3x, \dfrac{1}{2}x$	$5x, x^2$
$-6y, 2y, y$	$7x, 7y$

✔**Concept Check** True or false? The terms $-7xy$ and $-7yx$ are like terms. Explain.

A sum or difference of like terms can be simplified using the **distributive property.** Recall from Section 1.6 that the distributive property says that multiplication distributes over addition (and subtraction). Using variables, we can write the distributive property as follows:

$$(a + b)c = ac + bc.$$

Answer

6. 130 m

✔**Concept Check Answer**

true; the order of the variable factors does not make a difference

If we write the right side of the equation first, then the left side, we have the following:

Distributive Property

If a, b, and c are numbers, then

$$ac + bc = (a + b)c$$

Also,

$$ac - bc = (a - b)c$$

The distributive property guarantees that, no matter what number x is, $7x + 5x$ (for example) has the same value as $(7 + 5)x$, or $12x$. We then have that

$$7x + 5x = (7 + 5)x = 12x$$

This is an example of **combining like terms**. An algebraic expression is **simplified** when all like terms have been combined.

Example 7   Simplify each expression by combining like terms.

a. $3x + 2x$ **b.** $y - 7y$

Solution: We add or subtract like terms.

a. $3x + 2x = (3 + 2)x$
$$= 5x$$

$\overbrace{}^{\text{Understood 1}}$

b. $y - 7y = 1y - 7y$
$$= (1 - 7)y$$
$$= -6y$$

▦ Work Practice 7

Practice 7

Simplify each expression by combining like terms.
a. $8m - 11m$
b. $5a + a$

The commutative and associative properties of addition and multiplication can also help us simplify expressions. We presented these properties in Sections 1.3 and 1.6 and state them again using variables.

Properties of Addition and Multiplication

If a, b, and c are numbers, then

$a + b = b + a$ Commutative property of addition

$a \cdot b = b \cdot a$ Commutative property of multiplication

That is, the **order** of adding or multiplying two numbers can be changed without changing their sum or product.

$(a + b) + c = a + (b + c)$ Associative property of addition

$(a \cdot b) \cdot c = a \cdot (b \cdot c)$ Associative property of multiplication

That is, the **grouping** of numbers in addition or multiplication can be changed without changing their sum or product.

Helpful Hint

- Examples of these properties are

$$2 + 3 = 3 + 2$$ Commutative property of addition

$$7 \cdot 9 = 9 \cdot 7$$ Commutative property of multiplication

$$(1 + 8) + 10 = 1 + (8 + 10)$$ Associative property of addition

$$(4 \cdot 2) \cdot 3 = 4 \cdot (2 \cdot 3)$$ Associative property of multiplication

- These properties are not true for subtraction or division.

Practice 8

Simplify: $8m + 5 + m - 4$

Example 8   Simplify: $2y - 6 + 4y + 8$

Solution: We begin by writing subtraction as the addition of opposites.

$$2y - 6 + 4y + 8 = 2y + (-6) + 4y + 8$$
$$= 2y + 4y + (-6) + 8 \quad \text{Apply the commutative property of addition.}$$
$$= (2 + 4)y + (-6) + 8 \quad \text{Apply the distributive property.}$$
$$= 6y + 2 \quad \text{Simplify.}$$

■ Work Practice 8

Practice 9–12

Simplify each expression by combining like terms.

9. $7y + 11y - 8$
10. $2y - 6 + y + 7y$
11. $3.7x + 5 - 4.2x + 15$
12. $-9y + 2 - 4y - 8x$
    $+ 12 - x$

Examples   Simplify each expression by combining like terms.

9. $6x + 2x - 5 = 8x - 5$
10. $4x + 3 - 5x + 2x = 4x - 5x + 2x + 3$
$$= 1x + 3 \quad \text{or} \quad x + 3$$
11. $1.2y + 10 - 5.7y - 9 = 1.2y - 5.7y + 10 - 9$
$$= -4.5y + 1$$
12. $2x - 5 + 3y + 4x - 10y + 11 = 6x - 7y + 6$

■ Work Practice 9–12

Objective C Multiplying Expressions ▶

We can also use properties of numbers to multiply expressions such as $3(2x)$. By the associative property of multiplication, we can write the product $3(2x)$ as $(3 \cdot 2)x$, which simplifies to $6x$.

Practice 13–14

Multiply.
13. $7(8a)$
14. $-5(9x)$

Examples   Multiply.

13. $5(3y) = (5 \cdot 3)y$ Apply the associative property of multiplication.
$$= 15y \quad \text{Multiply.}$$

14. $-2(4x) = (-2 \cdot 4)x$ Apply the associative property of multiplication.
$$= -8x \quad \text{Multiply.}$$

■ Work Practice 13–14

We can use the distributive property to combine like terms, which we have done, and also to multiply expressions such as $2(3 + x)$. By the distributive property, we have that

$$2(3 + x) = 2 \cdot 3 + 2 \cdot x \quad \text{Apply the distributive property.}$$
$$= 6 + 2x \quad \text{Multiply.}$$

Answers

8. $9m + 1$ **9.** $18y - 8$ **10.** $10y - 6$
11. $-0.5x + 20$ **12.** $-13y - 9x + 14$
13. $56a$ **14.** $-45x$

Example 15   Use the distributive property to multiply: $6(x + 4)$

Solution: By the distributive property,

$$6(x + 4) = 6 \cdot x + 6 \cdot 4 \quad \text{Apply the distributive property.}$$
$$= 6x + 24 \quad \text{Multiply.}$$

■ Work Practice 15

Practice 15

Use the distributive property to multiply: $7(y + 2)$

✓**Concept Check**   What's wrong with the following?

$$8(a - b) = 8a - b$$

Example 16   Multiply: $-3(5a + 2)$

Solution: By the distributive property,

$$-3(5a + 2) = -3(5a) + (-3)(2) \quad \text{Apply the distributive property.}$$
$$= (-3 \cdot 5)a + (-6) \quad \text{Apply the associative property. Also, write } (-3)(2) \text{ as } -6.$$
$$= -15a - 6 \quad \text{Multiply.}$$

■ Work Practice 16

Practice 16
Multiply: $4(7a - 5)$

Objective D Simplifying Expressions ▶

Next, we will **simplify** expressions containing parentheses by first using the distributive property to multiply and then **combining** any like terms.

Example 17   Simplify: $2(3 + 7x) - 15$

Solution: First we use the distributive property to remove parentheses.

$$2(3 + 7x) - 15 = 2(3) + 2(7x) - 15 \quad \text{Apply the distributive property.}$$
$$= 6 + 14x - 15 \quad \text{Multiply.}$$
$$= 14x + (-9) \quad \text{or} \quad 14x - 9 \quad \text{Combine like terms.}$$

■ Work Practice 17

Practice 17

Simplify: $5(2y - 3) - 8$

Helpful Hint   2 is *not* distributed to the -15 since it is not within the parentheses.

Example 18   Simplify: $-2(x - 5) + 4(2x + 2)$

Solution: First we use the distributive property to remove parentheses.

$$-2(x - 5) + 4(2x + 2) = -2(x) - (-2)(5) + 4(2x) + 4(2) \quad \text{Apply the distributive property.}$$
$$= -2x + 10 + 8x + 8 \quad \text{Multiply.}$$
$$= 6x + 18 \quad \text{Combine like terms.}$$

■ Work Practice 18

Practice 18

Simplify:
$-7(x - 1) + 5(2x + 3)$

Answers
15. $7y + 14$ **16.** $28a - 20$
17. $10y - 23$ **18.** $3x + 22$

✓**Concept Check Answer**
did not distribute the 8

Objective E Finding Perimeter and Area

 Example 19 Find the perimeter of the triangle.

Solution: Recall that the perimeter of a figure is the distance around the figure. To find the perimeter, then, we find the sum of the lengths of the sides. We use the letter P to represent perimeter.

$$P = 2z + 3z + 5z$$
$$= 10z$$

The perimeter is $10z$ feet.

> **Helpful Hint**
> Don't forget to insert proper units.

■ Work Practice 19

Practice 19

Find the perimeter of the square.

$2x$
centimeters

Example 20 Finding the Area of a Basketball Court

Find the area of this YMCA basketball court.

$(2x - 6)$ ft

45 ft

Solution: Recall how to find the area of a rectangle. **Area = Length · Width**, or if A represents area, l represents length, and w represents width, we have $A = l \cdot w$.

$$A = l \cdot w$$
$$= 45(2x - 6) \quad \text{Let } l = 45 \text{ and } w = (2x - 6).$$
$$= 90x - 270 \quad \text{Multiply.}$$

The area is $(90x - 270)$ *square* feet.

■ Work Practice 20

Practice 20

Find the area of the rectangular garden.

$(12y + 9)$ yards

3 yards

> **Helpful Hint**
>
> Don't forget...
>
> | Area: | Perimeter: |
> |---|---|
> | • surface enclosed | • distance around |
> | • measured in square units | • measured in units |

Answers

19. $8x$ cm   **20.** $(36y + 27)$ sq yd

Vocabulary, Readiness & Video Check

Use the choices below to fill in each blank. Some choices may be used more than once.

numerical coefficient	combine like terms	like	term	variable	associative
constant	expression	unlike	distributive	commutative	

1. $14y^2 + 2x - 23$ is called a(n) _____ while $14y^2$, $2x$, and -23 are each called a(n) _____.
2. To multiply $3(-7x + 1)$, we use the _____ property.
3. To simplify an expression like $y + 7y$, we _____.
4. By the _____ properties, the *order* of adding or multiplying two numbers can be changed without changing their sum or product.
5. The term $5x$ is called a(n) _____ term while the term 7 is called a(n) _____ term.
6. The term z has an understood _____ of 1.
7. By the _____ properties, the grouping of numbers in addition or multiplication can be changed without changing their sum or product.
8. The terms $-x$ and $5x$ are _____ terms.
9. For the term $-3x^2y$, -3 is called the _____.
10. The terms $5x$ and $5y$ are _____ terms.

Identify each pair of terms as like terms or unlike terms.

11. $5x$ and $5y$

12. $-3a$ and $-3b$

13. x and $-2x$

14. $7y$ and y

15. $-5n$ and $6n^2$

16. $4m^2$ and $2m$

17. $8b$ and $-6b$

18. $12a$ and $-11a$

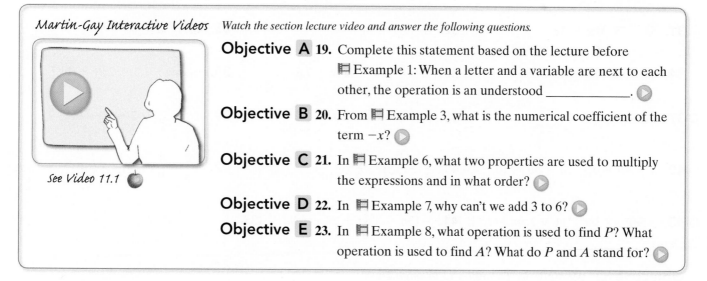

Martin-Gay Interactive Videos   *Watch the section lecture video and answer the following questions.*

Objective A 19. Complete this statement based on the lecture before Example 1: When a letter and a variable are next to each other, the operation is an understood _____.

Objective B 20. From Example 3, what is the numerical coefficient of the term $-x$?

Objective C 21. In Example 6, what two properties are used to multiply the expressions and in what order?

Objective D 22. In Example 7, why can't we add 3 to 6?

Objective E 23. In Example 8, what operation is used to find P? What operation is used to find A? What do P and A stand for?

See Video 11.1

11.1 Exercise Set MyMathLab® ▶

Objective A *Evaluate each expression when $x = -2$, $y = 5$, and $z = -3$. See Examples 1 through 6.*

1. $3 + 2z$

2. $7 + 3z$

3. $-y - z$

4. $-y - x$

5. $z - x + y$

6. $x + y - z$

7. $3x - z$

8. $y + 5z$

9. $8 - (5y - 7)$

10. $5 + (2x - 1)$

11. $y^3 - 4x$

12. $y^2 - 2z$

13. $\dfrac{6xy}{4}$

14. $\dfrac{8yz}{15}$

15. $\dfrac{2y - 2}{x}$

16. $\dfrac{6 + 3x}{z}$

17. $\dfrac{x + 2y}{2z}$

18. $\dfrac{2z - y}{3x}$

19. $\dfrac{5x}{y} - 10$

20. $7 - \dfrac{3y}{z}$

21. $\dfrac{xz}{y} + \dfrac{3}{10}$

22. $\dfrac{x}{yz} + \dfrac{31}{30}$

23. $|x| - |y| - 7.6$

24. $|z| - |y| - 12.7$

Objective B *Simplify each expression by combining like terms. See Examples 7 through 12.*

25. $3x + 5x$

26. $8y + 3y$

27. $5n - 9n$

28. $7z - 10z$

29. $4c + c - 7c$

30. $5b - 8b - b$

31. $5x - 7x + x - 3x$

32. $8y + y - 2y - y$

33. $4a + 3a + 6a - 8$

34. $5b - 4b + b - 15$

35. $1.7x + 3.4 - 2.6x + 7.8$

36. $-8.6y + 1.3 - 2.9y - 14.7$

37. $3x + 7 - x - 14$

38. $9x - 6 + x - 10$

39. $4x + 5y + 2 - y - 9x - 7$

40. $a + 4b + 3 - 7a - 5b - 10$

41. $\dfrac{5}{6} - \dfrac{7}{12}x - \dfrac{1}{3} - \dfrac{3}{10}x$

42. $-\dfrac{2}{5} + \dfrac{4}{9}y - \dfrac{4}{15} + \dfrac{1}{6}y$

43. $-5m - 2.3m + 11 + 2.5m - 15.1$ **44.** $-13n - 4.8n + 13 + 6.9n - 13.6$

Objective C *Multiply. See Examples 13 through 16.*

45. $6(5x)$

46. $4(4x)$

47. $-2(11y)$

48. $-3(21z)$

49. $-0.6(7a)$

50. $-0.4(9a)$

51. $\dfrac{2}{3}(-6a)$

52. $\dfrac{3}{4}(-8a)$

53. $2(y + 2)$

54. $3(x + 1)$

55. $5(3a - 8)$ **56.** $4(5y - 6)$ ▶**57.** $-4(3x + 7)$ **58.** $-8(8y + 10)$

59. $1.2(5x - 0.1)$ **60.** $3.1(7x - 0.3)$ **61.** $\frac{1}{2}(-8x - 3)$ **62.** $\frac{1}{5}(-20x - 7)$

Objective D *Simplify each expression. Use the distributive property to remove parentheses first. See Examples 17 and 18.*

63. $2(x + 4) - 17$ **64.** $5(6 + y) - 2$ **65.** $4(6n - 5) + 3n$

66. $3(5 - 2b) - 4b$ ▶**67.** $3 + 6(w + 2) + w$ **68.** $8z + 5(6 + z) + 20$

69. $-2(3x + 1) - 5(x - 2)$ **70.** $-3(5x - 2) - 2(3x + 1)$

Objective E *Find the perimeter of each figure. See Example 19.*

△**71.** 2y meters 6 meters 4y meters 3y meters 5y meters 16 meters

△**72.** 3x feet x feet 7 feet 5x feet 4x feet 12 feet

△**73.** 2a feet 2a feet 2a feet 6 feet 6 feet 5a feet

△**74.** 3z meters 1 meter 1 meter 5z meters

△**75.** Each side: $(-5x + 11)$ inches

△**76.** Each side: $(9y + 1)$ kilometers

Find the area of each rectangle. See Example 20.

△**77.** 9 inches 4y inches

△**78.** 5x centimeters 8 centimeters

△**79.** 32 kilometers $(x - 2)$ kilometers

△**80.** $(z - 6)$ meters 11 meters

▶△**81.** $(3y + 1)$ miles 20 miles

△**82.** 12 feet $(2x + 3)$ feet

Objectives A B C D E Mixed Practice *Solve. See Examples 1 through 20.*

83. Find the area of a regulation NCAA basketball court that is 94 feet long and 50 feet wide.

84. Find the area of a rectangular movie screen that is 50 feet long and 40 feet high.

85. A decorator wishes to put a wallpaper border around a rectangular room that measures 14 feet by 18 feet. Find the room's perimeter.

86. How much fencing will a rancher need for a rectangular cattle lot that measures 80 feet by 120 feet?

87. Find the perimeter of a triangular garden that measures 5 feet by x feet by $(2x + 1)$ feet.

88. Find the perimeter of a triangular picture frame that measures x inches by x inches by $(x - 14)$ inches.

89. How much interest will $3000 in a passbook savings account earn in 2 years at Money Bank, which pays 6% simple interest? Use $I = PRT$.

90. How much interest will a $12,000 certificate of deposit earn in 1 year at a rate of 8% simple interest? Use $I = PRT$.

91. Find the area of a circular braided rug with a radius of 5 feet. Use $A = \pi r^2$ and $\pi \approx 3.14$.

92. Mario's Pizza sells one 16″ cheese pizza or two 10″ cheese pizzas for $9.99. Which deal gives you more pizza? Use $A = \pi r^2$ with $\pi \approx 3.14$.

93. Convert Paris, France's, low temperature of $-5°C$ to Fahrenheit. Use $F = \dfrac{9}{5}C + 32$.

94. Convert Nome, Alaska's, 18°F high temperature to Celsius. Use $C = \dfrac{5}{9}(F - 32)$.

95. Find the volume of a box that measures 12 inches by 6 inches by 4 inches. Use $V = lwh$.

96. How many cubic meters does a space shuttle cargo compartment have if its dimensions are 8 meters long by 4 meters wide by 3 meters high? Use $V = lwh$.

97. Find the volume. Use $V = lwh$.

98. Find the volume. Use $V = lwh$.

Review

Perform each indicated operation. See Sections 10.2 and 10.3.

99. $-13 + 10$

100. $-15 + 23$

101. $-4 - (-12)$

102. $-7 - (-4)$

103. $-4 + 4$

104. $8 + (-8)$

Concept Extensions

If the expression on the left side of the equal sign is equivalent to the right, write "correct." If not, write "incorrect" and then write an expression that is equivalent to the left side. See the second Concept Check in this section.

105. $5(3x - 2) \stackrel{?}{=} 15x - 2$

106. $2(xy) \stackrel{?}{=} 2x \cdot 2y$

107. $7x - (x + 2) \stackrel{?}{=} 7x - x + 2$

108. $4(y - 3) + 11 \stackrel{?}{=} 4y - 3 + 11$

Review the commutative, associative, and distributive properties. Then identify which property allows us to write the equivalent expression on the right side of the equal sign.

109. $6(2x - 3) + 5 = 12x - 18 + 5$

110. $9 + 7x + (-2) = 7x + 9 + (-2)$

111. $-7 + (4 + y) = (-7 + 4) + y$

112. $(x + y) + 11 = 11 + (x + y)$

113. If x is a whole number, which expression is the largest: $2x, 5x,$ or $\frac{1}{3}x$? Explain your answer.

114. If x is a whole number, which expression is the smallest: $2x, 5x,$ or $\frac{1}{3}x$? Explain your answer.

Find the area of each figure.

△ **115.**

△ **116.**

To appraise the value of a large tree in landscaping, the trunk area is used. The trunk area A is calculated with the formula $A = 0.7854d^2$, where d is the diameter of the tree.

117. *Tane Mahuta* is the name of New Zealand's largest known living Kauri tree. Its trunk has a diameter of 173 inches. Use the formula to find the trunk area of this tree. Round your result to the nearest tenth. (*Source*: New Zealand Parks)

118. *Lost Monarch,* which was only discovered in 1998, is a coast redwood tree in northern California. It is located among other giant redwoods in "The Grove of Titans," in Jedediah Smith Redwoods State Park. This mammoth tree has confirmed diameter of 312 inches. Use the formula to find the trunk area of this tree. Round your result to the nearest tenth. (*Source:* Sierra Club)

Simplify.

119. $9684q - 686 - 4860q + 12,960$

120. $76(268x + 592) - 2960$

11.2 Solving Equations: The Addition Property

Objectives

A Determine Whether a Given Number Is a Solution of an Equation.

B Use the Addition Property of Equality to Solve Equations.

Frequently in this book we have written statements like $7 + 4 = 11$ or Area = length · width. Each of these statements is called an **equation.** An equation is of the form

expression = expression

An equation can be labeled as

$$\underset{\text{left side}}{\underbrace{x + 7}} = \underset{\text{right side}}{10}$$

equal sign

It is very important to know the difference between an **expression** and an **equation.** An equation contains an equal sign and an expression does not.

	Equations	Expressions	
equal signs	$7x = 6x + 4$	$7x - 6x + 4$	no equal signs
	$3(3y - 5) = 10y$	$y - 1 + 11y - 21$	

Objective A Determining Whether a Number Is a Solution

When an equation contains a variable, finding which values of the variable make an equation a true statement is called **solving** the equation for the variable. A **solution** of an equation is a value for the variable that makes the equation a true statement. For example, 2 is a solution of the equation $x + 5 = 7$ since replacing x with 2 results in the *true* statement $2 + 5 = 7$. Similarly, 3 is not a solution of $x + 5 = 7$ since replacing x with 3 results in the *false* statement $3 + 5 = 7$.

Practice 1

Determine whether 4 is a solution of the equation $3(y - 6) = 6$.

Example 1 Determine whether 6 is a solution of the equation $4(x - 3) = 12$.

Solution: We replace x with 6 in the equation.

$$4(x - 3) = 12$$
$$\downarrow$$
$$4(6 - 3) \overset{?}{=} 12 \quad \text{Replace } x \text{ with 6.}$$
$$4(3) \overset{?}{=} 12$$
$$12 \overset{?}{=} 12 \quad \text{True}$$

Since $12 = 12$ is a true statement, 6 *is* a solution of the equation.

■ Work Practice 1

Practice 2

Determine whether -2 is a solution of the equation $-4x - 3 = 5$.

Example 2 Determine whether -1 is a solution of the equation $3y + 1 = 3$.

Solution:

$$3y + 1 = 3$$
$$3(-1) + 1 \overset{?}{=} 3$$
$$-3 + 1 \overset{?}{=} 3$$
$$-2 \overset{?}{=} 3 \quad \text{False}$$

Since $-2 = 3$ is false, -1 is *not* a solution of the equation.

■ Work Practice 2

Answers

1. no **2.** yes

696

Objective B Using the Addition Property to Solve Equations

To solve an equation, we use properties of equality to write simpler equations, all equivalent to the original equation, until the final equation has the form

$x =$ **number**   or   **number** $= x$

Equivalent equations have the same solution, so the word "number" above represents the solution of the original equation. The first property of equality to help us write simpler, equivalent, equations is the **addition property of equality**.

> ### Addition Property of Equality
>
> Let a, b, and c represent numbers. Then
>
> | $a = b$ | Also, $a = b$ |
> |---|---|
> | and $a + c = b + c$ | and $a - c = b - c$ |
> | are equivalent equations. | are equivalent equations. |

In other words, the **same number** may be **added to or subtracted from both sides** of an equation without changing the solution of the equation.

A good way to visualize a true equation is to picture a balanced scale. Since it is balanced, each side of the scale weighs the same amount. Similarly, in a true equation the expressions on each side have the same value. Picturing our balanced scale, if we add the same weight to each side, the scale remains balanced.

Example 3   Solve: $x - 2 = 1$ for x.

Solution: To solve the equation for x, we need to rewrite the equation in the form $x =$ number. In other words, our goal is to get x alone on one side of the equation. To do so, we add 2 to both sides of the equation.

$$x - 2 = 1$$
$$x - 2 + 2 = 1 + 2 \quad \text{Add 2 to both sides of the equation.}$$
$$x + 0 = 3 \quad \text{Replace } -2 + 2 \text{ with } 0.$$
$$x = 3 \quad \text{Simplify by replacing } x + 0 \text{ with } x.$$

Check: To check, we replace x with 3 in the *original* equation.

$$x - 2 = 1 \quad \text{Original equation}$$
$$3 - 2 \overset{?}{=} 1 \quad \text{Replace } x \text{ with } 3.$$
$$1 \overset{?}{=} 1 \quad \text{True}$$

Since $1 = 1$ is a true statement, 3 is the solution of the equation.

■ Work Practice 3

Practice 3

Solve the equation for y: $y - 5 = -3$

Helpful Hint

Note that it is always a good idea to check the solution in the *original* equation to see that it makes the equation a true statement.

Answer

3. 2

Let's visualize how we used the addition property of equality to solve the equation in Example 3. Picture the original equation, $x - 2 = 1$, as a balanced scale. The left side of the equation has the same value as the right side.

If the same weight is added to each side of a scale, the scale remains balanced. Likewise, if the same number is added to each side of an equation, the left side continues to have the same value as the right side.

Copyright 2015 Pearson Education, Inc.

Practice 4

Solve: $-1 = z + 9$

Example 4 Solve: $-8 = x + 1$

Solution: To get x alone on one side of the equation, we subtract 1 from both sides of the equation.

$$-8 = x + 1$$
$$-8 - 1 = x + 1 - 1 \quad \text{Subtract 1 from both sides.}$$
$$-9 = x + 0 \quad \text{Replace } 1 - 1 \text{ with 0.}$$
$$-9 = x \quad \text{Simplify.}$$

Check:

$$-8 = x + 1$$
$$-8 \stackrel{?}{=} -9 + 1 \quad \text{Replace } x \text{ with } -9.$$
$$-8 \stackrel{?}{=} -8 \quad \text{True}$$

The solution is -9.

▦ Work Practice 4

Helpful Hint Remember that we can get the variable alone on either side of the equation. For example, the equations $x = 2$ and $2 = x$ both have the solution of 2.

Practice 5

Solve: $x - 2.6 = -1.8 - 5.9$

Example 5 Solve: $y - 1.2 = -3.2 - 6.6$

Solution: First we simplify the right side of the equation.

$$y - 1.2 = -3.2 - 6.6$$
$$y - 1.2 = -9.8$$

Next, we get y alone on the left side by adding 1.2 to both sides of the equation.

$$y - 1.2 + 1.2 = -9.8 + 1.2 \quad \text{Add 1.2 to both sides.}$$
$$y = -8.6 \quad \text{Simplify.}$$

Check to see that -8.6 is the solution.

▦ Work Practice 5

Answers

4. -10 5. -5.1

✓**Concept Check Answer**

subtract 2.1 from both sides

✓**Concept Check** What number should be added to or subtracted from both sides of the equation in order to solve the equation $-3.75 = y + 2.1$?

Example 6   Solve: $5x + 2 - 4x = 7 - 9$

Solution: First we simplify each side of the equation separately.

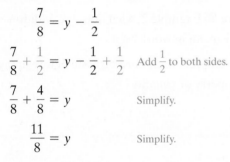

$$5x + 2 - 4x = 7 - 9$$
$$5x - 4x + 2 = 7 - 9$$
$$1x + 2 = -2$$

To get x alone on the left side, we subtract 2 from both sides.

$$1x + 2 - 2 = -2 - 2$$
$$1x = -4 \text{ or } x = -4$$

Check to verify that -4 is the solution.

■ **Work Practice 6**

Practice 6
Solve:
$-6y + 1 + 7y = 6 - 11$

Example 7   Solve: $\dfrac{7}{8} = y - \dfrac{1}{2}$

Solution: We use the addition property of equality to add $\dfrac{1}{2}$ to both sides.

$$\frac{7}{8} = y - \frac{1}{2}$$

$$\frac{7}{8} + \frac{1}{2} = y - \frac{1}{2} + \frac{1}{2} \quad \text{Add } \tfrac{1}{2} \text{ to both sides.}$$

$$\frac{7}{8} + \frac{4}{8} = y \qquad \text{Simplify.}$$

$$\frac{11}{8} = y \qquad \text{Simplify.}$$

Check to see that $\dfrac{11}{8}$ is the solution. (Although $\dfrac{11}{8} = 1\dfrac{3}{8}$, we will leave solutions as improper fractions.)

■ **Work Practice 7**

Practice 7

Solve: $\dfrac{2}{3} = x - \dfrac{4}{9}$

Example 8   Solve: $3(3x - 5) = 10x$

Solution: First we multiply on the left side to remove the parentheses.

$$3(3x - 5) = 10x$$
$$3 \cdot 3x - 3 \cdot 5 = 10x \quad \text{Use the distributive property.}$$
$$9x - 15 = 10x$$

Now we subtract $9x$ from both sides.

$$9x - 15 - 9x = 10x - 9x \qquad \text{Subtract } 9x \text{ from both sides.}$$
$$-15 = 1x \quad \text{or} \quad x = -15 \quad \text{Simplify.}$$

■ **Work Practice 8**

Practice 8

Solve: $13x = 4(3x - 1)$

Recall that the addition property of equality allows us to add or subtract the same number to or from both sides of an equation. Let's see how adding the same number to both sides of an equation also allows us to subtract the same number from both sides. To do so, let's add $(-c)$ to both sides of $a = b$. Then we have

$$a + (-c) = b + (-c)$$

which is the same as $a - c = b - c$, and there we have it.

Answers

6. -6 **7.** $\dfrac{10}{9}$ **8.** -4

Vocabulary, Readiness & Video Check

Use the choices below to fill in each blank.

equation   addition   simplifying
solving   equivalent   expression

1. The equations $x + 6 = 10$ and $x + 6 - 6 = 10 - 6$ are called _____ equations.

2. The difference between an equation and an expression is that a(n) _____ contains an equal sign, while a(n) _____ does not.

3. The process of writing $-3x + 10x$ as $7x$ is called _____ the expression.

4. For the equation $x - 1 = -21$, the process of finding that -20 is the solution is called _____ the equation.

5. By the _____ property of equality, $x = -2$ and $x + 7 = -2 + 7$ are equivalent equations.

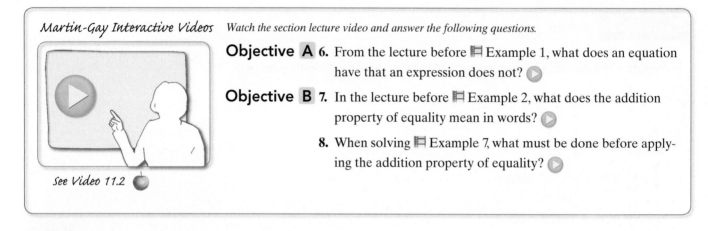

Martin-Gay Interactive Videos   Watch the section lecture video and answer the following questions.

Objective A 6. From the lecture before ⊞ Example 1, what does an equation have that an expression does not? ▶

Objective B 7. In the lecture before ⊞ Example 2, what does the addition property of equality mean in words? ▶

8. When solving ⊞ Example 7, what must be done before applying the addition property of equality? ▶

See Video 11.2 🍎

11.2 Exercise Set MyMathLab® ▶

Objective A *Decide whether the given number is a solution of the given equation. See Examples 1 and 2.*

1. Is 10 a solution of $x - 8 = 2$?

2. Is 9 a solution of $y - 2 = 7$?

▶ **3.** Is -5 a solution of $x + 12 = 17$?

4. Is -7 a solution of $a + 23 = -16$?

5. Is -8 a solution of $-9f = 64 - f$?

6. Is -6 a solution of $-3k = 12 - k$?

7. Is 3 a solution of $5(c - 5) = -10$?

8. Is 1 a solution of $2(b - 3) = 10$?

Objective B *Solve. Check each solution. See Examples 3 through 7.*

▶ **9.** $a + 5 = 23$

10. $f + 4 = -6$

11. $d - 9 = -17$

12. $s - 7 = 15$

▶ **13.** $7 = y - 2$

14. $1 = y + 7$

15. $-12 = x + 4$

16. $-10 = z - 15$

17. $x + \dfrac{1}{2} = \dfrac{7}{2}$

18. $x + \dfrac{1}{3} = \dfrac{4}{3}$

▶ 19. $y - \dfrac{3}{4} = -\dfrac{5}{8}$

20. $y - \dfrac{5}{6} = -\dfrac{11}{12}$

21. $x - 3 = -1 + 4$

22. $y - 8 = -5 - 1$

23. $-7 + 10 = m - 5$

24. $1 - 8 = n + 2$

25. $x - 0.6 = 4.7$

26. $y - 1.2 = 7.5$

▶ 27. $-2 - 3 = -4 + x$

28. $7 - (-10) = x - 5$

29. $y + 2.3 = -9.2 - 8.6$

30. $x + 4.7 = -7.5 - 3.4$

31. $-8x + 4 + 9x = -1 + 7$

32. $3x - 2x + 5 = 5 - 2$

33. $5 + (-12) = 5x - 7 - 4x$

34. $11 + (-15) = 6x - 4 - 5x$

▶ 35. $7x + 14 - 6x = -4 + (-10)$

36. $-10x + 11x + 5 = -9 + (-5)$

Solve. First multiply to remove parentheses. See Example 8.

▶ 37. $2(5x - 3) = 11x$

38. $6(3x + 1) = 19x$

39. $3y = 2(y + 12)$

40. $17x = 4(4x - 6)$

41. $21y = 5(4y - 6)$

42. $28z = 9(3z - 2)$

43. $-3(-4 - 2z) = 7z$

44. $-2(-1 - 3y) = 7y$

Review

Perform each indicated operation. See Section 10.4.

45. $\dfrac{-7}{-7}$

46. $\dfrac{4.2}{4.2}$

47. $\dfrac{1}{3} \cdot 3$

48. $\dfrac{1}{5} \cdot 5$

49. $-\dfrac{2}{3} \cdot -\dfrac{3}{2}$

50. $-\dfrac{7}{2} \cdot -\dfrac{2}{7}$

Concept Extensions

What number should be added to or subtracted from both sides of each equation in order to solve the equation?

51. $\dfrac{2}{3} + x = \dfrac{1}{12}$

52. $12.5 = -3.75 + x$

53. $-\dfrac{1}{7} = -\dfrac{4}{5} + x$

54. $9.1 = 5.9 + x$

55. In your own words, explain what is meant by the phrase "a number is a solution of an equation."

56. In your own words, explain how to check a possible solution of an equation.

Solve.

57. $x - 76{,}862 = 86{,}102$

58. $-968 + 432 = 86y - 508 - 85y$

A football team's total offense T is found by adding the total passing yardage P to the total rushing yardage R: T = P + R.

59. During the 2012 regular football season, the Pittsburgh Steelers' total offense was 5324 yards. The Steelers' passing yardage for the season was 3787 yards. How many yards did the Steelers gain by rushing during the season? (*Source:* National Football League)

60. During the 2012 regular football season, the New Orleans Saints' total offense was 6574 yards. The Saints' rushing yardage for the season was 1577 yards. How many yards did the Saints gain by passing during the season? (*Source:* National Football League)

In accounting, a company's annual net income I can be computed using the relation I = R − E, where R is the company's total revenues for the year and E is the company's total expenses for the year.

61. At the end of fiscal year 2013, Kohl's had a net income of $986,000,000. During the year, Kohl's had total expenses of $18,293,000,000. What were Kohl's total revenues for the year? (*Source:* Kohl's Corporation)

62. At the end of fiscal year 2013, Target had a net income of $2,999,000,000. During the year, Target had total expenses of $70,302,000,000. What were Target's total revenues for the year? (*Source:* Target Corporation)

11.3 Solving Equations: The Multiplication Property ▶

Objective

A Use the Multiplication Property to Solve Equations. ▶

Objective A Using the Multiplication Property to Solve Equations ▶

Although the addition property of equality is a powerful tool for helping us solve equations, it cannot help us solve all types of equations. For example, it cannot help us solve an equation such as $2x = 6$. To solve this equation, we use a second property of equality called the **multiplication property of equality**.

> **Multiplication Property of Equality**
>
> Let a, b, and c represent numbers and let $c \neq 0$. Then
>
> | $a = b$ | Also, $a = b$ |
> |---|---|
> | and $a \cdot c = b \cdot c$ | and $\dfrac{a}{c} = \dfrac{b}{c}$ |
> | are equivalent equations. | are equivalent equations. |

In other words, both sides of an equation may be multiplied or divided by the same nonzero number without changing the solution of the equation.

Picturing again our balanced scale, if we multiply or divide the weight on each side by the same nonzero number, the scale (or equation) remains balanced.

To solve $2x = 6$ for x, we use the multiplication property of equality to divide both sides of the equation by 2, and simplify as follows:

$$2x = 6$$
$$\frac{2x}{2} = \frac{6}{2} \quad \text{Divide both sides by 2.}$$
$$\frac{2}{2} \cdot x = 3$$
$$1 \cdot x = 3$$
$$x = 3$$

Example 1   Solve: $-5x = 15$

Solution: To get x by itself, we divide both sides by -5.

$$-5x = 15 \quad \text{Original equation}$$
$$\frac{-5x}{-5} = \frac{15}{-5}$$
$$\frac{-5}{-5} \cdot x = \frac{15}{-5} \quad \text{Divide both sides by } -5$$
$$1 \cdot x = -3 \quad \text{Simplify.}$$
$$x = -3$$

Check: To check, we replace x with -3 in the original equation.

$$-5x = 15 \quad \text{Original equation}$$
$$-5(-3) \stackrel{?}{=} 15 \quad \text{Let } x = -3.$$
$$15 \stackrel{?}{=} 15 \quad \text{True}$$

The solution is -3.

■ **Work Practice 1**

Example 2   Solve: $-8 = 2y$

Solution: To get y alone, we divide both sides of the equation by 2.

$$-8 = 2y$$
$$\frac{-8}{2} = \frac{2y}{2} \quad \text{Divide both sides by 2.}$$
$$-4 = 1 \cdot y \quad \text{or} \quad y = -4$$

Check to see that -4 is the solution.

■ **Work Practice 2**

Practice 1
Solve: $-3y = 18$

Practice 2
Solve: $-16 = 8x$

Answers
1. -6 **2.** -2

Practice 3

Solve: $-0.3y = -27$

Practice 4

Solve: $\frac{5}{7}b = 25$

Example 3 Solve: $-1.2x = -36$

Solution: We divide both sides of the equation by the numerical coefficient of x, which is -1.2.

$$-1.2x = -36$$

$$\frac{-1.2x}{-1.2} = \frac{-36}{-1.2}$$

$$1 \cdot x = 30$$

$$x = 30$$

Check to see that 30 is the solution.

■ Work Practice 3

Example 4 Solve: $\frac{3}{5}a = 9$

Solution: Recall that the product of a number and its reciprocal is 1. To get a alone then, we multiply both sides by $\frac{5}{3}$, the reciprocal of $\frac{3}{5}$.

$$\frac{3}{5}a = 9$$

$$\frac{5}{3} \cdot \frac{3}{5}a = \frac{5}{3} \cdot 9 \quad \text{Multiply both sides by } \frac{5}{3}.$$

$$1 \cdot a = \frac{5 \cdot \overset{3}{\cancel{9}}}{\cancel{3} \cdot 1} \quad \text{Multiply.}$$

$$a = 15 \quad \text{Simplify.}$$

Check: To check, we replace a with 15 in the original equation.

$$\frac{3}{5}a = 9 \quad \text{Original equation}$$

$$\frac{3}{5} \cdot 15 \overset{?}{=} 9 \quad \text{Replace } a \text{ with 15.}$$

$$\frac{3}{\cancel{5}} \cdot \frac{\overset{3}{\cancel{15}}}{1} \overset{?}{=} 9 \quad \text{Multiply.}$$

$$9 \overset{?}{=} 9 \quad \text{True}$$

Since $9 = 9$ is true, 15 is the solution of $\frac{3}{5}a = 9$.

■ Work Practice 4

Answers

3. 90 **4.** 35

Example 5 Solve: $-\dfrac{1}{4}x = \dfrac{1}{8}$

Solution: We multiply both sides of the equation by $-\dfrac{4}{1}$, the reciprocal of $-\dfrac{1}{4}$.

$$-\dfrac{1}{4}x = \dfrac{1}{8}$$

$$-\dfrac{4}{1}\cdot -\dfrac{1}{4}x = -\dfrac{4}{1}\cdot \dfrac{1}{8} \quad \text{Multiply both sides by } -\dfrac{4}{1}.$$

$$1\cdot x = -\dfrac{\overset{1}{\cancel{4}}\cdot 1}{1\cdot \underset{2}{\cancel{8}}} \quad \text{Multiply.}$$

$$x = -\dfrac{1}{2} \quad \text{Simplify.}$$

Check to see that $-\dfrac{1}{2}$ is the solution.

■ Work Practice 5

✔**Concept Check** Which operation is appropriate for solving each of the following equations, addition or division?
 a. $6 = -4x$ **b.** $6 = x - 4$

We often need to simplify one or both sides of an equation before applying the properties of equality to get the variable alone.

Example 6 Solve: $3y - 7y = 12$

Solution: First we combine like terms.

$$3y - 7y = 12$$
$$-4y = 12 \quad \text{Combine like terms.}$$
$$\dfrac{-4y}{-4} = \dfrac{12}{-4} \quad \text{Divide both sides by } -4.$$
$$y = -3 \quad \text{Simplify.}$$

Check: We replace y with -3.

$$3y - 7y = 12$$
$$3(-3) - 7(-3) \overset{?}{=} 12$$
$$-9 + 21 \overset{?}{=} 12$$
$$12 \overset{?}{=} 12 \quad \text{True}$$

The solution is -3.

■ Work Practice 6

Practice 5

Solve: $-\dfrac{7}{10}x = \dfrac{2}{5}$

Practice 6

Solve: $2m - 4m = 10$

Answers

5. $-\dfrac{4}{7}$ **6.** -5

✔**Concept Check Answers**
a. division **b.** addition

Practice 7

Solve: $-3a + 2a = -8 + 6$

Example 7   Solve: $-2z + z = 11 - 5$

Solution: We simplify both sides of the equation first.

$$-2z + z = 11 - 5$$
$$-1z = 6 \qquad \text{Combine like terms.}$$
$$\frac{-1z}{-1} = \frac{6}{-1} \qquad \text{Divide both sides by } -1.$$
$$z = -6 \qquad \text{Simplify.}$$

Check to see that -6 is the solution.

■ Work Practice 7

Practice 8

Solve: $-8 + 6 = \dfrac{a}{3}$

Example 8   Solve: $\dfrac{z}{-4} = 11 - 5$

Solution: Simplify the right side of the equation first.

$$\frac{z}{-4} = 11 - 5$$
$$\frac{z}{-4} = 6$$

Next, to get z alone, multiply both sides by -4.

$$-4 \cdot \frac{z}{-4} = -4 \cdot 6 \qquad \text{Multiply both sides by } -4.$$
$$\frac{-4}{-4} \cdot z = -4 \cdot 6$$
$$1z = -24 \quad \text{or} \quad z = -24$$

Check to see that -24 is the solution.

■ Work Practice 8

Answers

7. 2   **8.** -6

Vocabulary, Readiness & Video Check

Use the choices below to fill in each blank.

equation	multiplication	simplifying
solving	equivalent	expression

1. The equations $-3x = 51$ and $\dfrac{-3x}{-3} = \dfrac{51}{-3}$ are called _____ equations.

2. The difference between an equation and an expression is that a(n) _____ contains an equal sign, while a(n) _____ does not.

3. The process of writing $-3x + x$ as $2x$ is called _____ the expression.

4. For the equation $-5x = -20$, the process of finding that 4 is the solution is called _____ the equation.

5. By the _____ property of equality, $y = 8$ and $3 \cdot y = 3 \cdot 8$ are equivalent equations.

Martin-Gay Interactive Videos   *Watch the section lecture video and answer the following questions.*

Objective A **6.** In solving ⊟ Example 1, how is the multiplication property of
equality used? ▶

7. When solving ⊟ Example 5, what was done before applying
the multiplication property of equality? ▶

See Video 11.3 🍎

11.3 **Exercise Set** MyMathLab® ▶

Objective A *Solve. See Examples 1 through 5.*

1. $5x = 20$

2. $6y = 48$

▶ **3.** $-3z = 12$

4. $-2x = 26$

5. $0.4y = -12$

6. $0.8x = -8$

7. $2z = -34$

8. $7y = -21$

▶ **9.** $-0.3x = -15$

10. $-0.4z = -16$

11. $10 = \dfrac{2}{5}x$

12. $27 = \dfrac{3}{7}x$

13. $\dfrac{1}{6}y = -5$

14. $\dfrac{1}{8}y = -3$

15. $\dfrac{5}{6}x = \dfrac{5}{18}$

16. $\dfrac{4}{7}y = \dfrac{8}{21}$

17. $-\dfrac{2}{9}z = \dfrac{4}{27}$

18. $-\dfrac{3}{4}v = \dfrac{9}{14}$

Solve. First combine any like terms on each side of the equation. See Examples 6 and 7.

19. $2w - 12w = 40$

20. $-8y + y = 35$

21. $16 = 10t - 8t$

22. $100 = 15y - 5y$

23. $2z = 1.2 + 1.4$

24. $3x = 1.1 + 0.7$

25. $4 - 10 = -3z$

26. $12 - 20 = -4x$

Mixed Practice *Solve. See Examples 1 through 8.*

27. $-7x = 0$

28. $-20y = 0$

29. $0.4 = -8z$

30. $0.5 = -20x$

▶ **31.** $\dfrac{8}{5}t = -\dfrac{3}{8}$

32. $\dfrac{7}{4}r = -\dfrac{2}{7}$

33. $-\dfrac{3}{5}x = -\dfrac{6}{15}$

34. $-\dfrac{6}{7}y = -\dfrac{1}{14}$

35. $-3.6 = -0.9u + 0.3u$ **36.** $-5.4 = -1.4y + 1.3y$ ▶ **37.** $5 - 5 = 2x + 7x$

38. $12 + (-12) = 7x + 8x$ **39.** $-42 + 20 = -2x + 13x$ **40.** $-4y + 9y = -20 + 15$

41. $-3x - 3x = 50 - 2$ **42.** $5y - 9y = -14 + (-14)$ ▶ **43.** $23x - 25x = 7 - 9$

44. $6x - 8x = 12 - 22$ **45.** $\frac{1}{4}x - \frac{5}{8}x = 20 - 47$ **46.** $\frac{1}{2}x - \frac{4}{5}x = 10 - 19$

▶ **47.** $18 - 11 = \frac{x}{5}$ **48.** $14 - 9 = \frac{x}{12}$ **49.** $\frac{x}{-4} = 1 - (-6)$ **50.** $\frac{y}{-6} = 6 - (-1)$

Review

Evaluate each expression when $x = 5$. See Section 11.1.

51. $3x + 10$ **52.** $40x$ **53.** $\frac{x - 3}{2}$ **54.** $7x - 20$ **55.** $\frac{3x + 5}{x - 7}$ **56.** $\frac{2x - 1}{x - 8}$

The bar graph shows the number of acres burned by wildfires in the United States in recent years. Use the bar graph to answer Exercises 57 through 60. See Section 9.1.

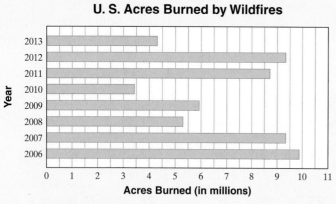

U. S. Acres Burned by Wildfires

Source: National Interagency Fire Center

57. During which year shown is the number of acres burned by wildfires the greatest?

58. During which year shown is the number of acres burned by wildfires the least?

59. Use the length of the bar to estimate the number of acres burned by wildfires in 2012.

60. Describe any trends shown in this graph.

Concept Extensions

What operation is appropriate for solving each equation: addition or division? See the Concept Check in this section.

61. $12 = x - 5$ **62.** $12 = -5x$ **63.** $-7x = 21$ **64.** $-7 + x = 21$

65. Why does the multiplication property of equality not allow us to divide both sides of an equation by zero?

66. Is the equation $-x = 6$ solved for the variable? Explain why or why not.

The equation $d = r \cdot t$ describes the relationship between distance d in miles, rate r in miles per hour, and time t in hours. If necessary, round answers to the nearest tenth.

67. The distance between New Orleans, Louisiana, and Memphis, Tennessee, by road is 390 miles. How long will it take to drive from New Orleans to Memphis if the driver maintains a speed of 60 miles per hour? (*Source: World Almanac*)

68. The distance between Boston, Massachusetts, and Milwaukee, Wisconsin, by road is 1050 miles. How long will it take to drive from Boston to Milwaukee if the driver maintains a speed of 55 miles per hour? (*Source: World Almanac*)

69. The distance between Cleveland, Ohio, and Indianapolis, Indiana, by road is 294 miles. At what speed should a driver drive if he or she would like to make the trip in 5 hours? (*Source: World Almanac*)

70. The distance between St. Louis, Missouri, and Minneapolis, Minnesota, by road is 552 miles. If it took 9 hours to drive from St. Louis to Minneapolis, what was the driver's average speed? (*Source: World Almanac*)

Solve.

71. $-0.025x = 91.2$

72. $3.6y = -1.259 - 3.277$

Solve.

73. $\dfrac{y}{72} = -86 - (-1029)$

74. $\dfrac{x}{-13} = 4^6 - 5^7$

75. $\dfrac{x}{-2} = 5^2 - |-10| - (-9)$

76. $\dfrac{y}{10} = (-8)^2 - |20| + (-2)^2$

Expressions and Equations

Answers

For the table below, identify each as an expression or an equation.

Expression or Equation	
1. $7x - 5y + 14$	
2. $7x = 35 + 14$	
3. $3(x - 2) = 5(x + 1) - 17$	
4. $-9(2x + 1) - 4(x - 2) + 14$	

1. _____

2. _____

3. _____

4. _____

Fill in each blank with "simplify" or "solve."

5. To _____ an expression, we combine any like terms.

5. _____

6. _____

6. To _____ an equation, we use the properties of equality to find any value of the variable that makes the equation a true statement.

7. _____

8. _____

Evaluate each expression when $x = -1$ and $y = 3$.

9. _____

7. $y - x$ **8.** $\dfrac{8y}{4x}$ **9.** $5x + 2y$ **10.** $\dfrac{y^2 + x}{2x}$

10. _____

11. _____

Simplify each expression by combining like terms.

12. _____

11. $7x + x$ **12.** $6y - 10y$

13. _____

13. $2a + 5a - 9a - 2$ **14.** $3x - y + 4 - 5x + 4y - 11$

14. _____

15. _____

Multiply and simplify if possible.

16. _____

15. $-2(4x + 7)$ **16.** $-3(2x - 10)$

17. $5(y + 2) - 20$

18. $12x + 3(x - 6) - 13$

△ **19.** Find the area.

Rectangle | 3 meters

$(4x - 2)$ meters

△ **20.** Find the perimeter.

Square

$5y$ inches

Solve and check.

21. $x + 7 = 20$

22. $-11 = x - 2$

23. $n - \dfrac{2}{5} = \dfrac{3}{10}$

24. $-7y = 0$

25. $12 = 11x - 14x$

26. $\dfrac{3}{5}x = 15$

27. $x - 1.2 = -4.5 + 2.3$

28. $8y + 7y = -45$

29. $6 - (-5) = x + 5$

30. $-0.2m = -1.6$

31. $-\dfrac{2}{3}n = \dfrac{6}{11}$

32. $11x = 55$

17. _____

18. _____

19. _____

20. _____

21. _____

22. _____

23. _____

24. _____

25. _____

26. _____

27. _____

28. _____

29. _____

30. _____

31. _____

32. _____

Objectives

A Solve Equations Using Addition and Multiplication Properties.

B Solve Equations Containing Parentheses.

C Write Numerical Sentences as Equations.

Objective A Solving Equations Using Addition and Multiplication Properties

We will now solve equations using more than one property of equality. To solve an equation such as $2x - 6 = 18$, we will first get the variable term $2x$ alone on one side of the equation.

Example 1 Solve: $2x - 6 = 18$

Solution: We start by adding 6 to both sides to get the variable term $2x$ alone.

$$2x - 6 = 18$$
$$2x - 6 + 6 = 18 + 6 \quad \text{Add 6 to both sides.}$$
$$2x = 24 \quad \text{Simplify.}$$

To finish solving, we divide both sides by 2.

$$\frac{2x}{2} = \frac{24}{2} \quad \text{Divide both sides by 2.}$$
$$1 \cdot x = 12 \quad \text{Simplify.}$$
$$x = 12$$

Check:

$$2x - 6 = 18$$
$$2(12) - 6 \stackrel{?}{=} 18 \quad \text{Replace } x \text{ with 12 and simplify.}$$
$$24 - 6 \stackrel{?}{=} 18$$
$$18 \stackrel{?}{=} 18 \quad \text{True}$$

The solution is 12.

■ Work Practice 1

Practice 1

Solve: $5y - 8 = 17$

Helpful Hint

Make sure you understand which property to use to solve an equation.

Addition
$$x + 2 = 10$$
To undo addition of 2, we subtract 2 from both sides.
$$x + 2 - 2 = 10 - 2 \quad \text{Use addition property of equality.}$$
$$x = 8$$

Check: $x + 2 = 10$
$$8 + 2 \stackrel{?}{=} 10$$
$$10 \stackrel{?}{=} 10 \quad \text{True}$$

Understood multiplication
$$2x = 10$$
To undo multiplication of 2, we divide both sides by 2.
$$\frac{2x}{2} = \frac{10}{2} \quad \text{Use multiplication property of equality.}$$
$$x = 5$$

Check: $2x = 10$
$$2 \cdot 5 \stackrel{?}{=} 10$$
$$10 \stackrel{?}{=} 10 \quad \text{True}$$

Answer

1. 5

Example 2   Solve: $17 - x + 3 = 15 - (-6)$

Solution: First we simplify each side of the equation.

$$17 - x + 3 = 15 - (-6)$$
$$20 - x = 21 \qquad \text{Combine like terms on each side of the equation.}$$

Next, we get the variable term alone on one side of the equation.

$$20 - x - 20 = 21 - 20 \qquad \text{Subtract 20 from both sides.}$$
$$-1x = 1 \qquad \text{Simplify. Recall that } -x \text{ means } -1x.$$
$$\frac{-1x}{-1} = \frac{1}{-1} \qquad \text{Divide both sides by } -1.$$
$$1 \cdot x = -1 \qquad \text{Simplify.}$$
$$x = -1$$

Check:

$$20 - x = 21$$
$$20 - (-1) \overset{?}{=} 21$$
$$21 \overset{?}{=} 21 \quad \text{True}$$

The solution is -1.

◼ **Work Practice 2**

Practice 2

Solve: $10 - y = 45$

Example 3   Solve: $1 = \frac{2}{3}x + 7$

Solution: Subtract 7 from both sides to get the variable term alone.

$$1 - 7 = \frac{2}{3}x + 7 - 7 \qquad \text{Subtract 7 from both sides.}$$
$$-6 = \frac{2}{3}x \qquad \text{Simplify.}$$
$$\frac{3}{2} \cdot -6 = \frac{3}{2} \cdot \frac{2}{3}x \qquad \text{Multiply both sides by } \frac{3}{2}.$$
$$\frac{3}{\cancel{2}} \cdot \frac{\overset{-3}{\cancel{-6}}}{1} = 1 \cdot x \qquad \text{Simplify.}$$
$$-9 = x \qquad \text{Simplify.}$$

Check to see that the solution is -9.

Helpful Hint   Don't forget that we can get the variable alone on either side of the equation.

◼ **Work Practice 3**

Practice 3

Solve: $11 = \frac{3}{4}y + 20$

If an equation contains variable terms on both sides, we use the addition property of equality to get all the variable terms on one side and all the constants, or numbers, on the other side.

Answers

2. -35 **3.** -12

Practice 4

Solve: $9x - 12 = x + 4$

Example 4 Solve: $3a - 6 = a + 4$

Solution:

$$3a - 6 = a + 4$$
$$3a - 6 + 6 = a + 4 + 6 \qquad \text{Add 6 to both sides.}$$
$$3a = a + 10 \qquad \text{Simplify.}$$
$$3a - a = a + 10 - a \qquad \text{Subtract } a \text{ from both sides.}$$
$$2a = 10 \qquad \text{Simplify.}$$
$$\frac{2a}{2} = \frac{10}{2} \qquad \text{Divide both sides by 2.}$$
$$a = 5 \qquad \text{Simplify.}$$

Check to see that the solution is 5.

◼ **Work Practice 4**

Practice 5

Solve: $8x + 4.2 = 10x - 11.6$

Example 5 Solve: $7x + 3.2 = 4x - 1.6$

Solution:

$$7x + 3.2 = 4x - 1.6$$
$$7x + 3.2 - 3.2 = 4x - 1.6 - 3.2 \qquad \text{Subtract 3.2 from both sides.}$$
$$7x = 4x - 4.8 \qquad \text{Simplify.}$$
$$7x - 4x = 4x - 4.8 - 4x \qquad \text{Subtract } 4x \text{ from both sides.}$$
$$3x = -4.8 \qquad \text{Simplify.}$$
$$\frac{3x}{3} = \frac{-4.8}{3} \qquad \text{Divide both sides by 3.}$$
$$x = -1.6 \qquad \text{Simplify.}$$

Check to see that -1.6 is the solution.

◼ **Work Practice 5**

Objective B Solving Equations Containing Parentheses ▶

If an equation contains parentheses, we must first use the distributive property to remove them.

Practice 6

Solve: $6(a - 5) = 7a - 13$

Example 6 Solve: $7(x - 2) = 9x - 6$

Solution: First we apply the distributive property.

$$\overset{\frown}{7(x - 2)} = 9x - 6$$
$$7x - 14 = 9x - 6 \qquad \text{Apply the distributive property.}$$

Next, we move variable terms to one side of the equation and constants to the other side.

$$7x - 14 - 9x = 9x - 6 - 9x \quad \text{Subtract } 9x \text{ from both sides.}$$
$$-2x - 14 = -6 \quad \text{Simplify.}$$
$$-2x - 14 + 14 = -6 + 14 \quad \text{Add 14 to both sides.}$$
$$-2x = 8 \quad \text{Simplify.}$$
$$\frac{-2x}{-2} = \frac{8}{-2} \quad \text{Divide both sides by } -2.$$
$$x = -4 \quad \text{Simplify.}$$

Check to see that -4 is the solution.

■ **Work Practice 6**

You may want to use the steps shown below to solve equations.

Steps for Solving an Equation

Step 1: If parentheses are present, use the distributive property.

Step 2: Combine any like terms on each side of the equation.

Step 3: Use the addition property of equality to rewrite the equation so that variable terms are on one side of the equation and constant terms are on the other side.

Step 4: Use the multiplication property of equality to divide both sides by the numerical coefficient of the variable to solve.

Step 5: Check the solution in the *original equation*.

Example 7   Solve: $3(2x - 6) + 6 = 0$

Solution:

$$3(2x - 6) + 6 = 0$$

Step 1: $6x - 18 + 6 = 0$ Apply the distributive property.

Step 2: $6x - 12 = 0$ Combine like terms on the left side of the equation.

Step 3: $6x - 12 + 12 = 0 + 12$ Add 12 to both sides.

$$6x = 12 \quad \text{Simplify.}$$

Step 4: $\dfrac{6x}{6} = \dfrac{12}{6}$ Divide both sides by 6.

$$x = 2 \quad \text{Simplify.}$$

Check:

Step 5: $3(2x - 6) + 6 = 0$

$$3(2 \cdot 2 - 6) + 6 \stackrel{?}{=} 0$$
$$3(4 - 6) + 6 \stackrel{?}{=} 0$$
$$3(-2) + 6 \stackrel{?}{=} 0$$
$$-6 + 6 \stackrel{?}{=} 0$$
$$0 \stackrel{?}{=} 0 \quad \text{True}$$

The solution is 2.

■ **Work Practice 7**

Practice 7

Solve: $4(2x - 3) + 4 = 0$

Answer

7. 1

Objective C Writing Sentences as Equations ▶

Next, we practice translating sentences into equations. Below are key words and phrases that translate to an equal sign:

Key Words or Phrases	Examples	Symbols
equals	3 equals 2 plus 1	$3 = 2 + 1$
gives	the quotient of 10 and -5 gives -2	$\dfrac{10}{-5} = -2$
is/was	17 minus 12 is 5	$17 - 12 = 5$
yields	11 plus 2 yields 13	$11 + 2 = 13$
amounts to	twice -15 amounts to -30	$2(-15) = -30$
is equal to	-24 is equal to 2 times -12	$-24 = 2(-12)$

Practice 8

Translate each sentence into an equation.

a. The difference of 110 and 80 is 30.

b. The product of 3 and the sum of -9 and 11 amounts to 6.

c. The quotient of 24 and -6 yields -4.

Answers

8. a. $110 - 80 = 30$

b. $3(-9 + 11) = 6$ **c.** $\dfrac{24}{-6} = -4$

Example 8 Translate each sentence into an equation.

a. The product of 7 and 6 is 42.

b. Twice the sum of 3 and 5 is equal to 16.

c. The quotient of -45 and 5 yields -9.

Solution:

a. In words: the product of 7 and 6   is   42

Translate: $7 \cdot 6$ $=$ 42

b. In words: twice   the sum of 3 and 5   is equal to   16

Translate: 2 $(3 + 5)$ $=$ 16

c. In words: the quotient of -45 and 5   yields -9

Translate: $\dfrac{-45}{5}$ $=$ -9

■ Work Practice 8

🖩 Calculator Explorations Checking Possible Solutions

A calculator can be used to check possible solutions of equations. To do this, replace the variable by the possible solution and evaluate each side of the equation separately. For example, to see whether 7 is a solution of the equation $52x = 15x + 259$, replace x with 7 and use your calculator to evaluate each side separately.

Equation: $52x = 15x + 259$

$52 \cdot 7 \overset{?}{=} 15 \cdot 7 + 259$ Replace x with 7.

Evaluate left side: 52 ⊠ 7 then = or ENTER.

Display: 364 .

Evaluate right side: 15 ⊠ 7 ⊞ 259 then =

or ENTER . Display: 364 .

Since the left side equals the right side, 7 is a solution of the equation $52x = 15x + 259$.

Use a calculator to determine whether the numbers given are solutions of each equation.

1. $76(x - 25) = -988$;   12

2. $-47x + 862 = -783$;   35

3. $x + 562 = 3x + 900$; -170

4. $55(x + 10) = 75x + 910$; -18

5. $29x - 1034 = 61x - 362$; -21

6. $-38x + 205 = 25x + 120$;   25

Vocabulary, Readiness & Video Check

Use the choices below to fill in each blank. Some choices may be used more than once.

addition	multiplication	combine like terms
$5(2x + 6) - 1 = 39$	$3x - 9 + x - 16$	distributive

1. An example of an expression is ——————— while an example of an equation is ———————.

2. To solve $\dfrac{x}{-7} = -10$, we use the ——————— property of equality.

3. To solve $x - 7 = -10$, we use the ——————— property of equality.

Use the order of the Steps for Solving an Equation in this section to answer Exercises 4 through 6.

4. To solve $9x - 6x = 10 + 6$, first ———————.

5. To solve $5(x - 1) = 25$, first use the ——————— property.

6. To solve $4x + 3 = 19$, first use the ——————— property of equality.

Martin-Gay Interactive Videos   *Watch the section lecture video and answer the following questions.*

See Video 11.4

Objective A 7. In ▣ Example 1, the number 1 is subtracted from the left side of the equation. What property tells us we must also subtract 1 from the right side? Why is it important to do the same thing to both sides? ▶

Objective B 8. From ▣ Example 3, what is the first step when solving an equation that contains parentheses? What property do we use to perform this step? ▶

Objective C 9. What word or phrase translates to "equals" in ▣ Example 5? In ▣ Example 6? ▶

11.4 Exercise Set MyMathLab® ▶

Objective A *Solve each equation. See Examples 1 through 5.*

1. $2x - 6 = 0$

2. $3y - 12 = 0$

3. $3n + 3.6 = 9.3$

4. $4z + 0.8 = 5.2$

5. $6 - n = 10$

6. $7 - y = 9$

7. $-\dfrac{2}{5}x + 19 = -21$

8. $-\dfrac{3}{7}y - 14 = 7$

9. $1.7 = 2y + 9.5$

10. $-5.1 = 3x + 2.4$

11. $2n + 8 = 0$

12. $8w + 40 = 0$

13. $3x - 7 = 4x + 5$

14. $7x - 1 = 8x + 4$

▶ **15.** $10x + 15 = 6x + 3$

16. $5x - 3 = 2x - 18$

17. $9 - 3x = 14 + 2x$

18. $4 - 7m = -3m + 4$

19. $-1.4x - 2 = -1.2x + 7$

20. $5.7y + 14 = 5.4y - 10$

21. $x + 20 + 2x = -10 - 2x - 15$

22. $2x + 10 + 3x = -12 - x - 20$

23. $40 + 4y - 16 = 13y - 12 - 3y$

24. $19x - 2 - 7x = 31 + 6x - 15$

Objective B *Solve each equation. See Examples 6 and 7.*

25. $-2(y + 4) = 2$

26. $-1(y + 3) = 10$

▶**27.** $3(x - 1) - 12 = 0$

28. $2(x + 5) + 8 = 0$

29. $35 - 17 = 3(x - 2)$

30. $22 - 42 = 4(x - 1)$

31. $2(y - 3) = y - 6$

32. $3(z + 2) = 5z + 6$

33. $2t - 1 = 3(t + 7)$

34. $-4 + 3c = 4(c + 2)$

35. $3(5c + 1) - 12 = 13c + 3$

36. $4(3t + 4) - 20 = 3 + 5t$

Mixed Practice (*Sections 11.1–11.4*) *Solve each equation. See Examples 1 through 7.*

37. $-4x = 44$

38. $-3x = 51$

39. $x + 9 = 2$

40. $y - 6 = -11$

41. $8 - b = 13$

42. $7 - z = 15$

43. $3r + 4 = 19$

44. $5m + 1 = 46$

45. $2x - 1 = -7$

46. $3t - 2 = -11$

47. $7 = 4c - 1$

48. $9 = 2b - 5$

49. $9a + 29 = -7$

50. $10 + 4v = -6$

51. $0 = 4x + 4$

52. $0 = 5y + 5$

53. $11(x - 2) = 22$

54. $5(a - 4) = 20$

▶**55.** $-7c + 1 = -20$

56. $-2b + 5 = -7$

57. $3(x - 5) = -7 - 11$

58. $4(x - 2) = -20 - 4$

59. $-5 + 7k = -13 + 8k$

60. $-7 + 9d = -17 + 10d$

61. $4x + 3 = 2x + 11$

62. $6y - 8 = 3y + 7$

63. $-8(n + 2) + 17 = -6n - 5$

64. $-10(x + 1) + 2 = -x + 10$

65. $\dfrac{3}{8}x + 14 = \dfrac{5}{8}x - 2$

66. $\dfrac{2}{7}x - 9 = \dfrac{5}{7}x - 15$

67. $10 + 5(z - 2) = -4z + 1$

68. $20 + 4(w - 5) = 5 - 2w$

69. $\dfrac{5}{8}a = \dfrac{1}{8}a + \dfrac{3}{4}$

70. $\dfrac{4}{9}a = \dfrac{1}{9}a + \dfrac{5}{6}$

71. $7(6 + w) = 6(w - 2)$

72. $6(5 + c) = 5(c - 4)$

73. $3 + 2(2n - 5) = 1$

74. $5 + 4(3x - 2) = 21$

75. $2(3z - 2) - 2(5 - 2z) = 4$

76. $2(3w + 7) - 4(5 - 2w) = 6$

77. $-20 - (-50) = \dfrac{x}{9}$

78. $-2 - 10 = \dfrac{z}{10}$

79. $12 + 5t = 6(t + 2)$ **80.** $4 + 3c = 2(c + 2)$ ▶**81.** $3(5c - 1) - 2 = 13c + 3$

82. $4(2t + 5) - 21 = 7t - 6$ **83.** $10 + 5(z - 2) = 4z + 1$ **84.** $14 + 4(w - 5) = 6 - 2w$

Objective C   *Write each sentence as an equation. See Example 8.*

85. The sum of -42 and 16 is -26.

86. The difference of -30 and 10 equals -40.

▶**87.** The product of -5 and -29 gives 145.

88. The quotient of -16 and 2 yields -8.

▶**89.** Three times the difference of -14 and 2 amounts to -48.

90. The product of -2 and the sum of 3 and 12 is -30.

91. The quotient of 100 and twice 50 is equal to 1.

92. Seventeen subtracted from -12 equals -29.

Review

The following bar graph shows the number of U.S. federal individual income tax returns that are filed electronically during the years shown (some years are projected). Use this graph to answer Exercises 93 through 96. See Section 9.1.

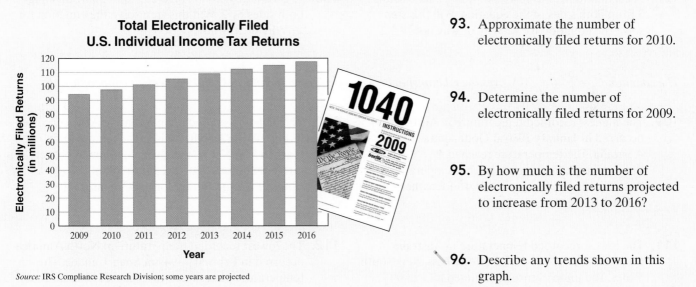

Total Electronically Filed U.S. Individual Income Tax Returns

Source: IRS Compliance Research Division; some years are projected

93. Approximate the number of electronically filed returns for 2010.

94. Determine the number of electronically filed returns for 2009.

95. By how much is the number of electronically filed returns projected to increase from 2013 to 2016?

96. Describe any trends shown in this graph.

Concept Extensions

Using the steps for solving an equation, choose the next operation for solving the given equation.

97. $2x - 5 = -7$
   a. Add 7 to both sides.
   b. Add 5 to both sides.
   c. Divide both sides by 2.

98. $3x + 2x = -x - 4$
   a. Add 4 to both sides.
   b. Subtract $2x$ from both sides.
   c. Add $3x$ and $2x$.

99. $-3x = -12$
    a. Divide both sides by -3.
    b. Add 12 to both sides.
    c. Add $3x$ to both sides.

100. $9 - 5x = 15$
    a. Divide both sides by -5.
    b. Subtract 15 from both sides.
    c. Subtract 9 from both sides.

A classmate shows you steps for solving an equation. The solution does not check, but the classmate is unable to find the error. For each set of steps, check the solution, find the error, and correct it.

101.
$$2(3x - 5) = 5x - 7$$
$$6x - 5 = 5x - 7$$
$$6x - 5 + 5 = 5x - 7 + 5$$
$$6x = 5x - 2$$
$$6x - 5x = 5x - 2 - 5x$$
$$x = -2$$

102.
$$37x + 1 = 9(4x - 7)$$
$$37x + 1 = 36x - 7$$
$$37x + 1 - 1 = 36x - 7 - 1$$
$$37x = 36x - 8$$
$$37x - 36x = 36x - 8 - 36x$$
$$x = -8$$

Solve.

103. $(-8)^2 + 3x = 5x + 4^3$

104. $3^2 \cdot x = (-9)^3$

105. $2^3(x + 4) = 3^2(x + 4)$

106. $x + 45^2 = 54^2$

107. A classmate tries to solve $3x = 39$ by subtracting 3 from both sides of the equation. Will this step solve the equation for x? Why or why not?

108. A classmate tries to solve $2 + x = 20$ by dividing both sides by 2. Will this step solve the equation for x? Why or why not?

The equation $C = \dfrac{5}{9}(F - 32)$ gives the relationship between Celsius temperatures C and Fahrenheit temperatures F.

109. The highest recorded temperature in Australia occurred in January 1960 at Oodnadatta, South Australia. The temperature reached 50.7°C. Use the given equation to convert this temperature to degrees Fahrenheit. (*Source:* World Weather Centre at Perth)

110. The highest recorded temperature in Africa occurred in July 1931 at Kebili, Tunisia. The temperature reached 55.0°C. Use the given equation to convert this temperature to degrees Fahrenheit. (*Source:* World Meteorological Organization)

111. The lowest recorded temperature in Australia occurred in June 1994 at Charlotte Pass, New South Wales. The temperature plummeted to -23.0°C. Use the given equation to convert this temperature to degrees Fahrenheit. (*Source:* World Weather Centre at Perth)

112. The lowest recorded temperature in North America occurred in February 1947 at Snag, Canada. The temperature plummeted to -63.0°C. Use the given equation to convert this temperature to degrees Fahrenheit. (*Source:* World Weather Centre at Perth)

11.5 Equations and Problem Solving

Objective A Writing Phrases as Algebraic Expressions

Now that we have practiced solving equations for a variable, we can extend considerably our problem-solving skills. We begin by writing phrases as algebraic expressions using the following key words and phrases as a guide:

Addition	Subtraction	Multiplication	Division	Equal Sign
sum	difference	product	quotient	equals
plus	minus	times	divided by	gives
added to	subtracted from	multiply	into	is/was
more than	less than	twice	per	yields
increased by	decreased by	of		amounts to
total	less	double		is equal to

Example 1 Write each phrase as an algebraic expression. Use x to represent "a number."

a. 7 increased by a number
b. 15 decreased by a number
c. the product of 2 and a number
d. the quotient of a number and 5
e. 2 subtracted from a number
f. the sum of 9 and twice a number

Solution:

a. In words:

7	increased by	a number
↓	↓	↓

Translate: $7 \quad + \quad x$

b. In words:

15	decreased by	a number
↓	↓	↓

Translate: $15 \quad - \quad x$

c. In words:

the product of

2	and	a number
↓	↓	↓

Translate: $2 \quad \cdot \quad x \qquad$ or $2x$

d. In words:

the quotient of

a number	and	5
↓	↓	↓

Translate: $x \quad \div \quad 5 \qquad$ or $\dfrac{x}{5}$

e. In words:

2	subtracted from	a number

Translate: $x \quad - \quad 2$

(Continued on next page)

Objectives

A Write Phrases as Algebraic Expressions.

B Write Sentences as Equations.

C Use Problem-Solving Steps to Solve Problems.

Practice 1

Write each phrase as an algebraic expression. Use x to represent "a number."

a. twice a number
b. 8 increased by a number
c. 10 minus a number
d. 10 subtracted from a number
e. the quotient of 6 and a number
f. the sum of 14 and triple a number

Answers
1. a. $2x$ **b.** $8 + x$ **c.** $10 - x$
d. $x - 10$ **e.** $\dfrac{6}{x}$ **f.** $14 + 3x$

721

f. In words: the sum of

9	and	twice a number
↓	↓	↓

Translate: 9 + 2x

■ Work Practice 1

Objective B Writing Sentences as Equations ▶

Now that we have practiced writing phrases as algebraic expressions, let's write sentences as equations. You may want to first study the key words and phrases chart to review some key words and phrases that translate to an equal sign.

Example 2   Write each sentence as an equation. Use x to represent "a number."

a. Nine increased by a number is 5.

b. Twice a number equals −10.

c. A number minus 6 amounts to 168.

d. Three times the sum of a number and 5 is −30.

e. The quotient of twice a number and 8 is equal to 2.

Solution:

a. In words:

nine	increased by	a number	is	5
↓	↓	↓	↓	↓

Translate: 9 + x = 5

b. In words:

twice a number	equals	−10
↓	↓	↓

Translate: 2x = −10

c. In words:

a number	minus	6	amounts to	168
↓	↓	↓	↓	↓

Translate: x − 6 = 168

d. In words:

three times	the sum of a number and 5	is	−30
↓	↓	↓	↓

Translate: 3 (x + 5) = −30

e. In words: the quotient of

twice a number	and	8	is equal to	2
↓	↓	↓	↓	↓

Translate: 2x ÷ 8 = 2

or $\frac{2x}{8} = 2$

■ Work Practice 2

Objective C Using Problem-Solving Steps to Solve Problems ▶

Our main purpose for studying arithmetic and algebra is to solve problems. The same problem-solving steps that have been used throughout this text are used in this section also. Those steps are next.

Problem-Solving Steps

1. UNDERSTAND the problem. During this step, become comfortable with the problem. Some ways of doing this are as follows:
   - Read and reread the problem.
   - Construct a drawing.
   - Propose a solution and check. Pay careful attention to how you check your proposed solution. This will help when writing an equation to model the problem.
   - Choose a variable to represent an unknown. Use this variable to represent any other unknowns.

2. TRANSLATE the problem into an equation.

3. SOLVE the equation.

4. INTERPRET the results: *Check* the proposed solution in the stated problem and *state* your conclusion.

The first problem that we solve consists of finding an unknown number.

Example 3 Finding an Unknown Number

Twice a number plus 3 is the same as the number minus 6. Find the unknown number.

Solution:

1. UNDERSTAND the problem. To do so, we read and reread the problem.

    Let's propose a solution to help us understand. Suppose the unknown number is 5. Twice this number plus 3 is $2 \cdot 5 + 3$ or 13. Is this the same as the number minus 6, or $5 - 6$, or -1? Since 13 is not the same as -1, we know that 5 is not the solution. However, remember that the purpose of proposing a solution is not to guess correctly, but to better understand the problem.

    Now let's choose a variable to represent the unknown. Let's let

    $x = $ unknown number

2. TRANSLATE the problem into an equation.

    In words:

twice a number	plus 3	is the same as	the number minus 6
↓	↓	↓	↓

    Translate: $2x$ $+ 3$ $=$ $x - 6$

3. SOLVE the equation. To solve the equation, we first subtract x from both sides.

    $2x + 3 = x - 6$

    $2x + 3 - x = x - 6 - x$

    $x + 3 = -6$ Simplify.

    $x + 3 - 3 = -6 - 3$ Subtract 3 from both sides.

    $x = -9$ Simplify.

    (Continued on next page)

Practice 3

Translate "The difference of a number and 2 equals 6 added to three times the number" into an equation and solve.

Answer

3. $x - 2 = 6 + 3x; -4$

4. INTERPRET the results. First, *check* the proposed solution in the stated problem. Twice "−9" is −18 and −18 + 3 is −15. This is equal to the number minus 6, or "−9" − 6, or −15. Then *state* your conclusion: The unknown number is −9.

■ Work Practice 3

✓**Concept Check** Suppose you have solved an equation involving perimeter to find the length of a rectangular table. Explain why you would want to recheck your math if you obtain the result of −5.

Practice 4

The distance by road from Cincinnati, Ohio, to Denver, Colorado, is 71 miles *less* than the distance from Denver to San Francisco, California. If the total of these two distances is 2399 miles, find the distance from Denver to San Francisco.

| Example 4 | Determining Distances |

The distance by road from Chicago, Illinois, to Los Angeles, California, is 1091 miles *more* than the distance from Chicago to Boston, Massachusetts. If the total of these two distances is 3017 miles, find the distance from Chicago to Boston. (*Source: World Almanac*)

Solution:

1. UNDERSTAND the problem. We read and reread the problem.

Let's propose and check a solution to help us better understand the problem. Suppose the distance from Chicago to Boston is 600 miles. Since the distance from Chicago to Los Angeles is 1091 *more* miles, then this distance is 600 + 1091 = 1691 miles. With these numbers, the total of the distances is 600 + 1691 = 2291 miles. This is less than the given total of 3017 miles, so we are incorrect. But not only do we have a better understanding of this exercise, we also know that the distance from Boston to Chicago is greater than 600 miles since this proposed solution led to a total too small. Now let's choose a variable to represent an unknown. Then we'll use this variable to represent any other unknown quantities. Let

$$x = \text{distance from Chicago to Boston}$$

Then

$$x + 1091 = \text{distance from Chicago to Los Angeles}$$

since that distance is 1091 more miles.

2. TRANSLATE the problem into an equation.

In words:	Chicago to Boston distance	+	Chicago to Los Angeles distance	=	total miles
	↓		↓		↓
Translate:	x	+	$x + 1091$	=	3017

3. SOLVE the equation:

$$x + x + 1091 = 3017$$
$$2x + 1091 = 3017 \qquad \text{Combine like terms.}$$
$$2x + 1091 - 1091 = 3017 - 1091 \qquad \text{Subtract 1091 from both sides.}$$
$$2x = 1926 \qquad \text{Simplify.}$$
$$\frac{2x}{2} = \frac{1926}{2} \qquad \text{Divide both sides by 2.}$$
$$x = 963 \qquad \text{Simplify.}$$

4. INTERPRET the results. First *check* the proposed solution in the stated problem. Since x represents the distance from Chicago to Boston, this is 963 miles. The distance from Chicago to Los Angeles is $x + 1091 = 963 + 1091 = 2054$ miles.

Answer

4. 1235 miles

✓**Concept Check Answer**

Length cannot be negative.

To check, notice that the total number of miles is $963 + 2054 = 3017$ miles, the given total of miles. Also, 2054 is 1091 more miles than 963, so the solution checks. Then, *state* your conclusion: The distance from Chicago to Boston is 963 miles.

■ Work Practice 4

Example 5   Calculating Separate Costs

A salesperson at an electronics store sold a computer system and software for $2100, receiving four times as much money for the computer system as for the software. Find the price of each.

Solution:

1. **UNDERSTAND** the problem. We read and re-read the problem. Then we choose a variable to represent an unknown. We use this variable to represent any other unknown quantities. We let

    $x =$ the software price

    $4x =$ the computer system price

2. **TRANSLATE** the problem into an equation.

In words:	software price	and	computer price	is	2100
	↓	↓	↓	↓	↓
Translate:	x	$+$	$4x$	$=$	2100

3. **SOLVE** the equation:

$$x + 4x = 2100$$
$$5x = 2100 \quad \text{Combine like terms.}$$
$$\frac{5x}{5} = \frac{2100}{5} \quad \text{Divide both sides by 5.}$$
$$x = 420 \quad \text{Simplify.}$$

4. **INTERPRET** the results. *Check* the proposed solution in the stated problem. The software sold for $420. The computer system sold for $4x = 4(\$420) = \$1680$. Since $\$420 + \$1680 = \$2100$, the total price, and $1680 is four times $420, the solution checks. *State* your conclusion: The software sold for $420, and the computer system sold for $1680.

■ Work Practice 5

Practice 5

A woman's $57,000 estate is to be divided so that her husband receives twice as much as her son. How much will each receive?

Answer

5. husband: $38,000; son: $19,000

Vocabulary, Readiness & Video Check

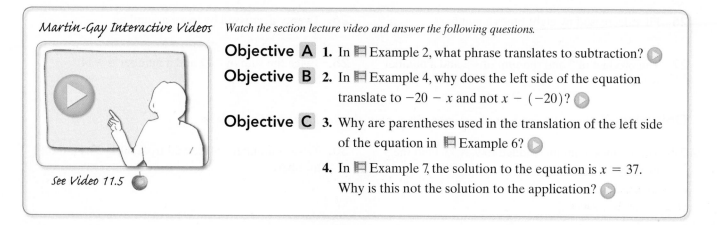

Martin-Gay Interactive Videos   Watch the section lecture video and answer the following questions.

Objective A 1. In ▣ Example 2, what phrase translates to subtraction? ▶

Objective B 2. In ▣ Example 4, why does the left side of the equation translate to $-20 - x$ and not $x - (-20)$? ▶

Objective C 3. Why are parentheses used in the translation of the left side of the equation in ▣ Example 6? ▶

4. In ▣ Example 7, the solution to the equation is $x = 37$. Why is this not the solution to the application? ▶

See Video 11.5 ●

11.5 Exercise Set MyMathLab®

Objective A Translating *Write each phrase as a variable expression. Use x to represent "a number." See Example 1.*

1. The sum of a number and five

2. Ten plus a number

3. The total of a number and eight

4. The difference of a number and five hundred

5. Twenty decreased by a number

6. A number less thirty

7. The product of 512 and a number

8. A number times twenty

9. A number divided by 2

10. The quotient of six and a number

11. The sum of seventeen, a number, and the product of five and the number

12. The difference of twice a number, and four

Objective B Translating *Write each sentence as an equation. Use x to represent "a number." See Example 2.*

13. A number added to −5 is −7.

14. Five subtracted from a number equals 10.

15. Three times a number yields 27.

16. The quotient of 8 and a number is −2.

17. A number subtracted from −20 amounts to 104.

18. Two added to twice a number gives −14.

Objectives A B Translating *Write each phrase as a variable expression or each sentence as an equation. Use x to represent "a number." See Examples 1 and 2.*

19. The product of five and a number

20. The quotient of twenty and a number, decreased by three

21. A number subtracted from 11

22. Twelve subtracted from a number

23. Twice a number gives 108.

24. Five times a number is equal to −75.

25. Fifty decreased by eight times a number

26. Twenty decreased by twice a number

27. The product of 5 and the sum of −3 and a number is −20.

28. Twice the sum of −17 and a number is −14.

Objective C *Translate each to an equation. Then solve the equation. See Example 3.*

29. Three times a number, added to 9, is 33. Find the number.

30. Twice a number, subtracted from 60, is 20. Find the number.

31. The sum of 3, 4, and a number amounts to 16. Find the number.

32. The sum of 7, 9, and a number is 40. Find the number.

33. The difference of a number and 3 is equal to the quotient of 10 and 5. Find the number.

34. Eight decreased by a number equals the quotient of 15 and 5. Find the number.

35. Thirty less a number is equal to the product of 3 and the sum of the number and 6. Find the number.

36. The product of a number and 3 is twice the sum of that number and 5. Find the number.

37. 40 subtracted from five times a number is 8 more than the number. Find the number.

38. Five times the sum of a number and 2 is 11 less than the number times 8. Find the number.

▶ 39. Three times the difference of a number and 5 amounts to the quotient of 108 and 12. Find the number.

40. Seven times the difference of a number and 1 gives the quotient of 70 and 10. Find the number.

41. The product of 4 and a number is the same as 30 less twice that same number. Find the number.

42. Twice a number equals 25 less triple that same number. Find the number.

Solve. For Exercises 43 and 44, the solutions have been started for you. See Examples 4 and 5.

43. Currently, Florida has 26 fewer electoral votes for president than California. If the total number of electoral votes for these two states is 84, find the number for each state. (*Source:* U.S. Electoral College)

Start the solution:

1. UNDERSTAND the problem. Reread it as many times as needed. Let's let

   x = number of electoral votes for California

   Then

   $x - 26$ = number of electoral votes for Florida

2. TRANSLATE into an equation. (Fill in the blanks below.)

$$\begin{array}{ccccc} \boxed{\text{votes for California}} & + & \boxed{\text{votes for Florida}} & = & 84 \\ \downarrow & & \downarrow & & \\ \underline{\quad\quad} & + & \underline{\quad\quad} & = & 84 \end{array}$$

   Now, you finish with

3. SOLVE the equation.
4. INTERPRET the results.

44. Currently, Ohio has twice the number of electoral votes for president as South Carolina. If the total number of electoral votes for these two states is 27, find the number for each state. (*Source:* U.S. Electoral College)

Start the solution:

1. UNDERSTAND the problem. Reread it as many times as needed. Let's let

   x = number of electoral votes for South Carolina

   Then

   $2x$ = number of electoral votes for Ohio

2. TRANSLATE into an equation. (Fill in the blanks below.)

$$\begin{array}{ccccc} \boxed{\text{votes for South Carolina}} & + & \boxed{\text{votes for Ohio}} & = & 27 \\ \downarrow & & \downarrow & & \\ \underline{\quad\quad} & + & \underline{\quad\quad} & = & 27 \end{array}$$

   Now, you finish with

3. SOLVE the equation.
4. INTERPRET the results.

45. A falcon, when diving, can travel five times as fast as a pheasant's top speed. If the total speed for these two birds is 222 miles per hour, find the fastest speed of the falcon and the fastest speed of the pheasant. (*Source: Fantastic Book of Comparisons*)

46. Norway has had three times as many rulers as Liechtenstein. If the total number of rulers for both countries is 56, find the number of rulers for Norway and the number for Liechtenstein.

Norway   Liechtenstein

47. The largest university (by enrollment) is Indira Gandhi National Open University in India, followed by Anadolu University in Turkey. If the enrollment in the Indian university is 1.5 million more students than the Turkish university and their combined enrollment is 5.5 million students, find the enrollment for each university. (*Source:* Wikipedia.org)

48. The average life expectancy for an elephant is 24 years longer than the life expectancy for a chimpanzee. If the total of these life expectancies is 130 years, find the life expectancy of each.

49. An Xbox 360 game system and several games are sold for $560. The cost of the Xbox 360 is 3 times as much as the cost of the games. Find the cost of the Xbox 360 and the cost of the games.

50. The two top-selling video games for 2012 were *Call of Duty: Black Ops II* and *Madden NFL 13*. The retail price of *Call of Duty: Black Ops II* is $30 more than the retail price of *Madden NFL 13*. If the total of these two prices is $88, find the price of each game. (*Source:* NPD Group, Amazon.com)

51. By air, the distance from New York City to London is 2001 miles *less* than the distance from Los Angeles to Tokyo. If the total of these two distances is 8939 miles, find the distance from Los Angeles to Tokyo.

52. By air, the distance from Melbourne, Australia, to Cairo, Egypt, is 2338 miles *more* than the distance from Madrid, Spain, to Bangkok, Thailand. If the total of these distances is 15,012 miles, find the distance from Madrid to Bangkok.

53. The two NCAA stadiums with the largest capacities are Michigan Stadium (Univ. of Michigan) and Beaver Stadium (Penn State). Michigan Stadium has a capacity of 3329 more than Beaver Stadium. If the combined capacity for the two stadiums is 216,473, find the capacity for each stadium. (*Source:* Wikipedia.org)

54. In 2020, China is projected to be the country with the greatest number of visiting tourists. This number is twice the number of tourists projected for Spain. If the total number of tourists for these two countries is projected to be 210 million, find the number projected for each. (*Source: The State of the World Atlas* by Dan Smith)

55. California contains the largest state population of native Americans. This population is three times the native American population of Washington state. If the total of these two populations is 412 thousand, find the native American population in each of these two states. (*Source:* U.S. Census Bureau)

56. In 2012, the number of Kohl's stores in California exceeded the number of Kohl's stores in Texas by 43 stores. If the combined number of Kohl's stores in California and Texas is 213, find the number of Kohl's stores in each state. (*Source:* Kohl's Corporation)

57. During the 2013 Super Bowl, the Baltimore Ravens scored 3 more points than the San Francisco 49ers. Together, both teams scored a total of 65 points. How many points did the 2013 champion Baltimore Ravens score during this game? (*Source:* NFL)

58. During the 2013 Men's NCAA Division I basketball championship game, the Michigan Wolverines scored 6 fewer points than the Louisville Cardinals. Together, both teams scored a total of 158 points. How many points did the 2013 champion Louisville Cardinals score during this game? (*Source:* National Collegiate Athletic Association)

59. In 2020, the shortage of nurses is projected to be 533,201 more nurses than the shortage in 2010. If the total number of nurse shortages for these two years is 1,083,631, find the nurse shortage for each year.

60. The percent shortage of nurses in 2019 is predicted to be three times the percent shortage in 2007. If the total percent shortages for these two years is 36%, find the percent shortage in 2007 and the percent shortage in 2019.

National Supply and Demand Projections for Full-Time Equivalent Registered Nurses: 2000 to 2020

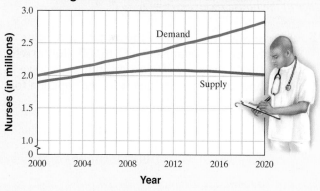

Source: Bureau of Health Professions, RN Supply and Demand Projections

61. In Germany, about twice as many cars are manufactured per day than in Spain. If the total number of these cars manufactured per day is 19,827, find the number manufactured in Spain and the number manufactured in Germany.

62. A Toyota Camry is traveling twice as fast as a Dodge truck. If their combined speed is 105 miles per hour, find the speed of the car and find the speed of the truck.

63. A biker sold his used mountain bike and accessories for $270. If he received five times as much money for the bike as he did for the accessories, find how much money he received for the bike.

64. A tractor and a plow attachment are worth $1200. The tractor is worth seven times as much money as the plow. Find the value of the tractor and the value of the plow.

65. In a 2014 report, the United States had the highest number of personal computers in use, followed by China. If the United States had 116 million more personal computers in use than China and the total number of computers for both countries was 506 million, find the number of personal computers in use for each country. (*Source of data:* Computer Industry Almanac Inc.)

66. Based on a 2014 report, the total number of personal computers in use for Russia and Italy was 99 million. If Russia had 9 million more personal computers than Italy, find the number of personal computers for each country. (*Source of data:* Computer Industry Almanac Inc.)

Review

Round each number to the given place value. See Section 1.5.

67. 586 to the nearest ten

68. 82 to the nearest ten

69. 1026 to the nearest hundred

70. 52,333 to the nearest thousand

71. 2986 to the nearest thousand

72. 101,552 to the nearest hundred

Concept Extensions

73. Solve Example 4 again, but this time let x be the distance from Chicago to Los Angeles. Did you get the same results? Explain why or why not.

74. Solve Exercise 43 again, but this time let x be the number of electoral votes for Florida. Did you get the same results? Explain why or why not.

In real estate, a house's selling price P is found by adding the real estate agent's commission C to the amount A that the seller of the house receives: $P = A + C$.

75. A house sold for $230,000. The owner's real estate agent received a commission of $13,800. How much did the seller receive? (*Hint:* Substitute the known values into the equation, then solve the equation for the remaining unknown.)

76. A homeowner plans to use a real estate agent to sell his house. He hopes to sell the house for $165,000 and keep $156,750 of that. If everything goes as he has planned, how much will his real estate agent receive as a commission?

In retailing, the retail price P of an item can be computed using the equation P = C + M, where C is the wholesale cost of the item and M is the amount of markup.

77. The retail price of a computer system is $999 after a markup of $450. What is the wholesale cost of the computer system? (*Hint:* Substitute the known values into the equation, then solve the equation for the remaining unknown.)

78. Slidell Feed and Seed sells a bag of cat food for $12. If the store paid $7 for the cat food, what is the markup on the cat food?

Chapter 11 Group Activity

Modeling Equation Solving with Addition and Subtraction

Sections 11.1–11.4

We can use positive counters ● and negative counters ● to help us model the equation-solving process. We also need to use an object that represents a variable. We use small slips of paper with the variable name written on them.

Recall that taking a ● and ● together creates a neutral or zero pair. After a neutral pair has been formed, it can be removed from or added to an equation model without changing the overall value. We also need to remember that we can add or remove the same number of positive or negative counters from both sides of an equation without changing the overall value.

We can represent the equation $x + 5 = 2$ as follows:

To get the variable by itself, we must remove 5 black counters from the left side of the model. To do so, we must add 5 negative counters to both sides of the model. Then we can remove neutral pairs: 5 from the left side and 2 from the right side (since there are only 2 black counters on the right side).

We are left with the following model, which represents the solution, $x = -3$.

Similarly, we can represent the equation $x - 4 = -6$ as follows:

To get the variable by itself, we must remove 4 red counters from both sides of the model

We are left with the following model, which represents the solution, $x = -2$.

Use the counter model to solve each equation.

1. $x - 3 = -7$ **2.** $x - 1 = -9$

3. $x + 2 = 8$ **4.** $x + 4 = 5$

5. $x + 8 = 3$ **6.** $x - 5 = -1$

7. $x - 2 = 1$ **8.** $x - 5 = 10$

9. $x + 3 = -7$ **10.** $x + 8 = -2$

Chapter 11 Vocabulary Check

Fill in each blank with one of the words or phrases listed below.

variable	addition	constant	algebraic expression	equation
terms	simplified	multiplication	evaluating the expression	solution
like	combined	numerical coefficient	distributive	

1. An algebraic expression is _____ when all like terms have been _____.

2. Terms that are exactly the same, except that they may have different numerical coefficients, are called _____ terms.

3. A letter used to represent a number is called a(n) _____.

4. A combination of operations on variables and numbers is called a(n) _____.

5. The addends of an algebraic expression are called the _____ of the expression.

6. The number factor of a variable term is called the _____.

7. Replacing a variable in an expression by a number and then finding the value of the expression is called _____ for the variable.

8. A term that is a number only is called a(n) _____.

9. A(n) _____ is of the form expression = expression.

10. A(n) _____ of an equation is a value for the variable that makes the equation a true statement.

11. To multiply $-3(2x + 1)$, we use the _____ property.

12. By the _____ property of equality, we may multiply or divide both sides of an equation by any nonzero number without changing the solution of the equation.

13. By the _____ property of equality, the same number may be added to or subtracted from both sides of an equation without changing the solution of the equation.

Helpful Hint

▶ Are you preparing for your test? Don't forget to take the Chapter 11 Test on page 740. Then check your answers at the back of the text and use the Chapter Test Prep Videos to see the fully worked-out solutions to any of the exercises you want to review.

11 Chapter Highlights

Definitions and Concepts	Examples
Section 11.1 Introduction to Variables	
A letter used to represent a number is called a **variable.** A combination of numbers, letters (variables), and operation symbols is called an **algebraic expression,** or **expression.**	x, y, z, a, b $3 + x, 7y, x^3 + y - 10$

Replacing a variable in an expression by a number and then finding the value of the expression is called **evaluating the expression.**	Evaluate: $2x + y$ when $x = 22$ and $y = 4$
	   $2x + y = 2 \cdot 22 + 4$ Replace x with 22 and y with 4.
	   $\qquad\qquad = 44 + 4$ Multiply.
	   $\qquad\qquad = 48$ Add.

The addends of an algebraic expression are called the **terms** of the expression.	$5x^2 + (-4x) + (-2)$ $\qquad\qquad\qquad\qquad$ 3 terms

Definitions and Concepts	Examples

Section 11.1   Introduction to Variables (continued)

The number factor of a variable term is called the **numerical coefficient.**

Term	Numerical Coefficient
$7x$	7
$-6y$	-6
x or $1x$	1

Terms that are exactly the same, except that they may have different numerical coefficients, are called **like terms.**

$$5x + 11x = (5 + 11)x = 16x$$
like terms
$$y - 6y = (1 - 6)y = -5y$$

An algebraic expression is **simplified** when all like terms have been **combined.**

Use the distributive property to multiply an algebraic expression within parentheses by a term. Once any like terms are then combined, the algebraic expression is simplified.

Simplify:
$$-4(x + 2) + 3(5x - 7)$$
$$= -4(x) + (-4)(2) + 3(5x) - 3(7)$$
$$= -4x + (-8) + 15x - (21)$$
$$= -4x + 15x + (-8) + (-21)$$
$$= 11x + (-29) \quad \text{or} \quad 11x - 29$$

Section 11.2   Solving Equations: The Addition Property

Addition Property of Equality

Let a, b, and c represent numbers. Then

$a = b$	Also, $a = b$
and $a + c = b + c$	and $a - c = b - c$
are equivalent equations.	are equivalent equations.

In other words, the same number may be added to or subtracted from both sides of an equation without changing the solution of the equation.

Solve for x:
$$x + 8 = 2 + (-1)$$
$$x + 8 = 1$$
$$x + 8 - 8 = 1 - 8 \qquad \text{Subtract 8 from both sides.}$$
$$x = -7 \qquad \text{Simplify.}$$
The solution is -7.

Section 11.3   Solving Equations: The Multiplication Property

Multiplication Property of Equality

Let a, b, and c represent numbers and let $c \neq 0$. then

$a = b$	Also, $a = b$
and $a \cdot c = b \cdot c$	and $\dfrac{a}{c} = \dfrac{b}{c}$
are equivalent equations.	are equivalent equations.

In other words, both sides of an equation may be multiplied or divided by the same nonzero number without changing the solution of the equation.

Solve: $-7x = 42$
$$\frac{-7x}{-7} = \frac{42}{-7} \qquad \text{Divide both sides by } -7.$$
$$x = -6 \qquad \text{Simplify.}$$

Solve: $\dfrac{2}{3}x = -10$

$$\frac{\overset{1}{\cancel{3}}}{\underset{1}{\cancel{2}}} \cdot \frac{\overset{1}{\cancel{2}}}{\underset{1}{\cancel{3}}}x = \frac{3}{2} \cdot -10 \qquad \text{Multiply both sides by } \frac{3}{2}.$$

$$x = -15 \qquad \text{Simplify.}$$

Definitions and Concepts	Examples

Section 11.4   Solving Equations Using Addition and Multiplication Properties

Steps for Solving an Equation

Step 1: If parentheses are present, use the distributive property.

Step 2: Combine any like terms on each side of the equation.

Step 3: Use the addition property of equality to rewrite the equation so that variable terms are on one side of the equation and constant terms are on the other side.

Step 4: Use the multiplication property of equality to divide both sides by the numerical coefficient of the variable to solve.

Step 5: Check the solution in the *original equation*.

Solve for x: $5(3x - 1) + 15 = -5$

Step 1: $15x - 5 + 15 = -5$ Apply the distributive property.

Step 2: $15x + 10 = -5$ Combine like terms.

Step 3: $15x + 10 - 10 = -5 - 10$ Subtract 10 from both sides.

$$15x = -15$$

Step 4: $\dfrac{15x}{15} = \dfrac{-15}{15}$ Divide both sides by 15.

$$x = -1$$

Step 5: Check to see that -1 is the solution.

Section 11.5   Equations and Problem Solving

Problem-Solving Steps

1. UNDERSTAND the problem. Some ways of doing this are

Read and reread the problem.
Construct a drawing.
Choose a variable to represent an unknown in the problem.

The incubation period for a golden eagle is three times the incubation period for a hummingbird. If the total of their incubation periods is 60 days, find the incubation period for each bird. (*Source: Wildlife Fact File,* International Masters Publishers)

1. UNDERSTAND the problem. Then choose a variable to represent an unknown. Let

$x =$ incubation period of a hummingbird

$3x =$ incubation period of a golden eagle

2. TRANSLATE the problem into an equation.

2. TRANSLATE.

incubation of hummingbird	+	incubation of golden eagle	is	60
↓		↓	↓	↓
x	$+$	$3x$	$=$	60

3. SOLVE the equation.

3. SOLVE:

$$x + 3x = 60$$
$$4x = 60$$
$$\frac{\overset{1}{\cancel{4}}x}{\underset{1}{\cancel{4}}} = \frac{60}{4}$$
$$x = 15$$

4. INTERPRET the results. *Check* the proposed solution in the stated problem and *state* your conclusion.

4. INTERPRET the solution in the stated problem. The incubation period for a hummingbird is 15 days. The incubation period for a golden eagle is $3x = 3 \cdot 15 = 45$ days.

Since 15 days + 45 days = 60 days and 45 is 3(15), the solution checks.

State your conclusion: The incubation period for a hummingbird is 15 days. The incubation period for a golden eagle is 45 days.

(11.1) *Evaluate each expression when $x = 5$, $y = 0$, and $z = -2$.*

1. $\dfrac{2x}{z}$

2. $4x - 3$

3. $\dfrac{x + 7}{y}$

4. $\dfrac{y}{5x}$

5. $x^3 - 2z$

6. $\dfrac{7 + x}{3z}$

△ **7.** Find the volume of a storage cube whose sides measure 2 feet. Use $V = s^3$.

2 feet

△ **8.** Find the volume of a wooden crate in the shape of a cube 4 feet on each side. Use $V = s^3$.

4 feet

9. Lamar deposited his $5000 bonus into an account paying 6% annual interest. How much interest will he earn in 6 years? Use $I = PRT$.

10. Jennifer Lewis borrowed $2000 from her grandmother and agreed to pay her 5% simple interest. How much interest will she owe after 3 years? Use $I = PRT$.

Simplify each expression by combining like terms.

11. $-6x - 9x$

12. $\dfrac{2}{3}x - \dfrac{9}{10}x$

13. $2y - 10 - 8y$

14. $8a + a - 7 - 15a$

15. $y + 3 - 9y - 1$

16. $1.7x - 3.2 + 2.9x - 8.7$

Multiply.

17. $-2(4y)$

18. $3(5y - 8)$

Simplify.

19. $7x + 3(x - 4) + x$

20. $4(x - 7) + 21$

21. $3(5a - 2) + 10(-2a + 1)$

22. $6y + 3 + 2(3y - 6)$

△ **23.** Find the area.

(2x − 1) yards

3
yards   Rectangle

△ **24.** Find the perimeter.

7y meters

Square

(11.2)

25. Is 4 a solution of $5(2 - x) = -10$?

26. Is 0 a solution of $6y + 2 = 23 + 4y$?

Solve.

27. $z - 5 = -7$

28. $x + 1 = 8$

29. $x + \dfrac{7}{8} = \dfrac{3}{8}$

30. $y + \dfrac{4}{11} = -\dfrac{2}{11}$

31. $n + 18 = 10 - (-2)$

32. $15 = 8x + 35 - 7x$

33. $m - 3.9 = -2.6$

34. $z - 4.6 = -2.2$

(11.3) *Solve.*

35. $-3y = -21$

36. $-8x = 72$

37. $-5n = -5$

38. $-3a = 15$

39. $\dfrac{2}{3}x = -\dfrac{8}{15}$

40. $-\dfrac{7}{8}y = 21$

41. $-1.2x = 144$

42. $-0.8y = -10.4$

43. $-5x = 100 - 120$

44. $18 - 30 = -4x$

(11.4) *Solve.*

45. $3x - 4 = 11$

46. $6y + 1 = 73$

47. $-\dfrac{5}{9}x + 23 = -12$

48. $-\dfrac{2}{3}x - 11 = \dfrac{2}{3}x - 55$

49. $6.8 + 4y = -2.2$

50. $-9.6 + 5y = -3.1$

51. $2x + 7 = 6x - 1$

52. $5x - 18 = -4x + 36$

53. $5(n - 3) = 7 + 3n$

54. $7(2 + x) = 4x - 1$

55. $2(4n - 11) + 8 = 5n + 4$

56. $3(5x - 6) + 9 = 13x + 7$

Write each sentence as an equation.

57. The difference of 20 and -8 is 28.

58. Nineteen subtracted from -2 amounts to -21.

59. The quotient of -75 and the sum of 5 and 20 is equal to -3.

60. The product of -5 and the sum of -2 and 6 yields -20.

(11.5) *Write each phrase as an algebraic expression. Use x to represent "a number."*

61. The quotient of 70 and a number

62. The difference of a number and 13

63. A number subtracted from 85.

64. Eleven added to twice a number

Write each sentence as an equation. Use x to represent "a number."

65. A number increased by 8 is 40.

66. Twelve subtracted from twice a number is 10.

Solve.

67. Five times a number subtracted from 40 is the same as three times the number. Find the number.

68. The product of a number and 3 is twice the difference of that number and 8. Find the number.

69. Bamboo and Pacific Kelp, a kind of seaweed, are two fast-growing plants. Bamboo grows twice as fast as kelp. If in one day both can grow a total of 54 inches, find how many inches each plant can grow in one day.

Bamboo   Kelp

70. In an election between the incumbent and a challenger, the incumbent received 11,206 more votes than the challenger. If a total of 18,298 votes were cast, find the number of votes for each candidate.

Mixed Review

Evaluate each expression when $x = 4$, $y = -3$, *and* $z = 5$.

71. $18 - (9 - 5x)$

72. $\dfrac{z}{100} + \dfrac{y}{10}$

Simplify.

73. $9x - 20x$

74. $-5(7x)$

75. $12x + 5(2x - 3) - 4$

76. $-7(x + 6) - 2(x - 5)$

Write each phrase as an algebraic expression. Use x to represent "a number."

77. Seventeen less than a number

78. Three times the sum of a number and five

Write each sentence as an equation using x to represent "a number."

79. The difference of a number and 3 is the quotient of the number and 4.

80. The product of a number and 6 is equal to the sum of the number and 2.

81. Is 3 a solution of $4y + 2 - 6y = 5 + 7$?

82. Is 7 a solution of $4(z - 8) + 12 = 8$?

Solve.

83. $c - 5 = -13 + 7$

84. $7x + 5 - 6x = -20$

85. $-7x + 3x = -50 - 2$

86. $-x + 8x = -38 - 4$

87. $14 - y = -3$

88. $7 - z = 0$

89. $9x + 12 - 8x = -6 + (-4)$

90. $-17x + 14 + 20x - 2x = 5 - (-3)$

91. $\dfrac{4}{9}x = -\dfrac{1}{3}$

92. $-\dfrac{5}{24}x = \dfrac{5}{6}$

93. $2y + 6y = 24 - 8$

94. $13x - 7x = -4 - 12$

95. $\dfrac{2}{3}x - 12 = -4$

96. $\dfrac{7}{8}x + 5 = -2$

97. $-5z + 3z - 7 = 8z - 7$

98. $4x - 3 + 6x = 5x - 3$

99. Three times a number added to twelve is 27. Find the number.

100. Twice the sum of a number and four is ten. Find the number.

Answers

1. Evaluate $\dfrac{3x - 5}{2y}$ when $x = 7$ and $y = -8$.

2. Simplify $7x - 5 - 12x + 10$ by combining like terms.

3. Multiply: $-2(3y + 7)$

4. Simplify: $5(3z + 2) - z - 18$

1. _____

2. _____

△ **5.** Write a product that represents the area of the rectangle. Then multiply.

4 meters

| Rectangle | $(3x - 1)$ meters |

3. _____

4. _____

5. _____

Solve.

6. _____

6. $x - 17 = -10$

7. $y + \dfrac{3}{4} = \dfrac{1}{4}$

7. _____

8. _____

8. $-4x = 48$

9. $-\dfrac{5}{8}x = -25$

9. _____

10. _____

10. $5x + 12 - 4x - 14 = 22$

11. $2 - c + 2c = 5$

11. _____

12. _____

12. $3x - 5 = -11$

13. $-4x + 7 = 15$

13. _____

740

14. $3.6 - 2x = -5.4$

15. $12 = 3(4 + 2y)$

16. $5x - 2 = x - 10$

17. $10y - 1 = 7y + 21$

18. $6 + 2(3n - 1) = 28$

19. $4(5x + 3) = 2(7x + 6)$

Solve.

△ **20.** A lawn is in the shape of a trapezoid with a height of 60 feet and bases of 70 feet and 130 feet. Find the area of the lawn. Use
$A = \dfrac{1}{2}h(B + b)$.

△ **21.** If the height of a triangular-shaped jib sail is 12 feet and its base is 5 feet, find the area of the sail. Use
$A = \dfrac{1}{2}bh$.

22. Translate the following phrases into mathematical expressions. Use x to represent "a number."
a. The product of a number and 17
b. Twice a number subtracted from 20

23. The difference of three times a number and five times the same number is 4. Find the number.

24. In a championship basketball game, Paula Zimmerman scored twice as many points as Maria Kaminsky. If the total number of points made by both women was 51, find how many points Paula scored.

25. In a 10-kilometer race, there are 112 more men entered than women. Find the number of women runners if the total number of runners in the race is 600.

14. _____

15. _____

16. _____

17. _____

18. _____

19. _____

20. _____

21. _____

22. a. _____

b. _____

23. _____

24. _____

25. _____

Answers

1. _____

2. _____

3. a. _____

b. _____

4. _____

5. _____

6. _____

7. _____

8. _____

9. _____

10. _____

11. _____

12. _____

13. a. _____

b. _____

c. _____

d. _____

14. _____

15. _____

16. _____

17. _____

18. _____

1. Multiply: 0.0531×16

2. Multiply: 0.0531×1000

3. Given the rectangle shown:

7 feet

5 feet

  a. Find the ratio of its width to its length.
  b. Find the ratio of its length to its perimeter.

4. Add: $\dfrac{5}{12} + \dfrac{2}{9}$

5. 12% of what number is 0.6?

6. Multiply: $\dfrac{7}{8} \cdot \dfrac{2}{3}$

7. What percent of 12 is 9?

8. Divide: $1\dfrac{4}{5} \div 2\dfrac{3}{10}$

9. Convert 3 pounds to ounces.

10. Round 23,781 to the nearest thousand.

11. Add 2400 ml to 8.9 L.

12. Round 0.02351 to the nearest thousandth.

13. Identify each figure as a line, a ray, a line segment, or an angle.

  a. C D

  b. E F

  c. M N O

  d. P T

14. Find the supplement of a 12° angle.

△ **15.** Find the diameter of the circle.

5 cm

16. Find the measure of the unknown angle.

92°

54°

?

△ **17.** Find the perimeter of the room shown below.

9 feet

10 feet

7 feet

15 feet

△ **18.** Find the area of the room in Exercise 17.

742

△ **19.** Find the area of the triangle.

8 centimeters

14 centimeters

20. Simplify: $\dfrac{(4 + \sqrt{4})^2}{\sqrt{100} - \sqrt{64}}$

21. Without a calculator or table:
    a. Determine which two whole numbers $\sqrt{78}$ is between.
    b. Use part **a** to approximate $\sqrt{78}$ to the nearest whole.

22. Divide: $0.1156 \div 0.02$

23. Mel Rose is a 6-foot-tall park ranger who needs to know the height of a particular tree. He measures the shadow of the tree to be 69 feet long when his own shadow is 9 feet long. Find the height of the tree.

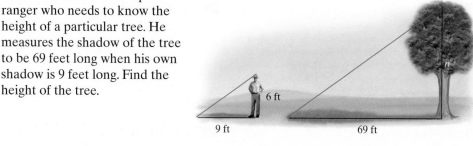

6 ft

9 ft   69 ft

24. What percent of 120 is 28.8?

25. The following bar graph shows the number of endangered species in 2013. Use this graph to answer the questions.

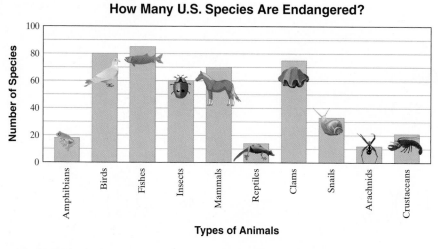

How Many U.S. Species Are Endangered?

Source: U.S. Fish and Wildlife Service

    a. Approximate the number of endangered species that are clams.
    b. Which category has the most endangered species?

26. Find the mean, median, and mode of 1, 7, 8, 10, 11, 11.

Insert < or > between each pair of numbers to make a true statement.

27. -2.9 -1 **28.** $\dfrac{2}{3}$ 0.66 **29.** -9 -11 **30.** $\dfrac{4}{5}$ $\dfrac{14}{17}$

19. _____

20. _____

21. a. _____

    b. _____

22. _____

23. _____

24. _____

25. a. _____

    b. _____

26. _____

27. _____

28. _____

29. _____

30. _____

31. _____

32. _____

33. _____

34. _____

35. _____

36. _____

37. _____

38. _____

39. _____

40. _____

41. _____

42. _____

43. _____

44. _____

45. _____

46. _____

47. _____

48. _____

49. _____

50. _____

Find each absolute value.

31. $|-2|$ **32.** $\left|\dfrac{5}{6}\right|$ **33.** $|1.2|$ **34.** $|0|$

Add.

35. $-5 + (-11)$ **36.** $-8.2 + 4.6$

37. $-\dfrac{3}{8} + \left(-\dfrac{1}{8}\right)$ **38.** $\dfrac{2}{5} + \left(-\dfrac{3}{10}\right)$

Subtract.

39. $8 - 15$ **40.** $4.6 - (-1.2)$

41. $-4 - (-5)$ **42.** $\dfrac{7}{10} - \dfrac{23}{24}$

Multiply.

43. $-2(-5)$ **44.** $-8(1.2)$

45. $\left(-\dfrac{1}{2}\right)\left(-\dfrac{2}{3}\right)$ **46.** $-2\dfrac{2}{9}\left(1\dfrac{4}{5}\right)$

47. Simplify: $(-3) \cdot |-5| - (-2) + 4^2$ **48.** Solve: $4x - 7.1 = 3x + 2.6$

49. Solve: $3(2x - 6) + 6 = 0$ **50.** Solve: $6(x - 5) = 4(x + 4) - 6$

Tables

Appendix

A

A.1 Addition Table and One Hundred Addition Facts ▶

+	0	1	2	3	4	5	6	7	8	9
0	0	1	2	3	4	5	6	7	8	9
1	1	2	3	4	5	6	7	8	9	10
2	2	3	4	5	6	7	8	9	10	11
3	3	4	5	6	7	8	9	10	11	12
4	4	5	6	7	8	9	10	11	12	13
5	5	6	7	8	9	10	11	12	13	14
6	6	7	8	9	10	11	12	13	14	15
7	7	8	9	10	11	12	13	14	15	16
8	8	9	10	11	12	13	14	15	16	17
9	9	10	11	12	13	14	15	16	17	18

One Hundred Addition Facts

Knowledge of the basic addition facts found above is an important prerequisite for a course in basic college mathematics. Study the table above and then perform the additions. Check your answers either by comparing them with those found in the back-of-the-book answer section or by using the table. Review any facts that you missed.

1. $1 + 4$ **2.** $5 + 6$ **3.** $2 + 3$ **4.** $7 + 8$ **5.** $3 + 9$ **6.** $6 + 1$

7. $4 + 4$ **8.** $0 + 6$ **9.** $9 + 5$ **10.** $8 + 2$ **11.** $5 + 7$ **12.** $3 + 2$

13. $5 + 5$ **14.** $1 + 1$ **15.** $8 + 1$ **16.** $6 + 6$ **17.** $2 + 9$ **18.** $3 + 5$

19. $9 + 9$ **20.** $5 + 2$ **21.** $6 + 4$ **22.** $0 + 0$ **23.** $1 + 9$ **24.** $3 + 7$

25. $9 + 8$ **26.** $0 + 8$ **27.** $4 + 9$ **28.** $3 + 0$ **29.** $7 + 5$ **30.** $8 + 9$

745

31. 9
   +7

32. 2
   +6

33. 4
   +3

34. 8
   +5

35. 3
   +1

36. 0
   +3

37. 7
   +1

38. 3
   +4

39. 8
   +0

40. 6
   +3

41. 2
   +4

42. 0
   +9

43. 8
   +8

44. 5
   +3

45. 3
   +6

46. 6
   +9

47. 4
   +8

48. 0
   +1

49. 2
   +5

50. 6
   +0

51. 2
   +0

52. 4
   +2

53. 8
   +3

54. 7
   +4

55. 1
   +7

56. 4
   +6

57. 0
   +5

58. 9
   +1

59. 8
   +6

60. 5
   +1

61. 6
   +7

62. 4
   +0

63. 1
   +6

64. 4
   +5

65. 0
   +7

66. 5
   +8

67. 7
   +6

68. 7
   +0

69. 4
   +1

70. 5
   +4

71. 0
   +4

72. 1
   +2

73. 7
   +9

74. 3
   +8

75. 7
   +7

76. 9
   +4

77. 1
   +0

78. 4
   +7

79. 2
   +2

80. 1
   +3

81. 2
   +8

82. 5
   +9

83. 6
   +2

84. 9
   +6

85. 5
   +0

86. 8
   +7

87. 7
   +3

88. 0
   +2

89. 9
   +2

90. 3
   +3

91. 9
   +3

92. 1
   +5

93. 2
   +7

94. 6
   +5

95. 7
   +2

96. 1
   +8

97. 6
   +8

98. 8
   +4

99. 9
   +0

100. 2
    +1

A.2 Multiplication Table and One Hundred Multiplication Facts

×	1	2	3	4	5	6	7	8	9
1	1	2	3	4	5	6	7	8	9
2	2	4	6	8	10	12	14	16	18
3	3	6	9	12	15	18	21	24	27
4	4	8	12	16	20	24	28	32	36
5	5	10	15	20	25	30	35	40	45
6	6	12	18	24	30	36	42	48	54
7	7	14	21	28	35	42	49	56	63
8	8	16	24	32	40	48	56	64	72
9	9	18	27	36	45	54	63	72	81

One Hundred Multiplication Facts

Knowledge of the basic multiplication facts found above is an important prerequisite for a course in basic college mathematics. Study the table above and then perform the multiplications. Check your answers either by comparing them with those found in the back-of-the-book answer section or by using the table. Review any facts that you missed.

1. $\begin{array}{r} 1 \\ \times 1 \\ \hline \end{array}$
2. $\begin{array}{r} 5 \\ \times 7 \\ \hline \end{array}$
3. $\begin{array}{r} 7 \\ \times 8 \\ \hline \end{array}$
4. $\begin{array}{r} 3 \\ \times 3 \\ \hline \end{array}$
5. $\begin{array}{r} 8 \\ \times 4 \\ \hline \end{array}$
6. $\begin{array}{r} 9 \\ \times 5 \\ \hline \end{array}$

7. $\begin{array}{r} 4 \\ \times 7 \\ \hline \end{array}$
8. $\begin{array}{r} 7 \\ \times 1 \\ \hline \end{array}$
9. $\begin{array}{r} 2 \\ \times 2 \\ \hline \end{array}$
10. $\begin{array}{r} 0 \\ \times 5 \\ \hline \end{array}$
11. $\begin{array}{r} 9 \\ \times 7 \\ \hline \end{array}$
12. $\begin{array}{r} 8 \\ \times 8 \\ \hline \end{array}$

13. $\begin{array}{r} 3 \\ \times 2 \\ \hline \end{array}$
14. $\begin{array}{r} 6 \\ \times 0 \\ \hline \end{array}$
15. $\begin{array}{r} 5 \\ \times 6 \\ \hline \end{array}$
16. $\begin{array}{r} 2 \\ \times 5 \\ \hline \end{array}$
17. $\begin{array}{r} 4 \\ \times 6 \\ \hline \end{array}$
18. $\begin{array}{r} 0 \\ \times 7 \\ \hline \end{array}$

19. $\begin{array}{r} 6 \\ \times 3 \\ \hline \end{array}$
20. $\begin{array}{r} 8 \\ \times 9 \\ \hline \end{array}$
21. $\begin{array}{r} 5 \\ \times 8 \\ \hline \end{array}$
22. $\begin{array}{r} 7 \\ \times 2 \\ \hline \end{array}$
23. $\begin{array}{r} 4 \\ \times 8 \\ \hline \end{array}$
24. $\begin{array}{r} 1 \\ \times 2 \\ \hline \end{array}$

25. $\begin{array}{r} 9 \\ \times 6 \\ \hline \end{array}$
26. $\begin{array}{r} 3 \\ \times 1 \\ \hline \end{array}$
27. $\begin{array}{r} 8 \\ \times 7 \\ \hline \end{array}$
28. $\begin{array}{r} 2 \\ \times 8 \\ \hline \end{array}$
29. $\begin{array}{r} 6 \\ \times 9 \\ \hline \end{array}$
30. $\begin{array}{r} 5 \\ \times 5 \\ \hline \end{array}$

31. $\begin{array}{r} 2 \\ \times 1 \\ \hline \end{array}$
32. $\begin{array}{r} 8 \\ \times 0 \\ \hline \end{array}$
33. $\begin{array}{r} 4 \\ \times 9 \\ \hline \end{array}$
34. $\begin{array}{r} 8 \\ \times 3 \\ \hline \end{array}$
35. $\begin{array}{r} 6 \\ \times 2 \\ \hline \end{array}$
36. $\begin{array}{r} 4 \\ \times 5 \\ \hline \end{array}$

37. 9
     ×4

38. 2
     ×9

39. 3
     ×4

40. 1
     ×6

41. 8
     ×6

42. 9
     ×8

43. 1
     ×8

44. 5
     ×1

45. 9
     ×0

46. 7
     ×4

47. 9
     ×3

48. 0
     ×3

49. 3
     ×5

50. 6
     ×8

51. 5
     ×9

52. 2
     ×6

53. 1
     ×0

54. 3
     ×9

55. 9
     ×9

56. 5
     ×4

57. 0
     ×6

58. 1
     ×9

59. 5
     ×0

60. 6
     ×1

61. 9
     ×2

62. 1
     ×7

63. 1
     ×3

64. 7
     ×3

65. 6
     ×6

66. 4
     ×0

67. 7
     ×9

68. 4
     ×3

69. 7
     ×5

70. 2
     ×0

71. 6
     ×7

72. 0
     ×8

73. 8
     ×5

74. 2
     ×4

75. 0
     ×1

76. 3
     ×8

77. 9
     ×1

78. 7
     ×0

79. 5
     ×3

80. 4
     ×4

81. 1
     ×5

82. 6
     ×5

83. 3
     ×0

84. 1
     ×4

85. 3
     ×7

86. 4
     ×2

87. 0
     ×2

88. 7
     ×7

89. 8
     ×2

90. 6
     ×4

91. 0
     ×0

92. 2
     ×7

93. 4
     ×1

94. 0
     ×4

95. 2
     ×3

96. 8
     ×1

97. 3
     ×6

98. 5
     ×2

99. 0
     ×9

100. 7
      ×6

Plane Figures Have Length and Width but No Thickness or Depth		
Name	Description	Figure
Polygon	Union of three or more coplanar line segments that intersect with each other only at each endpoint, with each endpoint shared by two segments.	
Triangle	Polygon with three sides (sum of measures of three angles is 180°).	
Scalene Triangle	Triangle with no sides of equal length.	
Isosceles Triangle	Triangle with two sides of equal length.	
Equilateral Triangle	Triangle with all sides of equal length.	
Right Triangle	Triangle that contains a right angle.	leg hypotenuse leg
Quadrilateral	Polygon with four sides (sum of measures of four angles is 360°).	
Trapezoid	Quadrilateral with exactly one pair of opposite sides parallel.	base leg parallel sides leg base
Isosceles Trapezoid	Trapezoid with legs of equal length.	
Parallelogram	Quadrilateral with both pairs of opposite sides parallel.	
Rhombus	Parallelogram with all sides of equal length.	
Rectangle	Parallelogram with four right angles.	

(*Continued*)

Plane Figures Have Length and Width but No Thickness or Depth (*continued*)

Name	Description	Figure
Square	Rectangle with all sides of equal length.	
Circle	All points in a plane the same distance from a fixed point called the **center.**	

Solid Figures Have Length, Width, and Height or Depth

Name	Description	Figure
Rectangular Solid	A solid with six sides, all of which are rectangles.	
Cube	A rectangular solid whose six sides are squares.	
Sphere	All points the same distance from a fixed point, called the **center.**	
Right Circular Cylinder	A cylinder having two circular bases that are perpendicular to its altitude.	
Right Circular Cone	A cone with a circular base that is perpendicular to its altitude.	

Table of Percents, Decimals, and Fraction Equivalents ▶

Percent	Decimal	Fraction
1%	0.01	$\frac{1}{100}$
5%	0.05	$\frac{1}{20}$
10%	0.1	$\frac{1}{10}$
12.5% or $12\frac{1}{2}$%	0.125	$\frac{1}{8}$
$16.\overline{6}$% or $16\frac{2}{3}$%	$0.1\overline{6}$	$\frac{1}{6}$
20%	0.2	$\frac{1}{5}$
25%	0.25	$\frac{1}{4}$
30%	0.3	$\frac{3}{10}$
$33.\overline{3}$% or $33\frac{1}{3}$%	$0.\overline{3}$	$\frac{1}{3}$
37.5% or $37\frac{1}{2}$%	$0.37\overline{5}$	$\frac{3}{8}$
40%	0.4	$\frac{2}{5}$
50%	0.5	$\frac{1}{2}$
60%	0.6	$\frac{3}{5}$
62.5% or $62\frac{1}{2}$%	0.625	$\frac{5}{8}$
$66.\overline{6}$% or $66\frac{2}{3}$%	$0.\overline{6}$	$\frac{2}{3}$
70%	0.7	$\frac{7}{10}$
75%	0.75	$\frac{3}{4}$
80%	0.8	$\frac{4}{5}$
$83.\overline{3}$% or $83\frac{1}{3}$%	$0.8\overline{3}$	$\frac{5}{6}$
87.5% or $87\frac{1}{2}$%	0.875	$\frac{7}{8}$
90%	0.9	$\frac{9}{10}$
100%	1.0	1
110%	1.1	$1\frac{1}{10}$
125%	1.25	$1\frac{1}{4}$
133.3% or $133\frac{1}{3}$%	$1.\overline{3}$	$1\frac{1}{3}$
150%	1.5	$1\frac{1}{2}$
$166.\overline{6}$% or $166\frac{2}{3}$%	$1.\overline{6}$	$1\frac{2}{3}$
175%	1.75	$1\frac{3}{4}$
200%	2.0	2

A.5 Table on Finding Common Percents of a Number

Common Percent Equivalences*	Shortcut Method for Finding Percent	Example
$1\% = 0.01 \left(\text{or } \frac{1}{100}\right)$	To find 1% of a number, multiply by 0.01. To do so, move the decimal point 2 places to the left.	1% of 210 is 2.10 or 2.1. 1% of 1500 is 15. 1% of 8.6 is 0.086.
$10\% = 0.1 \left(\text{or } \frac{1}{10}\right)$	To find 10% of a number, multiply by 0.1, or move the decimal point of the number one place to the left.	10% of 140 is 14. 10% of 30 is 3. 10% of 17.6 is 1.76.
$25\% = \frac{1}{4}$	To find 25% of a number, find $\frac{1}{4}$ of the number, or divide the number by 4.	25% of 20 is $\frac{20}{4}$ or 5. 25% of 8 is 2. 25% of 10 is $\frac{10}{4}$ or $2\frac{1}{2}$.
$50\% = \frac{1}{2}$	To find 50% of a number, find $\frac{1}{2}$ of the number, or divide the number by 2.	50% of 64 is $\frac{64}{2}$ or 32. 50% of 1000 is 500. 50% of 9 is $\frac{9}{2}$ or $4\frac{1}{2}$.
$100\% = 1$	To find 100% of a number, multiply the number by 1. In other words, 100% of a number is the number.	100% of 98 is 98. 100% of 1407 is 1407. 100% of 18.4 is 18.4.
$200\% = 2$	To find 200% of a number, multiply the number by 2.	200% of 31 is $31 \cdot 2$ or 62. 200% of 750 is 1500. 200% of 6.5 is 13.

*See Appendix A.4.

n	n^2	$\sqrt{n}$	n	n^2	$\sqrt{n}$
1	1	1.000	51	2601	7.141
2	4	1.414	52	2704	7.211
3	9	1.732	53	2809	7.280
4	16	2.000	54	2916	7.348
5	25	2.236	55	3025	7.416
6	36	2.449	56	3136	7.483
7	49	2.646	57	3249	7.550
8	64	2.828	58	3364	7.616
9	81	3.000	59	3481	7.681
10	100	3.162	60	3600	7.746
11	121	3.317	61	3721	7.810
12	144	3.464	62	3844	7.874
13	169	3.606	63	3969	7.937
14	196	3.742	64	4096	8.000
15	225	3.873	65	4225	8.062
16	256	4.000	66	4356	8.124
17	289	4.123	67	4489	8.185
18	324	4.243	68	4624	8.246
19	361	4.359	69	4761	8.307
20	400	4.472	70	4900	8.367
21	441	4.583	71	5041	8.426
22	484	4.690	72	5184	8.485
23	529	4.796	73	5329	8.544
24	576	4.899	74	5476	8.602
25	625	5.000	75	5625	8.660
26	676	5.099	76	5776	8.718
27	729	5.196	77	5929	8.775
28	784	5.292	78	6084	8.832
29	841	5.385	79	6241	8.888
30	900	5.477	80	6400	8.944
31	961	5.568	81	6561	9.000
32	1024	5.657	82	6724	9.055
33	1089	5.745	83	6889	9.110
34	1156	5.831	84	7056	9.165
35	1225	5.916	85	7225	9.220
36	1296	6.000	86	7396	9.274
37	1369	6.083	87	7569	9.327
38	1444	6.164	88	7744	9.381
39	1521	6.245	89	7921	9.434
40	1600	6.325	90	8100	9.487
41	1681	6.403	91	8281	9.539
42	1764	6.481	92	8464	9.592
43	1849	6.557	93	8649	9.644
44	1936	6.633	94	8836	9.695
45	2025	6.708	95	9025	9.747
46	2116	6.782	96	9216	9.798
47	2209	6.856	97	9409	9.849
48	2304	6.928	98	9604	9.899
49	2401	7.000	99	9801	9.950
50	2500	7.071	100	10,000	10.000

A.7 Compound Interest Table ▷

Compounded Annually

	5%	6%	7%	8%	9%	10%	11%	12%	13%	14%	15%	16%	17%	18%
1 year	1.05000	1.06000	1.07000	1.08000	1.09000	1.10000	1.11000	1.12000	1.13000	1.14000	1.15000	1.16000	1.17000	1.18000
5 years	1.27628	1.33823	1.40255	1.46933	1.53862	1.61051	1.68506	1.76234	1.84244	1.92541	2.01136	2.10034	2.19245	2.28776
10 years	1.62889	1.79085	1.96715	2.15892	2.36736	2.59374	2.83942	3.10585	3.39457	3.70722	4.04556	4.41144	4.80683	5.23384
15 years	2.07893	2.39656	2.75903	3.17217	3.64248	4.17725	4.78459	5.47357	6.25427	7.13794	8.13706	9.26552	10.53872	11.97375
20 years	2.65330	3.20714	3.86968	4.66096	5.60441	6.72750	8.06231	9.64629	11.52309	13.74349	16.36654	19.46076	23.10560	27.39303

Compounded Semiannually

	5%	6%	7%	8%	9%	10%	11%	12%	13%	14%	15%	16%	17%	18%
1 year	1.05063	1.06090	1.07123	1.08160	1.09203	1.10250	1.11303	1.12360	1.13423	1.14490	1.15563	1.16640	1.17723	1.18810
5 years	1.28008	1.34392	1.41060	1.48024	1.55297	1.62889	1.70814	1.79085	1.87714	1.96715	2.06103	2.15892	2.26098	2.36736
10 years	1.63862	1.80611	1.98979	2.19112	2.41171	2.65330	2.91776	3.20714	3.52365	3.86968	4.24785	4.66096	5.11205	5.60441
15 years	2.09757	2.42726	2.80679	3.24340	3.74532	4.32194	4.98395	5.74349	6.61437	7.61226	8.75496	10.06266	11.55825	13.26768
20 years	2.68506	3.26204	3.95926	4.80102	5.81636	7.03999	8.51331	10.28572	12.41607	14.97446	18.04424	21.72452	26.13302	31.40942

Compounded Quarterly

	5%	6%	7%	8%	9%	10%	11%	12%	13%	14%	15%	16%	17%	18%
1 year	1.05095	1.06136	1.07186	1.08243	1.09308	1.10381	1.11462	1.12551	1.13648	1.14752	1.15865	1.16986	1.18115	1.19252
5 years	1.28204	1.34686	1.41478	1.48595	1.56051	1.63862	1.72043	1.80611	1.89584	1.98979	2.08815	2.19112	2.29891	2.41171
10 years	1.64362	1.81402	2.00160	2.20804	2.43519	2.68506	2.95987	3.26204	3.59420	3.95926	4.36038	4.80102	5.28497	5.81636
15 years	2.10718	2.44322	2.83182	3.28103	3.80013	4.39979	5.09225	5.89160	6.81402	7.87809	9.10513	10.51963	12.14965	14.02741
20 years	2.70148	3.29066	4.00639	4.87544	5.93015	7.20957	8.76085	10.64089	12.91828	15.67574	19.01290	23.04980	27.93091	33.83010

Compounded Daily

	5%	6%	7%	8%	9%	10%	11%	12%	13%	14%	15%	16%	17%	18%
1 year	1.05127	1.06183	1.07250	1.08328	1.09416	1.10516	1.11626	1.12747	1.13880	1.15024	1.16180	1.17347	1.18526	1.19716
5 years	1.28400	1.34983	1.41902	1.49176	1.56823	1.64861	1.73311	1.82194	1.91532	2.01348	2.11667	2.22515	2.33918	2.45906
10 years	1.64866	1.82203	2.01362	2.22535	2.45933	2.71791	3.00367	3.31946	3.66845	4.05411	4.48031	4.95130	5.47178	6.04696
15 years	2.11689	2.45942	2.85736	3.31968	3.85678	4.48077	5.20569	6.04786	7.02625	8.16288	9.48335	11.01738	12.79950	14.86983
20 years	2.71810	3.31979	4.05466	4.95216	6.04831	7.38703	9.02202	11.01883	13.45751	16.43582	20.07316	24.51533	29.94039	36.56577

Unit Analysis

Let's suppose that you are traveling at a rate of 60 miles per hour. This means that

> in 1 hour, your distance traveled is 60 miles;
> in 2 hours, your distance traveled is $2 \cdot 60$ or 120 miles.

The rate 60 miles per hour can also be written as

$$60 \text{ mph}, \quad 60 \text{ miles per 1 hour}, \quad \text{or} \quad \frac{60 \text{ miles}}{1 \text{ hour}}.$$

We can use rates such as $\dfrac{60 \text{ miles}}{1 \text{ hour}}$ to help us solve problems. Before we do this, notice that if 60 miles are traveled in 1 hour, then in 1 hour a distance of 60 miles is traveled and we also have the rate $\dfrac{1 \text{ hour}}{60 \text{ miles}}$, if needed.

Let's practice writing the fraction form of rates once again. The rate of 30 drops of medicine per liter of saline solution can be written as:

$\dfrac{30 \text{ drops}}{1 \text{ liter}}$. If needed, we also have the rate $\dfrac{1 \text{ liter}}{30 \text{ drops}}$.

How do we know which version to use when solving a problem? As you will see in the examples, we use the rate that has *units to be divided out* in the *denominator* and *units asked to find* in the *numerator*.

Example 1 If you are traveling in a car at a rate of 60 miles per hour, how many miles do you travel in 21 hours?

Solution: Let's use a form of the given rate to help us answer this question. Since we are asked to find miles, we will multiply 21 hours by a version of the rate that has units to be divided out (hours) in the denominator and units asked to find (miles) in the numerator.

$$\begin{aligned}
\text{miles traveled in 21 hours} &= \frac{21 \text{ hours}}{1} \cdot \frac{60 \text{ miles}}{1 \text{ hour}} \\
&= \frac{21 \cdot 60}{1 \cdot 1} \text{miles} \\
&= 1260 \text{ miles}
\end{aligned}$$

You travel 1260 miles in 21 hours.

Helpful Hint

Pay careful attention to units and how they divide out.

Example 2 If you are traveling at the rate of 60 miles per hour, how many hours does it take to travel 900 miles in the car?

Solution: Let's use a form of the given rate again. For this example, we are asked to find hours, so we will write a rate so that units to be divided out (miles) are in the denominator and units needed in the answer (hours) are in the numerator.

$$\text{hours taken to travel 900 miles} = \frac{900 \text{ miles}}{1} \cdot \frac{1 \text{ hour}}{60 \text{ miles}}$$

$$= \frac{900 \cdot 1}{1 \cdot 60} \text{hours}$$

$$= 15 \text{ hours}$$

It takes 15 hours to travel 900 miles.

Helpful Hint

Remember: The rate you use depends on the exercise. A general rule is to use a factor that has units needed to divide out in the denominator and units asked to find in the numerator.

To solve some unit applications, it may be necessary to multiply by 2 or more rates. (More exercises involving unit conversions on length, weight, or mass can be found in the chapter on measurement, Chapter 7.)

Example 3 A telemarketer makes an average of 24 calls per hour. He is paid $0.50 per call. If he worked 25 hours last week, calculate his earnings (before deductions).

Solution: To calculate his earnings, notice we multiply by two different rates.

$$\text{earnings for 25 hours} = \frac{25 \text{ hours}}{1} \cdot \frac{24 \text{ calls}}{1 \text{ hour}} \cdot \frac{\$0.50}{1 \text{ call}}$$

$$= \frac{25 \cdot 24 \cdot \$0.50}{1 \cdot 1 \cdot 1}$$

$$= \$300$$

He made $300 last week.

B **Exercise Set** MyMathLab ▶

1. Suppose you make $16 per hour.
   a. Calculate your earnings (before deductions) for 40 hours.
   b. How many hours of work are required to earn $240 (before deductions)?

2. A lab technician needs a solution containing 20 drops of concentrated medicine per liter of saline solution.
   a. How many drops are needed for 17 liters of solution?
   b. How many drops are needed for 280 liters of saline solution?
   c. If you are told to make a solution using an entire container of 520 drops of medicine, how many liters of saline solution are needed?

3. A hummingbird's wings beat an average of 90 beats per second.
   a. How many beats occur during 14 seconds?
   b. How much time (in seconds) has elapsed after 5400 wing beats? How much time is your answer in minutes?
   c. How much time (in seconds) has elapsed after 8100 wing beats? How much time is your answer in minutes?

4. Noah Wyle (of *ER*) was once the highest paid TV drama actor per episode. He earned $400,000 per episode.
   a. How much money did he make for 12 episodes?
   b. How many episodes did he need to tape in order to earn $6,800,000?

5. Switzerland has an average of one hospital bed per 155 people (in population).
   a. Suppose there is a city in Switzerland with a population of 31,000. At this rate, how many hospital beds are there in this city?
   b. Suppose the only hospital in a city in Switzerland has 73 beds. Using the given rate, estimate the population of the city.

6. Niger has an average of one hospital bed per 2000 people (in population).
   a. Suppose there is a city in Niger with a population of 16,000. At this rate, how many hospital beds are there in this city?
   b. Suppose the only hospital in a city in Niger has 21 beds. Using the given rate, estimate the population of the city.

7. A recipe calls for 3 cups of water for every 2 cups of flour.
   a. If 9 cups of water are added, how many cups of flour are needed?
   b. If 20 cups of water are added, how many cups of flour are needed?
   c. If 15 cups of flour are added, how many cups of water are needed?

8. A recipe calls for 4 cups of sugar for every 3 cups of water.
   a. If 12 cups of sugar are added, how many cups of water are needed?
   b. If 14 cups of sugar are added, how many cups of water are needed?
   c. If 20 cups of water are added, how many cups of sugar are needed?

9. A computer technician on an assembly line makes $0.75 for her completion of each computer component. She can complete 26 components in an hour. If she worked 33 hours last week, what was her pay before deductions?

10. A cockroach is traveling at a rate of 3 miles per hour. Convert this speed to feet per second.

11. A lab technician needs a solution containing 15 drops of concentrated medicine per liter of saline solution.

    a. How many drops will be needed for 17.5 liters of solution?

    b. How many liters will be needed for 300 drops of concentrated medicine?

12. A car travels at a rate of 57 miles per hour.

    a. How many miles does the car travel in $5\frac{1}{2}$ hours?

    b. At this rate, how many hours does it take to travel 627 miles in the car?

13. A tortilla-making machine makes 16 tortillas per minute.

    a. How many tortillas can be made in 7.5 hours?

    b. How long will it take to make 200 tortillas?

14. During a flood watch in New Orleans, Louisiana, an average of 0.5 inch of rain fell per hour for 48 hours.

    a. How much water fell during the first 21 hours?

    b. At this given rate, how long did it take for 15.5 inches of rain to fall?

15. Ray Romano (of *Everybody Loves Raymond*) was once the world's highest paid TV actor. He earned $1.8 million per episode.

    a. At this rate, how much money did he earn for 6 episodes?

    b. How many episodes did he tape to earn $8.1 million?

16. Suppose a supporting actor of a television series makes $18,000 per episode.

    a. How much money does she make for 8 episodes?

    b. How many episodes did she perform in to earn $486,000?

    c. According to this actor's contract, it is possible to pay her for fractional portions of episodes. How much money does she make for 12.5 episodes?

17. A cheetah is the fastest known land animal and has been recorded traveling at speeds of 65 miles per hour. Suppose a particular cheetah is traveling at a speed of 45 miles per hour.

    a. How far has the cheetah traveled in $\frac{1}{4}$ hour?

    b. How long does it take the cheetah to travel 5 miles? Give your answer in hours and then in minutes.

18. The Gentoo penguin has a maximum speed of 27 kilometers per hour.

    a. Convert this to miles per hour. (Round to the nearest tenth.) (*Hint:* 1 mile $\approx$ 1.61 km)

    b. How far can the penguin travel in 10 minutes? Give your answer in miles.

Contents of Student Resources

Student Resources

Study Skills Builders

Attitude and Study Tips

Study Skills Builder 1

Have You Decided to Complete This Course Successfully?

Ask yourself if one of your current goals is to complete this course successfully.

If it is not a goal of yours, ask yourself why. One common reason is fear of failure. Amazingly enough, fear of failure alone can be strong enough to keep many of us from doing our best in any endeavor.

Another common reason is that you simply haven't taken the time to think about or write down your goals for this course. To help accomplish this, answer the questions below.

Exercises

1. Write down your goal(s) for this course.

2. Now list steps you will take to make sure your goal(s) in Exercise 1 are accomplished.

3. Rate your commitment to this course with a number between 1 and 5. Use the diagram below to help.

High Commitment		Average Commitment		Not Committed at All
5	4	3	2	1

4. If you have rated your personal commitment level (from the exercise above) as a 1, 2, or 3, list the reasons why this is so. Then determine whether it is possible to increase your commitment level to a 4 or 5.

Good luck, and don't forget that a positive attitude will make a big difference.

Study Skills Builder 2

Tips for Studying for an Exam

To prepare for an exam, try the following study techniques:

- Start the study process days before your exam.
- Make sure that you are up to date on your assignments.
- If there is a topic that you are unsure of, use one of the many resources that are available to you. For example,

  See your instructor.
  View a lecture video on the topic.
  Visit a learning resource center on campus.
  Read the textbook material and examples on the topic.

- Reread your notes and carefully review the Chapter Highlights at the end of any chapter.
- Work the review exercises at the end of the chapter.
- Find a quiet place to take the Chapter Test found at the end of the chapter. Do not use any resources when taking this sample test. This way, you will have a clear indication of how prepared you are for your exam. Check your answers and use the Chapter Test Prep Videos to make sure that you correct any missed exercises.

Good luck, and keep a positive attitude.

Exercises

Let's see how you did on your last exam.

1. How many days before your last exam did you start studying for that exam?

2. Were you up to date on your assignments at that time or did you need to catch up on assignments?

3. List the most helpful text supplement (if you used one).

4. List the most helpful campus supplement (if you used one).

5. List your process for preparing for a mathematics test.

6. Was this process helpful? In other words, were you satisfied with your performance on your exam?

7. If not, what changes can you make in your process that will make it more helpful to you?

Study Skills Builder 3

What to Do the Day of an Exam

Your first exam may be soon. On the day of an exam, don't forget to try the following:

- Allow yourself plenty of time to arrive.
- Read the directions on the test carefully.
- Read each problem carefully as you take your test. Make sure that you answer the question asked.
- Watch your time and pace yourself so that you may attempt each problem on your test.
- Check your work and answers.
- ***Do not turn your test in early.*** If you have extra time, spend it double-checking your work.

Good luck!

Exercises

Answer the following questions based on your most recent mathematics exam, whenever that was.

1. How soon before class did you arrive?

2. Did you read the directions on the test carefully?

3. Did you make sure you answered the question asked for each problem on the exam?

4. Were you able to attempt each problem on your exam?

5. If your answer to Exercise 4 is no, list reasons why.

6. Did you have extra time on your exam?

7. If your answer to Exercise 6 is yes, describe how you spent that extra time.

Study Skills Builder 4

Are You Satisfied with Your Performance on a Particular Quiz or Exam?

If not, don't forget to analyze your quiz or exam and look for common errors. Were most of your errors a result of:

- *Carelessness?* Did you turn in your quiz or exam before the allotted time expired? If so, resolve to use any extra time to check your work.

- *Running out of time?* Answer the questions you are sure of first. Then attempt the questions you are unsure of, and delay checking your work until all questions have been answered.

- *Not understanding a concept?* If so, review that concept and correct your work so that you make sure you understand it before the next quiz or the final exam.

- *Test conditions?* When studying for a quiz or exam, make sure you place yourself in conditions similar to test conditions. For example, before your next quiz or exam, take a sample test without the aid of your notes or text.

(For a sample test, see your instructor or use the Chapter Test at the end of each chapter.)

Exercises

1. Have you corrected all your previous quizzes and exams?

2. List any errors you have found common to two or more of your graded papers.

3. Is one of your common errors not understanding a concept? If so, are you making sure you understand all the concepts for the next quiz or exam?

4. Is one of your common errors making careless mistakes? If so, are you now taking all the time allotted to check over your work so that you can minimize the number of careless mistakes?

5. Are you satisfied with your grades thus far on quizzes and tests?

6. If your answer to Exercise 5 is no, are there any more suggestions you can make to your instructor or yourself to help? If so, list them here and share these with your instructor.

Study Skills Builder 5

How Are You Doing?

If you haven't done so yet, take a few moments and think about how you are doing in this course. Are you working toward your goal of successfully completing this course? Is your performance on homework, quizzes, and tests satisfactory? If not, you might want to see your instructor to see if he/she has any suggestions on how you can improve your performance. Reread Section 1.1 for ideas on places to get help with your mathematics course.

Exercises

Answer the following.

1. List any textbook supplements you are using to help you through this course.

2. List any campus resources you are using to help you through this course.

3. Write a short paragraph describing how you are doing in your mathematics course.

4. If improvement is needed, list ways that you can work toward improving your situation as described in Exercise 3.

Study Skills Builder 6

Are You Preparing for Your Final Exam?
To prepare for your final exam, try the following study techniques:

- Review the material that you will be responsible for on your exam. This includes material from your textbook, your notebook, and any handouts from your instructor.
- Review any formulas that you may need to memorize.
- Check to see if your instructor or mathematics department will be conducting a final exam review.
- Check with your instructor to see whether final exams from previous semesters/quarters are available to students for review.

- Use your previously taken exams as a practice final exam. To do so, rewrite the test questions in mixed order on blank sheets of paper. This will help you prepare for exam conditions.
- If you are unsure of a few concepts, see your instructor or visit a learning lab for assistance. Also, view the video segment of any troublesome sections.
- If you need further exercises to work, try the Cumulative Reviews at the end of the chapters.

Once again, good luck! I hope you are enjoying this textbook and your mathematics course.

Organizing Your Work

Study Skills Builder 7

Learning New Terms
Many of the terms used in this text may be new to you. It will be helpful to make a list of new mathematical terms and symbols as you encounter them and to review them frequently. Placing these new terms (including page references) on 3×5 index cards might help you later when you're preparing for a quiz.

Exercises

1. Name one way you might place a word and its definition on a 3×5 card.

2. How do new terms stand out in this text so that they can be found?

Study Skills Builder 8

Are You Organized?

Have you ever had trouble finding a completed assignment? When it's time to study for a test, are your notes neat and organized? Have you ever had trouble reading your own mathematics handwriting? (Be honest—I have.)

When any of these things happens, it's time to get organized. Here are a few suggestions:

- Write your notes and complete your homework assignments in a notebook with pockets (spiral or ring binder).
- Take class notes in this notebook, and then follow the notes with your completed homework assignment.
- When you receive graded papers or handouts, place them in the notebook pocket so that you will not lose them.
- Mark (possibly with an exclamation point) any note(s) that seem extra important to you.
- Mark (possibly with a question mark) any notes or homework that you are having trouble with.
- See your instructor or a math tutor to help you with the concepts or exercises that you are having trouble understanding.

- If you are having trouble reading your own handwriting, *slow down* and write your mathematics work clearly!

Exercises

1. Have you been completing your assignments on time?

2. Have you been correcting any exercises you may be having difficulty with?

3. If you are having trouble with a mathematical concept or correcting any homework exercises, have you visited your instructor, a tutor, or your campus math lab?

4. Are you taking lecture notes in your mathematics course? (By the way, these notes should include worked-out examples solved by your instructor.)

5. Is your mathematics course material (handouts, graded papers, lecture notes) organized?

6. If your answer to Exercise 5 is no, take a moment and review your course material. List at least two ways that you might better organize it.

Student Resources

Study Skills Builder 9

Organizing a Notebook

It's never too late to get organized. If you need ideas about organizing a notebook for your mathematics course, try some of these:

- Use a spiral or ring binder notebook with pockets and use it for mathematics only.
- Start each page by writing the book's section number you are working on at the top.
- When your instructor is lecturing, take notes. *Always* include any examples your instructor works for you.
- Place your worked-out homework exercises in your notebook immediately after the lecture notes from that section. This way, a section's worth of material is together.
- Homework exercises: Attempt and check all assigned homework.
- Place graded quizzes in the pockets of your notebook or a special section of your binder.

Exercises

Check your notebook organization by answering the following questions.

1. Do you have a spiral or ring binder notebook for your mathematics course only?

2. Have you ever had to flip through several sheets of notes and work in your mathematics notebook to determine what section's work you are in?

3. Are you now writing the textbook's section number at the top of each notebook page?

4. Have you ever lost or had trouble finding a graded quiz or test?

5. Are you now placing all your graded work in a dedicated place in your notebook?

6. Are you attempting all of your homework and placing all of your work in your notebook?

7. Are you checking and correcting your homework in your notebook? If not, why not?

8. Are you writing in your notebook the examples your instructor works for you in class?

Study Skills Builder 10

How Are Your Homework Assignments Going?

It is very important in mathematics to keep up with homework. Why? Many concepts build on each other. Often your understanding of a day's concepts depends on an understanding of the previous day's material.

Remember that completing your homework assignment involves a lot more than attempting a few of the problems assigned.

To complete a homework assignment, remember these four things:

- Attempt all of it.
- Check it.
- Correct it.
- If needed, ask questions about it.

Exercises

Take a moment and review your completed homework assignments. Answer the questions below based on this review.

1. Approximate the fraction of your homework you have attempted.

2. Approximate the fraction of your homework you have checked (if possible).

3. If you are able to check your homework, have you corrected it when errors have been found?

4. When working homework, if you do not understand a concept, what do you do?

MyMathLab and MathXL

Study Skills Builder 11

Tips for Turning In Your Homework on Time

It is very important to keep up with your mathematics homework assignments. Why? Many concepts in mathematics build upon each other.

Remember these 4 tips to help ensure your work is completed on time:

- Know the assignments and due dates set by your instructor.
- Do not wait until the last minute to submit your homework.
- Set a goal to submit your homework 6–8 hours before the scheduled due date in case you have unexpected technology trouble.
- Schedule enough time to complete each assignment.

Following the tips above will also help you avoid potentially losing points for late or missed assignments.

Exercises

Take a moment to consider your work on your homework assignments to date and answer the following questions:

1. What percentage of your assignments have you turned in on time?

2. Why might it be a good idea to submit your homework 6–8 hours before the scheduled deadline?

3. If you have missed submitting any homework by the due date, list some of the reasons why this occurred.

4. What steps do you plan to take in the future to ensure your homework is submitted on time?

Study Skills Builder 12

Tips for Doing Your Homework Online

Practice is one of the main keys to success in any mathematics course. Did you know that MyMathLab/MathXL provides you with **immediate feedback** for each exercise? If you are incorrect, you are given hints to work the exercise correctly. You have **unlimited practice opportunities** and can rework any exercises you have trouble with until you master them, and submit homework assignments unlimited times before the deadline.

Remember these success tips when doing your homework online:

- Attempt all assigned exercises.
- Write down (neatly) your step-by-step work for each exercise before entering your answer.
- Use the immediate feedback provided by the program to help you check and correct your work for each exercise.
- Rework any exercises you have trouble with until you master them.
- Work through your homework assignment as many times as necessary until you are satisfied.

Exercises

Take a moment to think about your homework assignments to date and answer the following:

1. Have you attempted all assigned exercises?

2. Of the exercises attempted, have you also written out your work before entering your answer—so that you can check it?

3. Are you familiar with how to enter answers using the MathXL player so that you avoid answer entry type errors?

4. List some ways the immediate feedback and practice supports have helped you with your homework. If you have not used these supports, how do you plan to use them with the success tips above on your next assignment?

Student Resources

Study Skills Builder 13

Organizing Your Work

Have you ever used any readily available paper (such as the back of a flyer, another course assignment, post-its, etc.) to work out homework exercises before entering the answer in MathXL? To save time, have you ever entered answers directly into MathXL without working the exercises on paper? When it's time to study, have you ever been unable to find your completed work or read and follow your own mathematics handwriting?

When any of these things happen, it's time to get organized. Here are some suggestions:

- Write your step-by-step work for each homework exercise, (neatly) on lined, loose-leaf paper and keep this in a 3-ring binder.

- Refer to your step-by-step work when you receive feedback that your answer is incorrect in MathXL. Double-check against the steps and hints provided by the program and correct your work accordingly.

- Keep your written homework with your class notes for that section.

- Identify any exercises you are having trouble with and ask questions about them.

- Keep all graded quizzes and tests in this binder as well to study later.

If you follow the suggestions above, you and your instructor or tutor will be able to follow your steps and correct any mistakes. You will also have a written copy of your work to refer to later to ask questions and study for tests.

Exercises

1. Why is it important that you write out your step-by-step work on homework exercises and keep a hard copy of all work submitted online?

2. If you have gotten an incorrect answer, are you able to follow your steps and find your error?

3. If you were asked today to review your previous homework assignments and first test, could you find them? If not, list some ways you might better organize your work.

Study Skills Builder 14

Getting Help with Your Homework Assignments

There are many helpful resources available to you through MathXL to help you work through any homework exercises you may have trouble with. It is important that you know what these resources are and know when and how to use them.

Let's review these features found in the homework exercises:

- **Help Me Solve This**—provides step-by-step help for the exercise you are working. You must work an additional exercise of the same type (without this help) before you can get credit for having worked it correctly.

- **View an Example**—allows you to view a correctly worked exercise similar to the one you are having trouble with. You can go back to your original exercise and work it on your own.

- **E-Book**—allows you to read examples from your text and find similar exercises.

- **Video**—your text author, Elayn Martin-Gay, works an exercise similar to the one you need help with. **Not all exercises have an accompanying video clip.

- **Ask My Instructor**—allows you to e-mail your instructor for help with an exercise.

Exercises

1. How does the "Help Me Solve This" feature work?

2. If the "View an Example" feature is used, is it necessary to work an additional problem before continuing the assignment?

3. When might be a good time to use the "Video" feature? Do all exercises have an accompanying video clip?

4. Which of the features above have you used? List those you found the most helpful to you.

5. If you haven't used the features discussed, list those you plan to try on your next homework assignment.

Study Skills Builder 15

Tips for Preparing for an Exam

Did you know that you can rework your previous homework assignments in MyMathLab and MathXL? This is a great way to prepare for tests. To do this, open a previous homework assignment and click "similar exercise." This will generate new exercises similar to the homework you have submitted. You can then rework the exercises and assignments until you feel confident that you understand them.

To prepare for an exam, follow these tips:

- Review your written work for your previous homework assignments along with your class notes.
- Identify any exercises or topics that you have questions on or have difficulty understanding.
- Rework your previous assignments in MyMathLab and MathXL until you fully understand them and can do them without help.
- Get help for any topics you feel unsure of or for which you have questions.

Exercises

1. Are your current homework assignments up to date and is your written work for them organized in a binder or notebook? If the answer is no, it's time to get organized. For tips on this, see Study Skills Builder 13—Organizing Your Work.

2. How many days in advance of an exam do you usually start studying?

3. List some ways you think that practicing previous homework assignments can help you prepare for your test.

4. List two or three resources you can use to get help for any topics you are unsure of or have questions on.

Good luck!

Study Skills Builder 16

How Well Do You Know the Resources Available to You in MyMathLab?

There are many helpful resources available to you in MyMathLab. Let's take a moment to locate and explore a few of them now. Go into your MyMathLab course, and visit the multimedia library, tools for success, and E-book.

Let's see what you found.

Exercises

1. List the resources available to you in the Multimedia Library.

2. List the resources available to you in the Tools for Success folder.

3. Where did you find the English/Spanish Audio Glossary?

4. Can you view videos from the E-Book?

5. Did you find any resources you did not know about? If so, which ones?

6. Which resources have you used most often or found most helpful?

Additional Help Inside and Outside Your Textbook

Study Skills Builder 17

How Well Do You Know Your Textbook?
The questions below will help determine whether you are familiar with your textbook. For additional information, see Section 1.1 in this text.

Exercises

1. What does the ● icon mean?

2. What does the ╲ icon mean?

3. What does the △ icon mean?

4. Where can you find a review for each chapter? What answers to this review can be found in the back of your text?

5. Each chapter contains an overview of the chapter along with examples. What is this feature called?

6. Each chapter contains a review of vocabulary. What is this feature called?

7. There are practice exercises that are contained in this text. What are they and how can they be used?

8. This text contains a student section in the back entitled Student Resources. List the contents of this section and how they might be helpful.

9. What exercise answers are available in this text? Where are they located?

Study Skills Builder 18

Are You Familiar with Your Textbook Supplements?
Below is a review of some of the student supplements available for additional study. Check to see if you are using the ones most helpful to you.

- Chapter Test Prep Videos. These videos provide video clip solutions to the Chapter Test exercises in this text. You will find this extremely useful when studying for tests or exams.
- Interactive DVD Lecture Series. These are keyed to each section of the text. The material is presented by me, Elayn Martin-Gay, and I have placed a ● by the exercises in the text that I have worked on the video.
- The *Student Solutions Manual*. This contains worked-out solutions to odd-numbered exercises as well as every exercise in the Integrated Reviews, Chapter Reviews, Chapter Tests, and Cumulative Reviews.
- Pearson Tutor Center. Mathematics questions may be phoned, faxed, or e-mailed to this center.
- MyMathLab is a text-specific online course. MathXL is an online homework, tutorial, and assessment system.

Take a moment and determine whether these are available to you.

As usual, your instructor is your best source of information.

Exercises

Let's see how you are doing with textbook supplements.

1. Name one way the Lecture Videos can be helpful to you.

2. Name one way the Chapter Test Prep Video can help you prepare for a chapter test.

3. List any textbook supplements that you have found useful.

4. Have you located and visited a learning resource lab located on your campus?

5. List the textbook supplements that are currently housed in your campus's learning resource lab.

Study Skills Builder 19

Are You Getting All the Mathematics Help That You Need?

Remember that, in addition to your instructor, there are many places to get help with your mathematics course. For example:

- This text has an accompanying video lesson for every section and the CD in this text contains worked-out solutions to every Chapter Test exercise.

- The back of the book contains answers to odd-numbered exercises and selected solutions.

- A *Student Solutions Manual* is available that contains worked-out solutions to odd-numbered exercises as well as solutions to every exercise in the Integrated Reviews, Chapter Reviews, Chapter Tests, and Cumulative Reviews.

- Don't forget to check with your instructor for other local resources available to you, such as a tutoring center.

Exercises

1. List items you find helpful in the text and all student supplements to this text.

2. List all the campus help that is available to you for this course.

3. List any help (besides the textbook) from Exercises 1 and 2 above that you are using.

4. List any help (besides the textbook) that you feel you should try.

5. Write a goal for yourself that includes trying everything you listed in Exercise 4 during the next week.

Bigger Picture—
Study Guide Outline

I. Some Operations on Sets of Numbers

   A. Whole Numbers

      1. Add or Subtract: (Sec. 1.3, 1.4)

$$\begin{array}{r} 14 \\ +\ 39 \\ \hline 53 \end{array} \qquad \begin{array}{r} 300 \\ -\ 27 \\ \hline 273 \end{array}$$

      2. Multiply or Divide: (Sec. 1.6, 1.7)

$$\begin{array}{r} 238 \\ \times\ 47 \\ \hline 1666 \\ 9520 \\ \hline 11{,}186 \end{array} \qquad \begin{array}{r} 127 \ \text{R2} \\ 7\overline{)891} \\ \underline{-7} \\ 19 \\ \underline{-14} \\ 51 \\ \underline{-49} \\ 2 \end{array}$$

      3. Exponent: (Sec. 1.9)

$$3^4 = \overbrace{3 \cdot 3 \cdot 3 \cdot 3}^{4\ \text{factors of } 3} = 81$$

      4. Square Root: (Sec. 1.9)

$$\sqrt{25} = 5\ because\ 5 \cdot 5 = 25\ \text{and 5 is a positive number.}$$

      5. Order of Operations: (Sec. 1.9)

$$\begin{aligned} 24 \div 3 \cdot 2 - (2 + 8) &= 24 \div 3 \cdot 2 - (10) && \text{Parentheses.} \\ &= 8 \cdot 2 - 10 && \text{Multiply or divide from left to right.} \\ &= 16 - 10 && \text{Multiply or divide from left to right.} \\ &= 6 && \text{Add or subtract from left to right.} \end{aligned}$$

   B. Fractions

      1. Simplify: Factor the numerator and denominator. Then divide out factors of 1 by dividing out common factors in the numerator and denominator. (Sec. 2.3)

$$\text{Simplify:}\ \frac{20}{28} = \frac{4 \cdot 5}{4 \cdot 7} = \frac{5}{7}$$

      2. Multiply: Numerator times numerator over denominator times denominator. (Sec. 2.4)

$$\frac{5}{9} \cdot \frac{2}{7} = \frac{10}{63}$$

      3. Divide: First fraction times the reciprocal of the second fraction. (Sec. 2.5)

$$\frac{2}{11} \div \frac{3}{4} = \frac{2}{11} \cdot \frac{4}{3} = \frac{8}{33}$$

      4. Add or Subtract: Must have same denominators. If not, find the LCD, and write each fraction as an equivalent fraction with the LCD as denominator. (Sec. 3.3)

$$\frac{2}{5} + \frac{1}{15} = \frac{2}{5} \cdot \frac{3}{3} + \frac{1}{15} = \frac{6}{15} + \frac{1}{15} = \frac{7}{15}$$

C. Decimals

1. **Add or Subtract:** Line up decimal points. (Sec. 4.3)

$$\begin{array}{r} 1.27 \\ + \; 0.6 \\ \hline 1.87 \end{array}$$

2. **Multiply:** (Sec. 4.4)

$$\begin{array}{r} 2.56 \\ \times \; 3.2 \\ \hline 512 \\ 7680 \\ \hline 8.192 \end{array}$$

2 decimal places

1 decimal place

$2 + 1 = 3$

3 decimal places

3. **Divide:** (Sec. 4.5) $8\overline{)5.6}\;^{0.7}$ $0.6\overline{)0.786}\;^{1.31}$

D. Signed Numbers

1. **Add:** (Sec. 10.2) $-5 + (-2) = -7$ Adding like signs.
   Add absolute value. Attach the common sign.

   $-5 + 2 = -3$ Adding unlike signs.
   Subtract absolute values. Attach the sign of
   the number with the larger absolute value.

2. **Subtract:** Add the first number to the opposite of the second number. (Sec. 10.3)

   $7 - 10 = 7 + (-10) = -3$

3. **Multiply or Divide:** Multiply or divide as usual. If the signs of the two numbers are the same, the answer is positive. If the signs of the two numbers are different, the answer is negative. (Sec. 10.4)

   $-5 \cdot 5 = -25$ $\dfrac{-32}{-8} = 4$

II. Solving Equations

A. Proportions: Set cross products equal to each other. Then solve. (Sec. 5.3)

$\dfrac{14}{3} = \dfrac{2}{n}$, or $14 \cdot n = 3 \cdot 2$, or $14 \cdot n = 6$, or $n = \dfrac{6}{14} = \dfrac{3}{7}$

B. Percent Problems

1. **Solved by Equations:** Remember that "of" means multiplication and "is" means equals. (Sec. 6.3)

   "12% of some number is 6" translates to

   $12\% \cdot n = 6$ or $0.12 \cdot n = 6$ or $n = \dfrac{6}{0.12}$ or $n = 50$

2. **Solved by Proportions:** Remember that percent, p, is identified by % or percent; base, b, usually appears after "of"; and amount, a, is the part compared to the whole. (Sec. 6.4)

   "12% of some number is 6" translates to

   $\dfrac{6}{b} = \dfrac{12}{100}$ or $6 \cdot 100 = b \cdot 12$ or $600 = b \cdot 12$ or $\dfrac{600}{12} = b$ or $50 = b$

C. Equations in General: Simplify both sides of the equation by removing parentheses and combining any like terms. Then use the Addition Property to write variable terms on one side, constants (or numbers) on the other side. Then use the Multiplication Property to solve for the variable by dividing both sides of the equation by the coefficient of the variable. (Sec. 11.4)

Solve: $2(x - 5) = 80$

$2x - 10 = 80$ Use the distributive property.

$2x - 10 + 10 = 80 + 10$ Add 10 to both sides.

$2x = 90$ Simplify.

$\dfrac{2x}{2} = \dfrac{90}{2}$ Divide both sides by 2.

$x = 45$ Simplify.

Study Guide

Practice Final Exam

1.

2.

3.

4.

5.

6.

7.

8.

9.

10.

11.

12.

13.

14.

15.

16.

17.

18.

19.

20.

21.

22.

Note: Exercises 1–41 review operations with nonnegative numbers. Simplify by performing the indicated operations.

1. $600 - 487$

2. $(2^4 - 5) \cdot 3$

3. $\dfrac{16}{3} \div \dfrac{3}{12}$

4. $\dfrac{11}{12} + \dfrac{3}{8} + \dfrac{5}{24}$

5. $\dfrac{64 \div 8 \cdot 2}{(\sqrt{9} - \sqrt{4})^2 + 1}$

6. $\begin{array}{r} 10.2 \\ \times\ 4.3 \\ \hline \end{array}$

7. $\dfrac{0.23 + 1.63}{0.3}$

8. $\begin{array}{r} 5\frac{1}{6} \\ -\ 3\frac{7}{8} \\ \hline \end{array}$

9. $3\dfrac{1}{3} \cdot 6\dfrac{3}{4}$

10. 126.9×100

11. $\left(\dfrac{3}{4}\right)^2 \div \left(\dfrac{2}{3} + \dfrac{5}{6}\right)$

12. Round 52,369 to the nearest thousand.

13. Round 34.8923 to the nearest tenth.

14. Write $\dfrac{16}{17}$ as a decimal. Round to the nearest thousandth.

15. Write 85% as a decimal.

16. Write 6.1 as a percent.

17. Write $\dfrac{3}{8}$ as a percent.

18. Write 0.2% as a fraction in simplest form.

19. Find the perimeter and the area of the rectangle below.

Rectangle — $\frac{2}{3}$ foot

1 foot

Write each ratio or rate as a fraction in simplest form.

20. 4500 trees to 6500 trees

21. 9 inches of rain in 30 days

22. Find the unit rate:
650 kilometers in 8 hours

23. Find each unit price and decide which is the better buy.

    Jelly:

    $1.49 for 16 ounces

    $2.39 for 24 ounces

24. Find the unknown number, n, in the proportion:

$$\frac{8}{n} = \frac{11}{6}$$

Solve.

25. Subtract 8.6 from 20.

26. A small airplane used $58\frac{3}{4}$ gallons of fuel on a $7\frac{1}{2}$-hour trip. How many gallons of fuel were used for each hour?

27. Jerome Grant worked 6 hours and packed 86 cartons of books. At this rate, how many cartons can he pack in 8 hours?

28. Twenty-nine cans of Sherwin-Williams paint cost $493. How much was each can?

29. 0.6% of what number is 7.5?

30. 567 is what percent of 756?

31. An alloy is 12% copper. How much copper is contained in 320 pounds of this alloy?

32. A $120 framed picture is on sale for 15% off. Find the discount and the sale price.

Convert.

33. 2.4 kg to grams

34. $2\frac{1}{2}$ gal to quarts

35. If 2 ft 9 in. of material is used to manufacture one scarf, how much material is needed for 6 scarves?

36. Find the complement of a 78° angle.

37. Find the measures of angles x, y, and z.

38. Find the measure of $\angle x$.

39. Given that the following triangles are similar, find the missing length n.

Find the mean, median, and mode of the list of numbers.

40. 26, 32, 42, 43, 49

23. _____

24. _____

25. _____

26. _____

27. _____

28. _____

29. _____

30. _____

31. _____

32. _____

33. _____

34. _____

35. _____

36. _____

37. _____

38. _____

39. _____

40. _____

41. _____

42. _____

43. _____

44. _____

45. _____

46. _____

47. _____

48. _____

49. _____

50. _____

51. _____

A professor measures the heights of the students in her class. The results are shown in the following histogram. Use this histogram to answer Exercise 41.

Student Heights

41. How many students are 5′7″ or shorter?

Note: Exercises 42–51 may contain signed numbers. Simplify by performing the indicated operations.

42. $(-5)^3 - 24 \div (-3)$

43. $-7 - (-19)$

44. $\dfrac{-3(-2) + 12}{-1(-4 - 5)}$

45. $\left(\dfrac{5}{9} - \dfrac{7}{9}\right)^2 + \left(-\dfrac{2}{9}\right)$

46. Evaluate: $\dfrac{3x - 5}{2y}$ when $x = 7$ and $y = -8$

47. Multiply, then simplify:
$5(3z + 2) - z - 18$

Solve.

48. $-\dfrac{5}{8}x = -25$

49. $5x + 12 - 4x - 14 = 22$

50. $3x - 5 = -11$

51. In a 10-kilometer race, there are 112 more men entered than women. Find the number of women runners if the total number of runners in the race is 600.

Answers to Selected Exercises

Chapter 1 The Whole Numbers

Section 1.2

Vocabulary, Readiness & Video Check **1.** whole **3.** words **5.** period **7.** hundreds **9.** 80,000

Exercise Set 1.2 **1.** tens **3.** thousands **5.** hundred-thousands **7.** millions **9.** three hundred fifty-four **11.** eight thousand, two hundred seventy-nine **13.** twenty-six thousand, nine hundred ninety **15.** two million, three hundred eighty-eight thousand **17.** twenty-four million, three hundred fifty thousand, one hundred eighty-five **19.** three hundred fifteen thousand, two hundred eighty-one **21.** two thousand, seven hundred seventeen **23.** thirty-two million, one hundred thousand **25.** fourteen thousand, four hundred thirty-three **27.** seventeen million **29.** 6587 **31.** 59,800 **33.** 13,601,011 **35.** 7,000,017 **37.** 260,997 **39.** 418 **41.** 16,732 **43.** $91,071,000 **45.** 101,000 **47.** 400 + 6 **49.** 3000 + 400 + 70 **51.** 80,000 + 700 + 70 + 4 **53.** 60,000 + 6000 + 40 + 9 **55.** 30,000,000 + 9,000,000 + 600,000 + 80,000 **57.** 5532; five thousand, five hundred thirty-two **59.** 5000 + 400 + 90 + 2 **61.** Mt. Washington **63.** National Gallery **65.** five million, sixty-five thousand **67.** 4 **69.** 9861 **71.** no; one hundred five **73.** answers may vary **75.** 34,000,000,000,000,000 **77.** Canton

Section 1.3

Calculator Explorations **1.** 134 **3.** 340 **5.** 2834

Vocabulary, Readiness & Video Check **1.** number **3.** sum; addend **5.** grouping; associative **7.** place; right; left **9.** increased by

Exercise Set 1.3 **1.** 36 **3.** 292 **5.** 49 **7.** 5399 **9.** 117 **11.** 512 **13.** 209,078 **15.** 25 **17.** 62 **19.** 212 **21.** 94 **23.** 910 **25.** 8273 **27.** 11,926 **29.** 1884 **31.** 16,717 **33.** 1110 **35.** 8999 **37.** 35,901 **39.** 632,389 **41.** 42 in. **43.** 25 ft **45.** 24 in. **47.** 8 yd **49.** 29 in. **51.** 44 m **53.** 2093 **55.** 266 **57.** 544 **59.** 3452 **61.** 21,141 thousand **63.** 6684 ft **65.** 340 ft **67.** 2425 ft **69.** 249,849 **71.** 1,474,711 vehicles **73.** 124 ft **75.** 3095 **77.** California **79.** 529 stores **81.** New York and Pennsylvania **83.** 6358 mi **85.** answers may vary **87.** answers may vary **89.** 1,044,473,765 **91.** correct **93.** incorrect: 530

Section 1.4

Calculator Explorations **1.** 770 **3.** 109 **5.** 8978

Vocabulary, Readiness & Video Check **1.** 0 **3.** minuend; subtrahend **5.** 0 **7.** 600 **9.** We cannot take 7 from 2 in the ones place, so we borrow one ten from the tens place and move it over to the ones place to give us 10 + 2 or 12.

Exercise Set 1.4 **1.** 44 **3.** 265 **5.** 135 **7.** 2254 **9.** 5545 **11.** 600 **13.** 25 **15.** 45 **17.** 146 **19.** 288 **21.** 168 **23.** 106 **25.** 447 **27.** 5723 **29.** 504 **31.** 89 **33.** 79 **35.** 39,914 **37.** 32,711 **39.** 5041 **41.** 31,213 **43.** 4 **45.** 20 **47.** 7 **49.** 72 **51.** 88 **53.** 264 pages **55.** 7 million sq km **57.** 264,000 sq mi **59.** 283,000 sq mi **61.** 6065 ft **63.** 28 ft **65.** 358 mi **67.** $619 **69.** 2630 thousand **71.** 100 dB **73.** 58 dB **75.** 346 **77.** 5920 sq ft **79.** Hartsfield-Jackson Atlanta International **81.** 14 million **83.** Jo; by 271 votes **85.** 1034 **87.** 9 **89.** 8518 **91.** 22,876 **93.** minuend: 48; subtrahend: 1 **95.** minuend: 70; subtrahend: 7 **97.** incorrect: 685 **99.** correct **101.** 5269 − 2385 = 2884 **103.** no; answers may vary **105.** no: 1089 more pages

Section 1.5

Vocabulary, Readiness & Video Check **1.** graph **3.** 70; 60 **5.** 3 is in the place value we're rounding to (tens), and the digit to the right of this place value is 5 or greater, so we need to add 1 to the 3. **7.** Each circled digit is to the right of the place value being rounded to and is used to determine whether or not we add 1 to the digit in the place value being rounded to.

Exercise Set 1.5 **1.** 420 **3.** 640 **5.** 2800 **7.** 500 **9.** 21,000 **11.** 34,000 **13.** 328,500 **15.** 36,000 **17.** 39,990 **19.** 30,000,000 **21.** 5280; 5300; 5000 **23.** 9440; 9400; 9000 **25.** 14,880; 14,900; 15,000 **27.** 28,000 **29.** 38,000 points **31.** $98,000,000,000 **33.** $5,022,000 **35.** $129,000,000,000 **37.** 130 **39.** 80 **41.** 5700 **43.** 300 **45.** 11,400 **47.** incorrect **49.** correct **51.** correct **53.** $3400 **55.** 900 mi **57.** 6000 ft **59.** 8,000,000,000 **61.** 3000 children **63.** $3,067,000,000; $3,070,000,000; $3,100,000,000 **65.** $2,910,000,000; $2,910,000,000; $2,900,000,000 **67.** 5723, for example **69. a.** 8550 **b.** 8649 **71.** answers may vary **73.** 140 m

Section 1.6

Calculator Explorations **1.** 3456 **3.** 15,322 **5.** 272,291

Vocabulary, Readiness & Video Check **1.** 0 **3.** product; factor **5.** grouping; associative **7.** length **9.** distributive property **11.** Think of the problem as 50 times 9 and then attach the two zeros from 900, or think of the problem as 5 times 9 and then attach the three zeros at the end of 50 and 900. Both approaches give us 45,000. **13.** Multiplication is also an application of addition since it is addition of the same addend.

Exercise Set 1.6   **1.** 24   **3.** 0   **5.** 0   **7.** 87   **9.** $6 \cdot 3 + 6 \cdot 8$ **11.** $4 \cdot 3 + 4 \cdot 9$ **13.** $20 \cdot 14 + 20 \cdot 6$ **15.** 512   **17.** 3678   **19.** 1662   **21.** 6444   **23.** 1157   **25.** 24,418   **27.** 24,786   **29.** 15,600   **31.** 0   **33.** 6400   **35.** 48,126   **37.** 142,506   **39.** 2,369,826   **41.** 64,790   **43.** 3,949,935   **45.** 800   **47.** 11,000   **49.** 74,060   **51.** 24,000   **53.** 45,000   **55.** 3,280,000   **57.** area: 63 sq m; perimeter: 32 m   **59.** area: 680 sq ft; perimeter: 114 ft   **61.** 240,000   **63.** 300,000   **65.** c   **67.** c   **69.** 880   **71.** 4200   **73.** 4480   **75.** 375 cal   **77.** \$3290   **79. a.** 20 boxes   **b.** 100 boxes   **c.** 2000 lb   **81.** 8800 sq ft   **83.** 56,000 sq ft   **85.** 5828 pixels   **87.** 2100 characters   **89.** 1360 cal   **91.** \$10, \$60; \$10, \$200; \$12, \$36; \$12, \$36; total cost: \$372   **93.** 1,440,000 tea bags   **95.** 135   **97.** 2144   **99.** 23   **101.** 15   **103.** $4 \cdot 7$ or $7 \cdot 4$ **105. a.** $5 + 5 + 5$ or $3 + 3 + 3 + 3 + 3$ **b.** answers may vary   **107.** 203   **109.** 42   **111.** answers may vary   **113.** 506 windows

$$\begin{array}{r} 203 \\ \times\ 14 \\ \hline 812 \\ 2030 \\ \hline 2842 \end{array} \qquad \begin{array}{r} 42 \\ \times\ 93 \\ \hline \end{array}$$

Section 1.7

Calculator Explorations   **1.** 53   **3.** 62   **5.** 261   **7.** 0

Vocabulary, Readiness & Video Check   **1.** quotient; dividend; divisor   **3.** 1   **5.** undefined   **7.** 0   **9.** $202 \cdot 102 + 15 = 20{,}619$ **11.** addition and division

Exercise Set 1.7   **1.** 6   **3.** 12   **5.** 0   **7.** 31   **9.** 1   **11.** 8   **13.** undefined   **15.** 1   **17.** 0   **19.** 9   **21.** 29   **23.** 74   **25.** 338   **27.** undefined   **29.** 9   **31.** 25   **33.** 68 R 3   **35.** 236 R 5   **37.** 38 R 1   **39.** 326 R 4   **41.** 13   **43.** 49   **45.** 97 R 8   **47.** 209 R 11   **49.** 506   **51.** 202 R 7   **53.** 54   **55.** 99 R 100   **57.** 202 R 15   **59.** 579 R 72   **61.** 17   **63.** 511 R 3   **65.** 2132 R 32   **67.** 6080   **69.** 23 R 2   **71.** 5 R 25   **73.** 20 R 2   **75.** 33 students   **77.** 165 lb   **79.** 310 yd   **81.** 89 bridges   **83.** 11 light poles   **85.** 5 mi   **87.** 1760 yd   **89.** 20   **91.** 387   **93.** 79   **95.** 74°   **97.** 9278   **99.** 15,288   **101.** 679   **103.** undefined   **105.** 9 R 12   **107.** c   **109.** b   **111.** 84   **113.** increase; answers may vary   **115.** no; answers may vary   **117.** 12 ft   **119.** answers may vary   **121.** 5 R 1

Integrated Review   **1.** 148   **2.** 6555   **3.** 1620   **4.** 562   **5.** 79   **6.** undefined   **7.** 9   **8.** 1   **9.** 0   **10.** 0   **11.** 0   **12.** 3   **13.** 2433   **14.** 9826   **15.** 213 R 3   **16.** 79,317   **17.** 27   **18.** 9   **19.** 138   **20.** 276   **21.** 1099 R 2   **22.** 111 R 1   **23.** 663 R 6   **24.** 1076 R 60   **25.** 1024   **26.** 9899   **27.** 30,603   **28.** 47,500   **29.** 65   **30.** 456   **31.** 6 R 8   **32.** 53   **33.** 183   **34.** 231   **35.** 9740; 9700; 10,000   **36.** 1430; 1400; 1000   **37.** 20,800; 20,800; 21,000   **38.** 432,200; 432,200; 432,000   **39.** perimeter: 24 ft; area: 36 sq ft   **40.** perimeter: 42 in.; area: 98 sq in.   **41.** 28 mi   **42.** 26 m   **43.** 24   **44.** 124   **45.** Lake Pontchartrain Bridge; 2175 ft   **46.** 730 qt

Section 1.8

Vocabulary, Readiness & Video Check   **1.** The George Washington Bridge has a length of 3500 feet.

Exercise Set 1.8   **1.** 49   **3.** 237   **5.** 42   **7.** 600   **9. a.** 400 ft   **b.** 9600 sq ft   **11.** \$15,500   **13.** 168 hr   **15.** 3500 ft   **17.** 141 yr   **19.** 372 billion bricks   **21.** 719 towns   **23.** \$21   **25.** 55 cal   **27.** 22 hot dogs   **29.** \$32,842,230   **31.** 1,215,051 people   **33.** 3987 mi   **35.** 13 paychecks   **37.** \$239   **39.** \$1045   **41.** b will be cheaper by \$3   **43.** Asia   **45.** 1053 million   **47.** 19 million   **49.** 798 million   **51.** \$14,754   **53.** 16,800 mg   **55. a.** 3750 sq ft   **b.** 375 sq ft   **c.** 3375 sq ft   **57.** \$2   **59.** answers may vary

Section 1.9

Calculator Explorations   **1.** 4096   **3.** 3125   **5.** 2048   **7.** 2526   **9.** 4295   **11.** 8

Vocabulary, Readiness & Video Check   **1.** base; exponent   **3.** addition   **5.** division   **7.** exponent; base   **9.** Because $8 \cdot 8 = 64$.   **11.** The area of a rectangle is length $\cdot$ width. A square is a special rectangle where length = width. Thus, the area of a square is side $\cdot$ side or $(\text{side})^2$.

Exercise Set 1.9   **1.** 4^3 **3.** 7^6 **5.** 12^3 **7.** $6^2 \cdot 5^3$ **9.** $9 \cdot 8^2$ **11.** $3 \cdot 2^4$ **13.** $3 \cdot 2^4 \cdot 5^5$ **15.** 64   **17.** 125   **19.** 32   **21.** 1   **23.** 7   **25.** 128   **27.** 256   **29.** 256   **31.** 729   **33.** 144   **35.** 100   **37.** 20   **39.** 729   **41.** 192   **43.** 162   **45.** 3   **47.** 8   **49.** 12   **51.** 4   **53.** 21   **55.** 7   **57.** 5   **59.** 16   **61.** 46   **63.** 8   **65.** 64   **67.** 83   **69.** 2   **71.** 48   **73.** 4   **75.** undefined   **77.** 59   **79.** 52   **81.** 44   **83.** 12   **85.** 21   **87.** 24   **89.** 28   **91.** 3   **93.** 25   **95.** 23   **97.** 13   **99.** area: 49 sq m; perimeter: 28 m   **101.** area: 529 sq mi; perimeter: 92 mi   **103.** true   **105.** false   **107.** $(2 + 3) \cdot 6 - 2$ **109.** $24 \div (3 \cdot 2) + 2 \cdot 5$ **111.** 1260 ft   **113.** 6,384,814   **115.** answers may vary; $(20 - 10) \cdot 5 \div 25 + 3$

Chapter 1 Vocabulary Check   **1.** whole numbers   **2.** perimeter   **3.** place value   **4.** exponent   **5.** area   **6.** square root   **7.** digits   **8.** average   **9.** divisor   **10.** dividend   **11.** quotient   **12.** factor   **13.** product   **14.** minuend   **15.** subtrahend   **16.** difference   **17.** addend   **18.** sum

Chapter 1 Review   **1.** tens   **2.** ten-millions   **3.** seven thousand, six hundred forty   **4.** forty-six million, two hundred thousand, one hundred twenty   **5.** $3000 + 100 + 50 + 8$ **6.** $400{,}000{,}000 + 3{,}000{,}000 + 200{,}000 + 20{,}000 + 5000$ **7.** 81,900   **8.** 6,304,000,000   **9.** 518,512,109   **10.** 184,177,220   **11.** Asia   **12.** Oceania/Australia   **13.** 63   **14.** 67   **15.** 48   **16.** 77   **17.** 956   **18.** 840   **19.** 7950   **20.** 7250   **21.** 4211   **22.** 1967   **23.** 1326   **24.** 886   **25.** 27,346   **26.** 39,300   **27.** 8032 mi   **28.** \$197,699   **29.** 276 ft   **30.** 66 km   **31.** 14   **32.** 34   **33.** 65   **34.** 304   **35.** 3914   **36.** 7908   **37.** 17,897   **38.** 34,658   **39.** 238,305   **40.** 249,795   **41.** 397 pages   **42.** \$25,626   **43.** May   **44.** August   **45.** \$110   **46.** \$240   **47.** 90   **48.** 50   **49.** 470   **50.** 500   **51.** 4800   **52.** 58,000

53. 50,000,000 **54.** 800,000 **55.** 126,000,000 **56.** 99,000 **57.** 7400 **58.** 4100 **59.** 2500 mi **60.** 900,000 **61.** 1911 **62.** 1396
63. 1410 **64.** 2898 **65.** 800 **66.** 900 **67.** 3696 **68.** 1694 **69.** 0 **70.** 0 **71.** 16,994 **72.** 8954 **73.** 113,634 **74.** 44,763
75. 411,426 **76.** 636,314 **77.** 375,000 **78.** 108,000 **79.** 12,000 **80.** 35,000 **81.** 5,100,000 **82.** 7,600,000 **83.** 1150 **84.** 4920
85. 108 **86.** 112 **87.** 24 g **88.** $152,340 **89.** 60 sq mi **90.** 500 sq cm **91.** 3 **92.** 4 **93.** 6 **94.** 7 **95.** 5 R 2 **96.** 4 R 2
97. undefined **98.** 0 **99.** 1 **100.** 10 **101.** 0 **102.** undefined **103.** 33 R 2 **104.** 19 R 7 **105.** 24 R 2 **106.** 35 R 15 **107.** 506 R 10
108. 907 R 40 **109.** 2793 R 140 **110.** 2012 R 60 **111.** 18 R 2 **112.** 21 R 2 **113.** 458 ft **114.** 13 mi **115.** 51 **116.** 59
117. 27 boxes **118.** $192 **119.** $5 billion **120.** 75¢ **121.** $898 **122.** 23,150 sq ft **123.** 49 **124.** 125 **125.** 45 **126.** 400 **127.** 13
128. 10 **129.** 15 **130.** 7 **131.** 12 **132.** 9 **133.** 42 **134.** 33 **135.** 9 **136.** 2 **137.** 1 **138.** 0 **139.** 6 **140.** 29 **141.** 40
142. 72 **143.** 5 **144.** 7 **145.** 49 sq m **146.** 9 sq in. **147.** 307 **148.** 682 **149.** 2169 **150.** 2516 **151.** 901 **152.** 1411 **153.** 458 R 8
154. 237 R 1 **155.** 70,848 **156.** 95,832 **157.** 1644 **158.** 8481 **159.** 740 **160.** 258,000 **161.** 2000 **162.** 40,000 **163.** thirty-six
thousand, nine hundred eleven **164.** one hundred fifty-four thousand, eight hundred sixty-three **165.** 70,943 **166.** 43,401 **167.** 64
168. 125 **169.** 12 **170.** 10 **171.** 12 **172.** 1 **173.** 2 **174.** 6 **175.** 4 **176.** 24 **177.** 24 **178.** 14 **179.** $4,205,000 **180.** $2,129,000
181. 53 full boxes with 18 left over **182.** $86

Chapter 1 Test **1.** eighty-two thousand, four hundred twenty-six **2.** 402,550 **3.** 141 **4.** 113 **5.** 14,880 **6.** 766 R 42 **7.** 200
8. 10 **9.** 0 **10.** undefined **11.** 33 **12.** 21 **13.** 8 **14.** 36 **15.** 5,698,000 **16.** 11,200,000 **17.** 52,000 **18.** 13,700 **19.** 1600
20. 92 **21.** 122 **22.** 1605 **23.** 7 R 2 **24.** $17 **25.** $126 **26.** 360 cal **27.** $7905 **28.** 20 cm; 25 sq cm **29.** 60 yd; 200 sq yd

Chapter 2 Multiplying and Dividing Fractions

Section 2.1

Vocabulary, Readiness & Video Check **1.** fraction; denominator; numerator **3.** improper; proper; mixed number **5.** The fraction is equal to 1. **7.** Each shape is divided into 3 equal parts. **9.** division

Exercise Set 2.1 **1.** numerator: 1; denominator: 2; proper **3.** numerator: 10; denominator: 3; improper **5.** numerator: 15; denominator: 15; improper **7.** 1 **9.** undefined **11.** 13 **13.** 0 **15.** undefined **17.** 16 **19.** $\frac{5}{6}$ **21.** $\frac{7}{12}$ **23.** $\frac{3}{7}$ **25.** $\frac{4}{9}$ **27.** $\frac{1}{6}$ **29.** $\frac{5}{8}$

31. **33.** **35.** **37.**

39. $\frac{42}{131}$ **41. a.** 89 **b.** $\frac{89}{131}$ **43.** $\frac{7}{44}$ **45.** $\frac{10}{19}$ **47.** $\frac{11}{31}$ **49.** $\frac{10}{31}$ **51. a.** $\frac{33}{50}$ **b.** 17 **c.** $\frac{17}{50}$ **53. a.** $\frac{21}{50}$ **b.** 29 **c.** $\frac{29}{50}$ **55. a.** $\frac{11}{4}$
b. $2\frac{3}{4}$ **57. a.** $\frac{23}{6}$ **b.** $3\frac{5}{6}$ **59. a.** $\frac{4}{3}$ **b.** $1\frac{1}{3}$ **61. a.** $\frac{11}{2}$ **b.** $5\frac{1}{2}$ **63.** $\frac{7}{3}$ **65.** $\frac{18}{5}$ **67.** $\frac{53}{8}$ **69.** $\frac{41}{15}$ **71.** $\frac{83}{7}$ **73.** $\frac{84}{13}$ **75.** $\frac{109}{24}$ **77.** $\frac{211}{12}$
79. $\frac{187}{20}$ **81.** $\frac{265}{107}$ **83.** $\frac{500}{3}$ **85.** $3\frac{2}{5}$ **87.** $4\frac{5}{8}$ **89.** $3\frac{2}{15}$ **91.** $2\frac{4}{21}$ **93.** 33 **95.** 15 **97.** $66\frac{2}{3}$ **99.** $10\frac{17}{23}$ **101.** $17\frac{13}{18}$ **103.** $1\frac{7}{175}$

105. $6\frac{65}{112}$ **107.** 9 **109.** 125 **111.** 7^5 **113.** $2^3 \cdot 3$ **115.** answers may vary **117.** $\frac{2}{3}$ **119.**

121. $\frac{18,000}{139,000}$ of employees **123.** $\frac{6}{36}$ of the countries

Section 2.2

Vocabulary, Readiness & Video Check **1.** prime factorization **3.** prime **5.** factors **7.** Because order doesn't matter when we multiply, so switching the order doesn't give us any new factors of 12. **9.** order; one

Exercise Set 2.2 **1.** 1, 2, 4, 8 **3.** 1, 5, 25 **5.** 1, 2, 4 **7.** 1, 2, 3, 6, 9, 18 **9.** 1, 29 **11.** 1, 2, 4, 5, 8, 10, 16, 20, 40, 80 **13.** 1, 2, 3, 4, 6, 12
15. 1, 2, 17, 34 **17.** prime **19.** composite **21.** prime **23.** composite **25.** prime **27.** composite **29.** prime **31.** composite
33. composite **35.** 2^5 **37.** $3 \cdot 5$ **39.** $2^3 \cdot 5$ **41.** $2^2 \cdot 3^2$ **43.** $3 \cdot 13$ **45.** $2^2 \cdot 3 \cdot 5$ **47.** $2 \cdot 5 \cdot 11$ **49.** $5 \cdot 17$ **51.** 2^7 **53.** $2 \cdot 7 \cdot 11$
55. $2^2 \cdot 3 \cdot 5^2$ **57.** $2^4 \cdot 3 \cdot 5$ **59.** $2^2 \cdot 3^2 \cdot 23$ **61.** $2 \cdot 3^2 \cdot 7^2$ **63.** $7^2 \cdot 13$ **65.** $3 \cdot 11$ **67.** $2 \cdot 7^2$ **69.** prime **71.** $3^3 \cdot 17$ **73.** prime

75. $2^2 \cdot 5^2 \cdot 7$ **77.** 4300 **79.** 7,660,000 **81.** 20,000 **83.** 2375 **85.** $\frac{1136}{2375}$ **87.** $2^2 \cdot 3^5 \cdot 5 \cdot 7$ **89.** answers may vary

91. answers may vary

Section 2.3

Calculator Explorations   1. $\dfrac{4}{7}$ **3.** $\dfrac{20}{27}$ **5.** $\dfrac{15}{8}$ **7.** $\dfrac{9}{2}$

Vocabulary, Readiness & Video Check   1. simplest form   **3.** cross products   **5.** 0   **7.** equivalent   **9.** $\dfrac{10}{24}$ is not in simplest form; $\dfrac{5}{12}$

Exercise Set 2.3   1. $\dfrac{1}{4}$ **3.** $\dfrac{2}{21}$ **5.** $\dfrac{7}{8}$ **7.** $\dfrac{2}{3}$ **9.** $\dfrac{7}{10}$ **11.** $\dfrac{7}{9}$ **13.** $\dfrac{3}{5}$ **15.** $\dfrac{27}{64}$ **17.** $\dfrac{5}{8}$ **19.** $\dfrac{5}{8}$ **21.** $\dfrac{14}{17}$ **23.** $\dfrac{3}{2}$ or $1\dfrac{1}{2}$ **25.** $\dfrac{3}{4}$ **27.** $\dfrac{5}{14}$

29. $\dfrac{3}{14}$ **31.** $\dfrac{11}{17}$ **33.** $\dfrac{3}{14}$ **35.** $\dfrac{7}{8}$ **37.** $\dfrac{3}{5}$ **39.** 14   **41.** equivalent   **43.** not equivalent   **45.** equivalent   **47.** equivalent

49. not equivalent   **51.** not equivalent   **53.** $\dfrac{1}{4}$ of a shift   **55.** $\dfrac{1}{2}$ mi   **57. a.** $\dfrac{8}{25}$ **b.** 34 states   **c.** $\dfrac{17}{25}$ **59.** $\dfrac{5}{12}$ of the wall   **61. a.** 30

b. $\dfrac{3}{5}$ **63.** $\dfrac{2}{5}$ of top-selling games   **65.** 364   **67.** 2322   **69.** 2520   **71.** answers may vary   **73.** $\dfrac{3}{5}$ **75.** $\dfrac{9}{25}$ **77.** $\dfrac{1}{25}$ **79.** $\dfrac{1}{10}$

81. answers may vary   **83.** $\dfrac{7}{100}$ **85.** answers may vary   **87.** 786, 222, 900, 1470   **89.** 6; answers may vary

Integrated Review   1. $\dfrac{3}{6}$ (or $\dfrac{1}{2}$ simplified)   **2.** $\dfrac{7}{4}$ or $1\dfrac{3}{4}$ **3.** $\dfrac{73}{85}$ **4.**

5. 1   **6.** 17   **7.** 0   **8.** undefined   **9.** $\dfrac{25}{8}$ **10.** $\dfrac{28}{5}$ **11.** $\dfrac{69}{7}$ **12.** $\dfrac{141}{7}$ **13.** $2\dfrac{6}{7}$ **14.** 5   **15.** $4\dfrac{7}{8}$ **16.** $8\dfrac{10}{11}$ **17.** 1, 5, 7, 35   **18.** 1, 2, 4, 5, 8, 10, 20, 40   **19.** composite   **20.** prime   **21.** $5\cdot 13$ **22.** $2\cdot 5\cdot 7$ **23.** $2^5\cdot 3$ **24.** $2^2\cdot 3\cdot 11$ **25.** $2^2\cdot 3^2\cdot 7$ **26.** prime   **27.** $3^2\cdot 5\cdot 7$ **28.** $3^2\cdot 7^2$ **29.** $2\cdot 11\cdot 13$ **30.** prime   **31.** $\dfrac{1}{7}$ **32.** $\dfrac{6}{5}$ or $1\dfrac{1}{5}$ **33.** $\dfrac{9}{19}$ **34.** $\dfrac{21}{55}$ **35.** $\dfrac{14}{15}$ **36.** $\dfrac{9}{10}$ **37.** $\dfrac{2}{5}$ **38.** $\dfrac{3}{8}$ **39.** $\dfrac{11}{14}$

40. $\dfrac{7}{11}$ **41.** not equivalent   **42.** equivalent   **43. a.** $\dfrac{1}{25}$ **b.** 48   **c.** $\dfrac{24}{25}$ **44. a.** $\dfrac{29}{108}$ **b.** 395   **c.** $\dfrac{79}{108}$

Section 2.4

Vocabulary, Readiness & Video Check   1. $\dfrac{a\cdot c}{b\cdot d}$ **3.** multiplication   **5.** There's a common factor of 2 in the numerator and denominator that can be divided out first.   **7.** radius is $\dfrac{1}{2}$ of diameter

Exercise Set 2.4   1. $\dfrac{2}{15}$ **3.** $\dfrac{6}{35}$ **5.** $\dfrac{9}{80}$ **7.** $\dfrac{5}{28}$ **9.** $\dfrac{12}{5}$ or $2\dfrac{2}{5}$ **11.** $\dfrac{1}{70}$ **13.** 0   **15.** $\dfrac{1}{110}$ **17.** $\dfrac{18}{55}$ **19.** $\dfrac{27}{80}$ **21.** $\dfrac{1}{56}$ **23.** $\dfrac{2}{105}$ **25.** 0

27. $\dfrac{1}{90}$ **29.** 8   **31.** 6   **33.** 20   **35.** 3   **37.** $\dfrac{5}{2}$ or $2\dfrac{1}{2}$ **39.** $\dfrac{1}{5}$ **41.** $\dfrac{5}{3}$ or $1\dfrac{2}{3}$ **43.** $\dfrac{2}{3}$ **45.** Exact: $\dfrac{77}{10}$ or $7\dfrac{7}{10}$; Estimate: 8   **47.** Exact: $\dfrac{836}{35}$ or $23\dfrac{31}{35}$; Estimate: 24   **49.** $\dfrac{25}{2}$ or $21\dfrac{1}{2}$ **51.** 15   **53.** 6   **55.** $\dfrac{45}{4}$ or $11\dfrac{1}{4}$ **57.** $\dfrac{49}{3}$ or $16\dfrac{1}{3}$ **59.** $\dfrac{1}{30}$ **61.** 0   **63.** $\dfrac{16}{5}$ or $3\dfrac{1}{5}$ **65.** $\dfrac{7}{2}$ or $3\dfrac{1}{2}$ **67.** $\dfrac{1}{8}$ **69.** $\dfrac{1}{56}$ **71.** $\dfrac{55}{3}$ or $18\dfrac{1}{3}$ **73.** 0   **75.** $\dfrac{208}{7}$ or $29\dfrac{5}{7}$ **77.** 50   **79.** 20   **81.** 128 freshmen   **83.** 117 million   **85.** 868 mi

87. $\dfrac{3}{16}$ in.   **89.** 30 gal   **91.** $\dfrac{17}{2}$ or $8\dfrac{1}{2}$ in.   **93.** $\dfrac{39}{2}$ or $19\dfrac{1}{2}$ in.   **95.** $\dfrac{2242}{625}$ or $3\dfrac{367}{625}$ sq in.   **97.** $\dfrac{1}{14}$ sq ft   **99.** $\dfrac{7}{2}$ or $3\dfrac{1}{2}$ sq yd   **101.** 3840 mi

103. 2400 mi   **105.** 206   **107.** 56 R 12   **109.** answers may vary   **111.** $3\dfrac{2}{3}\cdot 1\dfrac{1}{7}=\dfrac{11}{3}\cdot\dfrac{8}{7}=\dfrac{11\cdot 8}{3\cdot 7}=\dfrac{88}{21}$ or $4\dfrac{4}{21}$ **113.** b   **115.** a

117. 37 students   **119.** 664,950 Māori

Section 2.5

Vocabulary, Readiness & Video Check   1. reciprocals   **3.** $\dfrac{a\cdot d}{b\cdot c}$ **5.** $\dfrac{1}{n}$ **7.** Because we still have a division problem and we can't divide out common factors until we rewrite the division as a multiplication.

Exercise Set 2.5   1. $\dfrac{7}{4}$ **3.** 11   **5.** $\dfrac{1}{15}$ **7.** $\dfrac{7}{12}$ **9.** $\dfrac{4}{5}$ **11.** $\dfrac{16}{9}$ or $1\dfrac{7}{9}$ **13.** $\dfrac{18}{35}$ **15.** $\dfrac{3}{4}$ **17.** $\dfrac{1}{100}$ **19.** $\dfrac{1}{3}$ **21.** $\dfrac{5}{3}$ or $1\dfrac{2}{3}$ **23.** $\dfrac{35}{36}$ **25.** $\dfrac{14}{37}$

27. $\dfrac{8}{45}$ **29.** 1   **31.** undefined   **33.** 0   **35.** $\dfrac{7}{10}$ **37.** $\dfrac{1}{6}$ **39.** $\dfrac{40}{3}$ or $13\dfrac{1}{3}$ **41.** 5   **43.** $\dfrac{5}{28}$ **45.** $\dfrac{36}{35}$ or $1\dfrac{1}{35}$ **47.** $\dfrac{26}{51}$ **49.** 0

51. $\frac{17}{13}$ or $1\frac{4}{13}$ **53.** $\frac{35}{18}$ or $1\frac{17}{18}$ **55.** $\frac{19}{30}$ **57.** $\frac{1}{6}$ **59.** $\frac{121}{60}$ or $2\frac{1}{60}$ **61.** 96 **63.** $\frac{3}{4}$ **65.** undefined **67.** $\frac{11}{119}$ **69.** $\frac{35}{11}$ or $3\frac{2}{11}$ **71.** $\frac{9}{5}$ or $1\frac{4}{5}$

73. $3\frac{3}{16}$ miles **75.** $\frac{5}{6}$ Tbsp **77.** $\frac{19}{30}$ in. **79.** 20 lb **81.** $4\frac{2}{3}$ m **83.** $\frac{8}{35}$ **85.** $\frac{128}{51}$ or $2\frac{26}{51}$ **87.** $\frac{16}{15}$ or $1\frac{1}{15}$ **89.** $\frac{121}{400}$ **91.** 201 **93.** 196

95. 1569 **97.** $20\frac{2}{3} \div 10\frac{1}{2} = \frac{62}{3} \div \frac{21}{2} = \frac{62}{3} \cdot \frac{2}{21} = \frac{124}{63}$ or $1\frac{61}{63}$ **99.** c **101.** d **103.** 5 **105.** 634 aircraft **107.** answers may vary

Chapter 2 Vocabulary Check **1.** reciprocals **2.** composite number **3.** equivalent **4.** improper fraction **5.** prime number **6.** simplest form **7.** proper fraction **8.** mixed number **9.** numerator; denominator **10.** prime factorization **11.** undefined **12.** 0 **13.** cross products

Chapter 2 Review **1.** proper **2.** improper **3.** proper **4.** mixed number **5.** $\frac{2}{6}$ **6.** $\frac{4}{7}$ **7.** $\frac{7}{3}$ **8.** $\frac{13}{4}$ **9.** $\frac{11}{12}$ **10. a.** 108 **b.** $\frac{108}{131}$

11. $3\frac{3}{4}$ **12.** $45\frac{5}{6}$ **13.** 3 **14.** 5 **15.** $\frac{6}{5}$ **16.** $\frac{22}{21}$ **17.** $\frac{26}{9}$ **18.** $\frac{47}{12}$ **19.** composite **20.** prime **21.** 1, 2, 3, 6, 7, 14, 21, 42

22. 1, 2, 4, 5, 10, 20 **23.** $2^2 \cdot 17$ **24.** $2 \cdot 3^2 \cdot 5$ **25.** $5 \cdot 157$ **26.** $3 \cdot 5 \cdot 17$ **27.** $\frac{3}{7}$ **28.** $\frac{5}{9}$ **29.** $\frac{1}{3}$ **30.** $\frac{1}{2}$ **31.** $\frac{29}{32}$ **32.** $\frac{18}{23}$ **33.** 8

34. 6 **35.** $\frac{2}{3}$ of a foot **36.** $\frac{3}{5}$ of the cars **37.** no **38.** yes **39.** $\frac{3}{10}$ **40.** $\frac{5}{14}$ **41.** 9 **42.** $\frac{1}{2}$ **43.** $\frac{35}{8}$ or $4\frac{3}{8}$ **44.** $\frac{5}{2}$ or $2\frac{1}{2}$ **45.** $\frac{5}{3}$ or $1\frac{2}{3}$

46. $\frac{49}{3}$ or $16\frac{1}{3}$ **47.** Exact: $\frac{26}{5}$ or $5\frac{1}{5}$; Estimate: 6 **48.** Exact: $\frac{60}{11}$ or $5\frac{5}{11}$; Estimate: 8 **49.** $\frac{99}{4}$ or $24\frac{3}{4}$ **50.** $\frac{1}{6}$ **51.** $\frac{110}{3}$ or $36\frac{2}{3}$ g

52. $\frac{135}{4}$ or $33\frac{3}{4}$ in. **53.** $\frac{119}{80}$ or $1\frac{39}{80}$ sq in. **54.** $\frac{275}{8}$ or $34\frac{3}{8}$ sq m **55.** $\frac{1}{7}$ **56.** 8 **57.** $\frac{23}{14}$ **58.** $\frac{5}{17}$ **59.** 2 **60.** $\frac{15}{4}$ or $3\frac{3}{4}$ **61.** $\frac{5}{6}$

62. $\frac{8}{3}$ or $2\frac{2}{3}$ **63.** $\frac{21}{4}$ or $5\frac{1}{4}$ **64.** $\frac{121}{46}$ or $2\frac{29}{46}$ **65.** 22 mi **66.** $\frac{21}{20}$ or $1\frac{1}{20}$ mi **67.** proper **68.** improper **69.** mixed number

70. improper **71.** $31\frac{1}{4}$ **72.** 6 **73.** $\frac{95}{17}$ **74.** $\frac{47}{6}$ **75.** composite **76.** prime **77.** $2^2 \cdot 3^2 \cdot 5$ **78.** $2 \cdot 7^2$ **79.** $\frac{9}{10}$ **80.** $\frac{5}{7}$ **81.** $\frac{14}{15}$

82. $\frac{3}{5}$ **83.** $\frac{7}{12}$ **84.** $\frac{1}{4}$ **85.** 9 **86.** $\frac{27}{2}$ or $13\frac{1}{2}$ **87.** Exact: 10; Estimate: 8 **88.** Exact: $\frac{51}{4}$ or $12\frac{3}{4}$; Estimate: 12 **89.** $\frac{7}{3}$ or $2\frac{1}{3}$

90. $\frac{32}{5}$ or $6\frac{2}{5}$ **91.** $\frac{81}{2}$ or $40\frac{1}{2}$ sq ft **92.** $\frac{47}{61}$ in.

Chapter 2 Test **1.** $\frac{7}{16}$ **2.** $\frac{13}{5}$ **3.** $\frac{23}{3}$ **4.** $\frac{39}{11}$ **5.** $4\frac{3}{5}$ **6.** $18\frac{3}{4}$ **7.** $\frac{4}{35}$ **8.** $\frac{3}{5}$ **9.** not equivalent **10.** equivalent **11.** $2^2 \cdot 3 \cdot 7$

12. $3^2 \cdot 5 \cdot 11$ **13.** $\frac{4}{3}$ or $1\frac{1}{3}$ **14.** $\frac{4}{3}$ or $1\frac{1}{3}$ **15.** $\frac{1}{4}$ **16.** $\frac{16}{45}$ **17.** 16 **18.** $\frac{9}{2}$ or $4\frac{1}{2}$ **19.** $\frac{4}{11}$ **20.** 9 **21.** $\frac{64}{3}$ or $21\frac{1}{3}$ **22.** $\frac{45}{2}$ or $22\frac{1}{2}$

23. $\frac{18}{5}$ or $3\frac{3}{5}$ **24.** $\frac{20}{3}$ or $6\frac{2}{3}$ **25.** $\frac{34}{27}$ or $1\frac{7}{27}$ sq mi **26.** 24 mi **27.** $\frac{16,000}{3}$ or $5333\frac{1}{3}$ sq yd **28.** $90 per share

Cumulative Review **1.** hundred-thousands (Sec. 1.2, Ex. 1) **2.** two thousand, thirty-six **3.** 805 (Sec. 1.2, Ex. 9) **4.** 31 **5.** 184,046 (Sec. 1.3, Ex. 2) **6.** 39 **7.** 13 in. (Sec. 1.3, Ex. 5) **8.** 17 **9.** 14,440,060 (Sec. 1.3, Ex. 7) **10.** 5 **11.** 7321 (Sec. 1.4, Ex. 2) **12.** 64 **13. a.** Indonesia **b.** 331 (Sec. 1.3, Ex. 8) **14.** 25 R 5 **15.** 570 (Sec. 1.5, Ex. 1) **16.** 2400 **17.** 1800 (Sec. 1.5, Ex. 5) **18.** 300 **19. a.** 6 **b.** 0 **c.** 45 **d.** 0 (Sec. 1.6, Ex. 1) **20.** 20 **21. a.** $3 \cdot 4 + 3 \cdot 5$ **b.** $10 \cdot 6 + 10 \cdot 8$ **c.** $2 \cdot 7 + 2 \cdot 3$ (Sec. 1.6, Ex. 2) **22.** 180 **23. a.** 0 **b.** 0 **c.** 0 **d.** undefined (Sec. 1.7, Ex. 3) **24.** 154 sq mi **25.** 208 (Sec. 1.7, Ex. 5) **26.** 4014 **27.** 12 cards; 10 cards left over (Sec. 1.7, Ex. 11) **28.** 63 **29.** 40 ft (Sec. 1.8, Ex. 5) **30.** 16 **31.** 7^3 (Sec. 1.9, Ex. 1) **32.** 7^4 **33.** $3^4 \cdot 17^3$ (Sec. 1.9, Ex. 4)

34. $2^2 \cdot 3^4$ **35.** 7 (Sec. 1.9, Ex. 12) **36.** 0 **37.** $\frac{2}{5}$ (Sec. 2.1, Ex. 7) **38.** $2^2 \cdot 3 \cdot 13$ **39. a.** $\frac{38}{9}$ **b.** $\frac{19}{11}$ (Sec. 2.1, Ex. 17) **40.** $\frac{39}{5}$

41. 1, 2, 4, 5, 10, 20 (Sec. 2.2, Ex. 1) **42.** equivalent **43.** $\frac{7}{11}$ (Sec. 2.3, Ex. 2) **44.** $\frac{2}{3}$ **45.** $\frac{35}{12}$ or $2\frac{11}{12}$ (Sec. 2.4, Ex. 8)

46. $\frac{8}{3}$ or $2\frac{2}{3}$ **47.** $\frac{3}{1}$ or 3 (Sec. 2.5, Ex. 3) **48.** $\frac{1}{9}$ **49.** $\frac{5}{12}$ (Sec. 2.5, Ex. 6) **50.** $\frac{11}{56}$

Chapter 3 Adding and Subtracting Fractions

Section 3.1

Vocabulary, Readiness & Video Check **1.** like; unlike **3.** $\frac{a-c}{b}$ **5.** unlike **7.** like **9.** like **11.** unlike **13.** We can simplify by dividing out a common factor of 3 from the numerator and denominator to get $\frac{2}{3}$. **15.** $P = \frac{4}{20} + \frac{9}{20} + \frac{7}{20}$; 1 in.

Exercise Set 3.1   **1.** $\frac{3}{7}$ **3.** $\frac{1}{5}$ **5.** $\frac{2}{3}$ **7.** $\frac{7}{20}$ **9.** $\frac{1}{2}$ **11.** $\frac{13}{11}$ or $1\frac{2}{11}$ **13.** $\frac{7}{13}$ **15.** $\frac{2}{3}$ **17.** $\frac{6}{11}$ **19.** $\frac{3}{5}$ **21.** 1   **23.** $\frac{3}{4}$ **25.** $\frac{5}{6}$ **27.** $\frac{4}{5}$

29. $\frac{1}{90}$ **31.** $\frac{19}{33}$ **33.** $\frac{13}{21}$ **35.** $\frac{9}{10}$ **37.** 0   **39.** $\frac{3}{4}$ **41.** 1 in.   **43.** 2 m   **45.** $\frac{7}{10}$ mi   **47.** $\frac{3}{2}$ or $1\frac{1}{2}$ hr   **49.** $\frac{43}{100}$ **51.** $\frac{1}{10}$ **53.** $\frac{9}{20}$ **55.** $\frac{13}{50}$

57. $\frac{7}{25}$ **59.** $\frac{1}{50}$ **61.** $\frac{21}{25}$ **63.** $2\cdot5$ **65.** 2^3 **67.** $5\cdot11$ **69.** $\frac{5}{8}$ **71.** $\frac{8}{11}$ **73.** $\frac{2}{7}+\frac{9}{7}=\frac{11}{7}$ or $1\frac{4}{7}$ **75.** answers may vary

77. 1; answers may vary   **79.** $\frac{1}{4}$ mi

Section 3.2

Vocabulary, Readiness & Video Check   **1.** equivalent   **3.** multiple   **5.** Because 24 is a multiple of 8.

Exercise Set 3.2   **1.** 12   **3.** 45   **5.** 36   **7.** 72   **9.** 126   **11.** 75   **13.** 24   **15.** 42   **17.** 216   **19.** 150   **21.** 68   **23.** 588

25. 900   **27.** 1800   **29.** 363   **31.** 60   **33.** $\frac{20}{35}$ **35.** $\frac{14}{21}$ **37.** $\frac{15}{3}$ **39.** $\frac{15}{30}$ **41.** $\frac{30}{21}$ **43.** $\frac{21}{28}$ **45.** $\frac{30}{45}$ **47.** $\frac{36}{81}$ **49.** $\frac{90}{78}$ **51.** $\frac{56}{68}$

53. $\frac{86}{100},\frac{80}{100},\frac{58}{100},\frac{68}{100},\frac{12}{100},\frac{84}{100},\frac{68}{100},\frac{81}{100},\frac{90}{100},\frac{79}{100},\frac{74}{100},\frac{78}{100}$ **55.** drugs, health and beauty aids   **57.** $\frac{1}{2}$ **59.** $\frac{2}{5}$ **61.** $\frac{4}{9}$ **63.** 1

65. $\frac{814}{3630}$ **67.** answers may vary   **69.** a, b, and d

Section 3.3

Calculator Explorations   **1.** $\frac{37}{80}$ **3.** $\frac{95}{72}$ **5.** $\frac{394}{323}$

Vocabulary, Readiness & Video Check   **1.** equivalent; least common denominator   **3.** $\frac{15}{24};\frac{4}{24};\frac{19}{24}$ **5.** Multiplying by $\frac{3}{3}$ is

multiplying by a form of 1. Thus, the result is an equivalent fraction.   **7.** $\frac{34}{15}$ cm and $2\frac{4}{15}$ cm

Exercise Set 3.3   **1.** $\frac{5}{6}$ **3.** $\frac{5}{6}$ **5.** $\frac{8}{33}$ **7.** $\frac{9}{14}$ **9.** $\frac{3}{5}$ **11.** $\frac{13}{25}$ **13.** $\frac{53}{60}$ **15.** $\frac{1}{6}$ **17.** $\frac{67}{99}$ **19.** $\frac{98}{143}$ **21.** $\frac{13}{27}$ **23.** $\frac{75}{56}$ or $1\frac{19}{56}$

25. $\frac{19}{18}$ or $1\frac{1}{18}$ **27.** $\frac{19}{12}$ or $1\frac{7}{12}$ **29.** $\frac{11}{16}$ **31.** $\frac{17}{42}$ **33.** $\frac{33}{56}$ **35.** $\frac{37}{99}$ **37.** $\frac{1}{35}$ **39.** $\frac{11}{36}$ **41.** $\frac{1}{20}$ **43.** $\frac{1}{84}$ **45.** $\frac{9}{1000}$ **47.** $\frac{17}{99}$

49. $\frac{19}{36}$ **51.** $\frac{1}{5}$ **53.** $\frac{69}{280}$ **55.** $\frac{14}{9}$ or $1\frac{5}{9}$ **57.** $\frac{34}{15}$ or $2\frac{4}{15}$ cm   **59.** $\frac{17}{10}$ or $1\frac{7}{10}$ m   **61.** $\frac{7}{100}$ mph   **63.** $\frac{5}{8}$ in.   **65.** $\frac{31}{32}$ in.

67. $\frac{19}{100}$ of Girl Scout cookies   **69.** $\frac{19}{25}$ **71.** $\frac{17}{50}$ **73.** $\frac{81}{100}$ **75.** 5   **77.** $\frac{16}{29}$ **79.** $\frac{19}{3}$ or $6\frac{1}{3}$ **81.** $\frac{3}{5}+\frac{4}{5}=\frac{7}{5}$ or $1\frac{2}{5}$ **83.** $\frac{223}{540}$

85. $\frac{49}{44}$ or $1\frac{5}{44}$ **87.** answers may vary

Integrated Review   **1.** 30   **2.** 21   **3.** 14   **4.** 25   **5.** 100   **6.** 90   **7.** $\frac{9}{24}$ **8.** $\frac{28}{36}$ **9.** $\frac{10}{40}$ **10.** $\frac{12}{30}$ **11.** $\frac{55}{75}$ **12.** $\frac{40}{48}$ **13.** $\frac{1}{2}$ **14.** $\frac{2}{5}$

15. $\frac{7}{12}$ **16.** $\frac{13}{15}$ **17.** $\frac{3}{4}$ **18.** $\frac{2}{15}$ **19.** $\frac{17}{45}$ **20.** $\frac{19}{50}$ **21.** $\frac{37}{40}$ **22.** $\frac{11}{36}$ **23.** 0   **24.** $\frac{1}{17}$ **25.** $\frac{5}{33}$ **26.** $\frac{1}{42}$ **27.** $\frac{5}{18}$ **28.** $\frac{5}{13}$ **29.** $\frac{11}{18}$

30. $\frac{37}{50}$ **31.** $\frac{47}{30}$ or $1\frac{17}{30}$ **32.** $\frac{7}{30}$ **33.** $\frac{3}{5}$ **34.** $\frac{27}{20}$ or $1\frac{7}{20}$ **35.** $\frac{279}{350}$ **36.** $\frac{309}{350}$ **37.** $\frac{98}{5}$ or $19\frac{3}{5}$ **38.** $\frac{9}{250}$ **39.** $\frac{31}{3}$ or $10\frac{1}{3}$

40. $\frac{93}{64}$ or $1\frac{29}{64}$ **41.** $\frac{49}{54}$ **42.** $\frac{83}{48}$ or $1\frac{35}{48}$ **43.** $\frac{390}{101}$ or $3\frac{87}{101}$ **44.** $\frac{116}{5}$ or $23\frac{1}{5}$ **45.** $\frac{106}{135}$ **46.** $\frac{67}{224}$

Section 3.4

Vocabulary, Readiness & Video Check   **1.** mixed number   **3.** round   **5.** a   **7.** c   **9.** The fractional part of a mixed number should not be an improper fraction.   **11.** Because we need to borrow first.

Exercise Set 3.4   **1.** Exact: $6\frac{4}{5}$; Estimate: 7   **3.** Exact: $13\frac{11}{14}$; Estimate: 14   **5.** $17\frac{7}{25}$ **7.** $7\frac{5}{8}$ **9.** $7\frac{5}{24}$ **11.** $20\frac{1}{15}$ **13.** 19   **15.** $56\frac{53}{270}$

17. $13\frac{13}{24}$ **19.** $47\frac{53}{84}$ **21.** Exact: $2\frac{3}{5}$; Estimate: 3   **23.** Exact: $7\frac{5}{14}$; Estimate: 7   **25.** $\frac{24}{25}$ **27.** $2\frac{7}{15}$ **29.** $5\frac{11}{14}$ **31.** $23\frac{31}{72}$ **33.** $1\frac{4}{5}$

35. $1\frac{13}{15}$ **37.** $3\frac{5}{9}$ **39.** $15\frac{3}{4}$ **41.** $28\frac{7}{12}$ **43.** $15\frac{7}{8}$ **45.** 8 **47.** $17\frac{11}{12}$ **49.** $\frac{1}{16}$ in. **51.** no; she will be $\frac{1}{12}$ of a foot short **53.** $7\frac{13}{20}$ in.

55. $10\frac{1}{4}$ hr **57.** $2\frac{3}{8}$ hr **59.** $92\frac{99}{100}$ m **61.** $319\frac{1}{3}$ yd **63.** $9\frac{13}{30}$ min **65.** $1\frac{4}{5}$ min **67.** 7 mi **69.** $21\frac{5}{24}$ m **71.** 8 **73.** 25 **75.** 4

77. 167 **79.** 4 **81.** $9\frac{5}{8}$ **83.** a, b, c **85.** answers may vary **87.** Supreme is heavier by $\frac{1}{8}$ lb

Section 3.5

Vocabulary, Readiness & Video Check **1.** multiplication **3.** subtraction **5.** denominators; numerators **7.** We need to make sure we have the same denominator when adding and subtracting fractions; this is not necessary when multiplying and dividing fractions.

Exercise Set 3.5 **1.** > **3.** < **5.** < **7.** > **9.** > **11.** < **13.** > **15.** < **17.** $\frac{1}{16}$ **19.** $\frac{8}{125}$ **21.** $\frac{64}{343}$ **23.** $\frac{4}{81}$

25. $\frac{1}{6}$ **27.** $\frac{18}{125}$ **29.** $\frac{11}{15}$ **31.** $\frac{3}{35}$ **33.** $\frac{5}{9}$ **35.** $10\frac{4}{99}$ **37.** $\frac{1}{12}$ **39.** $\frac{9}{11}$ **41.** 0 **43.** 0 **45.** $\frac{2}{5}$ **47.** $\frac{2}{77}$ **49.** $\frac{17}{60}$ **51.** $\frac{5}{8}$ **53.** $\frac{1}{2}$

55. $\frac{29}{10}$ or $2\frac{9}{10}$ **57.** $\frac{27}{32}$ **59.** $\frac{1}{81}$ **61.** $\frac{5}{6}$ **63.** $\frac{3}{5}$ **65.** $\frac{1}{2}$ **67.** $\frac{19}{7}$ or $2\frac{5}{7}$ **69.** $\frac{9}{64}$ **71.** $\frac{3}{4}$ **73.** $\frac{13}{60}$ **75.** $\frac{13}{25}$ **77.** A **79.** M **81.** S

83. D **85.** M **87.** A **89.** no; answers may vary **91.** subtraction, multiplication, addition, division **93.** division, multiplication, subtraction, addition **95.** standard mail **97.** discretionary purchases

Section 3.6

Vocabulary, Readiness & Video Check **1.** To make sure we answer the question asked in the original problem.

Exercise Set 3.6 **1.** $\frac{1}{2} + \frac{1}{3}$ **3.** $20 \div 6\frac{2}{5}$ **5.** $\frac{15}{16} - \frac{5}{8}$ **7.** $\frac{21}{68} + \frac{7}{34}$ **9.** $8\frac{1}{3} \cdot \frac{7}{9}$ **11.** $3\frac{1}{3}$ c **13.** $12\frac{1}{2}$ in. **15.** $21\frac{1}{2}$ mi per gal

17. $1\frac{1}{2}$ yr **19.** $9\frac{2}{5}$ in. **21.** no; $\frac{1}{4}$ yd **23.** 5 pieces **25.** $\frac{9}{8}$ or $1\frac{1}{8}$ in. **27.** $3\frac{3}{4}$ c **29.** $11\frac{1}{4}$ sq in. **31.** $1\frac{29}{60}$ min **33.** $1\frac{13}{50}$ cu in.

35. 67 sheets **37. a.** yes **b.** 1 ft left over **39.** $2\frac{15}{16}$ lb **41.** area: $\frac{9}{128}$ sq in.; perimeter: $1\frac{1}{8}$ in. **43.** area: $\frac{25}{81}$ sq m; perimeter: $2\frac{2}{9}$ m

45. $4\frac{3}{4}$ ft **47.** $\frac{5}{26}$ ft **49.** 3 **51.** 81 **53.** 4 **55.** 30 **57.** 35 **59.** no; no; answers may vary **61.** $36\frac{44}{81}$ sq ft **63.** 68 customers

65. 22 hr

Chapter 3 Vocabulary Check

1. like **2.** least common multiple **3.** equivalent **4.** mixed number **5.** > **6.** <
7. least common denominator **8.** unlike **9.** exponent

Chapter 3 Review **1.** $\frac{10}{11}$ **2.** $\frac{3}{25}$ **3.** $\frac{2}{3}$ **4.** $\frac{1}{7}$ **5.** $\frac{3}{5}$ **6.** $\frac{3}{5}$ **7.** 1 **8.** 1 **9.** $\frac{19}{25}$ **10.** $\frac{16}{21}$ **11.** $\frac{3}{4}$ of his homework **12.** $\frac{3}{2}$ or $1\frac{1}{2}$ mi

13. 55 **14.** 60 **15.** 120 **16.** 80 **17.** 252 **18.** 72 **19.** $\frac{56}{64}$ **20.** $\frac{20}{30}$ **21.** $\frac{21}{33}$ **22.** $\frac{20}{26}$ **23.** $\frac{16}{60}$ **24.** $\frac{25}{60}$ **25.** $\frac{11}{18}$ **26.** $\frac{7}{15}$ **27.** $\frac{7}{26}$

28. $\frac{17}{36}$ **29.** $\frac{41}{42}$ **30.** $\frac{43}{72}$ **31.** $\frac{13}{45}$ **32.** $\frac{39}{70}$ **33.** $\frac{19}{9}$ or $2\frac{1}{9}$ m **34.** $\frac{3}{2}$ or $1\frac{1}{2}$ ft **35.** $\frac{1}{4}$ of a yd **36.** $\frac{7}{10}$ has been cleaned **37.** $45\frac{16}{21}$

38. 60 **39.** $32\frac{13}{22}$ **40.** $3\frac{19}{60}$ **41.** $111\frac{5}{18}$ **42.** $20\frac{7}{24}$ **43.** $5\frac{16}{35}$ **44.** $3\frac{4}{55}$ **45.** $7\frac{4}{5}$ in. **46.** $\frac{1}{40}$ oz **47.** 5 ft **48.** $11\frac{1}{6}$ ft **49.** <

50. > **51.** < **52.** > **53.** > **54.** > **55.** $\frac{9}{49}$ **56.** $\frac{64}{125}$ **57.** $\frac{9}{400}$ **58.** $\frac{9}{100}$ **59.** $\frac{8}{13}$ **60.** 2 **61.** $\frac{81}{196}$ **62.** $\frac{1}{27}$ **63.** $\frac{13}{18}$

64. $\frac{11}{15}$ **65.** $\frac{1}{7}$ **66.** $\frac{18}{5}$ or $3\frac{3}{5}$ **67.** $\frac{45}{28}$ or $1\frac{17}{28}$ **68.** $\frac{5}{6}$ **69.** $\frac{99}{56}$ or $1\frac{43}{56}$ **70.** $\frac{29}{110}$ **71.** $\frac{29}{54}$ **72.** $\frac{37}{60}$ **73.** 50 moons **74.** $15\frac{5}{8}$ acres

75. each measurement is $4\frac{1}{4}$ in. **76.** $\frac{7}{10}$ yd **77.** perimeter: $\frac{17}{11}$ or $1\frac{6}{11}$ mi; area: $\frac{3}{22}$ sq mi **78.** perimeter: $\frac{7}{3}$ or $2\frac{1}{3}$ m; area: $\frac{5}{16}$ sq m

79. 90 **80.** 60 **81.** $\frac{40}{48}$ **82.** $\frac{63}{72}$ **83.** $\frac{1}{6}$ **84.** $\frac{1}{5}$ **85.** $\frac{11}{12}$ **86.** $\frac{27}{55}$ **87.** $13\frac{5}{12}$ **88.** $12\frac{3}{8}$ **89.** $3\frac{16}{35}$ **90.** $8\frac{1}{21}$ **91.** $\frac{11}{25}$ **92.** $\frac{1}{8}$

93. $\frac{1}{144}$ **94.** $\frac{64}{27}$ or $2\frac{10}{27}$ **95.** $\frac{5}{17}$ **96.** $\frac{1}{12}$ **97.** < **98.** > **99.** $\frac{1}{2}$ hr **100.** $6\frac{7}{20}$ lb **101.** $44\frac{1}{2}$ yd **102.** $2\frac{2}{15}$ ft

103. $7\frac{1}{2}$ tablespoons **104.** $\frac{3}{8}$ gal

Chapter 3 Test   **1.** 60   **2.** 72   **3.** $<$ **4.** $<$ **5.** $\dfrac{8}{9}$ **6.** $\dfrac{2}{5}$ **7.** $\dfrac{13}{10}$ or $1\dfrac{3}{10}$ **8.** $\dfrac{8}{21}$ **9.** $\dfrac{13}{24}$ **10.** $\dfrac{2}{3}$ **11.** $\dfrac{67}{60}$ or $1\dfrac{7}{60}$ **12.** $\dfrac{7}{50}$

13. $\dfrac{3}{2}$ or $1\dfrac{1}{2}$ **14.** $14\dfrac{1}{40}$ **15.** $30\dfrac{13}{45}$ **16.** $1\dfrac{7}{24}$ **17.** $16\dfrac{8}{11}$ **18.** $\dfrac{5}{3}$ or $1\dfrac{2}{3}$ **19.** $\dfrac{16}{81}$ **20.** $\dfrac{9}{16}$ **21.** $\dfrac{153}{200}$ **22.** $\dfrac{3}{8}$ **23.** $\dfrac{11}{12}$ **24.** $3\dfrac{3}{4}$ ft

25. $7\dfrac{5}{6}$ gal   **26.** $\dfrac{23}{50}$ **27.** $\dfrac{13}{50}$ **28.** \$2820   **29.** perimeter: $\dfrac{10}{3}$ or $3\dfrac{1}{3}$ ft; area: $\dfrac{2}{3}$ sq ft   **30.** $\dfrac{5}{3}$ or $1\dfrac{2}{3}$ in.

Cumulative Review   **1.** eighty-five (Sec. 1.2, Ex. 4)   **2.** one hundred seven   **3.** one hundred twenty-six (Sec. 1.2, Ex. 5)   **4.** five thousand, twenty-six   **5.** 159 (Sec. 1.3, Ex. 1)   **6.** 19 in.   **7.** 514 (Sec. 1.4, Ex. 3)   **8.** 121 R 1   **9.** 278,000 (Sec. 1.5, Ex. 2)   **10.** 1, 2, 3, 5, 6, 10, 15, 30   **11.** 20,296 (Sec. 1.6, Ex. 4)   **12.** 0   **13. a.** 7   **b.** 12   **c.** 1   **d.** 1   **e.** 20   **f.** 1 (Sec. 1.7, Ex. 2)   **14.** 25

15. 1038 mi (Sec. 1.8, Ex. 1)   **16.** 11   **17.** 81 (Sec. 1.9, Ex. 5)   **18.** 125   **19.** 81 (Sec. 1.9, Ex. 7)   **20.** 1000   **21.** $\dfrac{4}{3}$ or $1\dfrac{1}{3}$ (Sec. 2.1, Ex. 15)

22. $\dfrac{11}{4}$ or $2\dfrac{3}{4}$ **23.** $\dfrac{5}{2}$ or $2\dfrac{1}{2}$ (Sec. 2.1, Ex. 16)   **24.** $\dfrac{14}{3}$ or $4\dfrac{2}{3}$ **25.** 3, 11, 17 are prime; 9, 26 are composite; (Sec. 2.2, Ex. 2)

26. 5   **27.** $2^2 \cdot 3^2 \cdot 5$ (Sec. 2.2, Ex. 4)   **28.** 62   **29.** $\dfrac{36}{13}$ or $2\dfrac{10}{13}$ (Sec. 2.3, Ex. 5)   **30.** $\dfrac{79}{8}$ **31.** equivalent (Sec. 2.3, Ex. 8)   **32.** $>$

33. $\dfrac{10}{33}$ (Sec. 2.4, Ex. 1)   **34.** $\dfrac{3}{2}$ or $1\dfrac{1}{2}$ **35.** $\dfrac{1}{8}$ (Sec. 2.4, Ex. 2)   **36.** 37   **37.** $\dfrac{11}{51}$ (Sec. 2.5, Ex. 9)   **38.** $\dfrac{25}{19}$ or $1\dfrac{6}{19}$

39. $\dfrac{51}{23}$ or $2\dfrac{5}{23}$ (Sec. 2.5, Ex. 10)   **40.** 16   **41.** $\dfrac{5}{8}$ (Sec. 3.1, Ex. 2)   **42.** $\dfrac{1}{5}$ **43.** 24 (Sec. 3.2, Ex. 1)   **44.** 35   **45.** 2 (Sec. 3.3, Ex. 4)

46. $\dfrac{25}{81}$ **47.** $4\dfrac{1}{3}$ (Sec. 3.4, Ex. 4)   **48.** $\dfrac{11}{100}$ **49.** $\dfrac{6}{13}$ (Sec. 3.5, Ex. 11)   **50.** $\dfrac{8}{175}$

Chapter 4 Decimals

Section 4.1

Vocabulary, Readiness & Video Check   **1.** words; standard form   **3.** and   **5.** tens   **7.** tenths   **9.** as "and"   **11.** Reading a decimal correctly gives you the correct place value, which tells you the denominator of your equivalent fraction.

Exercise Set 4.1   **1.** six and fifty-two hundredths   **3.** sixteen and twenty-three hundredths   **5.** two hundred five thousandths   **7.** one hundred sixty-seven and nine thousandths   **9.** two hundred and five thousandths   **11.** one hundred five and six tenths   **13.** two and forty-three hundredths   **15.** eighty-seven and ninety-seven hundredths   **17.** one hundred fifteen and six tenths   **19.** R. W. Financial; 321.42; Three hundred twenty-one and 42/100   **21.** Bell South; 59.68; Fifty-nine and 68/100

23. 6.5   **25.** 9.08   **27.** 705.625   **29.** 0.0046   **31.** 32.52   **33.** 1.3   **35.** $\dfrac{3}{10}$ **37.** $\dfrac{27}{100}$ **39.** $\dfrac{4}{5}$ **41.** $\dfrac{3}{20}$ **43.** $5\dfrac{47}{100}$ **45.** $\dfrac{6}{125}$ **47.** $7\dfrac{1}{125}$

49. $15\dfrac{401}{500}$ **51.** $\dfrac{601}{2000}$ **53.** $487\dfrac{8}{25}$ **55.** 0.6   **57.** 0.45   **59.** 3.7   **61.** 0.268   **63.** 0.09   **65.** 4.026   **67.** 0.028   **69.** 56.3   **71.** 0.43;

forty-three hundredths   **73.** 0.8; $\dfrac{8}{10}$ or $\dfrac{4}{5}$ **75.** seventy-seven thousandths; $\dfrac{77}{1000}$ **77.** 47,260   **79.** 47,000   **81.** answers may vary

83. twenty-six million, eight hundred forty-nine thousand, five hundred seventy-six hundred-billionths   **85.** 17.268

Section 4.2

Vocabulary, Readiness & Video Check   **1.** circumference   **3.** after   **5.** left to right

Exercise Set 4.2   **1.** $<$ **3.** $>$ **5.** $<$ **7.** $=$ **9.** $<$ **11.** $>$ **13.** 0.006, 0.0061, 0.06   **15.** 0.03, 0.042, 0.36   **17.** 1.01, 1.09, 1.1, 1.16   **19.** 20.905, 21.001, 21.03, 21.12   **21.** 0.6   **23.** 0.23   **25.** 0.594   **27.** 98,210   **29.** 12.3   **31.** 17.67   **33.** 0.5   **35.** 0.130   **37.** 3830   **39.** \$0.07   **41.** \$42,650   **43.** \$27   **45.** \$0.20   **47.** 0.3 cm   **49.** 1.43 hr   **51.** \$48   **53.** 106.5 people per sq mi   **55.** 24.623 hr   **57.** 0.5 min

59. 5766   **61.** 71   **63.** 243   **65.** b   **67.** a   **69.** 225.228; $225\dfrac{57}{250}$; Audi Sport North America   **71.** 225.228; 214.500; 201.265; 197.400

73. answers may vary   **75.** answers may vary   **77.** 0.26499, 0.25786   **79.** 0.10299, 0.1037, 0.1038, 0.9   **81.** \$3600 million

Section 4.3

Calculator Explorations   **1.** 328.742   **3.** 5.2414   **5.** 865.392

Vocabulary, Readiness & Video Check   **1.** 37.0   **3.** difference; minuend; subtrahend   **5.** false   **7.** Lining up the decimal points also lines up place values; to make sure we only add digits in the same place.   **9.** Check subtraction by addition.

Exercise Set 4.3   **1.** 3.5   **3.** 6.83   **5.** 0.094   **7.** 622.012   **9.** 583.09   **11.** Exact: 465.56; Estimate:
$$\begin{array}{r} 230 \\ + \ 230 \\ \hline 460 \end{array}$$

13. Exact: 115.123; Estimate: 100   **15.** 27.0578   **17.** 56.432   **19.** 6.5   **21.** 15.3   **23.** 598.23   **25.** Exact: 1.83; Estimate: $6 - 4 = 2$

$$\begin{array}{r} 6 \\ + \quad 9 \\ \hline 115 \end{array}$$

27. 861.6   **29.** 376.89   **31.** Exact: 876.6; Estimate: 1000   **33.** 194.4   **35.** 2.9988   **37.** 16.3   **39.** 88.028   **41.** 84.072   **43.** 243.17

$$\begin{array}{r} - \quad 100 \\ \hline 900 \end{array}$$

45. 56.83   **47.** 3.16   **49.** $7.52   **51.** $454.71   **53.** $0.14   **55.** 28.56 m   **57.** 9.14 in.   **59.** 196.3 mph   **61.** 64.8 degrees Fahrenheit
63. 763.035 mph   **65.** 14.46 in.   **67.** $0.03   **69.** 21.8 billion (21,800,000,000)   **71.** 326.0 in.   **73.** 67.44 ft   **75.** $1.294   **77.** 715.05 hr

79. Switzerland   **81.** 7.94 lb   **83.**

Country	Pounds of Chocolate per Person
Switzerland	26.24
Ireland	21.83
UK	20.94
Austria	19.40
Belgium	18.30

85. 46   **87.** 3870   **89.** $\frac{4}{9}$ **91.** incorrect;

$$\begin{array}{r} 9.200 \\ 8.630 \\ + \quad 4.005 \\ \hline 21.835 \end{array}$$

93. 6.08 in.   **95.** $1.20   **97.** 1 dime, 1 nickel, and 2 pennies; 3 nickels and 2 pennies; 1 dime and 7 pennies; 2 nickels and 7 pennies
99. answers may vary   **101.** answers may vary

Section 4.4

Vocabulary, Readiness & Video Check   **1.** sum   **3.** right; zeros   **5.** circumference   **7.** 3   **9.** 4   **11.** 8   **13.** Whether we placed the decimal point correctly in our product.   **15.** We used an approximation for π. The exact answer is 8π meters.

Exercise Set 4.4   **1.** 0.12   **3.** 0.6   **5.** 1.3   **7.** Exact: 22.26; Estimate: $5 \times 4 = 20$ **9.** 0.4032   **11.** Exact: 8.23854; Estimate:

$$\begin{array}{r} 1 \\ \times \quad 8 \\ \hline 8 \end{array}$$

13. 11.2746   **15.** 84.97593   **17.** 65   **19.** 0.65   **21.** 0.072   **23.** 709.3   **25.** 6046   **27.** 0.03762   **29.** 0.0492   **31.** 12.3   **33.** 1.29
35. 0.096   **37.** 0.5623   **39.** 43.274   **41.** 5,500,000,000   **43.** 97,800,000   **45.** 292,000   **47.** 8π m ≈ 25.12 m   **49.** 10π cm ≈ 31.4 cm
51. 18.2π yd ≈ 57.148 yd   **53.** $715.20   **55.** $9150   **57.** 24.8 g   **59.** 11.417 sq in.   **61.** 250π ft ≈ 785 ft   **63.** 135π m ≈ 423.9 m
65. 64.9605 in.   **67. a.** 62.8 m and 125.6 m   **b.** yes   **69.** 4.70 sq in.   **71.** 26   **73.** 36   **75.** 8   **77.** 9   **79.** 3.64   **81.** 3.56   **83.** 0.1105
85. 3,831,600 mi   **87.** answers may vary   **89.** answers may vary

Integrated Review   **1.** 2.57   **2.** 4.05   **3.** 8.9   **4.** 3.5   **5.** 0.16   **6.** 0.24   **7.** 11.06   **8.** 9.72   **9.** 4.8   **10.** 6.09   **11.** 75.56   **12.** 289.12
13. 25.026   **14.** 44.125   **15.** 82.7   **16.** 273.9   **17.** 280   **18.** 1600   **19.** 224.938   **20.** 145.079   **21.** 6   **22.** 6.2   **23.** 27.6092
24. 145.6312   **25.** 5.4   **26.** 17.74   **27.** 414.44   **28.** 1295.03   **29.** 116.81   **30.** 18.79   **31.** 156.2   **32.** 25.62   **33.** 5.62   **34.** 304.876
35. 114.66   **36.** 119.86   **37.** 0.000432   **38.** 0.000075   **39.** 0.0672   **40.** 0.0275   **41.** 862   **42.** 0.0293   **43.** 200 mi

Section 4.5

Calculator Explorations   **1.** not reasonable   **3.** reasonable

Vocabulary, Readiness & Video Check   **1.** quotient; divisor; dividend   **3.** left; zeros   **5.** 5.9   **7.** 0   **9.** 1   **11.** undefined
13. a whole number   **15.** We just need to know how to move the decimal point. 1000 has three zeros, so we move the decimal point in the decimal number three places to the left.   **17.** The fraction bar serves as a grouping symbol.

Exercise Set 4.5   **1.** 4.6   **3.** 0.094   **5.** 300   **7.** 5.8   **9.** Exact: 6.6; Estimate: $6\overline{)36}$ **11.** 0.413   **13.** 0.045   **15.** 7   **17.** 4.8   **19.** 2100

21. 30   **23.** 7000   **25.** Exact: 9.8; Estimate: $7\overline{)70}$ **27.** 9.6   **29.** 45   **31.** 54.592   **33.** 0.0055   **35.** 179   **37.** 23.87   **39.** 113.1

41. 0.54982   **43.** 2.687   **45.** 0.0129   **47.** 12.6   **49.** 1.31   **51.** 12.225   **53.** 0.045625   **55.** 11 qt   **57.** 202.1 lb   **59.** 5.1 m
61. 11.4 boxes   **63.** 24 tsp   **65.** 8 days   **67.** 290.3 mi   **69.** 122.8 mph   **71.** 18.1 points per game   **73.** 2.45   **75.** 0.66   **77.** 80.52

79. 14.7   **81.** 930.7   **83.** 571   **85.** 92.06   **87.** 144.4   **89.** $\frac{9}{10}$ **91.** $\frac{1}{20}$ **93.** 4.26   **95.** 1.578   **97.** 26.66   **99.** 904.29   **101.** c

103. b   **105.** 85.5   **107.** 8.6 ft   **109.** answers may vary   **111.** 65.2–82.6 knots   **113.** 319.64 m

Section 4.6

Vocabulary, Readiness & Video Check   **1.** false   **3.** true   **5.** We place a bar over just the repeating digits and only 6 repeats in our decimal answer.   **7.** $A = l \cdot w$; 0.248 sq yd

Exercise Set 4.6   **1.** 0.2   **3.** 0.68   **5.** 0.75   **7.** 0.08   **9.** 1.2   **11.** $0.91\overline{6}$ **13.** 0.425   **15.** 0.45   **17.** $0.\overline{3}$ **19.** 0.4375   **21.** $0.\overline{63}$
23. 5.85   **25.** 0.624   **27.** 0.33   **29.** 0.44   **31.** 0.6   **33.** 0.62   **35.** 0.38   **37.** 0.02   **39.** <   **41.** =   **43.** <   **45.** <   **47.** <   **49.** >

51. <   **53.** <   **55.** 0.32, 0.34, 0.35   **57.** 0.49, 0.491, 0.498   **59.** 0.73, $\frac{3}{4}$, 0.78   **61.** 0.412, 0.453, $\frac{4}{7}$ **63.** 5.23, $\frac{42}{8}$, 5.34   **65.** $\frac{17}{8}$, 2.37, $\frac{12}{5}$

67. 25.65 sq in.   **69.** 9.36 sq cm   **71.** 0.248 sq yd   **73.** 8   **75.** 72   **77.** $\frac{1}{81}$ **79.** $\frac{9}{25}$ **81.** $\frac{5}{2}$ **83.** = 1   **85.** > 1   **87.** < 1   **89.** 0.13

91. 8700 stations   **93.** answers may vary   **95.** answers may vary   **97.** 47.25   **99.** 3.37   **101.** 0.45

Chapter 4 Vocabulary Check   1. decimal   **2.** numerator; denominator   **3.** vertically   **4.** and   **5.** sum   **6.** circumference
7. standard form   **8.** circumference; diameter   **9.** difference   **10.** quotient   **11.** product   **12.** sum

Chapter 4 Review   1. tenths   **2.** hundred-thousandths   **3.** forty-five hundredths   **4.** three hundred forty-five hundred-
thousandths   **5.** one hundred nine and twenty-three hundredths   **6.** forty-six and seven thousandths   **7.** 2.15   **8.** 503.102
9. $\frac{4}{25}$ **10.** $12\frac{23}{1000}$ **11.** $1\frac{9}{2000}$ **12.** $25\frac{1}{4}$ **13.** 0.9   **14.** 0.25   **15.** 0.045   **16.** 26.1   **17.** >   **18.** =   **19.** 0.92, 8.09, 8.6   **20.** 0.09,
0.091, 0.1   **21.** 0.6   **22.** 0.94   **23.** $0.26   **24.** $12.46   **25.** $31,304   **26.** $10\frac{3}{4}$ **27.** 9.52   **28.** 2.7   **29.** 7.28   **30.** 26.007   **31.** 459.7
32. 100.278   **33.** 65.02   **34.** 189.98   **35.** 52.6 mi   **36.** $2.44   **37.** 22.2 in.   **38.** 38.9 ft   **39.** 18.5   **40.** 54.6   **41.** 72   **42.** 9345
43. 9.246   **44.** 3406.446   **45.** 14π m; 43.96 m   **46.** 63.8 mi   **47.** 887,000,000   **48.** 600,000   **49.** 0.087   **50.** 15.825   **51.** 70
52. 0.21   **53.** 8.059   **54.** 30.4   **55.** 0.0267   **56.** 9.3   **57.** 7.3 m   **58.** 45 mo   **59.** 16.94   **60.** 3.89   **61.** 129   **62.** 55   **63.** 0.81
64. 7.26   **65.** 0.8   **66.** 0.923   **67.** $2.\overline{3}$ or 2.333   **68.** $0.21\overline{6}$ or 0.217   **69.** =   **70.** =   **71.** <   **72.** <   **73.** 0.837, 0.839, $\frac{17}{20}$
74. $\frac{19}{12}$, 1.63, $\frac{18}{11}$ **75.** 6.9 sq ft   **76.** 5.46 sq in.   **77.** two hundred and thirty-two ten-thousandths   **78.** 16,025.014   **79.** $\frac{231}{100,000}$
80. 0.75, $\frac{6}{7}$, $\frac{8}{9}$ **81.** 0.07   **82.** 0.1125   **83.** 51.057   **84.** >   **85.** <   **86.** <   **87.** 42.90   **88.** 16.349   **89.** $123   **90.** $3646   **91.** 1.7
92. 2.49   **93.** 320.312   **94.** 148.74236   **95.** 8.128   **96.** 7.245   **97.** 4900   **98.** 23.904   **99.** 9600 sq ft   **100.** yes   **101.** 0.1024   **102.** 3.6

Chapter 4 Test   1. forty-five and ninety-two thousandths   **2.** 3000.059   **3.** 34.9   **4.** 0.862   **5.** <   **6.** $\frac{4}{9}$, 0.445, 0.454   **7.** $\frac{69}{200}$
8. $24\frac{73}{100}$ **9.** 0.65   **10.** $5.\overline{8}$ or 5.889   **11.** 0.941   **12.** 17.583   **13.** 11.4   **14.** 43.86   **15.** 56   **16.** 0.07755   **17.** 6.673   **18.** 12,690
19. 4.73   **20.** 0.363   **21.** 6.2   **22.** 4,583,000,000   **23.** 2.31 sq mi   **24.** 18π mi, 56.52 mi   **25. a.** 9904 sq ft   **b.** 198.08 oz   **26.** 54 mi

Cumulative Review   1. one hundred six million, fifty-two thousand, four hundred forty-seven (Sec. 1.2, Ex. 7)   **2.** 276,004
3. 14,440,060 (Sec 1.3, Ex. 7)   **4.** 288   **5.** 726 (Sec. 1.4, Ex. 4)   **6.** 200   **7.** 2300 (Sec. 1.5, Ex. 4)   **8.** 84   **9.** 57,600 megabytes
(Sec. 1.6, Ex. 11)   **10.** perimeter: 28 ft; area: 49 sq ft   **11.** 401 R 2 (Sec. 1.7, Ex. 8)   **12.** $\frac{21}{8}$ **13.** 47 (Sec. 1.9, Ex. 15)   **14.** $12\frac{4}{5}$
15. numerator: 3; denominator: 7 (Sec. 2.1, Ex. 1)   **16.** 9   **17.** $\frac{1}{10}$ (Sec. 2.3, Ex. 6)   **18.** 17   **19.** $\frac{15}{1}$ or 15 (Sec. 2.4, Ex. 9)   **20.** 13
21. $\frac{63}{16}$ (Sec. 2.5, Ex. 5)   **22.** 128   **23.** $\frac{15}{4}$ or $3\frac{3}{4}$ (Sec. 2.4, Ex. 10)   **24.** $9   **25.** $\frac{3}{20}$ (Sec. 2.5, Ex. 8)   **26.** $\frac{27}{20}$ or $1\frac{7}{20}$ **27.** $\frac{7}{9}$ (Sec. 3.1, Ex. 4)
28. $\frac{2}{5}$ **29.** $\frac{1}{4}$ (Sec. 3.1, Ex. 5)   **30.** $\frac{2}{5}$ **31.** $\frac{15}{20}$ (Sec. 3.2, Ex. 8)   **32.** $\frac{35}{45}$ **33.** $\frac{31}{30}$ or $1\frac{1}{30}$ (Sec. 3.3, Ex. 2)   **34.** $\frac{1}{90}$ **35.** $4\frac{7}{40}$ lb (Sec. 3.4, Ex. 7)
36. $27\frac{3}{4}$ lb   **37.** $\frac{1}{16}$ (Sec. 3.5, Ex. 3)   **38.** $\frac{49}{121}$ **39.** $\frac{3}{256}$ (Sec. 3.5, Ex. 5)   **40.** $\frac{2}{81}$ **41.** $\frac{43}{100}$ (Sec. 4.1, Ex. 8)   **42.** 0.75
43. > (Sec. 4.2, Ex. 1)   **44.** 5.06   **45.** 11.568 (Sec. 4.3, Ex. 4)   **46.** 75.329   **47.** 2370.2 (Sec. 4.4, Ex. 6)   **48.** 0.119
49. 768.05 (Sec. 4.4, Ex. 9)   **50.** 8.9

Chapter 5 Ratio and Proportion

Section 5.1

Vocabulary, Readiness & Video Check   1. true   **3.** false   **5.** true   **7.** true   **9.** We can use "to" as in 1 to 2, a colon as in 1 : 2, or
a fraction as in $\frac{1}{2}$.

Exercise Set 5.1   1. $\frac{11}{4}$ **3.** $\frac{23}{10}$ **5.** $\frac{151}{201}$ **7.** $\frac{2.8}{7.6}$ **9.** $\frac{5}{7\frac{1}{2}}$ **11.** $\frac{3\frac{3}{4}}{1\frac{2}{3}}$ **13.** $\frac{2}{3}$ **15.** $\frac{77}{100}$ **17.** $\frac{463}{821}$ **19.** $\frac{3}{4}$ **21.** $\frac{5}{12}$ **23.** $\frac{8}{25}$ **25.** $\frac{12}{7}$

27. $\frac{16}{23}$ **29.** $\frac{2}{5}$ **31.** $\frac{2}{7}$ **33.** $\frac{4}{1}$ **35.** $\frac{10}{29}$ **37.** $\frac{16}{51}$ **39.** $\frac{25}{144}$ **41.** $\frac{5}{4}$ **43.** $\frac{15}{1}$ **45.** $\frac{17}{40}$ **47.** $\frac{45}{659}$ **49.** $\frac{1}{3}$ **51.** $\frac{12}{151}$ **53.** $\frac{1}{49}$ **55.** 2.3

57. 0.15 **59.** the ratio of seven to nine **61.** the ratio of thirty to one **63.** no; answers may vary **65.** no; $\dfrac{3}{2}$ **67.** no; $\dfrac{2}{5}$ **69.** no; $\dfrac{9}{2}$
71. No, the shipment should not be refused. **73. a.** $\dfrac{6}{25}$ **b.** 38 states **c.** $\dfrac{6}{19}$

Section 5.2

Vocabulary, Readiness & Video Check **1.** unit **3.** division **5.** numerator; denominator **7.** We want a unit rate, which is a rate with a denominator of 1. A unit rate tells us how much of the first quantity ($) will occur in 1 of the second quantity (years).

Exercise Set 5.2 **1.** $\dfrac{1 \text{ shrub}}{3 \text{ ft}}$ **3.** $\dfrac{3 \text{ returns}}{20 \text{ sales}}$ **5.** $\dfrac{3 \text{ laser printers}}{14 \text{ computers}}$ **7.** $\dfrac{9 \text{ gal}}{2 \text{ acres}}$ **9.** $\dfrac{3 \text{ flight attendants}}{100 \text{ passengers}}$ **11.** $\dfrac{71 \text{ cal}}{2 \text{ fl oz}}$ **13.** 110 cal/oz
15. 75 riders/car **17.** 90 wingbeats/sec **19.** $50,000/yr **21.** 319,368 voters/senator **23.** 300 good/defective **25.** $6,874,600/player
27. 66 crayons/day **29.** $60,750/species **31.** $140,000/house **33.** 8.7 tornadoes/hour **35. a.** 31.25 computer boards/hr
b. 33.5 computer boards/hr **c.** Suellen **37. a.** 27.6 miles/gal **b.** 29.2 miles/gal **c.** the truck **39.** $11.50 per DVD
41. $0.17 per banana **43.** 8 oz: $0.149 per oz; 12 oz: $0.133 per oz; 12 oz **45.** 16 oz: $0.106 per oz; 6 oz: $0.115 per oz; 16 oz
47. 12 oz: $0.191 per oz; 8 oz: $0.186 per oz; 8 oz **49.** 100: $0.006 per napkin; 180: $0.005 per napkin; 180 napkins **51.** 10.2 **53.** 4.44
55. 1.9 **57.** 257; 19.2 **59.** 347; 21.6 **61.** 1.5 steps/ft **63.** answers may vary **65.** no; answers may vary

Integrated Review **1.** $\dfrac{9}{10}$ **2.** $\dfrac{9}{25}$ **3.** $\dfrac{43}{50}$ **4.** $\dfrac{8}{23}$ **5.** $\dfrac{173}{139}$ **6.** $\dfrac{6}{7}$ **7.** $\dfrac{7}{26}$ **8.** $\dfrac{20}{33}$ **9.** $\dfrac{2}{3}$ **10.** $\dfrac{1}{8}$ **11.** $\dfrac{2123}{1324}$ **12.** $\dfrac{1}{4}$ **13. a.** 4
b. $\dfrac{1}{3}$ **14.** $\dfrac{2}{3}$ **15.** $\dfrac{1 \text{ office}}{4 \text{ graduate assistants}}$ **16.** $\dfrac{2 \text{ lights}}{5 \text{ ft}}$ **17.** $\dfrac{2 \text{ senators}}{1 \text{ state}}$ **18.** $\dfrac{1 \text{ teacher}}{28 \text{ students}}$ **19.** $\dfrac{16 \text{ computers}}{25 \text{ households}}$ **20.** $\dfrac{9 \text{ students}}{2 \text{ computers}}$
21. 55 mi/hr **22.** 140 ft/sec **23.** 21 employees/fax line **24.** 17 phone calls/teenager **25.** 23 mi/gal **26.** 16 teachers/computer
27. 6 books/student **28.** 154 lb/adult **29.** 8 lb: $0.27 per lb; 18 lb: $0.28 per lb; 8 lb **30.** 100: $0.020 per plate; 500: $0.018 per plate;
500 paper plates **31.** 3 packs: $0.80 per pack; 8 packs: $0.75 per pack; 8 packs **32.** 4: $0.92 per battery; 10: $0.99 per battery; 4 batteries

Section 5.3

Vocabulary, Readiness & Video Check **1.** proportion; ratio **3.** true **5.** true **7.** false **9.** true **11.** equals or = **13.** a variable

Exercise Set 5.3 **1.** $\dfrac{10 \text{ diamonds}}{6 \text{ opals}} = \dfrac{5 \text{ diamonds}}{3 \text{ opals}}$ **3.** $\dfrac{3 \text{ printers}}{12 \text{ computers}} = \dfrac{1 \text{ printer}}{4 \text{ computers}}$ **5.** $\dfrac{6 \text{ eagles}}{58 \text{ sparrows}} = \dfrac{3 \text{ eagles}}{29 \text{ sparrows}}$
7. $\dfrac{2\frac{1}{4} \text{ cups flour}}{24 \text{ cookies}} = \dfrac{6\frac{3}{4} \text{ cups flour}}{72 \text{ cookies}}$ **9.** $\dfrac{22 \text{ vanilla wafers}}{1 \text{ cup cookie crumbs}} = \dfrac{55 \text{ vanilla wafers}}{2.5 \text{ cups cookie crumbs}}$ **11.** true **13.** false **15.** true **17.** true
19. false **21.** false **23.** true **25.** false **27.** true **29.** $\dfrac{8}{12} = \dfrac{4}{6}$; true **31.** $\dfrac{5}{2} = \dfrac{13}{5}$; false **33.** $\dfrac{1.8}{2} = \dfrac{4.5}{5}$; true **35.** $\dfrac{\frac{2}{3}}{\frac{1}{5}} = \dfrac{\frac{2}{5}}{\frac{1}{9}}$; false
37. 3 **39.** 9 **41.** 4 **43.** 3.2 **45.** 38.4 **47.** 25 **49.** 0.0025 **51.** 1 **53.** $\dfrac{9}{20}$ **55.** 12 **57.** $\dfrac{3}{4}$ **59.** $\dfrac{35}{18}$ **61.** < **63.** > **65.** <
67. $\dfrac{9}{3} = \dfrac{15}{5}; \dfrac{5}{15} = \dfrac{3}{9}; \dfrac{15}{9} = \dfrac{5}{3}$ **69.** $\dfrac{6}{1} = \dfrac{18}{3}; \dfrac{3}{18} = \dfrac{1}{6}; \dfrac{18}{6} = \dfrac{3}{1}$ **71.** possible answers: $\dfrac{d}{b} = \dfrac{c}{a}; \dfrac{a}{c} = \dfrac{b}{d}; \dfrac{b}{a} = \dfrac{d}{c}$ **73.** answers may vary
75. 14.9 **77.** 0.07 **79.** 3.163 **81.** 0 **83.** 1400 **85.** 252.5

Section 5.4

Vocabulary, Readiness & Video Check **1.** There are approximately 102.9 mg of cholesterol in a 5-ounce serving of lobster.

Exercise Set 5.4 **1.** 360 baskets **3.** 165 min **5.** 630 applications **7.** 23 ft **9.** 270 sq ft **11.** 25 gal **13.** 450 km **15.** 16 bags
17. 15 hits **19.** 27 people **21.** 18 applications **23.** 5 weeks **25.** $10\frac{2}{3}$ servings **27.** 37.5 sec **29. a.** 18 tsp **b.** 6 tbsp **31.** 6 people
33. 112 ft; 11-in. difference **35.** 102.9 mg **37.** 1248 feet; coincidentally, this is the actual height of the Empire State Building
39. 434 emergency room visits **41.** 113 million phones **43.** 2.4 c **45. a.** 0.1 gal **b.** 13 fl oz **47. a.** 2062.5 mg **b.** no **49.** $3 \cdot 5$
51. $2^2 \cdot 5$ **53.** $2^3 \cdot 5^2$ **55.** 2^5 **57.** 0.8 ml **59.** 1.25 ml **61.** $11 \approx 12$ or 1 dozen; $1.5 \times 8 = 12$; 12 cups of milk **63.** $4\frac{2}{3}$ ft
65. answers may vary

Chapter 5 Vocabulary Check **1.** ratio **2.** proportion **3.** unit rate **4.** unit price **5.** rate **6.** cross products **7.** equal **8.** not equal

Chapter 5 Review **1.** $\dfrac{23}{37}$ **2.** $\dfrac{14}{51}$ **3.** $\dfrac{5}{4}$ **4.** $\dfrac{11}{13}$ **5.** $\dfrac{7}{15}$ **6.** $\dfrac{17}{35}$ **7.** $\dfrac{18}{35}$ **8.** $\dfrac{35}{27}$ **9. a.** 1 **b.** $\dfrac{1}{25}$ **10. a.** 15 **b.** $\dfrac{3}{5}$ **11.** $\dfrac{1 \text{ stillborn birth}}{125 \text{ live births}}$
12. $\dfrac{3 \text{ professors}}{10 \text{ assistants}}$ **13.** $\dfrac{5 \text{ pages}}{2 \text{ min}}$ **14.** $\dfrac{4 \text{ computers}}{3 \text{ hr}}$ **15.** 52 mi/hr **16.** 15 ft/sec **17.** $6.96/CD **18.** $1\frac{1}{3}$ gal/acre **19.** $46.80/course

20. 13 bushels/tree   **21.** 8 oz: $0.124 per oz; 12 oz: $0.141 per oz; 8-oz size   **22.** 18 oz: $0.083 per oz; 28 oz: $0.085 per oz; 18-oz size   **23.** 16 oz: $0.037 per oz; 64 oz: $0.026 per oz; 128 oz: $0.018 per oz; 1-gal size

24. 12 oz: $0.049 per oz; 16 oz: $0.049 per oz; 32 oz: $0.037 per oz; 32-oz size   **25.** $\dfrac{20 \text{ men}}{14 \text{ women}} = \dfrac{10 \text{ men}}{7 \text{ women}}$ **26.** $\dfrac{50 \text{ tries}}{4 \text{ successes}} = \dfrac{25 \text{ tries}}{2 \text{ successes}}$ **27.** $\dfrac{16 \text{ sandwiches}}{8 \text{ players}} = \dfrac{2 \text{ sandwiches}}{1 \text{ player}}$

28. $\dfrac{12 \text{ tires}}{3 \text{ cars}} = \dfrac{4 \text{ tires}}{1 \text{ car}}$ **29.** no   **30.** yes   **31.** yes   **32.** no   **33.** 5   **34.** 15   **35.** 32.5   **36.** 5.625   **37.** 32   **38.** $6\dfrac{3}{4}$ **39.** 60   **40.** $7\dfrac{1}{5}$

41. 0.94   **42.** 0.36   **43.** $1\dfrac{1}{8}$ **44.** $1\dfrac{3}{7}$ **45.** 14 passes   **46.** 35 passes   **47.** 8 bags   **48.** 16 bags   **49.** no   **50.** 79 gal   **51.** $54,600

52. $1023.50   **53.** $40\dfrac{1}{2}$ ft   **54.** $8\dfrac{1}{4}$ in.   **55.** $\dfrac{3}{5}$ **56.** $\dfrac{4}{9}$ **57.** $\dfrac{1}{2}$ **58.** $\dfrac{5}{12}$ **59.** $\dfrac{1}{27}$ **60.** $\dfrac{1}{8}$ **61.** $\dfrac{1 \text{ teacher}}{9 \text{ students}}$ **62.** $\dfrac{1 \text{ nurse}}{4 \text{ patients}}$ **63.** 4 cups/person

64. 6 toys/child   **65.** 34 mi/hr   **66.** 2 gal/cow   **67.** 4 oz: $1.235 per oz; 8 oz: $1.248 per oz; 4-oz size   **68.** 12 oz: $0.054 per oz; 64 oz: $0.047 per oz; 64-oz size   **69.** $\dfrac{2 \text{ cups cookie dough}}{30 \text{ cookies}} = \dfrac{4 \text{ cups cookie dough}}{60 \text{ cookies}}$ **70.** $\dfrac{5 \text{ nickels}}{3 \text{ dollars}} = \dfrac{20 \text{ nickels}}{12 \text{ dollars}}$ **71.** yes   **72.** no

73. 1.6   **74.** 4.8   **75.** 84   **76.** 25   **77.** $609.50   **78.** $477   **79.** $171   **80.** $95

Chapter 5 Test   **1.** $\dfrac{15}{2}$ **2.** $\dfrac{9}{13}$ **3.** $\dfrac{7 \text{ men}}{1 \text{ woman}}$ **4.** $\dfrac{13 \text{ in.}}{10 \text{ days}}$ **5.** $\dfrac{43}{50}$ **6.** $\dfrac{47}{78}$ **7.** $\dfrac{197}{62}$ **8.** 81.25 km/hr   **9.** $\dfrac{2}{3}$ in./hr

10. 28 students/teacher   **11.** 9 in./sec   **12.** 8-oz size   **13.** 16-oz size   **14.** true   **15.** false   **16.** $\dfrac{25 \text{ computers}}{600 \text{ students}} = \dfrac{1 \text{ computer}}{24 \text{ students}}$

17. 5   **18.** $4\dfrac{4}{11}$ **19.** $\dfrac{7}{3}$ **20.** 8   **21.** $\dfrac{7}{8}$ **22.** $49\dfrac{1}{2}$ ft   **23.** $3\dfrac{3}{4}$ hr   **24.** $53\dfrac{1}{3}$ g   **25.** $114\dfrac{2}{3}$ cartons   **26.** 5925 adults

Cumulative Review   **1. a.** 3   **b.** 15   **c.** 0   **d.** 70 (Sec. 1.4, Ex. 1)   **2. a.** 0   **b.** 20   **c.** 0   **d.** 20   **3.** 249,000 (Sec. 1.5, Ex. 3)   **4.** 249,000   **5. a.** 200   **b.** 1230 (Sec. 1.6, Ex. 3)   **6.** 373 R 24   **7.** $6171 (Sec. 1.8, Ex. 3)   **8.** 16,591 ft   **9.** $3 \cdot 3 \cdot 5$ or $3^2 \cdot 5$ (Sec. 2.2, Ex. 3)   **10.** 8   **11.** $\dfrac{3}{5}$ (Sec. 2.3, Ex. 1)   **12.** 243   **13.** $\dfrac{6}{5}$ (Sec. 2.4, Ex. 5)   **14.** $15\dfrac{3}{8}$ **15.** $\dfrac{2}{5}$ (Sec. 2.4, Ex. 6)

16. $\dfrac{5}{54}$ **17.** $\dfrac{5}{7}$ (Sec. 3.1, Ex. 1)   **18.** $\dfrac{19}{30}$ **19.** $\dfrac{16}{13}$ or $1\dfrac{3}{13}$ (Sec. 3.1, Ex. 3)   **20.** $\dfrac{4}{5}$ **21.** 36 (Sec. 3.2, Ex. 2)   **22.** $\dfrac{49}{50}$ **23.** $\dfrac{12}{24}$ (Sec. 3.2, Ex. 9)

24. yes   **25.** $\dfrac{8}{33}$ (Sec. 3.3, Ex. 6)   **26.** $7\dfrac{47}{72}$ **27.** $\dfrac{1}{6}$ of an hour (Sec. 3.3, Ex. 9)   **28.** 27   **29.** $7\dfrac{17}{24}$ (Sec. 3.4, Ex. 1)   **30.** $\dfrac{16}{27}$

31. > (Sec. 3.5, Ex. 1)   **32.** 14,000,000   **33.** one and three tenths (Sec. 4.1, Ex. 1)   **34.** 0.075   **35.** 736.2 (Sec. 4.2, Ex. 5)

36. 736.236   **37.** 25.454 (Sec. 4.3, Ex. 1)   **38.** 681.24   **39.** 0.0849 (Sec. 4.4, Ex. 2)   **40.** 0.375   **41.** 0.125 (Sec. 4.5, Ex. 3)   **42.** $\dfrac{79}{10}$

43. 3.7 (Sec. 4.5, Ex. 12)   **44.** 3   **45.** $\dfrac{4}{9}, \dfrac{9}{20}$, 0.456 (Sec. 4.6, Ex. 9)   **46.** 140 m/sec   **47.** $\dfrac{2.6}{3.1}$ (Sec. 5.1, Ex. 2)   **48.** $\dfrac{1}{3}$

49. $\dfrac{1\frac{1}{2}}{7\frac{3}{4}}$ (Sec. 5.1, Ex. 3)   **50.** $\dfrac{1}{10}$

Chapter 6 Percent

Section 6.1

Vocabulary, Readiness & Video Check   **1.** Percent   **3.** percent   **5.** 0.01   **7.** $\dfrac{1}{100}$; 0.01

Exercise Set 6.1   **1.** 96%   **3. a.** 75%   **b.** 25%   **5.** football; 37%   **7.** 50%   **9.** 0.41   **11.** 0.06   **13.** 1.00 or 1   **15.** 0.736   **17.** 0.028   **19.** 0.006   **21.** 3.00 or 3   **23.** 0.3258   **25.** 0.38   **27.** 0.382   **29.** 0.45   **31.** 98%   **33.** 310%   **35.** 2900%   **37.** 0.3%   **39.** 22%   **41.** 530%   **43.** 5.6%   **45.** 33.28%   **47.** 300%   **49.** 70%   **51.** 77%   **53.** 28.1%   **55.** 8.1%   **57.** 0.25   **59.** 0.65   **61.** 0.9   **63.** b, d   **65.** 4%   **67.** personal care aides   **69.** 0.617   **71.** answers may vary

Section 6.2

Vocabulary, Readiness & Video Check   **1.** Percent   **3.** 100%   **5.** 13%   **7.** 87%   **9.** 1%   **11.** The fraction of $\dfrac{4}{100}$ can be simplified to $\dfrac{1}{25}$.   **13.** The difference is in how the percent symbol is replaced—for a decimal, replace % with the equivalent decimal form 0.01, and for a fraction, replace % with the equivalent fraction form $\dfrac{1}{100}$.

Exercise Set 6.2 **1.** $\frac{3}{25}$ **3.** $\frac{1}{25}$ **5.** $\frac{9}{200}$ **7.** $\frac{7}{4}$ or $1\frac{3}{4}$ **9.** $\frac{73}{100}$ **11.** $\frac{1}{8}$ **13.** $\frac{1}{16}$ **15.** $\frac{3}{50}$ **17.** $\frac{31}{300}$ **19.** $\frac{179}{800}$ **21.** 75% **23.** 70%

25. 40% **27.** 59% **29.** 34% **31.** $37\frac{1}{2}$% **33.** $31\frac{1}{4}$% **35.** 160% **37.** $77\frac{7}{9}$% **39.** 65% **41.** 250% **43.** 190% **45.** 63.64%

47. 26.67% **49.** 14.29% **51.** 91.67% **53.** 0.35, $\frac{7}{20}$; 20%, 0.2; 50%, $\frac{1}{2}$; 0.7, $\frac{7}{10}$; 37.5%, 0.375 **55.** 0.4, $\frac{2}{5}$; $23\frac{1}{2}$%, $\frac{47}{200}$; 80%, 0.8; 0.3333,

$\frac{1}{3}$; 87.5%, 0.875; 0.075, $\frac{3}{40}$ **57.** 2, 2; 280%, $2\frac{4}{5}$; 7.05, $7\frac{1}{20}$; 454%, 4.54 **59.** 0.67; $\frac{67}{100}$ **61.** 0.522; $\frac{261}{500}$ **63.** 80% **65.** 0.0875 **67.** 18%

69. 0.005; $\frac{1}{200}$ **71.** 0.142; $\frac{71}{500}$ **73.** 0.079; $\frac{79}{1000}$ **75.** 15 **77.** 10 **79.** 12 **81. a.** 52.9% **b.** 52.86% **83.** 107.8%

85. 65.79% **87.** 77% **89.** 75% **91.** 80% **93.** greater **95.** answers may vary **97.** 0.266; 26.6% **99.** 1.155; 115.5%

Section 6.3

Vocabulary, Readiness & Video Check **1.** is **3.** amount; base; percent **5.** greater **7.** percent: 42%; base: 50; amount: 21 **9.** percent: 125%; base: 86; amount: 107.5 **11.** "of" means multiplication; "is" means equal; "what" (or some equivalent) means the unknown number

Exercise Set 6.3 **1.** 18% · 81 = n **3.** 20% · n = 105 **5.** 0.6 = 40% · n **7.** n · 80 = 3.8 **9.** n = 9% · 43 **11.** n · 250 = 150 **13.** 3.5 **15.** 28.7 **17.** 10 **19.** 600 **21.** 110% **23.** 34% **25.** 1 **27.** 645 **29.** 500 **31.** 5.16% **33.** 25.2 **35.** 35% **37.** 35 **39.** 0.624 **41.** 0.5% **43.** 145 **45.** 63% **47.** 4% **49.** n = 30 **51.** n = $3\frac{7}{11}$ **53.** $\frac{17}{12} = \frac{n}{20}$ **55.** $\frac{8}{9} = \frac{14}{n}$ **57.** c **59.** b

61. Twenty percent of some number is eighteen and six tenths. **63.** b **65.** c **67.** c **69.** a **71.** a **73.** answers may vary **75.** 686.625 **77.** 12,285

Section 6.4

Vocabulary, Readiness & Video Check **1.** amount; base; percent **3.** amount **5.** amount: 12.6; base: 42; percent: 30 **7.** amount: 102; base: 510; percent: 20 **9.** 45 follows the word "of," so it is the base

Exercise Set 6.4 **1.** $\frac{a}{45} = \frac{98}{100}$ **3.** $\frac{a}{150} = \frac{4}{100}$ **5.** $\frac{14.3}{b} = \frac{26}{100}$ **7.** $\frac{84}{b} = \frac{35}{100}$ **9.** $\frac{70}{400} = \frac{p}{100}$ **11.** $\frac{8.2}{82} = \frac{p}{100}$ **13.** 26 **15.** 18.9

17. 600 **19.** 10 **21.** 120% **23.** 28% **25.** 37 **27.** 1.68 **29.** 1000 **31.** 210% **33.** 55.18 **33.** 45% **37.** 75 **39.** 0.864

41. 0.5% **43.** 140 **45.** 9.6 **47.** 113% **49.** $\frac{7}{8}$ **51.** $3\frac{2}{15}$ **53.** 0.7 **55.** 2.19 **57.** answers may vary **59.** no; a = 16 **61.** yes

63. answers may vary **65.** 12,011.2 **67.** 7270.6

Integrated Review **1.** 12% **2.** 68% **3.** 12.5% **4.** 250% **5.** 520% **6.** 800% **7.** 6% **8.** 44% **9.** 750% **10.** 325%
11. 3% **12.** 5% **13.** 0.65 **14.** 0.31 **15.** 0.08 **16.** 0.07 **17.** 1.42 **18.** 4 **19.** 0.029 **20.** 0.066 **21.** 0.03; $\frac{3}{100}$ **22.** 0.05; $\frac{1}{20}$
23. 0.0525; $\frac{21}{400}$ **24.** 0.1275; $\frac{51}{400}$ **25.** 0.38; $\frac{19}{50}$ **26.** 0.45; $\frac{9}{20}$ **27.** 0.123; $\frac{37}{300}$ **28.** 0.167; $\frac{1}{6}$ **29.** 8.4 **30.** 100 **31.** 250 **32.** 120%
33. 28% **34.** 76 **35.** 11 **36.** 130% **37.** 86% **38.** 37.8 **39.** 150 **40.** 62

Section 6.5

Vocabulary, Readiness & Video Check **1.** The price of the home was $175,000.

Exercise Set 6.5 **1.** 1600 bolts **3.** 8.8 lb **5.** 14% **7.** 91,800 businesses **9.** 45% **11.** 496 chairs; 5704 chairs **13.** 108,680 physician assistants **15.** 9880.36 thousand or approximately 9880 thousand **17.** 29% **19.** 50% **21.** 12.5% **23.** 29.2% **25.** $175.000 **27.** 31.2 hr **29.** increase: $867.87; new price: $20,153.87 **31.** 40 ft **33.** increase: $1210; tuition in 2013–2014: $9616 **35.** increase: 128,760 associate degrees; 2020–2021: 1,016,760 associate degrees **37.** 30; 60% **39.** 52; 80% **41.** 2; 25% **43.** 120; 75% **45.** 44% **47.** 1.3% **49.** 142.0% **51.** 9.0% **53.** 139.5% **55.** 18.8% **57.** 19.7% **59.** 54.3% **61.** 4.56 **63.** 11.18 **65.** 58.54

67. The increased number is double the original number. **69.** percent of increase = $\frac{30}{150}$ = 20% **71.** False; the percents are different.

Section 6.6

Vocabulary, Readiness & Video Check **1.** sales tax **3.** commission **5.** sale price **7.** We rewrite the percent as an equivalent decimal. **9.** Replace "amount of discount" in the second equation with "discount rate · original price": sale price = original price − (discount rate · original price).

Exercise Set 6.6 **1.** $7.50 **3.** $858.93 **5.** 7% **7. a.** $120 **b.** $130.20 **9.** $117; $1917 **11.** $485 **13.** 6% **15.** $16.10; $246.10 **17.** $53,176.04 **19.** 14% **21.** $4888.50 **23.** $185,500 **25.** $8.90; $80.10 **27.** $98.25; $98.25 **29.** $143.50; $266.50 **31.** $3255; $18,445 **33.** $45; $255 **35.** $27.45; $332.45 **37.** $3.08; $59.08 **39.** $7074 **41.** 8% **43.** 1200 **45.** 132 **47.** 16 **49.** d **51.** $4.00; $6.00; $8.00 **53.** $7.20; $10.80; $14.40 **55.** a discount of 60% is better; answers may vary **57.** $26,838.45

Section 6.7

Calculator Explorations   **1.** 1.56051   **3.** 8.06231   **5.** $634.49

Vocabulary, Readiness & Video Check   **1.** simple   **3.** Compound   **5.** Total amount   **7.** principal   **9.** The denominator is the total number of payments. We are asked to find the monthly payment for a 4-year loan, and since there are 48 months in 4 years, there are 48 total payments.

Exercise Set 6.7   **1.** $32   **3.** $73.60   **5.** $750   **7.** $33.75   **9.** $700   **11.** $101,562.50; $264,062.50   **13.** $5562.50   **15.** $14,280   **17.** $46,815.37   **19.** $2327.14   **21.** $58,163.65   **23.** 2915.75   **25.** $2938.66   **27.** $2971.89   **29.** $260.31   **31.** $637.26   **33.** 32 yd   **35.** 35 m   **37.** answers may vary   **39.** answers may vary

Chapter 6 Vocabulary Check   **1.** of   **2.** is   **3.** Percent   **4.** Compound interest   **5.** $\dfrac{\text{amount}}{\text{base}}$ **6.** 100%   **7.** 0.01   **8.** $\dfrac{1}{100}$ **9.** base; amount   **10.** Percent of decrease   **11.** Percent of increase   **12.** Sales tax   **13.** Total price   **14.** Commission   **15.** Amount of discount   **16.** Sale price

Chapter 6 Review   **1.** 37%   **2.** 77%   **3.** 0.83   **4.** 0.75   **5.** 0.735   **6.** 0.015   **7.** 1.25   **8.** 1.45   **9.** 0.005   **10.** 0.007   **11.** 2.00 or 2   **12.** 4.00 or 4   **13.** 0.2625   **14.** 0.8534   **15.** 260%   **16.** 102%   **17.** 35%   **18.** 5.5%   **19.** 72.5%   **20.** 25.2%   **21.** 7.6%   **22.** 8.5%   **23.** 71%   **24.** 65%   **25.** 400%   **26.** 900%   **27.** $\dfrac{1}{100}$ **28.** $\dfrac{1}{10}$ **29.** $\dfrac{1}{4}$ **30.** $\dfrac{17}{200}$ **31.** $\dfrac{51}{500}$ **32.** $\dfrac{1}{6}$ **33.** $\dfrac{1}{3}$ **34.** $1\dfrac{1}{10}$ **35.** 20%   **36.** 70%   **37.** $83\dfrac{1}{3}$%   **38.** 60%   **39.** 125%   **40.** $166\dfrac{2}{3}$%   **41.** 6.25%   **42.** 62.5%   **43.** 100,000   **44.** 8000   **45.** 23%   **46.** 114.5   **47.** 3000   **48.** 150%   **49.** 418   **50.** 300   **51.** 64.8   **52.** 180%   **53.** 110%   **54.** 165   **55.** 66%   **56.** 16%   **57.** 20.9%   **58.** 106.25%   **59.** $206,400   **60.** $13.23   **61.** $263.75   **62.** $1.15   **63.** $5000   **64.** $300.38   **65.** discount: $900; sale price: $2100   **66.** discount: $9; sale price: $81   **67.** $160   **68.** $325   **69.** $30,104.61   **70.** $17,506.54   **71.** $80.61   **72.** $32,830.10   **73.** 0.038   **74.** 0.245   **75.** 0.009   **76.** 54%   **77.** 9520%   **78.** 30%   **79.** $\dfrac{47}{100}$ **80.** $\dfrac{8}{125}$ **81.** $\dfrac{7}{125}$ **82.** $37\dfrac{1}{2}$%   **83.** $15\dfrac{5}{13}$%   **84.** 120%   **85.** 268.75   **86.** 110%   **87.** 708.48   **88.** 134%   **89.** 300%   **90.** 38.4   **91.** 560   **92.** 325%   **93.** 26%   **94.** $6786.50   **95.** $617.70   **96.** $3.45   **97.** 12.5%   **98.** $1491   **99.** $17,951.01   **100.** $11,687.50

Chapter 6 Test   **1.** 0.85   **2.** 5   **3.** 0.008   **4.** 5.6%   **5.** 610%   **6.** 39%   **7.** $\dfrac{6}{5}$ or $1\dfrac{1}{5}$ **8.** $\dfrac{77}{200}$ **9.** $\dfrac{1}{500}$ **10.** 55%   **11.** 37.5%   **12.** $155\dfrac{5}{9}$%   **13.** 33.6   **14.** 1250   **15.** 75%   **16.** 38.4 lb   **17.** $56,750   **18.** $358.43   **19.** 5%   **20.** discount: $18; sale price: $102   **21.** $395   **22.** 1%   **23.** $647.50   **24.** $2005.63   **25.** $427

Cumulative Review   **1.** 206 cases; 12 cans; yes (Sec. 1.8, Ex. 2)   **2.** 31,084   **3. a.** $4\dfrac{2}{7}$ **b.** $1\dfrac{1}{15}$ **c.** 14 (Sec. 2.1, Ex. 18)   **4. a.** $\dfrac{19}{7}$ **b.** $\dfrac{101}{10}$ **c.** $\dfrac{43}{8}$ **5.** $2 \cdot 2 \cdot 2 \cdot 2 \cdot 5$ or $2^4 \cdot 5$ (Sec. 2.2, Ex. 7)   **6.** 119 sq mi   **7.** $\dfrac{10}{27}$ (Sec. 2.3, Ex. 3)   **8.** 44   **9.** $\dfrac{23}{56}$ (Sec. 2.4, Ex. 4)   **10.** 76,500   **11.** $\dfrac{8}{11}$ (Sec. 2.5, Ex. 2)   **12.** $\dfrac{15}{4}$; $3\dfrac{3}{4}$ **13.** $\dfrac{4}{5}$ in. (Sec. 3.1, Ex. 6)   **14.** 50   **15.** 60 (Sec. 3.2, Ex. 4)   **16.** $\dfrac{1}{3}$ **17.** $\dfrac{2}{3}$ (Sec. 3.3, Ex. 1)   **18.** 340   **19.** $3\dfrac{5}{14}$ (Sec. 3.4, Ex. 5)   **20.** 33   **21.** $\dfrac{7}{16}$ (Sec. 3.5, Ex. 6)   **22.** $33\dfrac{27}{40}$ **23.** $\dfrac{2}{33}$ (Sec. 3.5, Ex. 8)   **24.** $6\dfrac{3}{8}$ **25.** 0.8 (Sec. 4.1, Ex. 13)   **26.** 0.09   **27.** 8.7 (Sec. 4.1, Ex. 14)   **28.** 0.0048   **29.** $3.18 (Sec. 4.2, Ex. 7)   **30.** 27.94   **31.** 829.6561 (Sec. 4.3, Ex. 2)   **32.** 1248.3   **33.** 18.408 (Sec. 4.4, Ex. 1)   **34.** 76,300   **35.** 0.7861 (Sec. 4.5, Ex. 8)   **36.** 1.276   **37.** 0.012 (Sec. 4.5, Ex. 9)   **38.** 50.65   **39.** 7.236 (Sec. 4.5, Ex. 11)   **40.** 0.191   **41.** 0.25 (Sec. 4.6, Ex. 1)   **42.** $0.\overline{5} \approx 0.556$ **43.** $\dfrac{5 \text{ nails}}{3 \text{ ft}}$ (Sec. 5.2, Ex. 1)   **44.** 23 miles/gal   **45.** no (Sec. 5.3, Ex. 3)   **46.** 18: 0.052 per tortilla; 24: 0.058 per tortilla; 18 tortillas is better buy.   **47.** $17\dfrac{1}{2}$ mi (Sec. 5.4, Ex. 1)   **48. a.** 0.07   **b.** 2   **c.** 0.005   **49.** $n = 25\% \cdot 0.008$ (Sec. 6.3, Ex. 3)   **50.** 37.5% or $37\dfrac{1}{2}$%

Chapter 7 Measurement

Section 7.1

Vocabulary, Readiness & Video Check   **1.** meter   **3.** yard   **5.** feet   **7.** feet   **9.** feet; Feet are the original units and we want them to divide out.   **11.** The sum of 21 yd 4 ft is correct, but is not in a good format since there is a yard in 4 feet. Convert 4 feet = 1 yd 1 ft and add again: 21 yd + 1 yd + 1 ft = 22 yd 1 ft.   **13.** 1.29 cm and 12.9 mm; These two different-unit lengths are equal.

Exercise Set 7.1   **1.** 5 ft   **3.** 36 ft   **5.** 8 mi   **7.** 102 in.   **9.** $3\dfrac{1}{3}$ yd   **11.** 33,792 ft   **13.** 4.5 yd   **15.** 0.25 ft   **17.** 13 yd 1 ft   **19.** 7 ft 1 in.   **21.** 1 mi 4720 ft   **23.** 62 in.   **25.** 26 ft   **27.** 84 in.   **29.** 11 ft 2 in.   **31.** 22 yd 1 ft   **33.** 6 ft 5 in.   **35.** 7 ft 6 in.   **37.** 14 ft 4 in.   **39.** 83 yd 1 ft   **41.** 6000 cm   **43.** 4 cm   **45.** 0.5 km   **47.** 1.7 m   **49.** 15 m   **51.** 42,000 cm   **53.** 7000 m   **55.** 83 mm   **57.** 0.201 dm

59. 40 mm   **61.** 8.94 m   **63.** 2.94 m or 2940 mm   **65.** 1.29 cm or 12.9 mm   **67.** 12.64 km or 12,640 m   **69.** 54.9 m   **71.** 1.55 km

73. $348\frac{2}{3}$; 12,552   **75.** $11\frac{2}{3}$; 420   **77.** 5000; 0.005; 500   **79.** 0.065; 65; 0.000065   **81.** 342,000; 342,000,000; 34,200,000   **83. a.** $213\frac{2}{3}$ yd

b. 7692 in.   **85.** 10 ft 6 in.   **87.** 5100 ft   **89.** 5.0 times   **91.** 13 ft 11 in.   **93.** 26.7 mm   **95.** 15 ft 9 in.   **97.** 3 ft 1 in.   **99.** 41.25 m or

4125 cm   **101.** 3.35 m   **103.** 2.13 m   **105.** $121\frac{1}{3}$ yd   **107.** 15 tiles   **109.** $\frac{21}{100}$ **111.** 0.13   **113.** 0.25   **115.** no   **117.** yes   **119.** no

121. Estimate: 13 yd   **123.** answers may vary; for example, $1\frac{1}{3}$ yd or 48 in.   **125.** answers may vary   **127.** 334.89 sq m

Section 7.2

Vocabulary, Readiness & Video Check   **1.** mass   **3.** gram   **5.** 2000   **7.** 2 lb   **9.** 2 tons   **11.** pounds; Pounds are the units we're converting to.   **13.** 3 places to the right; 4 g = 4000 mg

Exercise Set 7.2   **1.** 32 oz   **3.** 10,000 lb   **5.** 9 tons   **7.** $3\frac{3}{4}$ lb   **9.** $1\frac{3}{4}$ tons   **11.** 204 oz   **13.** 9800 lb   **15.** 76 oz   **17.** 1.5 tons

19. $\frac{1}{20}$ lb   **21.** 92 oz   **23.** 161 oz   **25.** 5 lb 9 oz   **27.** 53 lb 10 oz   **29.** 8 tons 750 lb   **31.** 3 tons 175 lb   **33.** 8 lb 11 oz   **35.** 31 lb 2 oz

37. 1 ton 700 lb   **39.** 0.5 kg   **41.** 4000 mg   **43.** 25,000 g   **45.** 0.048 g   **47.** 0.0063 kg   **49.** 15,140 mg   **51.** 6250 g   **53.** 350,000 cg

55. 13.5 mg   **57.** 5.815 g or 5815 mg   **59.** 1850 mg or 1.85 g   **61.** 1360 g or 1.36 kg   **63.** 13.52 kg   **65.** 2.125 kg   **67.** 200,000;

3,200,000   **69.** $\frac{269}{400}$ or 0.6725; 21,520   **71.** 0.5; 0.0005; 50   **73.** 21,000; 21,000,000; 2,100,000   **75.** 8.064 kg   **77.** 30 mg

79. 5 lb 8 oz   **81.** 35 lb 14 oz   **83.** 6 lb 15.4 oz   **85.** 144 mg   **87.** 6.12 kg   **89.** 130 lb   **91.** 211 lb   **93.** 0.16   **95.** 0.875   **97.** no

99. yes   **101.** no   **103.** answers may vary; for example, 250 mg or 0.25 g   **105.** true   **107.** answers may vary

Section 7.3

Vocabulary, Readiness & Video Check   **1.** capacity   **3.** fluid ounces   **5.** cups   **7.** quarts   **9.** 2 pt   **11.** 2 gal   **13.** 3 qt   **15.** 3 c

17. amount; unit   **19.** 3 places to the left; 5600 ml = 5.6 L

Exercise Set 7.3   **1.** 4 c   **3.** 16 pt   **5.** $3\frac{1}{2}$ gal   **7.** 5 pt   **9.** 8 c   **11.** $3\frac{3}{4}$ qt   **13.** $10\frac{1}{2}$ qt   **15.** 9 c   **17.** 23 qt   **19.** $\frac{1}{4}$ pt

21. 14 gal 2 qt   **23.** 4 gal 3 qt 1 pt   **25.** 22 pt   **27.** 13 gal 2 qt   **29.** 4 c 4 fl oz (or 2 pt 4 fl oz)   **31.** 1 gal 1 qt   **33.** 2 gal 3 qt 1 pt

35. 17 gal   **37.** 4 gal 3 qt   **39.** 5000 ml   **41.** 0.00016 kl   **43.** 5.6 L   **45.** 320 cl   **47.** 0.41 kl   **49.** 0.064 L   **51.** 160 L   **53.** 3600 ml

55. 19.3 L   **57.** 4.5 L or 4500 ml   **59.** 8410 ml or 8.41 L   **61.** 16,600 ml or 16.6 L   **63.** 3840 ml   **65.** 162.4 L   **67.** 336; 84; 168

69. $\frac{1}{4}$; 1; 2   **71.** 1.59 L   **73.** 18.954 L   **75.** 4.3 fl oz   **77.** yes   **79.** $0.316   **81.** $\frac{4}{5}$ **83.** $\frac{3}{5}$ **85.** $\frac{9}{10}$ **87.** no   **89.** no   **91.** less than;

answers may vary   **93.** answers may vary   **95.** 128 fl oz   **97.** 1.5 cc   **99.** 2.7 cc   **101.** 54 u or 0.54 cc   **103.** 86 u or 0.86 cc

Integrated Review   **1.** 3 ft   **2.** 2 mi   **3.** $6\frac{2}{3}$ yd   **4.** 19 ft   **5.** 11,088 ft   **6.** 38.4 in.   **7.** 3000 cm   **8.** 2.4 cm   **9.** 2 m   **10.** 18 m

11. 72 mm   **12.** 0.6 km   **13.** 15,000 lb   **14.** 5.5 tons   **15.** 136 oz   **16.** 4.5 lb   **17.** 6.5 lb   **18.** 80 oz   **19.** 28,000 g   **20.** 1.4 g

21. 0.0056 kg   **22.** 6000 g   **23.** 0.67 g   **24.** 0.0036 kg   **25.** 12 pt   **26.** 2.5 qt   **27.** 3.5 gal   **28.** 8.5 pt   **29.** 7 c   **30.** 6.5 gal

31. 7000 ml   **32.** 0.35 kl   **33.** 0.047 L   **34.** 970 L   **35.** 126 L   **36.** 0.075 L   **37.** 4 fl oz   **38.** 12 c

Section 7.4

Vocabulary, Readiness & Video Check   **1.** 1 L ≈ 0.26 gal or 3.79 L ≈ 1 gal

Exercise Set 7.4   **1.** 25.57 fl oz   **3.** 218.44 cm   **5.** 40 oz   **7.** 57.66 mi   **9.** 3.77 gal   **11.** 13.5 kg   **13.** 1.5; $1\frac{2}{3}$; 150; 60   **15.** 55; 5500;

180; 2160   **17.** 3.94 in.   **19.** 80.5 kph   **21.** 0.008 oz   **23.** yes   **25.** 2790 mi   **27.** 90 mm   **29.** 112.5 g   **31.** 104 mph   **33.** 26.24 ft

35. 3 mi   **37.** 8 fl oz   **39.** b   **41.** b   **43.** c   **45.** d   **47.** d   **49.** 29   **51.** 9   **53.** 5   **55.** 36   **57.** 2.13 sq m   **59.** 1.19 sq m

61. 1.69 sq m   **63.** 21.3 mg–25.56 mg   **65.** 800 sq m or 8606.72 sq ft

Section 7.5

Vocabulary, Readiness & Video Check   **1.** F = 1.8C + 32; 27

Exercise Set 7.5   **1.** 5°C   **3.** 40°C   **5.** 140°F   **7.** 239°F   **9.** 16.7°C   **11.** 61.2°C   **13.** 197.6°F   **15.** 61.3°F   **17.** 50°C   **19.** 80.6°F

21. 100°C   **23.** 37.9°C   **25.** 244.4°F   **27.** 462.2°C   **29.** 21.1°C   **31.** 12 in.   **33.** 12 cm   **35.** 8 ft 4 in.   **37.** false   **39.** true

41. ≈ 520,000,000°C   **43.** yes   **45.** no   **47.** no   **49.** yes   **51.** answers may vary

Section 7.6

Vocabulary, Readiness & Video Check   **1.** Energy   **3.** British Thermal Unit   **5.** 30 ft-lb   **7.** 60 ft-lb   **9.** 90 cal   **11.** 5 cal
13. U.S.; British; pound; Fahrenheit

Exercise Set 7.6   **1.** 1140 ft-lb   **3.** 3696 ft-lb   **5.** 425,000 ft-lb   **7.** 23,340 ft-lb   **9.** 778,000 ft-lb   **11.** 15,560,000 ft-lb
13. 26,553,140 ft-lb   **15.** 10,283 BTU   **17.** 805 cal   **19.** 750 cal   **21.** 1440 cal   **23.** 2.6 hr   **25.** 17.5 mi   **27.** 73   **29.** 163   **31.** 19
33. 92.925 ft-lb   **35.** 256 lb   **37.** answers may vary

Chapter 7 Vocabulary Check   **1.** Weight   **2.** Mass   **3.** meter   **4.** unit fractions   **5.** gram   **6.** Energy   **7.** British Thermal Unit
8. liter   **9.** calorie

Chapter 7 Review   **1.** 9 ft   **2.** 24 yd   **3.** 7920 ft   **4.** 18 in.   **5.** 17 yd 1 ft   **6.** 3 ft 10 in.   **7.** 4200 cm   **8.** 820 mm   **9.** 0.01218 m
10. 0.00231 km   **11.** 21 yd 1 ft   **12.** 7 ft 5 in.   **13.** 41 ft 3 in.   **14.** 3 ft 8 in.   **15.** 9.5 cm or 95 mm   **16.** 2.74 m or 274 cm   **17.** 34.48 m
18. 2.45 km   **19.** 169 yd 2 ft   **20.** 126 ft 8 in.   **21.** 108.5 km   **22.** 0.24 sq m   **23.** 4.125 lb   **24.** 4600 lb   **25.** 3 lb 4 oz   **26.** 5 tons 300 lb
27. 0.027 g   **28.** 40,000 g   **29.** 21 dag   **30.** 0.0003 dg   **31.** 3 lb 9 oz   **32.** 10 tons 800 lb   **33.** 2 tons 750 lb   **34.** 33 lb 8 oz
35. 21.5 mg   **36.** 0.6 kg or 600 g   **37.** 4 lb 4 oz   **38.** 9 tons 1075 lb   **39.** 7.85 kg   **40.** 1.1625 kg   **41.** 8 qt   **42.** 5 c   **43.** 7 pt   **44.** 72 c
45. 4 qt 1 pt   **46.** 3 gal 3 qt   **47.** 3800 ml   **48.** 0.042 dl   **49.** 1.4 kl   **50.** 3060 cl   **51.** 1 gal 1 qt   **52.** 7 gal   **53.** 736 ml or 0.736 L
54. 15.5 L or 15,500 ml   **55.** 2 gal 3 qt   **56.** 6 fl oz   **57.** 10.88 L   **58.** yes   **59.** 22.96 ft   **60.** 10.55 m   **61.** 4.55 gal   **62.** 8.27 qt
63. 425.25 g   **64.** 10.35 kg   **65.** 109 yd   **66.** 180.4 lb   **67.** 3.18 qt   **68.** 2.36 in.   **69.** 473°F   **70.** 320°F   **71.** 107.6°F   **72.** 34°C
73. 5.2°C   **74.** 26.7°C   **75.** 1.7°C   **76.** 329°F   **77.** 67.2 ft-lb   **78.** 136.5 ft-lb   **79.** 108,000 ft-lb   **80.** 9,336,000 ft-lb   **81.** 2600 BTU
82. 1125 cal   **83.** 15,120 cal   **84.** $2\frac{1}{3}$ hr   **85.** 13,200 ft   **86.** 75 in.   **87.** 4.5 mi   **88.** 10.75 ft   **89.** 4 tons 200 lb   **90.** 2 tons 300 lb
91. 500 cm   **92.** 0.000286 km   **93.** 1.4 g   **94.** 0.24 g   **95.** 27 qt   **96.** 21 qt   **97.** 17 c   **98.** 12.5 c   **99.** 186.8°F   **100.** 59°F
101. 11°C   **102.** 28°C   **103.** 9117 m or 9.117 km   **104.** 8.353 km or 8353 m   **105.** 13.9 L or 13,900 ml   **106.** 35.7 L or 35,700 ml
107. 8.1 g or 8100 mg   **108.** 1.6 g or 1600 mg   **109.** 50.4 kg   **110.** 12.8 kg   **111.** 7 gal 3 qt   **112.** 8 gal 2 qt   **113.** 38 dm or 3800 mm
114. 58 dm or 5800 mm   **115.** 2.25 tons   **116.** 2.25 tons

Chapter 7 Test   **1.** 23 ft 4 in.   **2.** 10 qt   **3.** 1.875 lb   **4.** 5600 lb   **5.** $4\frac{3}{4}$ gal   **6.** 0.04 g   **7.** 2400 g   **8.** 36 mm   **9.** 0.43 g   **10.** 830 ml
11. 1 gal 2 qt   **12.** 3 lb 13 oz   **13.** 8 ft 3 in.   **14.** 2 gal 3 qt   **15.** 66 mm or 6.6 cm   **16.** 2.256 km or 2256 m   **17.** 28.9°C   **18.** 54.7°F
19. 5.6 m   **20.** 4 gal 3 qt   **21.** 91.4 m   **22.** 16 ft 6 in.   **23.** 493 ft 6 in.   **24.** 150.368 m   **25.** 56.7°C   **26.** 105.8°F   **27.** 37,336 pt
28. 0.2 oz   **29.** 3.1 mi   **30.** 18.95 L   **31.** 679 ft-lb   **32.** 20,228,000 ft-lb   **33.** 900 cal

Cumulative Review   **1.** 2010 (Sec. 1.3, Ex. 4)   **2.** 1531   **3.** $3 \cdot 3 \cdot 3 \cdot 5 \cdot 7$ or $3^3 \cdot 5 \cdot 7$ (Sec. 2.2, Ex. 5)   **4.** 153 sq in.
5. 33 (Sec. 3.2, Ex. 7)   **6.** $\frac{10}{63}$ **7.** $5\frac{1}{15}$ (Sec. 3.4, Ex. 2)   **8.** $10\frac{1}{3}$ **9.** $\frac{1}{8}$ (Sec. 4.1, Ex.10)   **10.** $1\frac{1}{5}$ **11.** $105\frac{83}{1000}$ (Sec. 4.1, Ex. 12)
12. $\frac{8}{27}$ **13.** < (Sec. 4.2, Ex. 2)   **14.** 25   **15.** 67.69 (Sec. 4.3, Ex. 6)   **16.** 139.231   **17.** 4.21 (Sec. 4.4, Ex. 8)   **18.** 186,040
19. 0.0092 (Sec. 4.4, Ex. 10)   **20.** 5.8   **21.** 0.26 (Sec. 4.5, Ex. 2)   **22.** $\frac{21}{20}$ or $1\frac{1}{20}$ **23.** 2.1875 (Sec. 4.6, Ex. 4)   **24.** 7.3
25. $0.\overline{6}$ (Sec. 4.6, Ex. 2)   **26.** 2.16   **27.** $\frac{12}{17}$ (Sec. 5.1, Ex. 1)   **28.** $\frac{12}{23}$ **29.** 22.5 mi/gal (Sec. 5.2, Ex. 5)   **30.** $\frac{1}{4}$ **31.** $n = \frac{35}{6}$ or $5\frac{5}{6}$
(Sec. 5.3, Ex. 6)   **32.** 7.5 cups   **33.** 7 bags (Sec. 5.4, Ex. 3)   **34.** $\frac{23}{100}$ **35.** 0.23 (Sec. 6.1, Ex. 3)   **36.** 87.5% or $87\frac{1}{2}$%
37. 8.33% (Sec. 6.2, Ex. 9)   **38.** 24%   **39.** 14 (Sec. 6.3, Ex. 7)   **40.** 32 pints   **41.** $\frac{75}{30} = \frac{p}{100}$ (Sec. 6.4, Ex. 5)   **42.** 860 cm
43. 18% (Sec. 6.5, Ex. 6)   **44.** 6.5 tons   **45.** discount: $16.25; sale price: $48.75 (Sec. 6.6, Ex. 5)   **46.** 52 oz   **47.** $120 (Sec. 6.7, Ex. 1)
48. 15%   **49.** $4\frac{1}{2}$ tons (Sec. 7.2, Ex. 1)   **50.** 77°F

Chapter 8 Geometry
Section 8.1

Vocabulary, Readiness & Video Check   **1.** plane   **3.** Space   **5.** ray   **7.** straight   **9.** acute   **11.** Parallel; intersecting   **13.** degrees
15. vertical   **17.** $\angle WUV$, $\angle VUW$, $\angle U$, $\angle x$ **19.** $180° - 17° = 163°$

Exercise Set 8.1   **1.** line; line CD or line l or $\overleftrightarrow{CD}$ **3.** line segment; line segment MN or $\overline{MN}$ **5.** angle; $\angle GHI$ or $\angle IHG$ or $\angle H$
7. ray; ray UW or $\overrightarrow{UW}$ **9.** $\angle CPR$, $\angle RPC$ **11.** $\angle TPM$, $\angle MPT$ **13.** straight   **15.** right   **17.** obtuse   **19.** acute   **21.** 67°   **23.** 163°
25. 32°   **27.** 30°   **29.** $\angle MNP$ and $\angle RNO$; $\angle PNQ$ and $\angle QNR$ **31.** $\angle SPT$ and $\angle TPQ$; $\angle SPR$ and $\angle RPQ$; $\angle SPT$ and $\angle SPR$; $\angle TPQ$
and $\angle QPR$ **33.** 27°   **35.** 132°   **37.** $m\angle x = 30°$; $m\angle y = 150°$; $m\angle z = 30°$ **39.** $m\angle x = 77°$; $m\angle y = 103°$; $m\angle z = 77°$
41. $m\angle x = 100°$; $m\angle y = 80°$; $m\angle z = 100°$ **43.** $m\angle x = 134°$; $m\angle y = 46°$; $m\angle z = 134°$ **45.** $\angle ABC$ or $\angle CBA$ **47.** $\angle DBE$ or
$\angle EBD$ **49.** 15°   **51.** 50°   **53.** 65°   **55.** 95°   **57.** $\frac{9}{8}$ or $1\frac{1}{8}$ **59.** $\frac{7}{32}$ **61.** $\frac{5}{6}$ **63.** $\frac{4}{3}$ or $1\frac{1}{3}$ **65.** 54.8°   **67.** false; answers may vary
69. true   **71.** $m\angle a = 60°$; $m\angle b = 50°$; $m\angle c = 110°$; $m\angle d = 70°$; $m\angle e = 120°$ **73.** no; answers may vary   **75.** 45°; 45°

Section 8.2

Vocabulary, Readiness & Video Check **1.** Because the sum of the measures of the angles of a triangle equals 180°, each angle in an equilateral triangle must measure 60°.

Exercise Set 8.2 **1.** pentagon **3.** hexagon **5.** quadrilateral **7.** pentagon **9.** equilateral **11.** scalene; right **13.** isosceles **15.** 25° **17.** 13° **19.** 40° **21.** diameter **23.** rectangle **25.** parallelogram **27.** hypotenuse **29.** 14 m **31.** 14.5 cm **33.** 40.6 cm **35.** 84 in. **37.** cylinder **39.** rectangular solid **41.** cone **43.** cube **45.** rectangular solid **47.** sphere **49.** pyramid **51.** 14.8 in. **53.** 13 mi **55.** 72,368 mi **57.** 108 **59.** 12.56 **61.** true **63.** true **65.** false **67.** yes; answers may vary **69.** answers may vary

Section 8.3

Vocabulary, Readiness & Video Check **1.** perimeter **3.** π **5.** $\frac{22}{7}$(or 3.14); 3.14 $\left(\text{or } \frac{22}{7}\right)$ **7.** Opposite sides of a rectangle have the same measure, so we can just find the sum of the measures of all four sides.

Exercise Set 8.3 **1.** 64 ft **3.** 120 cm **5.** 21 in. **7.** 48 ft **9.** 42 in. **11.** 155 cm **13.** 21 ft **15.** 624 ft **17.** 346 yd **19.** 22 ft **21.** \$55 **23.** 72 in. **25.** 28 in. **27.** \$36.12 **29.** 96 m **31.** 66 ft **33.** 74 cm **35.** 17π cm; 53.38 cm **37.** 16π mi; 50.24 mi **39.** 26π m; 81.64 m **41.** 15π ft; 47.1 ft **43.** 12,560 ft **45.** 30.7 mi **47.** 14π cm $\approx$ 43.96 cm **49.** 40 mm **51.** 84 ft **53.** 23 **55.** 1 **57.** 6 **59.** 10 **61. a.** width: 30 yd; length: 40 yd **b.** 140 yd **63.** b **65. a.** 62.8 m; 125.6 m **b.** yes **67.** answers may vary **69.** 27.4 m **71.** 75.4 m **73.** 6.5 ft

Section 8.4

Vocabulary, Readiness & Video Check **1.** The formula for the area of a rectangle; We split the L-shaped figure into two rectangles, used the area formula twice to find the area of each, and then added these two areas.

Exercise Set 8.4 **1.** 7 sq m **3.** $9\frac{3}{4}$ sq yd **5.** 15 sq yd **7.** 2.25π sq in. $\approx$ 7.065 sq in. **9.** 17.64 sq ft **11.** 28 sq m **13.** 22 sq yd **15.** $36\frac{3}{4}$ sq ft **17.** $22\frac{1}{2}$ sq in. **19.** 25 sq cm **21.** 86 sq mi **23.** 24 sq cm **25.** 36π sq in. $\approx$ $113\frac{1}{7}$ sq in. **27.** 168 sqft **29.** 128,775 sq ft **31.** 1π sq cm $\approx$ 3.14 sq cm **33.** 128 sq in.; $\frac{8}{9}$ sq ft **35.** 510 sq in. **37.** 168 sq ft **39.** 9200 sq ft **41. a.** 381 sq ft **b.** 4 squares **43.** 14π in. $\approx$ 43.96 in. **45.** 25 ft **47.** $12\frac{3}{4}$ ft **49.** perimeter **51.** area **53.** area **55.** perimeter **57.** 12-in. pizza **59.** $1\frac{1}{3}$ sq ft; 192 sq in. **61.** 7.74 sq in. **63.** 7056π sq in. $\approx$ 22,155.84 sq in. **65.** 298.5 sq m **67. a.** width: 40 yd; length: 60 yd **b.** 2400 sq yd **69.** no; answers may vary

Section 8.5

Vocabulary, Readiness & Video Check **1.** volume **3.** cubic **5.** perimeter **7.** Exact answers are in terms of π, and approximate answers use an approximation for π.

Exercise Set 8.5 **1.** 72 cu in. **3.** 512 cu cm **5.** $12\frac{4}{7}$ cu yd **7.** $523\frac{17}{21}$ cu in. **9.** $28\frac{2}{7}$ cu in. **11.** 75 cu cm **13.** $2\frac{10}{27}$ cu in. **15.** 8.4 cu ft **17.** $10\frac{5}{6}$ cu in. **19.** 960 cu cm **21.** $\frac{1372}{3}\pi$ cu in. or $457\frac{1}{3}\pi$ cu in. **23.** $7\frac{1}{2}$ cu ft **25.** 288π cu ft **27.** 5.25π cu in. **29.** 7.96 cu m **31.** $12\frac{4}{7}$ cu cm **33.** 8.8 cu in. **35.** 10.648 cu in. **37.** 25 **39.** 9 **41.** 5 **43.** 20 **45.** 2093.33 cu m **47.** no; answers may vary **49.** 5.5 cu ft; 5.8 cu ft; (b) is larger **51.** $6\frac{2}{3}\pi$ cu in. $\approx$ 21 cu in.

Integrated Review **1.** 153°; 63° **2.** $m\angle x = 75°$; $m\angle y = 105°$; $m\angle z = 75°$ **3.** $m\angle x = 128°$; $m\angle y = 52°$; $m\angle z = 128°$ **4.** $m\angle x = 52°$ **5.** 4.6 in. **6.** $4\frac{1}{4}$ in. **7.** 20 m; 25 sq m **8.** 12 ft; 6 sq ft **9.** 10π cm $\approx$ 31.4 cm; 25π sq cm $\approx$ 78.5 sq cm **10.** 32 mi; 44 sq mi **11.** 54 cm; 143 sq cm **12.** 62 ft; 238 sq ft **13.** 64 cu in. **14.** 30.6 cu ft **15.** 400 cu cm **16.** $4\frac{1}{2}\pi$ cu mi $\approx$ $14\frac{1}{7}$ cu mi

Section 8.6

Calculator Explorations **1.** 32 **3.** 5.568 **5.** 9.849

Vocabulary, Readiness & Video Check **1.** 10 **3.** squaring **5.** 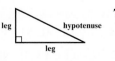 **7.** $\sqrt{49} = 7$ because 7^2 or $7 \cdot 7 = 49$.

9. The Pythagorean theorem works only for right triangles.

Exercise Set 8.6   **1.** 2   **3.** 11   **5.** $\frac{1}{9}$ **7.** $\frac{4}{8} = \frac{1}{2}$ **9.** 1.732   **11.** 3.873   **13.** 6.856   **15.** 5.099   **17.** 6, 7   **19.** 10, 11   **21.** 16   **23.** 9.592
25. $\frac{7}{12}$ **27.** 8.426   **29.** 13 in.   **31.** 6.633 cm   **33.** 52.802 m   **35.** 117 mm   **37.** 5   **39.** 12   **41.** 17.205   **43.** 44.822   **45.** 42.426
47. 1.732   **49.** 8.5   **51.** 141.42 yd   **53.** 25.0 ft   **55.** 340 ft   **57.** $n = 4$ **59.** $n = 45$ **61.** $n = 6$ **63.** 6   **65.** 10   **67.** answers may vary
69. yes   **71.** $\sqrt{80} - 6 \approx 2.94$ in.

Section 8.7

Vocabulary, Readiness & Video Check   **1.** false   **3.** true   **5.** false   **7.** $\angle M$ and $\angle Y$, $\angle N$ and $\angle X$, $\angle P$ and $\angle Z$, $\dfrac{p}{z} = \dfrac{m}{y} = \dfrac{n}{x}$
9. The ratios of corresponding sides are the same.

Exercise Set 8.7   **1.** congruent; SSS   **3.** not congruent   **5.** congruent; ASA   **7.** congruent; SAS   **9.** $\dfrac{2}{1}$ **11.** $\dfrac{3}{2}$ **13.** 4.5   **15.** 6
17. 5   **19.** 13.5   **21.** 17.5   **23.** 10   **25.** 28.125   **27.** 10   **29.** 520 ft   **31.** 500 ft   **33.** 60 ft   **35.** 14.4 ft   **37.** 52 neon tetras   **39.** 381 ft
41. 4.01   **43.** 1.23   **45.** $3\dfrac{8}{9}$ in.; no   **47.** 8.4   **49.** answers may vary   **51.** 200 ft, 300 ft, 425 ft

Chapter 8 Vocabulary Check   **1.** right triangle; hypotenuse; legs   **2.** line segment   **3.** complementary   **4.** line   **5.** perimeter
6. angle; vertex   **7.** Congruent   **8.** Area   **9.** ray   **10.** square root   **11.** transversal   **12.** straight   **13.** volume   **14.** vertical
15. adjacent   **16.** obtuse   **17.** right   **18.** acute   **19.** supplementary   **20.** Similar

Chapter 8 Review   **1.** right   **2.** straight   **3.** acute   **4.** obtuse   **5.** 65°   **6.** 75°   **7.** 58°   **8.** 98°   **9.** 90°   **10.** 25°   **11.** $\angle a$ and
$\angle b$; $\angle b$ and $\angle c$; $\angle c$ and $\angle d$; $\angle d$ and $\angle a$ **12.** $\angle x$ and $\angle w$; $\angle y$ and $\angle z$ **13.** $m\angle x = 100°$; $m\angle y = 80°$; $m\angle z = 80°$
14. $m\angle x = 155°$; $m\angle y = 155°$; $m\angle z = 25°$ **15.** $m\angle x = 53°$; $m\angle y = 53°$; $m\angle z = 127°$ **16.** $m\angle x = 42°$; $m\angle y = 42°$; $m\angle z = 138°$
17. 103°   **18.** 60°   **19.** 60°   **20.** 65°   **21.** $4\dfrac{1}{5}$ m   **22.** 7 ft   **23.** 9.5 m   **24.** $15\dfrac{1}{5}$ cm   **25.** cube   **26.** cylinder   **27.** pyramid
28. rectangular solid   **29.** 18 in.   **30.** 2.35 m   **31.** pentagon   **32.** hexagon   **33.** equilateral   **34.** isosceles, right   **35.** 89 m
36. 30.6 cm   **37.** 36 m   **38.** 90 ft   **39.** 32 ft   **40.** 440 ft   **41.** 5.338 in.   **42.** 31.4 yd   **43.** 240 sq ft   **44.** 140 sq m   **45.** 600 sq cm
46. 189 sq yd   **47.** 49π sq ft ≈ 153.86 sq ft   **48.** 82.81 sq m   **49.** 119 sq in.   **50.** 1248 sq cm   **51.** 144 sq m   **52.** 432 sq ft
53. 130 sq ft   **54.** $15\dfrac{5}{8}$ cu in.   **55.** 84 cu ft   **56.** $20{,}000\pi$ cu cm $\approx 62{,}800$ cu cm   **57.** $\dfrac{1}{6}\pi$ cu km $\approx \dfrac{11}{21}$ cu km   **58.** $2\dfrac{2}{3}$ cu ft
59. 307.72 cu in.   **60.** $7\dfrac{1}{2}$ cu ft   **61.** 0.5π cu ft or $\dfrac{1}{2}\pi$ cu ft   **62.** 8   **63.** 12   **64.** $\dfrac{2}{5}$ **65.** $\dfrac{1}{10}$ **66.** 13   **67.** 29   **68.** 10.7   **69.** 93
70. 127.3 ft   **71.** 88.2 ft   **72.** $37\dfrac{1}{2}$ **73.** $13\dfrac{1}{3}$ **74.** 17.4   **75.** 33 ft   **76.** $x = \dfrac{5}{6}$ in.; $y = 2\dfrac{1}{6}$ in.   **77.** 108°   **78.** 89°   **79.** 82°   **80.** 78°
81. 95°   **82.** 57°   **83.** 13 m   **84.** 12.6 cm   **85.** 22 dm   **86.** 27.3 in.   **87.** 194 ft   **88.** 1624 sq m   **89.** 9π sq m ≈ 28.26 sq m
90. $346\dfrac{1}{2}$ cu in.   **91.** 140 cu in.   **92.** 1260 cu ft   **93.** 28.728 cu ft   **94.** 1   **95.** 6   **96.** $\dfrac{4}{9}$ **97.** 86.6   **98.** 20.8   **99.** 48.1   **100.** 19.7
101. $6\dfrac{1}{2}$ **102.** 12

Chapter 8 Test   **1.** 12°   **2.** 56°   **3.** 57°   **4.** $m\angle x = 118°$; $m\angle y = 62°$; $m\angle z = 118°$ **5.** $m\angle x = 73°$; $m\angle y = 73°$; $m\angle z = 73°$
6. 6.2 m   **7.** $10\dfrac{1}{4}$ in.   **8.** 26°   **9.** circumference $= 18\pi$ in. ≈ 56.52 in.; area $= 81\pi$ sq in. ≈ 254.34 sq in.   **10.** perimeter $= 24.6$ yd;
area $= 37.1$ sq yd   **11.** perimeter $= 68$ in.; area $= 185$ sq in.   **12.** $62\dfrac{6}{7}$ cu in.   **13.** 30 cu ft   **14.** 7   **15.** 8.888   **16.** $\dfrac{8}{10} = \dfrac{4}{5}$ **17.** 16 in.
18. 18 cu ft   **19.** 62 ft; \$115.94   **20.** 5.66 cm   **21.** 198.08 oz   **22.** 7.5   **23.** approximately 69 ft

Cumulative Review   **1.** nineteen and five thousand twenty-three ten-thousandths (Sec. 4.1, Ex. 3)   **2.** $\dfrac{53}{66}$ **3.** 736.2 (Sec. 4.2, Ex. 5)
4. 700   **5.** 47.06 (Sec. 4.3, Ex. 3)   **6.** $\dfrac{20}{11}$ or $1\dfrac{9}{11}$ **7.** 76.8 (Sec. 4.4, Ex. 5)   **8.** $\dfrac{7}{66}$ **9.** 76,300 (Sec. 4.4, Ex. 7)   **10.** $\dfrac{23}{2}$ or $11\dfrac{1}{2}$
11. 38.6 (Sec. 4.5, Ex. 1)   **12.** 0.567   **13.** 3.7 (Sec. 4.5, Ex. 12)   **14.** $\dfrac{3}{5}$ or 0.6   **15.** > (Sec. 4.6, Ex. 7)   **16.** <   **17.** $\dfrac{26}{31}$ (Sec. 5.1, Ex. 5)
18. $\dfrac{16}{45}$ **19.** yes (Sec. 5.3, Ex. 2)   **20.** $\dfrac{35}{2}$ or $17\dfrac{1}{2}$ **21.** 25% (Sec. 6.1, Ex. 1)   **22.** 68%   **23.** $\dfrac{19}{1000}$ (Sec. 6.2, Ex. 2)   **24.** $\dfrac{13}{50}$ **25.** $\dfrac{5}{4}$ or
$1\dfrac{1}{4}$ (Sec. 6.2, Ex. 3)   **26.** $\dfrac{28}{5}$ or $5\dfrac{3}{5}$ **27.** 255 (Sec. 6.3, Ex. 8)   **28.** 15%   **29.** 52 (Sec. 6.4, Ex. 9)   **30.** 9   **31.** 775 freshmen (Sec. 6.5, Ex. 3)
32. \$2.25/sq ft   **33.** \$3210 (Sec. 6.6, Ex. 3)   **34.** 35 exercises   **35.** 96 in. (Sec. 7.1, Ex. 1)   **36.** 2 yd 2 ft 4 in.   **37.** 3200 g (Sec. 7.2, Ex. 7)
38. 0.07 m   **39.** 3 gal 3 qt (Sec. 7.3, Ex. 3)   **40.** 70,052   **41.** 84.2°F (Sec. 7.5, Ex. 2)   **42.** 12.5%   **43.** 50° (Sec. 8.2, Ex. 1)   **44.** 33 m
45. 28 in. (Sec. 8.3, Ex. 1)   **46.** 45 sq in.   **47.** $\dfrac{2}{5}$ (Sec. 8.6, Ex. 3)   **48.** $\dfrac{3}{4}$ **49.** $\dfrac{12}{19}$ (Sec. 8.7, Ex. 2)   **50.** $15\dfrac{5}{6}$

Chapter 9 Statistics and Probability

Section 9.1

Vocabulary, Readiness & Video Check **1.** bar **3.** line **5.** Count the number of symbols and multiply this number by how much each symbol stands for (from the key). **7.** bar graph

Exercise Set 9.1 **1.** Kansas **3.** 5.5 million or 5,500,000 acres **5.** South Dakota **7.** 4.0 million or 4,000,000 acres **9.** 66,000
11. 2006 **13.** 12,000 **15.** 66,000 wildfires/year **17.** September **19.** 76 **21.** $\frac{1}{38}$ **23.** Tokyo, Japan; about 34.8 million or
34,800,000 **25.** New York, U.S.; 21.6 million or 21,600,000 **27.** approximately 2 million or 2,000,000

29. **31.** **33.** 15 adults **35.** 61 adults **37.** 24 adults **39.** 12 adults
41. $\frac{9}{100}$ **43.** 20 to 44 **45.** 109 million **47.** 23 million **49.** answers may vary
51. $|$; 1 **53.** $\lvert\lvert\lvert\lvert\lvert\,\lvert\lvert\lvert$; 8 **55.** $\lvert\lvert\lvert\lvert\lvert\,\lvert$; 6 **57.** $\lvert\lvert\lvert\lvert\lvert\,\lvert$; 6 **59.** $\lvert\lvert$; 2
61. **63.** 50 **65.** 2011, 2013 **67.** 2006, 2008
69. 2006, 2008, 2012 **71.** 3.6 **73.** 6.2 **75.** 25%
77. 34% **79.** 83°F **81.** Sunday; 68°F
83. Tuesday; 13°F **85.** answers may vary

Section 9.2

Vocabulary, Readiness & Video Check **1.** circle **3.** 360 **5.** 100%

Exercise Set 9.2 **1.** parent or guardian's home **3.** $\frac{9}{35}$ **5.** $\frac{9}{16}$ **7.** Asia **9.** 37% **11.** 17,100,000 sq mi **13.** 2,850,000 sq mi
15. 55% **17.** nonfiction **19.** 31,400 books **21.** 27,632 books **23.** 25,120 books

25. **27.** **29.** $2^2 \times 5$ **31.** $2^3 \times 5$ **33.** 5×17 **35.** Pacific;
answers may vary **37.** 129,600,002 sq km
39. 55,542,858 sq km **41.** 602 respondents
43. 2276 respondents **45.** $\frac{301}{837}$
47. no; answers may vary

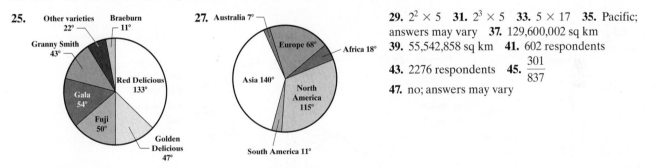

Integrated Review **1.** 700,000 **2.** 500,000 **3.** registered nurses, retail salespersons, home health aides **4.** food workers
5. Oroville Dam; 755 ft **6.** New Bullards Bar Dam; 635 ft **7.** 15 ft **8.** 4 dams **9.** Thursday and Saturday; 100°F **10.** Monday; 82°F
11. Sunday, Monday, and Tuesday **12.** Wednesday, Thursday, Friday, and Saturday **13.** 70 qt containers **14.** 52 qt containers
15. 2 qt containers **16.** 6 qt containers **17.** $\lvert\lvert$; 2 **18.** $\lvert$; 1 **19.** $\lvert\lvert\lvert$; 3 **20.** $\lvert\lvert\lvert\lvert\lvert\,\lvert$; 6 **21.** $\lvert\lvert\lvert\lvert\lvert$; 5

22.

Section 9.3

Vocabulary, Readiness & Video Check **1.** average **3.** mean (or average) **5.** grade point average **7.** Place the data numbers in numerical order (or verify that they already are).

Exercise Set 9.3 **1.** mean: 21; median: 23; no mode **3.** mean: 8.1; median: 8.2; mode: 8.2 **5.** mean: 0.5; median: 0.5; mode: 0.2 and 0.5
7. mean: 370.9; median: 313.5; no mode **9.** 1911.6 ft **11.** 1601 ft **13.** answers may vary **15.** 2.79 **17.** 3.64 **19.** 6.8
21. 6.9 **23.** 88.5 **25.** 73 **27.** 70 and 71 **29.** 9 rates **31.** $\frac{1}{3}$ **33.** $\frac{3}{5}$ **35.** $\frac{11}{15}$ **37.** 35, 35, 37, 43 **39.** yes; answers may vary

Section 9.4

Vocabulary, Readiness & Video Check   1. outcome   **3.** probability   **5.** 0   **7.** The number of outcomes equals the ending number of branches drawn.

Exercise Set 9.4   1.

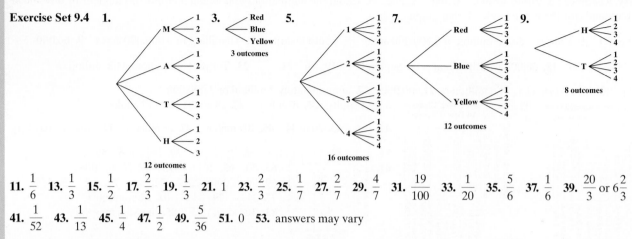

11. $\frac{1}{6}$ **13.** $\frac{1}{3}$ **15.** $\frac{1}{2}$ **17.** $\frac{2}{3}$ **19.** $\frac{1}{3}$ **21.** 1   **23.** $\frac{2}{3}$ **25.** $\frac{1}{7}$ **27.** $\frac{2}{7}$ **29.** $\frac{4}{7}$ **31.** $\frac{19}{100}$ **33.** $\frac{1}{20}$ **35.** $\frac{5}{6}$ **37.** $\frac{1}{6}$ **39.** $\frac{20}{3}$ or $6\frac{2}{3}$

41. $\frac{1}{52}$ **43.** $\frac{1}{13}$ **45.** $\frac{1}{4}$ **47.** $\frac{1}{2}$ **49.** $\frac{5}{36}$ **51.** 0   **53.** answers may vary

Chapter 9 Vocabulary Check   1. bar   **2.** mean   **3.** outcomes   **4.** pictograph   **5.** mode   **6.** line   **7.** median   **8.** tree diagram   **9.** experiment   **10.** circle   **11.** probability   **12.** histogram; class interval; class frequency

Chapter 9 Review   1. 1,250,000   **2.** 1,500,000   **3.** South   **4.** Northeast   **5.** South   **6.** Northeast, Midwest, West   **7.** 12%   **8.** 2012   **9.** 1992, 2002, 2012   **10.** answers may vary   **11.** 962   **12.** 927   **13.** 2000 and 2004   **14.** 1996 and 2000   **15.** 27   **16.** 120   **17.** 1 employee   **18.** 4 employees   **19.** 18 employees   **20.** 9 employees   **21.** ⊬⊬⊬; 5   **22.** |||; 3   **23.** ||||; 4

24.   **25.** mortgage payment   **26.** utilities   **27.** $1225   **28.** $700   **29.** $\frac{39}{160}$ **30.** $\frac{7}{40}$ **31.** 30   **32.** 12   **33.** 21

34. 1   **35.** mean: 17.8; median: 14; no mode   **36.** mean: 58.1; median: 60; mode: 45 and 86   **37.** mean: 24,500; median: 20,000; mode: 20,000   **38.** mean: 447.3; median: 420; mode: 400   **39.** 3.25   **40.** 2.57

46. $\frac{1}{6}$ **47.** $\frac{1}{6}$ **48.** $\frac{1}{5}$ **49.** $\frac{1}{5}$ **50.** $\frac{3}{5}$ **51.** $\frac{2}{5}$ **52.** $\frac{1}{2}$ **53.** $\frac{1}{2}$ **54.** mean: 74.4; median: 73; mode: none   **55.** mean: 48.8; median: 32;

mode: none   **56.** mean: 454; median: 463.5; mode: 500   **57.** mean: 619.17; median: 647.5; mode: 327   **58.** $\frac{1}{4}$ **59.** $\frac{3}{8}$ **60.** $\frac{1}{4}$ **61.** $\frac{1}{8}$

Chapter 9 Test **1.** $225 **2.** 3rd week; $350 **3.** $1100 **4.** June, August, September **5.** February; 3 cm **6.** March and November

7.
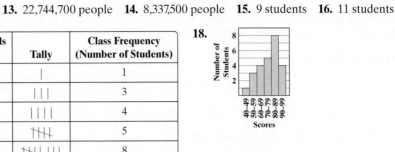
Most Common Blood Types by Percent in the Population

8. 3.0% **9.** 2004, 2005, 2007 **10.** 2003–2004, 2004–2005, 2006–2007, 2008–2009, 2010–2011

11. $\frac{17}{40}$ **12.** $\frac{31}{22}$ **13.** 22,744,700 people **14.** 8,337,500 people **15.** 9 students **16.** 11 students

17.

Class Intervals (Scores)	Tally	Class Frequency (Number of Students)
40–49	|	1
50–59	|||	3
60–69	||||	4
70–79	⊬⊢⊢	5
80–89	⊬⊢⊢ |||	8
90–99	||||	4

18. (bar graph: Number of Students vs Scores)

19. mean: 38.4; median: 42; no mode **20.** mean: 12.625; median: 12.5; mode: 12 and 16 **21.** 3.07

22.
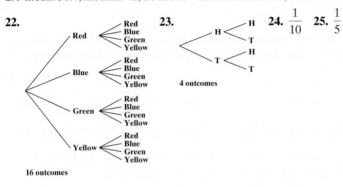
16 outcomes

23. (tree diagram) 4 outcomes **24.** $\frac{1}{10}$ **25.** $\frac{1}{5}$

Cumulative Review **1.** 28 (Sec. 1.9, Ex. 14) **2.** 12 **3.** $\frac{5}{18}$ (Sec. 2.3, Ex. 4) **4.** $\frac{7}{40}$ **5.** $8\frac{3}{10}$ (Sec. 3.4, Ex. 3) **6.** $\frac{34}{3}$ or $11\frac{1}{3}$

7. 8.4 sq ft (Sec. 4.6, Ex. 10) **8.** 10 m **9.** $\frac{3}{2}$ (Sec. 5.1, Ex. 4) **10.** $\frac{7}{18}$ **11.** $\frac{\$180}{1 \text{ week}}$ (Sec. 5.2, Ex. 2) **12.** $\frac{68 \text{ mi}}{1 \text{ hr}}$ or 68 mi/hr

13. $\frac{45 \text{ mi}}{2 \text{ gal}}$ (Sec. 5.2, Ex. 3) **14.** $\frac{39 \text{ files}}{2 \text{ hr}}$ **15.** yes (Sec. 5.3, Ex. 4) **16.** yes **17.** 22.4 cc (Sec. 5.4, Ex. 2) **18.** 262.5 mi

19. 0.046 (Sec. 6.1, Ex. 4) **20.** 0.0029 **21.** 1.9 (Sec. 6.1, Ex. 5) **22.** 4.52 **23.** $\frac{2}{5}$ (Sec. 6.2, Ex. 1) **24.** $\frac{27}{100}$ **25.** $\frac{1}{3}$ (Sec. 6.2, Ex. 4)

26. $\frac{107}{175}$ **27.** $5 = n \cdot 20$ (Sec. 6.3, Ex. 1) **28.** $\frac{5}{20} = \frac{p}{100}$ **29.** sales tax: $6.41; total price: $91.91 (Sec. 6.6, Ex. 1) **30.** $1610

31. $2400 (Sec. 6.7, Ex. 3) **32.** 33.75 **33.** $2\frac{1}{3}$ yd (Sec. 7.1, Ex. 2) **34.** 5000 pounds **35.** 4 lb 11 oz (Sec. 7.2, Ex. 5) **36.** 7

37. 0.0235 g (Sec. 7.2, Ex. 8) **38.** 1060 mm **39.** 3.21 L (Sec. 7.3, Ex. 6) **40.** 500 cm **41.** 59°F (Sec. 7.5, Ex. 1) **42.** 5°C **43.** 42° (Sec. 8.1, Ex. 4) **44.** 132° **45.** $\frac{1}{6}$ (Sec. 8.6, Ex. 2) **46.** $\frac{1}{5}$ **47.** 14 and 77 (Sec. 9.3, Ex. 5) **48.** 56 **49.** $\frac{1}{4}$ (Sec. 9.4, Ex. 3) **50.** $\frac{3}{5}$

Chapter 10 Signed Numbers

Section 10.1

Vocabulary, Readiness & Video Check **1.** integers **3.** inequality symbols **5.** is less than; is greater than **7.** absolute value **9.** number of feet a miner works underground **11.** negative **13.** opposite; opposite **15.** Eyre

Exercise Set 10.1 **1.** −1235 **3.** +14,433 **5.** +125 **7.** −13,215 **9.** −61,303,000 **11.** −160, −147; Guillermo **13.** −11

15. (number line, point at $-1\frac{1}{2}$) **17.** (number line, point at −4.7) **19.** < **21.** > **23.** > **25.** > **27.** < **29.** <

31. > **33.** < **35.** 9 **37.** 8 **39.** 0 **41.** 5 **43.** 8.1 **45.** $\frac{1}{2}$ **47.** $\frac{3}{8}$ **49.** 7.6 **51.** −5 **53.** 4 **55.** −23.6 **57.** $\frac{9}{16}$ **59.** 0.7

61. $-\frac{17}{18}$ **63.** 7 **65.** −20 **67.** −3 **69.** 8 **71.** 14 **73.** 29 **75.** 31; −31 **77.** −8.4; 8.4 **79.** Caspian Sea **81.** Lake Superior

83. iodine **85.** oxygen **87.** 13 **89.** 35 **91.** 360 **93.** $-|-8|, -|3|, 2^2, -(-5)$ **95.** $-|-6|, -|1|, -1, -(-6)$ **97.** $-10, -|-9|,$ $-(-2), |-12|, 5^2$ **99.** d **101.** 5 **103.** false **105.** true **107.** false **109.** answers may vary **111.** no; answers may vary

Section 10.2

Calculator Explorations   1. -159 **3.** 44   **5.** $-894{,}855$

Vocabulary, Readiness & Video Check   1. 0   **3.** a **5.** Negative; the numbers have different signs, so the sign of the sum is the same as the sign of the number with the larger absolute value, -9.

Exercise Set 10.2   1.

7. 35   **9.** -8 **11.** 0   **13.** 4   **15.** 2   **17.** -2 **19.** -11 **21.** 0   **23.** 7   **25.** -3 **27.** -9 **29.** -24 **31.** -57 **33.** 30   **35.** 20
37. -223 **39.** 151   **41.** -8.5 **43.** 4.6   **45.** $-\dfrac{8}{11}$ **47.** $-\dfrac{5}{6}$ **49.** $-\dfrac{7}{10}$ **51.** $5\dfrac{5}{8}$ **53.** -7 **55.** -24.6 **57.** 16   **59.** 13   **61.** 17
63. -10 **65.** $0 + (-215) + (-16) = -231$; 231 ft below the surface   **67.** Team 1: 7; Team 2: 6; winning team: Team 1   **69.** Park: -3;
Stanford: $+3$ **71.** Park: -5; Stanford: $+1$ **73.** $-\$39{,}000{,}000$ **75.** $\$2{,}054{,}000{,}000$ **77.** $2°C$ **79.** $\$13{,}609$ **81.** $-47°F$
83. -7679 m   **85.** 44   **87.** 72.01   **89.** 141   **91.** answers may vary   **93.** -3 **95.** -22 **97.** true   **99.** false   **101.** answers may vary

Section 10.3

Vocabulary, Readiness & Video Check   1. b   **3.** d   **5.** 0   **7.** 0   **9.** When subtracting, change the operation to addition and write
down the opposite of the second number. The second number is -8, therefore its opposite is 8.   **11.** The warmest temperature is
$263°F$ warmer than the coldest temperature.

Exercise Set 10.3   1. 0   **3.** 5   **5.** -5 **7.** 14   **9.** 3   **11.** -18 **13.** -14 **15.** -3.3 **17.** 4.02   **19.** $\dfrac{2}{5}$ **21.** $-\dfrac{3}{10}$ **23.** $\dfrac{5}{6}$ **25.** -38
27. -17 **29.** 13   **31.** -38 **33.** -11 **35.** -12.7 **37.** $\dfrac{3}{7}$ **39.** 2   **41.** 0   **43.** -1 **45.** -27 **47.** 40   **49.** -51 **51.** $15°F$ **53.** $4°F$
55. $263°F$ **57.** $-\$16$ **59.** $-10°C$ **61.** 154 ft   **63.** 69 ft   **65.** 652 ft   **67.** 144 ft   **69.** $1197°F$ **71.** $-\$534$ billion   **73.** 0   **75.** 8
77. 1058   **79.** answers may vary   **81.** 16   **83.** -20 **85.** -4 **87.** 0   **89.** -14 **91.** false   **93.** answers may vary

Integrated Review   1. $+29{,}028$ **2.** $-35{,}840$ **3.** -7 **4.** $+28{,}250$ **5.**   **6.** $>$ **7.** $>$ **8.** $<$
9. $>$ **10.** $<$ **11.** $>$ **12.** $>$ **13.** $<$ **14.** 1   **15.** 1   **16.** 0   **17.** -4 **18.** 8.6   **19.** 100.3   **20.** $\dfrac{3}{4}$ **21.** $-\dfrac{3}{4}$ **22.** -6 **23.** 13
24. -89.1 **25.** $\dfrac{2}{9}$ **26.** 5   **27.** -20.8 **28.** $\dfrac{2}{9}$ **29.** -2 **30.** $\dfrac{3}{4}$ **31.** -4 **32.** -9.6 **33.** -6 **34.** 12   **35.** 120   **36.** -3.44
37. $-\dfrac{11}{10}$ or $-1\dfrac{1}{10}$ **38.** 6   **39.** 1   **40.** -19 **41.** -12 **42.** -4 **43.** -44 **44.** -1.5 **45.** $191°F$ **46.** $219°F$

Section 10.4

Vocabulary, Readiness & Video Check   1. negative   **3.** positive   **5.** 0   **7.** undefined   **9.** multiplication

Exercise Set 10.4   1. 6   **3.** -36 **5.** -3.12 **7.** 0   **9.** $\dfrac{6}{35}$ **11.** -48 **13.** -8 **15.** 80   **17.** -150 **19.** 4   **21.** -25 **23.** -27 **25.** -8
27. 81   **29.** $\dfrac{25}{121}$ **31.** -4 **33.** -5 **35.** 8   **37.** 0   **39.** -13 **41.** -26 **43.** $\dfrac{25}{12}$ or $2\dfrac{1}{12}$ **45.** -5 **47.** -50 **49.** 3   **51.** -300
53. $-\dfrac{4}{5}$ **55.** 0   **57.** -125 **59.** -50 **61.** 43.4   **63.** $-\dfrac{32}{27}$ or $-1\dfrac{5}{27}$ **65.** $-\dfrac{3}{2}$ or $-1\dfrac{1}{2}$ **67.** -6 **69.** 252
71. $3 \cdot (-4) = 12$: a loss of 12 yd   **73.** $5 \cdot (-20) = -100$: a depth of 100 feet   **75.** $-210°C$ **77.** $-189°C$ **79.** $-\$896$ million
81. a. 393 condors   **b.** 15 condors per year   **83.** 300   **85.** 225   **87.** 109   **89.** 8   **91.** -29 **93.** -19 **95.** -8 **97.** false
99. true   **101.** negative   **103.** $\dfrac{-20}{41}, \dfrac{20}{-41}$ **105.** $(-5)^{17}, (-2)^{17}, (-2)^{12}, (-5)^{12}$ **107.** answers may vary

Section 10.5

Calculator Explorations   1. 48   **3.** -258

Vocabulary, Readiness & Video Check   1. division   **3.** average   **5.** subtraction   **7.** Finding the average is a good application of
both the order of operations and adding and dividing signed numbers.

Exercise Set 10.5   1. 3   **3.** 4   **5.** -7 **7.** -14 **9.** $-\dfrac{2}{9}$ **11.** -8 **13.** 4   **15.** 1   **17.** 4   **19.** -3 **21.** -55 **23.** -4 **25.** 16   **27.** 15
29. -3 **31.** -3 **33.** 4   **35.** 16   **37.** -27 **39.** 34   **41.** 0.65   **43.** -59 **45.** -7 **47.** $\dfrac{5}{36}$ **49.** -11 **51.** 36   **53.** -117 **55.** 30

57. −3 **59.** −2.21 **61.** −33 **63.** $-\dfrac{7}{5}$ **65.** −2 **67.** −19 **69.** −13.75°F **71.** no; answers may vary **73.** 4050 **75.** 45 **77.** 32 in.
79. 30 ft **81.** $2 \cdot (7 - 5) \cdot 3$ **83.** $-6 \cdot (10 - 4)$ **85.** answers may vary **87.** answers may vary **89.** 20,736

Chapter 10 Vocabulary Check 1. opposites **2.** signed **3.** absolute value **4.** integers **5.** additive inverse **6.** negative
7. positive **8.** inequality symbols **9.** average **10.** is less than; is greater than

Chapter 10 Review 1. −1435 **2.** +7562 **3.** [number line graph, −3.5 marked between −4 and −3; points at −2, 1, 3; scale −5 to 5] **4.** [number line graph, $-4\frac{1}{3}$ and −1.6 marked; scale −7 to 3] **5.** >

6. < **7.** > **8.** 12 **9.** 0 **10.** $\dfrac{7}{8}$ **11.** 12 **12.** $-\dfrac{1}{2}$ **13.** 3.9 **14.** −7 **15.** true **16.** true **17.** 2 **18.** 14 **19.** 4 **20.** 17 **21.** −23
22. −22 **23.** −2.1 **24.** −7 **25.** $\dfrac{4}{15}$ **26.** $-\dfrac{5}{9}$ **27.** −151 **28.** −606 **29.** −20°C **30.** −150 ft **31.** −15 **32.** −17 **33.** −8
34. −26 **35.** $-\dfrac{5}{4}$ or $-1\dfrac{1}{4}$ **36.** $-\dfrac{11}{10}$ or $-1\dfrac{1}{10}$ **37.** 20 **38.** 8 **39.** −4 **40.** −16 **41.** 1.2 **42.** 0.7 **43.** $-\dfrac{5}{18}$ **44.** $-\dfrac{3}{14}$
45. −$25 **46.** +692 ft **47.** 121 **48.** 81 **49.** −81 **50.** −121 **51.** $-\dfrac{64}{125}$ **52.** $-\dfrac{8}{27}$ **53.** 21 **54.** −18 **55.** −26 **56.** −19
57. −36 **58.** −40 **59.** $-\dfrac{7}{12}$ **60.** 6 **61.** undefined **62.** 0 **63.** 1.2 **64.** $\dfrac{4}{5}$ **65.** 15 **66.** −7 **67.** −16 **68.** −44 **69.** −14.8
70. 44 **71.** 7 **72.** 7 **73.** 39 **74.** −41 **75.** $-\dfrac{23}{9}$ **76.** $-\dfrac{89}{100}$ **77.** −2 **78.** −12 **79.** 16 **80.** −16 **81.** −15 **82.** −19 **83.** 48
84. −21 **85.** 2.1 **86.** $-\dfrac{1}{4}$ **87.** $\dfrac{6}{5}$ **88.** $-\dfrac{3}{2}$ **89.** $-\dfrac{1}{40}$ **90.** −1.35 **91.** −$9 **92.** 6°C **93.** 13,118 ft **94.** −27°C **95.** 2
96. 4 **97.** 3 **98.** 37 **99.** −5 **100.** −25 **101.** −20 **102.** 17

Chapter 10 Test 1. [number line graph, −6.5 marked, $-1\frac{2}{3}$ marked; points at −5, 2; scale −7 to 7] **2.** $-\dfrac{2}{3}$ **3.** 2.9 **4.** 98 **5.** −98 **6.** > **7.** > **8.** 3 **9.** −6
10. −10 **11.** 4 **12.** $-\dfrac{1}{2}$ **13.** 12 **14.** 65 **15.** 5 **16.** 12 **17.** −11 **18.** −46 **19.** −117 **20.** 128 **21.** 2 **22.** −5 **23.** $-\dfrac{14}{81}$
24. 19 **25.** −3 **26.** $-\dfrac{1}{45}$ **27.** $-\dfrac{9}{20}$ **28.** 88 ft below sea level **29.** 45 **30.** 31,642 **31.** 3820 ft below sea level **32.** −4

Cumulative Review 1. 78,875 (Sec. 1.6, Ex. 5) **2.** $\dfrac{25}{44}$ **3.** $\dfrac{4}{5}$ (Sec. 2.5, Ex. 7) **4.** 236 **5.** $\dfrac{17}{21}$ (Sec. 3.3, Ex. 3) **6.** $1\dfrac{11}{14}$ **7.** 48.26
(Sec. 4.1, Ex. 6) **8.** 0.08 **9.** 6.095 (Sec. 4.1, Ex. 7) **10.** 56,321 **11.** 3.432 (Sec. 4.3, Ex. 5) **12.** 13.5 **13.** 28.4405 (Sec. 4.5, Ex. 13)
14. 20 **15.** $\dfrac{4500 \text{ dollars}}{1 \text{ month}}$ or 4500 dollars/month (Sec. 5.2, Ex. 4) **16.** $\dfrac{60 \text{ miles}}{1 \text{ hour}}$ or 60 miles/hour **17.** $n = 2$ (Sec. 5.3, Ex. 5)
18. $n = \dfrac{48}{7}$ **19.** 46% (Sec. 6.1, Ex. 2) **20.** 80% **21.** 45% (Sec. 6.2, Ex. 6) **22.** 106% **23.** 150% (Sec. 6.2, Ex. 8) **24.** 500%
25. 200 (Sec. 6.3, Ex. 10) **26.** 242 **27.** $\dfrac{101}{200} = \dfrac{p}{100}$ (Sec. 6.4, Ex. 2) **28.** $101 = n \cdot 200$ **29.** $160 (Sec. 6.7, Ex. 2) **30.** $525
31. 9 ft 1 in. (Sec. 7.1, Ex. 6) **32.** 68 m **33.** 37°C (Sec. 7.5, Ex. 5) **34.** 720 cm **35.** 73° (Sec. 8.1, Ex. 5) **36.** 56° **37.** 60° (Sec. 8.2, Ex. 2)
38. 49° **39.** 5.1 sq mi (Sec. 8.4, Ex. 2) **40.** 25π sq m $\approx$ 78.5 sq m **41.** 36π cu in. $\approx 113\dfrac{1}{7}$ cu in. (Sec. 8.5, Ex. 2) **42.** 24 cu cm
43. 57 (Sec. 9.3, Ex. 3) **44.** 24.5 **45.** $\dfrac{1}{3}$ (Sec. 9.4, Ex. 4) **46.** −4.9 **47.** −23 (Sec. 10.2, Ex. 3) **48. a.** −1000 **b.** −1000
49. 82 (Sec. 10.5, Ex. 1) **50.** −4

Chapter 11 Introduction to Algebra
Section 11.1

Vocabulary, Readiness & Video Check 1. expression; term **3.** combine like terms **5.** variable; constant **7.** associative
9. numerical coefficient **11.** unlike **13.** like **15.** unlike **17.** like **19.** multiplication **21.** the distributive property first, then
the associative property of multiplication **23.** addition; multiplication; P = perimeter, A = area

Exercise Set 11.1   **1.** -3 **3.** -2 **5.** 4 **7.** -3 **9.** -10 **11.** 133 **13.** -15 **15.** -4 **17.** $-\dfrac{4}{3}$ or $-1\dfrac{1}{3}$ **19.** -12 **21.** $\dfrac{3}{2}$ or $1\dfrac{1}{2}$

23. -10.6 **25.** $8x$ **27.** $-4n$ **29.** $-2c$ **31.** $-4x$ **33.** $13a - 8$ **35.** $-0.9x + 11.2$ **37.** $2x - 7$ **39.** $-5x + 4y - 5$ **41.** $\dfrac{1}{2} - \dfrac{53}{60}x$

43. $-4.8m - 4.1$ **45.** $30x$ **47.** $-22y$ **49.** $-4.2a$ **51.** $-4a$ **53.** $2y + 4$ **55.** $15a - 40$ **57.** $-12x - 28$ **59.** $6x - 0.12$

61. $-4x - \dfrac{3}{2}$ **63.** $2x - 9$ **65.** $27n - 20$ **67.** $7w + 15$ **69.** $-11x + 8$ **71.** $(14y + 22)$ m   **73.** $(11a + 12)$ ft   **75.** $(-25x + 55)$ in.

77. $36y$ sq in.   **79.** $(32x - 64)$ sq km   **81.** $(60y + 20)$ sq mi   **83.** 4700 sq ft   **85.** 64 ft   **87.** $(3x + 6)$ ft   **89.** $\$360$ **91.** 78.5 sq ft

93. $23°$F   **95.** 288 cu in.   **97.** $91.2x$ cu in.   **99.** -3 **101.** 8 **103.** 0 **105.** incorrect; $15x - 10$ **107.** incorrect; $7x - x - 2$ or $6x - 2$

109. distributive   **111.** associative   **113.** answers may vary   **115.** $(20x + 16)$ sq mi   **117.** $23{,}506.2$ sq in.   **119.** $4824q + 12{,}274$

Section 11.2

Vocabulary, Readiness & Video Check   **1.** equivalent   **3.** simplifying   **5.** addition   **7.** We can add the same number to both sides of an equation or subtract the same number from both sides of an equation and we'll have an equivalent equation.

Exercise Set 11.2   **1.** yes   **3.** no   **5.** yes   **7.** yes   **9.** 18 **11.** -8 **13.** 9 **15.** -16 **17.** 3 **19.** $\dfrac{1}{8}$ **21.** 6 **23.** 8 **25.** 5.3 **27.** -1

29. -20.1 **31.** 2 **33.** 0 **35.** -28 **37.** -6 **39.** 24 **41.** -30 **43.** 12 **45.** 1 **47.** 1 **49.** 1 **51.** subtract $\dfrac{2}{3}$ from both sides

53. add $\dfrac{4}{5}$ to both sides   **55.** answers may vary   **57.** $162{,}964$ **59.** 1537 yd   **61.** $\$19{,}279{,}000{,}000$

Section 11.3

Vocabulary, Readiness & Video Check   **1.** equivalent   **3.** simplifying   **5.** multiplication   **7.** Simplified each side of the equation by combining like terms.

Exercise Set 11.3   **1.** 4 **3.** -4 **5.** -30 **7.** -17 **9.** 50 **11.** 25 **13.** -30 **15.** $\dfrac{1}{3}$ **17.** $-\dfrac{2}{3}$ **19.** -4 **21.** 8 **23.** 1.3 **25.** 2

27. 0 **29.** -0.05 **31.** $-\dfrac{15}{64}$ **33.** $\dfrac{2}{3}$ **35.** 6 **37.** 0 **39.** -2 **41.** -8 **43.** 1 **45.** 72 **47.** 35 **49.** -28 **51.** 25 **53.** 1 **55.** -10

57. 2006 **59.** 9.3 million acres or $9{,}300{,}000$ acres   **61.** addition   **63.** division   **65.** answers may vary   **67.** 6.5 hr   **69.** 58.8 mph
71. -3648 **73.** $67{,}896$ **75.** -48

Integrated Review   **1.** expression   **2.** equation   **3.** equation   **4.** expression   **5.** simplify   **6.** solve   **7.** 4 **8.** -6 **9.** 1 **10.** -4
11. $8x$ **12.** $-4y$ **13.** $-2a - 2$ **14.** $-2x + 3y - 7$ **15.** $-8x - 14$ **16.** $-6x + 30$ **17.** $5y - 10$ **18.** $15x - 31$ **19.** $(12x - 6)$ sq m

20. $20y$ in.   **21.** 13 **22.** -9 **23.** $\dfrac{7}{10}$ **24.** 0 **25.** -4 **26.** 25 **27.** -1 **28.** -3 **29.** 6 **30.** 8 **31.** $-\dfrac{9}{11}$ **32.** 5

Section 11.4

Calculator Explorations   **1.** yes   **3.** no   **5.** yes

Vocabulary, Readiness & Video Check   **1.** $3x - 9 + x - 16$; $5(2x + 6) - 1 = 39$ **3.** addition   **5.** distributive   **7.** the addition property of equality; to make sure we get an equivalent equation   **9.** gives; amounts to

Exercise Set 11.4   **1.** 3 **3.** 1.9 **5.** -4 **7.** 100 **9.** -3.9 **11.** -4 **13.** -12 **15.** -3 **17.** -1 **19.** -45 **21.** -9 **23.** 6 **25.** -5
27. 5 **29.** 8 **31.** 0 **33.** -22 **35.** 6 **37.** -11 **39.** -7 **41.** -5 **43.** 5 **45.** -3 **47.** 2 **49.** -4 **51.** -1 **53.** 4 **55.** 3 **57.** -1

59. 8 **61.** 4 **63.** 3 **65.** 64 **67.** $\dfrac{1}{9}$ **69.** $\dfrac{3}{2}$ **71.** -54 **73.** 2 **75.** $\dfrac{9}{5}$ **77.** 270 **79.** 0 **81.** 4 **83.** 1 **85.** $-42 + 16 = -26$

87. $-5(-29) = 145$ **89.** $3(-14 - 2) = -48$ **91.** $\dfrac{100}{2(50)} = 1$ **93.** 97 or 98 million returns   **95.** 9 million returns   **97.** b   **99.** a

101. $6x - 10 = 5x - 7$ **103.** 0 **105.** -4 **107.** no; answers may vary   **109.** $123.26°$F   **111.** $-9.4°$F
        $6x = 5x + 3$
          $x = 3$

Section 11.5

Vocabulary, Readiness & Video Check   **1.** decreased by   **3.** The phrase is "three times the difference of a number and 5." The "difference of a number and 5" translates to the expression $x - 5$, and in order to multiply 3 times this expression, we insert parentheses around the expression.

Exercise Set 11.5   **1.** $x + 5$ **3.** $x + 8$ **5.** $20 - x$ **7.** $512x$ **9.** $\dfrac{x}{2}$ **11.** $17 + x + 5x$ **13.** $-5 + x = -7$ **15.** $3x = 27$

17. $-20 - x = 104$ **19.** $5x$ **21.** $11 - x$ **23.** $2x = 108$ **25.** $50 - 8x$ **27.** $5(-3 + x) = -20$ **29.** $9 + 3x = 33$; 8

31. $3 + 4 + x = 16$; 9 **33.** $x - 3 = \dfrac{10}{5}$; 5 **35.** $30 - x = 3(x + 6)$; 3 **37.** $5x - 40 = x + 8$; 12 **39.** $3(x - 5) = \dfrac{108}{12}$; 8

41. $4x = 30 - 2x$; 5   **43.** California: 55 votes; Florida: 29 votes   **45.** falcon: 185 mph; pheasant: 37 mph   **47.** India: 3.5 million students; Turkey: 2.0 million students   **49.** Xbox: $420; games: $140   **51.** 5470 miles   **53.** Michigan Stadium: 109,901; Beaver Stadium: 106,572   **55.** California: 309 thousand; Washington: 103 thousand   **57.** 34 points   **59.** 2010: 275,215; 2020: 808,416   **61.** Spain: 6609 cars per day; Germany: 13,218 cars per day   **63.** $225   **65.** United States: 311 million; China: 195 million   **67.** 590   **69.** 1000   **71.** 3000   **73.** yes; answers may vary   **75.** $216,200   **77.** $549

Chapter 11 Vocabulary Check **1.** simplified; combined   **2.** like   **3.** variable   **4.** algebraic expression   **5.** terms   **6.** numerical coefficient   **7.** evaluating the expression   **8.** constant   **9.** equation   **10.** solution   **11.** distributive   **12.** multiplication   **13.** addition

Chapter 11 Review **1.** -5 **2.** 17   **3.** undefined   **4.** 0   **5.** 129   **6.** -2 **7.** 8 cu ft   **8.** 64 cu ft   **9.** $1800   **10.** $300
11. $-15x$ **12.** $-\dfrac{7}{30}x$ **13.** $-6y - 10$ **14.** $-6a - 7$ **15.** $-8y + 2$ **16.** $4.6x - 11.9$ **17.** $-8y$ **18.** $15y - 24$ **19.** $11x - 12$
20. $4x - 7$ **21.** $-5a + 4$ **22.** $12y - 9$ **23.** $(6x - 3)$ sq yd   **24.** $28y$ m   **25.** yes   **26.** no   **27.** -2 **28.** 7   **29.** $-\dfrac{1}{2}$ **30.** $-\dfrac{6}{11}$
31. -6 **32.** -20 **33.** 1.3   **34.** 2.4   **35.** 7   **36.** -9 **37.** 1   **38.** -5 **39.** $-\dfrac{4}{5}$ **40.** -24 **41.** -120 **42.** 13   **43.** 4   **44.** 3   **45.** 5
46. 12   **47.** 63   **48.** 33   **49.** -2.25 **50.** 1.3   **51.** 2   **52.** 6   **53.** 11   **54.** -5 **55.** 6   **56.** 8   **57.** $20 - (-8) = 28$
58. $-2 - 19 = -21$ **59.** $\dfrac{-75}{5 + 20} = -3$ **60.** $-5(-2 + 6) = -20$ **61.** $\dfrac{70}{x}$ **62.** $x - 13$ **63.** $85 - x$ **64.** $2x + 11$
65. $x + 8 = 40$ **66.** $2x - 12 = 10$ **67.** 5   **68.** -16 **69.** bamboos: 36 inches; kelp: 18 inches   **70.** incumbent: 14,752 votes; challenger: 3546 votes   **71.** 29   **72.** $-\dfrac{1}{4}$ **73.** $-11x$ **74.** $-35x$ **75.** $22x - 19$ **76.** $-9x - 32$ **77.** $x - 17$ **78.** $3(x + 5)$
79. $x - 3 = \dfrac{x}{4}$ **80.** $6x = x + 2$ **81.** no   **82.** yes   **83.** -1 **84.** -25 **85.** 13   **86.** -6 **87.** 17   **88.** 7   **89.** -22 **90.** -6
91. $-\dfrac{3}{4}$ **92.** -4 **93.** 2   **94.** $-\dfrac{8}{3}$ **95.** 12   **96.** -8 **97.** 0   **98.** 0   **99.** 5   **100.** 1

Chapter 11 Test **1.** -1 **2.** $-5x + 5$ **3.** $-6y - 14$ **4.** $14z - 8$ **5.** $4(3x - 1) = (12x - 4)$ sq m   **6.** 7   **7.** $-\dfrac{1}{2}$ **8.** -12
9. 40   **10.** 24   **11.** 3   **12.** -2 **13.** -2 **14.** 4.5   **15.** 0   **16.** -2 **17.** $\dfrac{22}{3}$ **18.** 4   **19.** 0   **20.** 6000 sq ft   **21.** 30 sq ft   **22. a.** $17x$
b. $20 - 2x$ **23.** -2 **24.** 34 points   **25.** 244 women

Cumulative Review **1.** 0.8496 (Sec. 4.4, Ex. 3)   **2.** 53.1   **3. a.** $\dfrac{5}{7}$ **b.** $\dfrac{7}{24}$ (Sec. 5.1, Ex. 8)   **4.** $\dfrac{23}{36}$ **5.** 5 (Sec. 6.3, Ex. 9)   **6.** $\dfrac{7}{12}$
7. 75% (Sec. 6.3, Ex. 11)   **8.** $\dfrac{18}{23}$ **9.** 48 oz (Sec. 7.2, Ex. 2)   **10.** 24,000   **11.** 11.3 L or 11,300 ml (Sec. 7.3, Ex. 8)   **12.** 0.024   **13. a.** line
b. line segment   **c.** angle   **d.** ray (Sec. 8.1, Ex. 1)   **14.** 168°   **15.** 10 cm (Sec. 8.2, Ex. 3)   **16.** 34°   **17.** 50 ft (Sec. 8.3, Ex. 6)
18. 132 sq ft   **19.** 56 sq cm (Sec. 8.4, Ex. 1)   **20.** 18   **21. a.** 8 and 9   **b.** 9 (Sec. 8.6, Ex. 5)   **22.** 5.78   **23.** 46 ft (Sec. 8.7, Ex. 4)
24. 24%   **25. a.** 75 clam species   **b.** fishes (Sec. 9.1, Ex. 3)   **26.** mean: 8; median: 9; mode: 11   **27.** < (Sec. 10.1, Ex. 6)
28. >   **29.** > (Sec. 10.1, Ex. 5)   **30.** <   **31.** 2 (Sec. 10.1, Ex. 10)   **32.** $\dfrac{5}{6}$ **33.** 1.2 (Sec. 10.1, Ex. 14)   **34.** 0   **35.** -16 (Sec. 10.2, Ex. 4)
36. -3.6 **37.** $-\dfrac{1}{2}$ (Sec. 10.2, Ex. 6)   **38.** $\dfrac{1}{10}$ **39.** -7 (Sec. 10.3, Ex. 6)   **40.** 5.8   **41.** 1 (Sec. 10.3, Ex. 7)   **42.** $-\dfrac{31}{120}$
43. 10 (Sec. 10.4, Ex. 2)   **44.** -9.6 **45.** $\dfrac{1}{3}$ (Sec. 10.4, Ex. 4)   **46.** -4 **47.** 3 (Sec. 10.5, Ex. 6)   **48.** 9.7   **49.** 2 (Sec. 11.4, Ex. 7)   **50.** 20

Appendices

A1: One Hundred Addition Facts **1.** 5   **3.** 5   **5.** 12   **7.** 8   **9.** 14   **11.** 12   **13.** 10   **15.** 9   **17.** 11   **19.** 18   **21.** 10   **23.** 10
25. 17   **27.** 13   **29.** 12   **31.** 16   **33.** 7   **35.** 4   **37.** 8   **39.** 8   **41.** 6   **43.** 16   **45.** 9   **47.** 12   **49.** 7   **51.** 2   **53.** 11   **55.** 8   **57.** 5
59. 14   **61.** 13   **63.** 7   **65.** 7   **67.** 13   **69.** 5   **71.** 4   **73.** 16   **75.** 14   **77.** 1   **79.** 4   **81.** 10   **83.** 8   **85.** 5   **87.** 10   **89.** 11   **91.** 12
93. 9   **95.** 9   **97.** 14   **99.** 9

A2: One Hundred Multiplication Facts **1.** 1   **3.** 56   **5.** 32   **7.** 28   **9.** 4   **11.** 63   **13.** 6   **15.** 30   **17.** 24   **19.** 18   **21.** 40
23. 32   **25.** 54   **27.** 56   **29.** 54   **31.** 2   **33.** 36   **35.** 12   **37.** 36   **39.** 12   **41.** 48   **43.** 8   **45.** 0   **47.** 27   **49.** 15   **51.** 45   **53.** 0
55. 81   **57.** 0   **59.** 0   **61.** 18   **63.** 3   **65.** 36   **67.** 63   **69.** 35   **71.** 42   **73.** 40   **75.** 0   **77.** 9   **79.** 15   **81.** 5   **83.** 0   **85.** 21
87. 0   **89.** 16   **91.** 0   **93.** 4   **95.** 6   **97.** 18   **99.** 0

Exercise Set B   **1. a.** \$640   **b.** 15 hr   **3. a.** 1260 beats   **b.** 60 sec or 1 min   **c.** 90 sec or $1\frac{1}{2}$ min   **5. a.** 200 beds   **b.** 11,315

7. a. 6 c   **b.** $13\frac{1}{3}$ c   **c.** $22\frac{1}{2}$ c   **9.** \$643.50   **11. a.** 262.5 drops   **b.** 20 liters   **13. a.** 7200 tortillas   **b.** 12.5 min   **15. a.** \$10.8 million

b. 4.5 episodes   **17. a.** $11\frac{1}{4}$ mi   **b.** $\frac{1}{9}$ hr or $6\frac{2}{3}$ min

Practice Final Exam   **1.** 113   **2.** 33   **3.** $\frac{64}{3}$ or $21\frac{1}{3}$ **4.** $\frac{3}{2}$ or $1\frac{1}{2}$ **5.** 8   **6.** 43.86   **7.** 6.2   **8.** $1\frac{7}{24}$ **9.** $22\frac{1}{2}$ **10.** 12,690   **11.** $\frac{3}{8}$

12. 52,000   **13.** 34.9   **14.** 0.941   **15.** 0.85   **16.** 610%   **17.** 37.5%   **18.** $\frac{1}{500}$ **19.** perimeter: $3\frac{1}{3}$ ft; area: $\frac{2}{3}$ sq ft   **20.** $\frac{9}{13}$

21. $\frac{3 \text{ in.}}{10 \text{ days}}$ **22.** 81.25 km/hr   **23.** 16-oz: \$0.0931/oz; 24-oz: \$0.0996/oz; 16-oz size   **24.** $\frac{48}{11}$ or $4\frac{4}{11}$ **25.** 11.4   **26.** $7\frac{5}{6}$ gal

27. $114\frac{2}{3}$ cartons   **28.** \$17   **29.** 1250   **30.** 75%   **31.** 38.4 lb   **32.** discount: \$18; sale price: \$102   **33.** 2400 g   **34.** 10 qt   **35.** 16 ft 6 in.

36. 12°   **37.** $m\angle x = 118°$; $m\angle y = 62°$; $m\angle z = 118°$ **38.** 26°   **39.** 7.5 or $7\frac{1}{2}$ **40.** mean: 38.4; median: 42; no mode   **41.** 11 students

42. -117 **43.** 12   **44.** 2   **45.** $-\frac{14}{81}$ **46.** -1 **47.** $14z - 8$ **48.** 40   **49.** 24   **50.** -2 **51.** 244 women

Solutions to Selected Exercises

Chapter 1

Exercise Set 1.2

1. The place value of the 5 in 657 is tens.

5. The place value of the 5 in 43,526,000 is hundred-thousands.

9. 354 is written as three hundred fifty-four.

13. 26,990 is written as twenty-six thousand, nine hundred ninety.

17. 24,350,185 is written as twenty-four million, three hundred fifty thousand, one hundred eighty-five.

21. 2717 is written as two thousand, seven hundred seventeen.

25. 14,433 is written as fourteen thousand, four hundred thirty-three.

29. Six thousand, five hundred eighty-seven in standard form is 6587.

33. Thirteen million, six hundred one thousand, eleven in standard form is 13,601,011.

37. Two hundred sixty thousand, nine hundred ninety-seven in standard form is 260,997.

41. Sixteen thousand, seven hundred thirty-two in standard form is 16,732.

45. One hundred one thousand in standard form is 101,000.

49. $3470 = 3000 + 400 + 70$

53. $66,049 = 60,000 + 6000 + 40 + 9$

57. 5532 is written as five thousand, five hundred thirty-two.

61. The tallest mountain in New England is Mt. Washington.

65. The number of visitors to the Vatican Museums was 5,065,000 which is written as five million, sixty-five thousand.

69. The largest number is 9861.

73. answers may vary

77. The Pro Football Hall of Fame is located in the town of Canton.

Exercise Set 1.3

1.
$$\begin{array}{r} 14 \\ + 22 \\ \hline 36 \end{array}$$

5.
$$\begin{array}{r} 12 \\ 13 \\ + 24 \\ \hline 49 \end{array}$$

9.
$$\begin{array}{r} 53 \\ + 64 \\ \hline 117 \end{array}$$

13.
$$\begin{array}{r} \overset{1\ \ 1\ 1}{22,781} \\ +186,297 \\ \hline 209,078 \end{array}$$

17.
$$\begin{array}{r} \overset{2}{6} \\ 21 \\ 14 \\ 9 \\ + 12 \\ \hline 62 \end{array}$$

21.
$$\begin{array}{r} \overset{1}{62} \\ 18 \\ + 14 \\ \hline 94 \end{array}$$

25.
$$\begin{array}{r} \overset{1\ 1\ 1}{7542} \\ 49 \\ + 682 \\ \hline 8273 \end{array}$$

29.
$$\begin{array}{r} \overset{2}{627} \\ 628 \\ + 629 \\ \hline 1884 \end{array}$$

33.
$$\begin{array}{r} \overset{1\ 1}{507} \\ 593 \\ + 10 \\ \hline 1110 \end{array}$$

37.
$$\begin{array}{r} \overset{1\ 1\ 2\ 2}{49} \\ 628 \\ 5\ 762 \\ +29,462 \\ \hline 35,901 \end{array}$$

41.
$$\begin{array}{r} \overset{2}{9} \\ 12 \\ 9 \\ + 12 \\ \hline 42 \end{array}$$
The perimeter is 42 inches.

45. Opposite sides of a rectangle have the same length.
$$\begin{array}{r} \overset{2}{4} \\ 4 \\ 8 \\ + 8 \\ \hline 24 \end{array}$$
The perimeter is 24 inches.

49. $8 + 3 + 5 + 7 + 5 + 1 = 8 + 1 + 3 + 7 + 5 + 5$
$$= 9 + 10 + 10$$
$$= 29$$
The perimeter is 29 inches.

53.
$$\begin{array}{r} \overset{1\ 1\ 1}{297} \\ + 1796 \\ \hline 2093 \end{array}$$

57. "Increased by" indicates addition.
$$\begin{array}{r} \overset{1}{452} \\ + 92 \\ \hline 544 \end{array}$$
452 increased by 92 is 544.

61.
$$\begin{array}{r} {\scriptstyle 11 \;\; 1} \\ 19{,}318 \\ +\,1823 \\ \hline 21{,}141 \end{array}$$

The population of Florida is projected to be 21,141 thousand in 2020.

65.
$$\begin{array}{r} {\scriptstyle 21} \\ 78 \\ 90 \\ 102 \\ +\,70 \\ \hline 340 \end{array}$$

He needs 340 feet of wiring.

69.
$$\begin{array}{r} {\scriptstyle 1 \;\;\; 1} \\ 172{,}251 \\ +\,77{,}598 \\ \hline 249{,}849 \end{array}$$

249,849 Harley-Davidson motorcycles were sold in 2012.

73. The sides of a square are all the same length.
$$\begin{array}{r} {\scriptstyle 1} \\ 31 \\ 31 \\ 31 \\ +\,31 \\ \hline 124 \end{array}$$

The perimeter of the board is 124 feet.

77. California has the most Target stores.

81. New York and Pennsylvania:
$$\begin{array}{r} {\scriptstyle 11} \\ 67 \\ +\,63 \\ \hline 130 \end{array}$$

Michigan and Ohio:
$$\begin{array}{r} {\scriptstyle 11} \\ 59 \\ +\,64 \\ \hline 123 \end{array}$$

New York and Pennsylvania have more Target stores.

85. answer may vary

89.
$$\begin{array}{r} {\scriptstyle 1111 \;\; 121 \; 1} \\ 56{,}468{,}980 \\ 1{,}236{,}785 \\ +\,986{,}768{,}000 \\ \hline 1{,}044{,}473{,}765 \end{array}$$

93.
$$\begin{array}{r} {\scriptstyle 22} \\ 14 \\ 173 \\ 86 \\ +\,257 \\ \hline 530 \end{array}$$

The given sum is incorrect; the correct sum is 530.

Exercise Set 1.4

1.
$$\begin{array}{r} 67 \\ -\,23 \\ \hline 44 \end{array}$$
Check:
$$\begin{array}{r} 44 \\ +\,23 \\ \hline 67 \end{array}$$

5.
$$\begin{array}{r} 167 \\ -\,32 \\ \hline 135 \end{array}$$
Check:
$$\begin{array}{r} 135 \\ +\,32 \\ \hline 167 \end{array}$$

9.
$$\begin{array}{r} 6998 \\ -\,1453 \\ \hline 5545 \end{array}$$
Check:
$$\begin{array}{r} 5545 \\ +\,1453 \\ \hline 6998 \end{array}$$

13.
$$\begin{array}{r} 62 \\ -\,37 \\ \hline 25 \end{array}$$
Check:
$$\begin{array}{r} {\scriptstyle 1} \\ 25 \\ +\,37 \\ \hline 62 \end{array}$$

17.
$$\begin{array}{r} 938 \\ -\,792 \\ \hline 146 \end{array}$$
Check:
$$\begin{array}{r} {\scriptstyle 1} \\ 146 \\ +\,792 \\ \hline 938 \end{array}$$

21.
$$\begin{array}{r} 600 \\ -\,432 \\ \hline 168 \end{array}$$
Check:
$$\begin{array}{r} {\scriptstyle 11} \\ 168 \\ +\,432 \\ \hline 600 \end{array}$$

25.
$$\begin{array}{r} 923 \\ -\,476 \\ \hline 447 \end{array}$$
Check:
$$\begin{array}{r} {\scriptstyle 11} \\ 447 \\ +\,476 \\ \hline 923 \end{array}$$

29.
$$\begin{array}{r} 533 \\ -\,29 \\ \hline 504 \end{array}$$
Check:
$$\begin{array}{r} {\scriptstyle 1} \\ 504 \\ +\,29 \\ \hline 533 \end{array}$$

33.
$$\begin{array}{r} 1983 \\ -\,1904 \\ \hline 79 \end{array}$$
Check:
$$\begin{array}{r} {\scriptstyle 1} \\ 79 \\ +\,1904 \\ \hline 1983 \end{array}$$

37.
$$\begin{array}{r} 50{,}000 \\ -\,17{,}289 \\ \hline 32{,}711 \end{array}$$
Check:
$$\begin{array}{r} {\scriptstyle 11 \;\, 11} \\ 32{,}711 \\ +\,17{,}289 \\ \hline 50{,}000 \end{array}$$

41.
$$\begin{array}{r} 51{,}111 \\ -\,19{,}898 \\ \hline 31{,}213 \end{array}$$
Check:
$$\begin{array}{r} {\scriptstyle 11 \;\, 11} \\ 31{,}213 \\ +\,19{,}898 \\ \hline 51{,}111 \end{array}$$

45.
$$\begin{array}{r} 41 \\ -\,21 \\ \hline 20 \end{array}$$

The difference of 41 and 21 is 20.

49.
$$\begin{array}{r} 108 \\ -\,36 \\ \hline 72 \end{array}$$

108 less 36 is 72.

53.
$$\begin{array}{r} 503 \\ -\,239 \\ \hline 264 \end{array}$$

She has 264 more pages to read.

57.
$$\begin{array}{r} 189{,}000 \\ +\,75{,}000 \\ \hline 264{,}000 \end{array}$$

The total land area drained by the Upper Mississippi and Lower Mississippi sub-basins is 264,000 square miles.

61.
$$\begin{array}{r} 20{,}320 \\ -\,14{,}255 \\ \hline 6065 \end{array}$$

The peak of Mt. McKinley is 6065 feet higher than the peak of Long's Peak.

65.
$$\begin{array}{r} 645 \\ -\ 287 \\ \hline 358 \end{array}$$

The distance between Hays, Kansas, and Denver, Colorado, is 358 miles.

69.
$$\begin{array}{r} 9129 \\ -\ 6499 \\ \hline 2630 \end{array}$$

The projected increase in the population of Arizona is 2630 thousand.

73.
$$\begin{array}{r} 88 \\ -\ 30 \\ \hline 58 \end{array}$$

Snoring is 58 dB louder than normal conversation.

77.
$$\begin{array}{r} 100{,}000 \\ -\ 94{,}080 \\ \hline 5920 \end{array}$$

The Dole Plantation maze is 5920 square feet larger than the Ruurlo maze.

81.
$$\begin{array}{r} 67 \\ -\ 53 \\ \hline 14 \end{array}$$

Chicago O'Hare International Airport has 14 million more passengers than Denver International Airport.

85.
$$\begin{array}{r} {}^{1\ 1} \\ 986 \\ +\ 48 \\ \hline 1034 \end{array}$$

89.
$$\begin{array}{r} 9000 \\ -\ 482 \\ \hline 8518 \end{array}$$

93. In $48 - 1$, 48 is the minuend and 1 is the subtrahend.

97.
$$\begin{array}{r} 741 \\ -\ 56 \\ \hline 685 \end{array}$$

The given answer is incorrect; the correct answer is 685.

101.
$$\begin{array}{r} 5269 \\ -2385 \\ \hline 2884 \end{array}$$

105.
$$\begin{array}{r} 289{,}462 \\ 369{,}477 \\ 218{,}287 \\ +121{,}685 \\ \hline 998{,}911 \end{array} \qquad \begin{array}{r} 1{,}000{,}000 \\ -998{,}911 \\ \hline 1089 \end{array}$$

No, they have not reached their goal. They have 1089 more pages to read.

Exercise Set 1.5

1. To round 423 to the nearest ten, observe that the digit in the ones place is 3. Since this digit is less than 5, we do not add 1 to the digit in the tens place. The number 423 rounded to the nearest ten is 420.

5. To round 2791 to the nearest hundred, observe that the digit in the tens place is 9. Since this digit is 5 or greater, we add 1 to the digit in the hundreds place. The number 2791 rounded to the nearest hundred is 2800.

9. To round 21,094 to the nearest thousand, observe that the digit in the hundreds place is 0. Since this digit is less than 5, we do not add 1 to the digit in the thousands place. The number 21,094 rounded to the nearest thousand is 21,000.

13. To round 328,495 to the nearest hundred, observe that the digit in the tens place is 9. Since this digit is 5 or greater, we add 1 to the digit in the hundreds place. The number 328,495 rounded to the nearest hundred is 328,500.

17. To round 39,994 to the nearest ten, observe that the digit in the ones place is 4. Since this digit is less than 5, we do not add 1 to the digit in the tens place. The number 39,994 rounded to the nearest ten is 39,990.

21. Estimate 5281 to a given place value by rounding it to that place value. 5281 rounded to the tens place is 5280, to the hundreds place is 5300, and to the thousands place is 5000.

25. Estimate 14,876 to a given place value by rounding it to that place value. 14,876 rounded to the tens place is 14,880, to the hundreds place is 14,900, and to the thousands place is 15,000.

29. To round 38,387 to the nearest thousand, observe that the digit in the hundreds place is 3. Since this digit is less than 5, we do not add 1 to the digit in the thousands place. Therefore, 38,387 points rounded to the nearest thousand is 38,000 points.

33. To round 5,021,850 to the nearest thousand, observe that the digit in the hundreds place is 8. Since this digit is 5 or greater, we add 1 to the digit in the thousands place. Therefore, $5,021,850 rounded to the nearest thousand is $5,022,000.

37.
$$\begin{array}{rll} 39 & \text{rounds to} & 40 \\ 45 & \text{rounds to} & 50 \\ 22 & \text{rounds to} & 20 \\ +\ 17 & \text{rounds to} & +\ 20 \\ \hline & & 130 \end{array}$$

41.
$$\begin{array}{rll} 1913 & \text{rounds to} & 1900 \\ 1886 & \text{rounds to} & 1900 \\ +\ 1925 & \text{rounds to} & +\ 1900 \\ \hline & & 5700 \end{array}$$

45.
$$\begin{array}{rll} 3995 & \text{rounds to} & 4000 \\ 2549 & \text{rounds to} & 2500 \\ +\ 4944 & \text{rounds to} & +\ 4900 \\ \hline & & 11{,}400 \end{array}$$

49. $229 + 443 + 606$ is approximately $230 + 440 + 610 = 1280$.

The answer of 1278 is correct.

53.
$$\begin{array}{rll} 899 & \text{rounds to} & 900 \\ 1499 & \text{rounds to} & 1500 \\ +\ 999 & \text{rounds to} & +\ 1000 \\ \hline & & 3400 \end{array}$$

The total cost is approximately $3400.

57.
$$\begin{array}{rll} 20{,}320 & \text{rounds to} & 20{,}000 \\ -\ 14{,}410 & \text{rounds to} & -\ 14{,}000 \\ \hline & & 6000 \end{array}$$

The difference in elevation is approximately 6000 feet.

61.
$$\begin{array}{rll} 1{,}128{,}030 & \text{rounds to} & 1{,}128{,}000 \\ -\ 1{,}125{,}209 & \text{rounds to} & -\ 1{,}125{,}000 \\ \hline & & 3000 \end{array}$$

The increase in enrollment was approximately 3000 children.

65. $2910 million written in standard form is $2,910,000,000.

$2,910,000,000 rounded to the nearest ten-million is $2,910,000,000.

$2,910,000,000 rounded to the nearest hundred-million is $2,900,000,000.

69. a. The smallest possible number that rounds to 8600 is 8550.

    b. The largest possible number that rounds to 8600 is 8649.

73. 54 rounds to 50

17 rounds to 20

$50 + 20 + 50 + 20 = 140$

The perimeter is approximately 140 meters.

Exercise Set 1.6

1. $1 \cdot 24 = 24$

5. $8 \cdot 0 \cdot 9 = 0$

9. $6(3 + 8) = 6 \cdot 3 + 6 \cdot 8$

13. $20(14 + 6) = 20 \cdot 14 + 20 \cdot 6$

17.
$$\begin{array}{r} \overset{1}{6}13 \\ \times \quad 6 \\ \hline 3678 \end{array}$$

21.
$$\begin{array}{r} \overset{4\,2}{1074} \\ \times \quad 6 \\ \hline 6444 \end{array}$$

25.
$$\begin{array}{r} 421 \\ \times \quad 58 \\ \hline 3\,368 \\ 21\,050 \\ \hline 24{,}418 \end{array}$$

29.
$$\begin{array}{r} 780 \\ \times \quad 20 \\ \hline 15{,}600 \end{array}$$

33. $(640)(1)(10) = (640)(10) = 6400$

37.
$$\begin{array}{r} 609 \\ \times \quad 234 \\ \hline 2\,436 \\ 18\,270 \\ 121\,800 \\ \hline 142{,}506 \end{array}$$

41.
$$\begin{array}{r} 589 \\ \times \quad 110 \\ \hline 5\,890 \\ 58\,900 \\ \hline 64{,}790 \end{array}$$

45. $8 \times 100 = 800$

49. $7406 \cdot 10 = 74{,}060$

53. $50 \cdot 900 = 45{,}000$

57. Area = (length)(width)

        = (9 meters)(7 meters)

        = 63 square meters

Perimeter = $(9 + 7 + 9 + 7)$ meters

           = 32 meters

61.
$$\begin{array}{r} 576 \quad \text{rounds to} \quad 600 \\ \times\, 354 \quad \text{rounds to} \quad \times \quad 400 \\ \hline 240{,}000 \end{array}$$

65. 38×42 is approximately 40×40, which is 1600.

The best estimate is c.

69. $80 \times 11 = (8 \times 10) \times 11$

             $= 8 \times (10 \times 11)$

             $= 8 \times 110$

             $= 880$

73.
$$\begin{array}{r} 2240 \\ \times \quad 2 \\ \hline 4480 \end{array}$$

77.
$$\begin{array}{r} 94 \\ \times \quad 35 \\ \hline 470 \\ 2820 \\ \hline 3290 \end{array}$$

The total cost is $3290.

81. Area = (length)(width)

       = (110 feet)(80 feet)

       = 8800 square feet

The area is 8800 square feet.

85.
$$\begin{array}{r} 94 \\ \times \quad 62 \\ \hline 188 \\ 5640 \\ \hline 5828 \end{array}$$

There are 5828 pixels on the screen.

89.
$$\begin{array}{r} 170 \\ \times \quad 8 \\ \hline 1360 \end{array}$$

There are 1360 calories in 8 ounces.

93. There are 60 minutes in one hour, so there are 24×60 minutes in one day.

$24 \times 60 \times 1000 = 24 \times 6 \times 10 \times 1000$

                       $= 144 \times 10{,}000$

                       $= 1{,}440{,}000$

They produce 1,440,000 tea bags in one day.

97.
$$\begin{array}{r} 134 \\ \times \quad 16 \\ \hline 804 \\ 1340 \\ \hline 2144 \end{array}$$

101.
$$\begin{array}{r} 19 \\ - \quad 4 \\ \hline 15 \end{array}$$

The difference of 19 and 4 is 15.

105. a. $3 \cdot 5 = 5 + 5 + 5 = 3 + 3 + 3 + 3 + 3$

     b. answers may vary

109. $42 \times 3 = 126$

$42 \times 90 = 3780$

The problem is 42×93.

113. On a side with 7 windows per row, there are $7 \times 23 = 161$ windows. On a side with 4 windows per row, there are $4 \times 23 = 92$ windows. $161 + 161 + 92 + 92 = 506$

There are 506 windows on the building.

Exercise Set 1.7

1. $54 \div 9 = 6$ because $6 \cdot 9 = 54$

5. $0 \div 8 = 0$ because $0 \cdot 8 = 0$

9. $\dfrac{18}{18} = 1$ because $1 \cdot 18 = 18$

13. $26 \div 0$ is undefined.

17. $0 \div 14 = 0$ because $0 \cdot 14 = 0$

21.
$$
\begin{array}{r}
29 \\
3\overline{)87} \\
-6 \\
\hline
27 \\
-27 \\
\hline
0
\end{array}
$$
Check: $3 \cdot 29 = 87$

25.
$$
\begin{array}{r}
338 \\
3\overline{)1014} \\
-9 \\
\hline
11 \\
-9 \\
\hline
24 \\
-24 \\
\hline
0
\end{array}
$$
Check: $3 \cdot 338 = 1014$

29.
$$
\begin{array}{r}
9 \\
7\overline{)63} \\
-63 \\
\hline
0
\end{array}
$$
Check: $7 \cdot 9 = 63$

33.
$$
\begin{array}{r}
68 \text{ R } 3 \\
7\overline{)479} \\
-42 \\
\hline
59 \\
-56 \\
\hline
3
\end{array}
$$
Check: $7 \cdot 68 + 3 = 479$

37.
$$
\begin{array}{r}
38 \text{ R } 1 \\
8\overline{)305} \\
-24 \\
\hline
65 \\
-64 \\
\hline
1
\end{array}
$$
Check: $8 \cdot 38 + 1 = 305$

41.
$$
\begin{array}{r}
13 \\
55\overline{)715} \\
-55 \\
\hline
165 \\
-165 \\
\hline
0
\end{array}
$$
Check: $55 \cdot 13 = 715$

45.
$$
\begin{array}{r}
97 \text{ R } 8 \\
97\overline{)9417} \\
-873 \\
\hline
687 \\
-679 \\
\hline
8
\end{array}
$$
Check: $97 \cdot 97 + 8 = 9417$

49.
$$
\begin{array}{r}
506 \\
13\overline{)6578} \\
-65 \\
\hline
07 \\
-0 \\
\hline
78 \\
-78 \\
\hline
0
\end{array}
$$
Check: $13 \cdot 506 = 6578$

53.
$$
\begin{array}{r}
54 \\
236\overline{)12744} \\
-1180 \\
\hline
944 \\
-944 \\
\hline
0
\end{array}
$$
Check: $236 \cdot 54 = 12{,}744$

57.
$$
\begin{array}{r}
202 \text{ R } 15 \\
102\overline{)20619} \\
-204 \\
\hline
21 \\
-0 \\
\hline
219 \\
-204 \\
\hline
15
\end{array}
$$
Check: $102 \cdot 202 + 15 = 20{,}619$

61.
$$
\begin{array}{r}
17 \\
7\overline{)119} \\
-7 \\
\hline
49 \\
-49 \\
\hline
0
\end{array}
$$

65.
$$
\begin{array}{r}
2132 \text{ R } 32 \\
40\overline{)85312} \\
-80 \\
\hline
53 \\
-40 \\
\hline
131 \\
-120 \\
\hline
112 \\
-80 \\
\hline
32
\end{array}
$$

69.
$$
\begin{array}{r}
23 \text{ R } 2 \\
5\overline{)117} \\
-10 \\
\hline
17 \\
-15 \\
\hline
2
\end{array}
$$
The quotient is 23 R 2.

73.
$$
\begin{array}{r}
20 \text{ R } 2 \\
3\overline{)62} \\
-6 \\
\hline
02 \\
-0 \\
\hline
2
\end{array}
$$
The quotient is 20 R 2.

77.
$$
\begin{array}{r}
165 \\
318\overline{)52470} \\
\underline{-318} \\
2067 \\
\underline{-1908} \\
1590 \\
\underline{-1590} \\
0
\end{array}
$$
The person weighs 165 pounds on Earth.

81.
$$
\begin{array}{r}
88 \text{ R } 1 \\
3\overline{)265} \\
\underline{-24} \\
25 \\
\underline{-24} \\
1
\end{array}
$$
There are 88 bridges every 3 miles over the 265 miles, plus the first bridge for a total of 89 bridges.

85.
$$
\begin{array}{r}
5 \\
5280\overline{)26400} \\
\underline{-26400} \\
0
\end{array}
$$
Broad Peak is 5 miles tall.

89.
$$
\begin{array}{r}
{}^{2} \\
10 \\
24 \\
35 \\
22 \\
17 \\
\underline{+12} \\
120
\end{array}
\qquad
\begin{array}{r}
20 \\
6\overline{)120} \\
\underline{-12} \\
0
\end{array}
$$
Average $= \dfrac{120}{6} = 20$

93.
$$
\begin{array}{r}
{}^{2} \\
86 \\
79 \\
81 \\
69 \\
\underline{+80} \\
395
\end{array}
\qquad
\begin{array}{r}
79 \\
5\overline{)395} \\
\underline{-35} \\
45 \\
\underline{-45} \\
0
\end{array}
$$
Average $= \dfrac{395}{5} = 79$

97.
$$
\begin{array}{r}
{}^{1\,1\,1} \\
82 \\
463 \\
29 \\
\underline{+8704} \\
9278
\end{array}
$$

101.
$$
\begin{array}{r}
722 \\
\underline{-\ 43} \\
679
\end{array}
$$

105.
$$
\begin{array}{r}
9 \text{ R } 12 \\
24\overline{)228} \\
\underline{-216} \\
12
\end{array}
$$

109. 200 divided by 20 is $200 \div 20$, which is choice b.

113. The average will increase; answers may vary.

117. Since Area $=$ length $\cdot$ width, length $= \dfrac{\text{Area}}{\text{width}}$.

length $= \dfrac{60 \text{ square feet}}{5 \text{ feet}} = \dfrac{60}{5}$ feet $= 12$ feet

The length is 12 feet.

121.
$$
\begin{array}{r}
26 \\
\underline{-\ 5} \\
21 \\
\underline{-\ 5} \\
16 \\
\underline{-\ 5} \\
11 \\
\underline{-\ 5} \\
6 \\
\underline{-\ 5} \\
1
\end{array}
$$
Thus $26 \div 5 = 5 \text{ R } 1$.

Exercise Set 1.8

1. 41 increased by 8 is what number

41 $+$ 8 $=$ what number
$$
\begin{array}{r}
41 \\
\underline{+\ 8} \\
49
\end{array}
$$

5. The total of 35 and 7 is what number

35 $+$ 7 $=$ what number
$$
\begin{array}{r}
35 \\
\underline{+\ 7} \\
42
\end{array}
$$

9. a. Perimeter is side $+$ side $+$ side $+$ side

Perimeter $=$ 120 $+$ 80 $+$ 120 $+$ 80
$= 400$

The perimeter is 400 feet.

b. Area is length times width

Area $=$ 120 $\times$ 80
$$
\begin{array}{r}
120 \\
\underline{\times\ 80} \\
9600
\end{array}
$$
The area is 9600 square feet.

13. Hours per week is hours per day times days per week

Hours per week $=$ 24 $\times$ 7
$$
\begin{array}{r}
24 \\
\underline{\times\ 7} \\
168
\end{array}
$$
There are 168 hours in a week.

17.

Difference in years	is	First State National Monument	minus	Yellowstone National Park
↓	↓	↓	↓	↓

Difference = 2013 − 1872

$$\begin{array}{r} 2013 \\ -1872 \\ \hline 141 \end{array}$$

Yellowstone National Park is 141 years older than First State National Monument.

21.

Total	is	number of Fairview	plus	number of Midway	plus	number of Riverside
↓	↓	↓	↓	↓	↓	↓

Total = 287 + 252 + 180

= 287 + 252 + 180 = 719

There are a total of 719 towns named Fairview, Midway, or Riverside.

25.

Calories in 1 ounce	is	calories	per	ounces
↓	↓	↓	↓	↓

Calories in 1 ounce = 165 ÷ 3

$$\begin{array}{r} 55 \\ 3\overline{)\,165} \\ -15 \\ \hline 15 \\ -15 \\ \hline 0 \end{array}$$

There are 55 calories in 1 ounce of canned tuna.

29.

Average	is	total sales	divided by	number of stores
↓	↓	↓	↓	↓

Average = 27,685,999,890 ÷ 843

$$\begin{array}{r} 32842230 \\ 843\overline{)\,27685999890} \\ -2529 \\ \hline 2395 \\ -1686 \\ \hline 7099 \\ -6744 \\ \hline 3559 \\ -3372 \\ \hline 1879 \\ -1686 \\ \hline 1938 \\ -1686 \\ \hline 2529 \\ -2529 \\ \hline 0 \end{array}$$

The average sales by each store was $32,842,230.

33.

Northern length	is	Southern length	plus	2054
↓	↓	↓	↓	↓

Length = 1933 + 2054

= 1933 + 2054 = 3987

The northern boundary of the conterminous United States is 3987 miles long.

37.

Total cost	is	number of sweaters	times	cost of sweater	plus	number of shirts	times	cost of shirt
↓	↓	↓	↓	↓	↓	↓	↓	↓

Total = 3 · 38 + 5 · 25

= 3·38 + 5·25 = 114 + 125 = 239

The total cost is $239.

41. Option a: a hamburger, onion rings, a candy bar, and a soda cost $4 + $3 + $2 + $1 = $10.

Option b: a hot dog, an apple, french fries, and a soda cost $3 + $1 + $2 + $1 = $7.

Thus option b is cheaper by $10 − $7, or $3.

45. The region with the most Internet users (Asia) had 1077 million users. The region with the fewest Internet users (Oceania/Australia) had 24 million users.

$$\begin{array}{r} 1077 \\ -24 \\ \hline 1053 \end{array}$$

Asia had 1053 million more users than Oceania/Australia.

49.

$$\begin{array}{r} 1077 \\ +\,519 \\ \hline 1596 \end{array} \qquad \begin{array}{r} 798 \\ 2\overline{)\,1596} \\ -14 \\ \hline 19 \\ -18 \\ \hline 16 \\ -16 \\ \hline 0 \end{array}$$

The average number of Internet users per region in these two world regions was 798 million.

53. There are 7 days in a week.

7 · 2400 = 16,800

No more than 16,800 milligrams of sodium should be consumed in one week.

57. 170,000,000 rounds to 200,000,000.

399,000,000 rounds to 400,000,000.

$$\begin{array}{r} 2 \\ 200000000\overline{)\,400000000} \\ -400000000 \\ \hline 0 \end{array}$$

The average revenue generated by each return receipt article was $2.

Exercise Set 1.9

1. $4 \cdot 4 \cdot 4 = 4^3$ **5.** $12 \cdot 12 \cdot 12 = 12^3$

9. $9 \cdot 8 \cdot 8 = 9 \cdot 8^2$

13. $3 \cdot 2 \cdot 2 \cdot 2 \cdot 2 \cdot 5 \cdot 5 \cdot 5 \cdot 5 \cdot 5 = 3 \cdot 2^4 \cdot 5^5$

17. $5^3 = 5 \cdot 5 \cdot 5 = 125$

21. $1^{10} = 1 \cdot 1 \cdot 1 \cdot 1 \cdot 1 \cdot 1 \cdot 1 \cdot 1 \cdot 1 \cdot 1 = 1$

25. $2^7 = 2 \cdot 2 \cdot 2 \cdot 2 \cdot 2 \cdot 2 \cdot 2 = 128$

29. $4^4 = 4 \cdot 4 \cdot 4 \cdot 4 = 256$

33. $12^2 = 12 \cdot 12 = 144$ **37.** $20^1 = 20$

41. $3 \cdot 2^6 = 3 \cdot 2 \cdot 2 \cdot 2 \cdot 2 \cdot 2 \cdot 2 = 192$

45. $\sqrt{9} = 3$ since $3 \cdot 3 = 9$

49. $\sqrt{144} = 12$ since $12 \cdot 12 = 144$

53. $15 + 3 \cdot 2 = 15 + 6 = 21$

57. $32 \div 4 - 3 = 8 - 3 = 5$

61. $6 \cdot 5 + 8 \cdot 2 = 30 + 16 = 46$

65. $(7 + 5^2) \div 4 \cdot 2^3 = (7 + 25) \div 4 \cdot 2^3$
$$= 32 \div 4 \cdot 2^3$$
$$= 32 \div 4 \cdot 8$$
$$= 8 \cdot 8$$
$$= 64$$

69. $\dfrac{18 + 6}{2^4 - 2^2} = \dfrac{24}{16 - 4} = \dfrac{24}{12} = 2$

73. $\dfrac{7(9 - 6) + 3}{3^2 - 3} = \dfrac{7(3) + 3}{9 - 3} = \dfrac{21 + 3}{6} = \dfrac{24}{6} = 4$

77. $2^4 \cdot 4 - (25 \div 5) = 2^4 \cdot 4 - 5$
$$= 16 \cdot 4 - 5$$
$$= 64 - 5$$
$$= 59$$

81. $(7 \cdot 5) + [9 \div (3 \div 3)] = (7 \cdot 5) + [9 \div (1)]$
$$= 35 + 9$$
$$= 44$$

85. $\dfrac{9^2 + 2^2 - 1^2}{8 \div 2 \cdot 3 \cdot 1 \div 3} = \dfrac{81 + 4 - 1}{4 \cdot 3 \cdot 1 \div 3}$
$$= \dfrac{85 - 1}{12 \cdot 1 \div 3}$$
$$= \dfrac{84}{12 \div 3}$$
$$= \dfrac{84}{4}$$
$$= 21$$

89. $4 \cdot \sqrt{49} - 0 \div \sqrt{100} = 4 \cdot 7 - 0 \div 10$
$$= 28 - 0 \div 10$$
$$= 28 - 0$$
$$= 28$$

93. $\sqrt{81} \div \sqrt{9} + 4^2 \cdot 2 - 10 = 9 \div 3 + 16 \cdot 2 - 10$
$$= 3 + 16 \cdot 2 - 10$$
$$= 3 + 32 - 10$$
$$= 35 - 10$$
$$= 25$$

97. $7^2 - \{18 - [40 \div (4 \cdot 2) + \sqrt{4}] + 5^2\}$
$$= 7^2 - \{18 - [40 \div 8 + \sqrt{4}] + 5^2\}$$
$$= 7^2 - \{18 - [40 \div 8 + 2] + 5^2\}$$
$$= 7^2 - \{18 - [5 + 2] + 5^2\}$$
$$= 7^2 - \{18 - 7 + 5^2\}$$
$$= 7^2 - \{18 - 7 + 25\}$$
$$= 7^2 - \{11 + 25\}$$
$$= 7^2 - 36$$
$$= 49 - 36$$
$$= 13$$

101. Area of a square $= (\text{side})^2$
$$= (23 \text{ miles})^2$$
$$= 529 \text{ square miles}$$

Perimeter $= (23 + 23 + 23 + 23)$ miles
$$= 92 \text{ miles}$$

105. $2^5 = 2 \cdot 2 \cdot 2 \cdot 2 \cdot 2$

The statement is false.

109. $24 \div (3 \cdot 2) + 2 \cdot 5 = 24 \div 6 + 2 \cdot 5 = 4 + 10 = 14$

113. $(7 + 2^4)^5 - (3^5 - 2^4)^2 = (7 + 16)^5 - (243 - 16)^2$
$$= 23^5 - 227^2$$
$$= 6{,}436{,}343 - 51{,}529$$
$$= 6{,}384{,}814$$

Chapter 1 Test

1. 82,426 is written as eighty-two thousand, four hundred twenty-six.

5.
$$\begin{array}{r} 496 \\ \times\ \ 30 \\ \hline 14{,}880 \end{array}$$

9. $0 \div 49 = 0$

13. $\dfrac{64 \div 8 \cdot 2}{(\sqrt{9} - \sqrt{4})^2 + 1} = \dfrac{8 \cdot 2}{(3 - 2)^2 + 1}$
$$= \dfrac{16}{1^2 + 1}$$
$$= \dfrac{16}{1 + 1}$$
$$= \dfrac{16}{2}$$
$$= 8$$

17. To round 52,369 to the nearest thousand, observe that the digit in the hundreds place is 3. Since this digit is less than 5, we do not add 1 to the digit in the thousands place. The number 52,369 rounded to the nearest thousand is 52,000.

21.
$$\begin{array}{r} \overset{1}{15} \\ +\ 107 \\ \hline 122 \end{array}$$

The sum of 15 and 107 is 122.

25.
$$\begin{array}{r} 725 \\ -599 \\ \hline 126 \end{array}$$

The more expensive refrigerator costs $126 more than the less expensive one.

29. Perimeter $= 20$ yards $+ 20$ yards $+ 10$ yards $+ 10$ yards
$$= 60 \text{ yards}$$

The perimeter is 60 yards.

Area $= (\text{length}) \cdot (\text{width})$
$$= (20 \text{ yards}) \cdot (10 \text{ yards})$$
$$= 200 \text{ square yards}$$

The area is 200 square yards.

Chapter 2

Exercise Set 2.1

1. In the fraction $\frac{1}{2}$, the numerator is 1 and the denominator is 2. Since 1 is less than 2, the fraction is proper.

5. In the fraction $\frac{15}{15}$, the numerator is 15 and the denominator is 15. Since 15 is greater than or equal to 15, the fraction is improper.

9. $\frac{5}{0}$ is undefined.   **13.** $\frac{0}{20} = 0$ **17.** $\frac{16}{1} = 16$

21. 7 of the 12 equal parts are shaded: $\frac{7}{12}$

25. 4 of the 9 equal parts are shaded: $\frac{4}{9}$

29. 5 of the 8 equal parts are shaded: $\frac{5}{8}$

33.

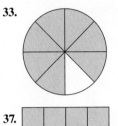

37.

41. a. $131 - 42 = 89$
89 students are not freshmen.

b. $\frac{89}{131}$ of the students are not freshmen.

45. $\frac{10}{19}$ of the named tropical storms turned into hurricanes in 2012.

49. $\frac{10}{31}$ of the class is sophomores.

53. a. $\frac{21}{50}$ of the marbles are blue

b. $50 - 21 = 29$
29 marbles are red.

c. $\frac{29}{50}$ of the marbles are red.

57. Each part is $\frac{1}{6}$ of a whole and there are 23 parts shaded, or 3 wholes and 5 more parts.

a. $\frac{23}{6}$

b. $3\frac{5}{6}$

61. Each part is $\frac{1}{2}$ of a whole and there are 11 parts shaded, or 5 wholes and 1 more part.

a. $\frac{11}{2}$

b. $5\frac{1}{2}$

65. $3\frac{3}{5} = \frac{5 \cdot 3 + 3}{5} = \frac{15 + 3}{5} = \frac{18}{5}$

69. $2\frac{11}{15} = \frac{15 \cdot 2 + 11}{15} = \frac{30 + 11}{15} = \frac{41}{15}$

73. $6\frac{6}{13} = \frac{13 \cdot 6 + 6}{13} = \frac{78 + 6}{13} = \frac{84}{13}$

77. $17\frac{7}{12} = \frac{12 \cdot 17 + 7}{12} = \frac{204 + 7}{12} = \frac{211}{12}$

81. $2\frac{51}{107} = \frac{107 \cdot 2 + 51}{107} = \frac{214 + 51}{107} = \frac{265}{107}$

85.
$$\begin{array}{r} 3\,\text{R}\,2 \\ 5\overline{)17} \\ -15 \\ \hline 2 \end{array}$$
$\frac{17}{5} = 3\frac{2}{5}$

89.
$$\begin{array}{r} 3\,\text{R}\,2 \\ 15\overline{)47} \\ -45 \\ \hline 2 \end{array}$$
$\frac{47}{15} = 3\frac{2}{15}$

93.
$$\begin{array}{r} 33 \\ 6\overline{)198} \\ -18 \\ \hline 18 \\ -18 \\ \hline 0 \end{array}$$
$\frac{198}{6} = 33$

97.
$$\begin{array}{r} 66\,\text{R}\,2 \\ 3\overline{)200} \\ -18 \\ \hline 20 \\ -18 \\ \hline 2 \end{array}$$
$\frac{200}{3} = 66\frac{2}{3}$

101.
$$\begin{array}{r} 17\,\text{R}\,13 \\ 18\overline{)319} \\ -18 \\ \hline 139 \\ -126 \\ \hline 13 \end{array}$$
$\frac{319}{18} = 17\frac{13}{18}$

105.
$$\begin{array}{r} 6\,\text{R}\,65 \\ 112\overline{)737} \\ -672 \\ \hline 65 \end{array}$$
$\frac{737}{112} = 6\frac{65}{112}$

109. $5^3 = 5 \cdot 5 \cdot 5 = 125$ **113.** $2 \cdot 2 \cdot 2 \cdot 3 = 2^3 \cdot 3$

117.

$\frac{2}{3}$ is larger than $\frac{1}{2}$.

121. $11{,}000 + 99{,}000 + 18{,}000 + 11{,}000 = 139{,}000$

$\frac{18{,}000}{139{,}000}$ of the employees work in the North American region.

Exercise Set 2.2

1. $1 \cdot 8 = 8$
$2 \cdot 4 = 8$
The factors of 8 are 1, 2, 4, and 8.

5. $1 \cdot 4 = 4$
$2 \cdot 2 = 4$
The factors of 4 are 1, 2, and 4.

9. $1 \cdot 29 = 29$
The factors of 29 are 1 and 29.

13. $1 \cdot 12 = 12$
$2 \cdot 6 = 12$
$3 \cdot 4 = 12$
The factors of 12 are 1, 2, 3, 4, 6, and 12.

17. 7 is prime, as its only factors are 1 and 7.

21. 23 is prime, since its only factors are 1 and 23.

25. 67 is prime, since its only factors are 1 and 67.

29. 31 is prime, since its only factors are 1 and 31.

33. 119 is composite, since its factors are 1, 7, 17, and 119.

37. $3\overline{)15}\,^{5}$

$15 = 3 \cdot 5$

41. $3\overline{)\,9}\,^{3}$
$2\overline{)18}$
$2\overline{)36}$

$36 = 2^2 \cdot 3^2$

45. $3\overline{)15}\,^{5}$
$2\overline{)30}$
$2\overline{)60}$

$60 = 2^2 \cdot 3 \cdot 5$

49. $5\overline{)85}\,^{17}$

$85 = 5 \cdot 17$

53. $7\overline{)\,77}\,^{11}$
$2\overline{)154}$

$154 = 2 \cdot 7 \cdot 11$

57. $3\overline{)\,15}\,^{5}$
$2\overline{)\,30}$
$2\overline{)\,60}$
$2\overline{)120}$
$2\overline{)240}$

$240 = 2^4 \cdot 3 \cdot 5$

61. $7\overline{)\,49}\,^{7}$
$3\overline{)147}$
$3\overline{)441}$
$2\overline{)882}$

$882 = 2 \cdot 3^2 \cdot 7^2$

65. $3\overline{)33}\,^{11}$

$33 = 3 \cdot 11$

69. 67 is prime, since its only factors are 1 and 67.

73. 97 is prime, since its only factors are 1 and 97.

77. To round 4267 to the nearest hundred, observe that the digit in the tens place is 6. Since this digit is 5 or greater, we add 1 to the digit in the hundreds place. The number 4267 rounded to the nearest hundred is 4300.

81. To round 19,764 to the nearest thousand, observe that the digit in the hundreds place is 7. Since this digit is 5 or greater, we add 1 to the digit in the thousands place. The number 19,764 rounded to the nearest thousand is 20,000.

85. $563 + 676 + 1136 = 2375$

There were a total of 2375 patents granted for the years shown.

$\dfrac{1136}{2375}$ of the patents were granted in 2012.

89. answers may vary

Exercise Set 2.3

1. $\dfrac{3}{12} = \dfrac{3}{3 \cdot 4} = \dfrac{1 \cdot 3}{4 \cdot 3} = \dfrac{1}{4}$

5. $\dfrac{14}{16} = \dfrac{2 \cdot 7}{2 \cdot 8} = \dfrac{7}{8}$

9. $\dfrac{35}{50} = \dfrac{5 \cdot 7}{5 \cdot 10} = \dfrac{7}{10}$

13. $\dfrac{24}{40} = \dfrac{8 \cdot 3}{8 \cdot 5} = \dfrac{3}{5}$

17. $\dfrac{25}{40} = \dfrac{5 \cdot 5}{5 \cdot 8} = \dfrac{5}{8}$

21. $\dfrac{56}{68} = \dfrac{4 \cdot 14}{4 \cdot 17} = \dfrac{14}{17}$

25. $\dfrac{90}{120} = \dfrac{30 \cdot 3}{30 \cdot 4} = \dfrac{3}{4}$

29. $\dfrac{66}{308} = \dfrac{22 \cdot 3}{22 \cdot 14} = \dfrac{3}{14}$

33. $\dfrac{75}{350} = \dfrac{25 \cdot 3}{25 \cdot 14} = \dfrac{3}{14}$

37. $\dfrac{288}{480} = \dfrac{96 \cdot 3}{96 \cdot 5} = \dfrac{3}{5}$

41. Equivalent, since the cross products are equal: $6 \cdot 4 = 24$ and $8 \cdot 3 = 24$

45. Equivalent, since the cross products are equal: $15 \cdot 6 = 90$ and $9 \cdot 10 = 90$

49. Not equivalent, since the cross products are not equal: $13 \cdot 12 = 156$ and $15 \cdot 10 = 150$

53. $\dfrac{2 \text{ hours}}{8 \text{ hours}} = \dfrac{2 \cdot 1}{2 \cdot 4} = \dfrac{1}{4}$

2 hours represents $\dfrac{1}{4}$ of a work shift.

57. a. $\dfrac{16 \text{ states}}{50 \text{ states}} = \dfrac{2 \cdot 8}{2 \cdot 25} = \dfrac{8}{25}$

$\dfrac{8}{25}$ of the states can claim at least one Ritz-Carlton hotel.

  b. $50 - 16 = 34$

34 states do not have a Ritz-Carlton hotel.

  c. $\dfrac{34}{50} = \dfrac{2 \cdot 17}{2 \cdot 25} = \dfrac{17}{25}$

$\dfrac{17}{25}$ of the states do not have a Ritz-Carlton hotel.

61. a. $50 - 20 = 30$

30 states do not have a Safeway-owned store.

  b. $\dfrac{30 \text{ states}}{50 \text{ states}} = \dfrac{3 \cdot 10}{5 \cdot 10} = \dfrac{3}{5}$

$\dfrac{3}{5}$ of the states do not have a Safeway-owned store.

65. $\begin{array}{r} 91 \\ \times\; 4 \\ \hline 364 \end{array}$

69. $\begin{array}{r} 72 \\ \times\; 35 \\ \hline 360 \\ 2160 \\ \hline 2520 \end{array}$

73. $\dfrac{3975}{6625} = \dfrac{3 \cdot 1325}{5 \cdot 1325} = \dfrac{3}{5}$

77. $3 + 1 = 4$ blood donors have an AB blood type.

$\dfrac{4 \text{ donors}}{100 \text{ donors}} = \dfrac{4 \cdot 1}{4 \cdot 25} = \dfrac{1}{25}$

$\dfrac{1}{25}$ of blood donors have an AB blood type.

81. answers may vary

85. answers may vary

89. The numbers are also divisible by 6; answers may vary

Exercise Set 2.4

1. $\dfrac{1}{3} \cdot \dfrac{2}{5} = \dfrac{1 \cdot 2}{3 \cdot 5} = \dfrac{2}{15}$

5. $\dfrac{3}{10} \cdot \dfrac{3}{8} = \dfrac{3 \cdot 3}{10 \cdot 8} = \dfrac{9}{80}$

9. $\dfrac{16}{5} \cdot \dfrac{3}{4} = \dfrac{16 \cdot 3}{5 \cdot 4} = \dfrac{4 \cdot 4 \cdot 3}{5 \cdot 4} = \dfrac{4 \cdot 3}{5} = \dfrac{12}{5}$ or $2\dfrac{2}{5}$

13. $0 \cdot \dfrac{8}{9} = 0$

17. $\dfrac{18}{20} \cdot \dfrac{36}{99} = \dfrac{18 \cdot 36}{20 \cdot 99} = \dfrac{9 \cdot 2 \cdot 4 \cdot 9}{4 \cdot 5 \cdot 9 \cdot 11} = \dfrac{2 \cdot 9}{5 \cdot 11} = \dfrac{18}{55}$

21. $\dfrac{11}{20} \cdot \dfrac{1}{7} \cdot \dfrac{5}{22} = \dfrac{11 \cdot 1 \cdot 5}{20 \cdot 7 \cdot 22}$

$= \dfrac{11 \cdot 1 \cdot 5}{5 \cdot 4 \cdot 7 \cdot 11 \cdot 2}$

$= \dfrac{1}{4 \cdot 7 \cdot 2}$

$= \dfrac{1}{56}$

25. $\dfrac{9}{20} \cdot 0 \cdot \dfrac{4}{19} = 0$

29. $7\dfrac{7}{8}$ rounds to 8.

33. $19\dfrac{11}{20}$ rounds to 20.

37. $\dfrac{5}{8} \cdot 4 = \dfrac{5}{8} \cdot \dfrac{4}{1} = \dfrac{5 \cdot 4}{8 \cdot 1} = \dfrac{5 \cdot 4}{2 \cdot 4 \cdot 1} = \dfrac{5}{2}$ or $2\dfrac{1}{2}$

41. $\dfrac{2}{5} \cdot 4\dfrac{1}{6} = \dfrac{2}{5} \cdot \dfrac{25}{6} = \dfrac{2 \cdot 25}{5 \cdot 6} = \dfrac{2 \cdot 5 \cdot 5}{5 \cdot 2 \cdot 3} = \dfrac{5}{3}$ or $1\dfrac{2}{3}$

45. Exact: $2\dfrac{1}{5} \cdot 3\dfrac{1}{2} = \dfrac{11}{5} \cdot \dfrac{7}{2} = \dfrac{11 \cdot 7}{5 \cdot 2} = \dfrac{77}{10}$ or $7\dfrac{7}{10}$

Estimate: $2\dfrac{1}{5}$ rounds to 2, $3\dfrac{1}{2}$ rounds to 4. $2 \cdot 4 = 8$, so the answer is reasonable.

49. $5 \cdot 2\dfrac{1}{2} = \dfrac{5}{1} \cdot \dfrac{5}{2} = \dfrac{5 \cdot 5}{1 \cdot 2} = \dfrac{25}{2}$ or $12\dfrac{1}{2}$

53. $\dfrac{3}{4} \cdot 16 \cdot \dfrac{1}{2} = \dfrac{3}{4} \cdot \dfrac{16}{1} \cdot \dfrac{1}{2}$

$= \dfrac{3 \cdot 16 \cdot 1}{4 \cdot 1 \cdot 2}$

$= \dfrac{3 \cdot 2 \cdot 2 \cdot 4 \cdot 1}{4 \cdot 1 \cdot 2}$

$= \dfrac{3 \cdot 2}{1}$

$= 6$

57. $3\dfrac{1}{2} \cdot 1\dfrac{3}{4} \cdot 2\dfrac{2}{3} = \dfrac{7}{2} \cdot \dfrac{7}{4} \cdot \dfrac{8}{3}$

$= \dfrac{7 \cdot 7 \cdot 8}{2 \cdot 4 \cdot 3}$

$= \dfrac{7 \cdot 7 \cdot 2 \cdot 4}{2 \cdot 4 \cdot 3}$

$= \dfrac{7 \cdot 7}{3}$

$= \dfrac{49}{3}$ or $16\dfrac{1}{3}$

61. $\dfrac{19}{37} \cdot 0 = 0$

65. $\dfrac{3}{2} \cdot \dfrac{7}{3} = \dfrac{3 \cdot 7}{2 \cdot 3} = \dfrac{7}{2}$ or $3\dfrac{1}{2}$

69. $\dfrac{7}{72} \cdot \dfrac{9}{49} = \dfrac{7 \cdot 9}{72 \cdot 49} = \dfrac{7 \cdot 9}{8 \cdot 9 \cdot 7 \cdot 7} = \dfrac{1}{8 \cdot 7} = \dfrac{1}{56}$

73. $9\dfrac{5}{7} \cdot 8\dfrac{1}{5} \cdot 0 = 0$

77. $\dfrac{1}{4}$ of $200 = \dfrac{1}{4} \cdot 200$

$= \dfrac{1}{4} \cdot \dfrac{200}{1}$

$= \dfrac{1 \cdot 200}{4 \cdot 1}$

$= \dfrac{1 \cdot 4 \cdot 50}{4 \cdot 1}$

$= \dfrac{50}{1}$

$= 50$

81. $\dfrac{4}{25}$ of $800 = \dfrac{4}{25} \cdot 800$

$= \dfrac{4}{25} \cdot \dfrac{800}{1}$

$= \dfrac{4 \cdot 800}{25 \cdot 1}$

$= \dfrac{4 \cdot 25 \cdot 32}{25 \cdot 1}$

$= \dfrac{4 \cdot 32}{1}$

$= 128$

128 of the students would be expected to major in business.

85. $\dfrac{2}{5}$ of $2170 = \dfrac{2}{5} \cdot 2170$

$= \dfrac{2}{5} \cdot \dfrac{2170}{1}$

$= \dfrac{2 \cdot 2170}{5 \cdot 1}$

$= \dfrac{2 \cdot 5 \cdot 434}{5 \cdot 1}$

$= \dfrac{2 \cdot 434}{1}$

$= 868$

He has hiked 868 miles of the trail.

89. $\dfrac{5}{6}$ of $36 = \dfrac{5}{6} \cdot 36$

$= \dfrac{5}{6} \cdot \dfrac{36}{1}$

$= \dfrac{5 \cdot 36}{6 \cdot 1}$

$= \dfrac{5 \cdot 6 \cdot 6}{6 \cdot 1}$

$= \dfrac{5 \cdot 6}{1}$

$= 30$

There are normally 30 gallons of solution in the vat.

93. $6 \cdot 3\frac{1}{4} = \frac{6}{1} \cdot \frac{13}{4} = \frac{6 \cdot 13}{1 \cdot 4} = \frac{2 \cdot 3 \cdot 13}{1 \cdot 2 \cdot 2} = \frac{3 \cdot 13}{1 \cdot 2} = \frac{39}{2}$ or $19\frac{1}{2}$

The sidewalk is $\frac{39}{2}$ or $19\frac{1}{2}$ inches wide.

97. Area $= $ length $\cdot$ width $= \frac{5}{14} \cdot \frac{1}{5} = \frac{5 \cdot 1}{14 \cdot 5} = \frac{1}{14}$

The area is $\frac{1}{14}$ square foot.

101. $\frac{8}{25} \cdot 12{,}000 = \frac{8}{25} \cdot \frac{12{,}000}{1}$

$= \frac{8 \cdot 12{,}000}{25 \cdot 1}$

$= \frac{8 \cdot 25 \cdot 480}{25 \cdot 1}$

$= \frac{8 \cdot 480}{1}$

$= 3840$

The family drove 3840 miles for work.

105.
$$\begin{array}{r} 206 \\ 8)\overline{1648} \\ -16 \\ \hline 04 \\ -\ 0 \\ \hline 48 \\ -48 \\ \hline 0 \end{array}$$

109. answers may vary

113. $3\frac{1}{5}$ rounds to 3

$4\frac{5}{8}$ rounds to 5

$3 \cdot 5 = 15$

The best estimate is b.

117. If $\frac{2}{3}$ of the students on the second bus are boys, then $\frac{1}{3}$ of the students on that bus are girls.

$\frac{3}{4}$ of $36 = \frac{3}{4} \cdot 36$

$= \frac{3}{4} \cdot \frac{36}{1}$

$= \frac{3 \cdot 36}{4 \cdot 1}$

$= \frac{3 \cdot 4 \cdot 9}{4 \cdot 1}$

$= \frac{3 \cdot 9}{1}$

$= 27$

27 students on the first bus are girls.

$\frac{1}{3}$ of $30 = \frac{1}{3} \cdot 30 = \frac{1}{3} \cdot \frac{30}{1} = \frac{1 \cdot 30}{3 \cdot 1} = \frac{1 \cdot 3 \cdot 10}{3 \cdot 1} = 10$

10 students on the second bus are girls.

$27 + 10 = 37$

There are 37 girls on the two buses.

Exercise Set 2.5

1. The reciprocal of $\frac{4}{7}$ is $\frac{7}{4}$.

5. The reciprocal of $15 = \frac{15}{1}$ is $\frac{1}{15}$.

9. $\frac{2}{3} \div \frac{5}{6} = \frac{2}{3} \cdot \frac{6}{5} = \frac{2 \cdot 6}{3 \cdot 5} = \frac{2 \cdot 3 \cdot 2}{3 \cdot 5} = \frac{2 \cdot 2}{5} = \frac{4}{5}$

13. $\frac{3}{7} \div \frac{5}{6} = \frac{3}{7} \cdot \frac{6}{5} = \frac{3 \cdot 6}{7 \cdot 5} = \frac{18}{35}$

17. $\frac{1}{10} \div \frac{10}{1} = \frac{1}{10} \cdot \frac{1}{10} = \frac{1 \cdot 1}{10 \cdot 10} = \frac{1}{100}$

21. $\frac{5}{8} \div \frac{3}{8} = \frac{5}{8} \cdot \frac{8}{3} = \frac{5 \cdot 8}{8 \cdot 3} = \frac{5}{3}$ or $1\frac{2}{3}$

25. $\frac{2}{37} \div \frac{1}{7} = \frac{2}{37} \cdot \frac{7}{1} = \frac{2 \cdot 7}{37 \cdot 1} = \frac{14}{37}$

29. $\frac{11}{12} \div \frac{11}{12} = \frac{11}{12} \cdot \frac{12}{11} = \frac{11 \cdot 12}{12 \cdot 11} = 1$

33. $0 \div \frac{7}{8} = 0 \cdot \frac{8}{7} = 0$

37. $\frac{2}{3} \div 4 = \frac{2}{3} \div \frac{4}{1} = \frac{2}{3} \cdot \frac{1}{4} = \frac{2 \cdot 1}{3 \cdot 4} = \frac{2 \cdot 1}{3 \cdot 2 \cdot 2} = \frac{1}{3 \cdot 2} = \frac{1}{6}$

41. $2\frac{1}{2} \div \frac{1}{2} = \frac{5}{2} \div \frac{1}{2} = \frac{5}{2} \cdot \frac{2}{1} = \frac{5 \cdot 2}{2 \cdot 1} = \frac{5}{1} = 5$

45. $3\frac{3}{7} \div 3\frac{1}{3} = \frac{24}{7} \div \frac{10}{3}$

$= \frac{24}{7} \cdot \frac{3}{10}$

$= \frac{24 \cdot 3}{7 \cdot 10}$

$= \frac{2 \cdot 12 \cdot 3}{7 \cdot 2 \cdot 5}$

$= \frac{12 \cdot 3}{7 \cdot 5}$

$= \frac{36}{35}$ or $1\frac{1}{35}$

49. $0 \div 15\frac{4}{7} = 0 \div \frac{109}{7} = 0 \cdot \frac{7}{109} = 0$

53. $1 \div \frac{18}{35} = 1 \cdot \frac{35}{18} = \frac{35}{18}$ or $1\frac{17}{18}$

57. $\frac{6}{15} \div \frac{12}{5} = \frac{6}{15} \cdot \frac{5}{12} = \frac{6 \cdot 5}{15 \cdot 12} = \frac{6 \cdot 5}{5 \cdot 3 \cdot 6 \cdot 2} = \frac{1}{3 \cdot 2} = \frac{1}{6}$

61. $12 \div \frac{1}{8} = 12 \cdot \frac{8}{1} = 12 \cdot 8 = 96$

65. $2\frac{3}{8} \div 0$ is undefined.

69. $4\frac{5}{11} \div 1\frac{2}{5} = \frac{49}{11} \div \frac{7}{5}$

$= \frac{49}{11} \cdot \frac{5}{7}$

$= \frac{49 \cdot 5}{11 \cdot 7}$

$= \frac{7 \cdot 7 \cdot 5}{11 \cdot 7}$

$= \frac{7 \cdot 5}{11}$

$= \frac{35}{11}$ or $3\frac{2}{11}$

73. $12\frac{3}{4} \div 4 = \frac{51}{4} \div \frac{4}{1} = \frac{51}{4} \cdot \frac{1}{4} = \frac{51}{16}$ or $3\frac{3}{16}$

The patient walked $3\frac{3}{16}$ miles per day.

77. $15\dfrac{1}{5} \div 24 = \dfrac{76}{5} \div \dfrac{24}{1}$

$\qquad = \dfrac{76}{5} \cdot \dfrac{1}{24}$

$\qquad = \dfrac{76 \cdot 1}{5 \cdot 24}$

$\qquad = \dfrac{4 \cdot 19 \cdot 1}{5 \cdot 4 \cdot 6}$

$\qquad = \dfrac{19 \cdot 1}{5 \cdot 6}$

$\qquad = \dfrac{19}{30}$

On average, $\dfrac{19}{30}$ inch of rain fell per hour.

81. $12 \div 2\dfrac{4}{7} = \dfrac{12}{1} \div \dfrac{18}{7}$

$\qquad = \dfrac{12}{1} \cdot \dfrac{7}{18}$

$\qquad = \dfrac{12 \cdot 7}{1 \cdot 18}$

$\qquad = \dfrac{6 \cdot 2 \cdot 7}{1 \cdot 6 \cdot 3}$

$\qquad = \dfrac{2 \cdot 7}{1 \cdot 3}$

$\qquad = \dfrac{14}{3} \text{ or } 4\dfrac{2}{3}$

The length of the rectangle is $4\dfrac{2}{3}$ meters.

85. $2\dfrac{2}{3} \div 1\dfrac{1}{16} = \dfrac{8}{3} \div \dfrac{17}{16} = \dfrac{8}{3} \cdot \dfrac{16}{17} = \dfrac{8 \cdot 16}{3 \cdot 17} = \dfrac{128}{51} \text{ or } 2\dfrac{26}{51}$.

89. $\dfrac{11}{20} \div \dfrac{20}{11} = \dfrac{11}{20} \cdot \dfrac{11}{20} = \dfrac{11 \cdot 11}{20 \cdot 20} = \dfrac{121}{400}$

93. $\begin{array}{r} 968 \\ -772 \\ \hline 196 \end{array}$

97. To divide mixed numbers, first write each mixed number as an improper fraction.

$20\dfrac{2}{3} \div 10\dfrac{1}{2} = \dfrac{62}{3} \div \dfrac{21}{2} = \dfrac{62}{3} \cdot \dfrac{2}{21} = \dfrac{62 \cdot 2}{3 \cdot 21} = \dfrac{124}{63} \text{ or } 1\dfrac{61}{63}$

101. $12\dfrac{2}{13}$ rounds to 12.

$3\dfrac{7}{8}$ rounds to 4.

$12 \div 4 = 3$

The best estimate is d.

105. $64 \div \dfrac{32}{317} = \dfrac{64}{1} \cdot \dfrac{317}{32} = \dfrac{2 \cdot 32 \cdot 317}{1 \cdot 32} = \dfrac{2 \cdot 317}{1} = 634$

The FedEx Express air fleet is 634 aircraft.

Chapter 2 Test

1. 7 of the 16 equal parts are shaded: $\dfrac{7}{16}$

5. $\begin{array}{r} 4 \text{ R } 3 \\ 5\overline{)23} \\ -20 \\ \hline 3 \end{array}$

$\dfrac{23}{5} = 4\dfrac{3}{5}$

9. Not equivalent, since the cross products are not equal: $7 \cdot 8 = 56$ and $11 \cdot 5 = 55$

13. $\dfrac{4}{4} \div \dfrac{3}{4} = \dfrac{4}{4} \cdot \dfrac{4}{3} = \dfrac{4 \cdot 4}{4 \cdot 3} = \dfrac{4}{3} \text{ or } 1\dfrac{1}{3}$

17. $8 \div \dfrac{1}{2} = \dfrac{8}{1} \cdot \dfrac{2}{1} = \dfrac{8 \cdot 2}{1 \cdot 1} = 16$

21. $\dfrac{16}{3} \div \dfrac{3}{12} = \dfrac{16}{3} \cdot \dfrac{12}{3}$

$\qquad = \dfrac{16 \cdot 12}{3 \cdot 3}$

$\qquad = \dfrac{16 \cdot 3 \cdot 4}{3 \cdot 3}$

$\qquad = \dfrac{16 \cdot 4}{3}$

$\qquad = \dfrac{64}{3} \text{ or } 21\dfrac{1}{3}$

25. $\dfrac{2}{3} \cdot 1\dfrac{8}{9} = \dfrac{2}{3} \cdot \dfrac{17}{9} = \dfrac{2 \cdot 17}{3 \cdot 9} = \dfrac{34}{27} \text{ or } 1\dfrac{7}{27}$

The area is $\dfrac{34}{27}$ or $1\dfrac{7}{27}$ square miles.

Chapter 3

Exercise Set 3.1

1. $\dfrac{1}{7} + \dfrac{2}{7} = \dfrac{1 + 2}{7} = \dfrac{3}{7}$

5. $\dfrac{2}{9} + \dfrac{4}{9} = \dfrac{2 + 4}{9} = \dfrac{6}{9} = \dfrac{3 \cdot 2}{3 \cdot 3} = \dfrac{2}{3}$

9. $\dfrac{3}{14} + \dfrac{4}{14} = \dfrac{3 + 4}{14} = \dfrac{7}{14} = \dfrac{7 \cdot 1}{7 \cdot 2} = \dfrac{1}{2}$

13. $\dfrac{4}{13} + \dfrac{2}{13} + \dfrac{1}{13} = \dfrac{4 + 2 + 1}{13} = \dfrac{7}{13}$

17. $\dfrac{10}{11} - \dfrac{4}{11} = \dfrac{10 - 4}{11} = \dfrac{6}{11}$

21. $\dfrac{7}{4} - \dfrac{3}{4} = \dfrac{7 - 3}{4} = \dfrac{4}{4} = 1$

25. $\dfrac{25}{12} - \dfrac{15}{12} = \dfrac{25 - 15}{12} = \dfrac{10}{12} = \dfrac{2 \cdot 5}{2 \cdot 6} = \dfrac{5}{6}$

29. $\dfrac{86}{90} - \dfrac{85}{90} = \dfrac{86 - 85}{90} = \dfrac{1}{90}$

33. $\dfrac{8}{21} + \dfrac{5}{21} = \dfrac{8 + 5}{21} = \dfrac{13}{21}$

37. $\dfrac{13}{28} - \dfrac{13}{28} = \dfrac{13 - 13}{28} = \dfrac{0}{28} = 0$

41. The perimeter is the distance around.

Perimeter $= \dfrac{4}{20} + \dfrac{7}{20} + \dfrac{9}{20}$

$\qquad = \dfrac{4 + 7 + 9}{20}$

$\qquad = \dfrac{20}{20}$

$\qquad = 1$

The perimeter is 1 inch.

45. To find the remaining amount of track to be inspected, subtract the $\dfrac{5}{20}$ mile that has already been inspected from the $\dfrac{19}{20}$ mile total that must be inspected.

$$\frac{19}{20} - \frac{5}{20} = \frac{19-5}{20} = \frac{14}{20} = \frac{2 \cdot 7}{2 \cdot 10} = \frac{7}{10}$$

$\frac{7}{10}$ of a mile of track remains to be inspected.

49. To find the part spent on eating and drinking, grooming, and sleeping, add the part spent eating and drinking $\left(\frac{4}{100}\right)$, the part spent grooming $\left(\frac{3}{100}\right)$, and the part spent sleeping $\left(\frac{36}{100}\right)$.

$$\frac{4}{100} + \frac{3}{100} + \frac{36}{100} = \frac{4+3+36}{100} = \frac{43}{100}$$

$\frac{43}{100}$ of a full-time college students' weekday is spent on eating and drinking, grooming, and sleeping.

53. To find the fraction of the total retail sales that were in the two categories, add the fraction that were in the health, beauty, and household essentials category $\left(\frac{5}{20}\right)$ and the fraction that were in the food and pet supplies category $\left(\frac{4}{20}\right)$.

$$\frac{5}{20} + \frac{4}{20} = \frac{5+4}{20} = \frac{9}{20}$$

$\frac{9}{20}$ of Target's total retail sales were in the categories of health, beauty, and household essentials and food and pet supplies.

57. North America takes up $\frac{16}{100}$ of the world's land area, while South America takes up $\frac{12}{100}$ of the land area.

$$\frac{16}{100} + \frac{12}{100} = \frac{16+12}{100} = \frac{28}{100} = \frac{4 \cdot 7}{4 \cdot 25} = \frac{7}{25}$$

$\frac{7}{25}$ of the world's land area is within North America and South America.

61. To find the fraction of U.S. theater screens that were digital, add the fraction that were digital 3-D $\left(\frac{17}{50}\right)$ and the fraction that were digital non-3-D $\left(\frac{25}{50}\right)$.

$$\frac{17}{50} + \frac{25}{50} = \frac{17+25}{50} = \frac{42}{50} = \frac{2 \cdot 21}{2 \cdot 25} = \frac{21}{25}$$

$\frac{21}{25}$ of U.S. theater screens were digital in 2012.

65.
$$\begin{array}{r} 2 \\ 2\overline{)4} \\ 2\overline{)8} \end{array}$$
$8 = 2 \cdot 2 \cdot 2 = 2^3$

69. $\frac{3}{8} + \frac{7}{8} - \frac{5}{8} = \frac{3+7-5}{8} = \frac{10-5}{8} = \frac{5}{8}$

73. $\frac{2}{7} + \frac{9}{7} = \frac{2+9}{7} = \frac{11}{7}$ or $1\frac{4}{7}$

77. $\frac{4}{100} + \frac{14}{100} + \frac{3}{100} + \frac{16}{100} + \frac{36}{100} + \frac{6}{100} + \frac{11}{100} + \frac{10}{100}$

$= \frac{4+14+3+16+36+6+11+10}{100}$

$= \frac{100}{100}$

$= 1$

answers may vary

Exercise Set 3.2

1. Multiples of 3: 3, 6, 9, $\boxed{12}$, 15, . . .
Multiples of 4: 4, 8, $\boxed{12}$, 16, . . .
LCM: 12

5. Multiples of 12: 12, 24, $\boxed{36}$, 48, . . .
Multiples of 18: 18, $\boxed{36}$, 54, . . .
LCM: 36

9. $18 = \boxed{2} \cdot \boxed{3 \cdot 3}$
$21 = 3 \cdot \boxed{7}$
LCM: $2 \cdot 3 \cdot 3 \cdot 7 = 126$

13. $8 = \boxed{2 \cdot 2 \cdot 2}$
$24 = 2 \cdot 2 \cdot 2 \cdot \boxed{3}$
LCM: $2 \cdot 2 \cdot 2 \cdot 3 = 24$

17. $8 = \boxed{2 \cdot 2 \cdot 2}$
$6 = 2 \cdot 3$
$27 = \boxed{3 \cdot 3 \cdot 3}$
LCM: $2 \cdot 2 \cdot 2 \cdot 3 \cdot 3 \cdot 3 = 216$

21. $34 = 2 \cdot \boxed{17}$
$68 = \boxed{2 \cdot 2} \cdot 17$
LCM: $2 \cdot 2 \cdot 17 = 68$

25. $30 = 2 \cdot 3 \cdot 5$
$36 = \boxed{2 \cdot 2} \cdot \boxed{3 \cdot 3}$
$50 = 2 \cdot \boxed{5 \cdot 5}$
LCM: $2 \cdot 2 \cdot 3 \cdot 3 \cdot 5 \cdot 5 = 900$

29. $11 = 11$
$33 = \boxed{3} \cdot 11$
$121 = \boxed{11 \cdot 11}$
LCM: $3 \cdot 11 \cdot 11 = 363$

33. $\frac{4}{7} = \frac{4 \cdot 5}{7 \cdot 5} = \frac{20}{35}$

37. $5 = \frac{5}{1} = \frac{5 \cdot 3}{1 \cdot 3} = \frac{15}{3}$

41. $\frac{10}{7} = \frac{10 \cdot 3}{7 \cdot 3} = \frac{30}{21}$

45. $\frac{2}{3} = \frac{2 \cdot 15}{3 \cdot 15} = \frac{30}{45}$

49. $\frac{15}{13} = \frac{15 \cdot 6}{13 \cdot 6} = \frac{90}{78}$

53. books and magazines: $\frac{43}{50} = \frac{43 \cdot 2}{50 \cdot 2} = \frac{86}{100}$

clothing and accessories: $\frac{4}{5} = \frac{4 \cdot 20}{5 \cdot 20} = \frac{80}{100}$

computer hardware: $\frac{29}{50} = \frac{29 \cdot 2}{50 \cdot 2} = \frac{58}{100}$

computer software: $\frac{17}{25} = \frac{17 \cdot 4}{25 \cdot 4} = \frac{68}{100}$

drugs, health and beauty aids: $\frac{3}{25} = \frac{3 \cdot 4}{25 \cdot 4} = \frac{12}{100}$

electronics and appliances: $\frac{21}{25} = \frac{21 \cdot 4}{25 \cdot 4} = \frac{84}{100}$

food, beer, and wine: $\frac{17}{25} = \frac{17 \cdot 4}{25 \cdot 4} = \frac{68}{100}$

home furnishings: $\frac{81}{100} = \frac{81}{100}$

music and videos: $\dfrac{9}{10} = \dfrac{9 \cdot 10}{10 \cdot 10} = \dfrac{90}{100}$

office equipment and supplies: $\dfrac{79}{100} = \dfrac{79}{100}$

sporting goods: $\dfrac{37}{50} = \dfrac{37 \cdot 2}{50 \cdot 2} = \dfrac{74}{100}$

toys, hobbies, and games: $\dfrac{39}{50} = \dfrac{39 \cdot 2}{50 \cdot 2} = \dfrac{78}{100}$

57. $\dfrac{7}{10} - \dfrac{2}{10} = \dfrac{7-2}{10} = \dfrac{5}{10} = \dfrac{5 \cdot 1}{5 \cdot 2} = \dfrac{1}{2}$

61. $\dfrac{23}{18} - \dfrac{15}{18} = \dfrac{23-15}{18} = \dfrac{8}{18} = \dfrac{2 \cdot 4}{2 \cdot 9} = \dfrac{4}{9}$

65. $\dfrac{37}{165} = \dfrac{37 \cdot 22}{165 \cdot 22} = \dfrac{814}{3630}$

69. a. $\dfrac{10}{15} = \dfrac{5 \cdot 2}{5 \cdot 3} = \dfrac{2}{3}$

   b. $\dfrac{40}{60} = \dfrac{20 \cdot 2}{20 \cdot 3} = \dfrac{2}{3}$

   c. $\dfrac{16}{20} = \dfrac{4 \cdot 4}{4 \cdot 5} = \dfrac{4}{5}$

   d. $\dfrac{200}{300} = \dfrac{100 \cdot 2}{100 \cdot 3} = \dfrac{2}{3}$

   a, b, and d are equivalent to $\dfrac{2}{3}$.

Exercise Set 3.3

1. The LCD of 3 and 6 is 6.

$\dfrac{2}{3} + \dfrac{1}{6} = \dfrac{2}{3} \cdot \dfrac{2}{2} + \dfrac{1}{6} = \dfrac{4}{6} + \dfrac{1}{6} = \dfrac{5}{6}$

5. The LCD of 11 and 33 is 33.

$\dfrac{2}{11} + \dfrac{2}{33} = \dfrac{2}{11} \cdot \dfrac{3}{3} + \dfrac{2}{33} = \dfrac{6}{33} + \dfrac{2}{33} = \dfrac{8}{33}$

9. The LCD of 35 and 7 is 35.

$\dfrac{11}{35} + \dfrac{2}{7} = \dfrac{11}{35} + \dfrac{2}{7} \cdot \dfrac{5}{5}$

$= \dfrac{11}{35} + \dfrac{10}{35}$

$= \dfrac{21}{35}$

$= \dfrac{7 \cdot 3}{7 \cdot 5}$

$= \dfrac{3}{5}$

13. The LCD of 15 and 12 is 60.

$\dfrac{7}{15} + \dfrac{5}{12} = \dfrac{7}{15} \cdot \dfrac{4}{4} + \dfrac{5}{12} \cdot \dfrac{5}{5}$

$= \dfrac{28}{60} + \dfrac{25}{60}$

$= \dfrac{53}{60}$

17. The LCD of 44 and 36 is 396.

$\dfrac{9}{44} + \dfrac{17}{36} = \dfrac{9}{44} \cdot \dfrac{9}{9} + \dfrac{17}{36} \cdot \dfrac{11}{11}$

$= \dfrac{81}{396} + \dfrac{187}{396}$

$= \dfrac{268}{396}$

$= \dfrac{4 \cdot 67}{4 \cdot 99}$

$= \dfrac{67}{99}$

21. The LCD of 3, 9, and 27 is 27.

$\dfrac{1}{3} + \dfrac{1}{9} + \dfrac{1}{27} = \dfrac{1}{3} \cdot \dfrac{9}{9} + \dfrac{1}{9} \cdot \dfrac{3}{3} + \dfrac{1}{27}$

$= \dfrac{9}{27} + \dfrac{3}{27} + \dfrac{1}{27}$

$= \dfrac{13}{27}$

25. The LCD of 36, 4, and 6 is 36.

$\dfrac{5}{36} + \dfrac{3}{4} + \dfrac{1}{6} = \dfrac{5}{36} + \dfrac{3}{4} \cdot \dfrac{9}{9} + \dfrac{1}{6} \cdot \dfrac{6}{6}$

$= \dfrac{5}{36} + \dfrac{27}{36} + \dfrac{6}{36}$

$= \dfrac{38}{36}$

$= \dfrac{2 \cdot 19}{2 \cdot 18}$

$= \dfrac{19}{18}$ or $1\dfrac{1}{18}$

29. The LCD of 8 and 16 is 16.

$\dfrac{7}{8} - \dfrac{3}{16} = \dfrac{7}{8} \cdot \dfrac{2}{2} - \dfrac{3}{16} = \dfrac{14}{16} - \dfrac{3}{16} = \dfrac{11}{16}$

33. The LCD of 7 and 8 is 56.

$\dfrac{5}{7} - \dfrac{1}{8} = \dfrac{5}{7} \cdot \dfrac{8}{8} - \dfrac{1}{8} \cdot \dfrac{7}{7} = \dfrac{40}{56} - \dfrac{7}{56} = \dfrac{33}{56}$

37. The LCD of 35 and 7 is 35.

$\dfrac{11}{35} - \dfrac{2}{7} = \dfrac{11}{35} - \dfrac{2}{7} \cdot \dfrac{5}{5} = \dfrac{11}{35} - \dfrac{10}{35} = \dfrac{1}{35}$

41. The LCD of 15 and 12 is 60.

$\dfrac{7}{15} - \dfrac{5}{12} = \dfrac{7}{15} \cdot \dfrac{4}{4} - \dfrac{5}{12} \cdot \dfrac{5}{5}$

$= \dfrac{28}{60} - \dfrac{25}{60}$

$= \dfrac{3}{60}$

$= \dfrac{3 \cdot 1}{3 \cdot 20}$

$= \dfrac{1}{20}$

45. The LCD of 100 and 1000 is 1000.

$$\frac{1}{100} - \frac{1}{1000} = \frac{1}{100} \cdot \frac{10}{10} - \frac{1}{1000}$$
$$= \frac{10}{1000} - \frac{1}{1000}$$
$$= \frac{9}{1000}$$

49. The LCD of 12 and 9 is 36.

$$\frac{5}{12} + \frac{1}{9} = \frac{5}{12} \cdot \frac{3}{3} + \frac{1}{9} \cdot \frac{4}{4} = \frac{15}{36} + \frac{4}{36} = \frac{19}{36}$$

53. The LCD of 28 and 40 is 280.

$$\frac{9}{28} - \frac{3}{40} = \frac{9}{28} \cdot \frac{10}{10} - \frac{3}{40} \cdot \frac{7}{7} = \frac{90}{280} - \frac{21}{280} = \frac{69}{280}$$

57. Add the lengths of the four sides. The LCD of 3 and 5 is 15.

$$\frac{4}{5} + \frac{1}{3} + \frac{4}{5} + \frac{1}{3} = \frac{4}{5} \cdot \frac{3}{3} + \frac{1}{3} \cdot \frac{5}{5} + \frac{4}{5} \cdot \frac{3}{3} + \frac{1}{3} \cdot \frac{5}{5}$$
$$= \frac{12}{15} + \frac{5}{15} + \frac{12}{15} + \frac{5}{15}$$
$$= \frac{34}{15} \text{ or } 2\frac{4}{15}$$

The perimeter is $\frac{34}{15}$ or $2\frac{4}{15}$ cm.

61. Subtract the speed of the three-toed sloth on the ground from its speed in trees.

The LCD of 10 and 100 is 100.

$$\frac{17}{100} - \frac{1}{10} = \frac{17}{100} - \frac{1}{10} \cdot \frac{10}{10} = \frac{17}{100} - \frac{10}{100} = \frac{7}{100}$$

A sloth can travel $\frac{7}{100}$ mile per hour faster in trees than it can on the ground.

65. Add the lengths shown.

The LCD of 16, 2, and 32 is 32.

$$\frac{5}{16} + \frac{1}{2} + \frac{5}{32} = \frac{5}{16} \cdot \frac{2}{2} + \frac{1}{2} \cdot \frac{16}{16} + \frac{5}{32}$$
$$= \frac{10}{32} + \frac{16}{32} + \frac{5}{32}$$
$$= \frac{31}{32}$$

The total length of the object is $\frac{31}{32}$ inch.

69. The Pacific Ocean takes up $\frac{1}{2}$ of the world's water area, while the Atlantic Ocean takes up $\frac{13}{50}$ of the water area. The LCD of 2 and 50 is 50.

$$\frac{1}{2} + \frac{13}{50} = \frac{1}{2} \cdot \frac{25}{25} + \frac{13}{50}$$
$$= \frac{25}{50} + \frac{13}{50}$$
$$= \frac{38}{50}$$
$$= \frac{2 \cdot 19}{2 \cdot 25}$$
$$= \frac{19}{25}$$

$\frac{19}{25}$ of the world's water area is accounted for by the Pacific Ocean and the Atlantic Ocean.

73. $1 - \frac{19}{100} = \frac{100}{100} - \frac{19}{100} = \frac{81}{100}$

$\frac{81}{100}$ of the areas maintained by the National Park Service are not designated as National Monuments.

77. $4 \div 7\frac{1}{4} = \frac{4}{1} \div \frac{29}{4} = \frac{4}{1} \cdot \frac{4}{29} = \frac{4 \cdot 4}{1 \cdot 29} = \frac{16}{29}$

81. a.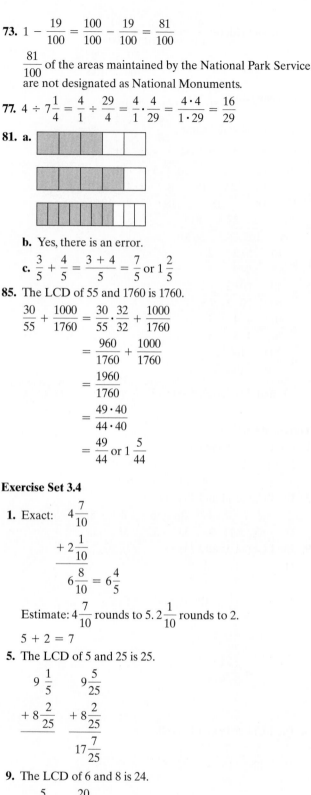

b. Yes, there is an error.

c. $\frac{3}{5} + \frac{4}{5} = \frac{3+4}{5} = \frac{7}{5}$ or $1\frac{2}{5}$

85. The LCD of 55 and 1760 is 1760.

$$\frac{30}{55} + \frac{1000}{1760} = \frac{30}{55} \cdot \frac{32}{32} + \frac{1000}{1760}$$
$$= \frac{960}{1760} + \frac{1000}{1760}$$
$$= \frac{1960}{1760}$$
$$= \frac{49 \cdot 40}{44 \cdot 40}$$
$$= \frac{49}{44} \text{ or } 1\frac{5}{44}$$

Exercise Set 3.4

1. Exact:

$$\begin{array}{r} 4\frac{7}{10} \\ + 2\frac{1}{10} \\ \hline 6\frac{8}{10} = 6\frac{4}{5} \end{array}$$

Estimate: $4\frac{7}{10}$ rounds to 5. $2\frac{1}{10}$ rounds to 2.

$5 + 2 = 7$

5. The LCD of 5 and 25 is 25.

$$\begin{array}{r} 9\frac{1}{5} \qquad 9\frac{5}{25} \\ + 8\frac{2}{25} \qquad + 8\frac{2}{25} \\ \hline 17\frac{7}{25} \end{array}$$

9. The LCD of 6 and 8 is 24.

$$\begin{array}{r} 1\frac{5}{6} \qquad 1\frac{20}{24} \\ + 5\frac{3}{8} \qquad + 5\frac{9}{24} \\ \hline 6\frac{29}{24} = 6 + 1\frac{5}{24} = 7\frac{5}{24} \end{array}$$

13. $11\dfrac{3}{5}$

$+7\dfrac{2}{5}$

$18\dfrac{5}{5} = 18 + 1 = 19$

17. The LCD of 8, 6, and 4 is 24.

$\begin{array}{ll} 3\dfrac{5}{8} & 3\dfrac{15}{24} \\[2mm] 2\dfrac{1}{6} & 2\dfrac{4}{24} \\[2mm] +7\dfrac{3}{4} & +7\dfrac{18}{24} \\[2mm] & 12\dfrac{37}{24} = 12 + 1\dfrac{13}{24} = 13\dfrac{13}{24} \end{array}$

21. Exact: $\quad 4\dfrac{7}{10}$

$-2\dfrac{1}{10}$

$2\dfrac{6}{10} = 2\dfrac{3}{5}$

Estimate: $4\dfrac{7}{10}$ rounds to 5. $2\dfrac{1}{10}$ rounds to 2.

$5 - 2 = 3$

25. The LCD of 5 and 25 is 25.

$\begin{array}{lll} 9\dfrac{1}{5} & 9\dfrac{5}{25} & 8\dfrac{30}{25} \\[2mm] -8\dfrac{6}{25} & -8\dfrac{6}{25} & -8\dfrac{6}{25} \\[2mm] & & \dfrac{24}{25} \end{array}$

29. The LCD of 7 and 14 is 14.

$\begin{array}{lll} 15\dfrac{4}{7} & 15\dfrac{8}{14} & 14\dfrac{22}{14} \\[2mm] -9\dfrac{11}{14} & -9\dfrac{11}{14} & -9\dfrac{11}{14} \\[2mm] & & 5\dfrac{11}{14} \end{array}$

33. $\begin{array}{ll} 10 & 9\dfrac{5}{5} \\[2mm] -8\dfrac{1}{5} & -8\dfrac{1}{5} \\[2mm] & 1\dfrac{4}{5} \end{array}$

37. $\begin{array}{ll} 6 & 5\dfrac{9}{9} \\[2mm] -2\dfrac{4}{9} & -2\dfrac{4}{9} \\[2mm] & 3\dfrac{5}{9} \end{array}$

41. The LCD of 6 and 12 is 12.

$\begin{array}{ll} 15\dfrac{1}{6} & 15\dfrac{2}{12} \\[2mm] +13\dfrac{5}{12} & +13\dfrac{5}{12} \\[2mm] & 28\dfrac{7}{12} \end{array}$

45. $5\dfrac{8}{9}$

$+2\dfrac{1}{9}$

$7\dfrac{9}{9} = 7 + 1 = 8$

49. Subtract the diameter for Eastern Bluebirds from the diameter for Mountain Bluebirds.

The LCD of 2 and 16 is 16.

$\begin{array}{ll} 1\dfrac{9}{16} & 1\dfrac{9}{16} \\[2mm] -1\dfrac{1}{2} & -1\dfrac{8}{16} \\[2mm] & \dfrac{1}{16} \end{array}$

The entrance holes for Mountain Bluebirds should be $\dfrac{1}{16}$ inch wider.

53. Subtract Yuma's average annual rainfall from Tucson's average annual rainfall. The LCD of 4 and 5 is 20.

$\begin{array}{lll} 11\dfrac{1}{4} & 11\dfrac{5}{20} & 10\dfrac{25}{20} \\[2mm] -3\dfrac{3}{5} & -3\dfrac{12}{20} & -3\dfrac{12}{20} \\[2mm] & & 7\dfrac{13}{20} \end{array}$

On average, Tucson gets $7\dfrac{13}{20}$ inches more rain annually than Yuma.

57. Subtract the time for a personal return from the time for a small business return.

The LCD of 2 and 8 is 8.

$\begin{array}{ll} 5\dfrac{7}{8} & 5\dfrac{7}{8} \\[2mm] -3\dfrac{1}{2} & -3\dfrac{4}{8} \\[2mm] & 2\dfrac{3}{8} \end{array}$

It takes him $2\dfrac{3}{8}$ hours longer to prepare a small business return.

61. Subtract the length of the Lacey V. Murrow Memorial Bridge from the length of the Evergreen Point Bridge.

$\begin{array}{ll} 2526 & 2525\dfrac{3}{3} \\[2mm] -2206\dfrac{2}{3} & -2206\dfrac{1}{3} \\[2mm] & 319\dfrac{1}{3} \end{array}$

The Evergreen Point Bridge is $319\dfrac{1}{3}$ yards longer than the Lacey V. Murrow Memorial Bridge.

65. Subtract the length of the August 21, 2017, eclipse from the length of the April 8, 2024, eclipse. The LCD of 3 and 15 is 15.

$$4\frac{7}{15} \qquad 4\frac{7}{15} \qquad 3\frac{22}{15}$$
$$-2\frac{2}{3} \qquad -2\frac{10}{15} \qquad -2\frac{10}{15}$$
$$\rule{2cm}{0.4pt} \qquad \rule{2cm}{0.4pt} \qquad 1\frac{12}{15} = 1\frac{4}{5}$$

The April 8, 2024, eclipse will be $1\frac{4}{5}$ minutes longer than the August 21, 2017, eclipse.

69. Add the lengths of the sides. The LCD of 3 and 8 is 24.

$$5\frac{1}{3} \qquad\quad 5\frac{8}{24}$$
$$5 \qquad\qquad 5$$
$$+7\frac{7}{8} \qquad +7\frac{21}{24}$$
$$\rule{2cm}{0.4pt}$$
$$20\frac{29}{24} = 20 + 1\frac{5}{24} = 21\frac{5}{24}$$

The perimeter is $21\frac{5}{24}$ meters.

73. $5^2 = 5 \cdot 5 = 25$

77.
$$2 + 3(8 \cdot 7 - 1) = 2 + 3(56 - 1)$$
$$= 2 + 3(55)$$
$$= 2 + 165$$
$$= 167$$

81. $9\frac{10}{16} = 9 + \frac{2 \cdot 5}{2 \cdot 8} = 9\frac{5}{8}$

85. answers may vary

Exercise Set 3.5

1. Since $7 > 6$, then $\frac{7}{9} > \frac{6}{9}$.

5. The LCD of 42 and 21 is 42.

$\frac{9}{42}$ has a denominator of 42.

$$\frac{5}{21} = \frac{5}{21} \cdot \frac{2}{2} = \frac{10}{42}$$

Since $9 < 10$, then $\frac{9}{42} < \frac{10}{42}$, so $\frac{9}{42} < \frac{5}{21}$.

9. The LCD of 4 and 3 is 12.

$$\frac{3}{4} = \frac{3}{4} \cdot \frac{3}{3} = \frac{9}{12}$$
$$\frac{2}{3} = \frac{2}{3} \cdot \frac{4}{4} = \frac{8}{12}$$

Since $9 > 8$, then $\frac{9}{12} > \frac{8}{12}$, so $\frac{3}{4} > \frac{2}{3}$.

13. The LCD of 10 and 11 is 110.

$$\frac{1}{10} = \frac{1}{10} \cdot \frac{11}{11} = \frac{11}{110}$$
$$\frac{1}{11} = \frac{1}{11} \cdot \frac{10}{10} = \frac{10}{110}$$

Since $11 > 10$, then $\frac{11}{110} > \frac{10}{110}$, so $\frac{1}{10} > \frac{1}{11}$.

17. $\left(\frac{1}{2}\right)^4 = \frac{1}{2} \cdot \frac{1}{2} \cdot \frac{1}{2} \cdot \frac{1}{2} = \frac{1}{16}$

21. $\left(\frac{4}{7}\right)^3 = \frac{4}{7} \cdot \frac{4}{7} \cdot \frac{4}{7} = \frac{64}{343}$

25.
$$\left(\frac{3}{4}\right)^2 \cdot \left(\frac{2}{3}\right)^3 = \left(\frac{3}{4} \cdot \frac{3}{4}\right) \cdot \left(\frac{2}{3} \cdot \frac{2}{3} \cdot \frac{2}{3}\right)$$
$$= \frac{3 \cdot 3 \cdot 2 \cdot 2 \cdot 2}{4 \cdot 4 \cdot 3 \cdot 3 \cdot 3}$$
$$= \frac{3 \cdot 3 \cdot 2 \cdot 2 \cdot 2}{2 \cdot 2 \cdot 2 \cdot 2 \cdot 3 \cdot 3 \cdot 3}$$
$$= \frac{1}{2 \cdot 3}$$
$$= \frac{1}{6}$$

29. The LCD of 15 and 5 is 15.

$$\frac{2}{15} + \frac{3}{5} = \frac{2}{15} + \frac{3}{5} \cdot \frac{3}{3} = \frac{2}{15} + \frac{9}{15} = \frac{11}{15}$$

33. $1 - \frac{4}{9} = \frac{9}{9} - \frac{4}{9} = \frac{5}{9}$

37. The LCD of 6 and 4 is 12.

$$\frac{5}{6} - \frac{3}{4} = \frac{5}{6} \cdot \frac{2}{2} - \frac{3}{4} \cdot \frac{3}{3} = \frac{10}{12} - \frac{9}{12} = \frac{1}{12}$$

41. $0 \cdot \frac{9}{10} = 0$

45. $\frac{20}{35} \cdot \frac{7}{10} = \frac{20 \cdot 7}{35 \cdot 10} = \frac{10 \cdot 2 \cdot 7}{7 \cdot 5 \cdot 10} = \frac{2}{5}$

49. $\frac{1}{5} + \frac{1}{3} \cdot \frac{1}{4} = \frac{1}{5} + \frac{1 \cdot 1}{3 \cdot 4} = \frac{1}{5} + \frac{1}{12}$

The LCD of 5 and 12 is 60.

$$\frac{1}{5} + \frac{1}{12} = \frac{1}{5} \cdot \frac{12}{12} + \frac{1}{12} \cdot \frac{5}{5} = \frac{12}{60} + \frac{5}{60} = \frac{17}{60}$$

53.
$$\frac{1}{5}\left(2\frac{5}{6} - \frac{1}{3}\right) = \frac{1}{5}\left(\frac{17}{6} - \frac{1}{3} \cdot \frac{2}{2}\right)$$
$$= \frac{1}{5}\left(\frac{17}{6} - \frac{2}{6}\right)$$
$$= \frac{1}{5}\left(\frac{15}{6}\right)$$
$$= \frac{1 \cdot 15}{5 \cdot 6}$$
$$= \frac{1 \cdot 5 \cdot 3}{5 \cdot 3 \cdot 2}$$
$$= \frac{1}{2}$$

57. $\left(\dfrac{3}{4}\right)^2 \div \left(\dfrac{3}{4} - \dfrac{1}{12}\right) = \left(\dfrac{3}{4}\right)^2 \div \left(\dfrac{3}{4} \cdot \dfrac{3}{3} - \dfrac{1}{12}\right)$

$= \left(\dfrac{3}{4}\right)^2 \div \left(\dfrac{9}{12} - \dfrac{1}{12}\right)$

$= \left(\dfrac{3}{4}\right)^2 \div \left(\dfrac{8}{12}\right)$

$= \left(\dfrac{3}{4}\right)^2 \div \left(\dfrac{4 \cdot 2}{4 \cdot 3}\right)$

$= \left(\dfrac{3}{4}\right)^2 \div \left(\dfrac{2}{3}\right)$

$= \left(\dfrac{3}{4} \cdot \dfrac{3}{4}\right) \div \left(\dfrac{2}{3}\right)$

$= \dfrac{9}{16} \div \dfrac{2}{3}$

$= \dfrac{9}{16} \cdot \dfrac{3}{2}$

$= \dfrac{9 \cdot 3}{16 \cdot 2}$

$= \dfrac{27}{32}$

61. $\dfrac{5}{9} \cdot \dfrac{1}{2} + \dfrac{2}{3} \cdot \dfrac{5}{6} = \dfrac{5 \cdot 1}{9 \cdot 2} + \dfrac{2 \cdot 5}{3 \cdot 6}$

$= \dfrac{5}{18} + \dfrac{10}{18}$

$= \dfrac{15}{18}$

$= \dfrac{3 \cdot 5}{3 \cdot 6}$

$= \dfrac{5}{6}$

65. $\dfrac{3}{13} \div \dfrac{9}{26} - \dfrac{7}{24} \cdot \dfrac{8}{14} = \dfrac{3}{13} \cdot \dfrac{26}{9} - \dfrac{7}{24} \cdot \dfrac{8}{14}$

$= \dfrac{3 \cdot 26}{13 \cdot 9} - \dfrac{7 \cdot 8}{24 \cdot 14}$

$= \dfrac{3 \cdot 13 \cdot 2}{13 \cdot 3 \cdot 3} - \dfrac{7 \cdot 8}{8 \cdot 3 \cdot 7 \cdot 2}$

$= \dfrac{2}{3} - \dfrac{1}{6}$

$= \dfrac{2}{3} \cdot \dfrac{2}{2} - \dfrac{1}{6}$

$= \dfrac{4}{6} - \dfrac{1}{6}$

$= \dfrac{3}{6}$

$= \dfrac{1}{2}$

69. $\left(\dfrac{3}{4} + \dfrac{1}{8}\right)^2 - \left(\dfrac{1}{2} + \dfrac{1}{8}\right) = \left(\dfrac{3}{4} \cdot \dfrac{2}{2} + \dfrac{1}{8}\right)^2 - \left(\dfrac{1}{2} \cdot \dfrac{4}{4} + \dfrac{1}{8}\right)$

$= \left(\dfrac{6}{8} + \dfrac{1}{8}\right)^2 - \left(\dfrac{4}{8} + \dfrac{1}{8}\right)$

$= \left(\dfrac{7}{8}\right)^2 - \left(\dfrac{5}{8}\right)$

$= \left(\dfrac{7}{8} \cdot \dfrac{7}{8}\right) - \dfrac{5}{8}$

$= \dfrac{49}{64} - \dfrac{5}{8}$

$= \dfrac{49}{64} - \dfrac{5}{8} \cdot \dfrac{8}{8}$

$= \dfrac{49}{64} - \dfrac{40}{64}$

$= \dfrac{9}{64}$

73. The average is the sum, divided by 3.

$\left(\dfrac{1}{5} + \dfrac{3}{10} + \dfrac{3}{20}\right) \div 3 = \left(\dfrac{1}{5} \cdot \dfrac{4}{4} + \dfrac{3}{10} \cdot \dfrac{2}{2} + \dfrac{3}{20}\right) \div 3$

$= \left(\dfrac{4}{20} + \dfrac{6}{20} + \dfrac{3}{20}\right) \div 3$

$= \dfrac{13}{20} \div 3$

$= \dfrac{13}{20} \cdot \dfrac{1}{3}$

$= \dfrac{13}{60}$

77. "Increased by" is most likely to translate to addition; A.

81. "Subtracted from" is most likely to translate to subtraction; S.

85. "Times" is most likely to translate to multiplication; M.

89. $\dfrac{2^3}{3} = \dfrac{2 \cdot 2 \cdot 2}{3} = \dfrac{8}{3}$

$\left(\dfrac{2}{3}\right)^3 = \dfrac{2}{3} \cdot \dfrac{2}{3} \cdot \dfrac{2}{3} = \dfrac{8}{27}$

$\dfrac{2^3}{3}$ and $\left(\dfrac{2}{3}\right)^3$ do not simplify to the same number.

answers may vary

93. The operations should be done in this order: division, multiplication, subtraction, addition

97. The LCD of 5 and 50 is 50.

$\dfrac{2}{5} = \dfrac{2}{5} \cdot \dfrac{10}{10} = \dfrac{20}{50}$

$\dfrac{13}{50} = \dfrac{13}{50}$

Since $20 > 13$, then $\dfrac{20}{50} > \dfrac{13}{50}$, so $\dfrac{2}{5} > \dfrac{13}{50}$.

The average college student spends more money on discretionary purchases than on room and board.

Exercise Set 3.6

1. "The sum" translates as addition.

$$\frac{1}{2} + \frac{1}{3}$$

5. "Subtract" indicates subtraction.

$$\frac{15}{16} - \frac{5}{8}$$

9. "product" indicates multiplication.

$$8\frac{1}{3} \cdot \frac{7}{9}$$

13.

Wall height	is	number of bricks	times	brick width	plus	layers of mortar	times	mortar width
↓	↓	↓	↓	↓	↓	↓	↓	↓
Height =		4	·	$2\frac{3}{4}$	+	3	·	$\frac{1}{2}$

$$= 4 \cdot \frac{11}{4} + 3 \cdot \frac{1}{2} = 11 + \frac{3}{2} = 11 + 1 + \frac{1}{2} = 12\frac{1}{2}$$

The wall is $12\frac{1}{2}$ inches tall.

17.

Bill life	is	$\frac{1}{20}$	times	coin life
↓	↓	↓	↓	↓
Life =		$\frac{1}{20}$	·	30

$$= \frac{30}{20} = \frac{10 \cdot 3}{10 \cdot 2} = \frac{3}{2} \text{ or } 1\frac{1}{2}$$

The life expectancy of circulating paper money is $1\frac{1}{2}$ years.

21.

Total needed	is	amount for large shirt	plus	number of small shirts	times	amount for small shirt
↓	↓	↓	↓	↓	↓	↓
Total =		$1\frac{1}{2}$	+	5	·	$\frac{3}{4}$

$$= 1\frac{1}{2} + 5 \cdot \frac{3}{4} = \frac{3}{2} + \frac{15}{4} = \frac{3}{2} \cdot \frac{2}{2} + \frac{15}{4} = \frac{6}{4} + \frac{15}{4}$$

$$= \frac{21}{4} \text{ or } 5\frac{1}{4}$$

The amount of cloth needed is $5\frac{1}{4}$ yards, so the 5-yard remnant is not enough. Another $\frac{1}{4}$ yard of material is required.

25.

Outer diameter	is	left thickness	plus	inner diameter	plus	right thickness
↓	↓	↓	↓	↓	↓	↓
outer diameter =		$\frac{3}{16}$	+	$\frac{3}{4}$	+	$\frac{3}{16}$

$$= \frac{3}{16} + \frac{3}{4} + \frac{3}{16} = \frac{3}{16} + \frac{3}{4} \cdot \frac{4}{4} + \frac{3}{16} = \frac{3}{16} + \frac{12}{16} + \frac{3}{16} = \frac{18}{16}$$

$$= \frac{2 \cdot 9}{2 \cdot 8} = \frac{9}{8} \text{ or } 1\frac{1}{8}$$

The outer diameter is $\frac{9}{8}$ or $1\frac{1}{8}$ inches.

29.

Area of photograph	is	length	times	width
↓	↓	↓	↓	↓
Area	=	$4\frac{1}{2}$	·	$2\frac{1}{2}$

$$= 4\frac{1}{2} \cdot 2\frac{1}{2} = \frac{9}{2} \cdot \frac{5}{2} = \frac{9 \cdot 5}{2 \cdot 2} = \frac{45}{4} \text{ or } 11\frac{1}{4}$$

The area of the photograph is $11\frac{1}{4}$ square inches.

33.

Volume	is	length	times	width	times	height
↓	↓	↓	↓	↓	↓	↓
Volume	=	$2\frac{4}{5}$	·	$1\frac{1}{2}$	·	$\frac{3}{10}$

$$= \frac{14}{5} \cdot \frac{3}{2} \cdot \frac{3}{10} = \frac{14 \cdot 3 \cdot 3}{5 \cdot 2 \cdot 10} = \frac{126}{100} = \frac{2 \cdot 63}{2 \cdot 50}$$

$$= \frac{63}{50} \text{ or } 1\frac{13}{50}$$

The volume of the box is $1\frac{13}{50}$ cubic inches.

37. a.

Length needed	is	number of pieces	times	length of each piece
↓	↓	↓	↓	↓
Length	=	12	·	$\frac{3}{4}$

$$= 12 \cdot \frac{3}{4} = \frac{12}{1} \cdot \frac{3}{4} = \frac{4 \cdot 3 \cdot 3}{1 \cdot 4} = \frac{3 \cdot 3}{1} = 9$$

Since he needs a total of 9 feet of tubing, the 10-foot piece is enough.

b. $10 - 9 = 1$

He will have 1 foot of tubing left over.

41.

Area	is	length	times	width
↓	↓	↓	↓	↓
Area	=	$\frac{3}{16}$	·	$\frac{3}{8}$

$$= \frac{3 \cdot 3}{16 \cdot 8} = \frac{9}{128}$$

The area is $\frac{9}{128}$ square inch.

Perimeter	is	two	times	length	plus	two	times	width
↓	↓	↓	↓	↓	↓	↓	↓	↓
Perimeter =		2	·	$\frac{3}{16}$	+	2	·	$\frac{3}{8}$

$$= \frac{2 \cdot 3}{16} + \frac{2 \cdot 3}{8} = \frac{6}{16} + \frac{6}{8} = \frac{6}{16} + \frac{12}{16} = \frac{18}{16} = \frac{9}{8} \text{ or } 1\frac{1}{8}$$

The perimeter is $1\frac{1}{8}$ inches.

45.

Length	is	$1\frac{9}{10}$	times	width
↓	↓	↓	↓	↓
Length	=	$1\frac{9}{10}$	·	$2\frac{1}{2}$

$$= 1\frac{9}{10} \cdot 2\frac{1}{2} = \frac{19}{10} \cdot \frac{5}{2} = \frac{19 \cdot 5}{10 \cdot 2} = \frac{19 \cdot 5}{5 \cdot 2 \cdot 2} = \frac{19}{2 \cdot 2}$$

$$= \frac{19}{4} \text{ or } 4\frac{3}{4}$$

The length of the flag is $4\frac{3}{4}$ feet.

49. $\sqrt{9} = 3$, because $3 \cdot 3 = 9$.

53. $8 \div 4 \cdot 2 = 2 \cdot 2 = 4$

57. $5 + 3[14 - (12 \div 3)] = 5 + 3[14 - 4]$

$$= 5 + 3[10]$$
$$= 5 + 30$$
$$= 35$$

61. The unmarked horizontal edge of the figure has length

$6\dfrac{7}{9}$ feet $- 3\dfrac{8}{9}$ feet.

$6\dfrac{7}{9} - 3\dfrac{8}{9} = 5\dfrac{16}{9} - 3\dfrac{8}{9} = 2\dfrac{8}{9}$

The unmarked vertical edge of the figure has length

$6\dfrac{2}{3}$ feet $- 4\dfrac{4}{9}$ feet.

$6\dfrac{2}{3} - 4\dfrac{4}{9} = 6\dfrac{6}{9} - 4\dfrac{4}{9} = 2\dfrac{2}{9}$

Area of smaller rectangle $= 2\dfrac{8}{9} \cdot 2\dfrac{2}{9}$

$$= \dfrac{26}{9} \cdot \dfrac{20}{9}$$
$$= \dfrac{26 \cdot 20}{9 \cdot 9}$$
$$= \dfrac{520}{81}$$
$$= 6\dfrac{34}{81}$$

Area of larger rectangle $= 4\dfrac{4}{9} \cdot 6\dfrac{7}{9}$

$$= \dfrac{40}{9} \cdot \dfrac{61}{9}$$
$$= \dfrac{40 \cdot 61}{9 \cdot 9}$$
$$= \dfrac{2440}{81}$$
$$= 30\dfrac{10}{81}$$

Total area $= 6\dfrac{34}{81} + 30\dfrac{10}{81}$

$$= 6 + 30 + \dfrac{34}{81} + \dfrac{10}{81}$$
$$= 36\dfrac{44}{81}$$

The area of the figure is $36\dfrac{44}{81}$ square feet.

65. There are 365 days in one year and 24 hours in one day.

3 years	=	3 years	$\cdot$	$\dfrac{365 \text{ days}}{1 \text{ year}}$	$\cdot$	$\dfrac{24 \text{ hours}}{1 \text{ day}}$	=	26,280 hours

$26{,}280 \cdot \dfrac{11}{13{,}140} = \dfrac{26{,}280}{1} \cdot \dfrac{11}{13{,}140} = \dfrac{2 \cdot 13{,}140 \cdot 11}{1 \cdot 13{,}140} = \dfrac{2 \cdot 11}{1} = 22$

Today it takes 22 hours to produce one million coins.

Chapter 3 Test

1. $4 = \boxed{2 \cdot 2}$

$15 = \boxed{3} \cdot \boxed{5}$

LCM: $2 \cdot 2 \cdot 3 \cdot 5 = 60$

5. $\dfrac{7}{9} + \dfrac{1}{9} = \dfrac{7 + 1}{9} = \dfrac{8}{9}$

9. The LCD of 8 and 3 is 24.

$\dfrac{7}{8} = \dfrac{7}{8} \cdot \dfrac{3}{3} = \dfrac{21}{24}$

$\dfrac{1}{3} = \dfrac{1}{3} \cdot \dfrac{8}{8} = \dfrac{8}{24}$

$\dfrac{7}{8} - \dfrac{1}{3} = \dfrac{21}{24} - \dfrac{8}{24} = \dfrac{13}{24}$

13. The LCD of 12, 8, and 24 is 24.

$\dfrac{11}{12} = \dfrac{11}{12} \cdot \dfrac{2}{2} = \dfrac{22}{24}$

$\dfrac{3}{8} = \dfrac{3}{8} \cdot \dfrac{3}{3} = \dfrac{9}{24}$

$\dfrac{11}{12} + \dfrac{3}{8} + \dfrac{5}{24} = \dfrac{22}{24} + \dfrac{9}{24} + \dfrac{5}{24}$

$$= \dfrac{36}{24}$$
$$= \dfrac{12 \cdot 3}{12 \cdot 2}$$
$$= \dfrac{3}{2} \text{ or } 1\dfrac{1}{2}$$

17.

$$\begin{array}{r} 19 \\ -2\dfrac{3}{11} \\ \hline \end{array} \qquad \begin{array}{r} 18\dfrac{11}{11} \\ -2\dfrac{3}{11} \\ \hline 16\dfrac{8}{11} \end{array}$$

21. $\left(\dfrac{4}{5}\right)^2 + \left(\dfrac{1}{2}\right)^3 = \left(\dfrac{4}{5} \cdot \dfrac{4}{5}\right) + \left(\dfrac{1}{2} \cdot \dfrac{1}{2} \cdot \dfrac{1}{2}\right)$

$$= \dfrac{16}{25} + \dfrac{1}{8}$$
$$= \dfrac{16}{25} \cdot \dfrac{8}{8} + \dfrac{1}{8} \cdot \dfrac{25}{25}$$
$$= \dfrac{128}{200} + \dfrac{25}{200}$$
$$= \dfrac{153}{200}$$

25. Divide the number of gallons of fuel by the number of hours.

$58\dfrac{3}{4} \div 7\dfrac{1}{2} = \dfrac{235}{4} \div \dfrac{15}{2} = \dfrac{235}{4} \cdot \dfrac{2}{15} = \dfrac{235 \cdot 2}{4 \cdot 15} = \dfrac{5 \cdot 47 \cdot 2}{2 \cdot 2 \cdot 5 \cdot 3}$

$$= \dfrac{47}{2 \cdot 3} = \dfrac{47}{6} \text{ or } 7\dfrac{5}{6}$$

$7\dfrac{5}{6}$ gallons of fuel were used each hour.

29. Perimeter $= 2 \cdot 1 + 2 \cdot \dfrac{2}{3}$

$\qquad\qquad = 2 + \dfrac{2}{1} \cdot \dfrac{2}{3}$

$\qquad\qquad = 2 + \dfrac{4}{3}$

$\qquad\qquad = 2 + 1 + \dfrac{1}{3}$

$\qquad\qquad = 3\dfrac{1}{3}$

The perimeter is $3\dfrac{1}{3}$ feet.

Area $= 1 \cdot \dfrac{2}{3} = \dfrac{2}{3}$

The area is $\dfrac{2}{3}$ square foot.

Chapter 4

Exercise Set 4.1

1. 6.52 in words is six and fifty-two hundredths.

5. 0.205 in words is two hundred five thousandths.

9. 200.005 in words is two hundred and five thousandths.

13. 2.43 in words is two and forty-three hundredths.

17. 115.6 in words is one hundred fifteen and six tenths.

21. The check should be paid to "Bell South," for the amount of "59.68," which is written in words as "Fifty-nine and $\dfrac{68}{100}$."

25. Nine and eight hundredths is 9.08.

29. Forty-six ten-thousandths is 0.0046.

33. One and three tenths is 1.3.

37. $0.27 = \dfrac{27}{100}$

41. $0.15 = \dfrac{15}{100} = \dfrac{3}{20}$

45. $0.048 = \dfrac{48}{1000} = \dfrac{6}{125}$

49. $15.802 = 15\dfrac{802}{1000} = 15\dfrac{401}{500}$

53. $487.32 = 487\dfrac{32}{100} = 487\dfrac{8}{25}$

57. $\dfrac{45}{100} = 0.45$ **61.** $\dfrac{268}{1000} = 0.268$

65. $\dfrac{4026}{1000} = 4.026$ **69.** $\dfrac{563}{10} = 56.3$

73. In standard form, eight tenths is 0.8, while as a fraction it is $\dfrac{8}{10}$ or $\dfrac{4}{5}$.

77. To round 47,261 to the nearest ten, observe that the digit in the ones place is 1. Since the digit is less than 5, we do not add 1 to the digit in the tens place. The number 47,261 rounded to the nearest ten is 47,260.

81. answers may vary

85. $17\dfrac{268}{1000} = 17.268$

Exercise Set 4.2

1. 0.15 0.16
$\qquad\uparrow\quad\uparrow$
$\qquad$5 < 6 so
$\qquad$0.15 < 0.16

5. 0.098   0.1
$\qquad\uparrow\qquad\uparrow$
$\qquad$0 < 1 so
$\qquad$0.098 < 0.1

9. 167.908 167.980
$\qquad\quad\uparrow\qquad\uparrow$
$\qquad\quad$0 < 8 so
$\quad$167.908 < 167.980

13. Smallest to largest: 0.006, 0.0061, 0.06

17. Smallest to largest: 1.01, 1.09, 1.1, 1.16

21. To round 0.57 to the nearest tenth, observe that the digit in the hundredths place is 7. Since this digit is 5 or greater, we add 1 to the digit in the tenths place. The number 0.57 rounded to the nearest tenth is 0.6.

25. To round 0.5942 to the nearest thousandth, observe that the digit in the ten-thousandths place is 2. Since this digit is less than 5, we do not add 1 to the digit in the thousandths place. The number 0.5942 rounded to the nearest thousandth is 0.594.

29. To round 12.342 to the nearest tenth, observe that the digit in the hundredths place is 4. Since this digit is less than 5, we do not add 1 to the digit in the tenths place. The number 12.342 rounded to the nearest tenth is 12.3.

33. To round 0.501 to the nearest tenth, observe that the digit in the hundredths place is 0. Since this digit is less than 5, we do not add 1 to the digit in the tenths place. The number 0.501 rounded to the nearest tenth is 0.5.

37. To round 3829.34 to the nearest ten, observe that the digit in the ones place is 9. Since this digit is 5 or greater, we add 1 to the digit in the tens place. The number 3829.34 rounded to the nearest ten is 3830.

41. To round 42,650.14 to the nearest one, observe that the digit in the tenths place is 1. Since this digit is less than 5, we do not add 1 to the digit in the ones place. The number 42,650.14 rounded to the nearest one is 42,650. The amount is $42,650.

45. To round 0.1992 to the nearest hundredth, observe that the digit in the thousandths place is 9. Since this digit is 5 or greater, we add 1 to the digit in the hundredths place. The number 0.1992 rounded to the nearest hundredth is 0.2. The amount is $0.20.

49. To round 1.4256 to the nearest hundredth, observe that the digit in the thousandths place is 5. Since this digit is 5 or greater, we add 1 to the digit in the hundredths place. The number 1.4256 rounded to the nearest hundredth is 1.43. The time is 1.43 hours.

53. To round 106.5157 to the nearest tenth, observe that the digit in the hundredths place is 1. Since this digit is less than 5, we do not add 1 to the digit in the tenths place. The number 106.5157 rounded to the nearest tenth is 106.5. The population density is 106.5 people per square mile.

57. To round 0.4667 to the nearest tenth, observe that the digit in the hundredths place is 6. Since this digit is 5 or greater, we add 1 to the digit in the tenths place. The number 0.4667 rounded to the nearest tenth is 0.5. The time is 0.5 minute.

61.
$$\begin{array}{r} 94 \\ -\ 23 \\ \hline 71 \end{array}$$

65. To round 2849.1738 to the nearest hundred, observe that the digit in the tens place is 4. Since this digit is less than 5, we do not add 1 to the digit in the hundreds place. The number 2849.1738 rounded to the nearest hundred is 2800, which is choice b.

69. Comparing each place value, we see that Audi Sport North America's 2010 average speed of 225.228 kilometers per hour is the fastest on the list.

$$225.228 = 225\frac{228}{1000} = 225\frac{57}{250}$$

73. answers may vary

77. 0.26499 and 0.25786 rounded to the nearest hundredths are 0.26, 0.26559 rounds to 0.27 and 0.25186 rounds to 0.25.

81. Round to the nearest hundred million, then add.

$$\begin{array}{r} 800 \\ 700 \\ 600 \\ 500 \\ 500 \\ +\ 500 \\ \hline 3600 \end{array}$$

The total amount of money is estimated as $3600 million.

Exercise Set 4.3

1.
$$\begin{array}{r} 1.3 \\ +\ 2.2 \\ \hline 3.5 \end{array}$$

5.
$$\begin{array}{r} 0.003 \\ +\ 0.091 \\ \hline 0.094 \end{array}$$

9.
$$\begin{array}{r} \overset{1}{4}90.00 \\ +\ 93.09 \\ \hline 583.09 \end{array}$$

13. Exact:
$$\begin{array}{r} \overset{1}{1}0\overset{11}{0}.009 \\ 6.080 \\ +\ 9.034 \\ \hline 115.123 \end{array}$$
Estimate:
$$\begin{array}{r} \overset{1}{1}00 \\ 6 \\ +\ 9 \\ \hline 115 \end{array}$$

17.
$$\begin{array}{r} \overset{1}{4}5.\overset{1}{0}23 \\ 3.006 \\ +\ 8.403 \\ \hline 56.432 \end{array}$$

21.
$$\begin{array}{r} 18.0 \\ -\ 2.7 \\ \hline 15.3 \end{array}$$
Check:
$$\begin{array}{r} \overset{1}{1}5.3 \\ +\ 2.7 \\ \hline 18.0 \end{array}$$

25. Exact:
$$\begin{array}{r} 5.90 \\ -\ 4.07 \\ \hline 1.83 \end{array}$$
Estimate:
$$\begin{array}{r} 6 \\ -\ 4 \\ \hline 2 \end{array}$$
Check:
$$\begin{array}{r} \overset{1}{1}.83 \\ +\ 4.07 \\ \hline 5.90 \end{array}$$

29.
$$\begin{array}{r} 500.34 \\ -\ 123.45 \\ \hline 376.89 \end{array}$$
Check:
$$\begin{array}{r} \overset{111\ 1}{3}76.89 \\ +\ 123.45 \\ \hline 500.34 \end{array}$$

33.
$$\begin{array}{r} 200.0 \\ -\ 5.6 \\ \hline 194.4 \end{array}$$
Check:
$$\begin{array}{r} \overset{111}{1}94.4 \\ +\ 5.6 \\ \hline 200.0 \end{array}$$

37.
$$\begin{array}{r} 23.0 \\ -\ 6.7 \\ \hline 16.3 \end{array}$$
Check:
$$\begin{array}{r} \overset{11}{1}6.3 \\ +\ 6.7 \\ \hline 23.0 \end{array}$$

41.
$$\begin{array}{r} 86.050 \\ -\ 1.978 \\ \hline 84.072 \end{array}$$

45.
$$\begin{array}{r} 150.00 \\ -\ 93.17 \\ \hline 56.83 \end{array}$$

49. Two $20 bills total $40.
$$\begin{array}{r} 40.00 \\ -\ 32.48 \\ \hline 7.52 \end{array}$$
Her change was $7.52.

53. The phrase "By how much did the price change" indicates that we should subtract the lower price from the higher price.
$$\begin{array}{r} 2.979 \\ -\ 2.839 \\ \hline 0.140 \end{array}$$
The price changed by $0.14.

57. To find the perimeter, we add the lengths of the sides. Opposite sides of a rectangle have the same length.
$$\begin{array}{r} 3.01 \\ 3.01 \\ 1.56 \\ +\ 1.56 \\ \hline 9.14 \end{array}$$
The perimeter is 9.14 inches.

61. Since the September 2013 average temperature is 2.5 degrees Fahrenheit above the 20th-century average for September, we subtract to find the 20th-century average for September.
$$\begin{array}{r} 67.3 \\ -\ 2.5 \\ \hline 64.8 \end{array}$$
The 20th-century average temperature for September is 64.8 degrees Fahrenheit.

65. To find the perimeter, we add the lengths of the sides. Opposite sides of a rectangle have the same length.
$$\begin{array}{r} 4.90 \\ 4.90 \\ 2.33 \\ +\ 2.33 \\ \hline 14.46 \end{array}$$
The perimeter is 14.46 inches.

69. The phrase "Find the decrease" indicates that we should subtract the monthly text messages in 2012 from the monthly text messages in 2011.
$$\begin{array}{r} 193.1 \\ -\ 171.3 \\ \hline 21.8 \end{array}$$
The number of text messages decreased by 21.8 billion (21,800,000,000) from 2011 to 2012.

73. Add the lengths of the sides to get the perimeter.
$$\begin{array}{r} \overset{11}{1}2.40 \\ 29.34 \\ +\ 25.70 \\ \hline 67.44 \end{array}$$
67.44 feet of border material is needed.

77. The phrase "Find the total" indicates that we should add the durations of the flights.

$$
\begin{array}{r}
\overset{2\ 1\ 2\ \ 1\ 1}{330.583} \\
94.567 \\
147.000 \\
+\ 142.900 \\
\hline
715.050
\end{array}
$$

James A. Lovell spent 715.05 hours in spaceflight.

81. The greatest chocolate consumption shown in the table is 26.24 pounds per year for Switzerland. The least chocolate consumption shown in the table is 18.30 pounds per year for Belgium.

$$
\begin{array}{r}
26.24 \\
-\ 18.30 \\
\hline
7.94
\end{array}
$$

The greatest chocolate consumption shown in the table is 7.94 pounds per person per year more than the least.

85.
$$
\begin{array}{r}
23 \\
\times\ 2 \\
\hline
46
\end{array}
$$

89. $\left(\dfrac{2}{3}\right)^2 = \dfrac{2}{3} \cdot \dfrac{2}{3} = \dfrac{2 \cdot 2}{3 \cdot 3} = \dfrac{4}{9}$

93. $10.68 - (2.3 + 2.3) = 10.68 - 4.60 = 6.08$
The unknown length is 6.08 inches.

97. 1 dime, 1 nickel, and 2 pennies:
$0.10 + 0.05 + 0.01 + 0.01 = 0.17$
3 nickels and 2 pennies:
$0.05 + 0.05 + 0.05 + 0.01 + 0.01 = 0.17$
1 dime and 7 pennies:
$0.10 + 0.01 + 0.01 + 0.01 + 0.01 + 0.01 + 0.01 + 0.01 = 0.17$
2 nickels and 7 pennies:
$0.05 + 0.05 + 0.01 + 0.01 + 0.01 + 0.01 + 0.01 + 0.01 + 0.01 = 0.17$

101. Answers may vary.

Exercise Set 4.4

1.
$$
\begin{array}{r}
0.2 \\
\times\ 0.6 \\
\hline
0.12
\end{array}
$$

5.
$$
\begin{array}{r}
0.26 \\
\times\ 5 \\
\hline
1.30 \text{ or } 1.3
\end{array}
$$

9.
$$
\begin{array}{r}
0.576 \\
\times\ 0.7 \\
\hline
0.4032
\end{array}
$$

13.
$$
\begin{array}{r}
490.2 \\
\times\ 0.023 \\
\hline
14706 \\
98040 \\
\hline
11.2746
\end{array}
$$

17. $6.5 \times 10 = 65$

21. $7.2 \times 0.01 = 0.072$

25. $6.046 \times 1000 = 6046$

29.
$$
\begin{array}{r}
0.123 \\
\times\ 0.4 \\
\hline
0.0492
\end{array}
$$

33.
$$
\begin{array}{r}
8.6 \\
\times\ 0.15 \\
\hline
430 \\
860 \\
\hline
1.290 \text{ or } 1.29
\end{array}
$$

37. $526.3 \times 0.001 = 0.5263$

41. 5.5 billion $= 5.5 \times 1$ billion
$= 5.5 \times 1,000,000,000$
$= 5,500,000,000$
The silos hold enough to make 5,500,000,000 bars.

45. 292 thousand $= 292 \times 1$ thousand
$= 292 \times 1000$
$= 292,000$
292,000 people visited the park each week.

49. Circumference $= \pi \cdot$ diameter
$C = \pi \cdot 10 = 10\pi$
$C \approx 10(3.14) = 31.4$
The circumference is 10π centimeters, which is approximately 31.4 centimeters.

53. Multiply his hourly wage by the number of hours worked.
$$
\begin{array}{r}
17.88 \\
\times\ 40 \\
\hline
715.20
\end{array}
$$
His pay for last week was $715.20.

57. Multiply the number of ounces by the number of grams of saturated fat in 1 ounce.
$$
\begin{array}{r}
6.2 \\
\times\ 4 \\
\hline
24.8
\end{array}
$$
There are 24.8 grams of saturated fat in 4 ounces of cream cheese.

61. Circumference $= \pi \cdot$ diameter
$C = \pi \cdot 250 = 250\pi$
$$
\begin{array}{r}
250 \\
\times\ 3.14 \\
\hline
1000 \\
2500 \\
75000 \\
\hline
785.00
\end{array}
$$
The circumference is 250π feet, which is approximately 785 feet.

65. Multiply her height in meters by the number of inches in 1 meter.
$$
\begin{array}{r}
39.37 \\
\times\ 1.65 \\
\hline
19685 \\
236220 \\
393700 \\
\hline
64.9605
\end{array}
$$
She is approximately 64.9605 inches tall.

69.
$$
\begin{array}{r}
3.01 \\
\times\ 1.56 \\
\hline
1806 \\
15050 \\
30100 \\
\hline
4.6956
\end{array}
$$
The area, rounded to the nearest hundredth, is 4.70 square inches.

73.
$$
\begin{array}{r}
36 \\
56\overline{)2016} \\
-168 \\
\hline
336 \\
-336 \\
\hline
0
\end{array}
$$

77. $\dfrac{24}{7} \div \dfrac{8}{21} = \dfrac{24}{7} \cdot \dfrac{21}{8} = \dfrac{24 \cdot 21}{7 \cdot 8} = \dfrac{3 \cdot 8 \cdot 3 \cdot 7}{7 \cdot 8} = \dfrac{3 \cdot 3}{1} = 9$

81.
```
   3.60
 − 0.04
   3.56
```

85.
```
    20.6
  × 1.86
    1236
   16480
   20600
  38.316
```
$38.316 \times 100,000 = 3,831,600$

Radio waves travel 3,831,600 miles in 20.6 seconds.

89. answers may vary

Exercise Set 4.5

1.
```
      4.6
  3)13.8
   − 12
     1 8
     1 8
       0
```

5. $0.06\overline{)18}$ becomes
```
      300
  6)1800
   −18
     000
```

9. Exact: $5.5\overline{)36.3}$ becomes
```
       6.6
  55)363.0
    − 330
      33 0
    − 33 0
         0
```
Estimate:
```
     6
  6)36
```

13.
```
       0.045
  12)0.540
    − 48
       60
     − 60
        0
```

17. $0.27\overline{)1.296}$ becomes
```
        4.8
  27)129.6
    − 108
      21 6
    − 21 6
         0
```

21. $0.6\overline{)18}$ becomes
```
       30.
  6)180.
   − 18
     00
```

25. Exact: $7.2\overline{)70.56}$ becomes
```
        9.8
  72)705.6
    − 648
      57 6
    − 57 6
         0
```
Estimate:
```
      10
  7)70
```

29. $0.027\overline{)1.215}$ becomes
```
        45
  27)1215
    − 108
      135
    − 135
        0
```

33. $3.78\overline{)0.02079}$ becomes
```
             0.0055
  378)2.0790
      − 1 890
        1890
      − 1890
          0
```

37. $0.023\overline{)0.549}$ becomes
```
         23.869 ≈ 23.87
  23)549.000
    −46
      89
     −69
      20 0
    −18 4
      1 60
    −1 38
       220
      −207
        13
```

41. $\dfrac{54.982}{100} = 0.54982$

45. $\dfrac{12.9}{1000} = 0.0129$

49. $\dfrac{13.1}{10} = 1.31$

53. $\dfrac{456.25}{10,000} = 0.045625$

57.
```
     202.14 ≈ 202.1
  39)7883.50
    −78
      08
     − 0
      83
     −78
       5 5
      −3 9
       1 60
      −1 56
          4
```
She needs to buy 202.1 pounds of fertilizer.

61. Divide the number of crayons by 64.

   11.40 rounded to the nearest tenth is 11.4 boxes.
```
  64)730.00
    −64
      90
     −64
      26 0
     −25 6
         40
```

65. From Exercise 63, we know that there are 24 teaspoons in 4 fluid ounces. Thus, there are 48 half teaspoons (0.5 tsp) or doses in 4 fluid ounces. To see how long the medicine will last if a dose is taken every 4 hours, there are $24 \div 4 = 6$ doses taken per day. $48 \text{ (doses)} \div 6 \text{ (per day)} = 8$ days. The medicine will last 8 days.

69. Divide the number of miles by the number of hours.

$$
\begin{array}{r}
122.79 \approx 122.8 \\
24\overline{)2947.10} \\
-24 \\
\hline
54 \\
-48 \\
\hline
67 \\
-48 \\
\hline
19\,1 \\
-16\,8 \\
\hline
2\,30 \\
-2\,16 \\
\hline
14
\end{array}
$$

The average speed was 122.8 miles per hour.

73. $0.7(6 - 2.5) = 0.7(3.5) = 2.45$

77. $30.03 + 5.1 \times 9.9 = 30.03 + 50.49 = 80.52$

81. $93.07 \div 10 \times 100 = 9.307 \times 100 = 930.7$

85. $5(20.6 - 2.06) - (0.8)^2 = 5(18.54) - (0.8)^2$
$= 5(18.54) - 0.64$
$= 92.70 - 0.64$
$= 92.06$

89. $0.9 = \dfrac{9}{10}$

93. $0.3\overline{)1.278}$ becomes
$$
\begin{array}{r}
4.26 \\
3\overline{)12.78} \\
-12 \\
\hline
0\,7 \\
-6 \\
\hline
18 \\
-18 \\
\hline
0
\end{array}
$$
$1.278 \div 0.3 = 4.26$

97.
$$
\begin{array}{r}
8.6 \\
\times\ 3.1 \\
\hline
86 \\
2580 \\
\hline
26.66
\end{array}
$$

101. 8.62×41.7 is approximately $9 \times 40 = 360$, which is choice c.

105. Add the numbers, then divide by 4.
$$
\begin{array}{r}
\overset{2}{8}6 \\
78 \\
91 \\
+\ 87 \\
\hline
342
\end{array}
$$
$$
\begin{array}{r}
85.5 \\
4\overline{)342.0} \\
-32 \\
\hline
22 \\
-20 \\
\hline
2\,0 \\
2\,0 \\
\hline
0
\end{array}
$$
The average is 85.5.

109. answers may vary

113. The rancher will need 4 times the perimeter of the corral.
$$
\begin{array}{r}
24.280 \\
24.280 \\
15.675 \\
+\ 15.675 \\
\hline
79.910
\end{array}
$$
The perimeter is 79.91 meters.
$$
\begin{array}{r}
79.91 \\
\times\quad 4 \\
\hline
319.64
\end{array}
$$
The rancher will need 319.64 meters of wire.

Exercise Set 4.6

1.
$$
\begin{array}{r}
0.2 \\
5\overline{)1.0} \\
-1\,0 \\
\hline
0
\end{array}
$$
$\dfrac{1}{5} = 0.2$

5.
$$
\begin{array}{r}
0.75 \\
4\overline{)3.00} \\
-2\,8 \\
\hline
20 \\
-20 \\
\hline
0
\end{array}
$$
$\dfrac{3}{4} = 0.75$

9.
$$
\begin{array}{r}
1.2 \\
5\overline{)6.0} \\
-5 \\
\hline
1\,0 \\
-1\,0 \\
\hline
0
\end{array}
$$
$\dfrac{6}{5} = 1.2$

13.
$$
\begin{array}{r}
0.425 \\
40\overline{)17.000} \\
-16\,0 \\
\hline
1\,00 \\
-80 \\
\hline
200 \\
200 \\
\hline
0
\end{array}
$$
$\dfrac{17}{40} = 0.425$

17.
$$
\begin{array}{r}
0.333\ldots \\
3\overline{)1.000} \\
-9 \\
\hline
10 \\
-9 \\
\hline
10 \\
9 \\
\hline
1
\end{array}
$$
$\dfrac{1}{3} = 0.\overline{3}$

21.
$$
\begin{array}{r}
0.636363\ldots \\
11\overline{)7.000000} \\
-6\,6 \\
\hline
40 \\
-33 \\
\hline
70 \\
-66 \\
\hline
40 \\
-33 \\
\hline
70 \\
-66 \\
\hline
40 \\
-33 \\
\hline
7
\end{array}
$$
$\dfrac{7}{11} = 0.\overline{63}$

25.
$$125)\overline{78.000}$$
$$\underline{-75\ 0}$$
$$3\ 00$$
$$\underline{-2\ 50}$$
$$500$$
$$\underline{-500}$$
$$0$$

$$\frac{78}{125} = 0.624$$

with quotient 0.624

29. $\frac{7}{16} = 0.4375 \approx 0.44$

33.
$$91)\overline{56.000}$$
$$\underline{-54\ 6}$$
$$1\ 40$$
$$\underline{-\ 91}$$
$$490$$
$$\underline{-455}$$
$$35$$

quotient $0.615 \approx 0.62$

$$\frac{56}{91} \approx 0.62$$

37.
$$50)\overline{1.00}$$
$$\underline{-1\ 00}$$
$$0$$

quotient 0.02

$$\frac{1}{50} = 0.02$$

41.
$$200)\overline{43.000}$$
$$\underline{-40\ 0}$$
$$3\ 00$$
$$\underline{-2\ 00}$$
$$1\ 000$$
$$\underline{-1\ 000}$$
$$0$$

quotient 0.215

$$0.215 = \frac{43}{200}$$

45.
$$6)\overline{5.000}$$
$$\underline{-4\ 8}$$
$$20$$
$$\underline{-18}$$
$$20$$
$$\underline{-18}$$
$$2$$

quotient $0.833\ldots$

$\frac{5}{6} = 0.8\overline{3}$ and $0.\overline{6} < 0.8\overline{3}$, so $0.\overline{6} < \frac{5}{6}$.

49.
$$9)\overline{1.00}$$
$$\underline{-9}$$
$$10$$
$$\underline{-9}$$
$$1$$

quotient 0.11

$\frac{1}{9} \approx 0.11$ and $0.11 > 0.1$, so $\frac{1}{9} > 0.1$

53.
$$64)\overline{456.000}$$
$$\underline{-448}$$
$$8\ 0$$
$$\underline{-6\ 4}$$
$$1\ 60$$
$$\underline{-1\ 28}$$
$$320$$
$$\underline{-320}$$
$$0$$

quotient 7.125

$\frac{456}{64} = 7.125$ and $7.123 < 7.125$, so $7.123 < \frac{456}{64}$

57. $0.49 = 0.490$

$0.49, 0.491, 0.498$

61. $\frac{4}{7} \approx 0.571$

$0.412, 0.453, \frac{4}{7}$

65. $\frac{12}{5} = 2.400$

$2.37 = 2.370$

$\frac{17}{8} = 2.125$

$\frac{17}{8}, 2.37, \frac{12}{5}$

69. Area $= \frac{1}{2} \cdot \text{base} \cdot \text{height}$

$$= \frac{1}{2} \cdot 5.2 \cdot 3.6$$

$$= 0.5 \cdot 5.2 \cdot 3.6$$

$$= 9.36$$

The area is 9.36 square centimeters.

73. $2^3 = 2 \cdot 2 \cdot 2 = 8$

77. $\left(\frac{1}{3}\right)^4 = \frac{1}{3} \cdot \frac{1}{3} \cdot \frac{1}{3} \cdot \frac{1}{3} = \frac{1}{81}$

81. $\left(\frac{2}{5}\right)\left(\frac{5}{2}\right)^2 = \frac{2}{5} \cdot \frac{5}{2} \cdot \frac{5}{2} = \frac{5}{2}$

85. $1.00001 > 1$

89. The fraction of stations with a news/talk format is $\frac{1953}{15{,}012}$.

$$15012)\overline{1953.000}$$

quotient $0.130 \approx 0.13$

$$\underline{-1501\ 2}$$
$$451\ 80$$
$$\underline{-450\ 36}$$
$$1\ 440$$
$$\underline{-\qquad 0}$$
$$0$$

$$\frac{1953}{15{,}012} \approx 0.13$$

93. answers may vary

97. $(9.6)(5) - \dfrac{3}{4} = (9.6)(5) - 0.75$

$= 48 - 0.75$

$= 47.25$

101. $\dfrac{3}{8}(5.9 - 4.7) = 0.375(5.9 - 4.7)$

$= 0.375(1.2)$

$= 0.45$

Chapter 4 Test

1. 45.092 in words is forty-five and ninety-two thousandths.

5. $25.0909 < 25.9090$

9.
```
      0.65
20)13.00
  -12 0
    1 00
   -1 00
       0
```
$\dfrac{13}{20} = 0.65$

13.
```
   20.0
  - 8.6
   11.4
```

17.
```
   6.6728 ≈ 6.673
7)46.7100
 -42
   4 7
  -4 2
    51
   -49
     20
    -14
     60
    -56
      4
```

21. $\dfrac{0.23 + 1.63}{0.3} = \dfrac{1.86}{0.3}$

$0.3\overline{)1.86}$ becomes $3\overline{)18.6}$
```
      6.2
 3)18.6
  -18
   06
   -6
    0
```
$\dfrac{0.23 + 1.63}{0.3} = 6.2$

25. a.
```
   123.8
 ×    80
  9904.0
```
The area of the lawn is 9904 square feet.

b. Multiply the number of square feet by the number of ounces per square foot.
```
   9904
 × 0.02
 198.08
```
198.08 ounces of insecticide are needed.

Chapter 5

Exercise Set 5.1

1. The ratio of 11 to 14 is $\dfrac{11}{14}$.

5. The ratio of 151 to 201 is $\dfrac{151}{201}$.

9. The ratio of 5 to $7\dfrac{1}{2}$ is $\dfrac{5}{7\frac{1}{2}}$.

13. The ratio of 16 to 24 is $\dfrac{16}{24} = \dfrac{8 \cdot 2}{8 \cdot 3} = \dfrac{2}{3}$.

17. The ratio of 4.63 to 8.21 is $\dfrac{4.63}{8.21} = \dfrac{4.63 \cdot 100}{8.21 \cdot 100} = \dfrac{463}{821}$.

21. The ratio of 10 hours to 24 hours is
$\dfrac{10 \text{ hours}}{24 \text{ hours}} = \dfrac{10}{24} = \dfrac{2 \cdot 5}{2 \cdot 12} = \dfrac{5}{12}$.

25. The ratio of 24 days to 14 days is
$\dfrac{24 \text{ days}}{14 \text{ days}} = \dfrac{24}{14} = \dfrac{2 \cdot 12}{2 \cdot 7} = \dfrac{12}{7}$.

29. The ratio of 8 inches to 20 inches is
$\dfrac{8 \text{ inches}}{20 \text{ inches}} = \dfrac{8}{20} = \dfrac{4 \cdot 2}{4 \cdot 5} = \dfrac{2}{5}$.

33. The ratio of $7\dfrac{3}{5}$ hours to $1\dfrac{9}{10}$ hours is

$\dfrac{7\frac{3}{5} \text{ hours}}{1\frac{9}{10} \text{ hours}} = \dfrac{7\frac{3}{5}}{1\frac{9}{10}}$

$= 7\dfrac{3}{5} \div 1\dfrac{9}{10}$

$= \dfrac{38}{5} \div \dfrac{19}{10}$

$= \dfrac{38}{5} \cdot \dfrac{10}{19}$

$= \dfrac{19 \cdot 2 \cdot 5 \cdot 2}{5 \cdot 19}$

$= \dfrac{2 \cdot 2}{1}$

$= \dfrac{4}{1}$

37. The perimeter of the mural is
$2L + 2W = 2(32 \text{ feet}) + 2(19 \text{ feet})$
$= 64 \text{ feet} + 38 \text{ feet}$
$= 102 \text{ feet}.$

The ratio of the width to the perimeter is
$\dfrac{32 \text{ feet}}{102 \text{ feet}} = \dfrac{32}{102} = \dfrac{2 \cdot 16}{2 \cdot 51} = \dfrac{16}{51}$.

41. The ratio of women to men is $\dfrac{125}{100} = \dfrac{25 \cdot 5}{25 \cdot 4} = \dfrac{5}{4}$.

45. The perimeter of the billboard is
$8 \text{ feet} + 15 \text{ feet} + 17 \text{ feet} = 40 \text{ feet}.$

The ratio of the longest side to the perimeter of the billboard is $\dfrac{17 \text{ feet}}{40 \text{ feet}} = \dfrac{17}{40}$.

49. The ratio of mountains over 14,000 feet tall found in Alaska to those found in Colorado is
$$\frac{19 \text{ mountains}}{57 \text{ mountains}} = \frac{19}{57} = \frac{19 \cdot 1}{19 \cdot 3} = \frac{1}{3}.$$

53. There is $50 - 49 = 1$ state without Target stores. The ratio of states without Target stores to states with Target stores is $\dfrac{1 \text{ state}}{49 \text{ states}} = \dfrac{1}{49}.$

57. $3.7\overline{)0.555}$ becomes $37\overline{)5.55}$
$$\begin{array}{r} 0.15 \\ 37\overline{)5.55} \\ -37 \\ \hline 1\,85 \\ -1\,85 \\ \hline 0 \end{array}$$

61. $30 : 1$ should be read as "the ratio of thirty to one."

65. $\dfrac{\$3}{\$2}$ is not in simplest form: $\dfrac{\$3}{\$2} = \dfrac{3}{2}.$

69. $4\dfrac{1}{2}$ is not in simplest form: $4\dfrac{1}{2} = \dfrac{9}{2}.$

73. a. The ratio of states with primary hand-held cell phone laws to total U.S. states is
$$\frac{12 \text{ states}}{50 \text{ states}} = \frac{12}{50} = \frac{2 \cdot 6}{2 \cdot 25} = \frac{6}{25}.$$

b. There are 50 states $-$ 12 states $=$ 38 states without primary hand-held cell phone laws.

c. The ratio of states with primary hand-held cell phone laws to states without such laws is
$$\frac{12 \text{ states}}{38 \text{ states}} = \frac{12}{38} = \frac{2 \cdot 6}{2 \cdot 19} = \frac{6}{19}.$$

Exercise Set 5.2

1. The rate of 5 shrubs every 15 feet is $\dfrac{5 \text{ shrubs}}{15 \text{ feet}} = \dfrac{1 \text{ shrub}}{3 \text{ feet}}.$

5. The rate of 6 laser printers for 28 computers is $\dfrac{6 \text{ laser printers}}{28 \text{ computers}} = \dfrac{3 \text{ laser printers}}{14 \text{ computers}}.$

9. The rate of 6 flight attendants for 200 passengers is $\dfrac{6 \text{ flight attendants}}{200 \text{ passengers}} = \dfrac{3 \text{ flight attendants}}{100 \text{ passengers}}.$

13. $\begin{array}{r} 110 \\ 3\overline{)330} \\ -3 \\ \hline 03 \\ -3 \\ \hline 0 \end{array}$

330 calories in a 3-ounce serving is $\dfrac{110 \text{ calories}}{1 \text{ ounce}}$ or 110 calories/ounce.

17. 1 minute $=$ 60 seconds
$$\begin{array}{r} 90 \\ 60\overline{)5400} \\ -540 \\ \hline 00 \\ -0 \\ \hline 0 \end{array}$$

$$\frac{5400 \text{ wingbeats}}{1 \text{ minute}} = \frac{5400 \text{ wingbeats}}{60 \text{ seconds}}$$
$$= \frac{90 \text{ wingbeats}}{1 \text{ second}}$$
$$= 90 \text{ wingbeats/second}$$

21. $\begin{array}{r} 319{,}368 \\ 2\overline{)638{,}736} \\ -6 \\ \hline 03 \\ -2 \\ \hline 18 \\ -18 \\ \hline 07 \\ -6 \\ \hline 13 \\ -12 \\ \hline 16 \\ -16 \\ \hline 0 \end{array}$

638,736 voters for 2 senators is $\dfrac{319{,}368 \text{ voters}}{1 \text{ senator}}$ or 319,368 voters/senator.

25. $\begin{array}{r} 6{,}874{,}600 \\ 20\overline{)137{,}492{,}000} \\ -120 \\ \hline 17\,4 \\ -16\,0 \\ \hline 1\,49 \\ -1\,40 \\ \hline 92 \\ -80 \\ \hline 12\,0 \\ -12\,0 \\ \hline 0 \end{array}$

$137,492,000 for 20 players is $\dfrac{\$6{,}874{,}600}{1 \text{ player}}$ or $6,874,600/player.

29. $\begin{array}{r} 60{,}750 \\ 400\overline{)24{,}300{,}000} \\ -24\,00 \\ \hline 300 \\ -0 \\ \hline 300\,0 \\ -280\,0 \\ \hline 20\,00 \\ -20\,00 \\ \hline 0 \end{array}$

$24,300,000 for 400 species is $\dfrac{\$60{,}750}{1 \text{ species}}$ or $60,750/species.

33. $\begin{array}{r} 8.70 \approx 8.7 \\ 24\overline{)209.00} \\ -192 \\ \hline 17\,0 \\ -16\,8 \\ \hline 20 \\ -0 \\ \hline 20 \end{array}$

209 tornadoes in 24 hours is about $\dfrac{8.7 \text{ tornadoes}}{1 \text{ hour}}$ or 8.7 tornadoes/hour.

37. a. $27.58 \approx 27.6$

$14.5\overline{)400}$ becomes $145\overline{)4000.00}$

$$
\begin{array}{r}
-290 \\
\hline
1100 \\
-1015 \\
\hline
85\,0 \\
-72\,5 \\
\hline
12\,50 \\
-11\,60 \\
\hline
90
\end{array}
$$

The unit rate for the car is approximately 27.6 miles/gallon.

b. $29.18 \approx 29.2$

$9.25\overline{)270}$ becomes $925\overline{)27000.00}$

$$
\begin{array}{r}
-1850 \\
\hline
8500 \\
-8325 \\
\hline
175\,0 \\
-92\,5 \\
\hline
82\,50 \\
-74\,00 \\
\hline
8\,50
\end{array}
$$

The unit rate for the truck is approximately 29.2 miles/gallon.

c. Since the unit rate of the truck is larger, the truck gets better gas mileage.

41. 0.17

$7\overline{)1.19}$

$$
\begin{array}{r}
-7 \\
\hline
49 \\
-49 \\
\hline
0
\end{array}
$$

The unit price is $0.17 per banana.

45. $0.1056 \approx 0.106$

$16\overline{)1.6900}$

$$
\begin{array}{r}
-16 \\
\hline
09 \\
-0 \\
\hline
90 \\
-80 \\
\hline
100 \\
-96 \\
\hline
4
\end{array}
$$

The 16-ounce size costs about $0.106 per ounce.

0.115

$6\overline{)0.690}$

$$
\begin{array}{r}
-6 \\
\hline
09 \\
-6 \\
\hline
30 \\
-30 \\
\hline
0
\end{array}
$$

The 6-ounce size costs $0.115 per ounce.

The 16-ounce size is the better buy.

49. $0.0059 \approx 0.006$

$100\overline{)0.5900}$

$$
\begin{array}{r}
-500 \\
\hline
900 \\
-900 \\
\hline
0
\end{array}
$$

The 100-count napkins cost about $0.006 per napkin.

$0.0051 \approx 0.005$

$180\overline{)0.9300}$

$$
\begin{array}{r}
-900 \\
\hline
300 \\
-180 \\
\hline
120
\end{array}
$$

The 180-count napkins cost about $0.005 per napkin. The 180-count package is the better buy.

53.
$$
\begin{array}{r}
3.7 \\
\times 1.2 \\
\hline
74 \\
370 \\
\hline
4.44
\end{array}
$$

57. $29{,}543 - 29{,}286 = 257$

$19.17 \approx 19.2$

$13.4\overline{)257}$ becomes $134\overline{)2570.00}$

$$
\begin{array}{r}
-134 \\
\hline
1230 \\
-1206 \\
\hline
24\,0 \\
-13\,4 \\
\hline
10\,60 \\
-9\,38 \\
\hline
1\,22
\end{array}
$$

There were 257 miles driven and the miles per gallon were 19.2.

61. $1.50 \approx 1.5$

$7759\overline{)11{,}674.00}$

$$
\begin{array}{r}
-7\,759 \\
\hline
3\,915\,0 \\
-3\,879\,5 \\
\hline
35\,50
\end{array}
$$

The unit rate is 1.5 steps/foot.

65. no; answers may vary

Exercise Set 5.3

1. $\dfrac{10 \text{ diamonds}}{6 \text{ opals}} = \dfrac{5 \text{ diamonds}}{3 \text{ opals}}$

5. $\dfrac{6 \text{ eagles}}{58 \text{ sparrows}} = \dfrac{3 \text{ eagles}}{29 \text{ sparrows}}$

9. $\dfrac{22 \text{ vanilla wafers}}{1 \text{ cup cookie crumbs}} = \dfrac{55 \text{ vanilla wafers}}{2.5 \text{ cups cookie crumbs}}$

13. $\dfrac{8}{6} \overset{?}{=} \dfrac{9}{7}$

$8 \cdot 7 \overset{?}{=} 9 \cdot 6$

$56 \neq 54$

false

17. $\dfrac{5}{8} \overset{?}{=} \dfrac{625}{1000}$

$5 \cdot 1000 \overset{?}{=} 8 \cdot 625$

$5000 = 5000$

true

21.
$$\frac{8}{10} \overset{?}{=} \frac{5.6}{0.7}$$
$$8(0.7) \overset{?}{=} 10(5.6)$$
$$5.6 \neq 56$$
false

25.
$$\frac{2\frac{2}{5}}{\frac{2}{3}} \overset{?}{=} \frac{1\frac{1}{9}}{\frac{1}{4}}$$
$$2\frac{2}{5} \cdot \frac{1}{4} \overset{?}{=} 1\frac{1}{9} \cdot \frac{2}{3}$$
$$\frac{12}{5} \cdot \frac{1}{4} \overset{?}{=} \frac{10}{9} \cdot \frac{2}{3}$$
$$\frac{3}{5} \neq \frac{20}{27}$$
false

29.
$$\frac{8}{12} \overset{?}{=} \frac{4}{6}$$
$$8 \cdot 6 \overset{?}{=} 12 \cdot 4$$
$$48 = 48$$
true

33.
$$\frac{1.8}{2} \overset{?}{=} \frac{4.5}{5}$$
$$1.8(5) \overset{?}{=} 4.5(2)$$
$$9 = 9$$
true

37.
$$\frac{n}{5} = \frac{6}{10}$$
$$10 \cdot n = 6 \cdot 5$$
$$10 \cdot n = 30$$
$$n = 3$$

41.
$$\frac{n}{8} = \frac{50}{100}$$
$$100 \cdot n = 8 \cdot 50$$
$$100 \cdot n = 400$$
$$n = 4$$

45.
$$\frac{24}{n} = \frac{60}{96}$$
$$24 \cdot 96 = 60 \cdot n$$
$$2304 = 60 \cdot n$$
$$38.4 = n$$

49.
$$\frac{0.05}{12} = \frac{n}{0.6}$$
$$0.05(0.6) = 12 \cdot n$$
$$0.03 = 12 \cdot n$$
$$0.0025 = n$$

53.
$$\frac{\frac{1}{3}}{\frac{3}{8}} = \frac{\frac{2}{5}}{n}$$
$$\frac{1}{3} \cdot n = \frac{3}{8} \cdot \frac{2}{5}$$
$$\frac{1}{3} \cdot n = \frac{3}{20}$$
$$n = \frac{3}{20} \div \frac{1}{3}$$
$$n = \frac{3}{20} \cdot \frac{3}{1}$$
$$n = \frac{9}{20}$$

57.
$$\frac{n}{1\frac{1}{5}} = \frac{4\frac{1}{6}}{6\frac{2}{3}}$$
$$\left(6\frac{2}{3}\right)n = \left(1\frac{1}{5}\right)\left(4\frac{1}{6}\right)$$
$$\frac{20}{3} \cdot n = \frac{6}{5} \cdot \frac{25}{6}$$
$$\frac{20}{3} \cdot n = \frac{5}{1}$$
$$n = \frac{5}{1} \div \frac{20}{3}$$
$$n = \frac{5}{1} \cdot \frac{3}{20}$$
$$n = \frac{3}{4}$$

61. 8.01   8.1
$$\uparrow \qquad \uparrow$$
$$0 \,<\, 1 \text{ so}$$
$$8.01 < 8.1$$

65. $5\frac{1}{3} = \frac{16}{3}$
$$6\frac{2}{3} = \frac{20}{3}$$
$$16 < 20, \text{ so } 5\frac{1}{3} < 6\frac{2}{3}$$

69.
$$\frac{6}{18} = \frac{1}{3}$$
$$\frac{6}{1} = \frac{18}{3}$$
$$\frac{3}{18} = \frac{1}{6}$$
$$\frac{18}{6} = \frac{3}{1}$$

73. answers may vary

77.
$$\frac{n}{5.2} = \frac{0.08}{6}$$
$$6 \cdot n = 5.2 \cdot 0.08$$
$$6 \cdot n = 0.416$$
$$n = \frac{0.416}{6}$$
$$n \approx 0.07$$

81.
$$\frac{n}{7} = \frac{0}{8}$$
$$8 \cdot n = 0 \cdot 7$$
$$8 \cdot n = 0$$
$$n = \frac{0}{8}$$
$$n = 0$$

85.
$$\frac{222}{1515} = \frac{37}{n}$$
$$222 \cdot n = 37 \cdot 1515$$
$$222 \cdot n = 56{,}055$$
$$n = \frac{56{,}055}{222}$$
$$n = 252.5$$

Exercise Set 5.4

1. Let n be the number of field goals (baskets) made.

$$\text{baskets} \rightarrow \frac{45}{100} = \frac{n}{800} \leftarrow \text{baskets} \atop \text{attempts}$$

$$45 \cdot 800 = 100 \cdot n$$
$$36{,}000 = 100 \cdot n$$
$$\frac{36{,}000}{100} = n$$
$$360 = n$$

He made 360 baskets.

5. Let n be the number of applications received.

$$\text{accepted} \rightarrow \frac{2}{7} = \frac{180}{n} \leftarrow \text{accepted} \atop \text{applied}$$

$$2 \cdot n = 7 \cdot 180$$
$$2 \cdot n = 1260$$
$$n = \frac{1260}{2}$$
$$n = 630$$

The school received 630 applications.

9. Let n be the number of square feet required.

$$\text{floor space} \rightarrow \frac{9}{1} = \frac{n}{30} \leftarrow \text{floor space} \atop \text{students}$$

$$9 \cdot 30 = 1 \cdot n$$
$$270 = n$$

30 students require 270 square feet of floor space.

13. Let n be the distance between Milan and Rome.

$$\text{kilometers} \rightarrow \frac{30}{1} = \frac{n}{15} \leftarrow \text{kilometers} \atop \text{cm on map}$$

$$30 \cdot 15 = 1 \cdot n$$
$$450 = n$$

Milan and Rome are 450 kilometers apart.

17. Let n be the number of hits the player is expected to get.

$$\text{hits} \rightarrow \frac{3}{8} = \frac{n}{40} \leftarrow \text{hits} \atop \text{at bats}$$

$$3 \cdot 40 = 8 \cdot n$$
$$120 = 8 \cdot n$$
$$\frac{120}{8} = n$$
$$15 = n$$

The player would be expected to get 15 hits.

21. Let n be the number of applications she should expect.

$$\text{applications} \rightarrow \frac{4}{3} = \frac{n}{14} \leftarrow \text{applications} \atop \text{ounces}$$

$$4 \cdot 14 = 3 \cdot n$$
$$56 = 3 \cdot n$$
$$\frac{56}{3} = n$$
$$18\frac{2}{3} = n$$

She should expect 18 applications from the 14-ounce bottle.

25. Let n be the number of servings he can make.

$$\text{milk} \rightarrow \frac{1\frac{1}{2}}{4} = \frac{4}{n} \leftarrow \text{milk} \atop \text{servings}$$

$$1\frac{1}{2} \cdot n = 4 \cdot 4$$
$$\frac{3}{2} \cdot n = 16$$
$$n = 16 \div \frac{3}{2}$$
$$n = \frac{32}{3} = 10\frac{2}{3}$$

He can make $10\frac{2}{3}$ servings.

29. a. Let n be the number of teaspoons of granules needed.

$$\text{water} \rightarrow \frac{25}{1} = \frac{450}{n} \leftarrow \text{water} \atop \text{granules}$$

$$25 \cdot n = 1 \cdot 450$$
$$25 \cdot n = 450$$
$$n = \frac{450}{25}$$
$$n = 18$$

18 teaspoons of granules are needed.

b. Let n be the number of tablespoons of granules needed.

$$\text{tsp} \rightarrow \frac{3}{1} = \frac{18}{n} \leftarrow \text{tsp} \atop \text{tbsp}$$

$$3 \cdot n = 1 \cdot 18$$
$$3 \cdot n = 18$$
$$n = \frac{18}{3}$$
$$n = 6$$

6 tablespoons of granules are needed.

33. Let n be the estimated head-to-toe height of the Statue of Liberty.

$$\text{height} \rightarrow \frac{n}{42} = \frac{5\frac{1}{3}}{2} \leftarrow \text{height} \atop \text{arm length}$$

$$n \cdot 2 = 42 \cdot 5\frac{1}{3}$$
$$2 \cdot n = 42 \cdot \frac{16}{3}$$
$$2 \cdot n = 224$$
$$n = \frac{224}{2}$$
$$n = 112$$

The estimated height is 112 feet.

$$112 - 111\frac{1}{12} = \frac{11}{12}$$

The difference is $\frac{11}{12}$ foot or 11 inches.

37. Let n be the estimated height of the Empire State Building.

$$\text{height} \rightarrow \frac{n}{102} = \frac{881}{72} \leftarrow \text{height} \atop \leftarrow \text{stories}$$

$$\text{stories} \rightarrow$$

$$n \cdot 72 = 102 \cdot 881$$
$$72 \cdot n = 89{,}862$$
$$n = \frac{89{,}862}{72}$$
$$n \approx 1248$$

The height of the Empire State Building is approximately 1248 feet.

41. Let n be the number of wireless phones expected to be smartphones.

$$\text{smartphones} \rightarrow \frac{n}{170} = \frac{2}{3} \leftarrow \text{smartphones} \atop \leftarrow \text{wireless phones}$$

$$\text{wireless phones} \rightarrow$$

$$n \cdot 3 = 170 \cdot 2$$
$$3 \cdot n = 340$$
$$n = \frac{340}{3}$$
$$n \approx 113$$

We would expect 113 million of the wireless phones sold to have been smartphones.

45. a. Let n be the number of gallons of oil needed.

$$\text{oil} \rightarrow \frac{n}{5} = \frac{1}{50} \leftarrow \text{oil} \atop \leftarrow \text{gas}$$

$$\text{gas} \rightarrow$$

$$n \cdot 50 = 5 \cdot 1$$
$$50 \cdot n = 5$$
$$n = \frac{5}{50}$$
$$n = \frac{1}{10} = 0.1$$

0.1 gallon of oil is needed.

b. $0.1 \text{ gallon} = 0.1(1 \text{ gallon})$
$$= 0.1(128 \text{ fluid ounces})$$
$$= 12.8 \text{ fluid ounces}$$

0.1 gallon is approximately 13 fluid ounces.

49. $3\overline{)15}$ 5

$$15 = 3 \cdot 5$$

53. $5\overline{)25}$
$2\overline{)50}$
$2\overline{)100}$
$2\overline{)200}$

$$200 = 2 \cdot 2 \cdot 2 \cdot 5 \cdot 5 = 2^3 \cdot 5^2$$

57. Let n be the number of ml.

$$\text{mg} \rightarrow \frac{15}{1} = \frac{12}{n} \leftarrow \text{mg} \atop \leftarrow \text{ml}$$

$$\text{ml} \rightarrow$$

$$15 \cdot n = 1 \cdot 12$$
$$15 \cdot n = 12$$
$$n = \frac{12}{15}$$
$$n = \frac{4}{5} = 0.8$$

0.8 ml of the medicine should be administered.

61. 11 muffins are approximately 1 dozen (12) muffins.

$$1.5 \cdot 8 = 12$$

Approximately 12 cups of milk will be needed.

65. answers may vary

Chapter 5 Test

1. $\dfrac{\$75}{\$10} = \dfrac{75}{10} = \dfrac{5 \cdot 15}{5 \cdot 2} = \dfrac{15}{2}$

5. $\dfrac{8.6}{10} = \dfrac{8.6 \cdot 10}{10 \cdot 10} = \dfrac{86}{100} = \dfrac{2 \cdot 43}{2 \cdot 50} = \dfrac{43}{50}$

9. $\dfrac{8 \text{ inches}}{12 \text{ hours}} = \dfrac{8}{12} \text{inch/hour} = \dfrac{2}{3} \text{inch/hour}$

13.
$$0.0931 \approx 0.093$$
$$16\overline{)1.4900}$$
$$\underline{-1\,44}$$
$$50$$
$$\underline{-48}$$
$$20$$
$$\underline{-16}$$
$$4$$

The 16-ounce size costs approximately \$0.093/ounce.

$$0.0995 \approx 0.100$$
$$24\overline{)2.3900}$$
$$\underline{-2\,16}$$
$$230$$
$$\underline{-216}$$
$$140$$
$$\underline{-120}$$
$$20$$

The 24-ounce size costs approximately \$0.100/ounce. The 16-ounce size is the better buy.

17. $\dfrac{n}{3} = \dfrac{15}{9}$

$$9 \cdot n = 3 \cdot 15$$
$$9 \cdot n = 45$$
$$n = \frac{45}{9}$$
$$n = 5$$

21. $\dfrac{n}{2\frac{5}{8}} = \dfrac{1\frac{1}{6}}{3\frac{1}{2}}$

$$\left(3\frac{1}{2}\right) \cdot n = \left(1\frac{1}{6}\right)\left(2\frac{5}{8}\right)$$
$$\frac{7}{2} \cdot n = \frac{7}{6} \cdot \frac{21}{8}$$
$$\frac{7}{2} \cdot n = \frac{49}{16}$$
$$n = \frac{49}{16} \div \frac{7}{2}$$
$$n = \frac{49}{16} \cdot \frac{2}{7}$$
$$n = \frac{7}{8}$$

25. Let n = the number of cartons.

$$\text{cartons} \to \frac{n}{8} = \frac{86}{6} \leftarrow \text{cartons}$$
$$\text{hours} \to 8 \quad 6 \leftarrow \text{hours}$$

$$6 \cdot n = 8 \cdot 86$$
$$6 \cdot n = 688$$
$$n = \frac{688}{6}$$
$$n = 114\frac{2}{3}$$

He can pack $114\frac{2}{3}$ cartons in 8 hours.

Chapter 6

Exercise Set 6.1

1. $\dfrac{96}{100} = 96\%$

5. The largest sector corresponds to football, so football was preferred by most adults.

$$\frac{37}{100} = 37\%$$

37% of adults prefer football.

9. $41\% = 41(0.01) = 0.41$

13. $100\% = 100(0.01) = 1.00$ or 1

17. $2.8\% = 2.8(0.01) = 0.028$

21. $300\% = 300(0.01) = 3.00$ or 3

25. $38\% = 38(0.01) = 0.38$ **29.** $45\% = 45(0.01) = 0.45$

33. $3.1 = 3.1(100\%) = 310\%$

37. $0.003 = 0.003(100\%) = 0.3\%$

41. $5.3 = 5.3(100\%) = 530\%$

45. $0.3328 = 0.3328(100\%) = 33.28\%$

49. $0.7 = 0.7(100\%) = 70\%$

53. $0.281 = 0.281(100\%) = 28.1\%$

57. $\dfrac{1}{4} = \dfrac{1}{4} \cdot \dfrac{25}{25} = \dfrac{25}{100} = 0.25$ **61.** $\dfrac{9}{10} = \dfrac{9}{10} \cdot \dfrac{10}{10} = \dfrac{90}{100} = 0.9$

65. If the percentages of the four blood types are added, the total will be 100%, since the blood types cover the whole population

$$45\% + 40\% + 11\% = (45 + 40 + 11)\% = 96\%$$
$$100\% - 96\% = (100 - 96)\% = 4\%$$

4% of the U.S. population has the blood type AB.

69. The percent change for biomedical engineers is 61.7%.

$$61.7\% = 61.7(0.01) = 0.617$$

Exercise Set 6.2

1. $12\% = 12 \cdot \dfrac{1}{100} = \dfrac{12}{100} = \dfrac{3 \cdot 4}{25 \cdot 4} = \dfrac{3}{25}$

5. $4.5\% = 4.5 \cdot \dfrac{1}{100}$

$$= \frac{4.5}{100}$$
$$= \frac{4.5}{100} \cdot \frac{10}{10}$$
$$= \frac{45}{1000}$$
$$= \frac{9 \cdot 5}{200 \cdot 5}$$
$$= \frac{9}{200}$$

9. $73\% = 73 \cdot \dfrac{1}{100} = \dfrac{73}{100}$

13. $6.25\% = 6.25 \cdot \dfrac{1}{100}$

$$= \frac{6.25}{100}$$
$$= \frac{6.25}{100} \cdot \frac{100}{100}$$
$$= \frac{625}{10,000}$$
$$= \frac{1 \cdot 625}{16 \cdot 625}$$
$$= \frac{1}{16}$$

17. $10\dfrac{1}{3}\% = \dfrac{31}{3}\% = \dfrac{31}{3} \cdot \dfrac{1}{100} = \dfrac{31}{300}$

21. $\dfrac{3}{4} = \dfrac{3}{4}(100\%) = \dfrac{300}{4}\% = 75\%$

25. $\dfrac{2}{5} = \dfrac{2}{5}(100\%) = \dfrac{200}{5}\% = 40\%$

29. $\dfrac{17}{50} = \dfrac{17}{50}(100\%) = \dfrac{1700}{50}\% = 34\%$

33. $\dfrac{5}{16} = \dfrac{5}{16}(100\%) = \dfrac{500}{16}\% = \dfrac{125}{4}\% = 31\dfrac{1}{4}\%$

37. $\dfrac{7}{9} = \dfrac{7}{9}(100\%) = \dfrac{700}{9}\% = 77\dfrac{7}{9}\%$

41. $2\dfrac{1}{2} = \dfrac{5}{2} = \dfrac{5}{2}(100\%) = \dfrac{500}{2}\% = 250\%$

45. $\dfrac{7}{11} = \dfrac{7}{11}(100\%) = \dfrac{700}{11}\% \approx 63.64\%$

$$\begin{array}{r} 63.636\ldots \approx 63.64 \\ 11\overline{)700.000} \\ \underline{-66} \\ 40 \\ \underline{-33} \\ 7\,0 \\ \underline{-6\,6} \\ 40 \\ \underline{-33} \\ 70 \\ \underline{-66} \\ 4 \end{array}$$

49. $\dfrac{1}{7} = \dfrac{1}{7}(100\%) = \dfrac{100}{7}\% \approx 14.29\%$

$$\begin{array}{r} 14.285 \approx 14.29 \\ 7\overline{)100.000} \\ \underline{-7} \\ 30 \\ \underline{-28} \\ 2\,0 \\ \underline{-1\,4} \\ 60 \\ \underline{-56} \\ 40 \\ \underline{-35} \\ 5 \end{array}$$

53.

Percent	Decimal	Fraction
35%	$35\% = 0.35$	$35\% = \dfrac{35}{100} = \dfrac{7}{20}$
$\dfrac{1}{5} = \dfrac{20}{100} = 20\%$	$\dfrac{1}{5} = \dfrac{2}{10} = 0.2$	$\dfrac{1}{5}$
$0.5 = 0.50 = 50\%$	0.5	$0.5 = \dfrac{5}{10} = \dfrac{1}{2}$
70%	$70\% = 0.7$	$70\% = \dfrac{70}{100} = \dfrac{7}{10}$
$\dfrac{3}{8} = \dfrac{375}{1000} = \dfrac{37.5}{100}$ $= 37.5\%$	$\dfrac{3}{8} = \dfrac{375}{1000} = 0.375$	$\dfrac{3}{8}$

57.

Percent	Decimal	Fraction
200%	$200\% = 2.00 = 2$	$200\% = \dfrac{200}{100} = 2$
$2.8 = \dfrac{28}{10} = \dfrac{280}{100}$ $= 280\%$	2.8	$2.8 = 2\dfrac{8}{10} = 2\dfrac{4}{5}$
705%	$705\% = 7.05$	$705\% = \dfrac{705}{100}$ $= 7\dfrac{5}{100} = 7\dfrac{1}{20}$
$4\dfrac{27}{50} = 4\dfrac{54}{100}$ $= \dfrac{454}{100} = 454\%$	$4\dfrac{27}{50} = 4\dfrac{54}{100} = 4.54$	$4\dfrac{27}{50}$

61. $52.2\% = 0.522$

$$52.2\% = \frac{52.2}{100} = \frac{522}{1000} = \frac{261}{500}$$

65. $8.75\% = 0.0875$

69. $0.5\% = 0.005$

$$0.5\% = \frac{0.5}{100} = \frac{50}{1000} = \frac{1}{200}$$

73. $7.9\% = 0.079$

$$7.9\% = \frac{7.9}{100} = \frac{79}{1000}$$

77. $8 \cdot n = 80$

$$\frac{8 \cdot n}{8} = \frac{80}{8}$$

$$n = 10$$

81. a. To round 52.8647 to the nearest tenth observe that the digit in the hundredths place is 6. Since this digit is 5 or greater, we add 1 to the digit in the tenths place. The number 52.8647 rounded to the nearest tenth is 52.9. Thus $52.8647\% \approx 52.9\%$.

b. To round 52.8647 to the nearest hundredth observe that the digit in the thousandths place is 4. Since this digit is less than 5, we do not add 1 to the digit in the hundredths place. The number 52.8647 rounded to the nearest hundredth is 52.86. Thus $52.8647\% \approx 52.86\%$.

85. $0.65794 = 65.794\%$

To round 65.794 to the nearest hundredth observe that the digit in the thousandths place is 4. Since this digit is less than 5, we do not add 1 to the digit in the hundredths place. The number 65.794 rounded to the nearest hundredth is 65.79. Thus, $0.65794 = 65.794\% \approx 65.79\%$.

89. 3 of the 4 equal parts are shaded.

$$\frac{3}{4} = \frac{3}{4} \cdot \frac{25}{25} = \frac{75}{100} = 75\%$$

75% of the figure is shaded.

93. A fraction written as a percent is greater than 100% when the numerator is <u>greater</u> than the denominator.

97.
$$\begin{array}{r} 0.2658 \approx 0.266 \\ 79\overline{)21.0000} \\ -15\,8 \\ \hline 5\,20 \\ -4\,74 \\ \hline 460 \\ -395 \\ \hline 650 \\ -632 \\ \hline 18 \end{array}$$

$$\frac{21}{79} \approx 0.266 \text{ or } 26.6\%$$

Exercise Set 6.3

1. 18% of 81 is what number?

$$18\% \cdot 81 = n$$

5. 0.6 is 40% of what number?

$$0.6 = 40\% \cdot n$$

9. What number is 9% of 43?

$$n = 9\% \cdot 43$$

13. $10\% \cdot 35 = n$
$0.10 \cdot 35 = n$
$3.5 = n$
10% of 35 is 3.5.

17. $1.2 = 12\% \cdot n$
$1.2 = 0.12 \cdot n$
$\dfrac{1.2}{0.12} = \dfrac{0.12 \cdot n}{0.12}$
$10 = n$
1.2 is 12% of 10.

21. $n \cdot 80 = 88$
$\dfrac{n \cdot 80}{80} = \dfrac{88}{80}$
$n = 1.1$
$n = 110\%$
88 is 110% of 80.

25. $0.1 = 10\% \cdot n$
$0.1 = 0.10 \cdot n$
$\dfrac{0.1}{0.1} = \dfrac{0.1 \cdot n}{0.1}$
$1 = n$
0.1 is 10% of 1.

29. $82.5 = 16\dfrac{1}{2}\% \cdot n$
$82.5 = 0.165 \cdot n$
$\dfrac{82.5}{0.165} = \dfrac{0.165 \cdot n}{0.165}$
$500 = n$
82.5 is $16\dfrac{1}{2}\%$ of 500.

33. $n = 42\% \cdot 60$
$n = 0.42 \cdot 60$
$n = 25.2$
25.2 is 42% of 60.

37. $120\% \cdot n = 42$
$1.20 \cdot n = 42$
$\dfrac{1.2 \cdot n}{1.2} = \dfrac{42}{1.2}$
$n = 35$
120% of 35 is 42.

41. $n \cdot 600 = 3$
$\dfrac{n \cdot 600}{600} = \dfrac{3}{600}$
$n = 0.005$
$n = 0.5\%$
0.5% of 600 is 3.

45. $1575 = n \cdot 2500$

$$\frac{1575}{2500} = \frac{n \cdot 2500}{2500}$$

$$0.63 = n$$

$$63\% = n$$

1575 is 63% of 2500.

49. $\dfrac{27}{n} = \dfrac{9}{10}$

$$27 \cdot 10 = 9 \cdot n$$

$$270 = 9n$$

$$\frac{270}{9} = n$$

$$30 = n$$

53. $\dfrac{17}{12} = \dfrac{n}{20}$

57. In the equation $5 \cdot n = 32$, the step that should be taken to find the value of n is to divide by 5, obtaining $n = \dfrac{32}{5}$, which is choice c.

61. $20\% \cdot n = 18.6$ in words is "twenty percent of some number is eighteen and six tenths."

65. Since 85 is less than 120, the percent is less than 100%; c.

69. Since 100% is 1, 100% of 45 is equal to 45; a.

73. answers may vary

77. $22{,}113 = 180\% \cdot n$

$$22{,}113 = 1.80 \cdot n$$

$$\frac{22{,}113}{1.8} = \frac{1.8 \cdot n}{1.8}$$

$$12{,}285 = n$$

22,113 is 180% of 12,285.

Exercise Set 6.4

1. 98%   of   45   is   what number?

   ↓   ↓   ↓

  percent   base   amount $= a$

$$\frac{a}{45} = \frac{98}{100}$$

5. 14.3   is   26%   of   what number?

   ↓   ↓   ↓

  amount   percent   base $= b$

$$\frac{14.3}{b} = \frac{26}{100}$$

9. What percent   of   400   is   70?

     ↓   ↓   ↓

  percent $= p$ base   amount

$$\frac{70}{400} = \frac{p}{100}$$

13. $\dfrac{a}{65} = \dfrac{40}{100}$ or $\dfrac{a}{65} = \dfrac{2}{5}$

$$a \cdot 5 = 65 \cdot 2$$

$$5a = 130$$

$$\frac{5a}{5} = \frac{130}{5}$$

$$a = 26$$

40% of 65 is 26.

17. $\dfrac{90}{b} = \dfrac{15}{100}$ or $\dfrac{90}{b} = \dfrac{3}{20}$

$$90 \cdot 20 = b \cdot 3$$

$$1800 = 3b$$

$$\frac{1800}{3} = \frac{3b}{3}$$

$$600 = b$$

15% of 600 is 90.

21. $\dfrac{42}{35} = \dfrac{p}{100}$ or $\dfrac{6}{5} = \dfrac{p}{100}$

$$6 \cdot 100 = 5 \cdot p$$

$$600 = 5p$$

$$\frac{600}{5} = \frac{5p}{5}$$

$$120 = p$$

42 is 120% of 35.

25. $\dfrac{3.7}{b} = \dfrac{10}{100}$ or $\dfrac{3.7}{b} = \dfrac{1}{10}$

$$3.7 \cdot 10 = b \cdot 1$$

$$37 = b$$

3.7 is 10% of 37.

29. $\dfrac{160}{b} = \dfrac{16}{100}$ or $\dfrac{160}{b} = \dfrac{4}{25}$

$$160 \cdot 25 = b \cdot 4$$

$$4000 = 4b$$

$$\frac{4000}{4} = \frac{4b}{4}$$

$$1000 = b$$

160 is 16% of 1000.

33. $\dfrac{a}{62} = \dfrac{89}{100}$

$$a \cdot 100 = 62 \cdot 89$$

$$100a = 5518$$

$$\frac{100a}{100} = \frac{5518}{100}$$

$$a = 55.18$$

55.18 is 89% of 62.

37. $\dfrac{105}{b} = \dfrac{140}{100}$ or $\dfrac{105}{b} = \dfrac{7}{5}$

$$105 \cdot 5 = b \cdot 7$$

$$525 = 7b$$

$$\frac{525}{7} = \frac{7b}{7}$$

$$75 = b$$

140% of 75 is 105.

41. $\dfrac{4}{800} = \dfrac{p}{100}$ or $\dfrac{1}{200} = \dfrac{p}{100}$

$$1 \cdot 100 = 200 \cdot p$$

$$100 = 200p$$

$$\frac{100}{200} = \frac{200p}{200}$$

$$0.5 = p$$

0.5% of 800 is 4.

45. $\dfrac{a}{48} = \dfrac{20}{100}$ or $\dfrac{a}{48} = \dfrac{1}{5}$

$a \cdot 5 = 48 \cdot 1$

$5a = 48$

$\dfrac{5a}{5} = \dfrac{48}{5}$

$a = 9.6$

20% of 48 is 9.6.

49. $\dfrac{11}{16} + \dfrac{3}{16} = \dfrac{11+3}{16} = \dfrac{14}{16} = \dfrac{7 \cdot 2}{8 \cdot 2} = \dfrac{7}{8}$

53. $\overset{1}{0.41}$

$\underline{+0.29}$

0.70

57. answers may vary

61. $\dfrac{p}{100} = \dfrac{13}{52}$

$\dfrac{25}{100} \overset{?}{=} \dfrac{13}{52}$

$\dfrac{1}{4} = \dfrac{1}{4}$ True

Yes, the percent is 25.

65. $\dfrac{a}{53,862} = \dfrac{22.3}{100}$

$a \cdot 100 = 22.3 \cdot 53,862$

$100a = 1,201,122.6$

$\dfrac{100a}{100} = \dfrac{1,201,122.6}{100}$

$a \approx 12,011.2$

Exercise Set 6.5

1. 24 is 1.5% of what number?

Method 1:

$24 = 1.5\% \cdot n$

$24 = 0.015n$

$\dfrac{24}{0.015} = \dfrac{0.015 \cdot n}{0.015}$

$1600 = n$

1600 bolts were inspected.

Method 2:

$\dfrac{24}{b} = \dfrac{1.5}{100}$

$24 \cdot 100 = 1.5 \cdot b$

$2400 = 1.5b$

$\dfrac{2400}{1.5} = \dfrac{1.5b}{1.5}$

$1600 = b$

1600 bolts were inspected.

5. 378 is what percent of 2700?

Method 1:

$378 = n \cdot 2700$

$\dfrac{378}{2700} = \dfrac{n \cdot 2700}{2700}$

$0.14 = n$

$14\% = n$

The student spent 14% of last semester's college costs on books.

Method 2:

$\dfrac{378}{2700} = \dfrac{p}{100}$

$378 \cdot 100 = 2700 \cdot p$

$37,800 = 2700p$

$\dfrac{37,800}{2700} = \dfrac{2700p}{2700}$

$14 = p$

The student spent 14% of last semester's college costs on books.

9. $12,780 is what percent of $28,400?

Method 1:

$12,780 = n \cdot 28,400$

$\dfrac{12,780}{28,400} = \dfrac{n \cdot 28,400}{28,400}$

$0.45 = n$

$45\% = n$

The reception venue is generally 45% of the average wedding budget.

Method 2:

$\dfrac{12,780}{28,400} = \dfrac{p}{100}$

$12,780 \cdot 100 = 28,400 \cdot p$

$1,278,000 = 28,400p$

$\dfrac{1,278,000}{28,400} = \dfrac{28,400p}{28,400}$

$45 = p$

The reception venue is generally 45% of the average wedding budget.

13. What number is 30% of 83,600?

Method 1:

$n = 30\% \cdot 83,600$

$n = 0.30 \cdot 83,600$

$n = 25,080$

The number of people employed as physician assistants is expected to be $83,600 + 25,080 = 108,680$.

Method 2:

$\dfrac{a}{83,600} = \dfrac{30}{100}$

$a \cdot 100 = 83,600 \cdot 30$

$100a = 2,508,000$

$\dfrac{100a}{100} = \dfrac{2,508,000}{100}$

$a = 25,080$

The number of people employed as physician assistants is expected to be $83,600 + 25,080 = 108,680$.

17. 38 is what percent of 131?

Method 1:

$38 = n \cdot 131$

$\dfrac{38}{131} = \dfrac{n \cdot 131}{131}$

$0.29 \approx n$

$29\% \approx n$

29% of the ski runs at Keystone ski area are rated intermediate.

Method 2:

$$\frac{38}{131} = \frac{p}{100}$$

$$38 \cdot 100 = 131 \cdot p$$

$$3800 = 131p$$

$$\frac{3800}{131} = \frac{131p}{131}$$

$$29 \approx p$$

29% of the ski runs at Keystone ski area are rated intermediate.

21. 10 is what percent of 80?

Method 1:

$$10 = n \cdot 80$$

$$\frac{10}{80} = \frac{n \cdot 80}{80}$$

$$0.125 = n$$

$$12.5\% = n$$

12.5% of the total calories come from fat.

Method 2:

$$\frac{10}{80} = \frac{p}{100}$$

$$10 \cdot 100 = 80 \cdot p$$

$$1000 = 80p$$

$$\frac{1000}{80} = \frac{80p}{80}$$

$$12.5 = p$$

12.5% of the total calories come from fat.

25. 26,250 is 15% of what number?

Method 1:

$$26{,}250 = 15\% \cdot n$$

$$26{,}250 = 0.15 \cdot n$$

$$\frac{26{,}250}{0.15} = \frac{0.15 \cdot n}{0.15}$$

$$175{,}000 = n$$

The price of the home was $175,000.

Method 2:

$$\frac{26{,}250}{b} = \frac{15}{100}$$

$$26{,}250 \cdot 100 = b \cdot 15$$

$$2{,}625{,}000 = 15b$$

$$\frac{2{,}625{,}000}{15} = \frac{15b}{15}$$

$$175{,}000 = b$$

The price of the home was $175,000.

29. What number is 4.5% of 19,286?

Method 1:

$$n = 4.5\% \cdot 19{,}286$$

$$n = 0.045 \cdot 19{,}286$$

$$n = 867.87$$

The price of the car will increase by $867.87. The new price of that model will be $19,286 + $867.87 = $20,153.87.

Method 2:

$$\frac{a}{19{,}286} = \frac{4.5}{100}$$

$$a \cdot 100 = 4.5 \cdot 19{,}286$$

$$100a = 86{,}787$$

$$a = \frac{86{,}787}{100}$$

$$a = 867.87$$

The price of the car will increase by $867.87. The new price of that model will be $19,286 + $867.87 = $20,153.87.

33. 14.4% of 8406 is what number?

Method 1:

$$14.4\%(8406) = n$$

$$0.144(8406) = n$$

$$1210.464 = n$$

$$1210 \approx n$$

The increase in tuition was $1210. The tuition for the 2013–2014 school year was $8406 + $1210 = $9616.

Method 2:

$$\frac{a}{8406} = \frac{14.4}{100}$$

$$a \cdot 100 = 14.4 \cdot 8406$$

$$100a = 121{,}046.4$$

$$\frac{100a}{100} = \frac{121{,}046.4}{100}$$

$$a = 1210.464$$

$$a \approx 1210$$

The increase in tuition was $1210. The tuition for the 2013–2014 school year was $8406 + $1210 = $9616.

37.

Original Amount	New Amount	Amount of Increase	Percent of Increase
50	80	$80 - 50 = 30$	$\frac{30}{50} = 0.6 = 60\%$

41.

Original Amount	New Amount	Amount of Decrease	Percent of Decrease
8	6	$8 - 6 = 2$	$\frac{2}{8} = 0.25 = 25\%$

45. percent of decrease $= \dfrac{\text{amount of decrease}}{\text{original amount}}$

$$= \frac{150 - 84}{150}$$

$$= \frac{66}{150}$$

$$= 0.44$$

The decrease in calories per cup is 44%.

49. percent of increase $= \dfrac{\text{amount of increase}}{\text{original amount}}$

$$= \frac{421 - 174}{174}$$

$$= \frac{247}{174}$$

$$\approx 1.420$$

The increase in the size of privately owned farms in the United States was about 142.0%.

53. percent of increase = $\dfrac{\text{amount of increase}}{\text{original amount}}$

$\phantom{\text{percent of increase}} = \dfrac{50.3 - 21}{21}$

$\phantom{\text{percent of increase}} = \dfrac{29.3}{21}$

$\phantom{\text{percent of increase}} \approx 1.395$

The increase in the number of people subscribing to high-speed internet via their cable provider was 139.5%.

57. percent of increase = $\dfrac{\text{amount of increase}}{\text{original amount}}$

$\phantom{\text{percent of increase}} = \dfrac{77,100 - 64,400}{64,400}$

$\phantom{\text{percent of increase}} = \dfrac{12,700}{64,400}$

$\phantom{\text{percent of increase}} \approx 0.197$

The percent of increase in the number of dieticians expected to be employed in the United States is 19.7%.

61.
$$\begin{array}{r} 0.12 \\ \times\ 38 \\ \hline 96 \\ 360 \\ \hline 4.56 \end{array}$$

65.
$$\begin{array}{r} 78.00 \\ -19.46 \\ \hline 58.54 \end{array}$$

69. percent of increase = $\dfrac{\text{amount of increase}}{\text{original amount}}$

$\phantom{\text{percent of increase}} = \dfrac{180 - 150}{150}$

$\phantom{\text{percent of increase}} = \dfrac{30}{150}$

$\phantom{\text{percent of increase}} = 0.20$

The percent of increase is 20%.

Exercise Set 6.6

1. sales tax = $5\% \cdot \$150 = 0.05 \cdot \$150 = \$7.50$

The sales tax is $7.50.

5. $\$335.30 = r \cdot \$4790$

$\dfrac{335.30}{4790} = r$

$0.07 = r$

The sales tax rate is 7%.

9. sales tax = $6.5\% \cdot \$1800 = 0.065 \cdot \$1800 = \$117$

total price = $\$1800 + \$117 = \$1917$

The sales tax is $117 and the total price of the bracelet is $1917.

13. $\$98.70 = r \cdot \$1645$

$\dfrac{98.70}{1645} = r$

$0.06 = r$

The sales tax rate is 6%.

17. commission = $4\% \cdot \$1,329,401 = 0.04 \cdot \$1,329,401$

$\phantom{\text{commission}} = \$53,176.04$

Her commission was $53,176.04.

21. commission = $1.5\% \cdot \$325,900 = 0.015 \cdot \$325,900$

$\phantom{\text{commission}} = \$4888.50$

His commission will be $4888.50.

	Original Price	Discount Rate	Amount of Discount	Sale Price
25.	$89	10%	10% · $89 = $8.90	$89 − $8.90 = $80.10
29.	$410	35%	35% · $410 = $143.50	$410 − $143.50 = $266.50

33. discount = $15\% \cdot \$300 = 0.15 \cdot \$300 = \$45$

sale price = $\$300 - \$45 = \$255$

The discount is $45 and the sale price is $255.

37.

Purchase Price	Tax Rate	Sales Tax	Total Price
$56	5.5%	5.5% · $56 = $3.08	$56 + $3.08 = $59.08

41.

Sale	Commission Rate	Commission
$17,900	$\dfrac{\$1432}{\$17,900} = 0.08 = 8\%$	$1432

45. $400 \cdot \dfrac{3}{100} \cdot 11 = 12 \cdot 11 = 132$

49. Round $68 to $70 and 9.5% to 10%.

$10\% \cdot \$70 = 0.10 \cdot \$70 = \$7$

$\$70 + \$7 = \$77$

The best estimate of the total price is $77; d.

53.

Bill Amount	10%	15%	20%
$72.17 ≈ $72.00	$7.20	$7.20 + $\dfrac{1}{2}$($7.20) = $7.20 + $3.60 = $10.80	2($7.20) = $14.40

57. $7.5\% \cdot \$24,966 = 0.075 \cdot \$24,966 = \$1872.45$

$\$24,966 + \$1872.45 = \$26,838.45$

The total price of the necklace is $26,838.45.

Exercise Set 6.7

1. simple interest = principal · rate · time

$\phantom{\text{simple interest}} = (\$200)(8\%)(2)$

$\phantom{\text{simple interest}} = (\$200)(0.08)(2) = \$32$

5. simple interest = principal · rate · time

$\phantom{\text{simple interest}} = (\$5000)(10\%)\left(1\dfrac{1}{2}\right)$

$\phantom{\text{simple interest}} = (\$5000)(0.10)(1.5) = \$750$

9. simple interest = principal · rate · time

$\phantom{\text{simple interest}} = (\$2500)(16\%)\left(\dfrac{21}{12}\right)$

$\phantom{\text{simple interest}} = (\$2500)(0.16)(1.75) = \$700$

13. simple interest = principal · rate · time

$\phantom{\text{simple interest}} = \$5000(9\%)\left(\dfrac{15}{12}\right)$

$\phantom{\text{simple interest}} = \$5000(0.09)(1.25) = \$562.5$

Total = $\$5000 + \$562.50 = \$5562.50$

17. $A = P\left(1 + \dfrac{r}{n}\right)^{n \cdot t}$

$ = 6150\left(1 + \dfrac{0.14}{2}\right)^{2 \cdot 15}$

$ = 6150(1.07)^{30}$

$ \approx 46{,}815.37$

The total amount is $46,815.37.

21. $A = P\left(1 + \dfrac{r}{n}\right)^{n \cdot t}$

$ = 10{,}000\left(1 + \dfrac{0.09}{2}\right)^{2 \cdot 20}$

$ = 10{,}000(1.045)^{40}$

$ \approx 58{,}163.65$

The total amount is $58,163.65.

25. $A = P\left(1 + \dfrac{r}{n}\right)^{n \cdot t}$

$ = 2000\left(1 + \dfrac{0.08}{1}\right)^{1 \cdot 5}$

$ = 2000(1.08)^{5}$

$ \approx 2938.66$

The total amount is $2938.66.

29. monthly payment $= \dfrac{\text{principal} + \text{simple interest}}{\text{number of payments}}$

$ = \dfrac{\$1500 + \$61.88}{6}$

$ = \dfrac{\$1561.88}{6}$

$ \approx \$260.31$

The monthly payment is $260.31.

33. perimeter $= 10 + 6 + 10 + 6 = 32$

The perimeter is 32 yards.

37. answers may vary

Chapter 6 Test

1. $85\% = 85(0.01) = 0.85$

5. $6.1 = 6.1(100\%) = 610\%$

9. $0.2\% = \dfrac{0.2}{100} = \dfrac{2}{1000} = \dfrac{1}{500}$

13. $n = 42\% \cdot 80$

$n = 0.42 \cdot 80$

$n = 33.6$

Therefore, 42% of 80 is 33.6.

17. 20% of what number is $11,350?

$0.20n = \$11{,}350$

$n = \dfrac{\$11{,}350}{0.20}$

$n = \$56{,}750$

The value is $56,750.

21. commission $= 4\% \cdot \$9875 = 0.04 \cdot \$9875 = \$395$

His commission is $395.

25. simple interest $=$ principal $\cdot$ rate $\cdot$ time

$ = (\$400)(13.5\%)\left(\dfrac{6}{12}\right)$

$ = (\$400)(0.135)(0.5)$

$ = \$27.00$

Total amount due the bank $= \$400 + \$27 = \$427$.

Chapter 7

Exercise Set 7.1

1. $60 \text{ in.} = \dfrac{60 \text{ in.}}{1} \cdot \dfrac{1 \text{ ft}}{12 \text{ in.}} = \dfrac{60}{12} \text{ ft} = 5 \text{ ft}$

5. $42{,}240 \text{ ft} = \dfrac{42{,}240 \text{ ft}}{1} \cdot \dfrac{1 \text{ mi}}{5280 \text{ ft}}$

$\phantom{42{,}240 \text{ ft}} = \dfrac{42{,}240}{5280} \text{ mi}$

$\phantom{42{,}240 \text{ ft}} = 8 \text{ mi}$

9. $10 \text{ ft} = \dfrac{10 \text{ ft}}{1} \cdot \dfrac{1 \text{ yd}}{3 \text{ ft}} = \dfrac{10}{3} \text{ yd} = 3\dfrac{1}{3} \text{ yd}$

13. $162 \text{ in.} = \dfrac{162 \text{ in.}}{1} \cdot \dfrac{1 \text{ yd}}{36 \text{ in.}}$

$\phantom{162 \text{ in.}} = \dfrac{162}{36} \text{ yd}$

$\phantom{162 \text{ in.}} = 4.5 \text{ yd}$

17. $40 \text{ ft} = \dfrac{40 \text{ ft}}{1} \cdot \dfrac{1 \text{ yd}}{3 \text{ ft}} = \dfrac{40}{3} \text{ yd}$

 13 yd 1 ft

$3\overline{)40}$

$\underline{-3}$

10

$\underline{-9}$

1

$40 \text{ ft} = 13 \text{ yd } 1 \text{ ft}$

21. $10{,}000 \text{ ft} = \dfrac{10{,}000 \text{ ft}}{1} \cdot \dfrac{1 \text{ mi}}{5280 \text{ ft}} = \dfrac{10{,}000}{5280} \text{ mi}$

 1 mi 4720 ft

$5280\overline{)10{,}000}$

$\underline{-5280}$

4720

$10{,}000 \text{ ft} = 1 \text{ mi } 4720 \text{ ft}$

25. $8 \text{ yd } 2 \text{ ft} = \dfrac{8 \text{ yd}}{1} \cdot \dfrac{3 \text{ ft}}{1 \text{ yd}} + 2 \text{ ft}$

$\phantom{8 \text{ yd } 2 \text{ ft}} = 24 \text{ ft} + 2 \text{ ft}$

$\phantom{8 \text{ yd } 2 \text{ ft}} = 26 \text{ ft}$

29. $3 \text{ ft } 10 \text{ in.} + 7 \text{ ft } 4 \text{ in.} = 10 \text{ ft } 14 \text{ in.}$

$\phantom{3 \text{ ft } 10 \text{ in.} + 7 \text{ ft } 4 \text{ in.}} = 10 \text{ ft} + 1 \text{ ft } 2 \text{ in.}$

$\phantom{3 \text{ ft } 10 \text{ in.} + 7 \text{ ft } 4 \text{ in.}} = 11 \text{ ft } 2 \text{ in.}$

33. $22 \text{ ft } 8 \text{ in.}$

$\underline{-16 \text{ ft } 3 \text{ in.}}$

$6 \text{ ft } 5 \text{ in.}$

37. $28 \text{ ft } 8 \text{ in.} \div 2 = 14 \text{ ft } 4 \text{ in.}$

41. $60 \text{ m} = \dfrac{60 \text{ m}}{1} \cdot \dfrac{100 \text{ cm}}{1 \text{ m}} = 6000 \text{ cm}$

45. $500 \text{ m} = \dfrac{500 \text{ m}}{1} \cdot \dfrac{1 \text{ km}}{1000 \text{ cm}} = \dfrac{500}{1000} \text{ km} = 0.5 \text{ km}$

49. $1500 \text{ cm} = \dfrac{1500 \text{ cm}}{1} \cdot \dfrac{1 \text{ m}}{100 \text{ cm}} = \dfrac{1500}{100} \text{ m} = 15 \text{ m}$

53. $7 \text{ km} = \dfrac{7 \text{ km}}{1} \cdot \dfrac{1000 \text{ m}}{1 \text{ km}} = 7000 \text{ m}$

57. $20.1 \text{ mm} = \dfrac{20.1 \text{ mm}}{1} \cdot \dfrac{1 \text{ dm}}{100 \text{ mm}}$

$= \dfrac{20.1}{100} \text{ dm}$

$= 0.201 \text{ dm}$

61.
$$\begin{array}{r} 8.60 \text{ m} \\ +0.34 \text{ m} \\ \hline 8.94 \text{ m} \end{array}$$

65.
$$\begin{array}{r} 24.8 \text{ mm} \\ -1.19 \text{ cm} \\ \hline \end{array} \quad \begin{array}{r} 24.8 \text{ mm} \\ -11.9 \text{ mm} \\ \hline 12.9 \text{ mm} \end{array} \text{ or } \begin{array}{r} 2.48 \text{ cm} \\ -1.19 \text{ cm} \\ \hline 1.29 \text{ cm} \end{array}$$

69. $18.3 \text{ m} \times 3 = 54.9 \text{ m}$

73.

	Yards	Feet	Inches
Chrysler Building in New York City	$348\frac{2}{3}$	1046	12,552

		Meters	Millimeters	Kilometers	Centimeters
77.	Length of elephant	5	5000	0.005	500
81.	Distance from London to Paris	342,000	342,000,000	342	34,200,000

85. $6 \text{ ft } 10 \text{ in.} + 3 \text{ ft } 8 \text{ in.} = 9 \text{ ft } 18 \text{ in.}$

$= 9 \text{ ft} + 1 \text{ ft } 6 \text{ in.}$

$= 10 \text{ ft } 6 \text{ in.}$

The bamboo is 10 ft 6 in. tall now.

89. $8 \text{ ft } 11 \text{ in.} = 96 \text{ in.} + 11 \text{ in.} = 107 \text{ in.}$

$21.5\overline{)107}$ becomes
$$\begin{array}{r} 4.97 \\ 215\overline{)1070.00} \\ -860 \\ \hline 210\,0 \\ -193\,5 \\ \hline 16\,50 \\ -15\,05 \\ \hline 1\,45 \end{array}$$

Robert was about 5.0 times the height of Chandra.

93.
$$\begin{array}{r} 80 \text{ mm} \\ -5.33 \text{ cm} \\ \hline \end{array} \quad \begin{array}{r} 80.0 \text{ mm} \\ -53.3 \text{ mm} \\ \hline 26.7 \text{ mm} \end{array}$$

The ice must be 26.7 mm thicker before skating is allowed.

97. $9 \text{ ft } 3 \text{ in.} \div 3 = 3 \text{ ft } 1 \text{ in.}$

Each cut piece will be 3 ft 1 in. long.

101.
$$\begin{array}{r} 3.35 \\ 20\overline{)67.00} \\ -60 \\ \hline 7\,0 \\ -6\,0 \\ \hline 1\,00 \\ -1\,00 \\ \hline 0 \end{array}$$

Each piece will be 3.35 meters long.

105. $2 \times 182 = 364$
$$\begin{array}{r} 121 \text{ R } 1 \\ 3\overline{)364} \\ -3 \\ \hline 06 \\ -6 \\ \hline 04 \\ -3 \\ \hline 1 \end{array}$$

Two of the trucks are $121\frac{1}{3}$ yards long.

109. $0.21 = \dfrac{21}{100}$

113. $\dfrac{1}{4} = \dfrac{1}{4} \cdot \dfrac{25}{25} = \dfrac{25}{100} = 0.25$

117. Yes, glass for a drinking glass being 2 millimeters thick is reasonable.

121. 5 yd 2 in. is close to 5 yd. 7 yd 30 in. is close to 7 yd 36 in. = 8 yd. Estimate: 5 yd + 8 yd = 13 yd.

125. answers may vary

Exercise Set 7.2

1. $2 \text{ lb} = \dfrac{2 \text{ lb}}{1} \cdot \dfrac{16 \text{ oz}}{1 \text{ lb}} = 2 \cdot 16 \text{ oz} = 32 \text{ oz}$

5. $18,000 \text{ lb} = \dfrac{18,000 \text{ lb}}{1} \cdot \dfrac{1 \text{ ton}}{2000 \text{ lb}}$

$= \dfrac{18,000}{2000} \text{ tons}$

$= 9 \text{ tons}$

9. $3500 \text{ lb} = \dfrac{3500 \text{ lb}}{1} \cdot \dfrac{1 \text{ ton}}{2000 \text{ lb}}$

$= \dfrac{3500}{2000} \text{ tons}$

$= \dfrac{7}{4} \text{ tons}$

$= 1\dfrac{3}{4} \text{ tons}$

13. $4.9 \text{ tons} = \dfrac{4.9 \text{ tons}}{1} \cdot \dfrac{2000 \text{ lb}}{1 \text{ ton}}$

$= 4.9 \cdot 2000 \text{ lb}$

$= 9800 \text{ lb}$

17. $2950 \text{ lb} = \dfrac{2950 \text{ lb}}{1} \cdot \dfrac{1 \text{ ton}}{2000 \text{ lb}}$

$= \dfrac{2950}{2000} \text{ tons}$

$= \dfrac{59}{40} \text{ tons}$

$\approx 1.5 \text{ tons}$

21. $5\dfrac{3}{4}$ lb $= \dfrac{23}{4}$ lb

$$= \dfrac{\frac{23}{4}\text{ lb}}{1} \cdot \dfrac{16\text{ oz}}{1\text{ lb}}$$

$$= \dfrac{23}{4} \cdot 16\text{ oz}$$

$$= 23 \cdot 4\text{ oz}$$

$$= 92\text{ oz}$$

25. $89\text{ oz} = \dfrac{89\text{ oz}}{1} \cdot \dfrac{1\text{ lb}}{16\text{ oz}} = \dfrac{89}{16}$ lb

$$
\begin{array}{r}
5\text{ lb }9\text{ oz} \\
16\overline{)89} \\
-80 \\
\hline
9
\end{array}
$$

$89\text{ oz} = 5\text{ lb }9\text{ oz}$

29. 3 tons 1820 lb $+ 4$ tons 930 lb

$= 7$ tons 2750 lb

$= 7$ tons $+ 1$ ton 750 lb

$= 8$ tons 750 lb

33.
$$
\begin{array}{ll}
12\text{ lb }4\text{ oz} & 11\text{ lb }20\text{ oz} \\
-\ 3\text{ lb }9\text{ oz} & -\ 3\text{ lb }\ 9\text{ oz} \\
\hline
& 8\text{ lb }11\text{ oz}
\end{array}
$$

37. 6 tons 1500 lb $\div 5 = \dfrac{6}{5}$ tons 300 lb

$$= 1\dfrac{1}{5}\text{ tons }300\text{ lb}$$

$$= 1\text{ ton} + \dfrac{2000\text{ lb}}{5} + 300\text{ lb}$$

$$= 1\text{ ton} + 400\text{ lb} + 300\text{ lb}$$

$$= 1\text{ ton }700\text{ lb}$$

41. $4\text{ g} = \dfrac{4\text{ g}}{1} \cdot \dfrac{1000\text{ mg}}{1\text{ g}} = 4 \cdot 1000\text{ mg} = 4000\text{ mg}$

45. $48\text{ mg} = \dfrac{48\text{ mg}}{1} \cdot \dfrac{1\text{ g}}{1000\text{ mg}} = \dfrac{48}{1000}\text{ g} = 0.048\text{ g}$

49. $15.14\text{ g} = \dfrac{15.14\text{ g}}{1} \cdot \dfrac{1000\text{ mg}}{1\text{ g}}$

$$= 15.14 \cdot 1000\text{ mg}$$

$$= 15{,}140\text{ mg}$$

53. $35\text{ hg} = \dfrac{35\text{ hg}}{1} \cdot \dfrac{10{,}000\text{ cg}}{1\text{ hg}}$

$$= 35 \cdot 10{,}000\text{ cg}$$

$$= 350{,}000\text{ cg}$$

57. $205\text{ mg} + 5.61\text{ g} = 0.205\text{ g} + 5.61\text{ g} = 5.815\text{ g}$

or

$205\text{ mg} + 5.61\text{ g} = 205\text{ mg} + 5610\text{ mg}$

$$= 5815\text{ mg}$$

61. $1.61\text{ kg} - 250\text{ g} = 1.61\text{ kg} - 0.250\text{ kg} = 1.36\text{ kg}$

or

$1.61\text{ kg} - 250\text{ g} = 1610\text{ g} - 250\text{ g} = 1360\text{ g}$

65. $17\text{ kg} \div 8 = \dfrac{17}{8}\text{ kg}$

$$
\begin{array}{r}
2.125 \\
8\overline{)17.000} \\
-16\ \ \ \ \ \ \\
\hline
1\,0\ \ \ \ \\
-8\ \ \ \ \\
\hline
20\ \ \\
-16\ \ \\
\hline
40 \\
-40 \\
\hline
0
\end{array}
$$

$17\text{ kg} \div 8 = 2.125\text{ kg}$

69.

Object	Tons	Pounds	Ounces
A 12-inch cube of osmium	$\dfrac{269}{400}$ or 0.6725	1345	21,520

73.

Object	Grams	Kilograms	Milligrams	Centigrams
A six-year-old boy	21,000	21	21,000,000	2,100,000

77. $0.09\text{ g} = \dfrac{0.09\text{ g}}{1} \cdot \dfrac{1000\text{ mg}}{1\text{ g}} = 0.09 \cdot 1000\text{ mg} = 90\text{ mg}$

$90\text{ mg} - 60\text{ mg} = 30\text{ mg}$

The extra-strength tablet contains 30 mg more medication.

81.
$$
\begin{array}{ll}
64\text{ lb }\ 8\text{ oz} & 63\text{ lb }24\text{ oz} \\
-28\text{ lb }10\text{ oz} & -28\text{ lb }10\text{ oz} \\
\hline
& 35\text{ lb }14\text{ oz}
\end{array}
$$

Carla's zucchini was 35 lb 14 oz lighter than the record weight.

85. $3 \times 16 = 48$

3 cartons contain 48 boxes of fruit.

$3\text{ mg} \times 48 = 144\text{ mg}$

3 cartons contain 144 mg of preservatives.

89. $3\text{ lb }4\text{ oz} \times 10 = 30\text{ lb }40\text{ oz}$

$$= 30\text{ lb} + 2\text{ lb }8\text{ oz}$$

$$= 32\text{ lb }8\text{ oz}$$

Each box weighs 32 lb 8 oz.

$32\text{ lb }8\text{ oz} \times 4 = 128\text{ lb }32\text{ oz}$

$$= 128\text{ lb} + 2\text{ lb}$$

$$= 130\text{ lb}$$

4 boxes of meat weigh 130 lb.

93. $\dfrac{4}{25} = \dfrac{4}{25} \cdot \dfrac{4}{4} = \dfrac{16}{100} = 0.16$

97. No, a pill containing 2 kg of medication is not reasonable.

101. No, a professor weighing less than 150 g is not reasonable.

105. True, a kilogram is 1000 grams.

Exercise Set 7.3

1. $32\text{ fl oz} = \dfrac{32\text{ fl oz}}{1} \cdot \dfrac{1\text{ c}}{8\text{ fl oz}} = \dfrac{32}{8}\text{ c} = 4\text{ c}$

5. $14\text{ qt} = \dfrac{14\text{ qt}}{1} \cdot \dfrac{1\text{ gal}}{4\text{ qt}} = \dfrac{14}{4}\text{ gal} = 3\dfrac{1}{2}\text{ gal}$

9. $2 \text{ qt} = \dfrac{2 \text{ qt}}{1} \cdot \dfrac{2 \text{ pt}}{1 \text{ qt}} \cdot \dfrac{2 \text{ c}}{1 \text{ pt}} = 2 \cdot 2 \cdot 2 \text{ c} = 8 \text{ c}$

13. $42 \text{ c} = \dfrac{42 \text{ c}}{1} \cdot \dfrac{1 \text{ qt}}{4 \text{ c}} = \dfrac{42}{4} \text{ qt} = 10\dfrac{1}{2} \text{ qt}$

17. $5 \text{ gal } 3 \text{ qt} = \dfrac{5 \text{ gal}}{1} \cdot \dfrac{4 \text{ qt}}{1 \text{ gal}} + 3 \text{ qt}$

$\qquad = 5 \cdot 4 \text{ qt} + 3 \text{ qt}$

$\qquad = 20 \text{ qt} + 3 \text{ qt}$

$\qquad = 23 \text{ qt}$

21. $58 \text{ qt} = 56 \text{ qt} + 2 \text{ qt}$

$\qquad = \dfrac{56 \text{ qt}}{1} \cdot \dfrac{1 \text{ gal}}{4 \text{ qt}} + 2 \text{ qt}$

$\qquad = \dfrac{56}{4} \text{ gal} + 2 \text{ qt}$

$\qquad = 14 \text{ gal } 2 \text{ qt}$

25. $2\dfrac{3}{4} \text{ gal} = \dfrac{11}{4} \text{ gal}$

$\qquad = \dfrac{\frac{11}{4} \text{ gal}}{1} \cdot \dfrac{4 \text{ qt}}{1 \text{ gal}} \cdot \dfrac{2 \text{ pt}}{1 \text{ qt}}$

$\qquad = \dfrac{11}{4} \cdot 4 \cdot 2 \text{ pt}$

$\qquad = 22 \text{ pt}$

29. $1 \text{ c } 5 \text{ fl oz} + 2 \text{ c } 7 \text{ fl oz} = 3 \text{ c } 12 \text{ fl oz}$

$\qquad\qquad\qquad\qquad\qquad = 3 \text{ c} + 1 \text{ c } 4 \text{ fl oz}$

$\qquad\qquad\qquad\qquad\qquad = 4 \text{ c } 4 \text{ fl oz (or } 2 \text{ pt } 4 \text{ fl oz)}$

33.

$$\begin{array}{ccc} 3 \text{ gal } 1 \text{ qt} & 2 \text{ gal } 5 \text{ qt} & 2 \text{ gal } 4 \text{ qt } 2 \text{ pt} \\ - \quad 1 \text{ qt } 1 \text{ pt} & - \quad 1 \text{ qt } 1 \text{ pt} & - \quad 1 \text{ qt } 1 \text{ pt} \\ \hline & & 2 \text{ gal } 3 \text{ qt } 1 \text{ pt} \end{array}$$

37. $9 \text{ gal } 2 \text{ qt} \div 2 = (8 \text{ gal } 4 \text{ qt} + 2 \text{ qt}) \div 2$

$\qquad\qquad\qquad\quad = 8 \text{ gal } 6 \text{ qt} \div 2$

$\qquad\qquad\qquad\quad = 4 \text{ gal } 3 \text{ qt}$

41. $0.16 \text{ L} = \dfrac{0.16 \text{ L}}{1} \cdot \dfrac{1 \text{ kl}}{1000 \text{ L}} = \dfrac{0.16}{1000} \text{ kl} = 0.00016 \text{ kl}$

45. $3.2 \text{ L} = \dfrac{3.2 \text{ L}}{1} \cdot \dfrac{100 \text{ cl}}{1 \text{ L}} = 3.2 \cdot 100 \text{ cl} = 320 \text{ cl}$

49. $64 \text{ ml} = \dfrac{64 \text{ ml}}{1} \cdot \dfrac{1 \text{ L}}{1000 \text{ ml}} = \dfrac{64}{1000} \text{ L} = 0.064 \text{ L}$

53. $3.6 \text{ L} = \dfrac{3.6 \text{ L}}{1} \cdot \dfrac{1000 \text{ ml}}{1 \text{ L}} = 3.6 \cdot 1000 \text{ ml} = 3600 \text{ ml}$

57. $2700 \text{ ml} + 1.8 \text{ L} = 2.7 \text{ L} + 1.8 \text{ L} = 4.5 \text{ L}$

or

$2700 \text{ ml} + 1.8 \text{ L} = 2700 \text{ ml} + 1800 \text{ ml} = 4500 \text{ ml}$

61. $17{,}500 \text{ ml} - 0.9 \text{ L} = 17{,}500 \text{ ml} - 900 \text{ ml}$

$\qquad\qquad\qquad\qquad\quad = 16{,}600 \text{ ml}$

or

$17{,}500 \text{ ml} - 0.9 \text{ L} = 17.5 \text{ L} - 0.9 \text{ L} = 16.6 \text{ L}$

65. $81.2 \text{ L} \div 0.5 = 81.2 \text{ L} \div \dfrac{1}{2}$

$\qquad\qquad\quad = 81.2 \text{ L} \cdot 2$

$\qquad\qquad\quad = 162.4 \text{ L}$

69.

Capacity	Cups	Gallons	Quarts	Pints
Your kidneys filter about this amount of blood every minute	4	$\dfrac{1}{4}$	1	2

73. $354 \text{ ml} + 18.6 \text{ L} = 0.354 \text{ L} + 18.6 \text{ L} = 18.954 \text{ L}$

There were 18.954 liters of gasoline in her tank.

77. $5 \text{ pt } 1 \text{ c} + 2 \text{ pt } 1 \text{ c} = 7 \text{ pt } 2 \text{ c}$

$\qquad\qquad\qquad\qquad\qquad = 7 \text{ pt} + 1 \text{ pt}$

$\qquad\qquad\qquad\qquad\qquad = 8 \text{ pt}$

$\qquad\qquad\qquad\qquad\qquad = \dfrac{8 \text{ pt}}{1} \cdot \dfrac{1 \text{ qt}}{2 \text{ pt}}$

$\qquad\qquad\qquad\qquad\qquad = \dfrac{8}{2} \text{ qt}$

$\qquad\qquad\qquad\qquad\qquad = 4 \text{ qt}$

$\qquad\qquad\qquad\qquad\qquad = \dfrac{4 \text{ qt}}{1} \cdot \dfrac{1 \text{ gal}}{4 \text{ qt}}$

$\qquad\qquad\qquad\qquad\qquad = \dfrac{4}{4} \text{ gal}$

$\qquad\qquad\qquad\qquad\qquad = 1 \text{ gal}$

Yes, the liquid can be poured into the container without causing it to overflow.

81. $\dfrac{20}{25} = \dfrac{4 \cdot 5}{5 \cdot 5} = \dfrac{4}{5}$

85. $\dfrac{72}{80} = \dfrac{8 \cdot 9}{8 \cdot 10} = \dfrac{9}{10}$

89. No, a tub filled with 3000 ml of hot water is not reasonable.

93. answers may vary

97. B indicates 1.5 cc.

101. B indicates 54 u or 0.54 cc.

Exercise Set 7.4

1. $756 \text{ ml} \approx \dfrac{756 \text{ ml}}{1} \cdot \dfrac{1 \text{ fl oz}}{29.57 \text{ ml}} \approx 25.57 \text{ fl oz}$

5. $1000 \text{ g} \approx \dfrac{1000 \text{ g}}{1} \cdot \dfrac{0.04 \text{ oz}}{1 \text{ g}} \approx 40 \text{ oz}$

9. $14.5 \text{ L} \approx \dfrac{14.5 \text{ L}}{1} \cdot \dfrac{0.26 \text{ gal}}{1 \text{ L}} \approx 3.77 \text{ gal}$

13.

	Meters	Yards	Centimeters	Feet	Inches
The height of a woman	1.5	$1\dfrac{2}{3}$	150	5	60

17. $10 \text{ cm} = \dfrac{10 \text{ cm}}{1} \cdot \dfrac{1 \text{ in.}}{2.54 \text{ cm}} \approx 3.94 \text{ in.}$

The balance beam is approximately 3.94 inches wide.

21. $200 \text{ mg} = 0.2 \text{ g} \approx \dfrac{0.2 \text{ g}}{1} \cdot \dfrac{0.04 \text{ oz}}{1 \text{ g}} \approx 0.008 \text{ oz}$

25. $4500 \text{ km} \approx \dfrac{4500 \text{ km}}{1} \cdot \dfrac{0.62 \text{ mi}}{1 \text{ km}} \approx 2790 \text{ mi}$

The trip is about 2790 miles.

29. $1.5 \text{ lb} - 1.25 \text{ lb} = 0.25 \text{ lb}$

$0.25 \text{ lb} \approx \dfrac{0.25 \text{ lb}}{1} \cdot \dfrac{0.45 \text{ kg}}{1 \text{ lb}} \cdot \dfrac{1000 \text{ g}}{1 \text{ kg}} \approx 112.5 \text{ g}$

The difference is approximately 112.5 grams.

33. $8 \text{ m} \approx \dfrac{8 \text{ m}}{1} \cdot \dfrac{3.28 \text{ ft}}{1 \text{ m}} \approx 26.24 \text{ ft}$

The base diameter is approximately 26.24 feet.

37. One dose every 4 hours results in $\dfrac{24}{4} = 6$ doses per day and

$6 \times 7 = 42$ doses per week.

$5 \text{ ml} \times 42 = 210 \text{ ml}$

$210 \text{ ml} \approx \dfrac{210 \text{ ml}}{1} \cdot \dfrac{1 \text{ fl oz}}{29.57 \text{ ml}} \approx 7.1 \text{ fl oz}$

8 fluid ounces of medicine should be purchased.

41. A liter has greater capacity than a quart; b.

45. An $8\frac{1}{2}$-ounce glass of water has a capacity of about

$250 \text{ ml} \left(\dfrac{1}{4} \text{L} \right)$; d.

49. $6 \cdot 4 + 5 \div 1 = 24 + 5 = 29$

53. $3 + 5(19 - 17) - 8 = 3 + 5(2) - 8$
$= 3 + 10 - 8$
$= 13 - 8$
$= 5$

57. $\text{BSA} = \sqrt{\dfrac{90 \times 182}{3600}} \approx 2.13$

The BSA is approximately 2.13 sq m.

61. $60 \text{ in.} = \dfrac{60 \text{ in.}}{1} \cdot \dfrac{2.54 \text{ cm}}{1 \text{ in.}} = 152.4 \text{ cm}$

$150 \text{ lb} \approx \dfrac{150 \text{ lb}}{1} \cdot \dfrac{0.45 \text{ kg}}{1 \text{ lb}} = 67.5 \text{ kg}$

$\text{BSA} \approx \sqrt{\dfrac{67.5 \times 152.4}{3600}} \approx 1.690$

The BSA is approximately 1.69 sq m.

65. $20 \text{ m} \times 40 \text{ m} = 800 \text{ sq m}$

$20 \text{ m} \approx \dfrac{20 \text{ m}}{1} \cdot \dfrac{3.28 \text{ ft}}{1 \text{ m}} \approx 65.6 \text{ ft}$

$40 \text{ m} \approx \dfrac{40 \text{ m}}{1} \cdot \dfrac{3.28 \text{ ft}}{1 \text{ m}} \approx 131.2 \text{ ft}$

$20 \text{ m} \times 40 \text{ m} \approx 65.6 \text{ ft} \times 131.2 \text{ ft} \approx 8606.72 \text{ sq ft}$

The area is 800 sq m or approximately 8606.72 sq ft.

Exercise Set 7.5

1. $C = \dfrac{5}{9}(F - 32)$

$= \dfrac{5}{9}(41 - 32)$

$= \dfrac{5}{9}(9)$

$= 5$

41°F is 5°C.

5. $F = \dfrac{9}{5}C + 32$

$= \dfrac{9}{5}(60) + 32$

$= 108 + 32$

$= 140$

60°C is 140°F.

9. $C = \dfrac{5}{9}(F - 32)$

$= \dfrac{5}{9}(62 - 32)$

$= \dfrac{5}{9}(30)$

≈ 16.7

62°F is 16.7°C.

13. $F = 1.8C + 32$
$= 1.8(92) + 32$
$= 165.6 + 32$
$= 197.6$

92°C is 197.6°F.

17. $C = \dfrac{5}{9}(F - 32)$

$= \dfrac{5}{9}(122 - 32)$

$= \dfrac{5}{9}(90)$

$= 50$

122°F is 50°C.

21. $C = \dfrac{5}{9}(F - 32)$

$= \dfrac{5}{9}(212 - 32)$

$= \dfrac{5}{9}(180)$

$= 100$

212°F is 100°C.

25. $F = 1.8C + 32$
$= 1.8(118) + 32$
$= 212.4 + 32$
$= 244.4$

118°C is 244.4°F.

29. $C = \dfrac{5}{9}(F - 32)$

$= \dfrac{5}{9}(70 - 32)$

$= \dfrac{5}{9}(38)$

≈ 21.1

70°F is 21.1°C.

33. $4 \text{ cm} + 3 \text{ cm} + 5 \text{ cm} = (4 + 3 + 5) \text{ cm} = 12 \text{ cm}$

37. False; the freezing point of water is 32°F.

41. $C = \dfrac{5}{9}(F - 32)$

$= \dfrac{5}{9}(936{,}000{,}000 - 32)$

$= \dfrac{5}{9}(935{,}999{,}968)$

$\approx 520{,}000{,}000$

936,000,000°F is approximately 520,000,000°C.

45. No, a fever of 40°F is not reasonable.

49. Yes, a fever of 40°C is reasonable.

Exercise Set 7.6

1. energy = 3 pounds · 380 feet = 1140 foot-pounds

5. energy = $\dfrac{2.5 \text{ tons}}{1} \cdot \dfrac{2000 \text{ pounds}}{1 \text{ ton}} \cdot 85 \text{ feet}$

= 425,000 foot-pounds

9. 1000 BTU = 1000 BTU · $\dfrac{778 \text{ foot-pounds}}{1 \text{ BTU}}$

= 778,000 foot-pounds

13. 34,130 BTU = 34,130 BTU · $\dfrac{778 \text{ foot-pounds}}{1 \text{ BTU}}$

= 26,553,140 foot-pounds

17. total hours = 1 · 7 = 7 hours

calories = 7 · 115 = 805 calories

21. total hours = $\dfrac{1}{3}$ · 6 = 2 hours

calories = 2 · 720 = 1440 calories

25. miles = $\dfrac{3500}{200}$ = 17.5 miles

29.
$$\begin{array}{r} {}^{7}\,{}^{10} \\ 1\,\cancel{8}\,\cancel{0} \\ -\ \ 17 \\ \hline 163 \end{array}$$

33. energy = 123.9 pounds · $\dfrac{9 \text{ inches}}{1} \cdot \dfrac{1 \text{ foot}}{12 \text{ inches}}$

= 92.925 foot-pounds

37. answers may vary

Chapter 7 Test

1.
$$\begin{array}{r} 23 \\ 12\overline{)280} \\ -\ 24 \\ \hline 40 \\ -\ 36 \\ \hline 4 \end{array}$$

280 inches = 23 feet 4 inches

5. 38 pt = $\dfrac{38 \text{ pt}}{1} \cdot \dfrac{1 \text{ qt}}{2 \text{ pt}} \cdot \dfrac{1 \text{ gal}}{4 \text{ qt}}$

= $\dfrac{38}{8}$ gal

= $\dfrac{19}{4}$ gal

= $4\dfrac{3}{4}$ gal

9. 4.3 dg = $\dfrac{4.3 \text{ dg}}{1} \cdot \dfrac{1 \text{ g}}{10 \text{ dg}} = \dfrac{4.3}{10}$ g = 0.43 g

13. 2 ft 9 in. × 3 = 6 ft 27 in.

= 6 ft + 2 ft 3 in.

= 8 ft 3 in.

17. C = $\dfrac{5}{9}$(F − 32)

= $\dfrac{5}{9}$(84 − 32)

= $\dfrac{5}{9}$(52)

≈ 28.9

84°F is 28.9°C.

21. 88 m + 340 cm = 88 m + 3.40 m = 91.4 m

The span is 91.4 meters.

25. C = $\dfrac{5}{9}$(F − 32)

= $\dfrac{5}{9}$(134 − 32)

= $\dfrac{5}{9}$(102)

≈ 56.7

134°F is 56.7°C.

29. 5 km ≈ $\dfrac{5 \text{ km}}{1} \cdot \dfrac{0.62 \text{ mi}}{1 \text{ km}}$ ≈ 3.1 mi

5 km is about 3.1 mi.

33. $\dfrac{180 \text{ cal}}{1 \text{ hour}} \cdot \dfrac{1 \text{ hour}}{1 \text{ day}} \cdot \dfrac{5 \text{ days}}{1}$ = 180 · 5 cal = 900 cal

She burns 900 calories.

Chapter 8

Exercise Set 8.1

1. The figure extends indefinitely in two directions. It is line CD, line l, or $\overleftrightarrow{CD}$.

5. The figure has two rays with a common endpoint. It is an angle, which can be named $\angle GHI$, $\angle IHG$, or $\angle H$.

9. Two other ways to name $\angle x$ are $\angle CPR$ and $\angle RPC$.

13. $\angle S$ is a straight angle.

17. $\angle Q$ measures between 90° and 180°. It is an obtuse angle.

21. The complement of an angle that measures 23° is an angle that measures 90° − 23° = 67°.

25. The complement of an angle that measures 58° is an angle that measures 90° − 58° = 32°.

29. 52° + 38° = 90°, so $\angle PNQ$ and $\angle QNR$ are complementary. 60° + 30° = 90°, so $\angle MNP$ and $\angle RNO$ are complementary.

33. $m\angle x = 74° - 47° = 27°$

37. $\angle x$ and the angle marked 150° are supplementary, so $m\angle x = 180° - 150° = 30°$. $\angle y$ and the angle marked 150° are vertical angles, so $m\angle y = 150°$. $\angle z$ and $\angle x$ are vertical angles so $m\angle z = m\angle x = 30°$.

41. $\angle x$ and the angle marked 80° are supplementary, so $m\angle x = 180° - 80° = 100°$. $\angle y$ and the angle marked 80° are alternate interior angles, so $m\angle y = 80°$. $\angle x$ and $\angle z$ are corresponding angles, so $m\angle z = m\angle x = 100°$.

45. $\angle x$ can also be named $\angle ABC$ or $\angle CBA$.

49. $m\angle ABC = 15°$

53. $m\angle DBA = m\angle DBC + m\angle CBA$

= 50° + 15°

= 65°

57. $\dfrac{7}{8} + \dfrac{1}{4} = \dfrac{7}{8} + \dfrac{2}{8} = \dfrac{9}{8}$ or $1\dfrac{1}{8}$

61. $3\dfrac{1}{3} - 2\dfrac{1}{2} = \dfrac{10}{3} - \dfrac{5}{2}$

$\qquad\qquad = \dfrac{20}{6} - \dfrac{15}{6}$

$\qquad\qquad = \dfrac{5}{6}$

65. The supplement of an angle that measures 125.2° is an angle with measure $180° - 125.2° = 54.8°$.

69. True; because 120° is less than 180°, it is possible to find the supplement of a 120° angle.

73. no; answers may vary

Exercise Set 8.2

1. The figure has five sides, so it is a pentagon.

5. The figure has four sides, so it is a quadrilateral.

9. All three sides of the triangle have the same length, therefore the triangle is equilateral.

13. Two sides of the triangle have the same length, therefore the triangle is isosceles.

17. $m\angle x = 180° - 95° - 72° = 13°$

21. Twice the radius of a circle is its diameter.

25. A quadrilateral with opposite sides parallel is a parallelogram.

29. $d = 2 \cdot r = 2 \cdot 7\,\text{m} = 14\,\text{m}$

33. $d = 2 \cdot r = 2 \cdot 20.3\,\text{cm} = 40.6\,\text{cm}$

37. The solid is a cylinder.

41. The solid is a cone.

45. The object has the shape of a rectangular solid.

49. The object has the shape of a pyramid.

53. $r = \dfrac{1}{2}d = \dfrac{1}{2} \cdot 26\,\text{miles} = 13\,\text{miles}$

57. $2(18) + 2(36) = 36 + 72 = 108$

61. True; since all four sides of a square are equal in length and a square is a parallelogram, a square is also a rhombus.

65. False; a pentagon has five sides, so it cannot be a quadrilateral, which has four sides.

69. answers may vary

Exercise Set 8.3

1. $P = 2 \cdot l + 2 \cdot w$

$\qquad = 2 \cdot 17\,\text{ft} + 2 \cdot 15\,\text{ft}$

$\qquad = 34\,\text{ft} + 30\,\text{ft}$

$\qquad = 64\,\text{ft}$

The perimeter is 64 feet.

5. $P = a + b + c$

$\qquad = 5\,\text{in.} + 7\,\text{in.} + 9\,\text{in.}$

$\qquad = 21\,\text{in.}$

The perimeter is 21 inches.

9. All sides of a regular triangle have the same length.

$P = a + b + c = 14\,\text{in.} + 14\,\text{in.} + 14\,\text{in.} = 42\,\text{in.}$

The perimeter is 42 inches.

13. Sum the lengths of the sides.

$P = 5\,\text{ft} + 3\,\text{ft} + 2\,\text{ft} + 7\,\text{ft} + 4\,\text{ft}$

$\qquad = 21\,\text{ft}$

The perimeter is 21 feet.

17. $P = 2 \cdot l + 2 \cdot w$

$\qquad = 2 \cdot 120\,\text{yd} + 2 \cdot 53\,\text{yd}$

$\qquad = 240\,\text{yd} + 106\,\text{yd}$

$\qquad = 346\,\text{yd}$

The perimeter of the football field is 346 yards.

21. The amount of stripping needed is 22 feet.

22 feet $\cdot$ \$2.50 per foot $=$ \$55

The total cost of the stripping is \$55.

25. $P = 4 \cdot s = 4 \cdot 7\,\text{in.} = 28\,\text{in.}$

The perimeter is 28 inches.

29. The unmarked vertical side must have length

$28\,\text{m} - 20\,\text{m} = 8\,\text{m}$.

The unmarked horizontal side must have length

$20\,\text{m} - 17\,\text{m} = 3\,\text{m}$.

Sum the lengths of the sides.

$P = 17\,\text{m} + 8\,\text{m} + 3\,\text{m} + 20\,\text{m} + 20\,\text{m} + 28\,\text{m}$

$\qquad = 96\,\text{m}$

The perimeter is 96 meters.

33. The unmarked vertical side must have length

$5\,\text{cm} + 14\,\text{cm} = 19\,\text{cm}$.

The unmarked horizontal side must have length

$18\,\text{cm} - 9\,\text{cm} = 9\,\text{cm}$.

Sum the lengths of the sides.

$P = 18\,\text{cm} + 19\,\text{cm} + 9\,\text{cm} + 14\,\text{cm} + 9\,\text{cm} + 5\,\text{cm}$

$\qquad = 74\,\text{cm}$

The perimeter is 74 centimeters.

37. $C = 2 \cdot \pi \cdot r$

$\qquad = 2 \cdot \pi \cdot 8\,\text{mi}$

$\qquad = 16\pi\,\text{mi}$

$\qquad \approx 50.24\,\text{mi}$

The circumference is exactly 16π miles, which is approximately 50.24 miles.

41. $\pi \cdot d = \pi \cdot 15\,\text{ft} = 15\pi\,\text{ft} \approx 47.1\,\text{ft}$

He needs 15π feet of netting or 47.1 feet.

45. Sum the lengths of the sides.

$P = 9\,\text{mi} + 6\,\text{mi} + 11\,\text{mi} + 4.7\,\text{mi}$

$\qquad = 30.7\,\text{mi}$

The perimeter is 30.7 miles.

49. The sides of a regular pentagon all have the same length. Sum the lengths of the sides.

$P = 8\,\text{mm} + 8\,\text{mm} + 8\,\text{mm} + 8\,\text{mm} + 8\,\text{mm}$

$\qquad = 40\,\text{mm}$

The perimeter is 40 millimeters.

53. $5 + 6 \cdot 3 = 5 + 18 = 23$

57. $72 \div (2 \cdot 6) = 72 \div 12 = 6$

61. a. The first age category that 8-year-old children fit into is "Under 9." Thus the minimum width is 30 yards, and the minimum length is 40 yards.

b. $P = 2 \cdot l + 2 \cdot w$

$\qquad = 2 \cdot 40\,\text{yd} + 2 \cdot 30\,\text{yd}$

$\qquad = 80\,\text{yd} + 60\,\text{yd}$

$\qquad = 140\,\text{yd}$

The perimeter of the field is 140 yards.

65. a. Smaller circle:

$C = 2 \cdot \pi \cdot r$

$= 2 \cdot \pi \cdot 10 \text{ m}$

$= 20\pi \text{ m}$

$\approx 62.8 \text{ m}$

Larger circle:

$C = 2 \cdot \pi \cdot r$

$= 2 \cdot \pi \cdot 20 \text{ m}$

$= 40\pi \text{ m}$

$\approx 125.6 \text{ m}$

b. Yes, when the radius of a circle is doubled, the circumference is also doubled.

69. Each of the three linear sides has length 6 meters. The length of the curved side is half of the circumference of a circle with diameter 6 meters, or

$\frac{1}{2}\pi \cdot d = \frac{1}{2}\pi \cdot 6 \text{ meters} = 3\pi \text{ meters} \approx 9.4 \text{ meters}$

$6 \text{ m} + 6 \text{ m} + 6 \text{ m} + 9.4 \text{ m} = 27.4 \text{ m}$

The perimeter of the window is 27.4 meters.

73. $P = 2 \cdot l + 2 \cdot w$

$31 \text{ ft} = 2 \cdot 9 \text{ ft} + 2 \cdot w$

$31 \text{ ft} = 18 \text{ ft} + 2 \cdot w$

$13 \text{ ft} = 2 \cdot w$

$6.5 \text{ ft} = w$

The width is 6.5 feet.

Exercise Set 8.4

1. $A = l \cdot w = 3.5 \text{ m} \cdot 2 \text{ m} = 7 \text{ sq m}$

5. $A = \frac{1}{2} \cdot b \cdot h = \frac{1}{2} \cdot 6 \text{ yd} \cdot 5 \text{ yd} = 15 \text{ sq yd}$

9. $A = s^2 = (4.2 \text{ ft})^2 = 17.64 \text{ sq ft}$

13. $A = \frac{1}{2}(b + B) \cdot h$

$= \frac{1}{2}(7 \text{ yd} + 4 \text{ yd}) \cdot 4 \text{ yd}$

$= \frac{1}{2}(11 \text{ yd}) \cdot 4 \text{ yd}$

$= 22 \text{ sq yd}$

17. $A = b \cdot h$

$= 5 \text{ in.} \cdot 4\frac{1}{2} \text{ in.}$

$= 5 \text{ in.} \cdot \frac{9}{2} \text{ in.}$

$= \frac{45}{2} \text{ sq in.}$

$= 22\frac{1}{2} \text{ sq in.}$

21.

Rectangle 1: $A = l \cdot w = 10 \text{ mi} \cdot 5 \text{ mi} = 50 \text{ sq mi}$

Rectangle 2: $A = l \cdot w = 12 \text{ mi} \cdot 3 \text{ mi} = 36 \text{ sq mi}$

The area of the figure is 50 sq mi $+$ 36 sq mi $=$ 86 sq mi.

25. $A = \pi r^2 = \pi(6 \text{ in.})^2 = 36\pi \text{ sq in.} \approx 113\frac{1}{7} \text{ sq in.}$

29. $A = l \cdot w = 505 \text{ ft} \cdot 255 \text{ ft} = 128{,}775 \text{ sq ft}$

The area of the flag is 128,775 square feet.

33. $A = l \cdot w = 16 \text{ in.} \cdot 8 \text{ in.} = 128 \text{ sq in.}$

$128 \text{ sq in.} \cdot \frac{1 \text{ sq ft}}{144 \text{ sq in.}} = \frac{128}{144} \text{ sq ft} = \frac{8}{9} \text{ sq ft}$

The side has area 128 square inches, which is $\frac{8}{9}$ square foot.

37. $A = l \cdot w = 7 \text{ ft} \cdot 6 \text{ ft} = 42 \text{ sq ft}$

$4 \cdot 42 \text{ sq ft} = 168 \text{ sq ft}$

Four panels have an area of 168 square feet.

41. a. $A = \frac{1}{2}(b + B) \cdot h$

$= \frac{1}{2}(25 \text{ ft} + 36 \text{ ft}) \cdot 12\frac{1}{2} \text{ ft}$

$= \frac{1}{2} \cdot 61 \text{ ft} \cdot 12\frac{1}{2} \text{ ft}$

$= 381\frac{1}{4} \text{ sq ft}$

To the nearest square foot, the area is 381 square feet.

b. Since each square covers 100 square feet, 4 squares of shingles need to be purchased.

45. Sum the lengths of the sides.

$3 \text{ ft} + 3\frac{1}{2} \text{ ft} + 6 \text{ ft} + 8\frac{1}{2} \text{ ft} + 4 \text{ ft} = 25 \text{ ft}$

The perimeter is 25 feet.

49. A fence goes around the edge of a yard, thus the situation involves perimeter.

53. Paint covers the surface of a wall, thus the situation involves area.

57. Note that the dimensions given are the diameters of the pizzas.

12-inch pizza:

$r = \frac{1}{2} \cdot d = \frac{1}{2} \cdot 12 \text{ in.} = 6 \text{ in.}$

$A = \pi \cdot r^2 = \pi(6 \text{ in.})^2 = 36\pi \text{ sq in.}$

Price per square inch $= \dfrac{\$10}{36\pi \text{ sq in.}} \approx \$0.0884$

8-inch pizzas:

$r = \frac{1}{2} \cdot d = \frac{1}{2} \cdot 8 \text{ in.} = 4 \text{ in.}$

$A = \pi \cdot r^2 = \pi(4 \text{ in.})^2 = 16\pi \text{ sq in.}$

$2 \cdot A = 2 \cdot 16\pi \text{ sq in.} = 32\pi \text{ sq in.}$

Price per square inch: $\dfrac{\$9}{32\pi \text{ sq in.}} \approx \$0.0895$

Since the price per square inch for the 12-inch pizza is less, the 12-inch pizza is the better deal.

61. The area of the shaded region is the area of the square minus the area of the circle.

Square: $A = s^2 = (6 \text{ in.})^2 = 36 \text{ sq in.}$

Circle: $r = \frac{1}{2} \cdot d = \frac{1}{2}(6 \text{ in.}) = 3 \text{ in.}$

$A = \pi \cdot r^2 = \pi(3 \text{ in.})^2 = 9\pi \text{ sq in.} \approx 28.26 \text{ sq in.}$

$36 \text{ sq in.} - 28.26 \text{ sq in.} = 7.74 \text{ sq in.}$

The shaded region has area of approximately 7.74 square inches.

65. The skating area is a rectangle with a half circle on each end.

Rectangle: $A = l \cdot w = 22 \text{ m} \cdot 10 \text{ m} = 220 \text{ sq m}$

Half circles: $A = 2 \cdot \dfrac{1}{2} \cdot \pi \cdot r^2$

$$= \pi(5 \text{ m})^2$$

$$= 25\pi \text{ sq m} \approx 78.5 \text{ sq m}$$

$220 \text{ sq m} + 78.5 \text{ sq m} = 298.5 \text{ sq m}$

The skating surface has area of 298.5 square meters.

69. no; answers may vary

Exercise Set 8.5

1. $V = l \cdot w \cdot h = 6 \text{ in.} \cdot 4 \text{ in.} \cdot 3 \text{ in.} = 72 \text{ cu in.}$

5. $V = \dfrac{1}{3} \pi \cdot r^2 \cdot h = \dfrac{1}{3} \pi \cdot (2 \text{ yd})^2 \cdot 3 \text{ yd} = 4\pi \text{ cu yd} \approx 12\dfrac{4}{7} \text{ cu yd}$

9. $V = \pi \cdot r^2 \cdot h$

$$= \pi(1 \text{ in.})^2 \cdot 9 \text{ in.}$$

$$= 9\pi \text{ cu in.}$$

$$\approx 28\dfrac{2}{7} \text{ cu in.}$$

13. $V = s^3 = \left(1\dfrac{1}{3} \text{ in.}\right)^3$

$$= \left(\dfrac{4}{3} \text{ in.}\right)^3$$

$$= \dfrac{64}{27} \text{ cu in.}$$

$$= 2\dfrac{10}{27} \text{ cu in.}$$

17. Note: $1\dfrac{3}{10} \text{ in.} = \dfrac{13}{10} \text{ in.}$

$$V = \dfrac{1}{3} s^2 h$$

$$= \dfrac{1}{3}(5 \text{ in.})^2 \cdot \dfrac{13}{10} \text{ in.}$$

$$= \dfrac{1}{3} \cdot 25 \cdot \dfrac{13}{10} \text{ cu in.}$$

$$= 10\dfrac{5}{6} \text{ cu in.}$$

21. $V = \dfrac{4}{3} \pi r^3$

$$= \dfrac{4}{3} \pi (7 \text{ in.})^3$$

$$= \dfrac{1372}{3} \pi \text{ cu in. or } 457\dfrac{1}{3} \pi \text{ cu in.}$$

25. $V = \dfrac{4}{3} \pi r^3$

$$= \dfrac{4}{3} \pi (6 \text{ ft})^3$$

$$= \dfrac{4}{3} \pi (216 \text{ cu ft})$$

$$= \dfrac{4}{3} \cdot 216 \cdot \pi \text{ cu ft}$$

$$= 288\pi \text{ cu ft}$$

29. $r = \dfrac{1}{2} \cdot d = \dfrac{1}{2} \cdot 2.6 \text{ m} = 1.3 \text{ m}$

$V = \pi \cdot r^2 \cdot h$

$$= \pi \cdot (1.3 \text{ m})^2 \cdot 1.5 \text{ m}$$

$$= \pi \cdot 1.69 \text{ sq m} \cdot 1.5 \text{ m}$$

$$= \pi \cdot 2.535 \text{ cu m}$$

$$\approx 7.96 \text{ cu m}$$

The volume of the cereal bowl was about 7.96 cubic meters.

33. $V = lwh = (2 \text{ in.})(2 \text{ in.})(2.2 \text{ in.}) = 8.8 \text{ cu in.}$

The volume of the Space Cube is 8.8 cubic inches.

37. $5^2 = 5 \cdot 5 = 25$

41. $1^2 + 2^2 = 1 \cdot 1 + 2 \cdot 2 = 1 + 4 = 5$

45. $r = \dfrac{1}{2} \cdot d = \dfrac{1}{2} \cdot 20 \text{ m} = 10 \text{ m}$

$V = \dfrac{1}{2} \cdot \dfrac{4}{3} \pi r^3$

$$= \dfrac{2}{3} \pi (10 \text{ m})^3$$

$$= \dfrac{2}{3} \pi \cdot 1000 \text{ cu m}$$

$$= \dfrac{2000}{3} \pi \text{ cu m}$$

$$\approx \dfrac{2000}{3} \cdot 3.14 \text{ cu m}$$

$$\approx 2093.33 \text{ cu m}$$

The volume of the dome is about 2093.33 cubic meters.

49. Kennel (a):

$$2\dfrac{1}{12} \text{ ft} \cdot 1\dfrac{8}{12} \text{ ft} \cdot 1\dfrac{7}{12} \text{ ft} = \left(\dfrac{25}{12} \cdot \dfrac{20}{12} \cdot \dfrac{19}{12}\right) \text{ cu ft}$$

$$= \dfrac{9500}{1728} \text{ cu ft}$$

$$= \dfrac{2375}{432} \text{ cu ft}$$

$$\approx 5.5 \text{ cu ft}$$

Kennel (b):

$$1\dfrac{1}{12} \text{ ft} \cdot 2 \text{ ft} \cdot 2\dfrac{8}{12} \text{ ft} = \left(\dfrac{13}{12} \cdot \dfrac{24}{12} \cdot \dfrac{32}{12}\right) \text{ cu ft}$$

$$= \dfrac{9984}{1728} \text{ cu ft}$$

$$= \dfrac{52}{9} \text{ cu ft}$$

$$\approx 5.8 \text{ cu ft}$$

Kennel (b) is larger.

Exercise Set 8.6

1. $\sqrt{4} = 2$ because $2^2 = 4$.

5. $\sqrt{\dfrac{1}{81}} = \dfrac{1}{9}$ because $\dfrac{1}{9} \cdot \dfrac{1}{9} = \dfrac{1}{81}$.

9. $\sqrt{3} \approx 1.732$

13. $\sqrt{47} \approx 6.856$

17. Since 38 is between $36 = 6 \cdot 6$ and $49 = 7 \cdot 7$, $\sqrt{38}$ is between 6 and 7; $\sqrt{38} \approx 6.16$.

21. $\sqrt{256} = 16$ because $16^2 = 256$.

25. $\sqrt{\dfrac{49}{144}} = \dfrac{7}{12}$ because $\left(\dfrac{7}{12}\right)^2 = \dfrac{7}{12} \cdot \dfrac{7}{12} = \dfrac{49}{144}$.

29. hypotenuse $= \sqrt{(\text{leg})^2 + (\text{other leg})^2}$

$$= \sqrt{(5)^2 + (12)^2}$$

$$= \sqrt{25 + 144}$$

$$= \sqrt{169}$$

$$= 13$$

The missing length is 13 inches.

33. hypotenuse $= \sqrt{(\text{leg})^2 + (\text{other leg})^2}$

$$= \sqrt{(22)^2 + (48)^2}$$

$$= \sqrt{484 + 2304}$$

$$= \sqrt{2788}$$

$$\approx 52.802$$

The missing length is approximately 52.802 meters.

37.

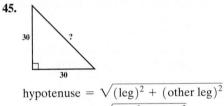

$$\text{hypotenuse} = \sqrt{(\text{leg})^2 + (\text{other leg})^2}$$
$$= \sqrt{(3)^2 + (4)^2}$$
$$= \sqrt{9 + 16}$$
$$= \sqrt{25}$$
$$= 5$$

The hypotenuse has length 5 units.

41.

$$\text{hypotenuse} = \sqrt{(\text{leg})^2 + (\text{other leg})^2}$$
$$= \sqrt{(10)^2 + (14)^2}$$
$$= \sqrt{100 + 196}$$
$$= \sqrt{296}$$
$$\approx 17.205$$

The hypotenuse is about 17.205 units.

45.

$$\text{hypotenuse} = \sqrt{(\text{leg})^2 + (\text{other leg})^2}$$
$$= \sqrt{(30)^2 + (30)^2}$$
$$= \sqrt{900 + 900}$$
$$= \sqrt{1800}$$
$$\approx 42.426$$

The hypotenuse is about 42.426 units.

49.

$$\text{hypotenuse} = \sqrt{(\text{leg})^2 + (\text{other leg})^2}$$
$$= \sqrt{(7.5)^2 + (4)^2}$$
$$= \sqrt{56.25 + 16}$$
$$= \sqrt{72.25}$$
$$= 8.5$$

The hypotenuse has length 8.5 units.

53. $\text{leg} = \sqrt{(\text{hypotenuse})^2 - (\text{other leg})^2}$
$$= \sqrt{(32)^2 - (20)^2}$$
$$= \sqrt{1024 - 400}$$
$$= \sqrt{624}$$
$$\approx 25.0$$

The tree is about 25 feet tall.

57. $\dfrac{n}{6} = \dfrac{2}{3}$
$$3 \cdot n = 2 \cdot 6$$
$$3n = 12$$
$$n = \dfrac{12}{3}$$
$$n = 4$$

61. $\dfrac{3}{n} = \dfrac{7}{14}$
$$3 \cdot 14 = 7 \cdot n$$
$$42 = 7 \cdot n$$
$$\dfrac{42}{7} = n$$
$$6 = n$$

65. From Exercise 19, $\sqrt{101}$ is between 10 and 11.
Since 101 is closer to 100 than it is to 121, $\sqrt{101}$ is approximately 10. $\sqrt{101} \approx 10.05$

69. $\sqrt{(25)^2 + (60)^2} = \sqrt{625 + 3600}$
$$= \sqrt{4225}$$
$$= 65$$

Since $\sqrt{(25)^2 + (60)^2} = 65$, the set forms the sides of a right triangle.

Exercise Set 8.7

1. The triangles are congruent by Side-Side-Side.

5. The triangles are congruent by Angle-Side-Angle.

9. $\dfrac{22}{11} = \dfrac{14}{7} = \dfrac{12}{6} = \dfrac{2}{1}$

The ratio of corresponding sides is $\dfrac{2}{1}$.

13. $\dfrac{n}{3} = \dfrac{9}{6}$
$$6 \cdot n = 3 \cdot 9$$
$$6 \cdot n = 27$$
$$n = \dfrac{27}{6}$$
$$n = 4.5$$

17. $\dfrac{n}{3.75} = \dfrac{12}{9}$
$$9 \cdot n = 12 \cdot 3.75$$
$$9 \cdot n = 45$$
$$n = \dfrac{45}{9}$$
$$n = 5$$

21. $\dfrac{n}{3.25} = \dfrac{17.5}{3.25}$
$$3.25 \cdot n = 17.5 \cdot 3.25$$
$$3.25 \cdot n = 56.875$$
$$n = \dfrac{56.875}{3.25}$$
$$n = 17.5$$

25. $\dfrac{n}{60} = \dfrac{15}{32}$
$$32 \cdot n = 60 \cdot 15$$
$$32 \cdot n = 900$$
$$n = \dfrac{900}{32}$$
$$n = 28.125$$

29. $\dfrac{n}{13} = \dfrac{80}{2}$
$$2 \cdot n = 80 \cdot 13$$
$$2 \cdot n = 1040$$
$$n = \dfrac{1040}{2}$$
$$n = 520$$

The observation deck is 520 feet high.

33.
$$\frac{n}{5} = \frac{48}{4}$$
$$\frac{n}{5} = \frac{12}{1}$$
$$1 \cdot n = 12 \cdot 5$$
$$n = 60$$
The tree is 60 feet tall.

37.
$$\frac{n}{55} = \frac{19}{20}$$
$$20 \cdot n = 55 \cdot 19$$
$$20 \cdot n = 1045$$
$$n = \frac{1045}{20}$$
$$n = 52.25$$
Pete would place 52 neon tetras in a 55-gallon tank.

41.
$$\begin{array}{r} \overset{1}{3.60} \\ + 0.41 \\ \hline 4.01 \end{array}$$

45.
$$\frac{n}{7 \text{ in.}} = \frac{5 \text{ in.}}{9 \text{ in.}}$$
$$\frac{n}{7} = \frac{5}{9}$$
$$7 \cdot 5 = 9 \cdot n$$
$$35 = 9 \cdot n$$
$$\frac{35}{9} = n$$
$$n = \frac{35}{9} \text{ or } 3\frac{8}{9} \text{ in.}$$

The new width is $3\frac{8}{9}$ in. Since the print area is now $3\frac{8}{9}$ in. by 5 in., it will not fit on a 3-by-5 inch index card. (Lengths are same but $3\frac{8}{9} > 3$.)

49. answers may vary

Chapter 8 Test

1. The complement of an angle that measures 78° is an angle that measures $90° - 78° = 12°$.

5. $\angle x$ and the angle marked 73° are vertical angles, so $m\angle x = 73°$. $\angle x$ and $\angle y$ are alternate interior angles, so $m\angle y = m\angle x = 73°$. $\angle x$ and $\angle z$ are corresponding angles, so $m\angle z = m\angle x = 73°$.

9. Circumference:
$$C = 2 \cdot \pi \cdot r$$
$$= 2 \cdot \pi \cdot 9 \text{ in.}$$
$$= 18\pi \text{ in.}$$
$$\approx 56.52 \text{ in.}$$
Area:
$$A = \pi r^2$$
$$= \pi (9 \text{ in.})^2$$
$$= 81\pi \text{ sq in.}$$
$$\approx 254.34 \text{ sq in.}$$

13. $V = l \cdot w \cdot h = 5 \text{ ft} \cdot 3 \text{ ft} \cdot 2 \text{ ft} = 30 \text{ cu ft}$

17. $P = 4 \cdot s = 4 \cdot 4 \text{ in.} = 16 \text{ in.}$
The perimeter of the photo is 16 inches.

21. First find the area of the lawn.
$$A = l \cdot w = 123.8 \text{ ft} \cdot 80 \text{ ft} = 9904 \text{ sq ft}$$
0.02 ounce per square foot is $\dfrac{0.02 \text{ oz}}{1 \text{ sq ft}}$.
$$\frac{0.02 \text{ oz}}{1 \text{ sq ft}} \cdot 9904 \text{ sq ft} = 198.08 \text{ oz}$$
Vivian needs to purchase 198.08 ounces of insecticide.

Chapter 9

Exercise Set 9.1

1. Kansas has the greatest number of wheat icons, so the greatest acreage of wheat was planted in the state of Kansas.

5. South Dakota is represented by 2.5 wheat icons, so the state of South Dakota planted about 2,500,000 acres of wheat in 2013.

9. The year 2012 has 5.5 flames and each flame represents 12,000 wildfires, so there were approximately $5.5(12,000) = 66,000$ wildfires in 2012.

13. 2010 has 5 flames and 2011 has 6 flames, which is 1 greater. Thus, the increase in the number of wildfires from 2010 to 2011 was about $1(12,000) = 12,000$ wildfires.

17. The longest bar corresponds to September, so the month in which most hurricanes made landfall is September.

21. Two of the 76 hurricanes that made landfall in August did so in 2008. The fraction is $\dfrac{2}{76} = \dfrac{1}{38}$.

25. The shortest bar corresponds to New York, U.S. The population is approximately 21.6 million or 21,600,000.

29.

Fiber Content of Selected Foods

33. The height of the bar for 100–149 miles per week is 15, so 15 of the adults drive 100–149 miles per week.

37. 15 of the adults drive 100–149 miles per week and 9 of the adults drive 150–199 miles per week, so $15 + 9 = 24$ of the adults drive 100–199 miles per week.

41. 9 of the 100 adults surveyed drive 150–199 miles per week, so the ratio is $\dfrac{9}{100}$.

45. According to the bar graph, the population of 20- to 44-year-olds is projected to be 109 million in 2020.

49. answers may vary

53.

Class Interval (Scores)	Tally	Class Frequency (Number of Games)
90–99	HHT III	8

57.

Class Interval (Account Balances)	Tally	Class Frequency (Number of People)
$200–$299	IIII I	6

61.

65. The dots for 2011 and 2013 are above the 50 level, so the total points scored in the Super Bowl were greater than 50 in 2011 and 2013.

69. The dots for 2006, 2008, and 2012 are below the 40 level, so the total points scored in the Super Bowl were fewer than 40 in 2006, 2008, and 2012.

73. 10% of 62 is $0.10 \cdot 62 = 6.2$

77. $\dfrac{17}{50} = \dfrac{17}{50} \cdot 100\% = \dfrac{17 \cdot 2 \cdot 50}{50}\% = 34\%$

81. The lowest point on the graph of low temperatures corresponds to Sunday. The low temperature on Sunday was 68°F.

85. answers may vary

Exercise Set 9.2

1. The largest sector corresponds to the category "parent or guardian's home," so most of the students live in a parent or guardian's home.

5. 180 of the students live in campus housing while 320 live in a parent or guardian's home.

$$\frac{180}{320} = \frac{9}{16}$$

The ratio is $\dfrac{9}{16}$.

9. $30\% + 7\% = 37\%$

37% of the land on Earth is accounted for by Europe and Asia.

13. Australia accounts for 5% of the land on Earth.

$$5\% \text{ of } 57,000,000 = 0.05 \cdot 57,000,000$$
$$= 2,850,000$$

Australia is 2,850,000 square miles.

17. The second-largest sector corresponds to nonfiction, so the second-largest category of books is nonfiction.

21. Children's fiction accounts for 22% of the books.

$$22\% \text{ of } 125,600 = 0.22 \cdot 125,600$$
$$= 27,632$$

The library has 27,632 children's fiction books.

25.

Type of Apple	Percent	Degrees in Sector
Red Delicious	37%	37% of 360° $= 0.37(360°) \approx 133°$
Golden Delicious	13%	13% of 360° $= 0.13(360°) \approx 47°$
Fuji	14%	14% of 360° $= 0.14(360°) \approx 50°$
Gala	15%	15% of 360° $= 0.15(360°) = 54°$
Granny Smith	12%	12% of 360° $= 0.12(360°) \approx 43°$
Other varieties	6%	6% of 360° $= 0.06(360°) \approx 22°$
Braeburn	3%	3% of 360° $= 0.03(360°) \approx 11°$

29. $20 = 2 \times 10 = 2 \times 2 \times 5 = 2^2 \times 5$

33. $85 = 5 \times 17$

37. Pacific Ocean:
$$49\% \cdot 264,489,800 = 0.49 \cdot 264,489,800$$
$$= 129,600,002 \text{ square kilometers}$$

41. $21.5\% \cdot 2800 = 0.215 \cdot 2800 = 602$

602 respondents spend $0–$15 online each month.

45. $\dfrac{\text{number of respondents who spend \$0–\$15}}{\text{number of respondents who spend \$15–\$175}} = \dfrac{602}{1674}$

$$= \frac{2 \cdot 301}{2 \cdot 837}$$

$$= \frac{301}{837}$$

Exercise Set 9.3

1. Mean: $\dfrac{15 + 23 + 24 + 18 + 25}{5} = \dfrac{105}{5} = 21$

Median: Write the numbers in order: 15, 18, 23, 24, 25

The middle number is 23.

Mode: There is no mode, since each number occurs once.

5. Mean:

$$\frac{0.5 + 0.2 + 0.2 + 0.6 + 0.3 + 1.3 + 0.8 + 0.1 + 0.5}{9}$$

$$= \frac{4.5}{9}$$

$$= 0.5$$

Median: Write the numbers in order:

0.1, 0.2, 0.2, 0.3, 0.5, 0.5, 0.6, 0.8, 1.3

The middle number is 0.5.

Mode: Since 0.2 and 0.5 occur twice, there are two modes, 0.2 and 0.5.

9. Mean:

$$\frac{2717 + 1972 + 1667 + 1614 + 1588}{5} = \frac{9558}{5} = 1911.6$$

The mean height is 1911.6 feet.

13. answers may vary

17. GPA $= \dfrac{4 \cdot 3 + 4 \cdot 3 + 4 \cdot 4 + 3 \cdot 3 + 2 \cdot 1}{3 + 3 + 4 + 3 + 1}$

$= \dfrac{51}{14}$

≈ 3.64

21. Mode: 6.9 since this number appears twice.

25. Mean: $\dfrac{\text{sum of 15 pulse rates}}{15} = \dfrac{1095}{15} = 73$

29. There are 9 rates lower than the mean. They are 66, 68, 71, 64, 71, 70, 65, 70, and 72

33. $\dfrac{18}{30} = \dfrac{3 \cdot 6}{5 \cdot 6} = \dfrac{3}{5}$

37. Since the mode is 35, 35 must occur at least twice in the set.

Since there is an odd number of numbers in the set, the median, 37, is in the set.

Let n be the remaining unknown number.

Mean: $\dfrac{35 + 35 + 37 + 40 + n}{5} = 38$

$\dfrac{147 + n}{5} = 38$

$5 \cdot \dfrac{147 + n}{5} = 5 \cdot 38$

$147 + n = 190$

$147 - 147 + n = 190 - 147$

$n = 43$

The missing numbers are 35, 35, 37, and 43.

Exercise Set 9.4

1.

12 outcomes

5.

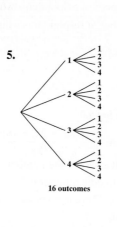

16 outcomes

9.

8 outcomes

13. A 1 or a 6 are two of the six possible outcomes. The probability is $\dfrac{2}{6} = \dfrac{1}{3}$.

17. Four of the six possible outcomes are numbers greater than 2. The probability is $\dfrac{4}{6} = \dfrac{2}{3}$.

21. A 1, a 2, or a 3 are three of three possible outcomes. The probability is $\dfrac{3}{3} = 1$.

25. One of the seven marbles is red. The probability is $\dfrac{1}{7}$.

29. Four of the seven marbles are either green or red. The probability is $\dfrac{4}{7}$.

33. The blood pressure did not change for 10 of the 200 people. The probability is $\dfrac{10}{200} = \dfrac{1}{20}$.

37. $\dfrac{1}{2} \cdot \dfrac{1}{3} = \dfrac{1 \cdot 1}{2 \cdot 3} = \dfrac{1}{6}$

41. One of the 52 cards is the king of hearts. The probability is $\dfrac{1}{52}$.

45. Thirteen of the 52 cards are hearts. The probability is $\dfrac{13}{52} = \dfrac{1}{4}$.

49.

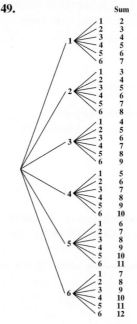

Five of the 36 sums are 6. The probability is $\dfrac{5}{36}$.

53. answers may vary

Chapter 9 Test

1. There are $4\dfrac{1}{2}$ dollar symbols for the second week.

Each dollar symbol corresponds to $50.

$4\dfrac{1}{2} \cdot \$50 = \dfrac{9}{2} \cdot \$50 = \dfrac{\$450}{2} = \$225$

$225 was collected during the second week.

5. The shortest bar corresponds to February. The normal monthly precipitation in February in Chicago is 3 centimeters.

9. The dots for 2004, 2005, and 2007 are above the 3 level, so the inflation rate was greater than 3% in 2004, 2005, and 2007.

13. 34.1% of the visitors in 2012 were from Canada.

34.1% of 66,700,000 $= 0.341 \times 66,700,000$

$= 22,744,700$

In 2012, 22,744,700 people visited the United States from Canada.

17.

Class Intervals (Scores)	Tally	Class Frequency (Number of Students)
40–49	|	1
50–59	|||	3
60–69	||||	4
70–79	�híⱦ	5
80–89	⫐⫐⫐ |||	8
90–99	||||	4

21.

Grade	Point Value	Credit Hours	$\left(\text{Point Value}\right)\cdot\left(\text{Credit Hours}\right)$
A	4	3	12
B	3	3	9
C	2	3	6
B	3	4	12
A	4	1	4
	Totals	14	43

$$\frac{43}{14} \approx 3.07$$

The grade point average is about 3.07.

25. A 3 or a 4 are two of the ten possible outcomes.

The probability is $\frac{2}{10} = \frac{1}{5}$.

Chapter 10

Exercise Set 10.1

1. If 0 represents ground level, then 1235 feet underground can be represented by -1235.

5. 125 degrees above zero Fahrenheit can be represented by $+125$.

9. A loss of \$61,303,000 can be represented by $-61,303,000$.

13. A decrease of 11 percent can be represented by -11.

17.

$$\begin{array}{c} -4.7 \\ \leftarrow\!\!\!\bullet\!\!+\!\!+\!\!+\!\!+\!\!+\!\!\diamond\!\!+\!\!+\!\!+\!\!+\!\!+\!\!+\!\!+\!\!+\!\!\diamond\!\!\rightarrow \\ -5\ -4\ -3\ -2\ -1\ \ 0\ \ 1\ \ 2\ \ 3\ \ 4\ \ 5 \end{array}$$

21. The graph of 4 is to the right of 0 on a number line so 4 is greater than 0, or $4 > 0$.

25. The graph of 0 is to the right of -3 on a number line, so 0 is greater than -3, or $0 > -3$.

29. The graph of $-1\frac{3}{4}$ is to the left of 0 on a number line, so $1\frac{3}{4}$ is less than 0, or $-1\frac{3}{4} < 0$.

33. The graph of $-2\frac{1}{2}$ is to the left of $-\frac{9}{10}$ on a number line, so $-2\frac{1}{2}$ is less than $-\frac{9}{10}$, or $-2\frac{1}{2} < -\frac{9}{10}$.

37. $|-8| = 8$ since -8 is 8 units from 0.

41. $|-5| = 5$ since -5 is 5 units from 0.

45. $\left|-\frac{1}{2}\right| = \frac{1}{2}$ since $-\frac{1}{2}$ is $\frac{1}{2}$ unit from 0.

49. $|7.6| = 7.6$ since 7.6 is 7.6 units from 0.

53. The opposite of -4 is $-(-4)$ or 4.

57. The opposite of $-\frac{9}{16}$ is $-\left(-\frac{9}{16}\right)$ or $\frac{9}{16}$.

61. The opposite of $\frac{17}{18}$ is $-\frac{17}{18}$.

65. $-|20| = -20$

69. $-(-8) = 8$

73. $-(-29) = 29$

77.

Number	Absolute Value of Number	Opposite of Number
-8.4	$|-8.4| = 8.4$	$-(-8.4) = 8.4$

81. The tallest bar on the graph corresponds to Lake Superior, so Lake Superior has the highest elevation.

85. The number on the graph closest to $-200°F$ is $-186°C$, which corresponds to the element oxygen.

89.
$$\begin{array}{r} 15 \\ + 20 \\ \hline 35 \end{array}$$

93.
$$2^2 = 2 \cdot 2 = 4$$
$$-|3| = -3$$
$$-(-5) = 5$$
$$-|-8| = -8$$
In order from least to greatest:
$$-|-8|, -|3|, 2^2, -(-5)$$

97. $-(-2) = 2$
$$5^2 = 5 \cdot 5 = 25$$
$$-10 = -10$$
$$-|-9| = -9$$
$$|-12| = 12$$
In order from least to greatest:
$$-10, -|-9|, -(-2), |-12|, 5^2$$

101. $-(-|-5|) = -(-5) = 5$

105. True; since every positive number is to the right of 0 on a number line and every negative number is to the left of 0, a positive number is always greater than a negative number.

109. answers may vary

Exercise Set 10.2

1.

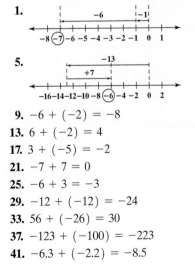

5.

9. $-6 + (-2) = -8$

13. $6 + (-2) = 4$

17. $3 + (-5) = -2$

21. $-7 + 7 = 0$

25. $-6 + 3 = -3$

29. $-12 + (-12) = -24$

33. $56 + (-26) = 30$

37. $-123 + (-100) = -223$

41. $-6.3 + (-2.2) = -8.5$

45. $-\dfrac{7}{11} + \left(-\dfrac{1}{11}\right) = -\dfrac{8}{11}$

49. $-\dfrac{4}{5} + \dfrac{1}{10} = -\dfrac{8}{10} + \dfrac{1}{10} = -\dfrac{7}{10}$

53. $-4 + 2 + (-5) = -2 + (-5) = -7$

57. $12 + (-4) + (-4) + 12 = 8 + (-4) + 12$
$$= 4 + 12$$
$$= 16$$

61. $-8 + 25 = 17$

65. $0 + (-215) + (-16) = -215 + (-16) = -231$
The diver is 231 feet below the surface.

69. Park:
$-1 + (-1) + 0 + (-1) + 0 + (+1) + 0 + (-1) + 0 = -3$
Stanford:
$+2 + (+1) + 0 + 0 + 0 + (-1) + 0 + (+1) + 0 = +3$
After 9 holes of golf, Park's score was -3 and Stanford's score was $+3$.

73. The height of the bar for 2012 is marked as -39, so the net income for Amazon.com, Inc. in 2012 was $-\$39,000,000$.

77. $-10 + 12 = 2$
The temperature at 11 p.m. was $2°C$.

81. $-55 + 8 = -47$
West Virginia's record low temperature is $-47°F$.

85. $44 - 0 = 44$

89.
$$\begin{array}{r} \overset{1}{\cancel{2}}\,\overset{9}{\cancel{\overset{10}{0}}}\,\overset{10}{\cancel{0}} \\ -\ \ 5\ \ 9 \\ \hline 1\ \ 4\ \ 1 \end{array}$$

93. Since the signs are different, the absolute values should be subtracted, rather than added.
$7 + (-10) = -3$

97. True; since two negative numbers have the same sign, the sum will have the common sign. That is, the sum will be negative.

101. answers may vary

Exercise Set 10.3

1. $-5 - (-5) = -5 + 5 = 0$

5. $3 - 8 = 3 + (-8) = -5$

9. $-5 - (-8) = -5 + 8 = 3$

13. $2 - 16 = 2 + (-16) = -14$

17. $3.62 - (-0.4) = 3.62 + 0.4 = 4.02$

21. $\dfrac{2}{5} - \dfrac{7}{10} = \dfrac{2}{5} + \left(-\dfrac{7}{10}\right) = \dfrac{4}{10} + \left(-\dfrac{7}{10}\right) = -\dfrac{3}{10}$

25. $-20 - 18 = -20 + (-18) = -38$

29. $2 - (-11) = 2 + 11 = 13$

33. $9 - 20 = 9 + (-20) = -11$

37. $\dfrac{4}{7} + \left(-\dfrac{1}{7}\right) = \dfrac{3}{7}$

41. $12 - 5 - 7 = 12 + (-5) + (-7)$
$$= 7 + (-7)$$
$$= 0$$

45. $-10 + (-5) - 12 = -10 + (-5) + (-12)$
$$= -15 + (-12)$$
$$= -27$$

49. $19 - 14 + (-6) + (-50) = 19 + (-14) + (-6) + (-50)$
$$= 5 + (-6) + (-50)$$
$$= -1 + (-50)$$
$$= -51$$

53. The two months with the lowest temperatures are January, $-10°F$, and December, $-6°F$.
$-6 - (-10) = -6 + 10 = 4$
The difference is $4°F$.

57. Represent checks with negative numbers and deposits with positive numbers.
$125 - 117 + 45 - 69 = 125 + (-117) + 45 + (-69)$
$$= 8 + 45 + (-69)$$
$$= 53 + (-69)$$
$$= -16$$
His balance is $-\$16$.

61. $-282 - (-436) = -282 + 436 = 154$
The difference in elevations is 154 feet.

65. The elevation of Lake Superior is 600 feet. The elevation of Lake Eyre is -52 feet.
$600 - (-52) = 600 + 52 = 652$
The difference in elevation is 652 feet.

69. $867 - (-330) = 867 + 330 = 1197$
The difference in temperatures is $1197°$ Fahrenheit.

73. $8 \cdot 0 = 0$

77.
$$\begin{array}{r} 23 \\ \times\ 46 \\ \hline 138 \\ 920 \\ \hline 1058 \end{array}$$

81. Since the signs of 9 and 7 are the same, the numbers should be added, not subtracted.
$9 - (-7) = 9 + 7 = 16$

85. $|-3| - |-7| = 3 - 7 = 3 + (-7) = -4$

89. $|-15| - |-29| = 15 - 29 = 15 + (-29) = -14$

93. answers may vary

Exercise Set 10.4

1. $-2(-3) = 6$

5. $(2.6)(-1.2) = -3.12$

9. $-\dfrac{3}{5}\left(-\dfrac{2}{7}\right) = \dfrac{6}{35}$

13. $(-1)(-2)(-4) = 2(-4) = -8$

17. $10(-5)(-1)(-3) = -50(-1)(-3)$
$$= 50(-3)$$
$$= -150$$

21. $-5^2 = -(5 \cdot 5) = -25$

25. $-2^3 = -(2 \cdot 2 \cdot 2) = -8$

29. $\left(-\dfrac{5}{11}\right)^2 = \left(-\dfrac{5}{11}\right)\left(-\dfrac{5}{11}\right) = \dfrac{25}{121}$

33. $\dfrac{-30}{6} = -5$ because $-5 \cdot 6 = -30$.

37. $\dfrac{0}{14} = 0$ because $0 \cdot 14 = 0$

41. $\dfrac{7.8}{-0.3} = -26$ because $-26 \cdot -0.3 = 7.8$

45. $\dfrac{100}{-20} = -5$ because $-5 \cdot -20 = 100$

49. $\dfrac{-12}{-4} = 3$ because $3 \cdot -4 = -12.$

53. $-\dfrac{8}{15} \div \dfrac{2}{3} = -\dfrac{8}{15} \cdot \dfrac{3}{2}$

$= -\dfrac{8 \cdot 3}{15 \cdot 2}$

$= -\dfrac{2 \cdot 4 \cdot 3}{3 \cdot 5 \cdot 2}$

$= -\dfrac{4}{5}$

57. $(-5)^3 = (-5)(-5)(-5) = 25(-5) = -125$

61. $-1(2)(7)(-3.1) = -2(7)(-3.1)$

$= -14(-3.1)$

$= 43.4$

65. $\dfrac{4}{3} \div \left(-\dfrac{8}{9}\right) = \dfrac{4}{3} \cdot \left(-\dfrac{9}{8}\right)$

$= -\dfrac{4 \cdot 9}{3 \cdot 8}$

$= -\dfrac{4 \cdot 3 \cdot 3}{3 \cdot 4 \cdot 2}$

$= -\dfrac{3}{2}$ or $-1\dfrac{1}{2}$

69.
$$\begin{array}{r} 42 \\ \times\ 6 \\ \hline 252 \end{array}$$
$-42(-6) = 252$

73. Moving down 20 feet can be represented by -20.
$5 \cdot (-20) = -100$
The diver reached a depth of 100 feet below the surface.

77. $-3 \cdot 63 = -189$
The melting point of argon is $-189°C.$

81. a. $420 - 27 = 420 + (-27) = 393$
The number of California condors changed by 393 from 1987 to 2013.

   b. 1987 to 2013 is 26 years.

   $\dfrac{393}{26} \approx 15$

   The average change was 15 condors per year.

85. $(3 \cdot 5)^2 = (15)^2 = 15 \cdot 15 = 225$

89. $12 \div 4 - 2 + 7 = 3 - 2 \div 7$

$= 1 + 7$

$= 8$

93. $-9 - 10 = -9 + (-10) = -19$

97. False; since the product of two numbers having the same sign is positive, the product of two negative numbers is always a positive number.

101. The product of an odd number of negative numbers is negative, so the product of seven negative numbers is negative.

105. The product of an odd number of negative numbers will be negative and the product of an even number of negative numbers will be positive. Also, $|-2| = 2 < 5 = |-5|$ and $-5 < -2$. In order from least to greatest, the numbers are $(-5)^{17}, (-2)^{17}, (-2)^{12}, (-5)^{12}.$

Exercise Set 10.5

1. $-1(-2) + 1 = 2 + 1 = 3$

5. $9 - 12 - 4 = -3 - 4 = -7$

9. $\dfrac{4}{9}\left(\dfrac{2}{10} - \dfrac{7}{10}\right) = \dfrac{4}{9}\left(-\dfrac{5}{10}\right)$

$= \dfrac{4}{9}\left(-\dfrac{1}{2}\right)$

$= -\dfrac{2}{9}$

13. $25 \div (-5) + \sqrt{81} = 25 \div (-5) + 9$

$= -5 + 9$

$= 4$

17. $\dfrac{24}{10 + (-4)} = \dfrac{24}{6} = 4$

21. $-19 - 12(3) = -19 - 36 = -55$

25. $[8 + (-4)]^2 = [4]^2 = 16$

29. $(3 - 12) \div 3 = (-9) \div 3 = -3$

33. $(5 - 9)^2 \div (4 - 2)^2 = (-4)^2 \div (2)^2$

$= 16 \div 4$

$= 4$

37. $(-12 - 20) \div 16 - 25 = (-32) \div 16 - 25$

$= -2 - 25$

$= -27$

41. $(0.2 - 0.7)(0.6 - 1.9) = (-0.5)(-1.3)$

$= 0.65$

45. $(-36 \div 6) - (4 \div 4) = (-6) - (1) = -7$

49. $(-5)^2 - 6^2 = 25 - 36 = -11$

53. $2(8 - 10)^2 - 5(1 - 6)^2 = 2(-2)^2 - 5(-5)^2$

$= 2(4) - 5(25)$

$= 8 - 125$

$= -117$

57. $\dfrac{(-7)(-3) - 4(3)}{3[7 \div (3 - 10)]} = \dfrac{21 - 12}{3[7 \div (-7)]}$

$= \dfrac{9}{3[-1]}$

$= \dfrac{9}{-3}$

$= -3$

61. $-3^2 + (-2)^3 - 4^2 = -9 + (-8) - 16$

$= -17 - 16$

$= -33$

65. $\dfrac{\text{sum of numbers}}{\text{number of numbers}}$

$= \dfrac{-10 + 8 + (-4) + 2 + 7 + (-5) + (-12)}{7}$

$= \dfrac{-2 + (-4) + 2 + 7 + (-5) + (-12)}{7}$

$= \dfrac{-6 + 2 + 7 + (-5) + (-12)}{7}$

$= \dfrac{-4 + 7 + (-5) + (-12)}{7}$

$= \dfrac{3 + (-5) + (-12)}{7}$

$= \dfrac{-2 + (-12)}{7}$

$= \dfrac{-14}{7}$

$= -2$

The average of the numbers is -2.

69. $\dfrac{\text{sum of temperatures}}{\text{number of temperatures}}$

$= \dfrac{-11 + (-14) + (-16) + (-14)}{4}$

$= \dfrac{-25 + (-16) + (-14)}{4}$

$= \dfrac{-41 + (-14)}{4}$

$= \dfrac{-55}{4}$

$= -13.75$

The average temperature for December through March is $-13.75°F$

73. $45 \cdot 90 = 4050$

77. $P = 4 \cdot s = 4 \cdot 8 \text{ in.} = 32 \text{ in.}$

The perimeter is 32 inches.

81. $2 \cdot (7 - 5) \cdot 3 = 2 \cdot 2 \cdot 3 = 4 \cdot 3 = 12$

85. answers may vary

89. $(-12)^4 = (-12)(-12)(-12)(-12) = 20{,}736$

Chapter 10 Test

1.

5. $-|-98| = -98$

9. $18 - 24 = 18 + (-24) = -6$

13. $-7 - (-19) = -7 + 19 = 12$

17. $-8 + 9 \div (-3) = -8 + (-3) = -11$

21. $\dfrac{-3(-2) + 12}{-1(-4 - 5)} = \dfrac{6 + 12}{-1(-9)} = \dfrac{18}{9} = 2$

25. $40 - 43 = 40 + (-43) = -3$

29. Checks and withdrawals are represented by negative numbers, while deposits are represented by positive numbers.

$129 - 79 - 40 + 35 = 50 - 40 + 35$

$= 10 + 35$

$= 45$

His balance is $45, which can be represented by the integer 45.

Chapter 11

Exercise Set 11.1

1. $3 + 2z = 3 + 2(-3)$

$= 3 + (-6)$

$= -3$

5. $z - x + y = -3 - (-2) + 5$

$= -3 + 2 + 5$

$= -1 + 5$

$= 4$

9. $8 - (5y - 7) = 8 - (5 \cdot 5 - 7)$

$= 8 - (25 - 7)$

$= 8 - 18$

$= -10$

13. $\dfrac{6xy}{4} = \dfrac{6(-2)(5)}{4} = \dfrac{-12(5)}{4} = \dfrac{-60}{4} = -15$

17. $\dfrac{x + 2y}{2z} = \dfrac{-2 + 2 \cdot 5}{2(-3)}$

$= \dfrac{-2 + 10}{-6}$

$= \dfrac{8}{-6}$

$= -\dfrac{4}{3} \text{ or } -1\dfrac{1}{3}$

21. $\dfrac{xz}{y} + \dfrac{3}{10} = \dfrac{-2(-3)}{5} + \dfrac{3}{10}$

$= \dfrac{6}{5} + \dfrac{3}{10}$

$= \dfrac{12}{10} + \dfrac{3}{10}$

$= \dfrac{15}{10}$

$= \dfrac{3}{2} \text{ or } 1\dfrac{1}{2}$

25. $3x + 5x = (3 + 5)x = 8x$

29. $4c + c - 7c = (4 + 1 - 7)c$

$= (5 - 7)c$

$= -2c$

33. $4a + 3a + 6a - 8 = (4 + 3 + 6)a - 8$

$= (7 + 6)a - 8$

$= 13a - 8$

37. $3x + 7 - x - 14 = 3x - x + 7 - 14$

$= (3 - 1)x + (7 - 14)$

$= 2x + (-7)$

$= 2x - 7$

41. $\dfrac{5}{6} - \dfrac{7}{12}x - \dfrac{1}{3} - \dfrac{3}{10}x = \dfrac{5}{6} - \dfrac{1}{3} - \dfrac{7}{12}x - \dfrac{3}{10}x$

$= \left(\dfrac{5}{6} - \dfrac{1}{3}\right) + \left(-\dfrac{7}{12} - \dfrac{3}{10}\right)x$

$= \left(\dfrac{5}{6} - \dfrac{2}{6}\right) + \left(-\dfrac{35}{60} - \dfrac{18}{60}\right)x$

$= \dfrac{3}{6} + \left(-\dfrac{53}{60}\right)x$

$= \dfrac{1}{2} - \dfrac{53}{60}x$

45. $6(5x) = (6 \cdot 5)x = 30x$

49. $-0.6(7a) = (-0.6 \cdot 7)a = -4.2a$

53. $2(y + 2) = 2 \cdot y + 2 \cdot 2 = 2y + 4$

57. $-4(3x + 7) = -4 \cdot 3x + (-4) \cdot 7$

$= (-4 \cdot 3)x + (-28)$

$= -12x - 28$

61. $\dfrac{1}{2}(-8x - 3) = \dfrac{1}{2}(-8x) - \dfrac{1}{2}(3)$

$= \left(\dfrac{1}{2} \cdot -8\right)x - \dfrac{3}{2}$

$= -4x - \dfrac{3}{2}$

65. $4(6n - 5) + 3n = 4 \cdot 6n - 4 \cdot 5 + 3n$

$= 24n - 20 + 3n$

$= 24n + 3n - 20$

$= (24 + 3)n - 20$

$= 27n - 20$

69. $-2(3x + 1) - 5(x - 2)$
$= -2 \cdot 3x + (-2) \cdot 1 - 5 \cdot x - (-5) \cdot 2$
$= (-2 \cdot 3)x + (-2) - 5x - (-10)$
$= -6x - 2 - 5x + 10$
$= (-6 - 5)x + (-2 + 10)$
$= -11x + 8$

73. $2a + 2a + 6 + 5a + 6 + 2a$
$= 2a + 2a + 5a + 2a + 6 + 6$
$= (2 + 2 + 5 + 2)a + 6 + 6$
$= 11a + 12$
The perimeter is $(11a + 12)$ feet.

77. Area $=$ (length) $\cdot$ (width)
$= (4y) \cdot (9)$
$= (4 \cdot 9)y$
$= 36y$
The area is $36y$ square inches.

81. Area $=$ (length) $\cdot$ (width)
$= (3y + 1) \cdot (20)$
$= 3y \cdot 20 + 1 \cdot 20$
$= (3 \cdot 20)y + 20$
$= 60y + 20$
The area is $(60y + 20)$ square miles.

85. Perimeter $= 2 \cdot$ (length) $+ 2 \cdot$ (width)
$= 2 \cdot (18) + 2 \cdot (14)$
$= 36 + 28$
$= 64$
The perimeter is 64 feet.

89. $I = PRT$
$= (3000) \cdot (0.06) \cdot (2)$
$= 360$
The account will earn $360 in interest.

93. $F = \dfrac{9}{5}C + 32$
$= \dfrac{9}{5}(-5) + 32$
$= -9 + 32$
$= 23$
The low temperature is 23°F.

97. Let $l = 7.6$, $w = 3$, and $h = 4x$.
$V = lwh = 7.6 \cdot 3 \cdot 4x = 91.2x$.
The volume is $91.2x$ cubic inches.

101. $-4 - (-12) = -4 + 12 = 8$

105. The given result is incorrect.
$5(3x - 2) = 5 \cdot 3x - 5 \cdot 2$
$= 15x - 10$

109. The parentheses have been removed by the distributive property.

113. answers may vary

117. Let $d = 173$.
$A = 0.7854d^2$
$= 0.7854(173)^2$
$= 0.7854 \cdot 29,929$
$\approx 23,506.2$
The trunk area is approximately 23,506.2 square inches.

Exercise Set 11.2

1. $x - 8 = 2$
$10 - 8 \stackrel{?}{=} 2$
$2 \stackrel{?}{=} 2$ True
Yes, 10 is a solution of the equation.

5. $-9f = 64 - f$
$-9(-8) \stackrel{?}{=} 64 - (-8)$
$72 \stackrel{?}{=} 64 + 8$
$72 \stackrel{?}{=} 72$ True
Yes, -8 is a solution of the equation.

9. $a + 5 = 23$
$a + 5 - 5 = 23 - 5$
$a = 18$
Check:
$a + 5 = 23$
$18 + 5 \stackrel{?}{=} 23$
$23 \stackrel{?}{=} 23$ True
The solution of the equation is 18.

13. $7 = y - 2$
$7 + 2 = y - 2 + 2$
$9 = y$
Check:
$7 = y - 2$
$7 \stackrel{?}{=} 9 - 2$
$7 \stackrel{?}{=} 7$ True
The solution of the equation is 9.

17. $x + \dfrac{1}{2} = \dfrac{7}{2}$
$x + \dfrac{1}{2} - \dfrac{1}{2} = \dfrac{7}{2} - \dfrac{1}{2}$
$x = \dfrac{6}{2}$
$x = 3$
Check:
$x + \dfrac{1}{2} = \dfrac{7}{2}$
$3 + \dfrac{1}{2} \stackrel{?}{=} \dfrac{7}{2}$
$\dfrac{6}{2} + \dfrac{1}{2} \stackrel{?}{=} \dfrac{7}{2}$
$\dfrac{7}{2} \stackrel{?}{=} \dfrac{7}{2}$ True
The solution of the equation is 3.

21. $x - 3 = -1 + 4$
$x - 3 = 3$
$x - 3 + 3 = 3 + 3$
$x = 6$
Check:
$x - 3 = -1 + 4$
$6 - 3 \stackrel{?}{=} -1 + 4$
$3 \stackrel{?}{=} 3$ True
The solution of the equation is 6.

25.
$$x - 0.6 = 4.7$$
$$x - 0.6 + 0.6 = 4.7 + 0.6$$
$$x = 5.3$$
Check:
$$x - 0.6 = 4.7$$
$$5.3 - 0.6 \overset{?}{=} 4.7$$
$$4.7 \overset{?}{=} 4.7 \quad \text{True}$$
The solution of the equation is 5.3.

29.
$$y + 2.3 = -9.2 - 8.6$$
$$y + 2.3 = -17.8$$
$$y + 2.3 - 2.3 = -17.8 - 2.3$$
$$y = -20.1$$
Check:
$$y + 2.3 = -9.2 - 8.6$$
$$-20.1 + 2.3 \overset{?}{=} -9.2 - 8.6$$
$$-17.8 \overset{?}{=} -17.8 \quad \text{True}$$
The solution of the equation is -20.1.

33.
$$5 + (-12) = 5x - 7 - 4x$$
$$5 + (-12) = 5x - 4x - 7$$
$$-7 = x - 7$$
$$-7 + 7 = x - 7 + 7$$
$$0 = x$$
Check:
$$5 + (-12) = 5x - 7 - 4x$$
$$5 + (-12) \overset{?}{=} 5(0) - 7 - 4(0)$$
$$5 + (-12) \overset{?}{=} 0 - 7 - 0$$
$$-7 \overset{?}{=} -7 \quad \text{True}$$
The solution of the equation is 0.

37.
$$2(5x - 3) = 11x$$
$$2 \cdot 5x - 2 \cdot 3 = 11x$$
$$10x - 6 = 11x$$
$$10x - 10x - 6 = 11x - 10x$$
$$-6 = x$$

41.
$$21y = 5(4y - 6)$$
$$21y = 5 \cdot 4y - 5 \cdot 6$$
$$21y = 20y - 30$$
$$21y - 20y = 20y - 20y - 30$$
$$y = -30$$

45. $\dfrac{-7}{-7} = 1$ because $1 \cdot -7 = -7$.

49. $-\dfrac{2}{3} \cdot -\dfrac{3}{2} = \dfrac{2 \cdot 3}{3 \cdot 2} = \dfrac{6}{6} = 1$

53. To solve $-\dfrac{1}{7} = -\dfrac{4}{5} + x$, $\dfrac{4}{5}$ should be added to both sides of the equation.

57.
$$x - 76{,}862 = 86{,}102$$
$$x - 76{,}862 + 76{,}862 = 86{,}102 + 76{,}862$$
$$x = 162{,}964$$

61. Use $I = R - E$ where $I = 986{,}000{,}000$ and $E = 18{,}293{,}000{,}000$.
$$I = R - E$$
$$986{,}000{,}000 = R - 18{,}293{,}000{,}000$$
$$986{,}000{,}000 + 18{,}293{,}000{,}000 = R - 18{,}293{,}000{,}000$$
$$+ \ 18{,}293{,}000{,}000$$
$$19{,}279{,}000{,}000 = R$$
Kohl's total revenues for the year were \$19,279,000,000.

Exercise Set 11.3

1.
$$5x = 20$$
$$\frac{5x}{5} = \frac{20}{5}$$
$$x = 4$$

5.
$$0.4y = -12$$
$$\frac{0.4y}{0.4} = \frac{-12}{0.4}$$
$$y = -30$$

9.
$$-0.3x = -15$$
$$\frac{-0.3x}{-0.3} = \frac{-15}{-0.3}$$
$$x = 50$$

13.
$$\frac{1}{6}y = -5$$
$$\frac{6}{1} \cdot \frac{1}{6}y = \frac{6}{1} \cdot -5$$
$$y = -30$$

17.
$$-\frac{2}{9}z = \frac{4}{27}$$
$$-\frac{9}{2} \cdot -\frac{2}{9}z = -\frac{9}{2} \cdot \frac{4}{27}$$
$$z = -\frac{2}{3}$$

21.
$$16 = 10t - 8t$$
$$16 = 2t$$
$$\frac{16}{2} = \frac{2t}{2}$$
$$8 = t$$

25.
$$4 - 10 = -3z$$
$$-6 = -3z$$
$$\frac{-6}{-3} = \frac{-3z}{-3}$$
$$2 = z$$

29.
$$0.4 = -8z$$
$$\frac{0.4}{-8} = \frac{-8z}{-8}$$
$$-0.05 = z$$

33.
$$-\frac{3}{5}x = -\frac{6}{15}$$
$$-\frac{5}{3} \cdot -\frac{3}{5}x = -\frac{5}{3} \cdot -\frac{6}{15}$$
$$x = \frac{2}{3}$$

37.
$$5 - 5 = 2x + 7x$$
$$0 = 9x$$
$$\frac{0}{9} = \frac{9x}{9}$$
$$0 = x$$

41.
$$-3x - 3x = 50 - 2$$
$$-6x = 48$$
$$\frac{-6x}{-6} = \frac{48}{-6}$$
$$x = -8$$

45. $\dfrac{1}{4}x - \dfrac{5}{8}x = 20 - 47$

$\dfrac{2}{8}x - \dfrac{5}{8}x = -27$

$-\dfrac{3}{8}x = -27$

$-\dfrac{8}{3} \cdot -\dfrac{3}{8}x = -\dfrac{8}{3} \cdot -27$

$x = 72$

49. $\dfrac{x}{-4} = 1 - (-6)$

$\dfrac{x}{-4} = 1 + 6$

$\dfrac{x}{-4} = 7$

$-4 \cdot \dfrac{x}{-4} = -4 \cdot 7$

$x = -28$

53. $\dfrac{x-3}{2} = \dfrac{5-3}{2} = \dfrac{2}{2} = 1$

57. The longest bar corresponds to 2006, so the number of acres burned by wildfires was greatest in 2006.

61. Addition should be used to solve the equation $12 = x - 5$. Specifically, 5 should be added to both sides of the equation.

65. answers may vary

69. Use $d = r \cdot t$, where $d = 294$ and $t = 5$.

$d = r \cdot t$

$294 = 5 \cdot r$

$\dfrac{294}{5} = \dfrac{5r}{5}$

$\dfrac{294}{5} = r$

The driver should drive at $\dfrac{294}{5} = 58.8$ miles per hour to make the drive in 5 hours.

73. $\dfrac{y}{72} = -86 - (-1029)$

$\dfrac{y}{72} = -86 + 1029$

$\dfrac{y}{72} = 943$

$72 \cdot \dfrac{y}{72} = 72 \cdot 943$

$y = 67{,}896$

Exercise Set 11.4

1. $2x - 6 = 0$

$2x - 6 + 6 = 0 + 6$

$2x = 6$

$\dfrac{2x}{2} = \dfrac{6}{2}$

$x = 3$

5. $6 - n = 10$

$6 - 6 - n = 10 - 6$

$-n = 4$

$\dfrac{-n}{-1} = \dfrac{4}{-1}$

$n = -4$

9. $1.7 = 2y + 9.5$

$1.7 - 9.5 = 2y + 9.5 - 9.5$

$-7.8 = 2y$

$\dfrac{-7.8}{2} = \dfrac{2y}{2}$

$-3.9 = y$

13. $3x - 7 = 4x + 5$

$3x - 7 - 5 = 4x + 5 - 5$

$3x - 12 = 4x$

$3x - 3x - 12 = 4x - 3x$

$-12 = x$

17. $9 - 3x = 14 + 2x$

$9 - 14 - 3x = 14 - 14 + 2x$

$-5 - 3x = 2x$

$-5 - 3x + 3x = 2x + 3x$

$-5 = 5x$

$\dfrac{-5}{5} = \dfrac{5x}{5}$

$-1 = x$

21. $x + 20 + 2x = -10 - 2x - 15$

$x + 2x + 20 = -10 - 15 - 2x$

$3x + 20 = -25 - 2x$

$3x + 2x + 20 = -25 - 2x + 2x$

$5x + 20 = -25$

$5x + 20 - 20 = -25 - 20$

$5x = -45$

$\dfrac{5x}{5} = \dfrac{-45}{5}$

$x = -9$

25. $-2(y + 4) = 2$

$-2y - 8 = 2$

$-2y - 8 + 8 = 2 + 8$

$-2y = 10$

$\dfrac{-2y}{-2} = \dfrac{10}{-2}$

$y = -5$

29. $35 - 17 = 3(x - 2)$

$18 = 3x - 6$

$18 + 6 = 3x - 6 + 6$

$24 = 3x$

$\dfrac{24}{3} = \dfrac{3x}{3}$

$8 = x$

33. $2t - 1 = 3(t + 7)$

$2t - 1 = 3t + 21$

$2t - 1 - 21 = 3t + 21 - 21$

$2t - 22 = 3t$

$2t - 2t - 22 = 3t - 2t$

$-22 = t$

37. $-4x = 44$

$\dfrac{-4x}{-4} = \dfrac{44}{-4}$

$x = -11$

41. $8 - b = 13$

$8 - 8 - b = 13 - 8$

$-b = 5$

$\dfrac{-b}{-1} = \dfrac{5}{-1}$

$b = -5$

45. $2x - 1 = -7$

$2x - 1 + 1 = -7 + 1$

$2x = -6$

$\dfrac{2x}{2} = \dfrac{-6}{2}$

$x = -3$

49.
$$9a + 29 = -7$$
$$9a + 29 - 29 = -7 - 29$$
$$9a = -36$$
$$\frac{9a}{9} = \frac{-36}{9}$$
$$a = -4$$

53.
$$11(x - 2) = 22$$
$$11x - 22 = 22$$
$$11x - 22 + 22 = 22 + 22$$
$$11x = 44$$
$$\frac{11x}{11} = \frac{44}{11}$$
$$x = 4$$

57.
$$3(x - 5) = -7 - 11$$
$$3x - 15 = -18$$
$$3x - 15 + 15 = -18 + 15$$
$$3x = -3$$
$$\frac{3x}{3} = \frac{-3}{3}$$
$$x = -1$$

61.
$$4x + 3 = 2x + 11$$
$$4x + 3 - 3 = 2x + 11 - 3$$
$$4x = 2x + 8$$
$$4x - 2x = 2x - 2x + 8$$
$$2x = 8$$
$$\frac{2x}{2} = \frac{8}{2}$$
$$x = 4$$

65.
$$\frac{3}{8}x + 14 = \frac{5}{8}x - 2$$
$$\frac{3}{8}x + 14 + 2 = \frac{5}{8}x - 2 + 2$$
$$\frac{3}{8}x + 16 = \frac{5}{8}x$$
$$\frac{3}{8}x - \frac{3}{8}x + 16 = \frac{5}{8}x - \frac{3}{8}x$$
$$16 = \frac{2}{8}x$$
$$16 = \frac{1}{4}x$$
$$4 \cdot 16 = 4 \cdot \frac{1}{4}x$$
$$64 = x$$

69.
$$\frac{5}{8}a = \frac{1}{8}a + \frac{3}{4}$$
$$\frac{5}{8}a - \frac{1}{8}a = \frac{1}{8}a - \frac{1}{8}a + \frac{3}{4}$$
$$\frac{4}{8}a = \frac{3}{4}$$
$$\frac{1}{2}a = \frac{3}{4}$$
$$2 \cdot \frac{1}{2}a = 2 \cdot \frac{3}{4}$$
$$a = \frac{3}{2}$$

73.
$$3 + 2(2n - 5) = 1$$
$$3 + 4n - 10 = 1$$
$$4n - 7 = 1$$
$$4n - 7 + 7 = 1 + 7$$
$$4n = 8$$
$$\frac{4n}{4} = \frac{8}{4}$$
$$n = 2$$

77.
$$-20 - (-50) = \frac{x}{9}$$
$$-20 + 50 = \frac{x}{9}$$
$$30 = \frac{x}{9}$$
$$9 \cdot 30 = 9 \cdot \frac{x}{9}$$
$$270 = x$$

81.
$$3(5c - 1) - 2 = 13c + 3$$
$$15c - 3 - 2 = 13c + 3$$
$$15c - 5 = 13c + 3$$
$$15c - 5 + 5 = 13c + 3 + 5$$
$$15c = 13c + 8$$
$$15c - 13c = 13c - 13c + 8$$
$$2c = 8$$
$$\frac{2c}{2} = \frac{8}{2}$$
$$c = 4$$

85. "The sum of -42 and 16 is -26" translates to $-42 + 16 = -26$.

89. "Three times the difference of -14 and 2 amounts to -48" translates to $3(-14 - 2) = -48$.

93. From the height of the bar, approximately 97 or 98 million returns were filed electronically in 2010.

97. The first step in solving $2x - 5 = -7$ is to add 5 to both sides, which is choice b.

101. The error is in the second line.
$$2(3x - 5) = 5x - 7$$
$$6x - 10 = 5x - 7$$
$$6x - 10 + 10 = 5x - 7 + 10$$
$$6x = 5x + 3$$
$$6x - 5x = 5x + 3 - 5x$$
$$x = 3$$

105.
$$2^3(x + 4) = 3^2(x + 4)$$
$$8(x + 4) = 9(x + 4)$$
$$8x + 32 = 9x + 36$$
$$8x + 32 - 36 = 9x + 36 - 36$$
$$8x - 4 = 9x$$
$$8x - 4 - 8x = 9x - 8x$$
$$-4 = x$$

109. Use $C = \frac{5}{9}(F - 32)$ with $C = 50.7$.
$$C = \frac{5}{9}(F - 32)$$
$$50.7 = \frac{5}{9}(F - 32)$$
$$\frac{9}{5}(50.7) = \frac{9}{5} \cdot \frac{5}{9}(F - 32)$$
$$91.26 = F - 32$$
$$91.26 + 32 = F - 32 + 32$$
$$123.26 = F$$
The temperature was $123.26°F$.

Exercise Set 11.5

1. "The sum of a number and five" is $x + 5$.

5. "Twenty decreased by a number" is $20 - x$.

9. "A number divided by 2" is $\dfrac{x}{2}$.

13. "A number added to -5 is -7" is $-5 + x = -7$.

17. "A number subtracted from -20 amounts to 104" is $-20 - x = 104$.

21. "A number subtracted from 11" is $11 - x$.

25. "Fifty decreased by eight times a number" is $50 - 8x$.

29. "Three times a number, added to 9, is 33" is $9 + 3x = 33$.

$$9 + 3x = 33$$
$$9 + 3x - 9 = 33 - 9$$
$$3x = 24$$
$$\frac{3x}{3} = \frac{24}{3}$$
$$x = 8$$

The number is 8.

33. "The difference of a number and 3 is equal to the quotient of 10 and 5" is $x - 3 = \dfrac{10}{5}$.

$$x - 3 = \frac{10}{5}$$
$$x - 3 = 2$$
$$x - 3 + 3 = 2 + 3$$
$$x = 5$$

The number is 5.

37. "40 subtracted from five times a number is 8 more than the number" is $5x - 40 = x + 8$.

$$5x - 40 = x + 8$$
$$5x - 40 + 40 = x + 8 + 40$$
$$5x = x + 48$$
$$5x - x = x - x + 48$$
$$4x = 48$$
$$\frac{4x}{4} = \frac{48}{4}$$
$$x = 12$$

The number is 12.

41. "The product of 4 and a number is the same as 30 less twice that same number" is $4x = 30 - 2x$.

$$4x = 30 - 2x$$
$$4x + 2x = 30 - 2x + 2x$$
$$6x = 30$$
$$\frac{6x}{6} = \frac{30}{6}$$
$$x = 5$$

The number is 5.

45. Let x be the fastest speed for the pheasant. Since the top speed of the falcon is five times the top speed of the pheasant, the top speed of the falcon is $5x$. Since the total of the speeds is 222 miles per hour, the sum of x and $5x$ is 222.

$$x + 5x = 222$$
$$6x = 222$$
$$\frac{6x}{6} = \frac{222}{6}$$
$$x = 37$$

The top speed of the pheasant is 37 miles per hour and the top speed of the falcon is $5 \cdot 37 = 185$ miles per hour.

49. Let x be the cost of the games. Since the cost of the Xbox 360 is three times the cost of the games, the cost of the Xbox 360 is $3x$. Since the total cost is $560, the sum of x and $3x$ is 560.

$$x + 3x = 560$$
$$4x = 560$$
$$\frac{4x}{4} = \frac{560}{4}$$
$$x = 140$$

The cost of the games is $140 and the cost of the Xbox 360 is $3 \cdot \$140 = \$420$.

53. Let x be the capacity of Beaver Stadium. Since the capacity of Michigan Stadium is 3329 more than the capacity of Beaver Stadium, the capacity of Michigan Stadium is $x + 3329$. Since the total capacity of the two stadiums is 216,473, the sum of x and $x + 3329$ is 216,473.

$$x + x + 3329 = 216{,}473$$
$$2x + 3329 = 216{,}473$$
$$2x + 3329 - 3329 = 216{,}473 - 3329$$
$$2x = 213{,}144$$
$$\frac{2x}{2} = \frac{213{,}144}{2}$$
$$x = 106{,}572$$

The capacity of Beaver Stadium is 106,572 and the capacity of Michigan Stadium is $106{,}572 + 3329 = 109{,}901$.

57. Let x be the number of points scored by the San Francisco 49ers. Since the Baltimore Ravens scored 3 more points than the 49ers, the Ravens scored $x + 3$ points. Since the total number of points scored was 65, the sum of x and $x + 3$ is 65.

$$x + x + 3 = 65$$
$$2x + 3 = 65$$
$$2x + 3 - 3 = 65 - 3$$
$$2x = 62$$
$$\frac{2x}{2} = \frac{62}{2}$$
$$x = 31$$

The San Francisco 49ers scored 31 points and the Baltimore Ravens scored $31 + 3 = 34$ points.

61. Let x be the number of cars manufactured per day in Spain. Since twice as many cars are manufactured per day in Germany, the number of cars manufactured per day in Germany is $2x$. Since the total number of cars manufactured in the two countries each day is 19,827, the sum of x and $2x$ is 19,827.

$$x + 2x = 19{,}827$$
$$3x = 19{,}827$$
$$\frac{3x}{3} = \frac{19{,}827}{3}$$
$$x = 6609$$

6609 cars are manufactured per day in Spain and $2 \cdot 6609 = 13{,}218$ cars are manufactured per day in Germany.

65. Let x be the number of personal computers, in millions, in use in China. Since the United States has 116 million more personal computers in use than China, the number of personal computers in use in the United States is $x + 116$. Since the total number of personal computers in use in both countries is 506 million, the sum of x and $x + 116$ is 506.

$$x + x + 116 = 506$$
$$2x + 116 = 506$$
$$2x + 116 - 116 = 506 - 116$$
$$2x = 390$$
$$\frac{2x}{2} = \frac{390}{2}$$
$$x = 195$$

The number of personal computers in use in China is 195 million and the number of personal computers in use in the United States is $195 + 116 = 311$ million.

69. 1026 rounded to the nearest hundred is 1000.

73. yes; answers may vary

77. Use $P = C + M$, where $P = 999$ and $M = 450$.

$$P = C + M$$
$$999 = C + 450$$
$$999 - 450 = C + 450 - 450$$
$$549 = C$$

The wholesale cost is $549.

Chapter 11 Test

1. $\dfrac{3x - 5}{2y} = \dfrac{3 \cdot 7 - 5}{2(-8)}$

$= \dfrac{21 - 5}{-16}$

$= \dfrac{16}{-16}$

$= -1$

5. Area $=$ length $\cdot$ width

$A = 4 \cdot (3x - 1)$
$= 4 \cdot 3x - 4 \cdot 1$
$= 12x - 4$

The area is $(12x - 4)$ square meters.

9. $-\dfrac{5}{8}x = -25$

$-\dfrac{8}{5} \cdot -\dfrac{5}{8}x = -\dfrac{8}{5} \cdot -25$

$x = 40$

13. $-4x + 7 = 15$
$-4x + 7 - 7 = 15 - 7$
$-4x = 8$
$\dfrac{-4x}{-4} = \dfrac{8}{-4}$
$x = -2$

17. $10y - 1 = 7y + 21$
$10y - 1 + 1 = 7y + 21 + 1$
$10y = 7y + 22$
$10y - 7y = 7y - 7y + 22$
$3y = 22$
$\dfrac{3y}{3} = \dfrac{22}{3}$
$y = \dfrac{22}{3}$

21. Use $A = \dfrac{1}{2}bh$, where $b = 5$ and $h = 12$.

$A = \dfrac{1}{2}bh = \dfrac{1}{2} \cdot 5 \cdot 12 = 30$

The area is 30 square feet.

25. Let x be the number of women runners entered. Then the number of men runners entered is $x + 112$, and the sum of these two quantities is 600.

$$x + x + 112 = 600$$
$$2x + 112 = 600$$
$$2x + 112 - 112 = 600 - 112$$
$$2x = 488$$
$$\frac{2x}{2} = \frac{488}{2}$$
$$x = 244$$

There were 244 women runners entered.

Appendices

A.1: One Hundred Addition Facts

1. $\begin{array}{r} 1 \\ +4 \\ \hline 5 \end{array}$

5. $\begin{array}{r} 3 \\ +9 \\ \hline 12 \end{array}$

9. $\begin{array}{r} 9 \\ +5 \\ \hline 14 \end{array}$

13. $\begin{array}{r} 5 \\ +5 \\ \hline 10 \end{array}$

17. $\begin{array}{r} 2 \\ +9 \\ \hline 11 \end{array}$

21. $\begin{array}{r} 6 \\ +4 \\ \hline 10 \end{array}$

25. $\begin{array}{r} 9 \\ +8 \\ \hline 17 \end{array}$

29. $\begin{array}{r} 7 \\ +5 \\ \hline 12 \end{array}$

33. $\begin{array}{r} 4 \\ +3 \\ \hline 7 \end{array}$

37. $\begin{array}{r} 7 \\ +1 \\ \hline 8 \end{array}$

41. $\begin{array}{r} 2 \\ +4 \\ \hline 6 \end{array}$

45. $\begin{array}{r} 3 \\ +6 \\ \hline 9 \end{array}$

49. $\begin{array}{r} 2 \\ +5 \\ \hline 7 \end{array}$

53. $\begin{array}{r} 8 \\ +3 \\ \hline 11 \end{array}$

57. $\begin{array}{r} 0 \\ +5 \\ \hline 5 \end{array}$

61. $\begin{array}{r} 6 \\ +7 \\ \hline 13 \end{array}$

65. $\begin{array}{r} 0 \\ +7 \\ \hline 7 \end{array}$

69. $\begin{array}{r} 4 \\ +1 \\ \hline 5 \end{array}$

73. $\begin{array}{r} 7 \\ +9 \\ \hline 16 \end{array}$

77. $\begin{array}{r} 1 \\ +0 \\ \hline 1 \end{array}$

81. $\begin{array}{r} 2 \\ +8 \\ \hline 10 \end{array}$

85. $\begin{array}{r} 5 \\ +0 \\ \hline 5 \end{array}$

89. $\begin{array}{r} 9 \\ +2 \\ \hline 11 \end{array}$

93. $\begin{array}{r} 2 \\ +7 \\ \hline 9 \end{array}$

97. $\begin{array}{r} 6 \\ +8 \\ \hline 14 \end{array}$

A.2: One Hundred Multiplication Facts

1. $\begin{array}{r} 1 \\ \times 1 \\ \hline 1 \end{array}$ **21.** $\begin{array}{r} 5 \\ \times 8 \\ \hline 40 \end{array}$ **41.** $\begin{array}{r} 8 \\ \times 6 \\ \hline 48 \end{array}$ **61.** $\begin{array}{r} 9 \\ \times 2 \\ \hline 18 \end{array}$ **81.** $\begin{array}{r} 1 \\ \times 5 \\ \hline 5 \end{array}$

5. $\begin{array}{r} 8 \\ \times 4 \\ \hline 32 \end{array}$ **25.** $\begin{array}{r} 9 \\ \times 6 \\ \hline 54 \end{array}$ **45.** $\begin{array}{r} 9 \\ \times 0 \\ \hline 0 \end{array}$ **65.** $\begin{array}{r} 6 \\ \times 6 \\ \hline 36 \end{array}$ **85.** $\begin{array}{r} 3 \\ \times 7 \\ \hline 21 \end{array}$

9. $\begin{array}{r} 2 \\ \times 2 \\ \hline 4 \end{array}$ **29.** $\begin{array}{r} 6 \\ \times 9 \\ \hline 54 \end{array}$ **49.** $\begin{array}{r} 3 \\ \times 5 \\ \hline 15 \end{array}$ **69.** $\begin{array}{r} 7 \\ \times 5 \\ \hline 35 \end{array}$ **89.** $\begin{array}{r} 8 \\ \times 2 \\ \hline 16 \end{array}$

13. $\begin{array}{r} 3 \\ \times 2 \\ \hline 6 \end{array}$ **33.** $\begin{array}{r} 4 \\ \times 9 \\ \hline 36 \end{array}$ **53.** $\begin{array}{r} 1 \\ \times 0 \\ \hline 0 \end{array}$ **73.** $\begin{array}{r} 8 \\ \times 5 \\ \hline 40 \end{array}$ **93.** $\begin{array}{r} 4 \\ \times 1 \\ \hline 4 \end{array}$

17. $\begin{array}{r} 4 \\ \times 6 \\ \hline 24 \end{array}$ **37.** $\begin{array}{r} 9 \\ \times 4 \\ \hline 36 \end{array}$ **57.** $\begin{array}{r} 0 \\ \times 6 \\ \hline 0 \end{array}$ **77.** $\begin{array}{r} 9 \\ \times 1 \\ \hline 9 \end{array}$ **97.** $\begin{array}{r} 3 \\ \times 6 \\ \hline 18 \end{array}$

Exercise Set B

1. a. $\dfrac{\$16}{1 \text{ hr}} \cdot \dfrac{40 \text{ hr}}{1} = \dfrac{\$16 \cdot 40}{1 \cdot 1} = \$640$

Before deductions, your earnings for 40 hours are $640.

b. $\dfrac{\$240}{1} \cdot \dfrac{1 \text{ hr}}{\$16} = \dfrac{240 \cdot 1 \text{ hr}}{1 \cdot 16} = 15 \text{ hr}$

It takes you 15 hours to earn $240 before deductions.

5. a. $\dfrac{1 \text{ bed}}{155 \text{ people}} \cdot \dfrac{31{,}000 \text{ people}}{1} = \dfrac{1 \text{ bed} \cdot 31{,}000}{155 \cdot 1}$

$= 200 \text{ beds}$

There would be 200 hospital beds in the city.

b. $\dfrac{155 \text{ people}}{1 \text{ bed}} \cdot \dfrac{73 \text{ beds}}{1} = \dfrac{155 \text{ people} \cdot 73}{1 \cdot 1}$

$= 11{,}315 \text{ people}$

The population is about 11,315 people.

9. $\dfrac{\$0.75}{1 \text{ component}} \cdot \dfrac{26 \text{ components}}{1 \text{ hr}} \cdot \dfrac{33 \text{ hr}}{1}$

$= \dfrac{\$0.75 \cdot 26 \cdot 33}{1 \cdot 1 \cdot 1}$

$= \$643.50$

Her pay before deductions was $643.50.

13. a. $\dfrac{16 \text{ tortillas}}{1 \text{ min}} \cdot \dfrac{60 \text{ min}}{1 \text{ hr}} \cdot \dfrac{7.5 \text{ hr}}{1}$

$= \dfrac{16 \text{ tortillas} \cdot 60 \cdot 7.5}{1 \cdot 1 \cdot 1}$

$= 7200 \text{ tortillas}$

The machine could make 7200 tortillas in 7.5 hours.

b. $\dfrac{200 \text{ tortillas}}{1} \cdot \dfrac{1 \text{ min}}{16 \text{ tortillas}} = \dfrac{200 \cdot 1 \text{ min}}{1 \cdot 16}$

$= 12\dfrac{1}{2} \text{ min or } 12.5 \text{ min}$

It takes the machine 12.5 minutes to make 200 tortillas.

17. a. $\dfrac{45 \text{ miles}}{1 \text{ hr}} \cdot \dfrac{\frac{1}{4} \text{ hr}}{1} = \dfrac{45 \text{ miles} \cdot \frac{1}{4}}{1 \cdot 1} = 11\dfrac{1}{4} \text{ miles}$

The cheetah has traveled $11\dfrac{1}{4}$ miles in $\dfrac{1}{4}$ hour.

b. $\dfrac{5 \text{ miles}}{1} \cdot \dfrac{1 \text{ hr}}{45 \text{ miles}} = \dfrac{5 \cdot 1 \text{ hr}}{1 \cdot 45} = \dfrac{1}{9} \text{ hr}$

$\dfrac{\frac{1}{9} \text{ hr}}{1} \cdot \dfrac{60 \text{ min}}{1 \text{ hr}} = \dfrac{\frac{1}{9} \cdot 60 \text{ min}}{1 \cdot 1} = \dfrac{60}{9} \text{ min} = 6\dfrac{2}{3} \text{ min}$

It takes the cheetah $\dfrac{1}{9}$ hour or $6\dfrac{2}{3}$ minutes to travel 5 miles.

Practice Final Exam

1. $\begin{array}{r} 600 \\ -\ 487 \\ \hline 113 \end{array}$

5. $\dfrac{64 \div 8 \cdot 2}{(\sqrt{9} - \sqrt{4})^2 + 1} = \dfrac{8 \cdot 2}{(3-2)^2 + 1} = \dfrac{16}{1^2 + 1}$

$= \dfrac{16}{1 + 1}$

$= \dfrac{16}{2}$

$= 8$

9. $3\dfrac{1}{3} \cdot 6\dfrac{3}{4} = \dfrac{10}{3} \cdot \dfrac{27}{4} = \dfrac{10 \cdot 27}{3 \cdot 4} = \dfrac{2 \cdot 5 \cdot 3 \cdot 9}{3 \cdot 2 \cdot 2}$

$= \dfrac{5 \cdot 9}{2}$

$= \dfrac{45}{2}$

$= 22\dfrac{1}{2}$

13. To round 34.8923 to the nearest tenth, observe that the digit in the hundredths place is 9. Since the digit is at least 5, we add 1 to the digit in the tenths place and drop all digits to the right of the tenths place. The number 34.8923 rounded to the nearest tenth is 34.9.

17. $\dfrac{3}{8} = \dfrac{3}{8}(100\%) = \dfrac{300}{8}\% = 37.5\%$

$$\begin{array}{r} 37.5 \\ 8\overline{)300.0} \\ \underline{-24} \\ 60 \\ \underline{-56} \\ 4\,0 \\ \underline{-4\,0} \\ 0 \end{array}$$

21. The rate of 9 inches in 30 days is $\dfrac{9\text{ inches}}{30\text{ days}} = \dfrac{3\text{ inches}}{10\text{ days}}$.

25. $\begin{array}{r} 20.0 \\ \underline{-8.6} \\ 11.4 \end{array}$

29. 0.6% of what number is 7.5?

Method 1:

$0.6\% \cdot x = 7.5$

$0.006x = 7.5$

$\dfrac{0.006x}{0.006} = \dfrac{7.5}{0.006}$

$x = 1250$

0.6% of 1250 is 7.5.

Method 2:

$\dfrac{7.5}{b} = \dfrac{0.6}{100}$

$7.5 \cdot 100 = 0.6 \cdot b$

$750 = 0.6b$

$\dfrac{750}{0.6} = \dfrac{0.6b}{0.6}$

$1250 = b$

0.6% of 1250 is 7.5.

33. $2.4\text{ kg} = \dfrac{2.4\text{ kg}}{1} \cdot \dfrac{1000\text{ g}}{1\text{ kg}} = 2400\text{ g}$

37. $\angle x$ and the angle marked 62° are supplementary, so $m\angle x = 180° - 62° = 118°$.

$\angle y$ and the angle marked 62° are vertical angles, so $m\angle y = 62°$.

$\angle x$ and $\angle z$ are vertical angles, so $m\angle z = m\angle x = 118°$.

41. 5 of the students are 5′0″ − 5′3″ and 6 of the students are 5′3″ − 5′7″, so 5 + 6 = 11 of the students are 5′7″ or shorter.

45. $\left(\dfrac{5}{9} - \dfrac{7}{9}\right)^2 + \left(-\dfrac{2}{9}\right) = \left(\dfrac{5-7}{9}\right)^2 + \left(-\dfrac{2}{9}\right)$

$= \left(-\dfrac{2}{9}\right)^2 + \left(-\dfrac{2}{9}\right)$

$= \left(-\dfrac{2}{9}\right)\cdot\left(-\dfrac{2}{9}\right) + \left(-\dfrac{2}{9}\right)$

$= \dfrac{4}{81} + \left(-\dfrac{2}{9}\cdot\dfrac{9}{9}\right)$

$= \dfrac{4}{81} + \left(-\dfrac{18}{81}\right)$

$= \dfrac{4-18}{81}$

$= -\dfrac{14}{81}$

49. $5x + 12 - 4x - 14 = 22$

$5x - 4x + 12 - 14 = 22$

$x - 2 = 22$

$x - 2 + 2 = 22 + 2$

$x = 24$

Index

Photo Credits

Geometric Formulas

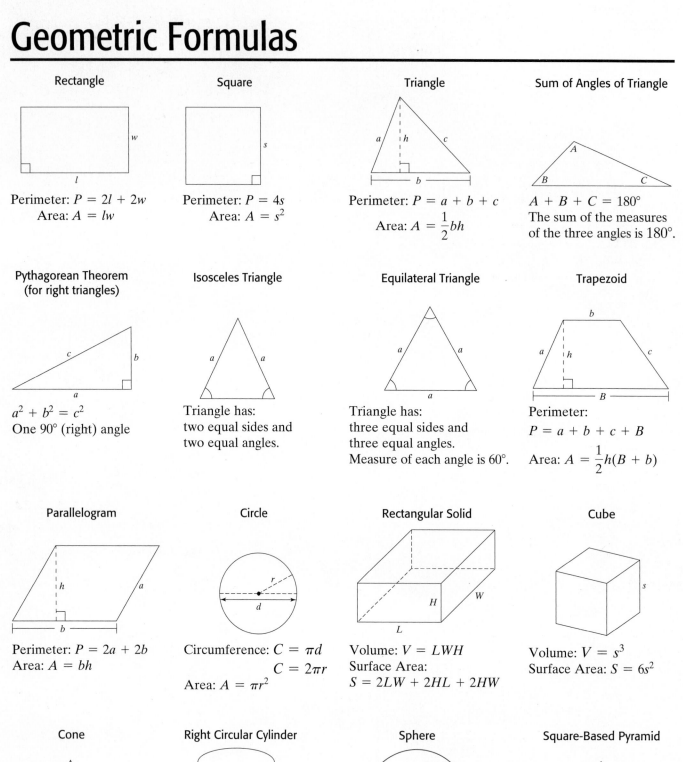

Rectangle

Perimeter: $P = 2l + 2w$
Area: $A = lw$

Square

Perimeter: $P = 4s$
Area: $A = s^2$

Triangle

Perimeter: $P = a + b + c$
Area: $A = \dfrac{1}{2}bh$

Sum of Angles of Triangle

$A + B + C = 180°$
The sum of the measures
of the three angles is $180°$.

Pythagorean Theorem (for right triangles)

$a^2 + b^2 = c^2$
One $90°$ (right) angle

Isosceles Triangle

Triangle has:
two equal sides and
two equal angles.

Equilateral Triangle

Triangle has:
three equal sides and
three equal angles.
Measure of each angle is $60°$.

Trapezoid

Perimeter:
$P = a + b + c + B$
Area: $A = \dfrac{1}{2}h(B + b)$

Parallelogram

Perimeter: $P = 2a + 2b$
Area: $A = bh$

Circle

Circumference: $C = \pi d$
$C = 2\pi r$
Area: $A = \pi r^2$

Rectangular Solid

Volume: $V = LWH$
Surface Area:
$S = 2LW + 2HL + 2HW$

Cube

Volume: $V = s^3$
Surface Area: $S = 6s^2$

Cone

Volume: $V = \dfrac{1}{3}\pi r^2 h$

Right Circular Cylinder

Volume: $V = \pi r^2 h$
Surface Area:
$S = 2\pi r^2 + 2\pi rh$

Sphere

Volume: $V = \dfrac{4}{3}\pi r^3$
Surface Area: $S = 4\pi r^2$

Square-Based Pyramid

Volume: $V = \dfrac{1}{3}\cdot s^2 \cdot h$